T0190026

Lecture Notes in Computer Science 10361

More information about this series at http://www.springer.com/series/7409

De-Shuang Huang · Vitoantonio Bevilacqua
Prashan Premaratne · Phalguni Gupta (Eds.)

Intelligent Computing Theories and Application

13th International Conference, ICIC 2017
Liverpool, UK, August 7–10, 2017
Proceedings, Part I

Editors
De-Shuang Huang
Tongji University
Shanghai
China

Prashan Premaratne
University of Wollongong
North Wollongong, NSW
Australia

Vitoantonio Bevilacqua
Politecnico di Bari
Bari
Italy

Phalguni Gupta
Indian Institute of Technology
Kanpur
India

ISSN 0302-9743 ISSN 1611-3349 (electronic)
Lecture Notes in Computer Science
ISBN 978-3-319-63308-4 ISBN 978-3-319-63309-1 (eBook)
DOI 10.1007/978-3-319-63309-1

Library of Congress Control Number: 2017946067

LNCS Sublibrary: SL3 – Information Systems and Applications, incl. Internet/Web, and HCI

Printed on acid-free paper

This Springer imprint is published by Springer Nature
The registered company is Springer International Publishing AG
The registered company address is: Gewerbestrasse 11, 6330 Cham, Switzerland

Preface

The International Conference on Intelligent Computing (ICIC) was started to provide an annual forum dedicated to the emerging and challenging topics in artificial intelligence, machine learning, pattern recognition, bioinformatics, and computational biology. It aims to bring together researchers and practitioners from both academia and industry to share ideas, problems, and solutions related to the multifaceted aspects of intelligent computing.

ICIC 2017, held in Liverpool, UK, August 7–10, 2017, constituted the 13th International Conference on Intelligent Computing. It built upon the success of ICIC 2016, ICIC 2015, ICIC 2014, ICIC 2013, ICIC 2012, ICIC 2011, ICIC 2010, ICIC 2009, ICIC 2008, ICIC 2007, ICIC 2006, and ICIC 2005 that were held in Lanzhou, Fuzhou, Taiyuan, Nanning, Huangshan, Zhengzhou, Changsha, China, Ulsan, Korea, Shanghai, Qingdao, Kunming, and Hefei, China, respectively.

This year, the conference concentrated mainly on the theories and methodologies as well as the emerging applications of intelligent computing. Its aim was to unify the picture of contemporary intelligent computing techniques as an integral concept that highlights the trends in advanced computational intelligence and bridges theoretical research with applications. Therefore, the theme for this conference was "Advanced Intelligent Computing Technology and Applications." Papers focused on this theme were solicited, addressing theories, methodologies, and applications in science and technology.

ICIC 2017 received 612 submissions from 21 countries and regions. All papers went through a rigorous peer-review procedure and each paper received at least three review reports. Based on the review reports, the Program Committee finally selected 212 high-quality papers for presentation at ICIC 2017, included in three volumes of proceedings published by Springer: two volumes of *Lecture Notes in Computer Science* (LNCS) and one volume of *Lecture Notes in Artificial Intelligence* (LNAI).

This volume of *Lecture Notes in Computer Science* (LNCS) includes 71 papers.

The organizers of ICIC 2017, including Tongji University and Liverpool John Moores University, UK, made an enormous effort to ensure the success of the conference. We hereby would like to thank the members of the Program Committee and the referees for their collective effort in reviewing and soliciting the papers. We would like to thank Alfred Hofmann, executive editor at Springer, for his frank and helpful advice and guidance throughout and for his continuous support in publishing the proceedings. In particular, we would like to thank all the authors for contributing their papers. Without the high-quality submissions from the authors, the success of the conference would not have been possible. Finally, we are especially grateful to the IEEE Computational Intelligence Society, the International Neural Network Society, and the National Science Foundation of China for their sponsorship.

May 2017

De-Shuang Huang
Vitoantonio Bevilacqua
Prashan Premaratne
Phalguni Gupta

ICIC 2017 Organization

General Co-chairs

De-Shuang Huang, China
Abir Hussain, UK

Program Committee Co-chairs

Kang-Hyun Jo, Korea
M. Michael Gromiha, India

Organizing Committee Co-chairs

Dhiya Al-Jumeily, UK
Tom Dowson, UK

Award Committee Co-chairs

Vitoantonio Bevilacqua, Italy
Phalguni Gupta, India

Tutorial Co-chairs

Juan Carlos Figueroa, Colombia
Valeriya Gribova, Russia

Publication Co-chairs

Kyungsook Han, Korea
Laurent Heutte, France

Workshop/Special Session Chair

Wenzheng Bao, China

Special Issue Chair

Ling Wang, China

International Liaison Chair

Prashan Premaratne, Australia

Publicity Co-chairs

Chun-Hou Zheng, China
Jair Cervantes Canales, Mexico
Huiyu Zhou, China

Exhibition Chair

Lin Zhu, China

Program Committee

Khalid Aamir
Mohd Helmy Abd Wahab
Abbas Amini
Vasily Aristarkhov
Waqas Haider Khan Bangyal
Shuui Bi
Hongmin Cai
Jair Cervantes
Pei-Chann Chang
Chen Chen
Shih-Hsin Chen
Weidong Chen
Wen-Sheng Chen
Xiyuan Chen
Jieren Cheng
Michal Choras
Angelo Ciaramella
Jose Alfredo Costa
Guojun Dai
Yanrui Ding
Ji-Xiang Du
Pufeng Du
Jianbo Fan
Paul Fergus
Juan Carlos Figueroa-Garcia
Liang Gao
Dunwei Gong
Yannis Goulermas
Michael Gromiha
Fei Han
Kyungsook Han
Tianyong Hao
Wei-Chiang Hong
Yuexian Hou
Saiful Islam
Chuleerat Jaruskulchai
Kang-Hyun Jo
Dah-Jing Jwo
Seeja K.R.
Sungshin Kim
Yong-Guk Kim
Yoshinori Kuno
Takashi Kuremoto
Xuguang Lan
Xinyi Le
Choong Ho Lee
Xiujuan Lei
Bo Li
Guoliang Li
Kang Li
Ming Li
Qiaotian Li
Shuai Li
Honghuang Lin
Bin Liu
Bingqiang Liu
Chunmei Liu
Shuo Liu
Xingwen Liu
Xiwei Liu
Yunxia Liu
Ahmad Lotfi
Jungang Lou
Yonggang Lu
Yingqin Luo
Jinwen Ma
Xiandong Meng
Kang Ning
Ben Niu
Gaoxiang Ouyang
Francesco Pappalardo
Young B. Park
Eros Pasero
Marzio Pennisi
Prashan Premaratne
Yuhua Qian
Daowen Qiu
Jiangning Song
Stefano Squartini
Zhixi Su
Shiliang Sun
Zhan-Li Sun
Jijun Tang
Joaquin Torres-Sospedra
Antonio Uva

Bing Wang
Jim Jing-Yan Wang
Shitong Wang
Xuesong Wang
Yong Wang
Yuanzhe Wang
Ze Wang
Wei Wei
Ka-Chun Wong
Hongjie Wu
QingXiang Wu
Yan Wu
Junfeng Xia
Shunren Xia
Xinzheng Xu
Wen Yu
Junqi Zhang
Rui Zhang
Shihua Zhang
Xiang Zhang
Zhihua Zhang
Dongbin Zhao
Xiaoguang Zhao
Huiru Zheng
Huiyu Zhou
Yongquan Zhou
Shanfeng Zhu
Quan Zou
Le Zhang

Additional Reviewers

Honghong Cheng
Jieting Wang
Xinyan Liang
Feijiang Li
Furong Lu
Mohd Farhan Md Fudzee
Husniza Husni
Rozaida Ghazali
Zarul Zaaba
Shuzlina Abdul-Rahman
Sasalak Tongkaw
Wan Hussain Wan Ishak
Mohamad Farhan Mohamad Mohsin
Masoud Mohammadian
Mario Diván
Nureize Arbaiy
Mohd Shamrie Sainin
Francesco Masulli
Nooraini Yusoff
Aida Mustapha
Nur Azzah Abu Bakar
Anang Hudaya Muhamad Amin
Sazalinsyah Razali
Norita Md Norwawi
Nasir Sulaiman
Ru-Ze Liang
Wan Hasrulnizzam Wan Mahmood
Azizul Azhar Ramli
Qian Guo
Mingjie Tang
Fang Jin
Muhammad Imran
Barnypok
Weizhi Li
Ri Hanafi
Jieyi Zhao
Zhichao Jiang
Y-H. Taguchi
Chenhao Xu
Shaofu Yang
Jian Xiao
Haoyong Yu
Prabakaran R.
Anusuya S.
A. Mary Thangakani
Fatih Adiguzel
Shanwen Zhu
Yuheng Wang
Xixun Zhu
Ambuj Srivastava
Xing He
Nagarajan Raju
Kumar Yugandhar
Sakthivel Ramasamy
Anoosha Paruchuri
Sherlyn
Ramakrishnan
Chandrasekaran
Akila Ranjith
Harihar Balasubramanian
Vimala A.
Farheen Siddiqui
Sameena Naaz
Parul Agarwal
Soniya Balram
Jaya Sudha
Deepa Anand
Yun Xue
Santo Motta
Nengfu Xie
Lei Liu
Nengfu Xie
Xiangyuan Lan
Rushi Lan
Jie Qin
Tomasz Andrysiak
Rafal Kozik
Adam Schmidt
Zhiqiang Liu
Liang Xiaobo
Zhiwei Feng
Xia Li
Chunxia Zhang
Jiaoyun Yang
Yi Xiong
Wenjun Shen
Xihao Hu
Xiangli Zhang
Xiaoyong Bian
Yan Xu

Lu-Chen Weng
Guan Xiao
Jianchao Fan
Ka-Chun Wong
Fei Wang
Beeno Cheong
Wei-Jie Yu
Meng Zhang
Jiao Zhang
Zhouhua Peng
Shankai Yan
Yueming Lyu
Xiaoping Wang
Cheng Liu
Jiecong Lin
Shan Gao
Xiu-Jun Gong
Lijun Quan
Haiou Li
Junyi Chen
Cheng Chen
Ukyo Liu
Weizhi Li
Xi Yuga
Bo Chen
Binbin Pan
Yaran Chen
Xiangtao Li
Fei Guo
Leyi Wei
Yungang Xu
Bin Qin
Yuanpeng Zhang
Wenlong Hang
Huo Xuan
Feng Cong
Hongguo Zhao
Xiongtao Zhang
Li Liu
Baohua Wang
Jianwei Yang
Naveed Anwer Butt
Seung Hoan Choi
Farid Garcia-Lamont
Sergio Ruiz
Josue Vicente Cervantes Bazan
Abd Ur Rehman
Ta Zhou
Wentao Fan
Xin Liu
Qing Lei
Ziyi Chen
Yewang Chen
Yan Chen
Jialin Peng
Sihai Yang
Hong-Bo Zhang
Xingguang Pan
Biqi Wang
Arturo
Lisbeth Rodríguez
Asdrubal Lopez-Chau
Xiaoli Li
Yuriy Orlov
Ken Wing-Kin Sung
Zhou Hufeng
Gonghong Wei
De-Shuang Huang
Dongbo Bu
Dariusz Plewczyński
Jialiang Yang
Musa Mhlanga
Shanshan Tuo
Dawei Li
Chee Hong Wong
Zhang Zhizhuo
Xiang Jinhai
Wada Youichiro
Shuai Cheng Li
Xuequn Shang
Ralf Jauch
Limsoon Wong
Edwin Cheung
Filippo Castiglione
Ping Zhang
Hong Zhu
Tianming Liang
Jiancong Fan
Guanying Wang
Shumei Zhang
Raymond Bond
Ginestra Bianconi
Antonino Staiano
Dong Shan
Xiangying Jiang
Hongjie Jia
Xiaopeng Hua
Sixiao Wei
Shumei Zhang
Francesco Camastra
Weikuan Jia
Huajuan Huang
Hui Li
Xinzheng Xu
Yan Qin
Jiayin Zhou
Xin Gao
Antonio Maratea
Wu Qingxiang
Xiaowei Zhang
Cunlu Xu
Yong Lin
Haizhou Wu
Ruxin Zhao
Alessio Ferone
Zheheng Jiang
Kun Zhang
Fei Chu
Chunyu Yang
Wei Dai
Gang Li
Guo Yanen
Hao Xin
Zhang Zanchao
Yonggang Wang
Yajun Zhang
Guimin Lin
Pengfei Li
Antony Lam
Hironobu Fujiyoshi
Yoshinori Kobayashi
Jin-Soo Kim
Erchao Li
Miao Rong

Yiping Liu
Nobutaka Shimada
Hae-Chul Choi
Irfan Mehmood
Yuyan Han
Jianhua Zhang
Biao Xu
Meirong Chen
Kazunori Onoguchi
Gongjing Chen
Yuntao Wei
Rina Su
Lu Xingjia
Casimiro Aday Curbelo Montanez
Mohammed Khalaf
Atsushi Yamashita
Hotaka Takizawa
Yasushi Mae
Hisato Fukuda
Shaohua Li
Keight Robert
Raghad Al-Shabandar
Xie Zhijun
Abir Hussain
Morihiro Hayashida
Ziding Zhang
Mengyao Li
Shen Chong
Wei Wang
Basma Abdulaimma
Hoshang Kolivand
Ala Al Kafri
Abbas, Rounaq
Kaihui Bian
Jiangning Song
Haya Alaskar
Francis, Hulya
Chuang Wu
Xinhua Tang
Hong Zhang
Sheng Zou
Cuci Quinstina
Bingbo Cui
Hongjie Wu
Wenrui Zhao
Xuan Wang
Junjun Jiang
Qiyan Sun
Zhuangguo Miao
Yuan Xu
Dengxin Dai
Zhiwu Huang
Xuetao Zhang
Shaoyi Du
Jihua Zhu
Bingxiang Xu
Zhixuan Wei
Hanfu Wang
Prashan Premaratne
Alessandro Naddeo
Jun Zhou
Bing Feng
Carlo Bianca
Raúl Montoliu
German Martín Mendoza-Silva
Ximo Torres
Bulent Ayhan
Qiang Liu
Xinle Liu
Jair Cervantes
Junming Zhang
Linting Guan
Guodong Zhao
Tao Yang
Zhenmin Zhang
Tao Wu
Yong Chen
Chien-Lung Chan
Liang-Chih Yu
Chin-Sheng Yang
Chin-Yuan Fan
Si-Woong Lee
Zhiwei Ji
Ke Zeng
Filipe Saraiva
Mario Malcangi
Francesco Ferracuti
José Alfredo Costa
Jacob Schrum
Shangxuan Tian
Yitian Zhao
Yonghuai Liu
Julien Leroy
Pierre Marighetto
Jheng-Long Wu
Falcon Rafael
Haytham Fayek
Yong Xu
Kozou Abdellah
Damiano Rossetti
Daniele Ferretti
Chengdong Li
Zhen Ni
Pavel Osinenko
Bakkiyaraj Ashok
Hector Menendez
Pandiri Venkatesh
Davendra Donald
Ming-Wei Li
Leo Chen
Xiaoli Wei
Chien-Yuan Lai
Yongquan Zhou
Ma Xiaotu
Yannan Bin
Wei-Chiang Hong
Seongho Kim
Xiang Zhang
Geethan Mendiz
Brendan Halloran
Kai Xu
Di Zhang
Sheng Ding
Jun Li
Lei Wang
Jin Gu
Wang, Xiaowo
Hao Lin
Yao Nie
Vincenzo Randazzo
Yujie Cai
Derek Wang
Gang Li
Fuhai Li
Junfeng Luo
Wei Lan

Chaowang Lan
Lasker Ershad Ali
Shuyi Zhang
Pengbo Bo
Abhijit Kundu
Jingkai Yang
Shamim Reza
Lee Kai Wah
Anass Nouri
Arridhana Ciptadi
Lun Li
Cheeyong Kim
Helen Hong
Guangchun Gao
Weilin Deng
Taeho Jo
Jiazhou Chen
Enhong Zhuo
Carlo Bianca
Gai-Ge Wang
Kwanggi Kim
Jiangling Song
Xiandong Xu
Guohui Zhang
Long Wen
Kunkun Peng
Qiqi Duan
Hui Wang
Shiying Sun
Yu Yongjia
Jianbo Fan
Lishuang Shen
Liming Xie
Geethan Mendiz
Ying Bi
Hong Wang
Juntao Liu
Yang Li
Enfeng Qi
Xiangyu Liu
Jia Liu
Hou Yingnan
Weihua Deng
Chuanfeng Li
Fuyi Li
Tomasz Talaska
Nathan Cannon
Bo Gao
Ting Yu
Kang Li
Li Zhang
Xingjia Lu
Qingfeng Li
Yong-Guk Kim
Kunikazu Kobayashi
Takashi Kuremoto
Shingo Mabu
Takaomi Hirata
Tuozhong Yao
Lvzhou Li
Fuchun Liu
Shenggen Zheng
Jingkai Yang
Chen Li
Sungshin Kim
Hansoo Lee
Jungwon Yu
Yongsheng Dong
Zhihua Cui
Duyu Liu
Timothy L. Bailey
Hongda Mao
Cai Xiao
Xin Bai
Qishui Zhong
Yansong Deng
Giansalvo Cirrincione
Shi Zhenwei
Shuhui Wang
Junning Gao
Ziye Wang
Hui Li
Chao Wu
Fuyuan Cao
Rui Hong
Wahyono Wahyono
Rafal Kozik
Yihua Zhou
Wenyan Wang
Juan Figueroa
Artem Lenskiy
Qiang Yan
Wenlong Xu
Yi Gu
Yi Kou
Chenhui Qiu
Tianyu Yang
Xiaoyi Yu
Rongjing Hu
Yuanyuan Liu
Zhang Haitao
Libing Shen
Dongliang Yu
Yanwu Zeng
Yuanyuan Wang
Giulia Russo
Emilio Mastriani
Di Tang
Yan Jiang
Hong-Guan Liu
Liyao Ma
Panpan Du
Hongguan Liu
Chenbin Liu
Dangdang Shao
Yunze Yang
Yang Chen
Yang Li
Heye Zhang
Austin Brockmeier
Hui Li
Xi Yang
Yan Wang
Lin Bai
Laksono Kurnianggoro
Ajmal Shahbaz
Yan Zhang
Xiaoyang Wang
Bo Wang
Sajjad Ahmed
Ke Zeng
Liang Mao
Yuan You
Qiuyang Liu
Shaojie He
Zehui Cao
Zhongpu Xia
Bin Ye

Zhi-Yu Shen
Alexander Filonenko
Zhenhu Liang
Jing Li
Yang Yang
Wenjun Xu
Yongjia Yu
Jyotsna Wassan
Vibha Patel
Saiful Islam
Ekram Khan
Angelo Ciaramella
Junyu Chen
Ziang Dong
Jingjing Fei
Meng Lei
Xuanfang Fei
Bing Zeng
Taifeng Li
Yan Wu
Paul Fergus
Jiulun Cai
Jingying Huang
Francesco Pappalardo
Xiaoguang Zhao
Qingfang Meng
Mohamed Alloghani
Ruihao Li
Zhengyu Yang
Xin Xie
Yifan Wu
Hong Zeng

Contents – Part I

Evolutionary Computation and Learning

Neural Networks

Information Security

Virtual Reality and Human-Computer Interaction

Business Intelligence and Multimedia Technology

Genetic Algorithms

Computational Intelligence and Security for Image Applications in Social Network

Neural Networks: Theory and Application

Evolutionary Computation and Learning

Gbest-Guided Covariance Matrix Adaptation Evolution Strategy for Large Scale Global Optimization

Zhang Fuxing[1], Zhang Tao[1,2], and Wang Rui[1(✉)]

[1] College of Information Systems and Management, National University of Defense Technology, Changsha 410073, People's Republic of China
ruiwangnudt@gmail.com
[2] State Key Laboratory of High Performance Computing, National University of Defense Technology, Changsha 410073, People's Republic of China

Abstract. The optimization of a large number of decision variables, so called large scale global optimization (LSGO) remains challenging for existing heuristics. Inspired by the concept of global best (gbest) guided strategy, this paper proposes a gbest-guided covariance matrix adaptation evolution strategy (GCMA-ES) where the gbest information is utilized in the search equation to guide the exploitation process. The GCMA-ES can take advantages from both the CMA-ES and the gbest-guided strategy. Its performance is demonstrated on the CEC 2010 LSGO benchmarks.

Keywords: Global best guided strategy · Evolution strategy · Large scale global optimization (LSGO) · Covariance matrix adaptation (CMA)

1 Introduction

Many real-world optimization problems [1, 2] can be formulated as LSGO problems in which the number of decision variables to be optimized is very large, e.g., 1000 or even more, see Eq. (1):

$$\min f(\mathrm{x}),\quad \mathrm{x} = (x_1, x_2, \ldots, x_n)$$
$$s.t.\begin{cases} x_{\min} \le x_i \le x_{\max} & i = 1, 2, \ldots, n \\ n \ge 1000, x_{\min} \in R, x_{\max} \in R \end{cases} \tag{1}$$

where $f(\mathrm{x})$ is the objective function to be optimized, x is the vector with n (usually, $n \geq 1000$) decision variables and each decision variable x_i is within the range [x_{min}, x_{max}]. Because of the high-dimensional characteristics and complex interactive relationship among different variables, traditional optimization techniques (e.g. Newton or Gradient based method) generally manifest poor global and/or local performance, by which high-quality solutions are difficult to be found.

D.-S. Huang et al. (Eds.): ICIC 2017, Part I, LNCS 10361, pp. 3–13, 2017.
DOI: 10.1007/978-3-319-63309-1_1

In order to effectively solve LSGO problems, researchers appear to heuristics [3–6]. For example, based on differential evolution, Yang [6, 7] proposed a self-adaptive differential evolution with neighborhood search (SaNSDE). Experimental results show that SaNSDE can greatly improve the performance of a pure DE on LSGO. However, it faces difficulties on non-separable and multimodal functions. In addition to differential evolution [5], evolution strategy (ES), decomposition based algorithm [8], particle swarm optimization (PSO) [9] have also been investigated to handle LSGO [4, 10]. By summarizing these algorithms, a common view is shared that the core of LSGO is to appropriately i) balance the exploration and exploitation, and ii) decompose separable and non-separable variables.

Amongst existing algorithms for LSGO, CMA-ES has gained considerable attentions in recent years due to its excellent ability in solving highly non-separable and multimodal functions [11]. The study [12] hybridizes the composite differential evolution (CODE) and covariance matrix adaptation evolution strategy (CMA-ES) to handle LSGO. The proposed algorithm (EBCC) is shown to perform well. In EBCC, the two constituent algorithms can learn from each other. By identifying the performance of the two constituent algorithms dynamically, EBCC assigned more function evaluations (FEs) to the better one, which aims to make full use of the number of FEs. In [8] a new decomposition based strategy (cooperative co-evolution (CC) framework [13]) was used in CMA-ES to solve LSGO problems. Moreover, some other variants of CMA-ES and their extended versions for constrained optimization problems have been discussed in [14]. Though CMA-ES has been regarded as one of the most efficient global optimizer, its exploitation performance is poor when the optimization problems feature complicated fitness landscapes which is often seen in real world optimization problems [15].

This study attempts to further explore the performance of CMA-ES on LSGO problems. To enhance the exploitation ability of CMA-ES, the gbest-guided strategy is incorporated into CMA-ES. The proposed algorithm, GCMA-ES, can utilize the strength of CMA-ES by adapting the step-size and updating the covariance matrix, and simultaneously update the solution through learning from the current best global solution. To verify the performance of GCMA-ES, it is compared to CMA-ES. Experimental results on the 2010 LSGO benchmark functions show that GCMA-ES performs well.

The remainder of this paper is organized as follows: Sect. 2 provides background knowledge related to CMA-ES and gbest-guided evolution algorithms. Section 3 describes the proposed method. In Sect. 4, the experimental setup and parameter setting are presented. Experiment results are demonstrated and discussed in Sect. 5. Section 6 concludes this paper and identifies future studies.

2 Background

2.1 Covariance Matrix Adaptation Evolution Strategy

CMA-ES is an improved version of ES [16]. In a standard ES, the mutation step-size works for all individuals in a global manner. There is no mutation step-size control in an individual level [18]. To solve this problem, CMA-ES is introduced by Hansen [16, 17], which contains two adaptation phases: the cumulative step-size adaptation and covariance matrix adaptation. Different from traditional ES, CMA-ES adopts the covariance matrix to adjust population distribution and makes full use of the step-size mechanism to guide the evolution path.

Typically, a CMA-ES includes the following components:

(1) Population with λ offspring individuals and μ parent individuals. Here, λ and μ can be computed by Eq. (2).

$$\begin{cases} \lambda = 4 + 3 * \ln(N) \\ \mu = \frac{\lambda}{2} \end{cases} \tag{2}$$

(2) Center of the selected individuals. μ parent individuals are selected and multiplied by a normalized weights array to update the search center $\langle x \rangle_\mu^{(G)}$.

$$\langle x \rangle_\mu^{(G)} = \frac{1}{\mu} \cdot \sum_{i \in I_{sel}^{(G)}} x_i^{(G)} \tag{3}$$

Then the new search center together with other related parameters are selected to generate λ offspring individuals by Eqs. (4) and (5).

$$x_k^{(G+1)} = \langle x \rangle_\mu^{(G)} + \sigma^{(G)} \cdot \underbrace{B^{(G)} \cdot D^{(G)} \cdot z_k^{(G+1)}}_{\sim N(0, C^{(G)})}, \; k = 1, \ldots, \lambda \tag{4}$$

$$\langle z \rangle_\mu^{(G+1)} = \frac{1}{\mu} \cdot \sum_{i \in I_{sel}^{(G+1)}} z_i^{(G+1)} \tag{5}$$

where Eq. (5) means how the search center of generation G + 1 with random vectors z_i is built.

(3) Covariance matrix. It is a symmetrical n*n matrix and is used to describe the distribution of updated population.

$$C^{(G)} = B^{(G)} \cdot D^{(G)} \cdot (B^{(G)} \cdot D^{(G)})^T \tag{6}$$

where Eq. (6) denotes how construct the covariance matrix C.

(4) Two adaptation mechanism: the cumulative step-size $\sigma^{(G)}$ adaptation and covariance matrix $C^{(G)}$ adaptation, which are described as follows:

First, the cumulative evolution path $p_{\sigma}^{(G+1)}$ is generated by Eq. (7) and is used to update the global step size σ of generation $G + 1$ by Eq. (8).

$$p_{\sigma}^{(G+1)} = (1 - c_{\sigma}) \cdot p_{\sigma}^{(G)} + \sqrt{c_{\sigma} \cdot (2 - c_{\sigma})} \cdot \sqrt{\mu} \cdot B^{(G)} \cdot \langle z \rangle_{\mu}^{(G+1)} \tag{7}$$

$$\sigma^{(G+1)} = \sigma^{(G)} \cdot \exp\left(\frac{1}{d_{\sigma}} \cdot \frac{\left\| p_{\sigma}^{(G+1)} \right\| - X_n}{X_n}\right) \tag{8}$$

Second, in order to adapt the covariance matrix C of generation $G + 1$, shown as Eq. (10), the cumulative evolution path $p_{c}^{(G+1)}$ is determined by Eq. (9).

$$p_{c}^{(G+1)} = (1 - c_{c}) \cdot p_{c}^{(G)} + \sqrt{c_{c} \cdot (2 - c_{c})} \cdot \sqrt{\mu} \cdot B^{(G)} \cdot D^{(G)} \cdot \langle z \rangle_{\mu}^{(G+1)} \tag{9}$$

$$C^{(G+1)} = (1 - c_{\text{cov}}) \cdot C^{(G)} + c_{\text{cov}} \cdot p_{c}^{(G+1)} \cdot (p_{c}^{(G+1)})^{T} \tag{10}$$

(5) Parameter specification. Other parameters are summarized in Table 1.

Table 1. Experimental setup parameters

Parameter	Description
N	Dimension of decision variables
$x_k^{(G+1)}$	Selected individuals of generation $G + 1$
Φ	Function values of all individuals.
$z_k^{(G+1)}$	The random vectors to generate offspring
$B^{(G)}$	Rotation matrix of generation G
$D^{(G)}$	Step size matrix ($n*n$ diagonal matrix)
c_{σ}	Determining the cumulation time of p_{σ}
c_c	Determining the cumulation time of p_c
d_{σ}	Damping parameter for change rate of $\sigma^{(G)}$
c_{cov}	Change rate of the covariance matrix C

2.2 Global Best Guided Evolution Strategy

The representative global best guided evolution strategy are reviewed in this section, which includes global best guided particle swarm optimization (GPSO) [19], global best guided differential evolution (GDE) [20], global best guided artificial bee colony (GABC) [21] and global best guided gravitational search algorithm (GGSA) [22].

(1) GPSO: The standard GPSO algorithm is a gbest topology-based stochastic algorithm in which the particles' behaviors are implemented by adjusting the velocity and/or step size of each particle. Each particle in a GPSO is influenced by a stochastic average of the best solutions found by all members of the population. Then the particle's search is centered around the mean value of the current best individual.
(2) GDE: The trail vector updates its position through learning from individuals selected randomly as well as the current global best individual found so far. By taking advantage of the information of the global best solution, more competitive candidate solutions are generated. It is shown that GDE is capable of balancing the exploration and exploitation.
(3) GABC: The standard ABC algorithm is good at exploration but perform poorly at exploitation, which results in a slow convergence speed and is easy to fall into a local optima. By incorporating the global information of present best solution into the search equation, the exploitation ability of the standard ABC is shown to be greatly improved.
(4) GGSA: The global best solution is used to accelerate the exploitation phase of the standard GSA which is inspired by the natural gravitational forces. By using the best solution, GGSA can effectively prevent particles from gathering together, and lead to a faster convergence speed.

Overall, (i) CMA-ES is one of the most efficient algorithms for solving high-dimensional optimization problems, (ii) and the global best guided evolution strategy has been widely demonstrated as effective in improving the exploitation ability of an ES. By combining the two strategies, it is expected to develop a more effective method for handling LSGO problems.

3 Description of GCMA-ES

Exploration and exploitation are two important processes for meta-heuristics [23]. In the original CMA-ES, both the cumulative step-size adaption and covariance matrix adaption are used to update the population, which ignores the information from the current global best individual. In this sense, the performance of the standard CMA-ES can be further improved by incorporating the global best guided strategy.

Specifically, the idea of GCMA-ES is that the current global best individual is recorded, updated and used to produce new offspring. The new search center, the covariance matrix, the global step size and the current global best individual are all considered to generate λ offspring individuals. GCMA-ES is obtained by adding the current global best individual to the generation loop stage of CMA-ES and Eq. (4) is modified as Eq. (11).

$$x_k^{(G+1)} = (1 - GGR) \cdot \left(\langle x \rangle_{\mu}^{(G)} + \sigma^{(G)} \cdot \underbrace{B^{(G)} \cdot D^{(G)} \cdot z_k^{(G+1)}}_{\sim N(0, C^{(G)})} \right) + GGR \cdot GB^{(G)} \quad (11)$$

where $GB^{(G)}$ denotes the global best individual of generation G. GGR stands for the ratio of constituting $x_k^{(G+1)}$ by $GB^{(G)}$. Other parameters are the same as Eq. (4). In Eq. (11), $GGR = 1$ and $GGR = 0$ means λ offspring individuals are completely generated by $GB^{(G)}$ and by Eq. (4), respectively. To examine the effect of GGR, four values are considered, i.e., 20%, 50%, 80% and a random value between 0 and 1.

Equation (12) is used to record and update the current global best individual $GB^{(G)}$.

$$GB^{(G+1)} = \begin{cases} Sol^{(G)} & \text{if } f(Sol^{(G)}) \le f(GB^{(G)}) \\ GB^{(G)} & \text{otherwise.} \end{cases} \tag{12}$$

where $Sol^{(G)}$ stands for the trial vector of generation G. $GB^{(G)}$ represents the global best individual of generation G. By Eq. (12), the better one between $Sol^{(G)}$ and $GB^{(G)}$ is selected as $GB^{(G+1)}$ in the next generation.

Overall,the flowchart of the gbest-guided covariance matrix adaptation evolution strategy is shown as Fig. 1. Its pseudo-code is presented in Algorithm 1.

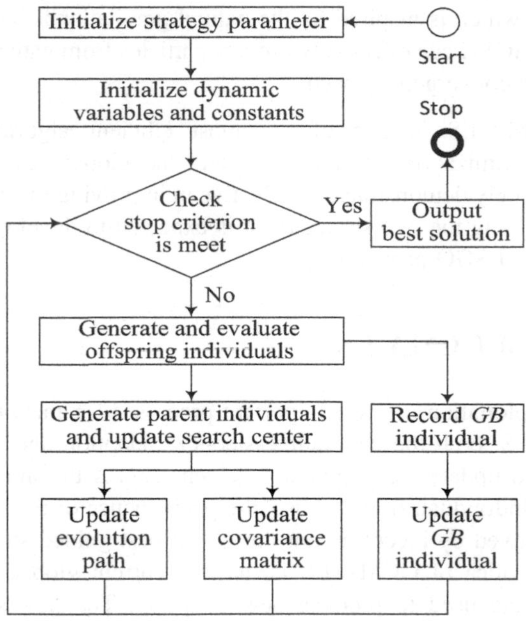

Fig. 1. Flowchart of the gbest-guided covariance matrix adaptation evolution strategy.

In Algorithm 1, $f(x)$ is the objective function to be optimized. n represents the dimensionality. Lb and Ub are the lower and upper bounds of decision variables. FEs_Max is the maximum number of function evaluations times.

There are five main steps in GCMA-ES. The first step is to initialize strategy parameters as well as dynamic variables and constants (line 1 to line 2). The second step is to generate and evaluate λ offspring individuals (line 4 to line 8). The third step

Algorithm 1 *(means, std, FEs)* = GCMA-ES *(f(x), n, Lb, Ub, FEs Max)*

1: Initialize strategy parameters λ, μ, c_σ, c_c, d_σ, c_{cov}
2: Initialize dynamic variables G, $\langle x \rangle_\mu$, σ, C, B, D, p_σ, p_c
3: **while** the stop criterion is not met **do**
4:     **for** i ← 1 to λ **do**
5:         Generate offspring individuals by Eq. (11)
6:         Evaluate fitness of the population
7:         FEs ← FEs+1
8:     **end for**
9:     Sort fitness of the λ offspring individuals by the ascending order
10:     Record and update the current global best individual *GB*
11:     Compute weighted mean to update search center by Eq. (3)
12:     Update evolution paths by Eq. (7) and Eq. (9)
13:     Update the covariance matrix by Eq. (10)
14:     Update the global step size σ by Eq. (8)
15:     Update other dynamic variables, e.g. B, D
16: **end while**
17: **return** optimization results

is to generate μ parent individuals and update the search center (line 9 to line 11). The fourth step is to update evolution paths (line 12). The fifth step is to update the covariance matrix, the global step size and other variables (line 13 to line 15).

4 Experimental Setup

To assess the performance of GCMA-ES, the CEC2010 LSGO benchmark functions with problem dimension D = 1000 are used. These test problems are classified into five categories [24].

In the experiment, CMA-ES [17] is compared to GCMA-ES so as to verify the effectiveness by incorporating the global best guided strategy. Specifically, we have realized four different GCMA-ES, denoted as GCMA-ES1 to GCMA-ES4. In these variants, parameter *GGR* is set as 20%, 50%, 80% and a random value between 0 and 1, respectively.

The maximum number of function evaluations is set as 5e + 5. For each test problem, the mean value and standard deviation of 31 independent runs are recorded. The population size for offspring and parent are 24 and 12, respectively. All the parameter settings with respect to CMA-ES is the same as introduced in [17].

5 Experiment Results

First the search behavior of GCMA-ES1 to 4 and CMA-ES is presented in Fig. 2. Due to the page limitation, only the results for F_7, F_{12}, F_{17}, F_{19} are shown. The x-axis and y-axis represent the function evaluation times and the mean function value of 31 independent runs, respectively. Note that the y-axis is in log scale.

From Fig. 2, it is observed that GCMA-ES1 to 4 in general converge faster than CMA-ES. Amongst the four variants, GCMA-ES1 and GCMA-ES2 is the best,

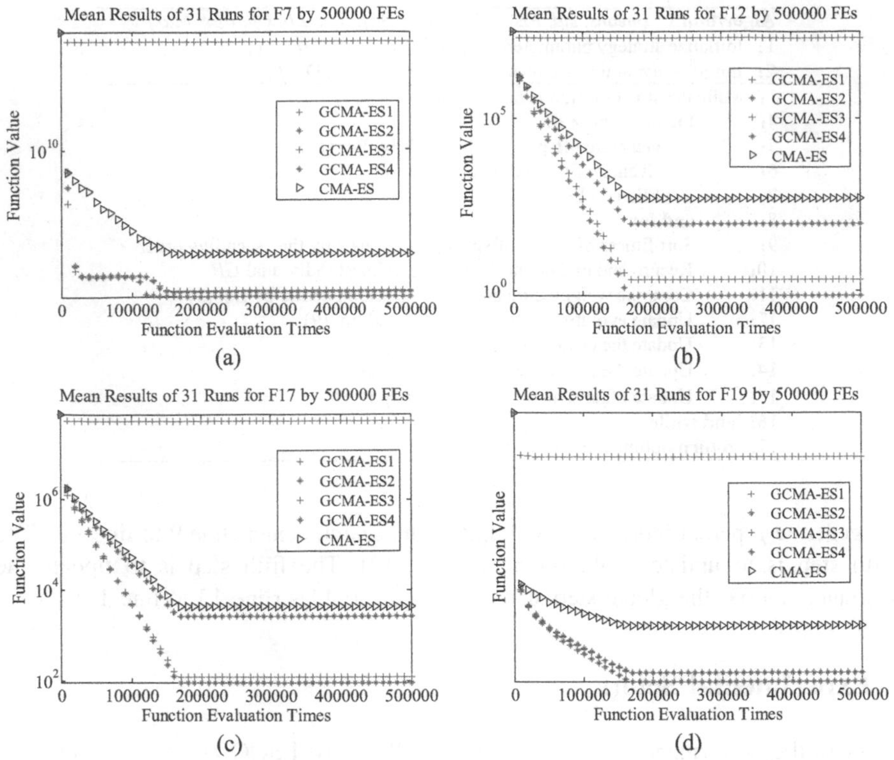

Fig. 2. Search behavior of algorithms (performance over 500000 function evaluation times) on test problems F_7, F_{12}, F_{17}, and F_{19}.

followed by GCMA-ES4, and GCMA-ES3 is the worst. This somehow indicates that *GGR* is crucial to algorithm performance.

In addition to the qualitative results, the mean results of the five competitor algorithms for all test problems are shown in Table 2. The Wilcoxon-ranksum two-sided statistical test at 95% confidence level is employed to examine whether the difference between CMA-ES and each of the GCMA-ES is significant. The symbol "+", "−" and "=" means that GCMA-ES is statistically better, worse or comparable to CMA-ES.

It is observed from Table 2 that GCMA-ES1 and GCMA-ES2 offer overall good performance, and are better than CMA-ES for a majority of problems. However, GCMA-ES3 and GCMA-ES4 show no obvious superiority over CMA-ES. Such results indicates that i) the global best guided strategy is beneficial to improve the performance of CMA-ES; and ii) the ratio of *GGR* plays an important role on the performance of GCMA-ES.

Overall, we can conclude that incorporating the global best information into the solution search equation is helpful in improving algorithm performance. However, choosing a suitable *GGR* for GCMA-ES is not easy.

Table 2. The obtained average optimal value by CMA-ES and GCMA-ES1 to 4. The symbol "+", "−" and "=" means that GCMA-ES is statistically better, worse or comparable to CMA-ES.

	CMA-ES	GCMA-ES1	GCMA-ES2	GCMA-ES3	GCMA-ES4
F_1	1.7014e + 07	1.1575e + 07 +	1.3538e + 07 +	1.7141e + 11 =	3.4696e + 07 −
F_2	4.8978e + 03	7.3791e + 03 −	7.6581e + 03 −	8.1007e + 03 −	7.5781e + 03 −
F_3	5.2032e-01	1.9349e + 01 −	1.9342e + 01 −	1.9481e + 01 −	1.9346e + 01 −
F_4	5.6037e + 11	1.9424e + 11 +	1.9257e + 11 +	3.2488e + 15 −	2.3393e + 11 +
F_5	6.6693e + 07	3.1541e + 08 −	4.2436e + 08 −	4.5769e + 08 −	4.2684e + 08 −
F_6	3.6681e + 04	1.9236e + 07 −	1.9233e + 07 −	1.9223e + 07 −	1.9239e + 07 −
F_7	1.1911e + 07	1.1206e + 06 +	7.9451e + 05 +	1.0795e + 13 −	1.0365e + 06 +
F_8	4.6729e + 07	5.5642e + 06 +	6.1992e + 06 +	3.4938e + 16 −	4.7041e + 07 =
F_9	1.7383e + 07	1.2280e + 07 +	1.5133e + 07 +	2.0812e + 11 −	3.7042e + 07 −
F_{10}	5.0555e + 03	7.6895e + 03 −	8.1093e + 03 −	8.5665e + 03 −	8.0472e + 03 −
F_{11}	1.1043e + 02	2.1252e + 02 −	2.1251e + 02 −	2.1406e + 02 −	2.1248e + 02 −
F_{12}	4.6681e + 02	2.1604e + 00 +	7.5411e-01 +	2.1077e + 07 −	8.5726e + 01 +
F_{13}	2.0947e + 04	1.7638e + 03 +	2.1690e + 03 +	5.7828e + 11 −	3.9320e + 04 −
F_{14}	1.7401e + 07	1.2375e + 07 +	1.5888e + 07 +	2.3628e + 11 −	3.9740e + 07 −
F_{15}	5.1497e + 03	7.7457e + 03 −	8.1760e + 03 −	8.4195e + 03 −	7.9362e + 03 −
F_{16}	2.8723e + 02	3.8608e + 02 −	3.8629e + 02 −	3.8922e + 02 −	3.8628e + 02 −
F_{17}	4.4922e + 03	1.2244e + 02 +	9.3125e + 01 +	5.1665e + 07 −	2.7379e + 03 +
F_{18}	4.9717e + 03	4.4007e + 03 +	3.9486e + 03 +	1.2936e + 12 −	1.1473e + 04 −
F_{19}	1.3660e + 06	2.6082e + 05 +	1.9595e + 05 +	5.1468e + 08 −	2.6684e + 05 +
F_{20}	9.7026e + 02	1.0580e + 03 −	1.0753e + 03 −	1.4580e + 12 −	1.0546e + 03 −
	+	11	11	0	5
	−	9	9	19	14
	=	0	0	1	1

6 Conclusions

Large scale global optimization problems remain challenging for heuristics. Inspired by the good performance of CMA-ES as well as the global best guided strategy, this study proposes to incorporate global best individual information into CMA-ES so as to improve the performance of CMA-ES. The proposed algorithm GCMA-ES is compared to CMA-ES on the CEC 2010 LSGO benchmarks. Experimental results show the in general GCAM-ES outperform CMA-ES. However, the introduced parameter *GGR* plays a crucial role on the performance of GCMA-ES.

With respect to future studies, first we would like to investigate how to design a self-adaptive strategy to control *GGR* during the search process. Second, we would like to incorporate the GCMA-ES into a cooperative co-evolutionary framework to better tackle LSGO problems. Third, it is also worth incorporating GCMA-ES into multi-objective evolutionary algorithms [25–29] to handle large-scale multi-objective problems.

Acknowledgements. The work is financially supported by the National Science Foundation of China (No. 71571187, 71201170, 61403404) and the Distinguished Natural Science Foundation of Hunan Province (2017JJ1001).

References

1. Zhang, Y., Zhang, T., Wang, R., Liu, Y., Guo, B.: Optimal operation of a smart residential microgrid based on model predictive control by considering uncertainties and storage impacts. J. Solar Energy. **122**, 1052–1065 (2015)
2. Zhang, Y., Liu, B., Zhang, T., Guo, B.: An intelligent control strategy of battery energy storage system for microgrid energy management under forecast uncertainties. J. Int. J. Electrochem. **9**, 4190–4204 (2014)
3. Mei, Y., Omidvar, M.N., Li, X., Yao, X.: Competitive divide-and-conquer algorithm for unconstrained large scale black-box optimization. J. ACM Trans. Math. Softw. **42**(2), Article 13 (2014)
4. Omidvar, M.N., Li, X., Mei, Y., Yao, X.: Cooperative co-evolution with differential grouping for large scale optimization. J. IEEE Trans. Evol. Comput. **18**(3), 378–393 (2014)
5. Sun, Y., Kirley, M., Halgamuge, S.K.: Extended differential grouping for large scale global optimization with direct and indirect variable interactions. In: Genetic and Evolutionary Computation Conference, pp. 313–320 (2015)
6. Yang, Z., Tang, K., Yao, X.: Self-adaptive differential evolution with neighborhood search. In: 2008 IEEE Congress on Evolutionary Computation (CEC), pp. 1110–1116 (2008)
7. Yang, Z., Tang, K., Yao, X.: Scalability of generalized adaptive differential evolution for large scale continuous optimization. J. Softw Comput. **15**(11), 2141–2155 (2011)
8. Liu, J.P.: CMA-ES and decomposition strategy for large scale continuous optimization problem. D. University of Science and Technology of China (2014)
9. Li, X., Yao, X.: Cooperatively coevolving particle swarms for large scale optimization. J. IEEE Trans. Evol. Comput. **16**(2), 210–224 (2012)
10. Mahdavi, S., Shiri, M.E., Rahnamayan, S.: Meta-heuristics in Large scale global continues optimization: a survey. J. Inf. Sci. **295**, 407–428 (2015)
11. Maesani, A., Iacca, G., Floreano, D.: Memetic viability evolution for constrained optimization. J. IEEE Trans. Evol. Comput. **20**(1), 125–144 (2016)
12. Niu, Y.Y.: The research on evolutionary algorithm ensembles for global numerical optimization problems. D. Central South University (2014)
13. Yang, Z.: Nature inspired real-valued optimization and applications. D. University of Science and Technology of China (2010)
14. Melo, V.V., Iacca, G.: A modified covariance matrix adaptation evolution strategy with adaptive penalty function and restart for constrained optimization. J. Expert Syst. Appl. **41**, 7077–7094 (2014)
15. Ghosh, S., Das, S., Roy, S., Minhazul Islam, S.K., Suganthan, P.N.: A differential covariance matrix adaptation evolutionary algorithm for real parameter optimization. J. Inf. Sci. **182**, 199–219 (2012)
16. Hansen, N., Ostermeier, A.: Completely derandomized self-adaptation in evolution strategies. J. Evol. Comput. **9**(2), 159–195 (2001)
17. Hansen, N., Müller, S.D., Koumoutsakos, P.: Reducing the time complexity of the derandomized evolution strategy with covariance matrix adaptation (CMA-ES). J. Evol. Comput. **11**(1), 1–18 (2003)

18. Shir, O.M., Emmerich, M., Bäck, T.: Self-adaptive niching CMA-ES with mahalanobis metric. In: 2007 IEEE Congress on Evolutionary Computation, pp. 820–827 (2007)
19. Eberhart, R., Shi, Y.: Particle swarm optimization: developments, applications and resources. In: IEEE Congress on Evolutionary Computation, pp. 81–86 (2001)
20. Mokan, M., Sharma, K., Sharma, H., Verma, C.: Gbest guided differential evolution. In: Industrial and Information Systems (ICIIS), pp. 1–6 (2014)
21. Jadhav, H.T., Roy, R.: Gbest guided artificial bee colony algorithm for environmental/economic dispatch considering wind power. J. Expert Syst. Appl. **40**, 6385–6399 (2013)
22. Mirjalili, S., Lewis, A.: Adaptive gbest-guided gravitational search algorithm. J. Neural Comput. Appl. **25**, 1569–1584 (2014)
23. Jadhav, H.T., Sharma, U., Patel, J., Roy, R.: Gbest guided artificial bee colony algorithm for emission minimization incorporating wind power. In: 11th IEEE International Conference on Environment and Electrical Engineering, pp. 1064–1069 (2012)
24. Tang, K., Li, X., Suganthan, P.N.: Benchmark functions for the cec 2010 special session and competition on large-scale global optimization. Technical report, Nature Inspired Computation and Applications Laboratory, USTC, China (2009). http://nical.ustc.edu.cn/cec10ss.php
25. Wang, R., Purshouse, R.C., Fleming, P.J.: Preference-inspired co-evolutionary algorithms for many objective optimization. J. IEEE Trans. Evol. Comput. **17**(4), 474–494 (2013)
26. Wang, R., Purshouse, R.C., Fleming, P.J.: Preference-inspired co-evolutionary algorithms using weight vectors. J. Eu. J. Oper. Res. **243**(2), 423–441 (2015)
27. Wang, R., Fleming, P.J., Purshouse, R.C.: General framework for localised multi-objective evolutionary algorithms. J. Inf. Sci. **258**, 29–53 (2014)
28. Wang, R., Zhang, Q.F., Zhang, T.: Decomposition based algorithms using Pareto adaptive scalarizing methods. J. IEEE Trans. Evol. Comput. **20**(6), 821–837 (2016)
29. Wang, Y., Wang, B.C., Li, H.X., Yen, G.G.: Incorporating objective function information into the feasibility rule for constrained evolutionary optimization. J. IEEE Trans. Cybern. **46**(12), 2938–2952 (2016)

A Fast Approximate Hypervolume Calculation Method by a Novel Decomposition Strategy

Weisen Tang, Hailin Liu(✉), and Lei Chen

Guangdong University of Technology, Guangzhou, China
hlliu@gdut.edu.cn

Abstract. In this paper, we present a new method to fast approximate the hypervolume measurement by improving the classical Monte Carlo sampling method. Hypervolume value can be used as a quality indicator or selection indicator for multiobjective evolutionary algorithms (MOEAs), and thus the efficiency of calculating this measurement is of crucial importance especially in the case of large sets or many dimensional objective spaces. To fast calculate hypervolume, we develop a new Monte Carlo sampling method by decreasing the amount of Monte Carlo sample points using a novel decomposition strategy in this paper. We first analyze the complexity of the proposed algorithm in theory, and then execute a series experiments to further test its efficiency. Both simulation experiments and theoretical analysis verify the effectiveness and efficiency of the proposed method.

Keywords: Hypervolume · MOEAs · Multiobjective optimization · Performance metrics

1 Introduction

In this paper we consider the problem of approximate hypervolume of the space dominated by a set of *d*–dimensional points. This hypervolume is often used as quality indicator in multi–objective evolutionary algorithms (MOEAs). Multi–objective evolutionary algorithms such as NSGA–II [1], MOEA/D [2, 3] have been successfully applied in various bi–objective and tri–objective optimization scenarios. However, they all appear to encounter difficulties when the number of objectives increases to more than three, which is known as many–objective optimization problems (MaOPs) [4].

As a consequence, researchers have tried to develop some alternative methods, and one of them is to use set quality measures [5], or quality indicators as the measurement to select the next generation population. Hypervolume [6] (or S–metric, Lebesgue) measure is one of the most popular quality indicator that can be fully sensitive to Pareto dominance and population diversity even when more than three objectives are involved. It was originally proposed and employed by Zitzler et al. [6] to quantitatively compare the outcomes of different EMO algorithms. In [6], the indicator was denoted as ‘size of the space covered’, and later also termed as ‘S–metric’, ‘hypervolume indicator’ [7], and hypervolume measure [8].

D.-S. Huang et al. (Eds.): ICIC 2017, Part I, LNCS 10361, pp. 14–25, 2017.
DOI: 10.1007/978-3-319-63309-1_2

It has been identified that the computational effort of computing hypervolume metric increases exponentially with the increasing number of objectives. Almost all the recent studies about hypervolume indicator are about how to fast calculate the hypervolume.

The "HV4D" algorithm [10, 11] is the most efficient in the case $d = 4$, see the computational results e.g. [12]. Above $d = 4$, the walking fish group (WFG) [13] and quick hypervolume (QHV) [12] are currently two of the most efficient with the computational experiments. Although both the complexity of them are exponential, they have acceptable runtime within $d = 14$.

Recently, some improved version about these two algorithms were proposed. Quick hypervolume was extended in computing the contributor of each point and speed it up with parallel computation [15]. Jaszkiewicz used a similar scheme as in the original quick hypervolume algorithm called QHV–II [14]. IWFG algorithm was proposed in [23] which used the adaptive slicing scheme, it processes less than a second that sets containing a thousand points in 10–13 objectives. On the other hand, Lacour [16] proposed another approach called 'HBDA' to calculate the computation of the hypervolume indicator. There are a couple of other methods that partly perfect the analysis and provement of computing runtime, e.g. [9, 17, 24].

In this paper, we mainly discuss not the exact hypervolume but the approximation. A first attempt in this direction presented in [18, 22]. The main idea is to estimate–by means of Monte Carlo simulation–the ranking of the individuals that is induced by the hypervolume indicator and not to determine the exact indicator values in the optimization problems. In [25] there are some performance analysis about this approximation method and hence proved the effectiveness of Monte Carlo simulation. This method is widely used such as HypE [19], SMS–EMOA [8], but few researchers give it a performance boost (e.g. [21] is one of them).

We design an new approximate hypervolume calculation method which greatly improve the approximate speed. The main idea is to decrease the number of Monte Carlo sample points under the precondition of retaining the precision by a novel decomposition strategy. This novel decomposition strategy is also an exact hypervolume calculation method and is similar to QHV but different. The accordingly complexity analyses will be given and this method will compare with the previous Monte Carlo simulation method [22] to show the faster calculating speed.

This paper is organized as follows. Section 2 briefly introduces the many–objective optimization problems and some basic concept of hypervolume. Section 3 gives a detail description of the proposed hypervolume calculation method. Some theoretical analysis proposes in Sect. 4. At last we conduct simulation experiments in 8, 10, 13 dimensional space respectively, and compare with the Monte Carlo simulation method in Sect. 5 to verify the effectiveness and efficiency of proposed method. Section 6 concludes this paper.

2 Problem Definition

Without loss of generality, the multi–objective optimization problem can be stated as follows

$$\begin{aligned} \min F(x) &= (f_1(x), f_2(x), \ldots, f_d(x))^T \\ s.t. x &\in \Omega. \end{aligned} \tag{1}$$

Where $\Omega \subset R^m$ is the decision space and m is the dimensionality of the decision variable x. $F : \Omega \to R^d$ consists d real–value objective functions and R^d is called the objective space. When $d \geq 4$, the problem (1) is regarded as a many–objective optimization problem.

Let $u = (u_1, \ldots, u_d)^T$ and $v = (v_1, \ldots, v_d) \in R^d$ are images of two solutions in the objective space, u is said to dominate v ($u \prec v$) if and only if $u_i \leq v_i$ for all $i = 1, \ldots, d$ and $u \neq v$. x^* is called Pareto optimal solution if there is no solution $x \in \Omega$ such that $F(x) \prec F(x^*)$. The set of all Pareto optimal solutions in Ω is denoted as $E(f, D)$ and we called that Pareto Front (PF) in the objective space.

According to the definition above, in the d–dimensional objective space, we define the dominated region namely hypervolume measure for given any set $\mathcal{F}$ of n points as follows

$$HV(\mathcal{F}) = VOL\left(\bigcup_{(x_1, x_2, \cdots, x_d) \in \mathcal{F}} [x_1, b_1] \times \ldots \times [x_d, b_d] \right) \tag{2}$$

where $(b_1, b_2, \cdots, b_d)$ is a bound of $\mathcal{F}$ which is dominated by all the solutions in $\mathcal{F}$ and $VOL(x) = \prod_{i=1}^{d} (x_i)$. Similarly, we define $VOL(x|b) = \prod_{i=1}^{d} (b_i - x_i)$ for convenience. Actually, hypervolume measure is the volume of the union of the boxes which is made up by the region between a point and the coordinate axis. This metric value is to be maximized, the larger measure of a population, the better it is.

3 The Proposed Fast Approximate Hypervolume Algorithm

In this section, we describe how our algorithm to fast calculate the approximate hypervolume indicator in high dimensions objective space. The main idea of our algorithm is that we use a novel decomposition strategy called "partial precision and partial approximation (PPPA)" to decrease the number of points that need to be sampled for hypervolume calculation.

3.1 A Calculating Exact Hypervolume Algorithm

First of all, we will introduce a pivot divide and conquer method which provides accurate calculation of hypervolume measure, and it works as follows:

1. Select a pivot point which matches certain definition or condition, calculate the volume of pivot point and add up the hypervolume measure;
2. To produce the sub–problems, we divide the space according to the pivot, classify other points into the possible space regions;
3. Recursively solve each of the sub–problems.

The methodology of this method is similar to QHV and QHV–II, but in a different way. The differences are mainly reflected in the splitting way and the pivot selecting way. Figure 1 illustrates the proposed method in an 3–dimension example for maximize objectives. In Fig. 1, each dimension d_i of pivot point would be segmented. The points which greater than the *Pivot* in one dimension will divide into two parts, one part get added to subset $\mathcal{Q}_i$, while the other part join the segmentation of next segmentation. The method is presented in Algorithm 1, where `F[i]` represents the individual i in population $\mathcal{F}$ and we denote as $\mathcal{F}^i$.

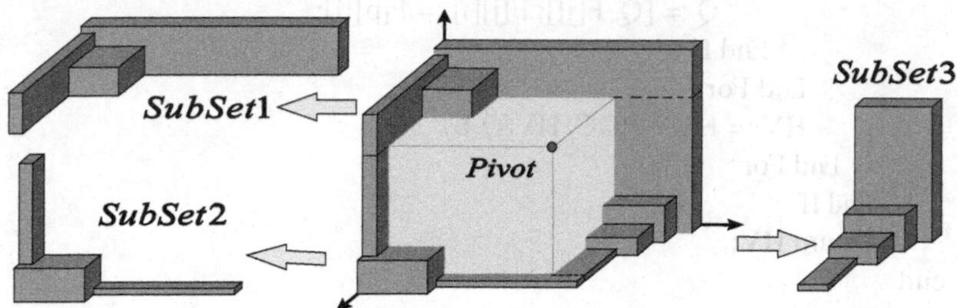

Fig. 1. Illustration of dividing a set in 3D case with Maximizing in three objectives relative to the origin.

The function `FindPivot(F,B)` in line 5 is to find the pivot. In Algorithm 1, the pivot is produced by the maximum–minimum method because this point is the mid-point of the set. That means $\mathcal{F}^p$ satisfies

$$\begin{aligned} g^{te}(\mathcal{F}^i) &= max\{\mathcal{F}_1^i, \mathcal{F}_2^i, \cdots, \mathcal{F}_d^i\} \\ g^{te}(\mathcal{F}^p) &= min\{g^{te}(\mathcal{F}^1), \cdots, g^{te}(\mathcal{F}^n)\}. \end{aligned} \tag{3}$$

In Algorithm 2, maximum volume point is chosen as pivot as it would reduce volume which needs to be approximate as larger as possible. In other words, $\mathcal{F}^p$ satisfies following equation

$$VOL(\mathcal{F}^p|b) = max\{VOL(\mathcal{F}^1|b), \cdots, VOL(\mathcal{F}^n|b)\}. \tag{4}$$

Algorithm 1. Pivot Divide to Calculate Hypervolume in d Dimension

```
PDC_HV(F,B)
  Input
    F: A set of n points;
    B: the point of computational boundary in Eq.(2);
  Output
    HV: the value of hypervolume measure for F;
  Begin
    If(n == 1)
      HV := VOL(F[1],B);        //Direct calculation of volume
    Else
      p := FindPivot(F,B);      //Find the pivot
      HV := VOL(F[p],B);
      For(i = 1 to d)
        Q := [];                //Initialize the subset of F
        For(j = 1 to n)
          If(F[j][i] < F[p][i])
            F[j][i] := B[i] – (F[p][i]- F[j][i]);
            Q := [Q, F[j]]; F[j][i] := F[p][i];
          End If
        End For
        HV := HV + PDC_HV(Q,B)
      End For
    End If
    Return HV;
end
```

3.2 The Proposed Approximate Hypervolume Method in High Dimension

We introduce a new approximate hypervolume method according to above description. Notice that the Algorithm 1 is also a good strategy to approximate hypervolume measure. After k layers recursion, d^k subsets can be obtained. Besides every recursion, this method would calculate a new volume of pivot and add to total volume of hypervolume. Therefore, we have the following formula

$$HV(\mathcal{F}) = \sum VOL(Pivot|b) + \sum_{i=1}^{d^k} HV(\mathcal{Q}_i) = V_p + V_{\mathcal{Q}} \tag{5}$$

Where $V_p = \sum VOL(Pivot|b)$ is the sum of the volume of all pivots, $\mathcal{Q}_i$ traverses all subsets which is obtained by the k–layer segmentation and $V_{\mathcal{Q}}$ is the sum of the hypervolume measure of all subsets. That is why we call this novel decomposition strategy "partial precision and partial approximation". Because after k layers recursion, Eq. (5) could be regarded as an equation with error term that approximate hypervolume measure. The former is the precise calculation of partial hypervolume and the latter is error term. Obviously, the former is easy to calculate for it is the multiplication of a series of number continuously, and the latter is difficult to calculate. Thus, a new method is proposed to approximate to the error term of Eq. (5).

In Eq. (5), k represents the layer of recursion that needs be presented. The larger the k value is, the more the layers of recursion is and the more accurate results are. But it also the more complex of computation because the more subsets will be generated. According to Eq. (5), the volume of this series of subsets have been reduced and then we can approximately compute them by the method as follows.

Algorithm 2. Partial Precision and Partial Approximation of Hypervolume

```
PPPA_HV(F,B,p,Iter)
  Input
    F and B is the Same as Algorithm 1;
    p: the number of sample points in unit volume;
    Iter: the layer of recursion;
  Output
    Same as Algorithm 1;
  Begin
    If(Iter <= k)
      Execute Algorithm 1 line 4 to 13;
      HV := HV + PPPA_HV(F,B,p,Iter+1);
    Else
      S := [];                    //Initialize the set of Sample
      For(i = 1 to n)
        Delete points in S which is dominated by F[i];
        s := p * VOL(F[i],B); //Sample size of F[i]
        Produce S' which Contains s points within F[i];
        S := [S, S'];
      End For
    End If
    Return |S|/p;
end
```

We now consider the Monte Carlo simulation method which is mentioned in [18, 19]. If the set $\mathcal{F}$ accounts for a small proportion in sample range $[0, b_1] \times \ldots \times [0, b_d]$, most of the sample points will tend to be beyond the set $\mathcal{F}$, and it will waste computation effort. The subsets $\mathcal{Q}_i$ obtained by Eq. (5) also have this situation. Therefore, Monte Carlo simulation has disadvantage in computing the above mentioned subsets. To overcome this, we develop a new method to compute the hypervolume measure of this class of subsets like $\mathcal{Q}_i$. The method associates the numbers of sample points with the size of hypervolume, and hence can reduce the numbers of sample points. It is described in Algorithm 2, where k is the default of max recursive layer, which often greater than 2, the parameter *Iter* is set as 1 and $|\mathcal{S}|$ represents the number of $\mathcal{S}$.

It is obvious that the complexity of the Algorithm 2 is highly related to the size of volume. The complexity of Algorithm 2 will be discussed in the next section. The essential principle of this method is the relationship of density and volume. As is well known, the mass equals the density times the volume. With density unchanged, we add points to the dominated region which is consisted by the set $\mathcal{Q}_i$ and the boundary b, until the dominated region is saturated. Then the hypervolume measure equals the numbers of total points in the dominated region divided by the density. It can be concluded that when the hypervolume measure of a set $\mathcal{F}$ satisfies $VOL(\mathcal{F}^i|b) < 1$, the algorithm would need less number of sample points, and computing efficiency can be increased.

4 Theoretical Analysis of the Proposed Algorithms

We prove that the complexity of Algorithm 2 is $O(dn\rho(V_{\mathcal{Q}})^{1+\varepsilon} + (dn)^k)$, where the definition of $V_{\mathcal{Q}}$ is similar to the last term (error term) of Eq. (5), ε is an arbitrarily small number and ρ is the sample points per unit volume. The greater the ρ, the more accurate the measured value will be, and vice versa.

Proposition 1. As described, the computational complexity of Algorithm 2 is $O(dn\rho(V_{\mathcal{Q}})^{1+\varepsilon} + (dn)^k)$.

Proof(of proposition). In general, let the sample boundary be $(b_1, \cdots, b_d) = (1, \cdots, 1)$ because arbitrary point set can be mapped in $[0, 1]^d$ by scaling. Let $\mathcal{F} = \{x^1, x^2, \cdots, x^n\}$, $V_i = VOL(x^i|b)$, $V_i < 1$, $i = 1, 2, \cdots, n$. Notice that in Algorithm 2 if we guarantee

$$\frac{S_1}{V_1} = \frac{S_2}{V_2} = \cdots = \frac{S_n}{V_n} = \rho$$

And then we uniformly generate $S_i (i = 1, 2, \cdots, n)$ sample points in the sample range $[x^i, b]^d$ and add into the set $\mathcal{S}$. So we just need to compare the dominated relationship between x^i and the sample points which in the $\mathcal{S}$, if the set $\mathcal{S}$ contains $|\mathcal{S}|$ sample points, the computational complexity is $O(|\mathcal{S}|d)$. Besides, because $|\mathcal{S}|/\rho$ converges to the measure value hv of hypervolume which means $||\mathcal{S}|/\rho - hv| < \varepsilon$. Therefore $|\mathcal{S}| < \rho(hv)^{1+\varepsilon}$ can be deduced. Such an operation have to be repeated n times, so the complexity of approximate calculation in Algorithm 2 is $O(dn\rho(hv)^{1+\varepsilon})$.

According to Algorithm 1, the cost of one time segmentation is $O(dn)$. A segmentation can generated d subsets. After k layers segmentation, d^k subsets are generated and the cost of segmentation is equal to $1 + d + \cdots + d^{k-1} = O((dn)^k)$.

We suppose that after segmentation the hypervolume measure of subset Q_i is hv_i, then the cost of calculate a subset Q_i is $O(dn\rho(hv_i)^{1+\varepsilon})$. Meanwhile the cost of calculate all subsets $Q_1, Q_2, \cdots, Q_{d^k}$ is

$$\sum_{i=1}^{d^k} O(dn\rho(hv_i)^{1+\varepsilon}) < O\left(dn\rho\left(\sum_{i=1}^{d^k}(hv_i)^{1+\varepsilon}\right)\right) < O(dn\rho(V_Q)^{1+\varepsilon}) \tag{6}$$

So the complexity of Algorithm 2 is $O(dn\rho(V_Q)^{1+\varepsilon} + (dn)^k)$.

5 Experimental Studies

In order to demonstrate the feasibility and effectiveness of the proposed method, we do a series of comparative experiments. To be specific, the proposed Algorithm 2 is compared to Monte Carlo simulation method in terms of speed and accuracy. We calculate each instances set 20 times from 8 to 13 dimension, instances include the following four instance types:

(C) Concave or so-called spherical instances;
(X) Convex instances;
(L) Linear instances;
(D) Degeneration instances;

Instances are obtained by drawing uniformly points from the open hypercube $(0, 1)^d$. Then each point z is modified as follows:

(C) $z_j \leftarrow \frac{z_j}{\sqrt{\sum_{k=1}^{d} z_k^2}}$, for each $j \in \{1, \ldots, d\}$, i.e. the component values of z are divided by their ℓ_2–norm.

(X) $z_j \leftarrow 1 - \frac{z_j}{\sqrt{\sum_{k=1}^{d} z_k^2}}$, for each $j \in \{1, \ldots, d\}$.

(L) $z_j \leftarrow \frac{z_j}{\sum_{k=1}^{d} z_k}$, for each $j \in \{1, \ldots, d\}$, i.e. the component values of z are divided by their ℓ_1–norm.

(D) $z_j \leftarrow \frac{1}{l}\sum_{k=1}^{l} z_k$, for each $j \in \{1, \ldots, l\}$, then $z_j \leftarrow \frac{z_j}{\sqrt{\sum_{k=1}^{d} z_k^2}}$, for each $j \in \{1, \ldots, d\}$, l is the degree of degradation, in experiment we let $l = \lceil \frac{d}{2} \rceil$.

We followed the suggestion of Russo and Francisco [12] to project uniformly distributed points on a hypersphere. In our study, 10, 000, 000 sample points per unit volume are used to ensure the accuracy, and the max segmentation recursive layer k is set 2. We use the following formula to calculate the accuracy

$$Accuracy = \frac{|hv^* - HV(\mathcal{F})|}{HV(\mathcal{F})}$$

Where $HV(\mathcal{F})$ is the exact hypervolume measure of the set $\mathcal{F}$ and hv^* is the approximation by proposed Algorithm 2 or Monte Carlo simulation. The results are shown in Fig. 2 and Table 1, Fig. 2 shows the running time comparison, and Table 1 shows the accuracy comparison for four kinds of different non-dominated sets.

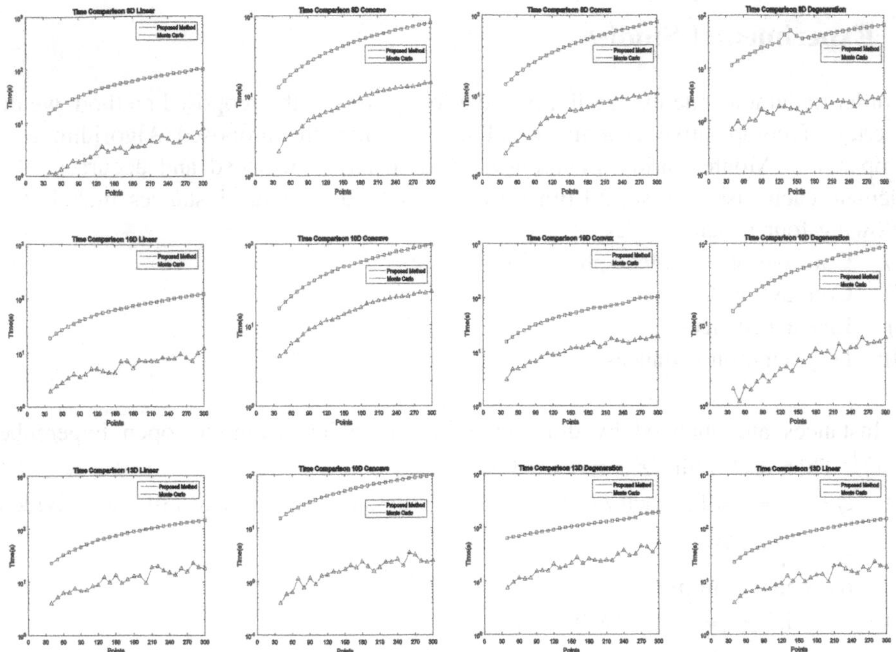

Fig. 2. The running time comparison for 8D, 10D, 13D with different instances set (seconds).

From those figures and table, we can conclude that our approximation algorithm outperforms Monte Carlo simulation in both speed and accuracy. The runtime of our algorithm is 10 times or more faster than Monte Carlo simulation, we only takes less than 10 s to perform a calculation and Monte Carlo simulation needs about 100 s or more. Besides, the accuracy of our algorithm is also far better than Monte Carlo. When the dimension is higher than 13, Monte Carlo simulation couldn't meet the requirement of accuracy, but our proposed algorithm is constant no matter what set it is.

We also have several disadvantages, As can be seen from the figure, our algorithm is not quite stable and the runtime would be fluctuate. Despite the fluctuation is in a certain range, it can't be predict perfectly. The reason is that our algorithm is influenced by the distribution and the number of the set. Obviously, this problem won't happen to Monte Carlo simulation.

Table 1. The accuracy comparison in different instances set(%).

Instances		Linear		Convex	
Algorithms		Proposed algorithm	Monte Carlo	Proposed algorithm	Monte Carlo
8D	100	5.617E-05	7.901E-05	1.034E-04	2.078E-04
	200	1.041E-04	7.350E-05	1.172E-04	2.336E-04
	300	4.763E-06	8.488E-05	3.945E-04	9.869E-05
10D	100	1.150E-04	1.051E-04	1.847E-04	1.013E-04
	200	1.074E-04	8.830E-05	7.575E-05	3.284E-04
	300	1.030E-04	1.084E-04	2.444E-04	2.764E-04
13D	100	1.653E-04	1.458E-04	1.412E-04	2.653E-01
	200	5.918E-06	1.364E-04	1.620E-04	5.381E-01
	300	4.447E-05	1.546E-04	1.029E-04	2.832E-01
Instances		Concave		Degeneration	
Algorithms		Proposed algorithm	Monte Carlo	Proposed algorithm	Monte Carlo
8D	100	1.341E-04	6.840E-04	4.200E-05	7.664E-03
	200	1.028E-04	7.542E-04	4.535E-05	3.335E-02
	300	1.601E-04	5.772E-04	1.618E-04	2.803E-02
10D	100	6.850E-05	1.537E-03	5.371E-05	1.740E-01
	200	2.349E-04	2.207E-03	6.587E-05	1.539E-01
	300	2.089E-04	1.490E-03	2.889E-04	3.316E-01
13D	100	5.474E-04	1.531E-01	8.870E-04	4.279E-02
	200	8.154E-05	2.601E-03	3.587E-05	1.539E-01
	300	4.089E-04	1.287E-01	5.933E-04	3.240E-01

6 Conclusion

In this paper, we propose a new method to approximate the hypervolume value, which can effectively decrease the running time of hypervolume indicator calculation in high dimension objective space. The performance of the proposed approximation hypervolume calculation method is verified by comparing it with the classical Monte Carlo sampling method on PFs of widely-used MaOPs test problems.

Acknowledgment. This work was supported in part by the National Natural Science Foundation of China under Grant 61673121, in part by the Projects of Science and Technology of Guangzhou under Grant 201508010008, and in part by the China Scholarship Council.

References

1. Deb, K.: Multi-Objective Optimization Using Evolutionary Algorithms, vol. 2(3), pp. 509. John Wiley & Sons, Inc., New York (2001)
2. Zhang, Q., Li, H.: MOEA/D: A multiobjective evolutionary algorithm based on decomposition. IEEE Trans. Evol. Comput. **11**(6), 712–731 (2007)

3. Liu, H., Gu, F., Zhang, Q.: Decomposition of a multiobjective optimization problem into a number of simple multiobjective subproblems. IEEE Trans. Evol. Comput. **18**(3), 450–455 (2014)
4. Schutze, O., Lara, A., Coello, C.A.C.: On the influence of the number of objectives on the hardness of a multiobjective optimization problem. IEEE Trans. Evol. Comput. **15**(4), 444–455 (2011)
5. Zitzler, E., Künzli, S.: Indicator-based selection in multiobjective search. In: Yao, X., et al. (eds.) PPSN 2004. LNCS, vol. 3242, pp. 832–842. Springer, Heidelberg (2004). doi:10.1007/978-3-540-30217-9_84
6. Zitzler, E., Thiele, L.: Multiobjective optimization using evolutionary algorithms — a comparative case study. In: Eiben, A.E., Bäck, T., Schoenauer, M., Schwefel, H.-P. (eds.) PPSN 1998. LNCS, vol. 1498, pp. 292–301. Springer, Heidelberg (1998). doi:10.1007/BFb0056872
7. Zitzler, E., Thiele, L., Laumanns, M., Fonseca, C.M., Grunert da Fonseca, V.: Performance assessment of multiobjective optimizers: an analysis and review. IEEE Trans. Evol. Comput. **7**(2), 117–132 (2003)
8. Beume, N., Naujoks, B.: SMS–MOA: Multiobjective selection based on dominated hypervolume. Eur. J. Oper. Res. **181**(3), 1653–1669 (2007)
9. Bringmann, K., Friedrich, T.: Approximating the volume of unions and intersections of high-dimensional geometric objects. In: Hong, S.-H., Nagamochi, H., Fukunaga, T. (eds.) ISAAC 2008. LNCS, vol. 5369, pp. 436–447. Springer, Heidelberg (2008). doi:10.1007/978-3-540-92182-0_40
10. Cox, W., While, L.: Improving and extending the HV4D algorithm for calculating hypervolume exactly. In: Kang, B.H., Bai, Q. (eds.) AI 2016. LNCS, vol. 9992, pp. 243–254. Springer, Cham (2016). doi:10.1007/978-3-319-50127-7_20
11. Guerreiro, A.P., Fonseca, C.M., Emmerich, M.T.: A fast dimension–sweep algorithm for the hypervolume indicator in four dimensions. In: Proceedings of the 24th Canadian Conference on Computational Geometry (CCCG 2012), pp. 77–82 (2012)
12. Russo, Francisco, A.P.: Quick hypervolume. IEEE Trans. Evol. Comput. **18**(4), 481–502 (2014)
13. While, L., Bradstreet, L., Barone, L.: A fast way of calculating exact hypervolumes. IEEE Trans. Evol. Comput. **16**(1), 86–95 (2012)
14. Jaszkiewicz: Improved quick hypervolume algorithm. In: Asilomar Conference on Signals, Systems and Computers (2017)
15. Russo, L.M.S., Francisco, A.P.: Extending quick hypervolume. J. Heuristics **22**(3), 245–271 (2016)
16. Lacour, R., Klamroth, K., Fonseca, C.M.: A box decomposition algorithm to compute the hypervolume indicator. Comput. Oper. Res. (2015)
17. Beume, N., Carlos, M.F.: On the complexity of computing the hypervolume indicator. IEEE Trans. Evol. Comput. **13**(5), 1075–1082 (2009)
18. Bader, J., Deb, K., Zitzler, E.: Faster hypervolume–based search using monte carlo sampling. In: Ehrgott, M., Naujoks, B., Stewart, T., Wallenius, J. (eds.) Conference on Multiple Criteria Decision Making (MCDM 2008). LNEMS, vol. 634, pp. 313–326. Springer, Heidelberg (2008)
19. Bader, J., Zitzler, E.: HypE: An algorithm for fast hypervolume–based many–objective optimization. Evol. Comput. **19**(1), 45–76 (2011)
20. Naujoks, B.: S–metric calculation by considering dominated hypervolume as Klee's measure problem. Evol. Comput. **17**(4), 477–492 (2009)
21. Bringmann, K., Friedrich, T.: Approximation quality of the hypervolume indicator. Artif. Intell. **195**(1), 265–290 (2013)

22. Bringmann, K., Friedrich, T.: Approximating the volume of unions and intersections of high-dimensional geometric objects. In: Hong, S.-H., Nagamochi, H., Fukunaga, T. (eds.) ISAAC 2008. LNCS, vol. 5369, pp. 436–447. Springer, Heidelberg (2008). doi:10.1007/978-3-540-92182-0_40
23. Cox, W., While, L.: Improving the IWFG algorithm for calculating incremental hypervolume. In: IEEE Congress on Evolutionary Computation IEEE, pp. 3969–3976 (2016)
24. Bringmann, K., Friedrich, T.: Parameterized average–case complexity of the hypervolume indicator. In: Conference on Genetic and Evolutionary Computation, pp. 575–582 (2013)
25. Nowak, K., Martens, M., Izzo, D.: Empirical performance of the approximation of the least hypervolume contributor. In: International Conference on Parallel Problem Solving From Nature, pp. 662–671 (2014)

Towards the Differentiation of Initial and Final Retention in Massive Open Online Courses

Raghad Al-Shabandar(✉), Abir Hussain, Andy Laws, Robert Keight, and Janet Lunn

Applied Computing Research Group, School of Computing and Mathematical Sciences, Liverpool John Moores University, Byrom Street, Liverpool L3 3AF, UK
R.N.AlShabandar@2013.ljmu.ac.uk,
{A.hussain,A.Laws,J.Lunn}@ljmu.ac.uk, R.Keight@2015.ljmu.ac.uk

Abstract. Following an accelerating pace of technological change, Massive Open Online Courses (MOOCs) have emerged as a popular educational delivery platform, leveraging ubiquitous connectivity and computing power to overcome longstanding geographical and financial barriers to education. Consequently, the demographic reach of education delivery is extended towards a global online audience, facilitating learning and development for a continually expanding portion of the world population. However, an extensive literature review indicates that the low completion rate is a major issue related to MOOCs. This is considered to be a lack of person to person interaction between instructors and learners on such courses and, the ability of tutors to monitor learners is impaired, often leading to learner withdrawals. To address this problem, learner drop out patterns across five courses offered by Harvard and MIT universities are investigated in this paper. Learning Analytics is applied to address key factors behind participant dropout events through the comparison of attrition during the first and last weeks of each course. The results show that the attrition of participants during the first week of the course is higher than during the last week, low percentages of learners' attrition are found prior to course closing dates. This could indicate that assessment fees may not represent a significant reason for learners' withdrawal. We introduce supervised machine learning algorithms for the analysis of learner retention and attrition within a MOOC platform. Results show that machine learning represents a viable direction for the predictive analysis of MOOCs outcomes, with the highest performances yielded by Boosted Tree classification for initial attrition and Neural Network based classification for final attrition.

Keywords: Machine learning (ML) · Massive Open Line Course (MOOC)

1 Introduction

With progress in Open Educational Resources (OER) advancing from an emerging field towards an increasingly important learning modality, Massive Open Online Courses (MOOCs) have seen dramatically increases in popularity over the last few years within the higher education sector [1]. The highest ranking universities have developed and delivered hundreds of courses, including HarvardX, Khan Academy, and Coursera [1]. MOOCs provide the same quality of learning as the traditional classroom without

D.-S. Huang et al. (Eds.): ICIC 2017, Part I, LNCS 10361, pp. 26–36, 2017.
DOI: 10.1007/978-3-319-63309-1_3

conventional time and geographical restrictions. As a result, learners are able to understand and learn courseware content at their own pace. Through the MOOC platform, learners are connected with an array of learning resources, including video lectures, regular assessments, and content in the form of pdf documents. Additionally, learners can interact with each other through participation in online discussion forums [2].

One of the distinctive features of MOOCs is their instant accessibility, coupled with the elimination of financial, geographical, and educational obstacles. Consequently, the proportion of participants engaging in such courses could increase quickly [1, 2]. For example, the number of participants has rapidly expanded in Harvard online courses, with 1.3 million unique learners engaged in online courses reported at the end of 2014,a low completion rate is the major issue related to MOOCs [2, 3]. Research investigations reveal on average that out of each one million participants in MOOCs, an overwhelming majority of them withdraw from MOOCs prior to completion [2]. Due to lack of face to face interaction between instructors and learners in such courses, it is understandably difficult for instructor's to maintain direct awareness of the reasons for individual learner withdrawals [4].

Learning Analytics (LA) is an emerging field of educational technology. LA approaches have demonstrated beneficial insight into the rate of attrition at an early stage. LA analysis, measures and abstracts comprehensive information about the learner from various aspects, including cognitive, social, and psychological facets to help the decision-maker to effectively reason about learner success and failures [5]. LA methods can provide course instructors further information about learner activity in a virtual environment and help them to tailor material to need of participants [5].

Machine learning is a space of techniques at the intersection of computer science, statistics, and mathematics, that has been subsequently adopted by researchers to predict student retention within virtual class environments [3].

Despite the large number of works reported in the literature for modelling student dropout rates, such models do not take into consideration the underlying factors that drive student withdrawals [4]. In this work, LA is therefore employed to analyse and address key factors behind participant dropout events, providing a window of opportunity in which to apply early stage intervention, thereby preventing such cases of withdrawal. It is hypothesised in this work that such withdrawal events are in fact largely preventable through the observation and analysis of learner behaviours over various time periods.

Machine learning (ML) represents a powerful data intensive approach which we apply within our proposed LA framework. ML is appropriate for the detection of potential patterns of learner attrition from course activity data through the examination of learning behaviour features over time [6]. Moreover, machine learning has the potential scope to infer the underlying emotional state of learners by discovering a latent pattern of learner behavior [1].

In this paper, supervised machine learning approaches will be presented to predict learner retention and attrition parameters in MOOCs platform. The performance of classifier models will be compared using a set of appropriate criteria.

2 Literature Review

MOOCs have attracted the attention of many researchers, with an aim to provide an advantage over traditional classroom environments. Much existing work focuses on participant attrition in MOOCs. In this section we will summarise the work of other researchers towards learner attrition in MOOCs.

The author in ref [3] applies supervised machine learning to predict the likelihood of learner dropout from MOOCs. Feature engineering over time was considered in order to obtain more accurate predication rates [3]. Other researchers emphasise forum posts as a prominent recourse of information for dropout analysis in MOOCs. In such works, the author in reference [7] adopts a sentiment analysis approach considering only forum post as the main criteria for analysis. The work considers the daily data of user forum posts and undertakes analysis in order to evaluate participant opinions regarding the quality of teaching, learning material, and peer-assessment. The results show a significant association between learner sentiment and attrition rate.

Although forum posts act as a major factor affecting attrition rates, it has been observed that around 5–10% of registrants participate in the discussion forums themselves [8]. Consequentially, the narrow focus on the forum post data imposes a critical limit on the generality of the approach, since other important factors such as behavioral activities are not accounted for [9].

The authors in reference [9] applies Support Vector Machines (SVM) and considers only click stream features. A set of features have been extracted from behavioral log data such as the number of times a student undertakes a particular quiz, the number of visits to the course home page, and length of the session [9].

The attrition phenomenon was described by [10] as a funnel of participation. The term funnel of participation emerges from the equivalent concept in marketing (marketing funnel). The funnel of participation approach attempts to describe learners' theoretical stages toward dropout from MOOCs according to four main stages. Such stages are defined as Awareness, Registration, Activity, and progress [10]. The author concludes that the fluctuation of learners behavioral activities leads to withdrawal from online courses.

Discussion threads are used to measure the negative behaviors of learners that lead to demotivate engagement within MOOCs platforms. Two kinds of features have been considered, namely click stream events and discussion threads [10]. Survival models have been developed by [2] for measuring the likelihood of attrition events. Survival model can be described as predictive models that apply logistic regression to infer the probability of learners' survival in the course over time [11] Additionally, feedforward neural networks have been implemented in [11] to predict completion rates in MOOCs, using student sentiments as input. In this case, only the behavioral attributes are used to measure the performance of learners.

3 Methodology

3.1 Data Description

The dataset used in this paper was obtained from Harvard University [12]. Harvard University collaborates with the Massachusetts Institute of Technology (MIT) to deliver high quality MOOCs. The click stream attribute is the main feature of dateset, which represents the number of events that correspond to the user interaction with courseware.

The Nchapters feature represents the number of chapters that participants proceed to read. The Explored feature is a binary discretisation of exploration learners. To become explorer, a participant should click more than half of the course content (chapter) [12]. Nplay_video feature represents specifically the number of events when the learner viewed a particular video. Viewed is also a binary discretised feature, which is encoded as 1 when the participants access the home page of assignments and related videos, or 0 otherwise [12].

The temporal features are an important features used to evaluate how learners activity change over time. The launch Date (course start date) attribute represents the date when course content available online, course wrap date (finish date) represents the date by certificates are issued [13]. There are two set of temporal attribute capture the user interaction activity with course, which are (start_time_DI, last_event_DI) [13]. ndays_act feature represents number of unique days when user interact with course [13]. The dataset also includes the demographic information of learners such as learners' educational levels, age, grade and sex. The final grade were computed by Course works (50%), 2 mid exam (25%) and final exam (25%). The learner must achieve 50% in final grade to be certified [13]. A brief description of dataset is explained in Table 1.

Table 1. Description features of HarvardX

Features	Description
User-Id LOE, YOB, Gande, Grade	Demographic feature of user including User_id, sex, date of birth, GPA and background
Launch Date, wrap date	Date feature describe start and end course date
Nevent nplay_video, Nchapters, nforum_post	Behavioural features including the number of click stream, play video event, interact with chapter
Viewed, Explored	Discrete features encoded as 1/0
Start-time _Di, Last event_DI, ndays_act	Date features describe start and end user interact with course. nday s_act (number of unique days)

3.2 Data Pre-processing

The Harvard University and MIT datasets used in this study captured 5 courses, classified into five types: Computer science, Electronic engineering, History, Chemistry, and Health. Due to the large size of date, we randomly sampled 700,000-log file entries representing the completed learners' activities on MOOCs, where each row represents a single user session. On inspection it was found that the Harvard dataset contains a large number of missing values inclusive of both behavioural and demographic features. To overcome this issue, Multivariate imputation by chained equations (MICE) has been applied [14]. MICE is capable of performing multiple imputations over a set of variables at single step regardless of the type of variables, making it a reasonable choice [14].

Data in the Harvard dataset does not match the normal distribution. Normality of data is a desirable property and may be required in the case of some classes of machine learning models [15]. To handle non normality issue, Box-Cox transformation was used, it is a member of the class of power transform functions, which are used for the efficient conversion of variables to a form of normality, the equalisation of variance, and to enhance the validity of tests for correlated variables [15]. Additionally, we scaled and centered the data through a z-score calculation. Furthermore, imbalanced classes are a notable concern in this dataset. As such, the procedure of Synthetic Minority Oversampling Technique (SMOTE) has been applied to equalise the class proportions through the generation of additional minority class examples [16]. In particular, SMOTE applies a kNN algorithm to interpolate new instances of each minority class through evaluation of its nearest neighbours according to some distance metric.

3.3 Experiments Introduction

The purpose of this study is to estimate the rate of learner dropout from MOOCs in the future. Five courses are considered in this study, provided by Harvard and MIT through the EDX platform in 2012–2013 [13]. The courses differ in both their structure and length. As such, the course material offered by Harvard was delivered on a weekly basis over 12–14 weeks, with MIT conversely releasing all materials at the launch date for each course [13]. Both HarvardX and MITx define successful certification of learners as the completion of weekly course works, followed by a pass mark for a final exam held at the end of the course [13].

The objective of this study is to estimate the learners dropout rate from future courses and additionally to identify the main reasons leading to learner withdrawal. A data-driven approach was used to describe patterns of activity drop off. The features considered comprise "ndays_act", which represents a number of unique days learners interact in the courseware, combined with temporal features. Importantly, there is no imposed limitation of time on learners' access to courseware content. Learners might enrol late in a given course; in addition, learners might withdraw from courses even prior to the completion date. Attrition was defined in terms of two main categories, namely initial and final attrition. A brief explanation of each category is provided below.

- Initial (in/out) state: The aim of drive initial (in/out state) feature examines the rate of participant dropout over the first week. Therefore, only learners who participated in the course since the first-week were considered. The date of learner first activity is compared with course start dates to determine learners who engaged since the beginning of course, and to examine if learners dropout from the course over the first week. The date of first activity compares with last activity if both activities happened in same first week and learners did not interact with course material. In this case, the learner state is defined as out (attrition), otherwise in (retention).

- Final (in/out) state: The aim of drive final (in/out) state feature is to evaluate the learners who enrol late and drop out from a course before the final exam date. In this case, only learners who enrolled after the course start were considered in order to explore if learners drop out of a course before the final exam data. The date of last activity was compared to the course end date. If last activity happened in the same period of course end date, the learner state is defined as out (attrition), otherwise in (retention).

3.4 Exploratory Data Analysis

Exploratory Data Analysis (EDA) was used as a precursor of modelling phase. The aim of EDA is to understand learners activity inuitively, in particular the percentage of withdrawal participants per individual course over time. To compare learner dropout rates over time, quantitative summaries were produced.

The information indicating the number of participants enrolled in courses since the beginning of each course lists in Table 2. In the "Health in Numbers" course, about 23,000 learner participants were enrolled; follow by "Computer Science" with 20,351 entrants. Furthermore, the table shows around 18,409 users participate in "Ancient Greek Hero", followed by 12,566 entrants in the "Circuits & Electronics" course [13]. The minority of learners enrolled in "Solid Chemistry".

Table 2. Numbers of In & Out learners over course first week

Code	Course title	Course acronym	No users	No in users	No out user
1	The Ancient Greek Hero	Ancient Hero	8,409	15,464	2945
4	Health in Numbers: Quantitative Methods in Clinical & Public Health Research	Health in Numbers	23,122	16,701	6421
9	Introduction to Solid State Chemistry	Solid Chemistry	3,094	2648	446
11	Circuits and Electronics	Circuits & Electronics	12,566	11,447	1119
13	Introduction to Computer Science and Programming	Computer Science	20,351	18,588	1763

The number of participants retained in courses following the actual course start dates list Table 3. The number of learners who register late in "Health in Numbers" course is set at 17,475, while the number of learners doubles in the "Computer Science" course. Registered late learners also remains less in both "Ancient Hero" and "Circuits & Electronics" courses. Figures 1 and 2 compare initial retention and attrition with final retention and attrition. It can be noticed that 30% of participants withdrew from "Health in

Numbers". Of the 23,122 entrants, 70% decided to continue to interact over the first week. Conversely, 92% of participant entrolled on the "Computer Science" course continued beyond the first week. Approximately 14% of learners withdrew from "Ancient Hero" course and 10% from the "Circuits & Electronics" within the first week, with last week drop offs of 3% and 2% respectively. An average of 5% and 3% of learners drop off from "Health in Numbers" course and "Computer Science" respectively over last week. In general, the number participant dropouts during the last week of the course are less than that experienced in the first week.

Table 3. Numbers of In & Out learners over course last week

Code	Course title	Course acronym	No users	No in users	No out user
1	The Ancient Greek Hero	Ancient Hero	1,374	11,075	299
4	Health in Numbers: Quantitative Methods in Clinical & Public Health Research	Health in Numbers	17,475	16,645	830
9	Introduction to Solid State Chemistry	Solid Chemistry	3,009	2845	158
11	Circuits and Electronics	Circuits & Electronics	9,523	9,341	182
13	Introduction to Computer Science and Programming	Computer Science	36,462	35,816	746

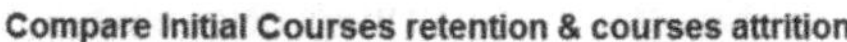

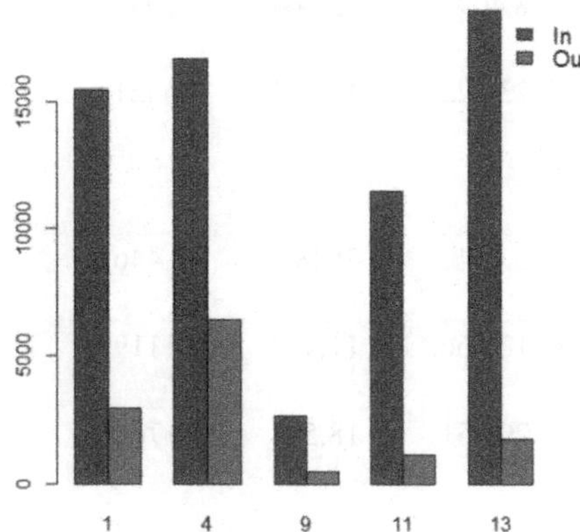

Fig. 1. Initial In/Out courses

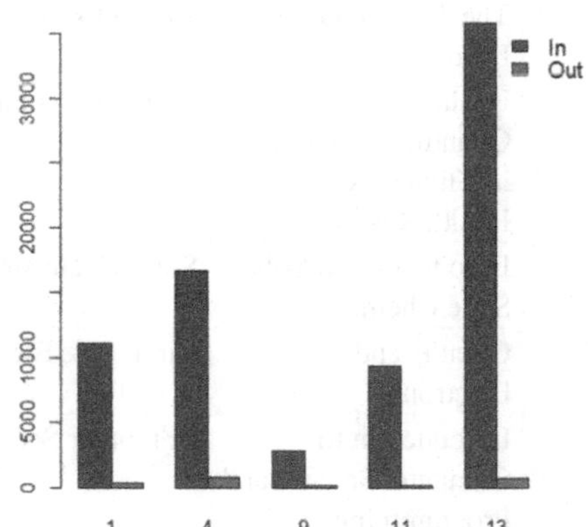

Fig. 2. Final In/Out courses

3.5 Experiments Setup

Two set of experiments are conducted in this study with the aim of predicting learner retention and attrition in MOOCs, over a different period of time. In both sets of experiments, similar the same courses are used to measure learner interaction with course syllabi over time. Only learners who interact with courseware content over the first week are considered in the first experiment. The prediction target is denoted as "Initial", comprising labels {in, out}, facilitating the prediction the of participant retention and attrition for each learner. In the second experiment, the learners who commence participation after course start dates and subsequently dropout prior to the final exam date were considered. The prediction target in this case is denoted "Final", again with possible labels values {in, out}.

Various Ensemble machine learning algorithms, including bagging and boosting techniques, are applied to the classification problem. Table 4 illustrates a brief description of the models used in this study. Ten-fold cross validation where five replicates are used to assess the performance of classifier models. Accordingly, 60% of original dataset were allocated to the cross-validation training set. A further 40% of the data was used as an external test dataset to validate generalization error for each model. The purpose of using Ensemble machine learning in our case study is to enhance the stability of the base classifiers, in particular to reduce the variance and decrease bias. Bootstrap aggregating (bagging) of weak classifiers into strong classifiers is achieved by randomly resampling the original training data of size m into a number of bootstrap samples, which retains the same size of the original dataset. New data points are then classified based on a voting procedure. Boosting leverages a multiplicity of weak base classifiers to form a strong classifier through the use of adaptive reweighting of data during training. Specifically, to obtain improved classification performance, a weight is assigned to each

Table 4. Brief description of ML models

Model	Description	Architecture	Type	Algorithm
NN	Feedforward Neural Network	Units 14-3-2	Nonlinear	Backpropagation
Treebag	Bagged CART	Ensemble DT using Bagging method	Nonlinear	Random subset Features Boot strap
Black boost	Boosted Tree	Ensemble DT using Boosting method	Nonlinear	Classical Gradient Boosting
Amdai	Adaptive Mixture Discriminant Analysis	Generalized Linear Model	Linear	Maximum Likelihood Estimation
GBM	Gradient Boosting Method	Ensemble DT using Boosting method	Nonlinear	Functional Gradient Descent
BagFDA	Bagged Flexible Discriminant Analysis	Ensemble FDA Bagging method	Linear	Maximum Likelihood Estimation
avNNet	Model Averaged Neural Network	Ensemble NN Begging method	Nonlinear	Backpropagation

data point, which is adjusted during the iterative learning process. The weight of data corresponding to misclassified samples increases while the weight of correctly classified sample decreases.

3.6 Result Evaluation and Discussion

A binary classification is performed, where retention is donated as the positive class while attrition is assigned to the negative class. Empirical results over both sets of experiments have been compared in terms of performance metrics comprising accuracy, specificity and sensitivity, precision, recall, and AUC. Tables 5 & 6 show the empirical results obtained for each classifier respectively.

Table 5. Empirical result for classification performance experiment 1

Model	Acc.	Sens.	Spec.	Precision	Recall	AUC
NN	0.8664	0.8580	0.9239	0.9873	0.8580	0.94082
Treebag	0.9464	0.9484	0.9324	0.9831	0.8321	0.98116
Blackboost	0.8085	0.8032	0.8451	0.9727	0.8032	0.897087
Amdai	0.7028	0.7046	0.6901	0.9400	0.7046	0.76550
GBM	0.9225	0.9239	0.9127	0.9865	0.9239	0.97676
bagfda	0.8957	0.9096	0.8000	0.9690	0.9096	0.9303
avNNet	0.8642	0.8535	0.9380	0.9896	0.8535	0.9606

Table 6. Empirical result for classification performance experiment 2

Model	Acc.	Sens.	Spec.	Precision	Recall	AUC
NN	0.89	0.9468	0.2464	0.9402	0.9468	0.7951
Treebag	0.70	0.6941	0.7971	0.9772	0.6941	0.8230
Blackboost	0.85	0.8920	0.4251	0.9510	0.8920	0.8397
Amdai	0.76	0.7728	0.6184	0.9621	0.7728	0.7888
GBM	0.71	0.7114	0.7778	0.9757	0.7114	0.8277
bagfda	0.66	0.6431	0.8744	0.9846	0.6431	0.8275
avNNet	0.72	0.7184	0.7536	0.9733	0.7184	0.8216

Bagged CART acquired the highest accuracy in experiment 1, with a value of 0.94%, while NN gives the best accuracy in experiment 2 where a value of 0.89% is obtained. There is a noticeable difference in accuracy for the boosting models, where GBM obtained higher accuracy than the Boosted Tree in experiment 1, achieving values of 0.92 and 0.80, respectively, while the Boosting tree classifier obtained better accuracy than GBM in experiment 2, yielding values of 0.85 and 0.71. A comparison of bagging models shows that BagFDA yielded slightly higher accuracy than the avNNet model with an average value of 0.89, whereas BagFDA showed the lowest accuracy in experiment 2, obtaining a value of 0.66%. In both sets of experiments, the linear classifier Amdai obtained the lowest average accuracy with values of 0.70 and 0.76, respectively.

Due to the number of learners who drop off from the course during the last week being much less than that of the first week, the True negative (specificity) results over all classifiers in experiment 1 are seen to be significantly higher than in those of experiment 2. In particular, models Treebag, avNNet, NN, and GBM obtained average values of 94%, 93%, 92%, and 91% respectively. Conversely, such models achieved worse specificity in experiment 2, with values of 79%, 75%, 30%, and 77% respectively. The linear model achieved a slightly higher specificity in experiment 1, with a value of 69%.

Receiver Operator Characteristic (ROC) and area Under Curve (AUC) were also considered. Figures 3 and 4 show ROC results for both sets of experiments. The curves are shown to converge to roughly the same semblance on the plot, indicating the similarity of performance across models in experiments 1 and 2, resulting in values around 90%, 80%, with the exception of the Amdai classifier where the lowest AUC values of both experiments were obtained, namely 76% and 78% respectively.

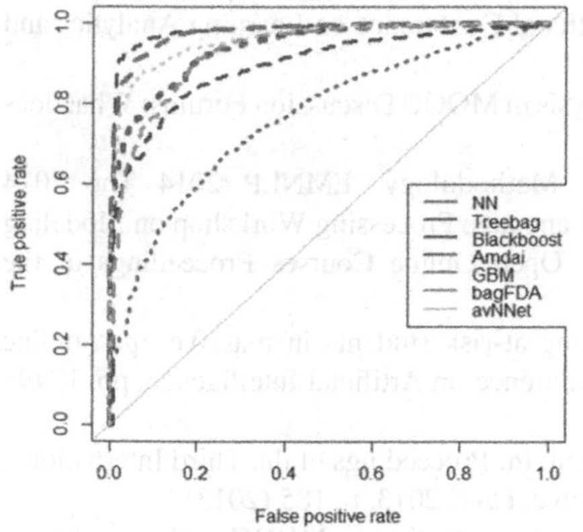

Fig. 3. Roc curve experiment 1

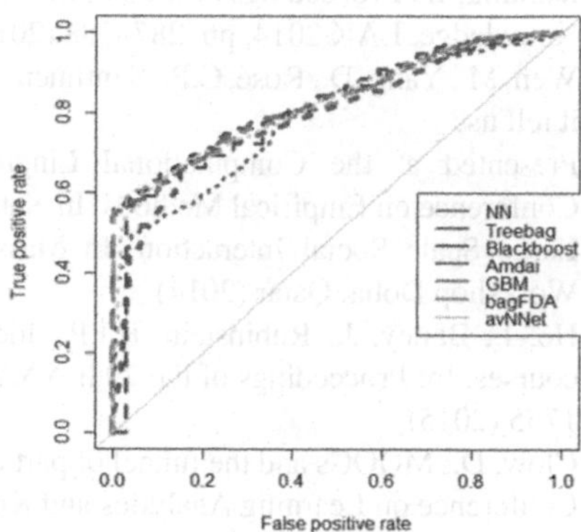

Fig. 4. Roc curve experiment 2

4 Conclusions

The principal focus of this study was to investigate the factors that affect learner dropout rates in MOOCs. Two sets of experiments have been conducted relating to different points of the course lifecycle. In the first experiment learners who enter into courses at the opening date, then subsequently withdraw during the first week were considered. The second experiment focuses on learners who enter after the commencement of courses, who then drop off prior to the final exam. We undertook EDA as prior step to enhance understanding of attrition correlates, indicating that factors such as exam fees are unlikely to constitute a key reason for withdrawal, since few participants attrited from the course during the last week. Machine learning is shown to be a valuable tool for predication of attrition and retention within MOOCs, Result reveal the ML models achieve high average performance across all metrics with range value 80%–90% in experiment 1 whereas, performance metrics fluctuated in experiment 2.

References

1. Qiu, J., Tang, J., Liu, T.X., Gong, J., Zhang, C., Zhang, Q., Xue, Y.: Modeling and predicting learning behavior in MOOCs. In: Proceedings of the Ninth ACM International Conference on Web Search and Data Mining, pp. 93–102 (2016)
2. Yang, D., Rose, C.P.: "Turn on, Tune in, Drop out": Anticipating student dropouts in Massive Open Online Courses. 1–8
3. Kloft, M., Stiehler, F., Zheng, Z., Pinkwart, N., Mattingly, K.D., Rice, M.C., Berge, Z.L.: Predicting MOOC dropout over weeks using machine learning methods. Knowl. Manag. E-Learn. **4**, 60–65 (2014)
4. United Kingdom: Dropout Rates Of Massive Open Online Courses: Behavioural Patterns Mooc Dropout and Completion: Existing Evaluations
5. Baker, R.S.J.D., Siemens, G.: Educational data mining and learning analytics. In: Cambridge Handbook of the Learning Sciences (2014)
6. Gašević, D., Rose, C., Siemens, G., Wolff, A., Zdrahal, Z.: Learning analytics and machine learning. In: Proceedings of the Fourth International Conference on Learning Analytics and Knowledge, LAK 2014, pp. 287–288 (2014)
7. Wen, M., Yang, D., Rosé, C.P.: Sentiment Analysis in MOOC Discussion Forums: What does it tell us?
8. Presented at the Computational Linguistics Methodology: EMNLP 2014 The 2014 Conference on Empirical Methods. In Natural Language Processing Workshop on Modeling Large Scale Social Interaction. In Massively Open Online Courses Proceedings of the Workshop Doha, Qatar (2014)
9. He, J., Bailey, J., Rubinstein, B.I.P.: Identifying at-risk students in massive open online courses. In: Proceedings of the 29th AAAI Conference on Artificial Intelligence, pp. 1749–1755 (2015)
10. Clow, D.: MOOCs and the funnel of participation. In: Proceedings of the Third International Conference on Learning Analytics and Knowledge, LAK 2013, p. 185 (2013)
11. Chaplot, D.S., Rhim, E., Kim, J.: Predicting student attrition in MOOCs using sentiment analysis and neural networks. In: Work. 17th International Confernce on Artificial Intelligence on Education, AIED-WS 2015, vol. 1432, pp. 7–12 (2015)
12. Ho, A.D., Chuang, I., Reich, J., Coleman, C.A., Whitehill, J., Northcutt, C.G., Williams, J.J., Hansen, J.D., Lopez, G., Petersen, R.: HarvardX and MITx: two years of open online courses fall 2012-summer 2014. SSRN Electron. J., 1–37 (2015)
13. Ho, A.D., Reich, J., Nesterko, S.O., Seaton, D.T., Mullaney, T., Waldo, J., Chuang, I.: HarvardX and MITx: the first year of open online courses, fall 2012-summer 2013. SSRN Electron. J., 1–33 (2014)
14. Groothuis-oudshoorn, K.: MICE: multivariate imputation by chained equations in R. J. Stat. Softw. **45**, 1–67 (2011)
15. Osborne, J.W.: Improving your data transformations: applying the box-cox transformation. Pract. Assess. Res. Eval. **15**, 1–9 (2010)
16. Chawla, N.V., Bowyer, K.W., Hall, L.O., Kegelmeyer, W.P.: SMOTE: synthetic minority over-sampling technique. J. Artif. Intell. Res. **16**, 321–357 (2002)

An Evolutionary Algorithm for Autonomous Agents with Spiking Neural Networks

Xianghong Lin(✉), Fanqi Shen, and Kun Liu

College of Computer Science and Engineering, Northwest Normal University, Lanzhou 730070, China
linxh@nwnu.edu.cn

Abstract. Inspired by the evolution of biological brains, the study of neurally-driven evolved autonomous agents has received more and more attention. In this paper, we propose an evolutionary algorithm for neurally-driven autonomous agents, each agent is controlled by a spiking neural network, and the network receives the sensory inputs and processes the motor outputs through the encoded spike information. The controlling spiking neural networks of autonomous agents are developed by the evolutionary algorithms that apply some of genetic operators and selection to a population of agents that undergo evolution. The corresponding food gathering experiment results show that the autonomous agents appear intelligent behaviours for the simulation environment. Additionally, the parameters of networks and agents play an important role in the evolutionary process.

Keywords: Spiking neural network · Evolutionary algorithm · Autonomous agent · Genetic operator

1 Introduction

Natural evolution discovered intelligent brains with billions of neurons and trillions of connections. In order to create an artificial evolutionary system, the researches proposed the approaches that combine evolutionary computation techniques with artificial neural networks [1]. The research of evolutionary neurally-driven autonomous agents is an important area, and still faces difficult conceptual and technical challenges from a neuroscience perspective [2]. The neurally-driven autonomous agents live a simulation environment and perform typical animate tasks, such as gathering food, evading predators, navigating path and seeking prey. Each autonomous agent include a "brain" that is controlled by an artificial neural network. Some of proposed approaches in evolutionary autonomous agents or robotics are to evolve the traditional artificial neural networks that encodes the neural information through the rate of neuronal spikes [3, 4]. By contrast, Motivated by recent advances in neurosciences, neural information is encoded in the brain through the precisely timed spike trains, not only through the neural firing rate [5]. Spiking neural networks (SNNs), the 3rd generation of artificial neural networks, are comprised of spiking neurons, which fire at certain points in time, thus sending an electric pulse that is regularly referred to as spike (action potential) [6]. The transmission of information in SNNs resemble the processing of the truly biological

D.-S. Huang et al. (Eds.): ICIC 2017, Part I, LNCS 10361, pp. 37–47, 2017.
DOI: 10.1007/978-3-319-63309-1_4

brain closely as far as possible. Thereby, SNNs are shown to be suitable tools for the processing of spatio-temporal information [7, 8].

In artificial intelligence, an evolutionary algorithm is a generic population-based metaheuristic optimization approach, which uses mechanisms inspired by biological evolution, such as reproduction, mutation, recombination, and selection [9]. Combination of evolutionary algorithms and spiking neuronal networks, recent years have witnessed a growing interest in the study for evolving SNNs [10]. The evolution of SNNs can be understood into two respects: the evolution of weights and structures, the weights and structures of SNNs are self-regulation without other interference. The most important feature of evolving SNNs is that it is capable of adapting to the dynamic changes of external environment. Wysoski et al. [11] proposed an evolving spiking neural network architecture that was designed as an audiovisual information processing system. Batllori et al. [12] presented an evolving SNNs method in which the robot "brain" was a SNN simulator whose parameters were tuned by a genetic algorithm, where fitness was assessed by the closeness to target output spike trains. Other researches have applied evolving SNNs as a classification method for spatio-temporal sequence data [13], or a clustering method for detecting overlapping clusters of irregular form [14].

In this paper, an evolutionary algorithm for neurally-driven autonomous agents is presented, in which the agents are controlled by the feedforward SNN model, and the weights of spiking neural controller are evolved by utilizing evolutionary operators. The experimental results confirmed that the population adaptability to environment visibly affected by the evolutionary operators. The structure of this paper is organized as follows: the SNN architecture and neuron model are described in Sect. 2. The evolutionary approach for the SNN autonomous agents is described in Sect. 3. The description of the experiment results is given in Sects. 4 and 5 provides the conclusion.

2 Spiking Neural Network and Neuron Model

2.1 Spiking Neural Network Architecture

The architecture of SNNs that control the autonomous agents is shown in Fig. 1(a). The three-layer feedforward SNNs are applied in this paper. The network is assumed to be fully connected, and all neurons in a layer are connected to all neurons in the subsequent layer. The feedforward SNNs contain three layers, including the input layer, hidden layer and output layer. The number of neurons in each layer is N_i, N_h and N_o respectively. In contrast to traditional feedforward artificial neural networks where two neurons are connected by one synapse only, the connection between two neurons in the SNNs consists of multiple synapses, where each synapse serves as a sub-connection that is associated with a different weight. The multiple synaptic connections between two neurons are shown in Fig. 1(b). In addition, the number of synaptic connections between two neurons is K, and each synapse transmit multiple spikes from a presynaptic neuron to a postsynaptic neuron.

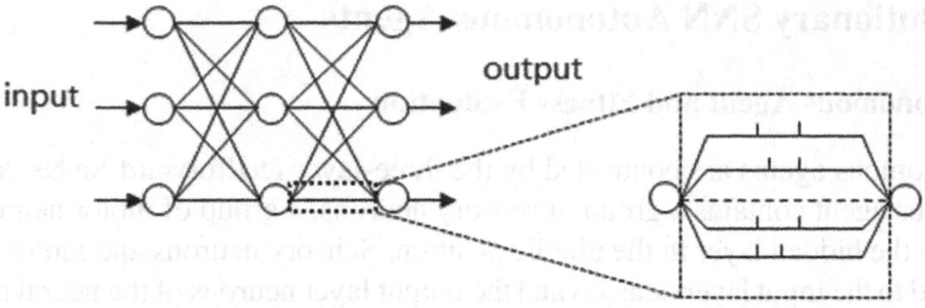

(a) Spiking neural network architecture. (b) Multiple synaptic connections.

Fig. 1. Three-layer feedforward architecture of spiking neural networks

2.2 Spike Response Model

Spiking neuron is the basic function unit for brain information processing. Spiking response model (SRM) [15] is employed in all simulations, and it expresses the variation of membrane potential at a given time for the input spike trains of all the presynaptic neurons. Assuming that a postsynaptic neuron has N presynaptic neurons, the *i*th presynaptic neuron fires a total of G_i spikes and the *g*th spike ($g \in [1,\ G_i]$) is fired at time t_i^g, and the weight of the *k*th synapse ($k \in [1, K]$) from the *i*th presynaptic neuron to the postsynaptic neuron is defined by w_i^k. The function expression of membrane potential u of the neuron at time t is given by:

$$u(t) = \eta\left(t - t^f\right) + \sum_{i=1}^{N} \sum_{k=1}^{K} \sum_{g=1}^{G_i} w_i^k \varepsilon\left(t - t_i^g\right) \tag{1}$$

Where t^f is the timing of the most recent output spike from neuron prior to time t.

The spike response function $\varepsilon(t)$ describes the relationship between input spikes of the presynaptic neuron and the internal state variable of the postsynaptic neuron, which is expressed as follow:

$$\varepsilon(t) = \begin{cases} \frac{t}{\tau} \exp\left(1 - \frac{t}{\tau}\right) & \text{when } t > 0 \\ 0 & \text{when } t \leq 0 \end{cases} \tag{2}$$

where τ is the time constant. In addition, $\eta(t)$ is the refractoriness function, which is mainly reflected the variation of membrane potential when the postsynaptic neuron was fired. The refractoriness function $\eta(t)$ can be expressed as follow:

$$\eta(t) = \begin{cases} -\theta \exp\left(-\frac{t}{\tau_R}\right) & \text{when } t > 0 \\ 0 & \text{when } t \leq 0 \end{cases} \tag{3}$$

Where θ is the neuron threshold, τ_R is the time constant of refractory period.

3 Evolutionary SNN Autonomous Agents

3.1 Autonomous Agent and Fitness Evaluation

The autonomous agents are controlled by the three-layer feedforward SNNs, and each autonomous agent contains a group of sensory neurons, a group of motor neurons, and neurons in the hidden layer in the middle position. Sensory neurons and motor neurons correspond to the input layer neurons and the output layer neurons of the neural network, and the hidden layer neurons receive the sensory spike inputs and transmit their firing spikes to the motor neurons. Autonomous agents only sense a range of surrounding environment in the process of movement, the sensory range is expressed as a circular area with radius of R_{agent}, and the sensory area is divided into several fan-shaped sub-regions, namely a plurality of direction of motion. In order to distinguish the environment information in each direction, each sub-region corresponds to a sensory neuron and a motor neuron. Figure 2(a) and (b) show the sensorimotor system of an autonomous agent, which includes 8 sensory neurons and 8 motor neurons respectively.

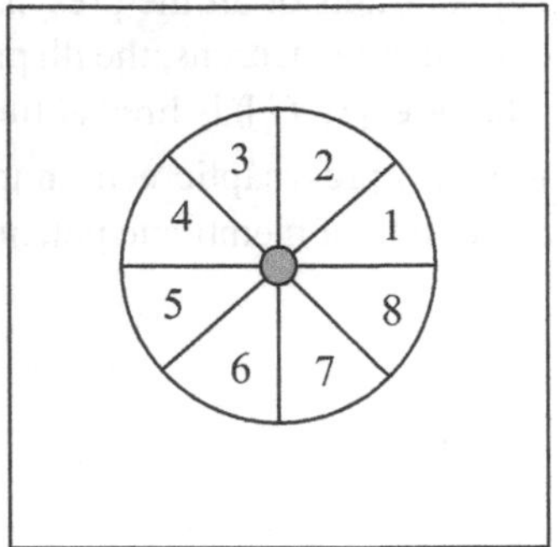

(a) Autonomousa gent.

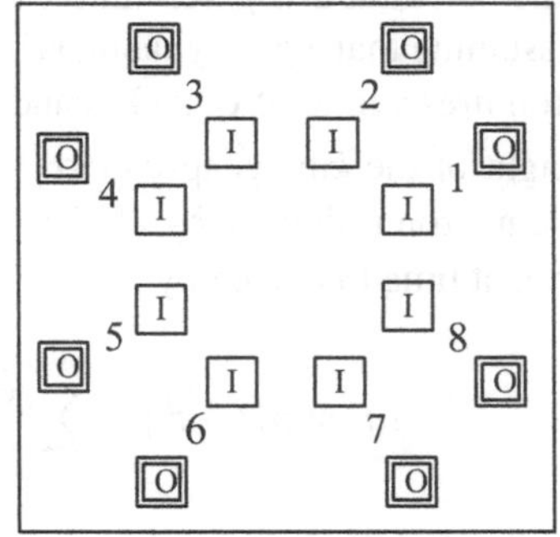

(b) Sensory inputs and motor outputs.

Fig. 2. The sensorimotor system of an autonomous agent

The autonomous agent lives the simulation environment of "food gathering" that can be seen as a 500 × 500 2D continuous field, and the adjacent edges connected to each other. There are 50 "foods" and 50 "poisons" in the simulation environment, they dispersed in the simulation environment randomly. If the autonomous agent feels the presence of substances (food or poison) in the sensory region, it will receive the specific frequency spike trains from environmental signals. In order to solve the contradiction between the food and poison signals, we employ the simple method of linear encoding to convert the environmental signal into the timing of spikes. In the simulation environment, the maximum number of steps T_{max} is 5000, the time window of each time step is 50 ms. At each time step, when a food is found in a fan-shaped sub-region, the corresponding sensory neuron fires a excitatory spike train with frequency of 12.5 Hz; when a poison is felt in a fan-shaped sub-region, the corresponding sensory neuron is converted into inhibitory neuron with a 12.5 Hz spike train; When there is no food and poison in a fan-shaped area, the input sensory neuron fires a frequency signal of 2.5 Hz and converts it into a spike train. For the motor neuron with the highest firing frequency of SNN in each time step, autonomous agents move a certain distance along the midline direction of this fan-shaped sub-regions, the distance is determined by the agent's speed S_{agent}.

The selection of adaptable fitness function is of great concern for evolutionary experiment. In the process of population evolution, we should select the parent used for genetic operators according to the fitness of individuals. Therefore, the concrete form of fitness function directly determines the evolution of the population. In the course of gathering food, what we will do as the following: if the autonomous agent gathers the food, fitness value increases; if the autonomous agent gathers the poison, the fitness value decreases correspondingly. Therefore, the fitness function of autonomous agents can be expressed as follow:

$$fitness = (F - P)/50 \tag{4}$$

Where F represents the total number of gathered foods, P stand for the total number of gathered poisons. Then, the range of fitness value is [–1, 1]. Autonomous agents should be capable to gather more foods as much as possible in their life cycle, and try to avoid the poisons, so as to have a higher fitness value.

3.2 Evolutionary Algorithm of SNNs

The evolutionary algorithm of SNNs is the key problem to construct and optimize the controlling networks of autonomous agents. The standard approach in evolutionary autonomous agents is to evolve neural networks for control by encoding the parameters of the network in the genome. In this paper, we explore the three-layer feedforward SNNs that is composed by spike response neuron model. Population initialization has completed their parameter setting, that is to say, neuron model and network structure of SNNs is fixed in the process of evolution. Here, the evolution of SNNs refers to the evolution of connection weights. The genome of multiple synaptic SNNs is represented by two three-dimensional weight matrices, one is the output weight matrix W_o from the hidden layer to the output layer, and another is the hidden weight matrix W_h from the input layer to the hidden layer.

The evolutionary algorithm represents time by a sequence of non-overlapping generations in the population. In the turnover from one generation to the next, pairs of individuals reproduce to form offspring. The core of the reproduction process is described as follows: (1) The crossover operator with probability P_{cross} is applied to form the offspring, in which swapping the columns of weight matrices of two parent individuals; (2) The mutation operator with probability P_{mut} is then applied. On the basis of SNNs weight matrix model, there are two mutation operators be designed: noise addition and weight replacement. Noise addition operator is that adding a random noise on original connection weight, noise value satisfies the Gaussian distribution G (0, 0.1). Weight replacement operator is that using a new weight to replace the original value, the method of generating new weight as same as the weight matrix initialization method. In addition, the elitism preserving strategy is applied in the evolution of SNNs. The evolutionary algorithm of SNNs based on weight matrix model is described in Table 1.

Table 1. The evolutionary algorithm of SNN autonomous agents

```
Initializing the size of population S_evo, and the evolutionary generation G_evo;
Setting the initialization parameters of spiking neural networks, generating initial population pop;
while !stopCondition() do //Termination condition for reaching the generation G_evo
    evaluate(pop); //Evaluation the autonomous agents population
    while the size of newPop < S_evo do
        parents←selection(pop); //Select operator, select parent individuals to participate in the crossover operation.
        offspring←crossOver(parents, P_cross); //Crossover operator for weight matrices
        offspring←mutation(offspring, P_mut); //Mutation operator for weight matrices
        newPop.add(offspring); //Add new individuals to the population
    end while
    pop←newPop;
    gen++;
end while //End the autonomous agents evolution
```

4 Experimental Results

In the evolutionary experiments, we need to set some basic parameters, including the parameters of neurons, the parameters of autonomous agents, and evolutionary parameters. Most of these parameters remained the same during the experiments. The number of synapses between two neurons is 5, and the synaptic weights is initialized as the uniform distribution in the interval [0, 0.2]. For each experiment, the results of evolutionary algorithm are averaged over 10 trials. Table 2 shows the basic parameter settings for evolving SNN autonomous agents.

In the basic benchmark experiment, the number of sensory neurons, hidden layer neurons and motor neurons in evolving SNNs are 16, 16 and 16, respectively. The sensory radius $R_{agent} = 50$, and movement speed of agents $S_{agent} = 5$. The mutation operator of SNN population adopts the noise addition operator with mutation probability of 0.1. As can be seen from the Fig. 3, the average fitness value and the optimal fitness value of the population are gradually increased. After about 8 generations of evolution, the average fitness value and the optimal fitness value of the population tend to be stable. In steady state, the optimal fitness value of the population is 0.66.

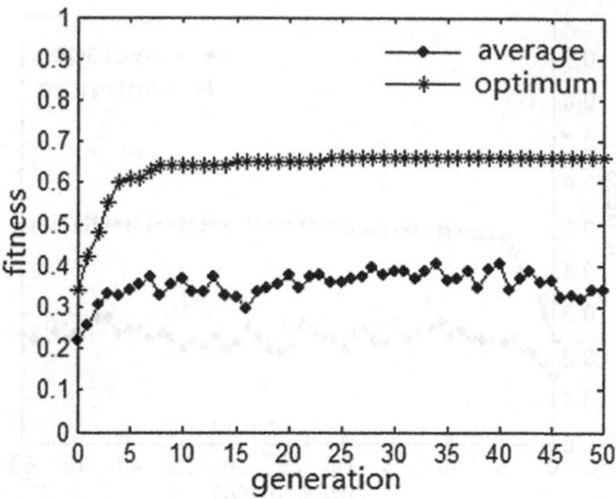

Fig. 3. The fitness of autonomous agents for basic benchmark networks.

4.1 Comparison of Different Network Scale

In the first comparison experiment, we investigate the intelligent behaviours of autonomous agent populations in different network scale. The small scale SNNs that control autonomous agents is $8 \times 8 \times 8$, that is, the number of sensory neurons, hidden layer neurons and motor neurons in population are 8, 8 and 8, respectively. The others parameters are same as shown in Table 2. From Fig. 4, we can see that after about 20 generations of evolution, the optimal fitness value of the population tend to be stable, that is 0.51, while the $16 \times 16 \times 16$ network population is 0.66 (seen as Fig. 3). Therefore, we can conclude that with the increase of the number of neurons and the complexity of the network structure, the autonomous agents perform better adaptability.

Table 2. The base parameter settings of the evolutionary algorithm in all experiments

Category	Parameter	Symbol	Values
Spike response model	The time constant of postsynaptic potential	τ	10 ms
	The time constant of refractory period	τ_R	50 ms
	The absolute refractory period	t_R	1 ms
	The neuronal threshold	θ	1
Autonomous agent	The radius of sensory region	R_{agent}	50
	Movement speed	S_{agent}	5
	The number of sensory neurons	N_{i}	16
	The number of hidden neurons	N_{h}	16
	The number of motor neurons	N_{o}	16
Evolutionary algorithm	The size of population	S_{evo}	20
	Evolutionary generation	G_{evo}	50
	Crossover probability	P_{corss}	0.5
	Mutation probability	P_{mut}	0.1

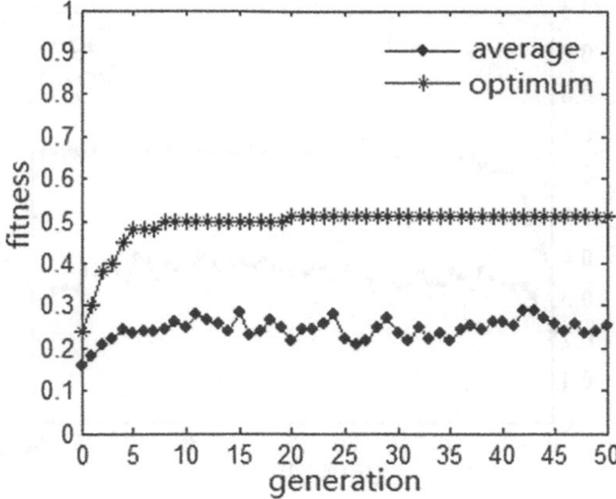

Fig. 4. The fitness of autonomous agents for $8 \times 8 \times 8$ network scale.

4.2 Comparison of Mutation Operators

Then, we investigate the influence of two kinds of mutation operators, noise addition and weight replacement, on the evolution of SNN autonomous agents. Setting the network scale is $16 \times 16 \times 16$, the others parameters are same as previous. Figure 5 shows the optimal fitness value of autonomous agent population evolved through the weight replacement operator is 0.61. The average fitness values are in a similar level for two kinds of mutation operators. The result from this experiment is that: either noise addition or weight replacement, the implementation methods are different, but the effect is basically consistent for evolutionary algorithm of SNNs.

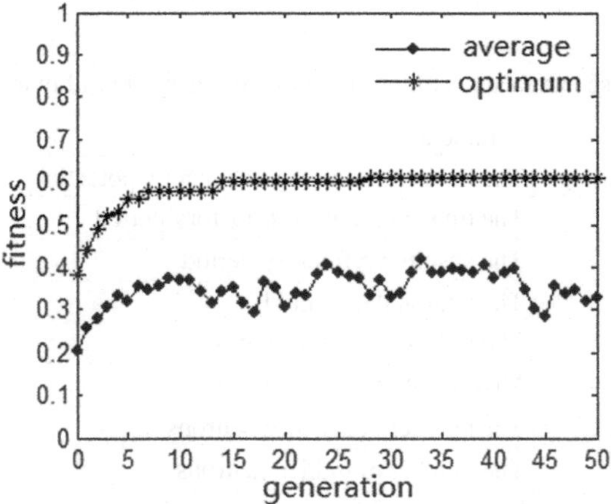

Fig. 5. The fitness of autonomous agents for weight replacement mutation operator

4.3 Comparison of Sensory Radius

And then, we investigate the performance of autonomous agents with the different sensory radius. From Fig. 6 we note that the optimal value is 0.72 in the population with sensory radius $R_{agent} = 30$, it is more than the optimal value of the population with sensory radius $R_{agent} = 50$ (seen as Fig. 3, the optimal value is 0.66). Comparing with the two populations for the different sensory radius, the autonomous agent population with smaller sensory radius maintains a higher level of average fitness. Therefore, the sensory radius of autonomous agents have an important effect on population fitness. In fact, the bigger the sensory radius is, the more material (foods or poisons) can be sensed, but if it beyond the reasonable value range, the larger sensory radius would reduce the fitness of autonomous agents.

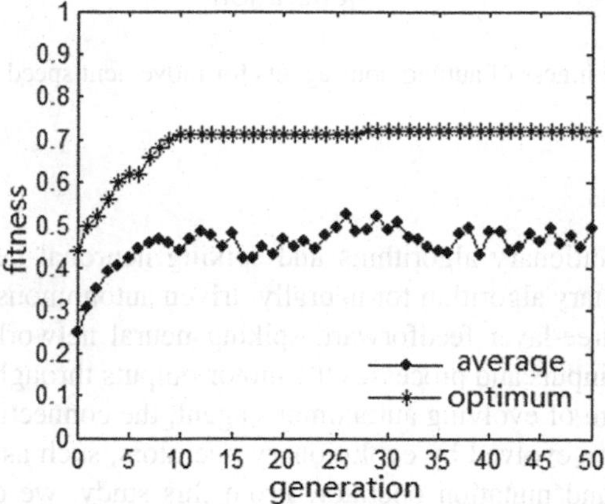

Fig. 6. The fitness of autonomous agents for sensory radius $R_{agent} = 30$.

4.4 Comparison of Agent Speed

Finally, we investigate the autonomous agents with different movement speed in the "food gathering" simulated environment. According to the experimental results, it can be found that whether the speed of movement will affect the performance of autonomous agents. From Fig. 7 we can note that the optimal value is 0.52 in autonomous agent population with the movement speed $S_{agent} = 3$, while in basic benchmark SNN population the value is 0.66 for $S_{agent} = 5$. Comparing the experiment results, we conclude that the movement speed of autonomous agents can produce a remarkable effect to the population's fitness. Within a certain scope, the faster movement speed means the higher fitness.

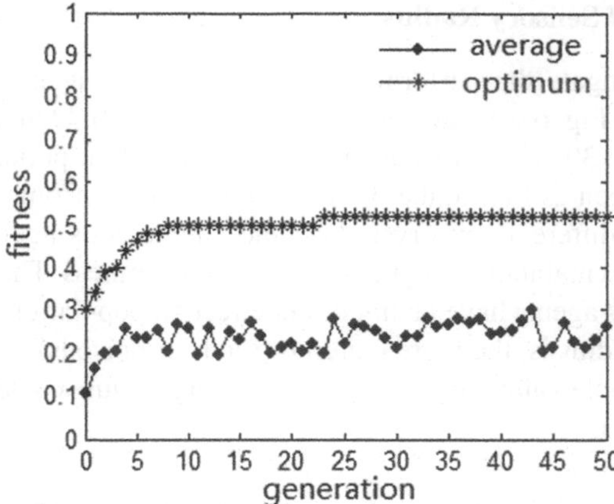

Fig. 7. The fitness of autonomous agents for movement speed $S_{agent} = 3$.

5 Conclusions

Combination of evolutionary algorithms and spiking neuronal networks, this paper proposes an evolutionary algorithm for neurally-driven autonomous agents, each agent is controlled by a three-layer feedforward spiking neural network, and the network receives the sensory inputs and processes the motor outputs through the encoded spike trains. On the problem of evolving autonomous agent, the connection weight matrices of neural networks are evolved by evolutionary operators, such as selection operator, crossover operator, and mutation operator. From this study, we draw the following conclusions: (1) the fitness of spiking neural network population increases gradually in the process of evolution; (2) the individual's network structure is more complex, the autonomous agent shows the higher fitness; (3) the parameters of autonomous agents have an important influence on their fitness in the simulation environment.

Acknowledgement. This research is supported by the Natural Science Foundation of Gansu Province of China under Grant No. 1506RJZA127, and the Scientific Research Project of Universities of Gansu Province under Grant No. 2015A-013.

References

1. Tettamanzi, A.G.B., Tomassini, M.: Soft Computing: Integrating Evolutionary, Neural, and Fuzzy Systems. Springer Science and Business Media, Heidelberg (2013)
2. Keinan, A., Sandbank, B., Hilgetag, C.C., et al.: Axiomatic scalable neurocontroller analysis via the Shapley value. Artif. Life **12**(3), 333–352 (2006)
3. Floreano, D., Mondada, F.: Evolutionary neurocontrollers for autonomous mobile robots. Neural Netw. **11**(7), 1461–1478 (1998)
4. Bredeche, N., Montanier, J.M., Liu, W., et al.: Environment-driven distributed evolutionary adaptation in a population of autonomous robotic agents. Math. Comput. Model. Dyn. Syst. **18**(1), 101–129 (2012)

5. Seth, A.K.: Neural coding: rate and time codes work together. Curr. Biol. **25**(3), R110–R113 (2015)
6. Ghosh-Dastidar, S., Adeli, H.: Spiking neural networks. Int. J. Neural Syst. **19**(4), 295–308 (2009)
7. Rekabdar, B., Nicolescu, M., Nicolescu, M., et al.: A scale and translation invariant approach for early classification of spatio-temporal patterns using spiking neural networks. Neural Process. Lett. **43**(2), 327–343 (2016)
8. Rekabdar, B., Nicolescu, M., Nicolescu, M., et al.: Using patterns of firing neurons in spiking neural networks for learning and early recognition of spatio-temporal patterns. Neural Comput. Appl., 1–17 (2016)
9. Črepinšek, M., Liu, S.H., Mernik, M.: Exploration and exploitation in evolutionary algorithms: A survey. ACM Comput. Surv. **45**(3), 35 (2013)
10. Schliebs, S., Kasabov, N.: Evolving spiking neural network—a survey. Evolving Syst. **4**(2), 87–98 (2013)
11. Wysoski, S.G., Benuskova, L., Kasabov, N.: Evolving spiking neural networks for audiovisual information processing. Neural Netw. **23**(7), 819–835 (2010)
12. Batllori, R., Laramee, C.B., Land, W., et al.: Evolving spiking neural networks for robot control. Procedia Comput. Sci. **6**, 329–334 (2011)
13. Kasabov, N., Scott, N.M., Tu, E., et al.: Evolving spatio-temporal data machines based on the NeuCube neuromorphic framework: design methodology and selected applications. Neural Netw. **78**, 1–14 (2016)
14. Bodyanskiy, Y., Dolotov, A., Vynokurova, O.: Evolving spiking wavelet-neuro-fuzzy self-learning system. Appl. Soft Comput. **14**, 252–258 (2014)
15. Gerstner, W., Kistler, W.M.: Spiking Neuron Models: Single Neurons, Populations, Plasticity. Cambridge University Press, Cambridge (2002)

Sparse Reconstruction with Bat Algorithm and Orthogonal Matching Pursuit

Chunmei Zhang, Xingjuan Cai(✉), and Zhentao Shi

Complex System and Computational Intelligence Laboratory, Taiyuan University of Science and Technology, Taiyuan 030024, Shanxi, China
Zhzhchunmei@163.com, Xingjuancai@163.com, sophia_szt@tyust.edu.cn

Abstract. Sparse reconstruction problem is a typical research topic in compressed sensing theory and applications, and many algorithms are designed to solve it. As one of the most well-known reconstruction algorithm, orthogonal matching pursuit (OMP) is widely used in many applications. However, due to the limited global search capability, it is often fallen into the local optimal. In this paper, bat algorithm, a new population-based swarm intelligent algorithm, is incorporated into the methodology of OMP to increase the global search capability. To test the performance, two different signals are employed to compare, and simulation results show our algorithm increases the performance significantly.

Keywords: Sparse reconstruction · Bat algorithm · Orthogonal matching pursuit

1 Introduction

With the arrival of large data era, amount number of data (including signals, images, videos, etc.) is required to be transmitted and stored. However, the traditional Nyquist sampling theorem causes a great waste of resources. In order to solve this problems, D. David and T. Terence proposed the theory of compressed sensing (CS) [1], and it has been successfully applied in imaging system, image fusion, target recognition and target tracking.

A signal is compressible or sparse means it is projected from a high-dimension space to low-dimension space by a measurement matrix [2]. The total process includes three different parts: sparse representation [3], signal compression [4] and signal reconstruction [5–7]. In this paper, we only focus on the signal reconstruction.

Typical reconstruction algorithms are divided into three categories: greedy algorithm based on local search (matching pursuit algorithm(MP) [8] and orthogonal matching pursuit algorithm(OMP) [8]), convex relaxation algorithm based on linear programming (base pursuit algorithm (BP) [9] and base pursuit denoising algorithm (BPDN) [9]), and some other algorithms (gradient projection algorithm (GPSR) [10] and iterative hard threshold algorithm (IHT) [11]). As one of the most well-known reconstruction algorithm, orthogonal matching pursuit is widely used in many applications.

D.-S. Huang et al. (Eds.): ICIC 2017, Part I, LNCS 10361, pp. 48–56, 2017.
DOI: 10.1007/978-3-319-63309-1_5

T. Tony [12] designed the iterative stopping rule to improve the performance of OMP. N.B. Karahanoglu [13] added the A^* strategy to search more sparse solution. W. Jian [14] proposed generate orthogonal matching pursuit algorithm (gOMP) which generate multi index set in per iterative cause smaller number of iterative than OMP.

However, due to the limited global search capability, it is often fallen into the local optimal. To solve this problem, some bio-inspired algorithms are introduced to optimize the performance of OMP. J. Wu [15] proposed the evolutionary algorithm to optimal the residual by iterative. L. Li [16] proposed a multi-objective algorithm which combined NSGA-2 and IST local research to reconstruction signals.

Bio-inspired computation is an umbrella for all population-based stochastic algorithms [17]. Due to the different background, there are many algorithms designed, such as artificial bee colony algorithm [18, 19], bat algorithm [20–22], social emotional optimisation algorithm [23], cuckoo search [24, 25], fruit fly optimization [26], firefly algorithm [27, 28], artificial plant optimization algorithm [29]. In this paper, we try to introduce bat algorithm into OMP.

The rest of this paper is organized as follows. In Sect. 2, the detail about the CS model and OMP algorithm are presented, while Sect. 3 provides the BA-OMP algorithm. The experiments are performed in Sect. 4. In Sect. 5 concludes the paper.

2 Sparse Reconstruction

2.1 Main Framework

Suppose signal $\alpha(\alpha \in R^N)$ with length N, it is the linear combination of a set of vectors:

$$\alpha = \sum_{i=1}^{N} \theta_i \varphi_i = \Psi\Theta \tag{1}$$

where $\Psi = \{\varphi_i | i = 1, 2, \cdots, N\}$ is called the sparse transform base matrix. If $\|\Theta\|_0 = K(K \ll N)$ is satisfied, the signal α is called $K-$ Sparse, and K is called sparsity.

To compress the signal α, a measurement matrix $\Phi \in R^{M \times N}(M \ll N)$ (independent of matrix Ψ) is incorporated with the following manner:

$$y = \Phi\alpha = \Phi\Psi\Theta = A_S\Theta \tag{2}$$

where the A_S is called sensing matrix.

Now, we want to reconstruct the signal α with the compressed signal y, and the reconstructed signal is closely to the original signal, in other words, the reconstruction problem can be modeled as a constraint optimal problem:

$$\begin{cases} \min\limits_{\Theta} \|\Theta\|_0 \\ s.t. y = \Phi\Psi\Theta \end{cases} \tag{3}$$

2.2 Orthogonal Matching Pursuit Algorithm

In Reference [12], authors provide a clear description for OMP as follows:

> "OMP is an iterative greedy algorithm that selects at each step the column of A_S which is most correlated with the current residuals. This column is then added into the set of selected columns. The algorithm updates the residuals by projecting the observation y onto the linear subspace spanned by the columns that have already been selected and the algorithm then iterate. Compared with other alternative methods, a major advantage of the OMP is its simplicity and fast implementation."

The main flow of the OMP algorithm is as follows (Table 1):

Table 1. Procedure of OMP algorithm

Procedure of OMP algorithm
Step1 Import: Import original signal α, measurement matrix Φ, sparse transform base matrix Ψ, sparsity K and compressed rate $C_r = M / N$. **Step2 Initialization:** the residual $re = y$, index set $\beta_K = \phi$, $t = 1$. **Step3 Projection:** Find the index β $$\beta = \arg \max_{j=1,\ldots,N} \left|\langle re_{t-1}, A_S \rangle\right|$$ **Step4 Update:** Update the index set $\beta_K = \beta_K \cup \{\beta\}$, the sensing matrix $A_s[:,\beta] = \phi$. **Step5 Update support set:** Get the support set J_K from A_S by using the index set β_K. **Step6 The least-square method:** The reconstruction signal $\bar{\alpha}$ is solved by least-square method $\bar{\alpha} = (J_K{}^T J_K)^{-1} J_K{}^T y$. **Step7 Update: Update residual** $re = y - A_S \bar{\alpha}$, $t = t + 1$. **Step9 Judge:** If $t > K$, stop iteration, else go to Step 3.

To solve the problem (3), the greedy method and other nonlinear optimal method is employed, but as we all know, it is most get sub-optimal solutions. In step 4–7 is always need long iteration time. Hence, this paper attempted to employ a bio-inspired approach, namely Bat algorithm to deal with the signal reconstruction problems.

3 Proposed Approach

Bat algorithm is a biological heuristic algorithm based on bats foraging behavior, the bats transmit biological waves and their echoes to identify objects in a specific region, when they receive the waves with their ears, they calculate the reflective pulse to measure the distance of prey and go catch it.

There are three characters for bat algorithm: (1) the echolocation is used to detect prey distances and avoid obstacles; (2) the pulse rate *r*and loudness *A*, are adjusted dynamically; (3) the loudness is defined as a constant value varies from a large A_0 to a minimum $A_{\min}$.

Suppose the search region is $[x_{\min}, x_{\max}]$, for bat *i*, the initial position x_i^0 is generated randomly as follows:

$$x_i^0 = x_{\min} + rand_1 \times (x_{\max} - x_{\min}) \tag{4}$$

where $rand_1$ is a random number with uniform distribution varying from [0,1].

In iteration *t*, the bat *i* is updated as follows:

$$f_i = f_{\min} + \beta \times (f_{\max} - f_{\min}) \tag{5}$$

$$v_i^{t+1} = v_i^t + (x_i^t - x_{best}) \times f_i \tag{6}$$

$$x_i^{t+1} = x_i^t + v_i^{t+1} \tag{7}$$

where f_i is frequency, β is a random number varies from 0 to 1.

To improve the exploitation, some bats will be perform the following local perturbation:

$$x_{new} = x_{best} + \varepsilon \bar{A}^i \tag{8}$$

where ε is a random number varies from -1 to 1, $\bar{A}^i$ is the average loudness at generation *t* of all bats.

If the position x_i^{t+1} is better than x_i^t, the loudness A_i^{t+1} and pulse rate r_i^{t+1} are also updated as follows:

$$A_i^{t+1} = \lambda A_i^t \tag{9}$$

$$r_i^{t+1} = r_i^0[1 - \exp(-\gamma t)] \tag{10}$$

where the λ and γ is constants.

To improve the quality of the support set J_K listed in Step 5 of OMP algorithm. In this paper, bat algorithm is used to find the optimal index set. This hybrid algorithm is called BA-OMP algorithm, and the flowchart is listed in Table 2.

Table 2. Procedure of BA-OMP algorithm

Procedure of BA-OMP
Step1 Import: Import original signal α, measurement matrix Φ, sparse transform base matrix Ψ, sparsity K and compressed rate $C_r = M/N$.
Step2 Initialization: Import the pulse emission loudness A, pulse emission rate r. Generate the initial position x_i^0 and velocity v_i^0 randomly.
Step3 Compression: Compressed the original signal α to signal y by using sensing matrix A_S.
Step4 Calculate fitness: Choose current position to be index set β_K and to update support set J_K. Calculate frequency-domain signal $\bar{\alpha}$ by the least square method. Then get the time-domain signal $\hat{\alpha}$ by sparse transform base Ψ. Calculate fitness as follows: $$fitness = \frac{\lVert A_S\hat{\alpha} - y\rVert_2^2}{\lVert y\rVert_2^2} \quad (11)$$
Step5 Update: Update position and velocity by using formula (6), (7).
Step6 Local disturbance: Generate a random number $rand1$. If $rand1 \geq r(t)$, then update position using formula (8). Generate a random number $rand2$. If $rand2 \geq A(t)$ and $f(x(t+1)) < f(x(t))$, then accept the current position and velocity and update $A(t)$ and $r(t)$ by formula (9), (10).
Step7 Update: Update the best position to be index set β_K.
Step 8 Judge: Whether the iterative conditions are satisfied. If not, go to Step5.
Step9 Update support set: Get the support set J_K from A_S by using the index set β_K.
Step10 The least square method: Calculate frequency signal by least square method in formula (12). The recovery time-domain signal $\hat{\alpha}$ in formula (13). $$\bar{\alpha} = (J_K{}^T J_K)^{-1} J_K{}^T y \quad (12)$$ $$\hat{\alpha} = \Psi^{-1}\bar{\alpha} \quad (13)$$
Step11 Output: Output the reconstruction Error: $$Error = \lVert\hat{\alpha} - \alpha\rVert_2^2 / \lVert\alpha\rVert_2^2 \quad (14)$$

4 Performance Evolution

To demonstrate the performance of BA-OMP, the following two ground-truth signals are chosen for compare:

$$f(x_1) = 0.1 \cdot \cos(2\pi \cdot f_1 \cdot Ts \cdot ts) + 0.2 \cdot \cos(2\pi \cdot f_2 \cdot Ts \cdot ts) + 0.3 \cdot \cos(2\pi \cdot f_3 \cdot Ts \cdot ts) + 0.4 \cdot \cos(2\pi \cdot f_4 \cdot Ts \cdot ts) \quad (15)$$

$$f(x_2) = \sin(11\pi \cdot f_1 \cdot Ts \cdot ts) + \cos(17\pi \cdot f_2 \cdot Ts \cdot ts) - 0.4\sin(6\pi \cdot f_3 \cdot Ts \cdot ts) \quad (16)$$

where sampling frequency $f_s = 800$, sampling interval $ts = 1/f_s$, sampling sequence $Ts = 1 : N$, signal frequency were $f_1 = 50$ Hz, $f_2 = 100$ Hz, $f_3 = 200$ Hz, $f_4 = 400$ Hz respectively.

The measurement matrix Φ is a Gaussian random matrix with N(0,1), the length of the signal is $N = 256$, population is set to 100, the maximum generation is $15 \cdot K$, the dimension $M = 4 \cdot K$ (K is sparsity of signal), and each signal is run 100 times. The reconstruction error is the average difference between the recovery signal $\hat{\alpha}$ and ground-truth signal α, and computing as follows:

$$Error = \sum_{i=1}^{N} (\alpha_i - \hat{\alpha}_i)^2 / \sum_{i=1}^{N} \alpha_i^2 \tag{17}$$

where α_i presents the ith value of ground-truth and $\hat{\alpha}_i$ presents the ith value of recovery signal. N is the length of signal.

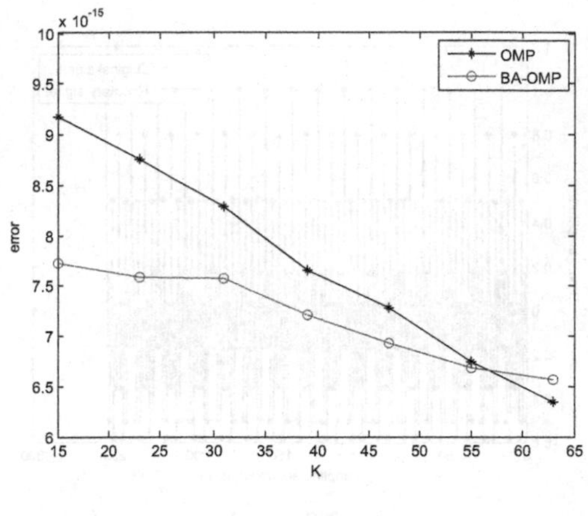

(a) Signal 1

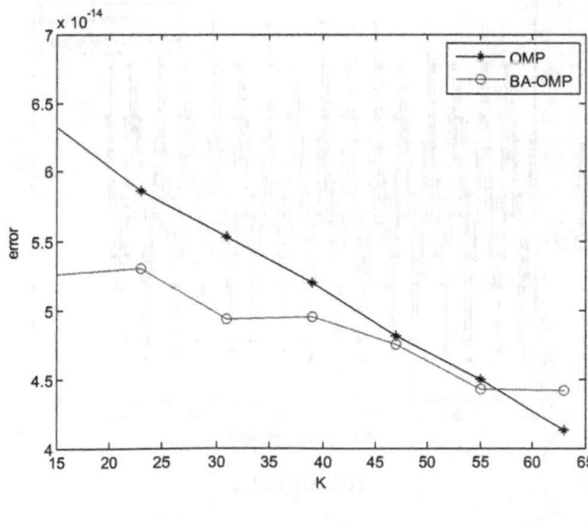

(b) Signal 2

Fig. 1. Dynamic comparison for reconstruction errors v.s. sparsity K

Tables 3 and 4 are the comparison results of reconstruction errors for mentioned above two signals. There are eight different settings for sparsity K: 7, 15, 23, 31, 39, 47, 55, 63. To provide a deep insight, we also plot them with Figs. 1 and 2. Simulation results show BA-OMP is suit for a proper sparsity from 15 to 55.

Table 3. Reconstruction error with change in sparisity of Signal 1

K	7	15	23	31
OMP	0.1010	9.1775e–15	8.7502e–15	8.2812e–15
BA-OMP	0.24352	**7.7225e–15**	**7.5835e–15**	**7.5766e–15**
K	39	47	55	63
OMP	7.6509e–15	7.2854e–15	6.7483e–15	6.3493e–15
BA-OMP	**7.2126e–15**	**6.9283e–15**	**6.6827e–15**	6.5700e–15

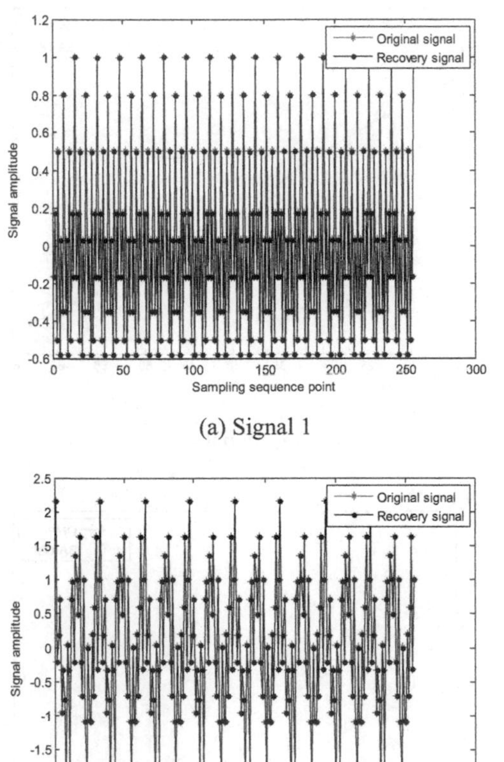

(a) Signal 1

(b) Signal 2

Fig. 2. Dynamic comparison for signal amplitude v.s. sampling sequence point

Table 4. Reconstruction error with change in sparisity of Signal 2

K	7	15	23	31
OMP	0.0787	6.3182e–14	5.8665e–14	5.5271e–14
BA-OMP	0.34805	**5.2594e–14**	**5.3006e–14**	**4.9362e–14**
K	39	47	55	63
OMP	5.2001e–14	4.8128e–14	4.4946e–14	4.1234e–14
BA-OMP	**7.2126e–15**	**6.9283e–15**	**6.6827e-15**	6.5700e–15

5 Conclusion

Orthogonal matching pursuit only finds a local optima in some cases. To improve the performance, in this paper, bat algorithm is incorporated into the methodology of OMP, simulation results show it is superior to the standard version of OMP.

Acknowledgments. This research is supported by the Natural Science Foundation of Shanxi Province under No. 201601D011045.

References

1. Donoho, D.L.: Compressed sensing. IEEE Trans. Inf. Theor. **52**(4), 1289–1306 (2006)
2. Candès, E.J.: Compressive sampling. Marta Sanz Solé **17**(2), 1433–1452 (2007)
3. Candès, E.J., Tao, T.: Decoding by linear programming. IEEE Trans. Inf. Theor. **51**(12), 4203–4215 (2005)
4. Candes, E.J., Tao, T.: Near-optimal signal recovery from random projections: universal encoding strategies? IEEE Trans. Inf. Theor. **52**(12), 5406–5425 (2006)
5. Donoho, D.L., Elad, M.: Optimally sparse representation in general (nonorthogonal) dictionaries via $\ell 1$ minimization. Proc. Natl. Acad. Sci. **100**(5), 2197–2202 (2003)
6. Candès, E.J., Plan, Y.: Near-ideal model selection by l1 minimization. Annal. Stat. **37**(5A), 2145–2177 (2008)
7. Lee, H., Battle, A., Raina, R., Ng, A.Y.: Efficient sparse coding algorithms. Adv. Neural. Inf. Process. Syst. **19**, 801–808 (2007)
8. Tropp, J.A., Gilbert, A.C.: Signal recovery from random measurements Via orthogonal matching pursuit. IEEE Trans. Inf. Theor. **53**(12), 4655–4666 (2007)
9. Chen, S.S., Donoho, D.L., Saunders, M.A.: Atomic decomposition by basis pursuit. Siam Rev. **43**(1), 129–159 (2006)
10. Figueiredo, M.A.T., Nowak, R.D., Wright, S.J.: Gradient projection for sparse reconstruction: application to compressed sensing and other inverse problems. IEEE J. Sel. Top. Sig. Process. **1**(4), 586–597 (2007)
11. Blumensath, T., Davies, M.E.: Iterative hard thresholding for compressed sensing. Appl. Comput. Harmonic Anal. **27**(3), 265–274 (2009)
12. Cai, T.T., Wang, L.: Orthogonal matching pursuit for sparse signal recovery with noise. IEEE Trans. Inf. Theor. **57**(7), 4680–4688 (2011)
13. Karahanoglu, N.B., Erdogan, H.: A* orthogonal matching pursuit: Best-first search for compressed sensing signal recovery. Digit. Sign. Proc. **22**(4), 555–568 (2012)

14. Jian, W., Kwon, S., Shim, B.: Generalized orthogonal matching pursuit. IEEE Trans. Signal Process. **60**(12), 6202–6216 (2012)
15. Wu, J., Liu, F., Jiao, L.C.: Compressive sensing SAR image reconstruction based on Bayesian framework and evolutionary computation. IEEE Trans. Image Process. **20**(7), 04–11 (2011)
16. Li, L., Yao, X., Stolkin, R.: An evolutionary multiobjective approach to sparse reconstruction. IEEE Trans. Evol. Comput. **18**(6), 827–845 (2014)
17. Xiao, R., Zhang, Y., Huang, Z.: Emergent computation of complex systems: a comprehensive review. Int. J. Bio-inspired Comput. **7**(2), 75–97 (2015)
18. Sun, H., Wang, K., Zhao, J., Yu, X.: Artificial bee colony algorithm with improved special centre. Int. J. Comput. Sci. Math. **7**(6), 548–553 (2016)
19. Lu, Y., Li, R.X., Li, S.M.: Artificial bee colony with bidirectional search. Int. J. Comput. Sci. Math. **7**(6), 586–593 (2016)
20. Cai, X., Wang, L., Kang, Q.: Bat algorithm with Gaussian walk. Int. J. Bio-Inspired Comput. **6**(3), 166–174 (2014)
21. Cai, X., Gao, X.Z., Xue, Y.: Improved bat algorithm with optimal forage strategy and random disturbance strategy. Int. J. Bio-Inspired Comput. **8**(4), 205–214 (2016)
22. Xue, F., Cai, Y., Cao, Y.: Optimal parameter settings for bat algorithm. Int. J. Bio-Inspired Comput. **7**(2), 125–128 (2015)
23. Guo, Z.L., Wang, S.W., Yue, X.Z.: Enhanced social emotional optimisation algorithm with elite multi-parent crossover. Int. J. Comput. Sci. Math. **7**(6), 568–574 (2016)
24. Yang, X.S., Deb, S.: Cuckoo search via levy flights. In: Proceedings of World Congress on Nature and Biologically Inspired Computing, pp. 210–214 (2009)
25. Cui, Z., Sun, B., Wang, G., Xue, Y.: A novel oriented cuckoo search algorithm to improve DV-hop performance for cyber-physical systems. J. Parallel Distrib. Comput. **103**, 42–52 (2017)
26. Zhang, Y.W., Wu, J.T., Guo, X., Li, G.N.: Optimising web service composition based on differential fruit fly optimisation algorithm. Int. J. Comput. Sci. Math. **7**(1), 87–101 (2016)
27. Wang, H., Wang, W.J., Zhou, X.Y.: Firefly algorithm with neighborhood attraction. Inf. Sci. **382**, 374–387 (2017)
28. Wang, H., Wang, W.J., Sun, H.: Firefly algorithm with random attraction. Int. J. Bio-Inspired Comput. **8**(1), 33–41 (2016)
29. Cui, Z., Fan, S., Zeng, J.: Artificial plant optimisation algorithm with three-period photosynthesis. Int. J. Bio-Inspired Comput. **5**(2), 133–139 (2013)

Data Transfer and Extension for Mining Big Meteorological Data

Tianwen Huang[1] and Fei Jiao[2(✉)]

[1] Zhaoqing Meteorological Bureau, Zhaoqing 526040, China
[2] Education Technology and Computer Center, Zhaoqing University, Zhaoqing 526040, China
huangtw_zq@126.com

Abstract. It is necessary for mining meteorological big data to build a machine learning model by using historical data to predict the future meteorological elements. This work is significant and has a technical challenge. However, the maintained data of the small cities and the medium cities are very limited due to historical reasons. It is adverse to build an accurate forecast model. Aiming at this problem, a temperature forecast method based on transfer learning technique is proposed. It extends the data of the target city by transferring the data from related cities. It builds a forecast model based on the extended dataset, and then solves the problem of the insufficient samples in machine learning. In this experiment, the temperature sequence of Gaoyao weather station in Zhaoqing area is extended according to the yearly average temperature from 1884 to 1997 of Hongkong. It is corrected by Macau data. Temperature trend of Zhaoqing area is modeled by the time power function and the least square method. The fitting curves and the regression function of the temperature change are obtained. The forecasting model is tested by the actual temperature data of 2014, 2015 and 2016. The results support the effectiveness of the proposed method and they also justify the superiority of applying data transfer to temperature forecast.

Keywords: Big data · Transfer learning · Temperature forecast

1 Introduction

How to predict the future temperature or other meteorological elements by using machine learning techniques is an essential problem in the area of mining meteorological big data. Transfer learning is a relatively new term in the field of machine learning. The goal of transfer learning is to use a priori knowledge and solve the new problem faster and better [1]. Machine learning technique is effective in a large variety of tasks, such as the classification, the detection, and the recommendation [2–5], but it usually faces the challenge of data sparseness when it is applied to meteorological big data. As the training data of some meteorological data is very rare, it is more difficult to get sufficient data for data mining. Due to historical reasons, the meteorological data can be very limited, especially in the small cities and the medium cities. There were only about 200 meteorological stations in China in 1949. Then more and more meteorological stations were built. There are about 2400 stations now. Before 1949, many meteorological

D.-S. Huang et al. (Eds.): ICIC 2017, Part I, LNCS 10361, pp. 57–66, 2017.
DOI: 10.1007/978-3-319-63309-1_6

stations were forced to stop working for some reasons, such as the war, the unrest and the economic depression. Some weather stations have existed for a long time, but the number of meteorological records is quite limited.

Machine learning requires a large amount of training data for each area. A lot of research of machine learning cannot be carried out if there are not enough labeled data. It is very wasteful if these training data whose distribution is different are completely discarded. How to use these data reasonably is the main problem which should be solved by transfer learning. The previous studies accumulated a lot of experience on transfer learning [6]. The meteorological data of a certain area may be short, but the data in different but related regions of the world are massive. So a temperature forecast method based on transfer learning is proposed in this paper. It solves the problem of insufficient samples in machine learning by extending the data of the target city through transferring data from related cities, and it also builds a forecast model based on the extended dataset. According to the monthly mean temperature data of Hongkong whose geographical position is close to Zhaoqing, the temperature data of Zhaoqing city is extended and expanded. On this basis, combined with the time power function and the least square curve fitting, this paper constructs a prediction model to solve the problem of insufficient samples in machine learning.

At present, the application of data mining in meteorology is mainly focused on weather forecast, climatic prediction and meteorological disasters prediction. Data mining method for time series data [7, 8] can also be used in the long term forecast of temperature. Review the history, enhancement of the greenhouse effect has aggravated the occurrence frequency and negative effects of El Nino to a certain extent. Temperature has a close relationship with human society and the ecological system. So it is necessary to study the temperature data of Gaoyao weather station in Zhaoqing area.

Due to their own characteristics, the meteorological data have some strong temporal and spatial correlation characteristics. So it is an effective method to improve the forecast level by having a spatial-temporal correlation analysis of the meteorological data. In data mining, valuable information about the analysis and research of the time series are provided by predecessors. The fuzzy time series model is introduced into the short-range weather forecast by Wang Yong-di [9]. On the research of air temperature, the predecessors did a lot of work and they provided the basis data [10–12]. The application of data mining method in the meteorological field is worth popularizing. Especially in the era of big data, weather services are constantly expanding and the meteorological department needs to keep the data in a constant growth. Then it is necessary and feasible to use mathematical methods to analyze and predict these data. In this paper, the results before and after introducing data transfer and expansion were compared. The conclusion shows that data transfer can improve the accuracy of temperature forecast and it also can be utilized to other time series prediction.

2 Data Transfer and Extension

This section presents data transfer and extension approach which can solve the problem of training data scarcity by introducing related data from different sources. It consists of two parts, data source selection and data processing. The goal is to obtain meaningful data while reduce the error or noise introduced for facilitating the target forecasting task.

2.1 Data Source

It is essential to establish a reliable model to forecast the target field data by using a small number of labeled training samples. How to draw a representative sample set as a training set is an important issue to be considered [13]. The premise of data mining is that the amount of the data should be large enough to be excavated. The longer the data sequence is, the more accurate the model can reflect the authenticity of the climate statistics. The effect of noise will be smaller if the data sequence is long enough. The Gaoyao weather observation station was built in China in 1954. It only has about 60 years' observation data. The actual condition of air temperature in Gaoyao is shown in Fig. 1. Through collecting China meteorological data, it is found that the time of the temperature data in Hongkong is relatively long. There are 114 years of temperature data from 1884 to 1997. But in the war of resistance against Japan and the liberation war, there are irresistible factors. Hongkong is lack of some observation records of temperature data. The discontinuity of the sequence brings great difficulties to the research work. Macao has 97 years of complete temperature data from 1901 to 1997 and its geographic position and climate are similar to that of Hongkong. To ensure the completeness of the data, temperature data of Hongkong is corrected by that of Macao and then the data of Zhaoqing is corrected by that of Hongkong. 131 years of temperature time series from 1884 to 2014 in Zhaoqing are produced.

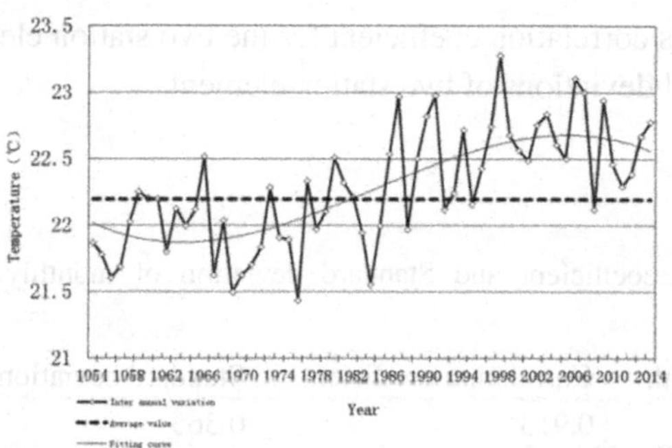

Fig. 1. The actual condition of air temperature in Gaoyao

2.2 Data Preprocessing

The raw data are preprocessed by using the difference correction method. One meteorological element of one meteorological observation station maybe varies greatly, while the difference between two meteorological observation stations varies relatively small. So it is suitable for the use of difference correction method. The station with a longer time and more complete series can be utilized to revise another station that has a shorter time or incomplete series. Temperature, humidity, air pressure, wind speed and other factors have the characteristics that the difference values between two meteorological observations are stable.

X denotes the station whose data sequence is long and uniform as the base station and it has *N* years' data. *Y* denotes the station as the corrected station and it has *n* years' data ($n < N$). Let $\bar{X}_n$ and $\bar{Y}_n$ be *n* years' average value of observation records in two stations respectively. Let X_i and Y_i be the observation value of one year. Then the average difference of *n* years is:

$$d_n = \bar{Y}_n - \bar{X}_n \tag{1}$$

So the difference correction is obtained as follows:

$$Y_i = d_n + X_i \tag{2}$$

The theoretical basis of the difference correction method is: under the influence of atmospheric circulation, the yearly variety of the meteorological elements of two meteorological observation stations is consistent. When the distance between the two stations increases, the consistency of the meteorological elements between them will not exist, the correction will not have the meaning. To solve this problem, we should select two stations whose geographical location and climate are similar as much as possible. Furthermore, adequacy criterion of a revised formula is required.

$$r > \frac{1\sigma_x}{2\sigma_y} \tag{3}$$

In the Eq. (3), *r* is correlation coefficient for the two station elements, σ_x and σ_y are respectively standard deviations of two station elements.

Table 1. Correlation coefficient and Standard deviation of monthly average temperature correction

Month	Correlation coefficient	Standard deviation
1	0.913	0.365
2	0.958	0.405
3	0.902	0.423
4	0.894	0.489
5	0.804	0.581
6	0.718	0.487
7	0.725	0.455
8	0.798	0.430
9	0.707	0.421
10	0.737	0.334
11	0.878	0.417
12	0.925	0.401

In this paper, the correction coefficients are between 0.7–0.9, basically meet the appropriate standards. Firstly, the monthly average temperature in Hongkong is revised by that in Macao, and then using the monthly average temperature in Hongkong to revise the monthly average temperature of Gaoyao, a total of 131 years (1884–2014) of monthly average temperature time series can be obtained. It increases the amount of the data, so the impact of noise is smaller and the prediction of the temperature change trend is more accurate. The test results of the correction are shown in Table 1.

3 Forecast Model Based on Data Transfer and Expansion

Forecast model can be built in many ways. Visualization of temperature data is an intuitive way that helps find out the characteristics of the time series. In our forecast system, Zhaoqing annual average temperature is processed by 10-year moving average and the resulting scatter plots can be a preliminary estimate of the approximate form of the curve equation. The least square regression method was used to get the temperature variation trend equation, and the significance of the regression equation was tested. The method is applied in all kinds of fields, and it is considered that the moving average can be realized quickly [14]. Through the experiment, it is found that the fitting degree of the temperature change trend equation is the highest after 10-year moving average.

Forecast model based on data transfer and expansion mainly uses time power function and least square fitting. On the short-range weather forecast, especially the provincial and provincial station, mainly rely on statistical analysis methods. Among them, the multiple regression analysis method is widely used. The parameter estimation usually uses the least square parameter estimation method. Referring to the relevant literature, we found that the application of partial least squares (PLS), many predecessors have done a lot of research work [15–17]. The method is applicable to the correlation between the independent variables of the regression model. In this paper, only one independent variable is studied. It is *t*, which represents the time: year. So, least square fitting (LSF) can be utilized to meet the requirements [18].

The most common method to determine the parameters in the fitting curve is the least square method. The basic principle is to minimize the square sum of deviations between the predicted values and the measured values and the actual observed values. The change trend equation of temperature is generally considered as the time power function. Its general form is: $T = b_0 + b_1 t + b_2 t^2 + \ldots + b_m t^m$. The original nonlinear regression equation can be transformed into multiple linear regressions: $y = b_0 + b_1 x_1 + b_2 x_2 + \ldots + b_m x_m$. Multivariate linear regression analysis is widely used in industry, agriculture, medicine, social investigation, biological information processing and other fields as an effective data processing method. Multiple linear regression forecast is a quantitative analysis method that the establishment of multiple linear regression model is used to study the relationship between a dependent variable and multiple independent variables by using historical data [19–22]. Following the method of previous studies, the F-distribution test is used.

4 Results and Discussions

The meteorological department usually defines that in the Gregorian calendar, spring is from March to May, summer is from June to August, autumn is from September to November and winter is from December to February of the next year. Migrated by Temperature data of Hongkong which is revised by that of Macao, the temperature data of Gaoyao weather station in Zhaoqing city is extended.

The prediction model is built so as to solve the problem of lack of samples in machine learning. Table 2 shows the experimental results of the influence of the equation fitting before data migration and after data migration. Among them, r is the correlation coefficient of equation fitting and S is the standard deviation. Before migration, the r and S of year's average temperature equation fitting are 0.70 and 0.32. After migration, the r and S change to be 0.98 and 0.09. If the absolute value of R is above 0.8, equation fitting is generally considered to have a strong correlation. Between 0.3 and 0.8, there is a weak correlation while below 0.3, there is no correlation. It can be found that the fitting degree of the equation is obviously improved after data migration, especially in spring.

Table 2. Effect of data migration on fitting equation

Average temperature equation fitting	Before migration r/S	After migration r/S
Year	0.70/0.32	0.98/0.09
Spring	0.21/0.76	0.92/0.22
Summer	0.57/0.35	0.93/0.14
Autumn	0.65/0.55	0.92/0.18
Winter	0.09/3.90	0.92/0.18

With the time power function and the least square method, the annual average temperature change curve and the monthly mean temperature of every season and every month changing curve are obtained. For space limitation, here we only discuss the annual mean temperature change trend equation. Every r/S of the n order polynomial fitting equation is obtained by LSF. When $n = 5$, the fitting effect is the best. So, temperature time series after 10-year moving average can be thought as the 5 order polynomial fitting equation:

$$T = b_0 + b_1t + b_2t^2 + b_3t^3 + b_4t^4 + b_5t^5 \tag{4}$$

Among them, t represents time (year), T represents the corresponding annual average temperature. The coefficients $b_0, b_1, b_2, b_3, b_4, b_5$ are: -1.23×10^5, 2.05×10^2, -9.74×10^{-2}, -7.22×10^{-6}, 1.64×10^{-8} and -3.16×10^{-12}. r is 0.98 and S is 0.09. The fitting curve of the equation is shown in Fig. 2.

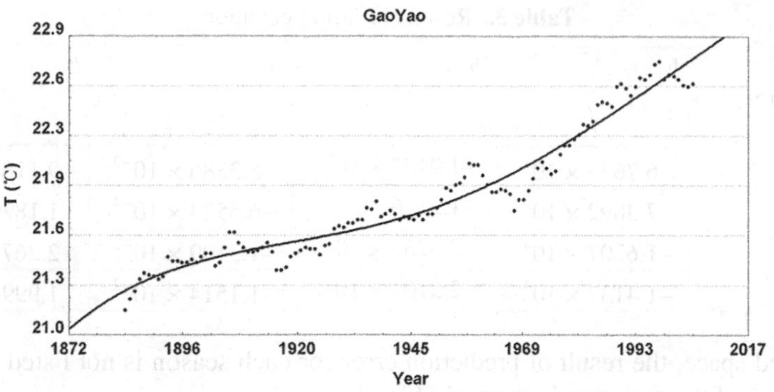

Fig. 2. The change trend of annual average temperature in Zhaoqing

As a test, $t = 2014$, $t = 2015$ and $t = 2016$ are respectively input into the equation. 22. 98°C, 23.00°C and 23.02°C can be obtained as the values of T. The actual annual average temperature of 2014, 2015 and 2016 in Zhaoqing are 22.78°C, 23.4°C and 22.5°C. But, the prediction error of the original fitting equation is great before introducing data migration. So the fitting effect of the equation after introducing data migration is better. Suppose the difference between the real data and the solution of the fitting equation which obtained before migration is *d_before*. Suppose the difference between the real data and the solution of the fitting equation which obtained after migration is *d_after*. To test the fitting effect, *d_before* and *d_after* are compared. The result is shown in Fig. 3.

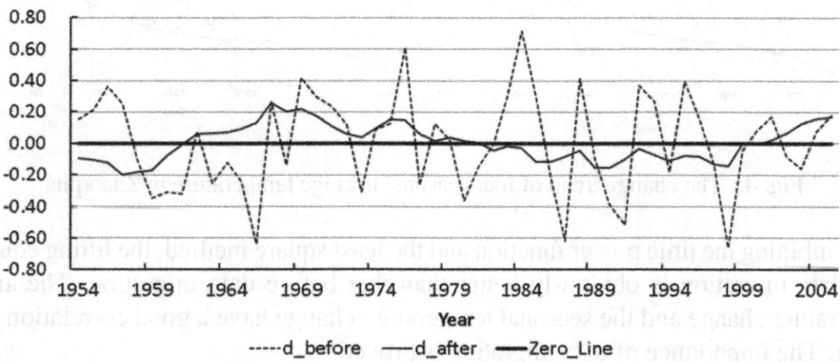

Fig. 3. Effect observation of data migration on equation fitting

After further experiments, T represents the average temperature of four seasons. The fitting effect of the equation is good when they are thought as the 3 order polynomial fitting equation. $T = b_0 + b_1t + b_2t^2 + b_3t^3$, then the results of fitting equation are shown as Table 3.

Table 3. Results of fitting equation

Average temperature period	b_0	b_1	b_2	b_3
Spring	6.7673×10^{3}	-1.0402×10	5.3386×10^{-3}	-9.1186×10^{-7}
Summer	-7.3092×10^{2}	1.2176	-6.5533×10^{-4}	1.1820×10^{-7}
Autumn	-1.6207×10^{4}	2.5263×10	-1.3109×10^{-2}	2.2678×10^{-6}
Winter	-1.4137×10^{4}	2.2107×10	-1.1514×10^{-2}	1.9992×10^{-6}

Limited space, the result of prediction error for each season is not listed here. The fitting curve of the equation is shown in Fig. 4.

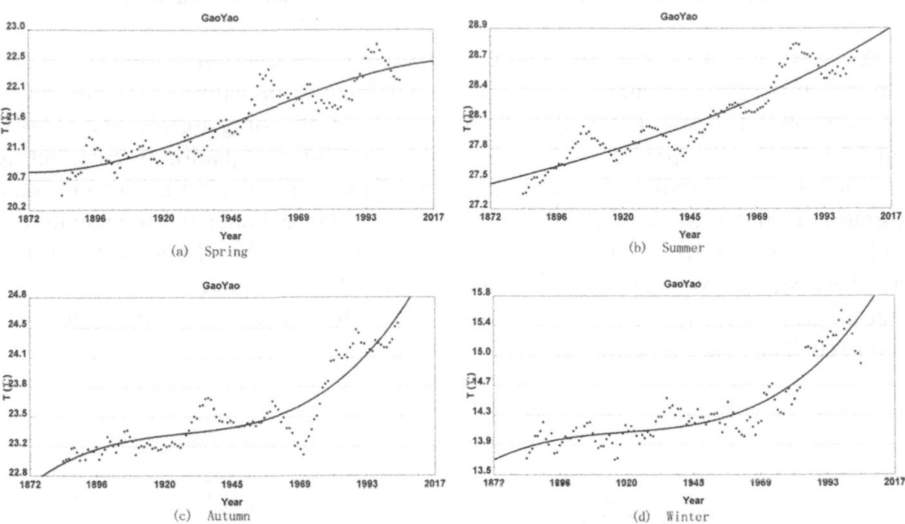

Fig. 4. The change trend of four seasons' average temperature in Zhaoqing

Combining the time power function and the least square method, the fitting equation after data migration is obviously better than that before data migration. The annual temperature change and the seasonal temperature change have a good correlation coefficient. The importance of data migration is proved.

5 Conclusion

In the era of meteorological big data, the problem of lacking training data in machine learning is solved by incorporating the idea of transfer learning. After using data transfer, the period of the meteorological element is extended. Then the time power function and the least square method are utilized to deal with the data after extension. The changing curve fitting is implemented and the regression equation of the time series change trend is obtained. The experiments on real meteorological data show that: (1) The proposed method can effectively obtain the reliable weather forecast model and the regression results obtained are superior. (2) Data transfer and expansion can significantly improve the regression model, and they can improve the accuracy of the temperature forecast. (3) The prediction model was tested with actual data of 2014 and 2015. The results demonstrate that the proposed method is effective and it makes progress of the application of data migration in temperature forecast.

The results of this study have a reference value for the medium and long term temperature prediction of the small or medium cities and towns. They are also valuable to study the future climate change, agricultural plan. In the future, we will study how to apply data transfer and expansion into the forecast of other meteorological elements, such as humidity, rainfall, evaporation, and so on.

Acknowledgments. This research was supported by Science and technology research project of Guangdong Meteorological Bureau (Grant No.2016B51), Science and technology research project of Zhaoqing Meteorological Bureau (Grant No.201609), Science and technology innovation project of Zhaoqing (Grant No.201624030904).

References

1. Pan, S., Yang, Q.: A survey on transfer learning. IEEE Trans. Knowl. Data Eng. **22**(10), 1345–1359 (2010)
2. Wang, M., Li, W., Wang, X.: Transferring a generic pedestrian detector towards specific scenes. In: Proceedings of 2012 IEEE Conference on Computer Vision and Pattern Recognition, Providence, RI, USA, pp. 3274–3281, IEEE (2012)
3. Zhang, Z., Jin, X., Li, L., et al.: Multi-domain active learning for recommendation. In: Proceeding of 30th AAAI Conference on Artificial Intelligence (2016)
4. Wan, C., Jin, X., Ding, G., et al.: Gaussian cardinality restricted boltzmann machines. In: Proceeding of 29th AAAI Conference on Artificial Intelligence (2015)
5. Zhang, S., Jin, X., Shen, D., et al.: Short text classification by detecting information path. In: Proceedings of the 22nd ACM International Conference on Information and Knowledge Management (2013)
6. Wang, X., Wang, M., Li, W.: Scene-specific pedestrian detection for static video surveillance. IEEE Trans on PAMI **36**(2), 361–374 (2014)
7. Jin, X., Lu, Y., Shi, C.: Distribution discovery: Local analysis of temporal rules. PAKDD **2002**, 469–480 (2002)
8. Jin, X., Lu, Y., Shi, C.: Similarity measure based on partial information of time series. In: Proceedings of 8th ACM SIGKDD International Conference on Knowledge Discovery and Data Mining (2002)
9. Wang, Y.: Application of fuzzy time series model in short-term climate prediction. J. Nanjing Univ. Inf. Sci. Technol. **4**, 316–320 (2012). Chinese
10. Haylock, M., Peterson, T., Alves, L., et al.: Trends in total and extreme South American rainfall in 1960–2000 and links with sea surface temperature. J. Clim. **19**(8), 1490–1512 (2006)
11. Thorne, P., Willett, K., Allsn, R., et al.: Guiding the creation of a comprehensive surface temperature resource for twenty-first-century climate science. Bull. Amer. Meteor. Soc. **92**, ES40–ES47 (2001)

12. Muller, R., Rohde, R., Jacobsen, R., et al.: A new estimate of the average earth surface land temperature spanning 1753 to 2011. Geoinfor. Geosta. Overview (2013). doi:10.4172/gigs.1000101
13. He, Q., Zhao, X., Shi, Z.: Minimal consistent subset for hyper surface classification method. J. Pattern Recogn. Artif. Intell. **22**(1), 95–108 (2008). doi:10.1142/S0218001408006132
14. Xiong, B., Yin, Z.: Fast non-localmeans for image de-noising on moving average and modified weight function. J. Image Graph. **17**(5), 628–634 (2012). Chinese
15. Kruger, U., Xie, L.: Partial least squares. In: Statistical monitoring of complex multivariate processes: With applications in industrial process control, pp. 375–409 (2012)
16. Cleophas, T., Zwinderman, A.: Partial least squares, pp. 197–213. Machine learning in medicine. Springer, Netherlands (2013)
17. Abdi, H.: Partial least squares regression and projection on latent structure regression (PLS Regression). Wiley Interdisc. Rev. Comput. Stat. **2**(1), 97–106 (2010)
18. Qu, F., Meng, X.: Source localization using TDOA and FDOA measurements based on constrained total least squares algorithm. J. Electron. Inf. Technol. **36**(5), 1075–1081 (2014). Chinese
19. Seghouane, A.: New AIC corrected variants for multivariate linear regression model selection. IEEE Trans. Aerosp. Electron. Syst. **47**(2), 1154–1165 (2011)
20. Aköz, O., Karsligil, M.: Severity detection of traffic accidents at intersections based on vehicle motion analysis and ultiphase linear regression. In: Proceedings of the 13th International IEEE Conference on Intelligent Transportation Systems. Piscataway, pp. 474–479, IEEE (2010)
21. Nandi, A., Yu, C., Bohannon, P., Ramakrishnan, R.: Data cube materialization and mining over Map Reduce. IEEE Trans. Knowl. Data Eng. **24**(20), 1747–1759 (2012)
22. Wang, Z., Agrawal, D., Tan, K.: COSAC: A framework for combinatorial statistical analysis on cloud. IEEE Trans. Knowl. Data Eng. **25**(9), 2010–2023 (2013)

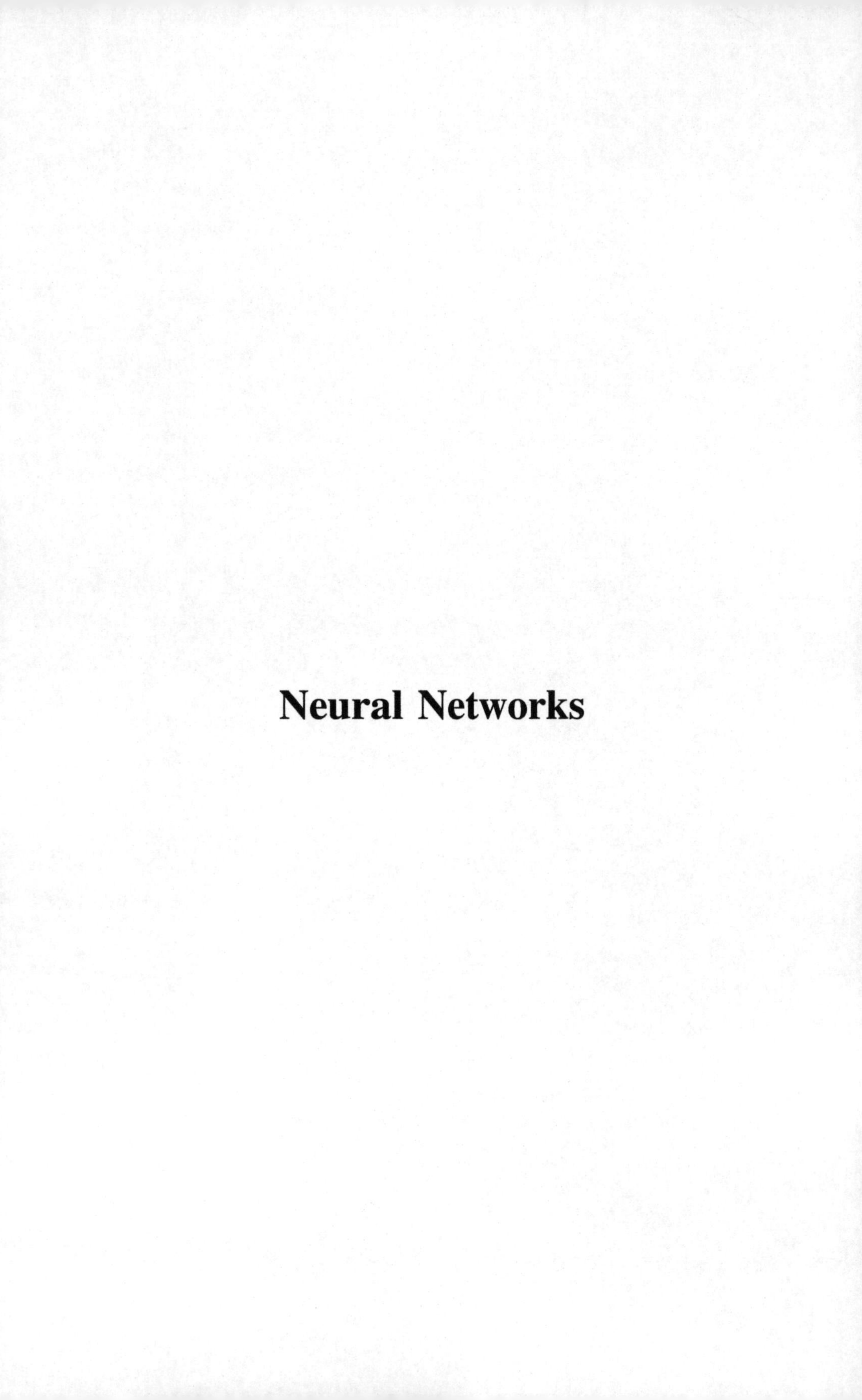

Neural Networks

A Review of Image Recognition with Deep Convolutional Neural Network

Qing Liu, Ningyu Zhang, Wenzhu Yang(✉), Sile Wang, Zhenchao Cui, Xiangyang Chen, and Liping Chen

School of Computer Science and Technology, Hebei University, Baoding 071002, People's Republic of China
wenzhuyang@163.com

Abstract. Image recognition technology is widely used in industry, space military, medicine and agriculture. At present, most of the image recognition methods use artificial feature extraction which is not only laborious, time consuming, but also difficult to do. Deep convolutional neural network is becoming a research hotspot in recent years. It has successfully applied to character recognition, face recognition, and so on. The traditional deep convolutional neural network still has some defaults when dealing with large-scale images and high-resolution complex images. So many research works are rolling ahead to improve the network to make it more efficient and robust. Firstly, the principle of the traditional convolutional neural network was briefly introduced. Then, the improvements on convolutional layer, pooling layer, activation function of convolutional neural network in recent years were summarized. Its applications in image recognition were also presented. Finally, the challenges in convolutional neural network research were analyzed and our recent works ware introduced.

Keywords: Convolutional neural network · Image recognition · Convolutional layer · Pooling layer · Activation function

1 Introduction

Image recognition means using computer technology for image processing, analysis and understanding. With the rapid development of computer technology, the application of image recognition technology is also prevalent in people's daily work and life. Banks use handwritten character recognition systems to identify the figure of checks. Fingerprint recognition and face recognition systems are embedded in computers and mobile phones. The application of image recognition technology can be seen everywhere in our life.

Early image recognition system mainly uses Scale-invariant Feature Transform [1] and Histogram of Oriented Gradients [2] to extract the image features, and then entering these image features to the classifier. At last, the classifier outputs the recognition results of these images. However, due to the diversity of image recognition problems, these image recognition systems are basically for specific identification tasks, and have poor generalization ability. For example, when the handwritten digital

D.-S. Huang et al. (Eds.): ICIC 2017, Part I, LNCS 10361, pp. 69–80, 2017.
DOI: 10.1007/978-3-319-63309-1_7

recognition methods are used to recognition faces or other images, it does not show good performance. So most of the recognition system requires a lot of algorithm research to get a breakthrough in the performance of specific identification problems, such as improving the recognition rate, the training speed, etc. Therefore, it is necessary to find a more generic method that can get better recognition results in different identification problems.

LeNet [3] is a classical model of Convolution Neural Network (CNN) proposed by LeCun. When applied to various different image recognition tasks, LeNet has obtain good results [4]. LeNet is one of the representatives of the universal image recognition system.

The traditional deep CNN still has some defaults when dealing with large-scale images and high-resolution complex images. So many research works are rolling ahead to improve the network to make it more efficient and robust. In the following sections, we introduce the basic working principle of deep CNN and its application in image recognition. At the same time, we discussed the improvements of deep CNN in different aspects. Firstly, we made a brief introduction of Traditional deep Convolution Neural Network. Then, we make a summary of some improvements on network structure and training algorithm of deep CNN such as convolutional layer, pooling layer, activation function and so on. Next, we discussed two typical applications of deep CNN including character recognition and face identification. Finally, we analyze the main problems need to be solved and the main research direction in the future.

2 Traditional Deep Convolution Neural Networks

Deep Convolution Neural Network [3–5] (deep CNN) is a kind of Artificial Neural Network with more than three layers. It is an efficient method developed in recent years and has attracted wide attention in the field of Artificial Intelligence. Taking LeNet-5 as an example, as shown in Fig. 1, it is a hierarchical neural network with convolutional layers alternated with pooling layers, followed by some fully connected layers.

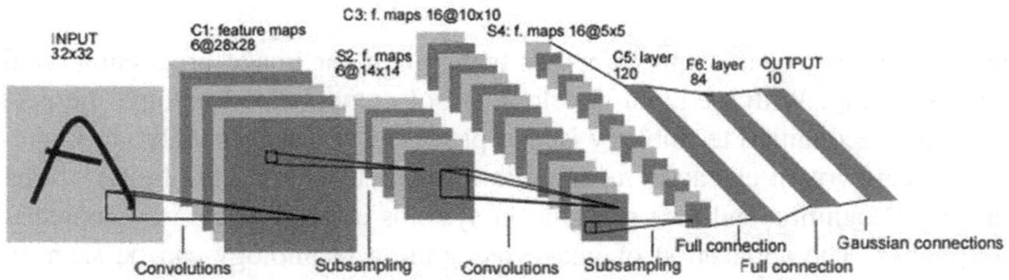

Fig. 1. The architecture of LeNet-5 network

Convolutional layer is a feature extraction layer. It utilizes various kernels to convolve the whole image as well as the former feature map to get new features. It followed by an activation function to decide whether the extracted feature can be passed to next layer. Denoting the activation functions as f(·), the feature map can be computed as follows.

$$x_j^l = f\left(\sum\nolimits_{i \in R_j} x_i^{l-1} * W_{ij}^l + b_j^l\right) \tag{1}$$

Here x_j^l is the output at i-th feature map of the l-th layer, * is the 2D discrete convolution operator, W_{ij}^l is a trainable filter (kernel) and b_j^l is a trainable bias parameter.

Generally, a pooling layer follows a convolutional layer to reduce the feature dimension by down sampling the features extracted by convolutional layers. The pooling operation not only reduces the complexity of the convolutional layers, but also restrains the phenomenon of over-fitting. Meanwhile, it enhances the tolerance of features to minor distortions and rotations, and increases the performance and robustness of deep CNN. Denoting the pooling function as *down*(·), the feature map after pooling can be calculated as follows.

$$x_j^l = f(down(x_i^{l-1}) * w_j^l + b_j^l) \tag{2}$$

3 Improvement of Deep CNN

In the past few years, deep learning in many pattern recognition problems (e.g., computer vision, image recognition) has made remarkable achievements. Among them, the deep CNN is particularly eye-catching. In order to make the deep CNN become more efficient and robust, many researchers have proposed effective improvement measures for deep CNN. We introduce the major improvements of deep CNN in two aspects: improvements on network architectures and improvements on training algorithms.

3.1 Improvements on Network Architectures

(1) Convolutional Layer

In general, traditional convolutional neuron network is composed of alternated convolutional layers with pooling layers. The filters in convolutional layer for feature extraction are always linear. It followed by nonlinear activation functions. The convolutional layer can be described as follows.

$$f_{i,j,k} = \max(w_k^T x_{i,j}, 0) \tag{3}$$

Here using RELU as activation function, and the *(i, j)* stands for the pixel index in the image, x_{ij} is the input patch centered at location *(i, j)*, and *k* is used to index the channels of the feature map.

As show in Fig. 2(a), when the instances of the latent concepts are linearly separable, the linear convolution can extract these features. The linear convolutional layer uses a linear filter to convolute the feature maps.

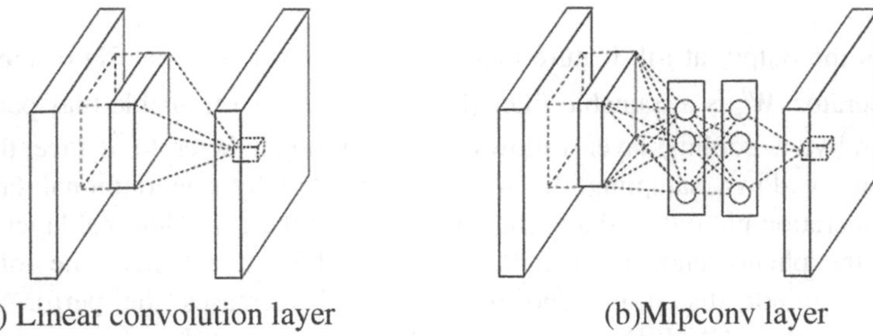

Fig. 2. The comparison of linear convolution layer and mlpconv layer

However, representations that achieve good abstraction are generally highly non-linear functions of the input data. In classic CNN, this can be compensated by using utilizing a super complete cluster of filters to cover all kinds of the latent concepts. Nevertheless, it needs too much parameters to train, and training time is relatively long.

Lin *et al.* [6] proposed Network in Network (NIN), which is the improvement of CNN. As show in Fig. 2(b), they designed more complex operations in each local receptive field, and utilized Multilayer Perceptron (MLP) to replace the GLM to convolve over the input. This new type of layer is called Multilayer Perceptron Convolution (Mlpconv). The calculation performed by Mlpconv layer is shown as follows.

$$\begin{aligned} f_{i,j,k_1}^{1} &= \max\left(w_{k_1}^{1T} x_{i,j} + b_{k_1}, 0\right), \\ &\vdots \\ f_{i,j,k_n}^{n} &= \max\left(w_{k_n}^{n\,T} f_{i,j}^{n-1} + b_{k_n}, 0\right). \end{aligned} \tag{4}$$

Here n is the number of layers in the multilayer perceptron.

(2) Weights Initialization

The initialization of CNN is mainly to initialize the filter weight and bias of the convolution layers. Weights initialization is to assign all connection weights in the network to an initial value.

Sparse Auto Encoder (SAE) [7] is an unsupervised learning model that recodes the input data by making the final output value is approximately equal to the original input values. In this way, SAE can obtain to the feature of the data.

Using the SAE carries on the pre-training of the CNN, the purpose of the pre-training is to obtain a set of filters that are accorded with the statistic characteristics of the dataset and whose have good initial values [8, 9]. As shown in Fig. 3, SAE is used to pre-train the filter W of CNN. If the first convolution layer requires n filters of size m × m, then we cut n patches of size m × m randomly from input image, and sent these patches into input layers of SAE. The SAE weights of hidden layer is what we want. Through this pre-training and the supervised training of CNN, network will be the fastest speed to achieve the optimal weight and the training time will be shorter [9].

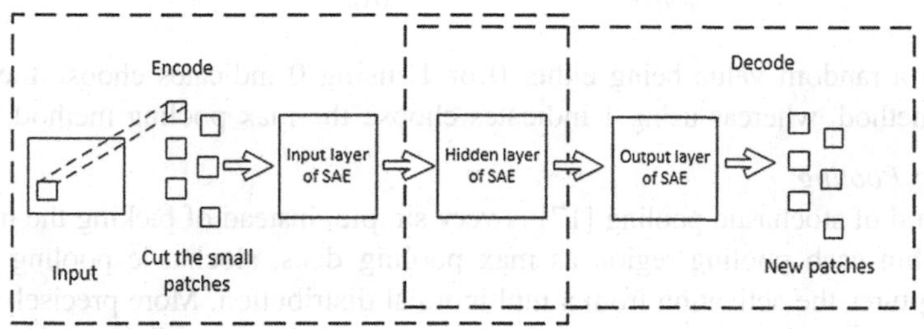

Fig. 3. Pre-training framework of SAE

3.2 Improvements on Training Algorithms

(1) Pooling Layers
The pooling layer is also called the down sampling layer, which take features of the convolutional layer to sample, reduces the computational burden of the network, and decrease the connection parameters between the convolutional layers. In this section, we describe a few of pooling approaches used in CNN.

Lp Pooling
Lp pooling [10, 11] is biologically inspired by the research of how complex cells deal with information a theoretical analysis of this method is that the Lp pooling compared with max pooling has better generalization [12, 13]. There have two special cases of Lp pooling, P = 1 corresponds to a simple Gaussian averaging, whereas P = ∞ corresponds to mix pooling. L2 pooling can be formulated as follows.

$$O = \left(\sum\sum I(i,j)^P \times G(i,j)\right)^{1/P} \tag{5}$$

Here G is a Gaussian kernel, I is the input feature map, and O is the output feature map.

Mixed Pooling
The max pooling and average pooling methods have highly computational efficiency, so they are frequently used in pooling layers of CNN. These two kinds of pooling methods showed good classification accuracy in some datasets. However, both the max pooling and average pooling operators have their own drawbacks. The max pooling only thinks about the maximum element and ignores the others element in the pooling areas. Sometimes, this operator will lead to an unacceptable result. About average pooling, it considers the average of all the elements within the pooling field. This method will consider all noises and the characteristic of the feature map will be reducing.

Therefore, Yu *et al.* [14] proposed mixed pooling operation, which is inspired by the Dropout [15] and DropConnect [16], and this method is combination of max pooling and average pooling. The pooled output can be calculated as follows.

$$y_{k,i,j} = \lambda \max_{(p,q)\in R_{ij}} a_{m,n,k} + (1-\lambda)\frac{1}{|R_{ij}|}\sum\nolimits_{(m,n)\epsilon R_{ij}} a_{m,n,k} \tag{6}$$

Here λ is a random value being either 0 or 1, using 0 indicates choose the average pooling method, whereas using 1 indicates choose the max pooling method.

Stochastic Pooling

The method of stochastic pooling [17] is very simple, instead of picking the maximum value within each pooling region as max pooling does, stochastic pooling just randomly captures the activation from a multinomial distribution. More precisely, we first compute the probabilities p for each pooling region j by normalizing the activations within the region.

$$p_i = \frac{a_i}{\sum_{k\in R_j} a_k} \tag{7}$$

Then we sample from the multinomial distribution based on p to pick a location l within the region. The pooled activation is then simply a_l.

$$s_j = a_l \; where \; l \sim P\left(p_1, \ldots, p_{|R_j|}\right) \tag{8}$$

(1) Spatial Pyramid Pooling

Spatial pyramid pooling (SPP) [18] can generate a fixed-length representation regardless of the input size. To adopt the deep neural network for images of arbitrary sizes, we replace the last pooling layer with SPP. Take a two-layer network as an example, the size of input image is arbitrary and the output layer has 21 neurons, in other words, when the size of input feature map is arbitrary, we need extract 21 features and obtain a 21D feature vector.

Figure 4 [18] illustrates the SPP method, we use three different scales to divide the input image which size is arbitrary; we can obtain 21 pieces (16 + 4 + 1 = 21) and extract a feature from each piece. Thus, we get the 21D feature vector.

(2) Activation Function

In the artificial network, the function of the input and output of neuron is called the activation function. The activation function [19] can be derivable practically anywhere. Therefore, in theory all the continuous functions can be used as activation functions. At present, the common activation function is piecewise linear function and exponential nonlinear function.

ReLU

In traditional convolutional neural network, Sigmoid is the most commonly used activation function. But it also has some shortcomings: when we solve the gradient during the back propagation, if the derivative of the Sigmoid is in the saturation region,

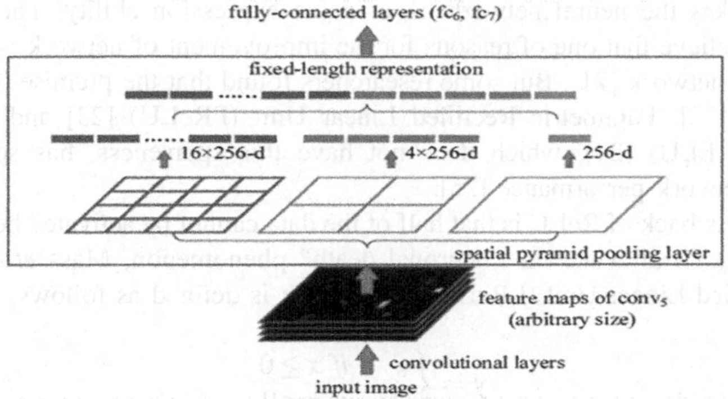

Fig. 4. A network structure with a spatial pyramid pooling layer

it will be infinitely close to 0. That is, the gradient will be small. At this point, the parameters of network will be difficult to get effective training, this phenomenon is the so-called "vanishing gradient" problem.

Compared with the traditional Sigmoid, Rectified Linear Unit (ReLU) [20] can effectively alleviate vanishing gradient problem. It will train CNN directly in a supervised manner without relying on unsupervised layer-by-layer pre-training. ReLU is defined as follows.

$$y = \begin{cases} x & if\ x \geq 0 \\ 0 & if\ x < 0 \end{cases} \tag{9}$$

As shown in Fig. 5(a), it can be found that when x $\geq$ 0, the derivative is 1, the gradient remains unchanged, and the vanishing gradient problem is also alleviated. However, when x < 0, the function is 0, the weight cannot be updated, there is "neuronal death" phenomenon.

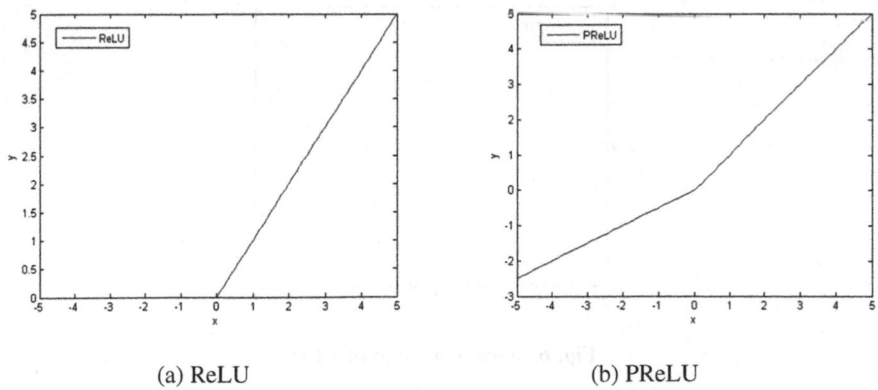

(a) ReLU (b) PReLU

Fig. 5. The comparison of ReLU and PReLU

ReLU makes the neural network have sparse expression ability. There are some researchers believe that one of reasons for the improvement of network is the sparsity of the neural network [21]. But some researchers found that the premise of network is not sparsity [22]. Parametric Rectified Linear Unit (PReLU) [23] and Exponential Linear Unit (ELU) [24], which does not have this sparseness, has some help in improving network performance [25].

A chief drawback of ReLU is that half of the data cannot be activated because of the zero gradient. To alleviate the "neuronal death" phenomenon, Mass *et al.* proposed Leaky Rectified Linear Unit (LReLU) [26], which is defined as follows.

$$\mathrm{y} = \begin{cases} x & if\ x \geq 0 \\ \lambda x & if\ x < 0 \end{cases} \tag{10}$$

Here λ is a predefined parameter in range (0, 1).

PReLU is an improved version of ReLU and LReLU. PReLU is defined as follows.

$$\mathrm{y} = \begin{cases} x & if\ x \geq 0 \\ \lambda_i x & if\ x < 0 \end{cases} \tag{11}$$

As shown in Fig. 5(b), compared with LReLU, the negative half-axis slope λ_i of PReLU is a trainable gain parameter. Compared with ReLU, PReLU converges faster.

ELU

ELU combines the advantages of Sigmoid and ReLU, with left soft saturation, which is defined as follows.

$$\mathrm{y} = \begin{cases} x & if\ x \geq 0 \\ \lambda(e^x - 1) & if\ x < 0 \end{cases} \tag{12}$$

As shown in Fig. 6, when x ≥ 0, ELU can mitigate the vanishing gradient, whereas, when x < 0, ELU is more robust to input changes and noise.

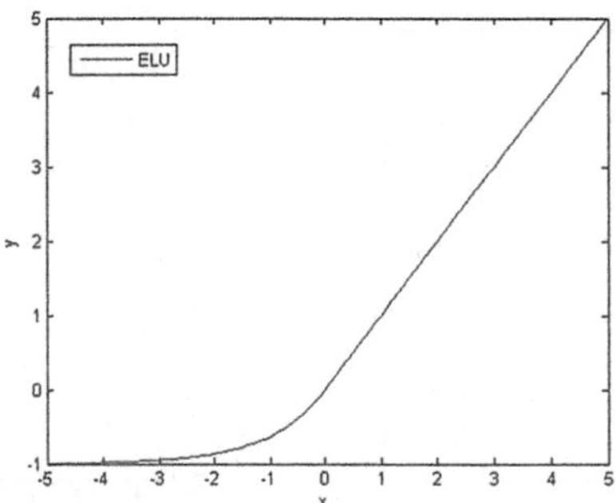

Fig. 6. Function curve of ELU

4 Several Applications of Convolution Neural Network in Image Recognition

In recent years, there have been many successful examples of applying CNN to image recognition tasks. In this section, we will character recognition and face identification.

4.1 Character Recognition

In 1990, when LeCun *et al.* [27] researched the handwritten digital recognition, they proposed applying the back propagation algorithm to deep CNN to train the parameters of the network. This makes deep CNN show better recognition accuracy on the MNIST [28]. Then, Sermanet *et al.* [29] applied deep CNN to digit classification of house numbers. The proposed deep CNN used two feature extraction layers and two nonlinear classifiers, and Lp pooling. They trained and tested the network on SVHN. Some Chinese researchers use deep CNN to identify Chinese characters [30]. The trained deep CNN can automatically learn and identify similar characters. Zhao et al. [31] improved LeNet-5 and applied it to the identification of license plate characters.

4.2 Face Recognition

Face recognition is a very important research direction of image recognition. Deep CNN has achieved good effect in face recognition.

DeepFace [32] is an image recognition system proposed by Facebook research team using deep CNN to identify human face. They trained the deep CNN with 4 million face images and got the accuracy of experimentation similar to human recognition in LFW. DeepFace also introduced an improved convolution structure that is locally connected. The network is trained at each spatial location with a separate convolutional kernel. But there are also some drawbacks: parameter inflating.

The Chinese University of Hong Kong proposed the DeepID series [33–36]. The first generation of DeepID [33] includes 4 convolution layers, and the softmax function is used as loss function. DeepID2 [34] has been greatly improved comparing with DeepID. The main reason is adding a verification signal based on DeepID. DeepID2+ [35] employs the supervisory signal at each layer of CNN, and uses a deeper network structure and more training data. DeepID3 [36] has 2 very deep neural network, it references the VGG net [37] and GoolgeNet [38], and introduces Inception layer, but the results found the recognition rate of DeepID3 is almost the same with DeepID2+.

FaceNet [39] is a face recognition system based on CNN, which is put forward by the Google team in CVPR 2015, the system is mainly composed of 22 layers of deep convolutional neural network and a large number of face data and ternary loss function. FaceNet achieved a new accuracy record of 99.63%.

5 Our Research Work

Using machine vision technology to analyze the growth status of the crop is a deeply integrated topic of computer technology with agricultural engineering technology. The key scientific problem in it is the recognition of high-resolution complex image. The background of the crop growth status image is very complex, and the status of the crop growth is difficult to be described by specific features. This makes it difficult to recognize the image by these pattern recognition algorithms based on manually extracted features. So this makes it practicability to use the feature-learning based image recognition methods to resolve the above-mentioned problem. Though deep convolutional neural network has great advantage in low resolution image recognition, it still has some unresolved problems when dealing with high resolution image, including too many layers to construct the network, too large number of parameters to train, and so on. We will thoroughly study the deep convolutional neural network for high resolution image recognition, specifically including:

- A layer-adaptive multi-scale convolution layer model will be studied by analyzing the learning mechanism of the convolution layer models.
- A Top-N based multi-channel pooling model will be discussed by analyzing the feature pooling methods.
- A flexible activation function which can avoid gradient vanishing and neuron death will be designed by analyzing the activation model of the neuron.
- The balancing method of the network size will also be considered by quantizing the problem scale and the network depth.

6 Conclusion

Image recognition technology has been widely used in our daily work and life. Early image recognition systems mainly use the manual way to extract the specified features of the image, and its generalization ability is poor. The feature recognition method based on feature learning is highly versatile and has high recognition precision. In particular, deep convolution neural network is the state-of-the-art model for image recognition. However, the traditional deep convolutional neural network still has many problems to be resolved. These problems involve the design of new convolution layer, the design of new pool layer, the design of new activation function, the balance of network scale and so on. The deep convolutional neural network has been improved in some aspects, but how to construct an efficient deep convolutional neural network is still an unresolved problem. Especially when dealing with high-resolution images it shows performance degradation, difficult to train and other problems. Therefore, improvements on deep convolutional neural network will not stop before it becomes perfect.

Acknowledgments. The authors thank The Natural Science Foundation of Hebei Provence for their financial support (F2015201033), the Natural Science Foundation of Hebei Provence for their financial support (F2017201069). The authors also thank Information Technology Center of Hebei University for providing the high performance computing platform.

References

1. Lowe, D.G.: Object recognition from local scale-invariant features. In: The Proceedings of the Seventh IEEE International Conference on Computer Vision, p. 1150 (1999). doi:10.1109/ICCV.1999.790410
2. Dalal, N., Triggs, B.: Histograms of oriented gradients for human detection. In: IEEE Computer Society Conference on Computer Vision and Pattern Recognition, vol. 1, pp. 886–893. IEEE (2005). doi:10.1109/CVPR.2005.177
3. Lecun, Y., Boser, B., Denker, J.S., et al.: Backpropagation applied to handwritten zip code recognition. Neural Comput. **1**(4), 541–551 (1989). doi:10.1162/neco.1989.1.4.541
4. Yu, K., Jia, L., Chen, Y., Xu, W.: Deep learning: yesterday, today, and tomorrow. J. Comput. Res. Dev. **50**(9), 1799–1804 (2013). doi:10.7544/issn1000-1239.2013.20131180
5. LeCun, Y., Bottou, L., Bengio, Y., et al.: Gradient-based learning applied to document recognition. Proc. IEEE **86**(11), 2278–2324 (1998). doi:10.1109/5.726791
6. Lin, M., Chen, Q., Yan, S.: Network in network. arXiv preprint arXiv:1312.4400, 2013
7. Baldi, P., Zhiqin, L.: Complex-valued autoencoders. Neural Netw. **33**(3), 136–147 (2012). doi:10.1016/j.neunet.2012.04.011
8. Zhang, W., Xu, Y., Ni, J., et al.: Image target recognitions method based on multi-scale block convolutional neural network. J. Comput. Appl. 1033–1038 (2016). doi:10.11772/j.issn.1001-9081.2016.04.1033
9. Wang, G., Xu, J.: Fast feature representation method based on multi-level pyramid convolution neural network. Appl. Res. Comput. **32**(8), 2492–2495 (2015). doi:10.3969/j.issn.1001-3695.2015.08.061
10. Simoncelli, E.P., Heeger, D.J.: A model of neuronal responses in visual area MT. Vis. Res. (1998). doi:10.1016/S0042-6989(97)00183-1
11. Hyvärinen, A., Köster, U.: Complex cell pooling and the statistics of natural images. Netw. Comput. Neural Syst. **18**(2), 81–100 (2007). doi:10.1080/09548980701418942
12. Bruna, J., Szlam, A., Lecun, Y.: Signal recovery from pooling representations. In: ICML (2014)
13. Gulcehre, C., Cho, K., Pascanu, R., Bengio, Y.: Learned-norm pooling for deep feedforward and recurrent neural networks. In: Calders, T., Esposito, F., Hüllermeier, E., Meo, R. (eds.) ECML PKDD 2014. LNCS, vol. 8724, pp. 530–546. Springer, Heidelberg (2014). doi:10.1007/978-3-662-44848-9_34
14. Yu, D., Wang, H., Chen, P., Wei, Z.: Mixed pooling for convolutional neural networks. Rough Sets Knowl. Technol. (2014). doi:10.1007/978-3-319-11740-9_34
15. Hinton, G.E., Srivastava, N., Krizhevsky, A., Sutskever, I., Salakhutdinov, R.R.: Improving neural networks by preventing co-adaptation of feature detectors. arXiv preprint arXiv:1207.0580 (2012)
16. Wan, L., Zeiler, M., Zhang, S., Cun, Y.L., Fergus, R.: Regularization of neural networks using dropconnect. In: ICML (2013)
17. Zeiler, M.D., Fergus, R.: Stochastic pooling for regularization of deep convolutional neural networks. CoRR (2013)
18. He, K., Zhang, X., Ren, S., Sun, J.: Spatial pyramid pooling in deep convolutional networks for visual recognition. In: Fleet, D., Pajdla, T., Schiele, B., Tuytelaars, T. (eds.) ECCV 2014. LNCS, vol. 8691, pp. 346–361. Springer, Cham (2014). doi:10.1007/978-3-319-10578-9_23
19. Gulcehre, C., Moczulski, M., Denil, M., et al.: Noisy activation functions. In: International Conference on Machine Learning (2016)
20. Nair, V., Hinton, G.E.: Rectified linear units improve restricted Boltzmann machines Vinod Nair. In: ICML, pp. 807–814 (2014)

21. Glorot, X., Bordes, A., Bengio, Y.: Deep sparse rectifier neural networks. In: International Conference on Artificial Intelligence and Statistics (2011)
22. Xu, B., Wang, N., Chen, T., Li, M.: Empirical evaluation of rectified activations in convolutional network. arXiv preprint arXiv:1505.00853 (2015)
23. He, K., Zhang, X., Ren, S., Sun, J.: Delving deep into rectifiers: surpassing human-level performance on ImageNet classification. arXiv preprint arXiv:1502.01852 (2015). doi:10.1109/iccv.2015.123
24. Clevert, D.-A., Unterthiner, T., Hochreiter, S.: Fast and accurate deep network learning by exponential linear units (ELUs). arXiv preprint arXiv:1511.07289 (2015)
25. Li, Y., Fan, C., Li, Y., et al.: Improving deep neural network with multiple parametric exponential linear units. arXiv preprint arXiv:1606.00305 (2016)
26. Maas, A.L., Hannun, A.Y., Ng, A.Y.: Rectifier nonlinearities improve neural network acoustic models. In: ICML (2013)
27. LeCun, Y., Denker, J.S., Henderson, D., et al.: Handwritten digit recognition with a back-propagation network. In: Advances in Neural Information Processing Systems, Colorado, USA, pp. 396–404 (1994)
28. LeCun, Y., Cortes, C.: MNIST handwritten digit database (2010). http://yann.lecun.com/esdb/mnist
29. Sermanet, P., Chintala, S., Lecun, Y.: Convolutional neural networks applied to house numbers digit classification. In: International Conference on Pattern Recognition, pp. 3288–3291. IEEE (2012)
30. Yang, Z., Tao, D., Zhang, S., et al.: Similar handwritten chinese character recognition based on deep neural networks with big data. J. Commun. **35**(9), 184–189 (2014). doi:10.33969/.j.issn.1000-436x.2014.09.019
31. Zhao, Z., Yang, S., Ma, Z.: Lincese plate character recongnition based on convolutional neural network LeNet-5. J. Syst. Simul. **22**(3), 638–641 (2010). doi:10.16182/j.cnki.joss.2010.03.040
32. Taigman, Y., Yang, M., Ranzato, M.A., et al.: Deepface: closing the gap to human-level performance in face verification. In: Proceedings of IEEE Conference on Computer Vision and Pattern Recognition (CVPR), pp. 1701–1708 (2014). doi:10.1109/cvpr.2014.220
33. Sun, Y., Wang, X., Tang, X.: Deep learning face representation from predicting 10,000 classes. In: 2014 IEEE Conference on Computer Vision and Pattern Recognition (CVPR), pp. 1891–1898. IEEE (2014). doi:10.1109/cvpr.2014.244
34. Sun, Y., Chen, Y., Wang, X., et al.: Deep learning face representation by joint identification-verification. In: Advances in Neural Information Processing Systems, pp. 1988–1996 (2014)
35. Sun, Y., Wang, X., Tang, X.: Deeply learned face representations are sparse, selective, and robust. arXiv preprint arXiv:1412.1265 (2014). doi:10.1109/cvpr.2015.7298907
36. Sun, Y., Liang, D., Wang, X., et al.: DeepID3: face recognition with very deep neural networks. arXiv preprint arXiv:1502.00873 (2015)
37. Simonyan, K., Zisserman, A.: Very deep convolutional networks for large-scale image recognition. arXiv preprint arXiv:1409.1556 (2014)
38. Szegedy, C., Liu, W., Jia, Y., et al.: Going deeper with convolutions. arXiv preprint arXiv:1409.4842 (2014). doi:10.1109/cvpr.2015.7298594
39. Schroff, F., Kalenichenko, D., Philbin, J.: FaceNet: a unified embedding for face recognition and clustering. arXiv preprint arXiv:1503.03832 (2015). doi:10.1109/cvpr.2015.7298682

Enhance a Deep Neural Network Model for Twitter Sentiment Analysis by Incorporating User Behavioral Information

Ahmed Sulaiman M. Alharbi[1,2(✉)] and Elise DeDoncker[1]

[1] Department of Computer Science, Western Michigan University, Kalamazoo, MI, USA
{ahmedsulaiman.alharbi,elise.dedoncker}@wmich.edu

[2] Department of Computer Science, Taibah University, Medina, Kingdom of Saudi Arabia

Abstract. Sentiment analysis on social media such as Twitter has become a very important and challenging task. Most existing sentiment classification methods for social media detect the sentiment polarity primarily based on the textual content and neglect of other information. Therefore, in this paper, we propose a neural network model that incorporates user behavioral information with a given document (tweet). The neural network used in this paper is Convolutional Neural Network (CNN). The system is evaluated on two datasets provided by SemEval. The proposed model outperforms the baselines. That means going beyond the content of tweets benefits sentiment classification; providing the classifier with a deep understanding of the task.

Keywords: Sentiment analysis · Social media · Neural network · Deep learning

1 Introduction

Twitter[1] gives us access to the unprompted views of a wide set of users on particular products or events. The expressions of sentiment about organizations, products, events and people have proven extremely useful for marketing and social studies [1].

Unlike conventional texts with many words that help gather sufficient statistics, the texts in social media, especially Twitter, consist only of a few characters. Moreover, tweets may have abbreviations or acronyms that infrequently appear in conventional documents. Therefore, applying traditional methods to such settings will not provide us with acceptable performance.

To address this issue, we propose a neural network based model that is going beyond the content of a given document (tweet). It takes into account, besides a given tweet, the behavior of a user who wrote that tweet. Accordingly, the main research question here is: is there a relationship between users' behaviors and their posts? And if such a relationship exists, can it be used to enhance sentiment analysis performance?

[1] http://twitter.com/.

D.-S. Huang et al. (Eds.): ICIC 2017, Part I, LNCS 10361, pp. 81–88, 2017.
DOI: 10.1007/978-3-319-63309-1_8

The proposed model classifies tweets into positive and negative categories based on proposed set of features that enhance the classification performance. These features help the model to understand a user's behavior (see Sect. 3.2).

The remainder of this paper is structured as follows. We start with a background in Sect. 2. Next, Sect. 3 presents a description of our proposed system. Section 4 shows the experiments and results. Finally, we conclude our work and discuss future work in Sect. 5.

2 Related Work

Most existing sentiment classification methods for social media focus on document-level classification; for instance, the ConSent "Context-based Sentiment Analysis" algorithm is created by [2]. It has two phases, learning and detection. A set of key terms and context terms from a training set are produced in the learning phase. The detection phase is where the classification task takes place. It scans all tweets searching for the key and context terms, and uses the classifier to identify sentiments in the document. As it is clear, ConSent algorithm focuses mainly on the tweet content.

However, there are other proposed platforms use some assistance features such as the emotional state of a tweet's writer and relationships among users. The main goal of such platform is clustering users. In [3], the authors propose an emotional aware clustering approach to group tweets based on eight primary emotions. These emotions are acceptance, fear, anger, joy, anticipation, sadness, disgust and surprise. Their proposed model relies on using an existing dictionary (WordNet). However, [4] points out that WordNet is not a very reliable source since it introduces too much noise.

Going beyond the content of a document benefits sentiment classification because it is providing the classifier with a deep understanding of the task. To investigate its usefulness, [5] develops a multidimensional framework in order to analyze the spatial, temporal and sentiment aspects of tweets discussed the same topic. The authors point out that the combination of the sentiment aspects with the temporal and spatial dimension allow them to infer interesting insights about topics.

3 Methodology

3.1 Deep Learning Architecture

The proposed model for sentiment analysis consists of the convolutional neural network. The system architecture is presented in Fig. 1. The model is implemented using the Weka[2] library.

The main components of our network: input, convolutional, pooling, activations and softmax layers. The input layer consists of: word embeddings and a list of features. Regarding the word embedding, it may be randomly initialized or pre-trained. For the purpose of this work, we utilize the publicly available word2vec

[2] http://www.cs.waikato.ac.nz/ml/weka/.

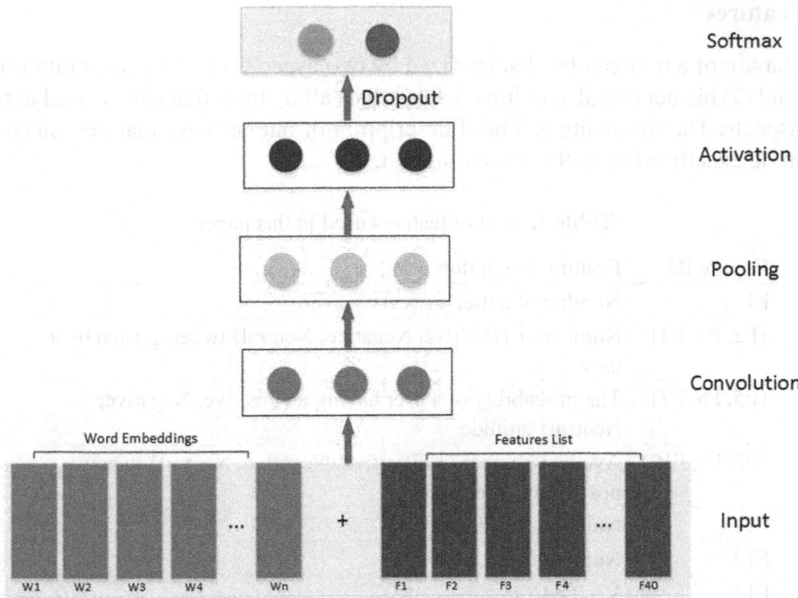

Fig. 1. The architecture of the proposed model

embeddings [6]. We pre-train the 200-dimensional word embeddings on each dataset. Therefore, each tweet is tokenized and each generated token is mapped to a distributional feature representation, known as word embedding. Moreover, the features that describe the writer of that tweet are appended to the generated vector and then fed into the next layer. Next is convolutional layer; its main goal is to extract patterns.

In order to allow the learning of non-linear decision boundaries, a non-linear activation function is located at the activations layer. There are a number of common choices of activation functions used with neural networks; for example, sigmoid (or logistic), hyperbolic tangent tanh, and a rectified linear (ReLU) functions. In our model, we use ReLU because as it is pointed out by several studies; such as [7], that ReLU speeds up the training and produces more accurate results than other activation functions.

The softmax layer is an activation function whose output is the probability distribution over labels. In this 2-classes task, given an input representation vector v a softmax operation is computed as follows:

$$softmax(v, l) = \frac{\exp(v_l)}{\sum_1^L \exp(\widehat{v}_l)} \tag{1}$$

Where L is the number of sentiment classes and $\widehat{v}_l$ is the predicted probability of sentiment class l. The reason for adding the dropout layer before the softmax layer is to prevent overfitting.

3.2 Features

The behavior of a user can be characterized by two aspects: (1) the personality traits of a user and (2) his/her social activities. Table 1 lists all features that can be used to reveal these aspects. The following is a brief description of one of these features, since most of them do exactly what as the names suggest.

Table 1. List of features used in this paper

Feature ID	Feature description
F1	Number of a user's tweets
{F2, F3, F4}	Number of {Positive, Negative, Neutral} tweets posted by a user
{F5, F6, F7}	The probability of a user having a {Positive, Negative, Neutral} attitude
{F8, F9, F10}	Average number of {Positive, Negative, Neutral} tweets posted by a user
F11	Number of followers
F12	Number of friends
F13	Verified
F14, F15	Number of adjectives, and their average
F16, F17	Number of nouns, and their average
F18, F19	Number of adverbs and their average
F20, F21	Number of verbs, and their average
F22, F23	Number of hashtags, and their average
F24, F25	Number of mentions, and their average
F26, F27	Number of URLs, and their average
F28, F29	Number of emoticons, and their average
F30, F31	Number of question marks, and their average
F32, F33	Number of exclamation marks, and their average
F34, F35	Number of words per tweet, and their average
F36, F73	Number of positive words in bing liu lexicon, and their average
F38, F39	Number of negative words in bing liu lexicon and their average
F40	Number of retweets

Number of Positive, Negative and Neutral Tweets Posted by a User and Their Averages. In our experiments, the frequencies of tweets based on three aspects, positive, negative and neutral are used. To achieve this, tweets from each user in the dataset are collected; this provides us with a huge number of unlabeled tweets. Labeling all these tweets manually is not an easy task to do and it will be a very time consuming and costly. To solve this issue we adopt the best solution used by the researchers; the SentiStrength algorithm. It was developed by [8], and it uses a scoring range from −5 (very negative) to +5 (very positive). The SentiStrength outputs two integers, one for each label. Accordingly, we create an application that receives the SentiStrength's output and computed as follow.

$$SS(t_i) = \begin{cases} Positive \;\; if \; SS(t_i)_{pos} > 1 \; and \; SS(t_i)_{pos} > SS(t_i)_{neg} \\ Negative \;\; if \; SS(t_i)_{neg} > 1 \; and \; SS(t_i)_{neg} > SS(t_i)_{pos} \\ Neutral \;\; if \; SS(t_i)_{pos} = SS(t_i)_{neg} = 1 \; or \; SS(t_i)_{pos} = SS(t_i)_{neg} \end{cases} \tag{2}$$

For each user $u \in U$, the application calculates $SS(t_i)$. Where $SS(t_i)$ is a function that takes as input a tweet $t_i \in T_u$; [T_u has all tweets retrieved from user u's timeline], and returns its label based on SentiStrength algorithm. $SS(t_i)_{pos}$ and $SS(t_i)_{neg}$ are scores of both positive and negative sentiment strengths, respectively.

4 Experiments and Results

4.1 Datasets

In this paper, we train our model on two datasets. Both of them are Twitter datasets published by SemEval-2016[3]. They are annotated manually; positive and negative labels. For each dataset, a list of users who posted the tweets is created. Then, tweets in timelines of each listed user along with their public profiles data are retrieved. Table 2 shows statistical information of the datasets used in our experiments.

Table 2. Labels and users distribution in datasets

	SemEval-2016_1	SemEval-2016_2
Number of tweets	3,694	1,122
Number of positive tweets	3,054	832
Number of negative tweets	643	290
Number of users	3,536	2,198
Number of tweets from users timelines	774,244	491,902
Average number of tweets per user	218,96	223,80

4.2 Result

All experimental results are reported using 10-fold cross validation. The performance of our model is compared with the following baseline methods on both datasets; Naive Bayes[4] (NB) and Support Vector Machines (SVM).

Since many features are included in our experiments, we find it suitable classify features into sets. In addition, using sets provides us with more clarity of which set has more influence on the performance of the model. The sets are shown in Table 3.

[3] http://alt.qcri.org/semeval2016/.

[4] It is a probabilistic algorithm that uses Bayes rule.

Table 3. The feature sets with different combinations.

Set no.	Features used
1	F5, F6 and F7
2	F1 to F13
3	F2, F3 and F4
4	Word embedding, F5, F6 and F7
5	Word embedding
6	All features (F1 to F40)
7	All features (F1 to F40) + Word embedding
8	All features (F1 to F40) of users with more than 200 tweets in their timelines
9	Same as set no. 8 except for here we eliminate outer and extreme values

Table 4 shows the accuracy of all classification methods. Generally, the proposed CNN classifier keeps its performance at a steady trend, unlike SVM and NB classifiers.

Table 4. Accuracy rates of CNN, SVM and NB

Set no.	1	2	3	4	5	6	7	8	9
Accuracy									
CNN	82.63	82.85	82.93	88.71	87.08	83.90	88.46	86.94	83.55
SVM	82.58	82.58	82.58	86.84	86.67	82.58	86.75	86.39	82.46
NB	82.63	48.31	83.62	85.16	85.16	59.38	70.40	84.43	75.31

The effect of the unbalance dataset is observed clearly on NB and SVM. One possible reason the unbalance dataset does not has a big effect on CNN classifier performance is the way the weights are calculated in this type of classifier.

It can be observed that the best classification performance in accuracy comes from CNN and SVM. However, CNN has higher recall and precision than SVM across all sets. This means using accuracy as an only measurement for a classifier performance is

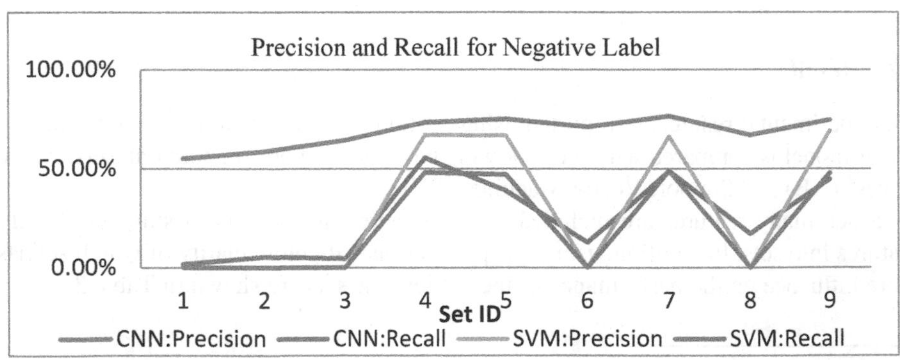

Fig. 2. Precision and reacall for the negative label of CNN and SVM.

not recommended. Figure 2 makes this observation clear. Although the number of negative instances in both datasets is lower than the positive, the proposed CNN model renders a good and stable performance.

The poor performance of NB is not surprising since NB relies on the assumption of conditional independence among the features, something that is clearly not valid here. All the features used in the experiments have some degree of inter-dependency.

Finally, we notice that the best accuracy comes from CNN classifier which is 88.71 with set number 4. This supports our claim of the efficiency of utilizing information beyond the content of a given tweet since the features F5, F6 and F7 are computed by retrieving the timeline tweets of the user.

5 Conclusions

In this paper, we presented a sentiment analysis model developed by combing a list of features. We proposed the architecture of a Convolutional Neural Network (CNN) that takes into account not only the text but also user behavior. Our evaluation results demonstrated the efficiency of the model in social media setting.

Our model outperformed the baseline methods in accuracy, recall and precision. Also, the proposed model was affected less by unbalance dataset issue.

This work further suggests interesting directions for future work. For example, it would be interesting to investigate the contributions of the produced list of features for non-binary sentiment classification tasks. In future work, we also plan to explore other learning neural network based models, such as Recurrent Neural Network (RNN), Long Short-Term Memory (LSTM), gated feedback RNN for sentiment analysis.

References

1. Thelwall, M., Buckley, K., Paltoglou, G.: Sentiment in Twitter events. J. Am. Soc. Inf. Sci. Technol. **62**, 406–418 (2011)
2. Katz, G., Ofek, N., Shapira, B.: ConSent: context-based sentiment analysis. Knowl. Based Syst. **84**, 162–178 (2015)
3. Tsagkalidou, K., Koutsonikola, V., Vakali, A., Kafetsios, K.: Emotional aware clustering on micro-blogging sources. In: D'Mello, S., Graesser, A., Schuller, B., Martin, J.-C. (eds.) ACII 2011. LNCS, vol. 6974, pp. 387–396. Springer, Heidelberg (2011). doi: 10.1007/978-3-642-24600-5_42
4. Mudinas, A., Zhang, D., Levene, M.: Combining lexicon and learning based approaches for concept-level sentiment analysis. In: Proceedings of the First International Workshop on Issues of Sentiment Discovery and Opinion Mining - WISDOM 2012, pp. 1–8 (2012)
5. Davidov, D., Tsur, O., Rappoport, A.: Enhanced sentiment learning using Twitter hashtags and smileys. In: Proceedings of the 23rd International Conference on Computational Linguistics Posters, pp. 241–249 (2010)
6. Mikolov, T., Chen, K., Corrado, G., Dean, J.: Distributed representations of words and phrases and their compositionality. In: Proceedings of the 26th International Conference on Neural Information Processing Systems, pp. 3111–3119 (2013)

7. Severyn, A., Moschitti, A.: Twitter sentiment analysis with deep convolutional neural networks. In: Proceedings of the 38th International ACM SIGIR Conference on Research and Development in Information Retrieval - SIGIR 2015, pp. 959–962 (2015)
8. Thelwall, M., Buckley, K., Paltoglou, G., Cai, D., Kappas, A.: Sentiment in short strength detection informal text. J. Am. Soc. Inf. Sci. Technol. **6**, 2544–2558 (2010)

Topological Structure Analysis of Developmental Spiking Neural Networks

Xianghong Lin(✉), Ying Li, and Jichang Zhao

School of Computer Science and Engineering, Northwest Normal University, Lanzhou 730070, China
linxh@nwnu.edu.cn

Abstract. The complex network structure of biological brains is obtained through the developmental processes. Type and complexity of network structure directly reflect the ability of the network to deal with information processing. In this paper, we propose a developmental method for creating recurrent spiking neural networks based on genetic regulatory network model. This research investigates the developmental process of spiking neural networks, and analyzes the network structure in the different parameter settings, such as the number of regulatory nodes, the weights scale of genetic networks, and the developmental scale. The experimental results show that the developmental spiking neural networks have the similar topological characteristics as biological networks, namely scale-free and small-world properties.

Keywords: Spiking neural network · Developmental process · Recurrent genetic regulatory network · Scale-free · Small-world

1 Introduction

In recent years, inspired by evolution and development of biological neural system, spiking neural networks (SNNs) have been paid more and more attention [1, 2]. The brain-inspired SNNs are highly nonlinear complex networks connected by a large number of neurons. They have powerful parallel processing, distributed storage, adaptive learning, a high degree of fault-tolerant capability. In most approaches of the SNNs evolution, the network weights are evolved and the network topologies are predefined by subjective experience of designer and repeated experiments. Such network topologies may not guarantee the optimal network performance, and its application and development is limited to some extent. For evolution of the SNNs topologies, the researchers proposed many developmental approaches to solve above problem. According to different expression of genotypes, developmental methods are divided into two categories: one is grammatical method which including graph generation system [3, 4], cellular encoding [5] and so on; the other is genetic regulatory networks (GRNs) [6–8]. In grammatical method, a set of grammatical rules for expressing the developmental process are used to represent genotypes and the SNNs are generated by repeated grammar rules. Ahmadizar [9] proposed an evolutionary-based algorithm that is developed to simultaneously evolve the topology. GRNs are a dynamic network model based on DNA or

D.-S. Huang et al. (Eds.): ICIC 2017, Part I, LNCS 10361, pp. 89–100, 2017.
DOI: 10.1007/978-3-319-63309-1_9

RNA in biological cells, and they regulate cell division, differentiation and migration through gene expression. Federici et al. [10] applied a recurrent neural network to express a GRN and used this model to develop a neural network with fault tolerance. By using indirect coding, some methods for gradually developing phenotype from genotype are investigated. These methods have a advantage that a gene is expressed multiple times at different developmental stages and can be rapidly searched for SNNs in smaller gene spaces. In fact, GRN is an ideal model for controlling the development of evolutionary neural networks at the genetic level.

In this paper, we propose a recurrent GRNs model for generating developmental recurrent SNNs. By analyzing the influence of some parameters on topology of developmental SNNs, we found that the developmental SNNs have similar structural characteristics to biological networks, that is, scale-free and small-world properties.

2 Recurrent Genetic Regulatory Networks Model

GRNs are capable of modelling regulatory networks for gene transcription, simulating and analyzing for specific species or tissue structure in the whole process of gene expression. GRNs transform abstract process of biological differentiation into a man-made control framework system by computer science, bioscience and other technical means, to further explore and understand essence of life and manner of transmitting information in organisms. In the process of modeling GRNs, we should consider not only positive gene expression to generate network structures, but also reverse effect of gene regulation on the structures. Thus, recurrent GRN models can be approximated to exhibit biological characteristics such as interactivity, complexity, randomness, high dimensionality and so forth.

The structural diagram of a recurrent GRN is shown in Fig. 1. The recurrent GRNs contain three layers, including the input layer, regulation layer and output layer, and the number of neurons in each layer is N_I, N_R and N_O respectively. N_R is determined by the genes in the genome, N_I and N_O is determined by the corresponding developmental model. In the network, all nodes of each layer are fully connected feedforward, that is, each input node can directly send information to all regulatory nodes, and all regulatory nodes also can directly send the information that needs to be delivered to each output node. However, the regulatory nodes represent genes in the genome and play a role in gene regulation, in which each node is recurrently connected and the connection degree k is 8.

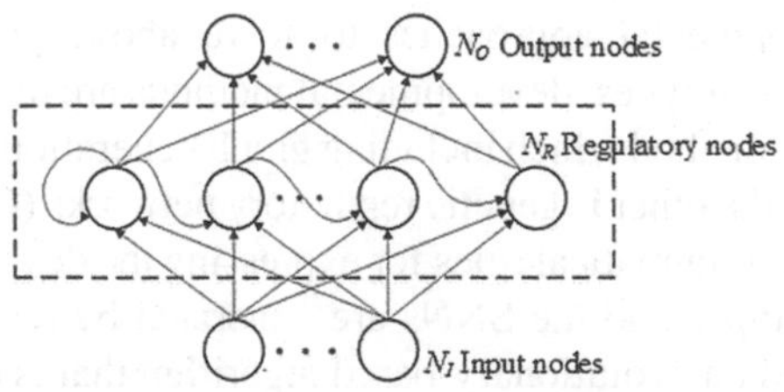

Fig. 1. The structure of a recurrent genetic regulatory network model.

In this paper, we utilize weight matrices [11] that represent the mutual regulation of genes to model the recurrent GRNs. The recurrent GRNs can be transformed as follows: the initial gene expression state is obtained after the initialization of the weight, where n-dimensional space vector $v(t)$ represents the expression state of n genes at time t, and each element of $v(t)$ corresponds to the expression level of each gene at time t. The regulation intensity between the two genes is represented by w, which is a real value in the interval $[0, W]$, where W represents weight scale. In a GRN with the weight matrix, a single gene expression state is discrete and irrelevant. At this time, all gene expression states can be updated at the next moment. The activation $A_i(t+1)$ of gene i at time $t+1$ can be expressed by:

$$A_i(t+1) = \sigma\left(\sum_{j=1}^{n} w_{ij}v_j(t) - \theta_i\right) = \sigma\left(r_i(t) - \theta_i\right), \tag{1}$$

where $r_i(t)$ represents total gene input of gene i at time t, w_{ij} represents a regulation weight of gene j to gene i, and $v_j(t)$ represents the activation of gene j at time t. $\sigma(x)$ is the sigmoid function, given by

$$\sigma(x) = \frac{1}{1+e^{-x}}. \tag{2}$$

θ_i is the activation threshold of node i, given by

$$\theta_i = \sum_{j=1}^{n} w_{ij}\delta, \tag{3}$$

where δ is weight bias (by default, $\delta = 0.5$).

3 Developmental Method of Spiking Neural Networks

Development of a specific SNN is controlled by a recurrent GRN. Considering convenience of description, during the developmental process, at time t, the activation of the i-th input node, the activation of the i-th regulatory node and the activation of the i-th output node are expressed as $R_i(t)$, $i = 1, 2, \ldots, N_R$, $I_i(t)$, $i = 1, 2, \ldots, N_I$ and $O_i(t)$, $i = 1, 2, \ldots, N_O$, respectively. The output nodes activation is divided into two parts: one part as a qualitative feature is utilized to compare with a given threshold, and the other part convert into specific parameter values within given range $[P_{min}, P_{max}]$, which is represented as $P_{min} + (P_{max} - P_{min})O_i(t)$.

3.1 The SNN Developmental Procedure

The procedure of generating a SNN from a GRN is shown in Fig. 2. The initial GRN represents an embryonic cell that can develop into a nervous system, and activation of its nodes is initialized randomly, namely in an interval [0, 1]. The state of the network is synchronized in a discrete time step, each of which represents a single cell division or differentiation event. The modeling process of a SNN is adjusted by the activation of output nodes. Firstly, an embryonic cell are divided into two different types of daughter cells by the cell division, secondly, these daughter cells are divided into again sub-cells continuously, then, all daughter cells form a cell division tree until without cell division, finally they are differentiated into the spiking neurons. Among them, the connections between neurons are generated by spiking neurons, and the neural network is formed by the connection of neurons.

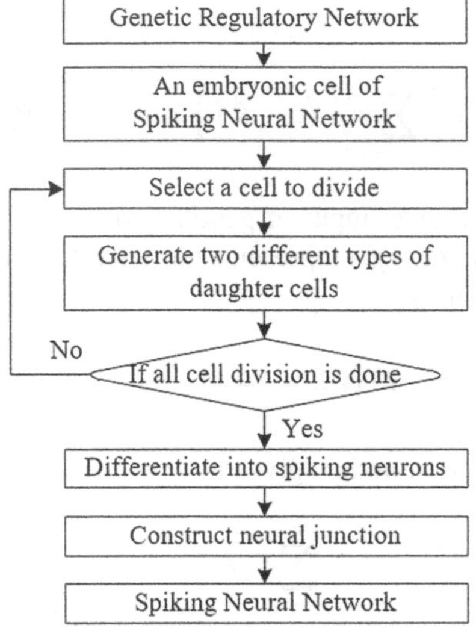

Fig. 2. Developmental procedure of a spiking neural network.

3.2 The Developmental Control Processes

In a GRN, the regulation of output nodes activation is divided into three parts, where first part controls cell division, second part determines parameters to generate spiking neuron, and third part regulates the connection between the neurons and determines the connection properties.

(1) **Generation of cell division tree.** Suppose that there are 3 output nodes to control the development of SNN, in which the activation of all nodes is evenly within the interval [0, 1]. Thus, the activation of the output node $O_1^d(t)$ indicates possibility of cell division whether the cell is continue to split or has been divided into final neuron. By introducing the dividing thresholds θ_d, if $O_1^d(t) \leq \theta_d$, the cells will differentiate into specific spiking neurons, otherwise the cells will divide. To simulate this phenomenon, the dividing threshold has dynamic changes in the developmental process and it is expressed as $\theta_d = 0.01e^{\lambda d}$, where d represents the current cell depth in the cell division tree and λ represents developmental scale that is used

to control developmental scale of the spiking neural network. The activation of output node $O_2^d(t)$ indicates direction of cell division whether it is horizontal or vertical division, where if $O_2^d(t) \leq 0.5$, the cells will divide into two daughter cells of left and right distribution on horizontal direction, otherwise the cells will divide into two sub-cells of the upper and lower distribution on vertical direction. The activation of output node $O_3^d(t)$ indicates distribution of the daughter cells, which it determines the mother cell for symmetrical or asymmetric division, where if $O_3^d(t) \leq 0.5$, the mother cell will symmetrically divide and the two daughter cells have the asymmetric division gene with the same activation, otherwise the mother cell will asymmetrically divide and the activation of the asymmetric division gene are also different in generating two sub-cells. However, asymmetric division only affects the asymmetric division gene activation in sub-cells, the rest of the regulatory gene activation is the same as the mother cell.

(2) **Generation of spiking neuron.** The spiking neurons are formed by all terminal node cells at the end of cell division, and the cells that have never been divided. In developmental SNNs, the number of hidden neurons M_H depends on the cells in the cell division tree, M_I and M_O represent the number of input and output layers neurons respectively, where specific values are determined by the size of the network and the practical problems. Thus, the features and properties of the generated spiking hidden neurons are encoded by activation of output nodes $O_i^n(t)$, $i = 1, 2, \ldots, 5$, and their specific number depends on complexity of target spiking neurons, where $O_1^n(t) \sim O_5^n(t)$ represent leak reversal potential, membrane time constant, threshold potential, reset potential and absolute refractory period of the hidden neurons, respectively.

(3) **Construction of neural connections.** The formation of neural connections indicates generation of SNN models. The establishment of neural connections and the characteristics are also encoded by output nodes, the output node activation is expressed by $O_i^c(t)$, $i = 1, 2, \ldots, 7$. Each output node has its own function and coding object in the process of neural junction, where $O_1^c(t)$ represents connection property of the neuron interface, namely the connection between the hidden neuron and the input/output neurons; $O_2^c(t)$ and $O_3^c(t)$, as the branch connection attribute, represent the recurrent connection between the hidden neurons and their own; $O_4^c(t)$ represents presynaptic weight maker, $O_5^c(t)$ denotes postsynaptic weight marker, all of which are synaptic weight of spiking neurons and $O_6^c(t)$ and $O_7^c(t)$ is the presynaptic and postsynaptic maker delay, respectively.

The connections between the hidden neurons are established according to the following rules: firstly, we find a target neuron that connects the source neuron, that is, starting from the source neuron, we can find out an ancestor node according to the reverse order of the cell division by $dO_2^c(t)$ times, where d is the depth of the terminal neuron in entire cell division tree; secondly, the ancestor node is regarded as the root node and the target neuron is searched according to certain rules to find the target neurons: if $2O_c^3(t) < 1$, target neuron is sought in the direction of the left/next daughter cells and $O_c^3(t) = 2O_c^3(t) - 1$, otherwise, it is directed toward the direction of the opposite daughter cells and $O_c^3(t) = 2O_c^3(t)$; finally, each daughter cell repeats the above rules until the terminal node cells are reached, and the terminal node cell is the target neuron.

When the cell division tree is projected in a certain way, the space position between neurons was transformed into plane position distribution and we can find the terminal node neuron and its adjacent neurons in target neuron, namely the target neurons group. The terminal neurons establish the connection with all members of the target neurons group.

4 Structural Analysis of Developmental Neural Networks

The experiments of GRNs reflect the scale-free and small-world properties in a simple single nuclear microbial or a complex intelligent organisms. These two properties can effectively improve the robustness, reduce the failure rate and enhance information processing and transmission rate in the networks. A series of experiments analyzed the topological structure of the developmental neural networks by scale-free and small-world properties.

In order to study the effect of the control parameters in the recurrent GRNs, we select the number of regulatory nodes, the weight scale in the GRNs and the developmental scale as variables in the developmental process, and the rest of the parameters to do the basic settings, shown in Table 1 for details. However, the number of input and output neurons of developmental SNNs are generally determined by the actual problem and the size of these networks. Here, we focus on the research of topological structure in SNNs.

Table 1. Basic parameter setting of genetic regulatory networks

Variable	Description	Value
N_I	Number of input nodes	2
N_R	Number of regulatory nodes	32
N_O	Number of output nodes	15
W	Weight scale of GRNs	3.0
λ	Developmental scale	0.5

4.1 Structural Properties of Complex Networks

Scale-free property [12] mainly reflects average degree and degree distribution in neural network structure. Average degree k represents that a neuron is connected to k neurons on average, that is to say, the average degree of a node is k in light of the perspective of graph theory. Degree distribution is more specific to describe dynamic diversity of network structure than average degree.

In the biological neural network for gene replication, transcription, expression, protein production, metabolism and other activities, degree distribution $P(k)$ of the network obey power law as follows:

$$P(k) \sim k^{-\gamma},\ \gamma > 1 \tag{4}$$

where k is the connection degree between nodes in the network and γ is an exponential parameter greater than one.

Small-world property [13, 14] is mainly composed of characteristic path length L in the neural network, that is, average typical distance (minimum number of edges connecting two random nodes) between all nodes in the network, and node cluster coefficient C_i. The path length L as follows:

$$L = \left(\frac{1}{N(N-1)} \sum_{i,j \in V} \frac{1}{d_{ij}} \right)^{-1}, \quad (i \neq j), \tag{5}$$

where N is total number of nodes, V is set of all nodes in the network, d_{ij} is the typical distance between node i and node j, and the typical distance between nodes is defined as $d_{ij} = +\infty$.

The cluster coefficient C_i reflects integration level of the network. The cluster coefficients average C of all nodes is defined as the whole cluster coefficients, and its expression:

$$C = \frac{1}{N} \sum_{i \in V} C_i = \frac{1}{N} \sum_{i \in V} \frac{e_i}{k_i(k_i - 1)}, \tag{6}$$

where k_i is degree of node i, e_i is number of directional connections within the target neuron group, and C_i represents the connection relation within the target neuron group that terminal neuron i is connected.

According to the above definition, to determine whether a network has the small-world property, it is necessary to calculate its specific path length L and cluster coefficient C, and compare it with a random network with same nodes number and average degree. If $L/L_{random} \geq 1$ and $C/C_{random} \gg 1$, the network has a small-world property.

4.2 GRN Parameter Analysis in Developmental Process

In this section, the effect of the above two parameters on the number, average degree, degree distribution, characteristic path length as well as cluster coefficient average of hidden neuron in developmental SNNs were investigated.

Firstly, the influence of the size of control nodes on the scale-free property of developmental neural networks is analyzed. The scale of regulatory nodes N_R is increased from 8 to 56 with an interval of 8 and the other parameters take the benchmark values. Figure 3 shows the scale-free property of developmental spiking neural networks with different regulatory neurons numbers of GRNs. Figure 3(a) describes the effect that the regulatory nodes number on the amount and average degree of hidden neurons. As shown in the figure, the hidden neurons number varies on a parabola going up with the regulatory nodes number. Such as, when N_R is 16, that of hidden neurons is 247.97 and when N_R is 48, that of hidden neurons is 237.39. With the number of regulatory nodes gradually increases, average degree of hidden neurons is gradually decreases, as can be seen from this figure. As when N_R is 16, average degree of hidden neurons is 13.15 and average degree of hidden neurons is 11.74 when $N_R = 48$. Figure 3(b) indicates the degree distribution for N_R of 200 SNNs are 8, 32 and 56. The degree distribution curve of the network peaked at a certain point, but the basic trend was declining. Overall, the change of the number of control nodes has a weak effect on the curve shape of degree distribution in the networks.

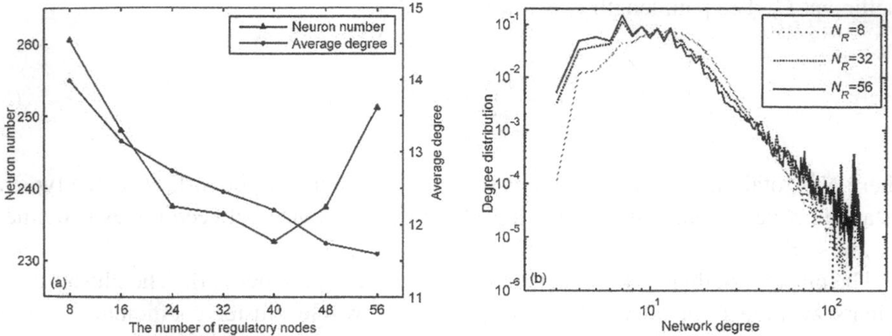

Fig. 3. Effect of number of regulatory neurons on the scale-free property in the developmental spiking neural networks. (a) Effects on the number and average degree of hidden neurons. (b) The degree distribution with different numbers of regulatory nodes.

Then, the influence of weight scale W of GRNs on the scale-free property of developmental SNNs is analyzed. The value of W is increased from 1.0 to 7.0 with an interval of 1.0, and the other parameters remain unchanged. Figure 4 shows the scale-free property of the developmental SNNs with different weight scales. Figure 4(a) shows the effect on the number and average degree of hidden neurons in different weight scales. From Fig. 4(a) we can see that the number of hidden neurons decreases at first, then increases when the weight scale increases gradually, so the suitable weight scale for the developmental SNNs is 3. Such as, the number of hidden neurons is 236.38 when $W = 3.0$, and when it is 5.0, the number of hidden neurons is 241.64. The effects of the value of W on the average degree of hidden neurons is also represented in the Fig. 4(a). When W is gradually increased, the average degree of neurons in the hidden layer is gradually decreased. For example, the average degree of the hidden neurons is 12.45 when $W = 3.0$ and the corresponding values is 10.68 when $W = 5.0$. Figure 4(b) shows the degree distribution for 200 SNNs with weight scales 1.0, 4.0 and 7.0. The degree distribution curve of the network reaches a maximum at a certain point, but the basic trend is declining. In general, the change of the regulatory nodes number has a weak effect on the curve shape of the degree distribution in the SNNs.

Finally, the developmental neural networks topological structures whether they have the small-world property are analyzed, where the number of regulatory nodes and the weight scale are taken as the benchmark value. Figure 5 shows the distribution of L/L_{random} and C/C_{random} in 200 SNNs, where the number of regulatory nodes and the synaptic weight are 32 and 3.0 respectively. For above networks, the value of L/L_{random}

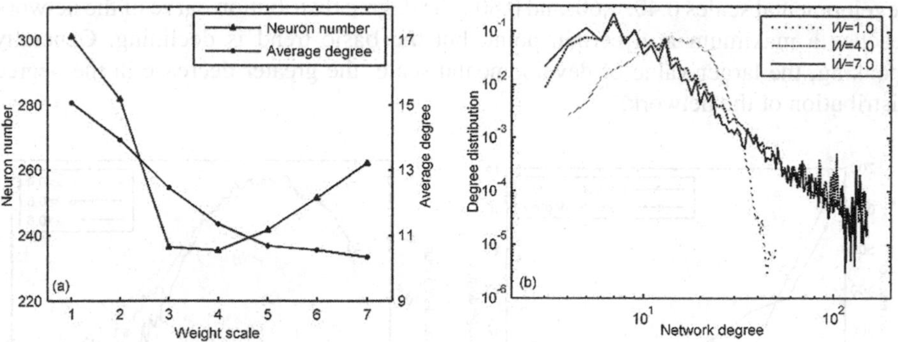

Fig. 4. Effect of weight scale of GRNs on the scale-free property in the developmental spiking neural networks. (a) Effects on the number and average degree of hidden neurons. (b) The degree distribution with different weight scales.

and C/C_{random} are 1.29 and 10.27, respectively. Thus, the developmental neural networks have obvious small-world properties.

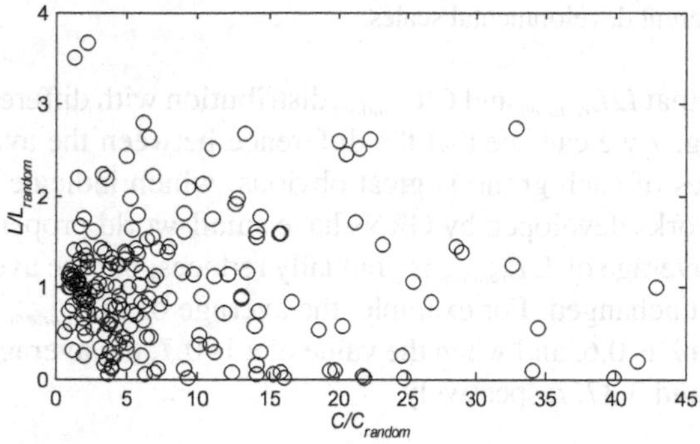

Fig. 5. L/L_{random} and C/C_{random} distribution of spiking neural network.

4.3 Developmental Scale Analysis in Developmental Process

Developmental scale λ is another important parameter for topological analysis of developmental SNNs. Mainly, influence of λ on the developmental SNNs is analyzed in this section. The value of λ is increased from 0.40 to 0.80 with an interval of 0.05 and the other parameters are consistent with the reference values.

Figure 6 shows the scale-free property of the developmental SNNs with different developmental scales. Figure 6(a) shows variation on the number and average degree of the neurons in developmental neural networks. With the value of λ gradually increases, the number and average degree of hidden neurons both decreases gradually. For example, the number and average degree of hidden neurons are 236.38 and 12.45 respectively when $\lambda = 0.50$; the corresponding values are 52.56 and 10.80 respectively when $\lambda = 0.70$. Figure 6(b) shows the degree distribution for 200 SNNs with developmental scales 0.40, 0.60, and 0.80. The degree distribution curve of the network reaches a maximum at a certain point, but the basic trend is declining. Generally speaking, the larger value of developmental scale, the greater decrease in the degree distribution of the network.

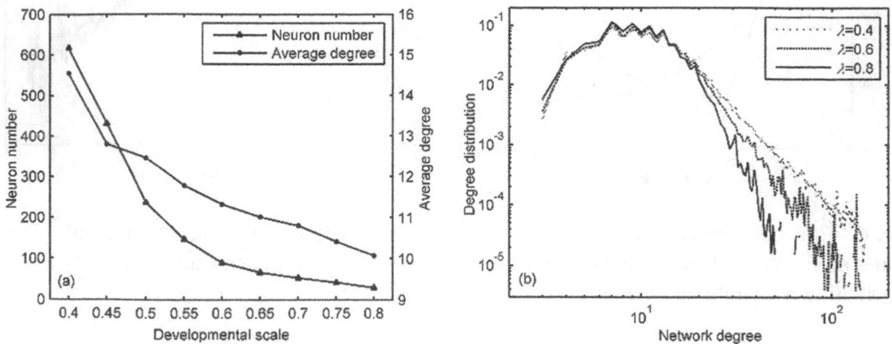

Fig. 6. Effect of developmental scales on the scale-free property in the developmental spiking neural networks. (a) Effects on the number and average degree of hidden neurons. (b) The degree distribution with different developmental scales.

Figure 7 shows that L/L_{random} and C/C_{random} distribution with different developmental scales. Through Fig. 7 we can see that the difference between the average of L/L_{random} and C/C_{random} values of each group is great obvious, which indicate that the developmental neural networks developed by GRNs have small-world property. With λ gradually increases, the average of L/L_{random} is gradually reduces, but the average of C/C_{random} remain essentially unchanged. For example, the average of L/L_{random} and C/C_{random} are 1.13 and 4.87 when λ is 0.6, and when the value of λ is 0.7, the average of L/L_{random} and C/C_{random} are 1.19 and 3.37, respectively.

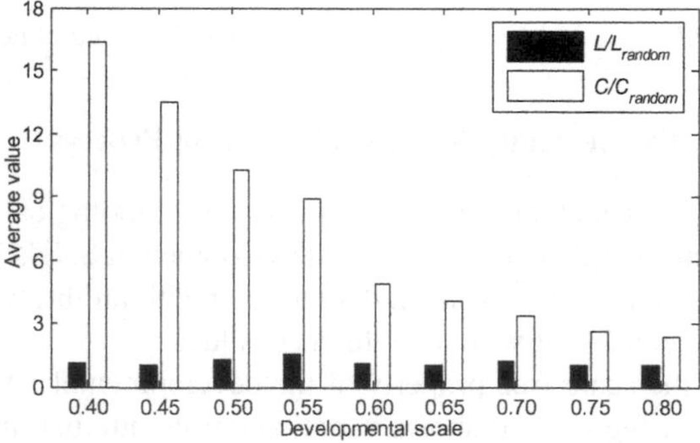

Fig. 7. The average of L/L_{random} and C/C_{random} distribution with different developmental scales.

5 Conclusions

SNNs are comprised of spiking neurons, which are shown to be suitable tools for the processing of spatio-temporal information. Based on the dynamic gene expression of GRN model, we propose a developmental method for creating recurrent SNNs, in which the neurons are connected in a recurrent manner. The influence of different parameters on the developmental SNN topology is analyzed. The experimental results show that the networks have structural features, namely, scale-free and small-world properties, similar to those of biological networks. In addition, the SNNs generated by the GRNs, especially the gene fragment replication and disproportionation, plays an important role in the performance of network structure and dynamic characteristics. In the future research, we will further investigate the spike firing properties for the spiking neural networks generated by the proposed developmental method.

Acknowledgement. This research is supported by the National Natural Science Foundation of China under Grant No. 61165002, and the Natural Science Foundation of Gansu Province of China under Grant No. 1506RJZA127, and the Scientific Research Project of Universities of Gansu Province under Grant No. 2015A-013.

References

1. Kasabov, N.K.: NeuCube: a spiking neural network architecture for mapping, learning and understanding of spatio-temporal brain data. Neural Netw. **52**, 62–76 (2014)
2. Diehl, P.U., Zarrella, G., Cassidy, A., et al.: Conversion of artificial recurrent neural networks to spiking neural networks for low-power neuromorphic hardware. In: Proceedings of the IEEE International Conference on Rebooting Computing, pp. 1–8 (2016)
3. Kitano, H.: Designing neural networks using genetic algorithms with graph generation system. Complex Syst. **4**(4), 461–476 (1990)
4. Homma, N., Aoki, T. Higuchi, T.: Evolutionary graph generation system with transmigration capability for arithmetic circuit design. In: Proceedings of the 2001 IEEE International Symposium on Circuits and Systems, vol. 5, pp. 171–174 (2001)
5. Gruau F.: Neural network synthesis using cellular encoding and the genetic algorithm. Ph.D. thesis, Université Claude Bernard-Lyon (1994)
6. Longabaugh, W.J., Davidson, E.H., Bolouri, H.: Computational representation of developmental genetic regulatory networks. Dev. Biol. **283**, 1–16 (2005)
7. Martín, M., Organista, M.F., de Celis, J.F.: Structure of developmental gene regulatory networks from the perspective of cell fate-determining genes. Transcription **7**(1), 32–37 (2016)
8. Lones, M.A.: Computing with Artificial Gene Regulatory Networks: Evolutionary Algorithms in Gene Regulatory Network Research. Wiley, Hoboken (2016)
9. Ahmadizar, F., Soltanian, K., AkhlaghianTab, F., et al.: Artificial neural network development by means of a novel combination of grammatical evolution and genetic algorithm. Eng. Appl. Artif. Intell. **39**, 1–13 (2015)
10. Federici, D.: A regenerating spiking neural network. Neural Netw. Off. J. Int. Neural Netw. Soc. **18**(5–6), 746–754 (2005)
11. Weaver, D.C., Workman, C.T., Stormo, G.D.: Modeling regulatory networks with weight matrices. In: Proceedings of Pacific Symposium on Biocomputing, vol. 4, pp. 112–123 (1999)

12. Fazekas, I, Porvázsnyik, B.: Scale-free property for degrees and weights in a preferential attachment random graph model. J. Probab. Stat. (2013). Article ID 707960
13. Suzuki, T., Okazawa, M., Ohkura, K.: Small-world property evaluated by exchanging network topology. Int. J. Mod. Phys. C **26**(11), 1550122 (2015)
14. Mehlhorn, H., Schreiber, F.: Small-world property. In: Dubitzky, W., Wolkenhauer, O., Cho, K.-H., Yokota, H. (eds.) Encyclopedia of Systems Biology, pp. 1957–1959. Springer, New York (2013). doi:10.1007/978-1-4419-9863-7_2

A Unified Deep Neural Network for Scene Text Detection

Yixin Li and Jinwen Ma(✉)

LMAM, Department of Information Science, School of Mathematical Sciences, Peking University, Beijing 100871, China
liyixin@spku.edu.cn, jwma@math.pku.edu.cn

Abstract. Scene text detection is important and valuable for text recognition in natural scenes, but it is still a very challenging problem. In this paper, we propose a unified deep neural network for scene text detection, which is composed of a Fully Convolutional Network (FCN) for text saliency map generation and a Bounding box Regression Network (BRN) for text bounding boxes prediction. The FCN is trained with a hybrid loss function based on two types of pixel-wise ground truth masks while the unified neural network is fine-tuned with a multi-task loss function. Additionally, the post-processing procedures including scoring the predicted bounding boxes by the saliency map and eliminating the redundant boxes via the Non-Maximum Suppression (NMS) method are applied to improve the final text detection results. It is demonstrated by the experimental results on ICDAR2013 benchmark that our proposed unified deep neural network can achieve good performance of text detection and process images at 5 fps, being faster than most of the existing text detection methods.

Keywords: Scene text detection · Fully Convolutional Network · Deep learning

1 Introduction

As a series of abstract symbols for human communication, text carries rich contextual and semantic information. Reading and understanding text in a natural image plays an important role in many computer vision tasks, such as image matching, robot navigation and human-computer interaction. Since the background can strongly affect the result of text recognition, it is necessary to detect or localize text lines in natural images before text recognition. Thus, text detection has become a popular research topic in text recognition and computer vision.

Although Optical Character Recognition (OCR) is considered to be a powerful character recognition tool for the scanned images, it is still rather difficult to detect and recognize text in natural scenes. In fact, we can spot and localize text instantly just by a glance of the scene, even if the text is written in a language we do not know. However, real-time text detection in natural scene can be a difficult task for a computer, due to the diversity of scene text, complexity of background and low quality of natural images [1]. Actually, text can be in different colors, fonts, sizes, and orientations even in a single natural image. On the other hand, some common objects in natural scene with certain textures like fences, trees and traffic signs, can be easily confused with text, and produce

D.-S. Huang et al. (Eds.): ICIC 2017, Part I, LNCS 10361, pp. 101–112, 2017.
DOI: 10.1007/978-3-319-63309-1_10

false positive samples to the text detection system. Moreover, natural images with low resolution, non-uniform illumination, and partial occlusion are not uncommon, which can be another challenge to the detection system.

To tackle these difficulties, a variety of methods and algorithms have been established from different aspects. The primary focus of these approaches is generally to learn an effective and robust feature representation of text. It is an extremely difficult task, due to the variation of text and interference from the environment. Conventional methods treat text as regions with certain texture, which are sensitive to the uneven lighting conditions and background interferences causing by human defined rules and handcraft features. Deep learning based methods are more robust under the supervised training on large amounts of labeled data. Most of the deep learning based methods predict a saliency map so that they are unable to generate bounding boxes directly. In this case a post-processing algorithm is needed to generate the final detection results.

In this paper, we propose a deep learning architecture for scene text detection which is composed of a Fully Convolutional Network (FCN) [2] and a Bounding Box Regression Network (BRN). The unified deep neural network is trained to generate a pixel-wise saliency map and predict the locations of candidate bounding boxes simultaneously. Then we score the generated bounding boxes by the text saliency map, and the non-maximum suppression (NMS) method [3] is adopted to filter the overlapping bounding boxes. It is demonstrated by the experiments on ICDAR2013 dataset [4] that our proposed unified network can achieve good text detection performance. Moreover, it can process 5 images per second, being faster than most of the existing scene text detection methods.

The rest of this paper is organized as follows. We briefly review the related text detection methods in Sect. 2. The methodology of our scene text detection approach is introduced in Sect. 3. Section 4 summarizes the experiment results and comparisons. Finally, a brief conclusion is given in Sect. 5.

2 Related Works

In recent years, with more and more attention to text detection, many effective algorithms and strategies have been established to extract and locate text in natural images. Most of current text detection methods mostly have a step-wise pipeline. They extract letter or word candidates or generate a text saliency map from the input image, then group the word candidates or regions with high response in saliency map into text lines, some of the approaches filter the text lines using an offline trained classifier to achieve higher precision.

Epshtein et al. [5] proposed the stroke width transform (SWT) to extract letter candidates and then merge them into text lines. Maximally Stable Extremal Regions (MSER) [6] was proposed in the early 2000s as a kind of affine invariant regions, and was brought into text detection task in 2010s [7]. MSER-based text detection algorithms [7, 8], taking MSERs as letter candidates, achieved stat-of-the-art performance on ICDAR2013 [4] benchmark, USTB_STAR [9] in particular, even won the ICDAR2013 robust reading competition [4]. Sun et al. [10] proposed the Color-enhanced Extremal Regions (CER) method which has been one of the most powerful text detection approaches without any deep learning technology. But these kinds of methods extract letter candidates with human defined rules and filter them with handcraft features, followed by many parameters which are very difficult to optimize.

In the past few years, deep learning, especially convolution neural networks, has been widely used in almost all the computer vision tasks, and text detection is no exception. Because of the powerful generalization abilities and the supervised training on big data, deep learning based text detection methods beat the conventional methods to the top of benchmarks. Also, the parameters in deep neural networks can be learned automatically. In the Text-spotter [11], a CNN filter was utilized to perform a sliding window type search and produce a saliency map to predict text regions. Zhang [12] and Yao [13] considered text detection as a semantic segmentation task, applied a FCN to generate a pixel-wise saliency map and then utilized a graph partition or classification algorithm to localize text lines. Although the deep neural networks can perform a feed-forward procedure rather fast after the training phase is completed, these methods can hardly be real-time due to the complex post-processing procedures.

3 Methodology

3.1 Overview

In this section, we describe our unified deep learning neural network architecture for scene text detection in detail. As shown in Fig. 1, our proposed text detection pipeline has two parts: a single deep neural network that can simultaneously generates text saliency map and locates the candidate bounding boxes, and an extremely simple post-processing procedure that filters the overlapping bounding boxes.

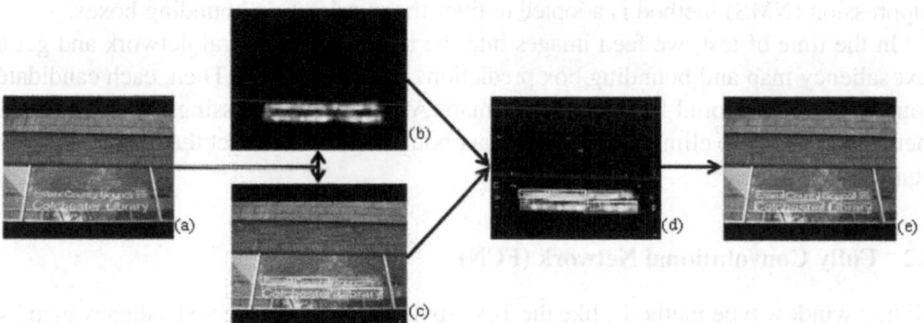

Fig. 1. The pipeline of the proposed unified text detection method. (a) Original image; (b), (c) Text saliency map and bounding boxes generated by the network; (d) Candidate bounding boxes on text saliency map; € Final detection result.

For the network part, we respectively train two networks: a Fully Convolutional Network (FCN) to predict each pixel and a small sized network sharing feature with the FCN called bounding box regression network (BRN) to learn the locations of bounding boxes. Then, these two networks sharing the layers before 'pool5' layer of the FCN are fine-tuned with a multi-task loss function together. The architecture of our unified deep neural network is shown in Fig. 2.

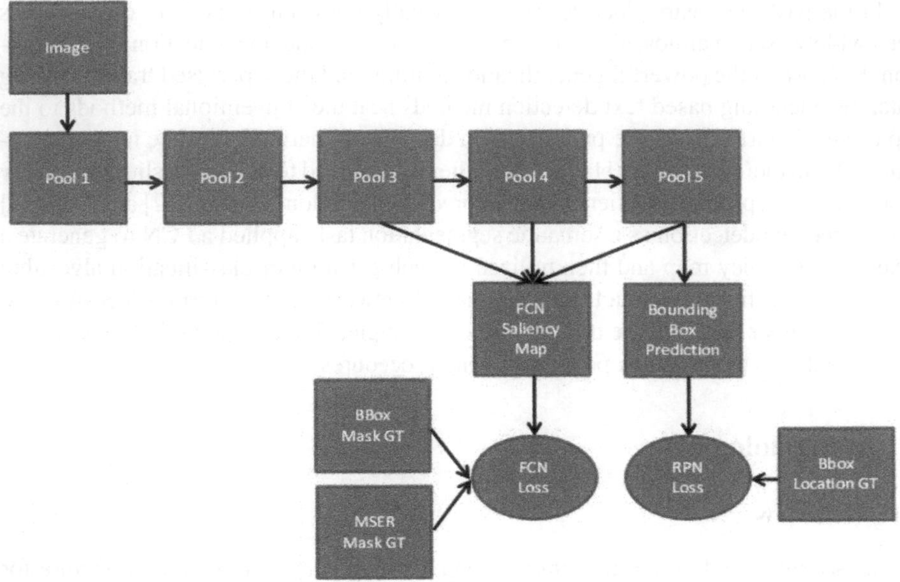

Fig. 2. The architecture of our proposed unified deep neural network.

As for the post-processing part, we use the saliency map generated by the FCN to score the candidate bounding boxes predicted by the BRN. Then, the non-maximum suppression (NMS) method is adopted to filter the overlapping bounding boxes.

In the time of test, we feed images into the unified deep neural network and get a text saliency map and bounding box predictions simultaneously. Then, each candidate bounding box is scored by the saliency map. At the post-processing stage, the NMS method is applied to eliminate the redundant bounding boxes to get the final text detection results.

3.2 Fully Convolutional Network (FCN)

Sliding window type methods, like the Text-spotter [11], generate text saliency map by classification confidence of each window. Each value of the confidence map only considers the feature within a limited region. Some objects like tree leaves, traffic signs and piano keyboards, with parts very similar to text, can be easily misclassified to text and produce false positives.

Actually, Fully Convolutional Network (FCN) [2] was recently proposed to solve the semantic segmentation problem, and achieved state-of-the-art performance on PASCAL VOC dataset. By the means of deconvolution and upsampling, different feature layers from a classification network are able to merge into one single feature map and perform a pixel-wise prediction. Since FCN uses different feature layers including shallow layers with low level feature and deep layers that carry high level semantic information, it considers both local and global information, thus it can 'see the bigger picture' then a sliding window type neural network.

Inspired by the idea of text detection via semantic segmentation given in [12, 13], we use FCN to classify each pixel to text or non-text. We modify VGG-16 [14] to our FCN architecture by removing the fully connected layers and adding deconvolution layers after pooling stages. In our FCN architecture, the 3rd, 4th and 5th pooling layers are fused into one feature map by adding deconvolution layers and upsampling by 8,16 and 32 times using bilinear interpolation, resulting in the prediction feature map with the same size of input image. The FCN is trained on a log softmax loss function to perform the pixel-wise prediction.

In order to train the FCN, we need a pixel-wise image label for each training image, but most available text detection datasets [4, 15, 16] provide ground truth by bounding box parameters. The most straightforward way is to set pixels within ground truth bounding boxes to positive and pixels outside text bounding boxes to negative like the second row of Fig. 3. But this may lead to a so called 'sticking' problem [13]: when multiple text lines are close to each other, we may not be able to separate them from the saliency map predicted by FCN.

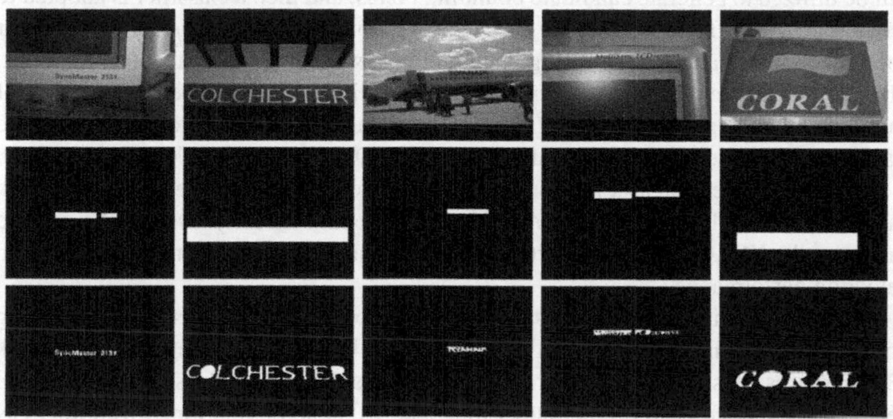

Fig. 3. Two types of ground truth mask. Original images are in the first row, the second and third rows contain bounding box ground truth masks and MSER ground truth masks.

To tackle this problem, we utilize the Maximally Stable Extremal Region (MSER) [6] ground truth mask and we train the FCN with a hybrid loss function. Since text components carry rich edge information and generally have significant color contrast with backgrounds, many text detection approaches (e.g. [7, 9, 10, 17]) extract the MSERs or Ers as letter candidates. In fact, we extract the MSERs within the bounding box ground truth, and only set the pixels in the MSERs as positive. As shown in the third row of Fig. 3, the contour and shape of letters are well described by the MSER ground truth mask. Thus, we can define the hybrid loss function as follows:

$$\mathcal{L}_{FCN}(M, M_{bbox}, M_{mser}) \\ = \lambda_{bbox} \cdot \mathcal{L}_{\log softmax}(M, M_{bbox}) + \lambda_{mser} \cdot \mathcal{L}_{\log softmax}(M, M_{mser}), \quad (1)$$

where M is the output prediction map of the FCN, and M_{bbox} and M_{mser} are the bounding box and MSER ground truth map, respectively. Accordingly, the FCN can learn the region of text lines by minimizing the first part of loss function, and learn the contour and shape of letter by minimizing the last part of loss function. In our experiments, we treat the loss functions of two types of ground truth mask equally, thus the parameters λ_{bbox} and λ_{mser} are both set to 0.5.

We modify the VGG-16 pre-trained on the ImageNet dataset to our FCN model, train the FCN by minimizing the hybrid loss function defined in (1) by the stochastic gradient descent. During the training phase, we inherit the most of the used parameters including the learning rate with 1e−4, the weight decay with 0.9 and the dropout with 0.5 from the FCN model proposed in [2].

3.3 Bounding Box Regression Network (BRN)

R-CNN based object detection approaches treat the detection as a classification problem. A region proposal algorithm, selective search (SS) or deep neural network for instance, can be utilized to generate candidate bounding boxes, and then a classifier is adopted to score each bounding box. Finally, the bounding boxes are filtered according to their confidence scores and the remaining bounding boxes become the final detection result. This kind of framework is inefficient and hard to optimize because the great number of candidate bounding boxes causes a mass of redundant calculation and each part of the framework must be trained individually. The YOLO (You Only Look Once) approach [18] treats the detection as a regression problem directly and a unified neural network can be trained to learn the locations of bounding boxes directly. Due to this simple structure, the YOLO approach can run in real-time, and in the meantime, achieve a high detection performance.

Our main idea is to train a regression network to learn the locations of text bounding boxes directly. This regression network we called bounding box regression network (BRN) shares the same layers before 'pool5' layer as the FCN and a small sized neural network is added to perform the regression. The BRN can be fine-tuned with the FCN together.

Instead of using fully connected layers in the YOLO structure, we use convolution layers in BRN to maintain the spatial information. Inspired by Darknet-19 [19], we apply 1×1 convolution layers to reduce the number of the feature map tunnels, and batch normalization layers to accelerate the convergence and avoid the overfitting.

Like the YOLO model, our proposed BRN can be regarded as dividing $32n \times 32n$ input image to $n \times n$ even grid. If a text bounding box region overlaps with a grid cell, then this cell is responsible for detecting the nearest text bounding box. The YOLO network learns 4 location coordinates and a confidence score for each bounding box in a grid cell, and learns C class probabilities. Since we just have one class of text, it is unnecessary to learn the class probability. So, we only train BRN to learn 4 location parameters, and the confidence score can be calculated from the FCN saliency map. Thus, our BRN output map is $12 \times 12 \times 4$ (for a 384×384 image).

The four location coordinates are relative distance from four bounding box edges to the center of the grid cell. We use a Sigmoid function to convert the coordinates to 0 to 1. The L2 loss we use weights each grid cell equally, but we want grid cells that overlap with text bounding box to be more important and outweigh the cells with no text region in it. Therefore, we introduce two parameters to control the weights between text cells and non-text cells. The loss function is designed as follows:

$$\begin{aligned} \mathcal{L}_{BRN}(C, \widehat{C}) = \lambda_{text} \cdot \sum_{i=1}^{n} \sum_{j=1}^{n} I_{text}^{ij} \left\| C_{ij} - \widehat{C}_{ij} \right\|_{L^2}^{2} \\ + \lambda_{non\text{-}text} \cdot \sum_{i=1}^{n} \sum_{j=1}^{n} (1 - I_{text}^{ij}) \left\| C_{ij} - \widehat{C}_{ij} \right\|_{L^2}^{2} \end{aligned} \tag{2}$$

where C and $\widehat{C}$ are the output of BRN and the coordinates ground truth, I_{text}^{ij} denotes if a text bounding box region overlaps with the grid cell ij. And the two parameters λ_{text} and $\lambda_{non-text}$ are set to 10 and 1 in our experiments. And we set the learning rate to 1e−5 to avoid gradient explosion.

3.4 Joint Training of FCN and BRN

Our idea is to generate text saliency map and predict the locations of text bounding boxes in the same time by a single deep neural network. So we dock the FCN and BRN by connecting the 'pool5' layer and sharing the feature layers before. We then fine-tune the whole unified network architecture by optimizing a multi-task loss function combined with the loss functions of FCN and BRN:

$$\begin{aligned} &\mathcal{L}(M, M_{bbox}, M_{mser}, C, \widehat{C}) \\ &\quad = \lambda_{FCN} \cdot \mathcal{L}_{FCN}(M, M_{bbox}, M_{mser}) + \lambda_{BRN} \cdot \mathcal{L}_{BRN}(C, \widehat{C}) \end{aligned} \tag{3}$$

In our experiments, we set λ_{FCN} and λ_{BRN} to be 10 and 1, respectively. We gradually reduce the learning rate from 1e−4 to 1e−5 to avoid the gradient explosion.

3.5 Non-Maximum Suppression (NMS)

Our BRN output map divides the input image into even grid cells which are 32×32 patches of the input image. That is, the BRN predicts one bounding box for every grid cell, but most of them are redundant. For instance, the BRN generates a $12 \times 12 \times 4$ map for a 384×384 input image, which is 144 bounding boxes predicted by the BRN. We score these bounding boxes by the saliency map from the FCN, and then apply the non-maximum suppression (NMS) algorithm to eliminate the redundant bounding boxes and get the final detection result.

In the scoring procedure, we would like the bounding box with more text content and less non-text content gets a higher score. That is, the bounding box should contain as much pixel value of saliency map as possible. Thus, we define the confidence score as follows:

$$score(R) = \sum_{(i,j)\in R} map_{ij},$$

where map_{ij} is the pixel-wise value of text saliency map generated by the FCN.

Then, we apply the non-maximum suppression (NMS) algorithm to filter the redundant bounding boxes by their confidence scores. That is, if the overlap ratio of two bounding boxes is higher than a given threshold, we remove the one with lower confidence score. The post-processing procedure only takes about 0.03 s on a 3.0 Hz CPU.

4 Experiments

In this section, we begin to introduce the datasets and present the experiment results and comparisons of our proposed method on them. Moreover, the running time and limitations of our proposed method are also discussed.

4.1 Datasets

All the images used to train our network are harvested from ICDAR2013 [4], ICDAR2015 [15], and COCO-Text [16] with data augmentation.

ICDAR2013. The 'robust reading' competition held by International Conference on Document Analysis and Recognition (ICDAR) every two years provides a dataset and a benchmark on scene text detection. In fact, ICDAR2013 [4] contains about 500 images with text annotations for training and testing. We apply flip and rotation to make our network capable to detect multi-orientated text layout.

ICDAR2015. ICDAR2015 'robust reading' competition [15] is divided into four different tasks. While task 2 'Focused Scene Text' is same as ICDAR2013 which is well-captured by camera and focus on horizontal text lines, task 4 'Incidental Scene Text' is introduced as a new task. This dataset is captured by wearable device without prior knowledge of the whereabouts of text lines. ICDAR2015 dataset is more close to the real word scenario so it is much more difficult than the previous ICDAR2013 dataset. We apply random crop to augment the data.

COCO-Text. COCO (Common objects in content) is a huge image dataset for numerous computer vision tasks, such as object detection, semantic segmentation and human key point detection. With text instance labeling on COCO dataset, it can be used for text detection task, and this subset of COCO is called COCO-Text [16]. We filter the text instances and remove the ones with low quality and we take 12172 images for training and 5641 images for validation.

4.2 FCN Predictions

We train three FCNs by different kinds of ground truth mask individually to evaluate the performance on pixel-wise predictions. We call these three FCNs trained by bounding box ground truth mask, MSER ground truth mask, and hybrid ground truth mask using the loss function defined by (1) BboxNet, MserNet and HybridNet.

All three FCNs can predict the text lines in the most scenarios as shown in the first row of Fig. 4. But when two text lines are very close, the prediction of BboxNet tends to be a blob of high response which makes us unable to separate the text lines. For example, the 'Key' and 'West' cannot be separated from the BboxNet saliency map in the second row of Fig. 4. There are still some problems when we train the FCN only by MSER ground truth mask. The most severe one is that, when the stroke width of the text is relatively large and the text fills up the nearby region, the MserNet might sometimes get the background and text regions mixed up like the 'animal' in the last row of Fig. 4. While the HybridNet is capable of separating the close text lines, and in the meantime, it is unlikely to be confused by the background and text regions.

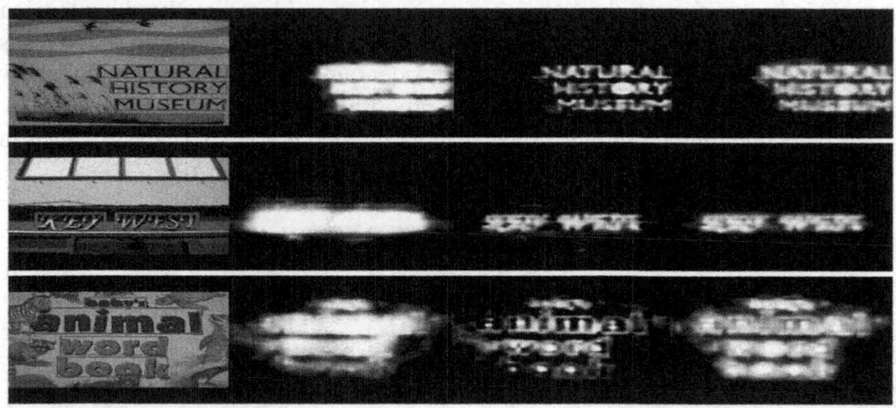

Fig. 4. Saliency maps produced by BboxNet, MserNet and HybridNet.

4.3 Experiment Results

The BRN is trained to learn four parameters of horizontal bounding boxes. So we test our proposed text detection system on ICDAR2013 dataset and pass the ICDAR2015 and COCO-Text which has images with multi-oriented text instances. Our method achieves 0.79 precision, 0.73 recall and 0.76 F-measure on ICDAR2013 benchmark. The evaluation details can be found in the official document of ICDAR2013 robust reading competition [4].

The most of the top methods on the ICDAR2013 leaderboard have not been described in academic papers to date, so we only list our proposed method along with other methods which are published in Table 1.

Table 1. Test results on ICDAR2013 dataset

Methods	Precision	Recall	F-measure	Time/s
CER [10]	0.87	0.86	0.86	
FCN_Megvii [13]	0.89	0.80	0.84	0.62
Proposed	0.79	0.73	0.76	0.19
USTB_TextStar [9]	0.88	0.66	0.76	0.80
Text-Spotter [11]	0.88	0.65	0.74	>1[a]
CASIA [8]	0.85	0.63	0.72	
I2R NUS [4]	0.73	0.66	0.69	
TH-TextLoc [4]	0.70	0.65	0.67	
TD-Mixture [20]	0.69	0.66	0.67	7.20
SWT [5]	0.73	0.60	0.66	>3[a]

[a]Evaluated by our experiments

It can be found that our precision is much lower than most of other methods, since our proposed network treats the detection as a regression problem directly without region proposals, the locations of predicted bounding boxes might be inaccurate which leads to a low precision performance. On the bright side, our proposed method is extremely fast for no redundant calculation from region proposals.

The detection samples in Fig. 5 show that our method is able to handle text lines in different colors, fonts and scales and performs well in several scenarios.

Fig. 5. Detection samples of the proposed unified deep learning neural network.

4.4 Running Time

The framework of the proposed method is very simple. On a 384 × 384 image, it takes only 0.12 s on average to generate a text saliency map and bounding box predictions running on a single GTX1080 GPU without batch input. Actually, it should be at least twice faster when running on a GTX Titan X GPU. The post-processing procedure of NMS can be efficiently done on a single CPU in 0.03 s. Our whole text detection system can process 5 images per second but it is still far from real-time. The deep learning platform we use in our experiments is MatConvNet [21].

4.5 Limitations

Since the FCN is sensitive to several certain scenarios including uneven illumination and blur, our proposed text detection system might fail under these certain conditions. Light spots causing by the reflection of light source split the text when they appear in the middle of text lines. Moreover, the severe blur on text regions makes the FCN difficult to extract the feature representation from the text regions.

Our unified deep neural network directly learns the locations of text bounding boxes as a regression problem. Therefore, the bounding boxes predicted by the BRN are not so accurate in comparison with the other classification based methods, which leads to a decrease on precision. Moreover, we only train the BRN to learn horizontal rectangle bounding boxes for text lines, but text in real life images can be in different layouts like multiple orientations or circles. Our system might fail when it comes to these scenarios.

5 Conclusion

We have established a unified deep neural network architecture for scene text detection. A Fully Convolutional Network (FCN) is trained to predict the text saliency map in a pixel-wise style with a hybrid loss function to overcome the 'sticking' problem. Moreover, the Bounding box Regression Network (BRN) sharing the feature layers with the FCN is directly trained with the locations of indexed bounding boxes. The unified network is fine-tuned in an end-to-end manner by a multi-task loss function. For scene text detection, we input the natural images to the unified network to generate a text saliency map and predict the locations of candidate bounding boxes at the meantime. We then score each bounding box by the saliency map and use the non-maximum suppression (NMS) algorithm to eliminate the redundant bounding boxes. It is demonstrated by the experimental results on ICDAR2013 benchmark that our unified network can achieve 0.76 F-measure and run at 5 fps on a GPU, which is faster than most of the other existing text detection methods.

Acknowledgements. This work was supported by the Natural Science Foundation of China for Grant 61171138.

References

1. Zhu, Y., Yao, C., Bai, X.: Scene text detection and recognition: recent advances and future trends. Front. Comput. Sci. **10**(1), 19–36 (2016)
2. Long, J., Shelhamer, E., Darrell, T.: Fully convolutional networks for semantic segmentation. IEEE Trans. Pattern Anal. Mach. Intell. **PP**(99), 640–651 (2014)
3. Neubeck, A., Gool, L.V.: Efficient non-maximum suppression. In: International Conference on Pattern Recognition, pp. 850–855. DBLP (2006)

4. Karatzas, D., Shafait, F., Uchida, S., et al.: ICDAR 2013 robust reading competition. In: International Conference on Document Analysis and Recognition, pp. 1484–1493. IEEE Computer Society (2013)
5. Epshtein, B., Ofek, E., Wexler, Y.: Detecting text in natural scenes with stroke width transform. In: Computer Vision and Pattern Recognition, pp. 2963–2970. IEEE (2010)
6. Matas, J., Chum, O., Urban, M., et al.: Robust wide-baseline stereo from maximally stable extremal regions. Image Vis. Comput. **22**(10), 761–767 (2004)
7. Neumann, L., Matas, J.: A method for text localization and recognition in real-world images. In: Kimmel, R., Klette, R., Sugimoto, A. (eds.) ACCV 2010. LNCS, vol. 6494, pp. 770–783. Springer, Heidelberg (2011). doi:10.1007/978-3-642-19318-7_60
8. Shi, C., Wang, C., Xiao, B., et al.: Scene text detection using graph model built upon maximally stable extremal regions. Pattern Recogn. Lett. **34**(2), 107–116 (2013)
9. Yin, X.C., Yin, X., Huang, K., et al.: Robust text detection in natural scene images. IEEE Trans. Pattern Anal. Mach. Intell. **36**(5), 970–983 (2014)
10. Sun, L., Huo, Q., Jia, W., et al.: A robust approach for text detection from natural scene images. Pattern Recogn. **48**(9), 2906–2920 (2015)
11. Jaderberg, M., Vedaldi, A., Zisserman, A.: Deep features for text spotting. In: Fleet, D., Pajdla, T., Schiele, B., Tuytelaars, T. (eds.) ECCV 2014. LNCS, vol. 8692, pp. 512–528. Springer, Cham (2014). doi:10.1007/978-3-319-10593-2_34
12. Zhang, Z., Zhang, C., Shen, W., et al.: Multi-oriented text detection with fully convolutional networks. In: Proceedings of the IEEE Conference on Computer Vision and Pattern Recognition, pp. 4159–4167 (2016)
13. Yao, C., Bai, X., Sang, N., et al.: Scene text detection via holistic, multi-channel prediction. arXiv preprint arXiv:1606.09002 (2016)
14. Simonyan, K., Zisserman, A.: Very deep convolutional networks for large-scale image recognition. arXiv preprint arXiv:1409.1556 (2014)
15. Karatzas, D., Gomez-Bigorda, L., Nicolaou, A., et al.: ICDAR 2015 competition on robust reading. In: 2015 13th International Conference on Document Analysis and Recognition (ICDAR), pp. 1156–1160. IEEE (2015)
16. Veit, A., Matera, T., Neumann, L., et al.: Coco-text: dataset and benchmark for text detection and recognition in natural images. arXiv preprint arXiv:1601.07140 (2016)
17. Huang, W., Qiao, Yu., Tang, X.: Robust scene text detection with convolution neural network induced MSER trees. In: Fleet, D., Pajdla, T., Schiele, B., Tuytelaars, T. (eds.) ECCV 2014. LNCS, vol. 8692, pp. 497–511. Springer, Cham (2014). doi:10.1007/978-3-319-10593-2_33
18. Redmon, J., Divvala, S., Girshick, R., et al.: You only look once: unified, real-time object detection. In: Proceedings of the IEEE Conference on Computer Vision and Pattern Recognition, pp. 779–788 (2016)
19. Redmon, J., Farhadi, A.: YOLO9000: better, faster, stronger. arXiv preprint arXiv: 1612.08242 (2016)
20. Yao, C., Bai, X., Liu, W., et al.: Detecting texts of arbitrary orientations in natural images. In: 2012 IEEE Conference on Computer Vision and Pattern Recognition (CVPR), pp. 1083–1090. IEEE (2012)
21. Vedaldi, A., Lenc, K.: Matconvnet: convolutional neural networks for matlab. In: Proceedings of the 23rd ACM International Conference on Multimedia, pp. 689–692. ACM (2015)

Exploring the New Application of Morphological Neural Networks

Bin Sun[1(✉)] and Naiqin Feng[1,2]

[1] School of Information Engineering,
Zhengzhou University of Industrial Technology, Zhengzhou 451150, China
2937603937@qq.com

[2] College of Computer and Information Engineering,
Henan Normal University, Xinxiang 453007, China

Abstract. The traditional artificial neural networks can simulate the psychological phenomenon of "implicit learning", but can't simulate the cognitive phenomenon of "one-trial learning". In this paper we took advantage of morphological associative memory networks to realize the simulation of "one-trial learning" for the first time. Theoretical analysis and simulation experiments show that the method of morphological associative memory networks is a higher effective machine learning method, and can very well simulate the cognitive phenomenon of "one-trial learning", therefore, it will provide a new experimental tool for the study of intelligent science and cognitive science.

Keywords: One-trial learning · Cognitive psychology · Simulation · Morphological neural networks · Morphological associative memories

1 Introduction

In real life, learning and memory ability of some people is very strong, even can be "having a photographic memory". So efficient way of learning, in the cognitive psychology, is known as "one-trial learning" [1], of course, can also be called "one-trial memory". Obviously, "one-trial learning" is the human dream of learning methods and learning efficiency. However, learning and memory have occurred in the brain, and can't be seen directly by us. Accordingly till now, the working mechanisms of human brain for their own learning and memory, is not entirely clear.

There are many ways to learn and memorize. In addition to "one-trial learning" and "one-trial memory", there are "implicit learning" [2] and "explicit learning" [3]. In order to explore the mysteries of human learning and memory, neuroscientists, brain scientists, psychologists, computer scientists and artificial intelligence scholars have devoted a lot of effort [4–6]. Literature research shows that people, with the help of the traditional artificial neural networks and machine learning, can successfully simulate the "implicit learning", "explicit learning" and other psychological phenomena. However, it is a pity that so far, people have not been able to simulate the cognitive phenomenon of "one-trial learning". Therefore, some psychologists pointed out that cognitive psychology is challenging the traditional artificial neural networks [7].

D.-S. Huang et al. (Eds.): ICIC 2017, Part I, LNCS 10361, pp. 113–120, 2017.
DOI: 10.1007/978-3-319-63309-1_11

Why can't the traditional artificial neural networks be able to do a task of "one-trial learning"? Is there a possibility to solve the problem? This paper tried to explore the problem and give some preliminary results, in order to throw out a minnow to catch a whale, and initiate resonating for studying the machine learning method [8] of "one-trial learning" or "one-trial memory", also provide a new idea for the research and development of intelligent science, cognitive science and machine learning.

2 Limitations of Traditional Artificial Neural Networks

The drawbacks of traditional artificial neural networks are evident to anyone.

(1) **Lower learning efficiency**
A typical traditional neural network uses each sample of the sample set to learn one by one, very time-consuming. For the complex network with the large sample set, multi layers and multi neurons, the trained process is very long. Sometimes it may take hours or even days, and hundreds and hundreds of iterations of learning, in order to achieve a certain precision of training. Obviously, it is not realistic to use this kind of neural network to complete the task of "one-trial learning" or "one-trial memory".

(2) **Larger learning error**
The traditional neural network method is applied to learn from each sample, and it is often occurred to learn the new and forget the old, not only long training time and many times of iteration, but also larger training error. Naturally, it is not conducive to the completion of the task of "one-trial learning".

(3) **Poor interpretability**
The traditional artificial neural networks are often compared to the "black box", and their working process is difficult to observe and explain. This disadvantage of traditional neural networks is extremely detrimental to explore the learning mechanism.

(4) **Weak generalization ability**
The learning effect of the traditional neural networks is related to the selection of the training set. For unknown samples that have never been studied, the actual outputs of the network may have a large gap with the target outputs.
In order to overcome these shortcomings of traditional neural networks, people have made a lot of efforts, including neural network ensembles, improved learning algorithms, extracting rules from neural networks, and so on. However, these efforts are not enough to qualify for the task of "one-trial learning".

3 Analysis of Morphological Neural Networks

In 1990, Davidson and Ritter put forward the theory of morphological neural networks (MNN) [9]. In 1998, Ritter et al. applied MNN to the simulation of associative memories, and proposed morphological associative memories (MAM) [10], or Real MAM (RMAM). After that, people presented many MAM models, such as fuzzy morphological associative memories (Fuzzy MAM, FMAM [11]), enhanced fuzzy morphological associative memories based on the empirical kernel map (Enhanced FMAM, EFMAM [12]), unified framework of morphological associative memories [13], and some other new methods of MAMs [14–17].

The research shows that MNNs are a new kind of artificial neural networks, and their network structure and learning algorithm are qualitatively different from those of traditional neural networks. It is a two layer network, the input layer and the output layer. The connection weight matrix (memory matrix W_{XY}/M_{XY}) between two layers of neurons is not obtained by dynamic learning or training, but through a calculation ("one-trial learning").

3.1 Stronger Nonlinear Neural Computation

The traditional artificial neural network models are determined by the network topology, node characteristics and learning rules. The basic algebraic system used in these models is the set of real numbers **R** together with the operations of addition and multiplication and the laws governing these operations. This algebraic system, known as a ring, is commonly denoted by (**R**, +, ×). The two basic equations governing the theory of computation in the standard neural network model are:

$$\tau_i(t+1) = \sum\nolimits_{j=1}^{n} a_j(t) + w_{ij} \tag{1}$$

$$a_i(t+1) = f(\tau_i(t+1) - \theta_i) \tag{2}$$

Where $a_j(t)$ denotes the value of the jth neuron at time t, n represents the number of neurons in the network, w_{ij} the synaptic connectivity value between the ith neuron and the jth neuron, $\tau_i(t+1)$ the next total input effect on the ith neuron, θ_i a threshold, and f the next state function which usually introduces a nonlinearity into the network.

Different from the traditional neural network model, the basic calculations in the MNN model are based on the algebraic lattice structure (**R**, ∧, ∨, +), wherein the symbols ∧ and ∨ represent the binary operations of minimum and maximum, respectively. The nonlinear of MNNs is stronger than that of traditional neural network. The basic computational model of MNNs is as follows:

$$\tau_i(t+1) = \wedge_{j=1}^{n} a_j(t) + w_{ij} \tag{3}$$

$$a_i(t+1) = f(\tau_i(t+1) - \theta_i) \tag{4}$$

or

$$\tau_i(t+1) = \vee_{j=1}^{n} a_j(t) + w_{ij} \tag{5}$$

$$a_i(t+1) = f(\tau_i(t+1) - \theta_i) \tag{6}$$

Let X and Y are a pair of associative memory matrices, the X is the input mode matrix of dimension $n \times p$, and the Y is the output mode matrix of dimension $m \times p$, using the unified framework of MAMs for the representation, then the minimal product associative memory that contacts X and Y is defined by

$$\boldsymbol{W}_{XY} = \boldsymbol{Y} \bar{\wedge} \boldsymbol{X}' \tag{7}$$

W_{XY} is matrix of dimension $m \times n$, and its i, jth entry is defined by

$$w_{ij} = \wedge_{k=1}^{p} y_{ik} - x_{jk} = (y_{i1} - x_{j1}) \wedge \cdots \wedge (y_{ip} - x_{jp}). \tag{8}$$

The maximum product memory contacting X and Y is given by

$$\boldsymbol{M}_{XY} = \boldsymbol{Y} \bar{\vee} \boldsymbol{X}' \tag{9}$$

M_{XY} is matrix of dimension $m \times n$, and its i, jth entry is defined by

$$m_{ij} = \vee_{k=1}^{p} y_{ik} - x_{jk} = (y_{i1} - x_{j1}) \vee \cdots \vee (y_{ip} - x_{jp}). \tag{10}$$

When the input pattern matrix X is provided to the network, the associative output matrix Y_0 is calculated by

$$\boldsymbol{Y}_0 = \boldsymbol{W}_{XY} \overset{+}{\vee} \boldsymbol{X} \tag{11}$$

Y_0 is matrix of dimension $m \times p$, and the i, jth entry is defined by

$$y_{ij} = \vee_{k=1}^{n} w_{ik} + x_{kj} = (w_{i1} + x_{1j}) \vee \cdots \vee (w_{in} + x_{nj}) \tag{12}$$

or

$$\boldsymbol{Y}_0 = \boldsymbol{M}_{XY} \overset{+}{\wedge} \boldsymbol{X} \tag{13}$$

and the i, jth entry is defined by

$$y_{ij} = \wedge_{k=1}^{n} m_{ik} + x_{kj} = (m_{i1} + x_{1j}) \wedge \cdots \wedge (m_{in} + x_{nj}). \tag{14}$$

When Y_0 = Y, W_{XY} and M_{XY} are called the complete recall memories for (X, Y).

3.2 Better Associative Memory Performance

To sum up, MAMs have four main advantages.

(1) Unlimited storage capacity
The memories of MAMs are established by the associative matrix between input and output. RMAM, for example, its memory is stored in the memory matrix W_{XY} or M_{XY}. The size of the memory matrix is $m \times n$, where m and n are the

dimension of the output and input pattern vectors, respectively. The recall output of the mode **y**, is the result of morphological operations between W_{XY} (or M_{XY}) and input mode **x**. That is to say, the size of W_{XY} and M_{XY} is determined by $m \times n$, and it has nothing to do with p, the number of pattern pairs. As long as m and n are limited, there is no limit on p, the number of mode pairs required to remember, and can even exceed s, the storage capacity of the computer.

(2) One step recall memory
MAMs calculate and converge in one step, there is no iterative process. Therefore there is no convergence problem of traditional neural networks. This advantage of morphological associative memories is very important for the realization of simulating "one-trial learning".

(3) Good memory performance
Auto RMAM, Auto FMAM and Auto LEMAM, for perfect inputs, are capable of giving a guarantee of complete recall memory. Hetero RMAM, Hetero FMAM and Hetero LEMAM, for complete inputs, not unconditionally guarantee to recall completely, but in a certain range of noises and satisfying the completely recall theorem, they also can do perfect recall. Such a good memory and recall performance is an important condition to complete the task of "one-trial learning".

(4) Robust noise performance
MAMs have strong noise robustness. For example, W_{XY} of RMAM, A_{XY} of FMAM, and T_{XY} of LEMAM can resist strong corrosion noises; and M_{XY} of RMAM, B_{XY} of FMAM, V_{XY} of LEMAM can resist strong expansion noises.
From the above analysis, we can see that the advantages of MAMs are compatible with the need to fulfill the task of "one-trial learning", therefore, MAMs have the ability to do this task.

4 Simulation Experiments of "One-Trial Learning"

Here, two experiments were designed to simulate "one-trial learning".

Experiment 1. "one-trial learning" to remember phone numbers.

In Former Soviet Union there was a "Memory Superman" – Yuri Novikov. He had an amazing learning efficiency and super memory for remembering telephone numbers. In order to simulate his ability of learning and memory, we designed the following experiment.

There were 16 phone numbers of 7 digits, arranged into a matrix X as follows:

$$X = \begin{bmatrix} 3325016 & 2022222 & 5115235 & 6760054 \\ 3047901 & 2021927 & 5523798 & 6750395 \\ 3051229 & 2143888 & 5069733 & 6058999 \\ 3150705 & 2836311 & 5083333 & 6543928 \end{bmatrix}$$

Present the matrix X quickly to a tested learner ("Memory Superman"), no more than 1 s. Then present the X_1, a damaged version of X, to the learner. The damaged positions are represented by -1:

$$X_1 = \begin{bmatrix} 3325016 & -1 & -1 & 6760054 \\ -1 & -1 & 5523798 & 6750395 \\ 3051229 & -1 & 5069733 & -1 \\ 3150705 & 2836311 & -1 & -1 \end{bmatrix}$$

After that, the learner tried to recover the lost phone numbers. As a result, the learer in the test filled the missing phone numbers quickly and correctly.

The memory matrix W_{XX} of auto RMAM (ARMAM) can successfully simulate the experiment. Let Y = X, and use formula (7) and (8), we can calculate and obtain the memory matrix W_{XX} as follows:

$$W_{XX} = X \bar{\wedge} X' = \begin{bmatrix} 0 & -408563 & -121666 & -814089 \\ -277115 & 0 & -121961 & -814384 \\ -701055 & -691396 & 0 & -692423 \\ -216126 & -440465 & 13600 & 0 \end{bmatrix}.$$

Using formulas (11) and (12), after the recall or calculation, we can get

$$W_{XX} \overset{+}{\vee} X_1 = = \begin{bmatrix} 3325016 & 2022222 & 5115235 & 6760054 \\ 3047901 & 2021927 & 5523798 & 6750395 \\ 3051229 & 2143888 & 5069733 & 6058999 \\ 3150705 & 2836311 & 5083333 & 6543928 \end{bmatrix} = X$$

That is, we achieved a simulation of "one-trial learning" or " one-trial memory". Here, we give only the results of the memory and recall, but for the specific computing process, please check them by readers self.

Experiment 2. Learning words through images.

In the early childhood education, teachers often use the image-reading-teaching-method, in order to make the children to establish the connection between an image and its word, and some children learn quickly. To simulate this situation, was designed the following experiment.

There are 5 groups of digitized images of letters and fruits: (A, Apple), (B, Banana), (C, Cherry), (S, Strawberry), (W, Watermelon), as shown in Fig. 1.

Fig. 1. Pattern pairs of letters and fruits

The simulation experiment was carried out on the MATLAB-6.5 platform. In this case, X = {A, B, C, S, W}, Y = {Apple, Banana, Cherry, Strawberry, Watermelon}. The letter images in X and the fruit images in Y are 60 × 60 Boolean images, which are used as input and output pattern sets, respectively. By using the reverse RMAM method [15], we made the hetero association of X → Y ($n = m = 3600$). First of all, a "one-trial learning" was done to form a memory matrix A_{XY}; and then, a recall was done to obtain the remembered result. The experimental results show that all the telephone numbers are remembered and a complete recall memory is achieved through only one "one-trial learning". Namely, for any given letter in X, the learner can immediately recall the corresponding fruit image in Y. This experiment is a good simulation for the phenomenon of "one-trial learning" or "one-trial memory" in the process of children's learning with flashcards.

5 Conclusions

MNNs are a new method of machine learning. This paper took advantage of this method tentatively to study the cognitive phenomenon of human "one-trial learning" and to simulate the cognitive phenomenon successfully, therefor we opened a new application area, at the same time, also explored a new way and provided a new tool for revealing the cognitive mechanism of "one-trial learning".

Acknowledgement. This work was supported in part by the science and technology research project of Zhengzhou city (Grant No. 153PKJGG153).

References

1. Brosgole, L., Contino, A.F., Hansen, K.H.: What is one-trial learning? Psychon. Sci. **15**(2), 89–90 (1969)
2. Sariyska, R., Lachmann, B., Markett, S.: Individual differences in implicit learning abilities and impulsive behavior in the context of Internet addiction and Internet Gaming Disorder under the consideration of gender. Addict. Behav. Rep. **5**, 19–28 (2017)
3. Ziori, E., Dienes, Z.: The time course of implicit and explicit concept learning. Conscious. Cogn. **21**(1), 204–216 (2012)
4. Zhan, Y., Luo, Y.Z., Deng, X.F.: Spatiotemporal prediction of continuous daily PM2.5 concentrations across China using a spatially explicit machine learning algorithm. Atmos. Environ. **155**, 129–139 (2017)
5. Barber, T.A.: Discrimination of shape and size sues by day-old chicks in two one-trial learning tasks. Behav. Process. **124**, 10–14 (2016)
6. Feng, N.Q., Qin, L.J., Wang, X.F.: Morphological associative memories applied to the implicit learning. J. Henan Normal Univ. (Nat. Sci. Ed.) **41**(3), 156–159 (2013)
7. Yu, J.Y.: Cognitive psychology and neural networks. In: Cao, C.G., Zhou, Z.H. (eds.) Neural Networks and their Applications, pp. 406–445. Tsinghua University Press, Beijing (2004)

8. Taffese, W.Z., Sistonen, E.: Machine learning for durability and service-life assessment of reinforced concrete structures: recent advances and future directions. Autom. Constr. **77**, 1–14 (2017)
9. Davidson, J.L., Hummer, F.: Morphology neural networks: an introduction with applications. Circuits Syst. Signal Process. **12**(2), 177–210 (1993)
10. Ritter, G.X., Sussner, P., Dia-de-Leon, J.L.: Morphological associative memories. IEEE Trans. Neural Netw. **9**(2), 281–292 (1998)
11. Wang, M., Wang, S.T., Wu, X.J.: Initial results on fuzzy morphological associative memories. Acta Electronica Sinica **31**(5), 690–693 (2003)
12. Wang, M., Chen, S.C.: Enhanced FMAM based on empirical kernel map. IEEE Trans. Neural Netw. **16**(3), 557–564 (2005)
13. Feng, N.Q., Liu, C.H., Zhang, C.P.: Research on the framework of morphological associative memories. Chin. J. Comput. **33**(1), 157–166 (2010)
14. Feng, N.Q., Tian, Y., Wang, X.F.: Logarithmic and exponential morphological associative memories. J. Software **26**(7), 1662–1674 (2015)
15. Feng, N.Q., Tian, Y., Feng, G.H.: An effective method of hetero associative morphological memories. J. Chin. Comput. Syst. **36**(10), 2374–2378 (2015)
16. Feng, N.Q., Wang, X.F., Mao, W.T.: Heteroassociative morphological memories based on four-dimensional storage. Neurocomputing **116**, 76–86 (2013)
17. Feng, N., Yao, Y.: No rounding reverse fuzzy morphological associative memories. Neural Netw. World **6**, 571–587 (2016)

An Improved Evolutionary Random Neural Networks Based on Particle Swarm Optimization and Input-to-Output Sensitivity

Qing-Hua Ling[1,2(✉)], Yu-Qing Song[1], Fei Han[1], and Hu Lu[1]

[1] School of Computer Science and Communication Engineering, Jiangsu University, Zhenjiang 212013, Jiangsu, China
lingee_2000@163.com

[2] School of Computer Science and Engineering, Jiangsu University of Science and Technology, Zhenjiang 212013, Jiangsu, China

Abstract. Extreme learning machine (ELM) for random single-hidden-layer feedforward neural networks (SLFN) has been widely applied in many fields because of its fast learning speed and good generalization performance. Since ELM randomly selects the input weights and hidden biases, it typically requires high number of hidden neurons and thus decreases its convergence performance. To overcome the deficiency of the traditional ELM, an improved ELM based on particle swarm optimization (PSO) and input-to-output sensitivity information is proposed in this study. In the improved ELM, PSO encoding the input-to-output sensitivity information of the SLFN is used to optimize the input weights and hidden biases. The improved ELM could obtain better generalization performance and improve the conditioning of the SLFN by decreasing the input-to-output sensitivity of the network. Experiment results on the classification problems verify the improved performance of the proposed ELM.

1 Introduction

Traditional gradient based learning methods for single-hidden-layer feedforward neural networks (SLFN) are apt to converge to local minima and time consuming, which have a negative effect on their applications [1, 2]. To overcome the defects of gradient based learning algorithms, random vector functional link networks (RVFL) was proposed [3] where actual values of the weights from the input layer to hidden layer can be randomly generated in a suitable domain without training in the learning stage [4]. The independently developed method, SLFN with random weights (RWSLFN) in [5] without direct links between the inputs and outputs, belongs to the family of RVFL. RWSLFN has the potential of achieving better generalization performance because of its simple network structure, and it also requires less computational cost than those with the direct links [6].

Extreme learning machine (ELM) [7], an effective learning algorithm for RWSLFN, has been widely employed to solve problems in diverse domains, which randomly chooses the input weights and hidden biases and analytically determines the

D.-S. Huang et al. (Eds.): ICIC 2017, Part I, LNCS 10361, pp. 121–127, 2017.
DOI: 10.1007/978-3-319-63309-1_12

output weights of the SLFN by Moore-Penrose (MP) generalized inverse method. Compared with gradient based learning algorithm, ELM obtains better generalization performance with thousands of times faster speed. However, owing to randomly selecting input weights and hidden biases, ELM tends to require more hidden neurons and ill-condition SLFN is established, so the over-fitting and uncertainty performance still remain to be solved.

To avoid the defects of the traditional ELM, many improved ELM were proposed. In [8], an evolutionary ELM (E-ELM) was proposed which used the differential evolutionary algorithm to select the input weights and MP generalized inverse to analytically determine the output weights. The E-ELM could achieve good generalization performance with much more compact networks. In [9], particle swarm optimization (PSO) was used to optimize the input weights and hidden biases of the SLFN to solve some prediction problems, which mainly encoded the boundary conditions into PSO to improve the performance of ELM. In [10], an improved ELM optimized by PSO (IPSO-ELM) was proposed, which used an improved PSO to select the input weights and hidden biases and MP generalized inverse method analytically determine the output weights. The IPSO-ELM has improved generalization performance and the conditioning of the SLFN.

In this paper, we also use PSO to optimize the input weights and hidden biases of SLFN to improve the performance of the traditional ELM. Different from IPSO-ELM, PSO encodes the input-to-output sensitivity information of the SLFN to select optimal input weights and hidden biases in this study. According to [11], low input-to-output sensitivity may improve the generalization performance of the SLFN with high probability. In the proposed method, PSO could prefer to select those input weights and hidden biases resulting into low input-to-output sensitivity. The proposed ELM optimized with PSO and input-to-output sensitivity in this study is called PSOIOS-ELM.

2 Preliminaries

2.1 Extreme Learning Machine

In [7], a learning algorithm for SLFN called ELM was proposed to solve the problems caused by gradient-based learning algorithms. ELM randomly chose the input weights and hidden biases, and analytically determined the output weights of SLFN. ELM has much better generalization performance with much faster learning speed than gradient-based algorithms [12].

For N arbitrary distinct samples, (x_i, t_i), where $x_i = [x_{i1}, x_{i2}, \ldots, x_{in}]^T \in R^n$, $t_i = [t_{i1}, t_{i2}, \ldots, t_{im}]^T \in R^m$, a SLFN with NH hidden neurons can approximate these N samples with zero error. This means that

$$Hwo = T \tag{1}$$

where

$$H(wh_1,\ldots,wh_{N_H},b_1,\ldots,b_{N_H},x_1,\ldots,x_N)$$

$$=\begin{bmatrix} g(wh_1\cdot x_1+b_1) & \cdots & g(wh_{N_H}\cdot x_1+b_{N_H}) \\ \vdots & \cdots & \vdots \\ g(wh_1\cdot x_N+b_1 & \ldots & g(wh_{N_H}\cdot x_N+b_{N_H}) \end{bmatrix}_{N\times N_H},\ wo=\begin{bmatrix} wo_1^T \\ \vdots \\ wo_{N_H}^T \end{bmatrix}_{N_H\times m},\ T=\begin{bmatrix} t_1^T \\ \vdots \\ t_N^T \end{bmatrix}_{N\times m}.$$

$wh_i=[wh_{i1},wh_{i2},\ldots,wh_{in}]^T$ is the weight vector connecting the *i*-th hidden neuron to the input neurons, $wo_i=[wo_{i1},wo_{i2},\cdots wo_{im}]^T$ is the weight vector connecting the *i*-th hidden neuron to the output neurons, b_i is the bias of the *i*-th hidden neuron, and $g(\cdot)$ is the activation function of hidden neurons.

Thus, to determine the output weights is to find the least square (LS) solution to the given linear system. The minimum norm LS solution to the Eq. (1) is

$$wo=H^+T \tag{2}$$

where H^+ is the MP generalized inverse of matrix H. The minimum norm LS solution is unique and has the smallest norm among all the LS solutions. ELM tends to obtain good generalization performance [7, 12]. Since the solution is obtained by a analytical method and all the parameters of SLFN need not be adjusted, ELM converges much faster than gradient-based algorithms.

2.2 Particle Swarm Optimization

PSO is an evolutionary computation technique in search of the best solution by simulating the movement of birds in a flock [13]. The population of the birds is called swarm, and the members of the population are particles. Each particle represents a possible solution to the optimization problem. During each iteration, each particle flies independently in its own direction which is guided by its own previous best position as well as the global best position of all the particles. Assume that the dimension of the search space is *R*, and the swarm is S = (X_1, X_2, X_3, …, X_{Np}); each particle represents a position in R dimension space. The previous best position of the *i*-th particle is called pbest which is expressed as Pi = (p_{i1}, p_{i2}, …, p_{iR}). The best position of the all particles is called gbest which is expressed as Pg = (p_{g1}, p_{g2}, …, p_{gR}). The velocity of the i-th particle is expressed as V_i = (v_{i1}, v_{i2}, …, v_{iR}). According to [14], the adaptive PSO was described as:

$$V_i(t+1)=W(t)\times V_i(t)+c_1\times rand()\times(P_i(t)-X_i(t))+c_2\times rand()\times(P_g(t)-X_i(t)) \tag{3}$$

$$X_i(t+1)=X_i(t)+V_i(t+1) \tag{4}$$

where c_1, c_2 are the acceleration constants with positive values; *rand()* is a random number ranged from 0 to 1; *W(t)* is the inertia weight to keep a balance between global search and local search.

3 The Proposed Extreme Learning Machine

In this study, in the optimization process of the input weights and hidden biases, the input-to-output sensitivity of the SLFN is partly considered. The proposed ELM improves not only the generalization performance but also the condition performance of the SLFN. The detailed steps of the PSOIOS-ELM method are described as follows:

Step 1: The data set is divided into training and testing datasets, and the training data set is further divided into training and validation datasets.

Step 2: The swarm is randomly initialized. The components in each particle represents the candidate input weights and hidden biases of the SLFN. All components in the particle are randomly initialized within the range of [1,1].

Step3: Each particle's fitness value is calculated. According to the input weights and hidden biases defined by each particle, within the training data set, the corresponding output weighs are calculated according to Eq. (2). Then, the fitness of each particle is calculated as the root mean squared error (RMSE) on the validation set.

Step4: The pbest of each particle as well as the gbest of all particles is updated. Since low input-to-output sensitivity may provide good robustness of the SLFN, the input-to-output sensitivity of the network along with the RMSE on the validation set are considered for updating the pbests of all particles and the gbest of the swarm as follows:

$$P_{ib} = \begin{cases} P_i & (f(P_{ib}) > f(P_i)) \quad or \quad (f(P_{ib}) < f(P_i) < f(P_{ib}) + \lambda f(P_{ib}) \quad and \quad |sen_{P_i}| < |sen_{P_g}|) \\ P_{ib} & else \end{cases} \tag{5}$$

$$P_g = \begin{cases} P_i & (f(P_g) > f(P_i)) \quad or \quad (f(P_g) < f(P_i) < f(P_g) + \lambda f(P_g) \quad and \quad |sen_{P_i}| < |sen_{P_g}|) \\ P_g & else \end{cases} \tag{6}$$

where $f(P_i)$, $f(P_{ib})$ and $f(P_g)$ are the corresponding fitness for the i-th particle, the pbest of the *i*-th particle and gbest of all particles, respectively. sen_{P_i}, $sen_{P_{ib}}$ and sen_{P_g} are the corresponding input-to-output sensitivity value of the SLFN represented by the *i*-th particle, the pbest of the *i*-th particle and gbest of all particles, respectively. $\lambda > 0$ is a tolerance rate.

Since the input-to-output sensitivity of the SLFN is determined by the derivative of the hidden and output neurons, it is defined in this study as follows:

$$sen(\,) = \sum_{j=1}^{N}\left(\sum_{i=1}^{N_H}\left|g'(wh_i^T x_j + b_i)\right| \times \left|\sum_{k=1}^{N_H}\sum_{l=1}^{m} wo_{kl}\right|\right) \tag{7}$$

Step 5: Each particle updates its position according to Eqs. (3) and (4), and the new population is generated. When the component of the particle is beyond the constrained range of [−1,1], the direction of the corresponding velocity component is changed as its opposite direction.

Step 6: The above optimization process is repeated until the goal is met or the maximum optimization epochs are completed. Thus the ELM with the optimal input weights is obtained, and then the optimal ELM is applied to the testing data.

4 Experiment Results and Discussion

In this section, the PSOIOS-ELM is compared with the traditional ELM, PSO-ELM [9], E-ELM [8], and IPSO-ELM [10] on three benchmark classification data including Diabetes, Satellite image and Image segmentation. The parameters in all algorithms in all experiments are determined by trial and error. For the optimization algorithm in the PSOIOS-ELM, IPSO-ELM, PSO-ELM and E-ELM, the maximum optimization epochs both are 20, and the population size both are 200. For the PSOIOS-ELM, IPSO-ELM and PSO-ELM, the initial inertial weight and the final inertial weight are selected as 1.2 and 0.4 in all experiments. The tolerance rate in the PSOIOS-ELM is set as 0.02 in all cases. All the results shown in this paper are the mean values of 50 trials.

The training, testing and validation datasets are randomly regenerated at each trial of simulations according to [10] for all the algorithms. The corresponding performance of five algorithms on three classification problems is listed in Tables 1, 2 and 3. From Tables 1, 2 and 3, the IPSOIOS-ELM achieve higher testing accuracy than other ELM in all cases but on the Satellite image data. Among all evolutionary ELM, the IPSOIOS-ELM obtains the highest testing accuracies on all data. The traditional ELM requires the most hidden nodes in all cases, which results into largest norm values of the output weights, condition values of the hidden output matrix and input-to-output sensitivity values. The input-to-output sensitivity values in the IPSOIOS-ELM are comparatively low almost in all cases. Moreover, the IPSOIOS-ELM also achieves the low norm values of the output weights and low condition values of the hidden output matrix as other evolutionary ELM.

Figure 1 depicts the input-to-output sensitivity curves of the different ELM on the three benchmark classification problems. From Fig. 1, the ELM obtains the largest input-to-output sensitivity values on all data, which is mainly owing to much more hidden nodes. Although the E-ELM achieves the smallest input-to-output sensitivity values nearly in all cases, its stability needs to improve. The input-to-output sensitivity of the SLFNs trained by the three PSO-based ELM are relatively close. Since decreasing the input-to-output sensitivity is not the only target of the PSOIOS-ELM, it does not obtain the least sensitivity values in all cases.

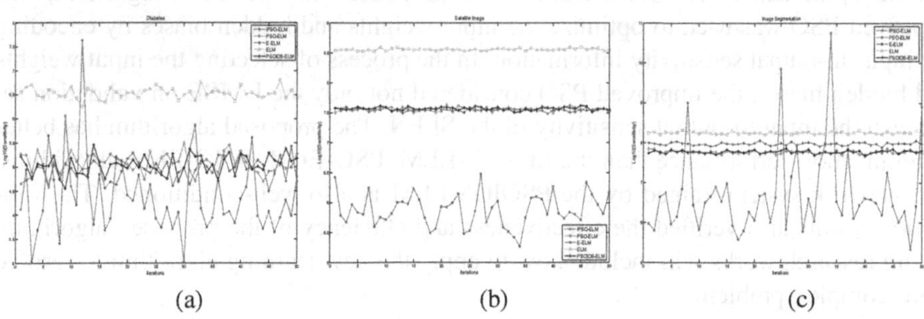

(a) (b) (c)

Fig. 1. The input-to-output sensitivity value of the SLFN trained by the five ELMs (a) Diabetes (b) Satellite image (c) Segmentation image

Table 1. Performance of five algorithms on Diabetes classification problem

Algorithms	Cpu time	Accuracy(%)		Hidden neurons	Norm	Condition	Sensitivity
	Training	Training	Testing ± Std.				
ELM	0.0061 s	81.45	75.19 ± 0.0282	30	382.5409	7.6265e + 3	3.5550e + 6
E-ELM	12.2287 s	76.98	75.51 ± 0.0273	10	62.5096	1.2576e + 3	9.3036e + 5
PSO-ELM	12.4323 s	77.15	76.38 ± 0.0251	10	232.9444	3.1850e + 3	1.3060e + 5
IPSO-ELM	12.1810 s	77.70	76.40 ± 0.0241	10	72.4941	1.0556e + 3	9.7827e + 4
PSOIOS-ELM	12.0105 s	77.90	76.45 ± 0.0230	10	104.9829	1.9605e + 3	1.1778e + 5

Table 2. Performance of five algorithms on Satellite image classification problem

Algorithms	Cpu time	Accuracy(%)		Hidden neurons	Norm	Condition	Sensitivity
	Training	Training	Testing ± Std.				
ELM	2.0034 s	89.50	88.12 ± 0.0103	300	28.6327	3235.5	3.6492e + 8
E-ELM	3.111e + 3 s	88.39	87.25 ± 0.0098	90	2.4371	129.0319	8.7247e + 5
PSO-ELM	3.213e + 3 s	88.21	87.18 ± 0.0106	90	15.3416	808.9574	3.5924e + 7
IPSO-ELM	3.105e + 3 s	88.03	87.30 ± 0.0115	90	14.7202	630.1026	3.4896e + 7
PSOIOS-ELM	2.9801e + 3 s	88.12	87.35 ± 0.0103	90	14.5742	820.7504	3.5173e + 7

Table 3. Performance of five algorithms on Image segmentation classification problem

Algorithms	Cpu time	Accuracy(%)		Hidden neurons	Norm	Condition	Sensitivity
	Training	Training	Testing ± Std.				
ELM	0.4116 s	96.46	92.16 ± 0.0053	150	163.8552	1.2811e + 4	1.7864e + 8
E-ELM	1.1111e + 3 s	95.61	94.51 ± 0.0125	100	52.6075	1.0254e + 4	1.3205e + 9
PSO-ELM	1.3231e + 3	95.94	94.12 ± 0.0130	100	126.0788	7.6494e + 3	5.8952e + 7
IPSO-ELM	1.2731e + 3 s	95.33	94.62 ± 0.0123	100	94.0883	5.1130e + 3	9.4836e + 7
PSOIOS-ELM	1.2563e + 3 s	95.90	94.70 ± 0.0116	100	90.1446	5.0251e + 3	6.0025e + 7

5 Conclusions

In this paper, an improved evolutionary extreme learning machine based on particle swarm optimization (PSOIOS-ELM) was proposed. In the new algorithm, the improved PSO was used to optimize the input weights and hidden biases by encoding the input-to-output sensitivity information. In the process of selecting the input weights and hidden biases, the improved PSO considered not only the RMSE on validation set but also the input-to-output sensitivity of the SLFN. The proposed algorithm has better generalization performance than the ELM, E-ELM, PSO-ELM, IPSO-ELM algorithms. The network system trained by the PSOIOS-ELM is also well-conditioned. The simulation results also verified the effectiveness and efficiency of the proposed algorithm. Future research works will include how to apply the new learning algorithm to resolve more complex problem.

Acknowledgements. This work was supported by the National Natural Science Foundation of China (Nos. 61572241 and 61271385), the Foundation of the Peak of Six Talents of Jiangsu

Province (No. 2015-DZXX-024), and the Fifth "333 High Level Talented Person Cultivating Project" of Jiangsu Province (No. (2016) III-0845).

References

1. Huang, D.S., Du, J.X.: A constructive hybrid structure optimization methodology for radial basis probabilistic neural networks. IEEE Trans. Neural Netw. **19**(12), 2099–2115 (2008)
2. Huang, D.S.: A constructive approach for finding arbitrary roots of polynomials by neural networks. IEEE Trans. Neural Netw. **15**(2), 477–491 (2004)
3. Pao, Y.H., Phillips, S.M., Sobajic, D.J.: Neural-net computing and the intelligent control of systems. Int. J. Control **56**(2), 263–289 (1992)
4. Zhang, L., Suganthan, P.N.: A comprehensive evaluation of random vector functional link networks. Inf. Sci. **367–368**, 1094–1105 (2016)
5. Schmidt, W.F., Kraaijveld, M., Duin, R.P.: Feedforward neural networks with random weights. In: Proceedings of the 11th International Conference on Pattern Recognition (IAPR 1992), pp. 1–4 (1992)
6. Ling, Q.H., Song, Y.Q., Han, F., et al.: An improved ensemble of random vector functional link networks based on particle swarm optimization with double optimization strategy. PLoS ONE **11**(11), e0165803 (2016)
7. Huang, G.B., Zhu, Q.Y., Siew, C.K.: Extreme learning machine: a new learning scheme of feedforward neural networks. In: Proceedings of the 2004 International Joint Conference on Neural Networks (IJCNN 2004), pp. 985–990 (2004)
8. Zhu, Q.Y., Qin, A.K., Suganthan, P.N., Huang, G.-B.: Evolutionary extreme learning machine. Pattern Recogn. **38**(10), 1759–1763 (2005)
9. Xu, Y., Shu, Y.: Evolutionary extreme learning machine – based on particle swarm optimization. In: Wang, J., Yi, Z., Zurada, J.M., Lu, B.-L., Yin, H. (eds.) ISNN 2006. LNCS, vol. 3971, pp. 644–652. Springer, Heidelberg (2006). doi:10.1007/11759966_95
10. Han, F., Yao, H.F., Ling, Q.H.: An improved evolutionary extreme learning machine based on particle swarm optimization. Neurocomputing. **116**, 87–93 (2013)
11. Han, F., Ling, Q.H., Huang, D.S.: Modified constrained learning algorithms incorporating additional functional constraints into neural network. Inf. Sci. **178**(3), 907–919 (2008)
12. Han, F., Huang, D.S.: Improved extreme learning machine for function approximation by encoding a priori information. Neurocomputing. **69**(16–18), 2369–2373 (2006)
13. Eberhart, R.C., Kennedy, J.: Particle swarm optimization. In: Proceeding of the IEEE International Conference on Neural Network, Perth, Australia, pp. 1942–1948 (1995)
14. Shi, Y., Eberhart, R.C.: A modified particle swarm optimizer. In: Proceedings of the IEEE International Conference on Evolutionary Computation, pp. 69–73 (1998)

Nature Inspired Computing and Optimization

Evolutionary Metaphor of Genetic Encoding for Self-organizable Robots

Kiwon Yeom(✉)

Department of Intelligent Robotics, Youngsan University, Yangsan, Korea
pragman@gmail.com

Abstract. Computational models of multi-agent system are widely used for studying a variety of morphogenic processes, which include cell signaling, cell-cell interactions, pattern formation, and cell sorting during tissue self-assembly. This article describes a very simple genetic encoding and developmental system designed for self-organization of multi-cellular agents. The morphogenetic evolution system is guided by gene expression (genetic encoding) and cellular differentiation. Computer simulations show that the method can generate arbitrary 3D shape.

1 Introduction

In general, biological organisms start their development from a single egg cell and form a perfect shape through cell differentiation [3]. Before each differentiation, a copy of the genome is symmetrically made and divided from the origin. The embryo is gradually shaped not only by the fact that different gene products from but also by the physical interactions between the cell components, and between the cells and their environments. The processes allow cells to self-bridge and highlight the discrepancy between the amount of information encoded in the genome and the contribution of self-organizational processes originating in the complex structure of the organism.

The power of such evolutionary processes might be improved if the encoding is biased toward locations of the search space, which are more likely to contain good solutions, or if the encoding and the genetic operators generate fitness values that are better suited for search by evolutionary algorithms.

In this research below we use a simple genetic encoding approach and dynamic developmental system for multi-cellular agents' self-construction. It is based on gene expression principle and cellular differentiation of previous.

A short review and a new classification of developmental systems in evolvable system are given in Sect. 2. We describe general problems and assumptions in Sect. 3, and present the model of development and morphogenetic system in Sect. 4. Section 5 investigates its capacity to evolve 2D structures of various complexities. In Sect. 6 we describe evolutionary model. We discuss the biological implications with experimental results in Sect. 7 and draw the conclusions and outline of future work in Sect. 9.

D.-S. Huang et al. (Eds.): ICIC 2017, Part I, LNCS 10361, pp. 131–139, 2017.
DOI: 10.1007/978-3-319-63309-1_13

2 Related Works

Reference [10] describes a robot that moves along and manipulates passive cubic blocks. The system's creators focus on the software design of the robot, but do not consider automatic control of the robot. Everist et al. in [17] discuss a two-dimensional self-assembly system using self-mobile pucks to assemble passive struts. Reference [11] introduces the inter-robot communication method with cubic blocks. Reference [12] describes a very simple approach for manipulation of circular objects. Reference [18] proposes a polymer foam based hardware system for building terrestrial robots.

Many other works address an emergent system with identical mobile units which are required to organize and rearrange themselves autonomously into a given or desired configuration with a distributed mechanism [1–9]. Another related area is self-assembly which is programmed to form a user-specified shape. [14] presents hardware design for cube based self-assembly in three dimensional spaces.

Reference [10, 13] developed a evolutional model for convex and non-convex shapes using 2D cellular automata. They applied the grammar tree as developmental instructions at each node. Reference [12] described an evolvable computational organism using Cartesian Genetic Programming. Their goal is to develop a cell division mechanism which allows a cell to generate multi-cellular organisms.

3 General Problems and Assumptions

The goal is to design a three dimensional construction system by self-organizable swarm robot as follows: Firstly, specified number of simple cubic robots are deployed into the workspace. Secondly, the robots is to construct three dimensional structures. A user-specified shape map is given as the only input information and robots build the given structure by themselves. To this cause, a set of rules is established so that robots successfully complete their construction task. It describes for robots to follow which will move to appropriate direction, and bind with each other.

We assume the three dimensional space is a kind of weightless or micro gravity environment and the agent can move freely in any direction in the environment without any limitation. Because dealing with gravity would make a variety of considerations complicated, we leave this issue for future work.

4 Activity Rules

Once the partial structure is built, agents are assumed to be able to move along its surface in any direction and attach to other agents which are fastened at sites. There is no centralized control, and coordination is implicitly controlled via the structure being built. Agents are cubic blocks which are the same as modular robots. The reason we use cubic block is for convenience in a Cartesian coordinate system and the easy control of their movements (generally 6 directions only). Moving agents are able to communicate with physically attached neighbor agents, to share the coordinate system, and to decide whether attachment at a site is possible or not. Each agent stores the shape map into its static memory and we assume the memory of agent is enough. Agents can store the coordinate information such as the location at which it is [12].

In this paper we assume that robots are able to attach exactly with automatic aligning and connecting mechanism. The assembled agents plays a role as a reference coordinate system to keep their position and to provide information of distance and direction when they move.

The clustering (binding each other) is based on a hierarchical method which is most commonly used unsupervised clustering method that does not require the use of information of the known classes. However, to overcome the shortcoming of the previous work of the author [12], we devised a self-organizing map based varying weight algorithm.

5 Self-organizable Swarm Robots

In this paper, self-organizable swarm robots consist of a number of independent, simple, and moving robots. In order to construct any user-specified 3D shape, each robot is randomly flocked and proceed construction behavior. Then a flocked set is maintained by repeatedly computing the fitness of swarm. Figure 1 roughly shows the scheme of the constructional process. The gray color cube block presents that the location is already taken by another agent when another agent tries to take the position.

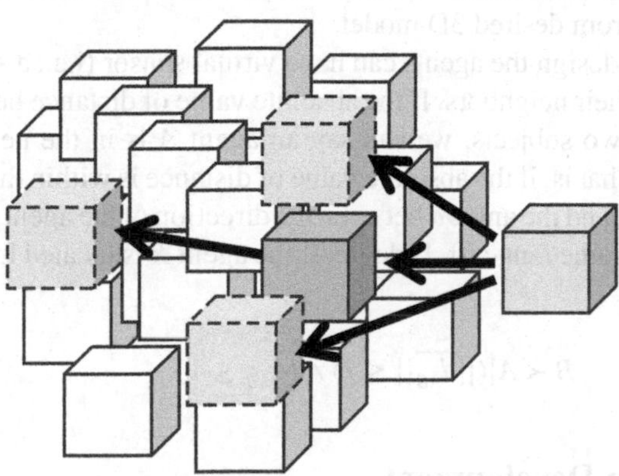

Fig. 1. Clustering process by comparison of fitness.

The set of possible actions after an agent collides with other agent block as shown in Fig. 1 is listed in Table 1. A collective construction of structures can be generated at a location relative to the spatial coordinates of the construction element which the agent has collided with.

Table 1. Description of agent's activity for construction

Action	Flag	Parameter
Move to the forward	↑	$\vec{d}_{forward}$
Move to the backward	→	$\vec{d}_{backward}$
Move to the left	←	$\vec{d}_{leftt}$
Move to the right	↓	$\vec{d}_{right}$
Move to above	∩	$\vec{d}_{above}$
Move to below	∪	$\vec{d}_{below}$
Stay here	⊙	$\vec{d}_{here}$

The movement and construction policy is stated by a direction vector using $\vec{d}$. There is a set of direction vectors such as $\vec{d}_{forward}$, $\vec{d}_{backward}$, $\vec{d}_{leftt}$, $\vec{d}_{right}$, $\vec{d}_{above}$, $\vec{d}_{below}$, and $\vec{d}_{here}$. The absolute values of these direction vectors are normalized to the size of the construction elementary cube. Therefore, it has the same size as a cube, so that distance of a vector $\vec{d}$ points from the center of one construction elementary cube exactly to its immediate neighbor cube's center.

There are two types of maintenance methods. One is that the agent may adhere to the construction elementary cube it has collided with. Another method is to find other candidate locations to combine within a relatively near distance from the construction elementary cube. The information of the position of the construction elementary cube is arbitrarily given from desired 3D model.

In this work, we design the agents can have virtual sensor (i.e., a sort of pheromone sensor) to perceive their neighbors. If the absolute value of distance between two agents satisfies following two subjects, we can say an agent *A* is in the neighborhood of an agent *B* vice versa. That is, if the absolute value of distance is within the radius *r* of agent *A′* perceptive sensor, and the angle α between the direction of the agent *A* and the location of agent *B* is within some range [0, 1] degrees, the agent *B* is located in perceptive range of agent *A*.

$$B \prec A|(||\overrightarrow{d_{AB}}|| \leq r) \wedge (\alpha_{AB} \leq 1) \quad (1)$$

6 Construction Development

In this paper the proposed developmental model, which is the phenotype of the user-specified structure, is represented by a three dimensional array of the cube blocks (cells) which each element of array is associated to a location information of the cells on a grid. The reason we choose a three dimensional array and a cubic-shaped block for the evolutionary phenotype is that it is easier to map into computational array of grids which represents the information of three dimensional target shape. Of course, one must consider sphere-like cells which are closer to the natural cells. However, to adopt such a circular cell model, we have to consider physical environments and physics among cells as well. Therefore, in this paper, we only focus on the cubic-shaped cells and remain the other as future work. The development starts with a single bunch cell placed in the middle of the grid, and unfolds in development phases.

Usually the cells show different responses to signal of different type in nature. This variety means that the cells contain several diffusers, which distribute a chemical signal, and a receiver. In addition, when the cells receive a specific stimulus, and if the cells contain any response mechanism to the stimulus, it shows high intensive response such as propagating chemical elements like pheromone. In this paper, the communication mechanism between agents is based on an analogy with such chemical process for communication between live cells. The first step of signaling begins from distributing a specific substance via diffuser placed in cells. This is another inspiration to model the way of cellular interaction in this work.

There are several diffusers in a computational cell (agent or robot) for the different type of signals. If a cell receives a particular signal from outside, the concentration level of that signal type in the cell always increases towards the maximum value. We refer to the concentration level of signal solute of signal type *st* in cell i^{th} as Concentration Level of Signal Solute ($CLSS_i^{st}$) and the diffusers of type *dt* in cells j^{th} as Diffuser Type (DT_j^{dt}). However, when there are cells which their signal characteristic have not been initialized via diffusers, the signaling process tries to set the signal for them. For this reason, each signal in each cell has a flag which indicates if it is initialized (or valid). When $DT_i^{st} = 1$, the signal of type *st* in cell *i* is initialized, otherwise $DT_i^{st} = 0$. Signals of each type are processed independently, without interactions among them, as if they were in different chemical layers (see pseudo code). It can be re-written as following

$$CLSS_i^{st} = \begin{cases} h(high) & if\ DT_i^{st} = 1 \\ l(low) & Otherwise \end{cases} \tag{2}$$

The expression phase assigns a function to each cell by matching the signal intensities inside that cell with the entries of an expression table PE stored in the genetic code (Fig. 2, bottom right). Figure 2 shows a phenotype expression table with *n* entries and $CLSS = 4$ types of signals. The table contains both the information of intensities of the signals and the function to express in case of match for each entry. The intensity of signal *CLSS* in the entry *j* of the table is noted by expression $CLSS_j^s$. A cell *i* is said to match an entry *j* of the expression table when the distance between i^{th} and j^{th} cells is the smallest among all entries of the expression table.

$$Distance = \sum_{s=1}^{S} HD(CLSS_i^s - CLSS_j^s) \tag{3}$$

The Distance HD is the Hamming distance.

The genetic code contains the expression table PE, and the location of the diffusers (see Fig. 2). The genetic code therefore affects the pattern of diffusion and the expression rules of the cells in the cellular network. Therefore each entry of the expression table is 16 bits (4 signals coded on 4 bits each). It should be noted that the functions in this system are not encoded, so that it could not be evolved in this environment. The locations

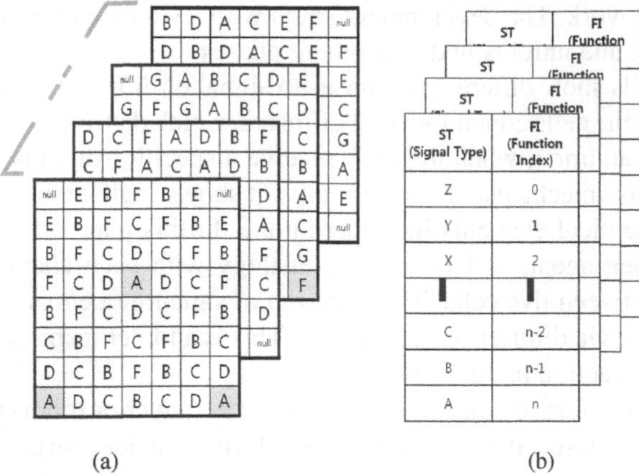

Fig. 2. The snapshots of the signaling phase with one type of signal and several diffusers (a) and (b) shows expression tables for functionality mapping.

of the diffusers are stored as a general coordinate system with pairs of X and Y with two bits which indicate the type of the diffuser. A population of genetic strings is randomly initialized and evolved using a standard genetic algorithm [15].

7 Evolutionary Model Based on Genetic Encoding

The cell association mechanism presented in the previous chapter allows a swarm of agents to get together in three-dimensional spaces. The construction process of the target structure can be achieved by supplying a user-defined 3D structure towards which the swarm is supposed to orientate its corresponding position in the structure.

The weights w_1 to w_8 and maximum values for volume V_{max} are basic properties for construction. Swarms have a set of m policies and it determines the constructional behaviors of a swarm. Each constructional policy represented by CP_i consists of a combination of conditions

$$CP_i = cCE_{i0} \wedge cCE_{il} \wedge \cdots \wedge cCE_{in} \tag{4}$$

where if a construction elementary cube (agent) is found at a position by collision with a moving agent, a condition cCE_{ij} is fulfilled. The behavior policy aP_i contains a specific action (one of the lists in Table 1), which the agent should have to do, and a direction vector as the action's parameter. Once a collision occurs, each constructional policy CP_i is tested and if applicable its consequent action is executed. If an agent has more than one behavior policy, the corresponding actions are executed in the order of their genotypical appearance. In total a swarm genotype $sgen_{total}$ is described following as.

$$sgen_{total} = w_1, \cdots, w_8, V_{max}, CP_1, \cdots, CP_m, aP_1, \cdots, aP_{m,} \tag{5}$$

As usual evolution process in genetic algorithm, we randomly select two parts of $sgen_{total}$ as the crossover masks for the recombination of the swarm genotypes. If there exist any dependencies within the agent's genotype, it is divided into three parts to conserve them.

The offspring's number of policies is limited to the smallest order of the ancestors' policy sets. Therefore the average size of the swarm policy sets decreases in the course of evolution. We choose the smallest one to lessen the computational consumption, and [16] describes that a small set of rules makes a good performance corresponding to evolution.

For two arbitrary genotypes gen_i and gen_j, a crossover mask cm_{ij} which has the shorter genotype's length is generated. gen_0 inherits genotype elements whose entry in cCE_{ij} is 0 from gen_i, if it is 1 from gen_j's surplus of rules is not considered. Mutation is applied on all genotypes of a new generation. Conditions and the action parameter of each behavior policy also undergo the mutation process separately.

As the initial mutation process, the mutation operator checks whether the given value should be altered or not according to a mutation value. If the decision is made in favor of change, an update value, which is smaller than a given mutation value, is chosen and changed by adding to or subtracting from the original value. Whenever the resulting values are out of a given interval [e.g., lower bound, upper bound], they are trimmed to the boundary. This trimming process ensures that the evolved parameters are meaningful and the mutation value defines the procedure's optimal effect. The next generations' members are chosen by means of fitness evaluation.

8 Experimental Results

"In-silico" experiments (see example in Fig. 3) have proved the effectiveness of the method in evolving any kind of shape, of any complexity (in terms of number of cells, number of colors, etc.). Using the complexity of structures as a metaphor for the complexity of organism, the proposed approach is able to establish the potential applications such as generating a complex biological systems or any artificial construction of architectures.

The effectiveness of the proposed method essentially depends on the features of the swarm model of development and evolutionary process. In particular it is related to the presence of a homogeneous distribution of construction elementary cells, which keeps the shape "plastic" throughout evolutionary development and allows artificial evolution process to meet its ends.

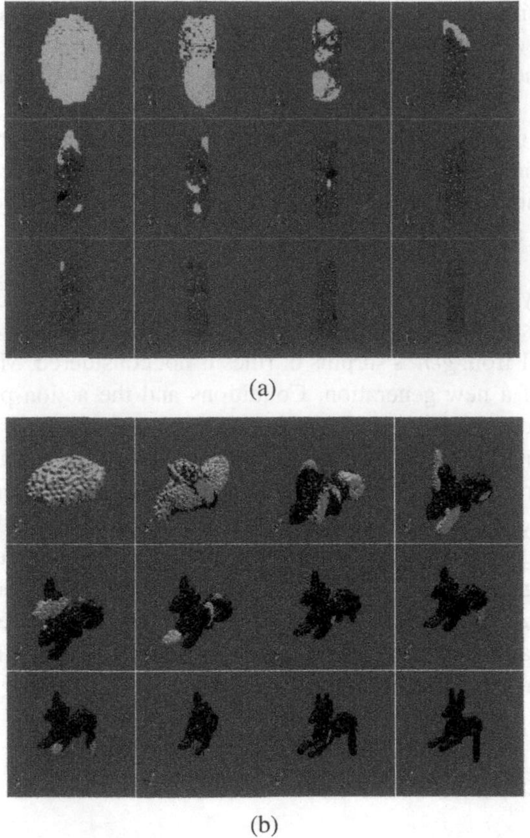

(a)

(b)

Fig. 3. (a) Example of developmental process encoded by a genome, evolved in 30,000 generations (top view of the shape). (b) visualization of generated three dimensional shape.

9 Conclusions

In this work, we argued that biological ideas are useful and applicable to the field of evolutionary systems. Using developmental process of the cells that include growth, differentiation, and agglomerate, we described agents that are capable of changing their shape according to the user-input or environment variation.

Acknowledgments. This work was supported by a 2017 research grant from Youngsan University, Republic of Korea.

References

1. Floyd, S., Heiskanen, T., Taylor, T.W., Mann, G.E., Ray, W.H.: Polymerization of olefins through heterogeneous catalysis VI. Effect of particle heat and mass transfer on polymerization behavior and polymer properties. J. Appl. Polym. Sci. **33**, 1021–1065 (1987)
2. Kosek, J., Stepanek, F., Novak, A., Grof, Z., Marek, M.: Multi-scale modelling of growing polymer particles in heterogeneous catalytic reactors. In: Gani, R., Jłrgensen, S.B. (eds.) Proceedings of the European Symposium on Computer Aided Process Engineering 11 (ESCAPE - 2011), pp. 171–176. Elsevier, Amsterdam (2001)
3. Grof, Z., Kosek, J., Marek, M., Adler, P.M.: Modeling of morphogenesis of polyolefin particles: catalyst fragmentation. AIChE J. **49**, 1002–1013 (2003)
4. Grof, Z., Kosek, J., Marek, M.: Principles of the morphogenesis of polyolefin particles. Ind. Eng. Chem. Res. **44**, 2389–2404 (2005)
5. Tsuji, Y., Kawaguchi, T., Tanaka, T.: Discrete particle simulation of two-dimensional fluidized bed. Powder Technol. **77**, 79–87 (1993)
6. Young, R.J., Lovell, P.A.: Introduction to Polymers, 2nd edn. Chapman & Hall, London (1995)
7. Bay, J.S., Unsal, C.: Spatial self-organization in large populations of mobile robots. In: IEEE Symposium on Intelligent Control, Columbus (1994)
8. Coore, D.: Botanical computing: a developmental approach to generating interconnect topologies on an amorphous computer. Ph.D. thesis. MIT (1999)
9. Day, S.J., Lawrence, P.A.: Morphogens: measuring dimensions: the regulation of size and shape. Development **127**, 2977–2987 (2000)
10. Nagpal, R.: Programmable self-assembly using biologically-inspired multirobot control. In: ACM Joint Conference on Autonomous Agents and Multiagent Systems, Bologna I (2002)
11. Yeom, K.: Morphological approach for autonomous and adaptive system: the construction of three-dimensional artificial model based on self-reconfigurable modular agents. Neurocomputing **148**, 100–111 (2015)
12. Yeom, K., You, B.-J.: Three-dimensional construction based on self-reconfigurable modular robots. Int. J. Comput. Sci. Eng. **11**(4), 368–379 (2015)
13. White, P., Zykov, V., Bongard, J., Lipson, H.: Three dimensional stochastic reconfiguration of modular robots. In: Proceedings of Robotics Science and Systems, pp. 161–168 (2005)
14. Gara, A., Blumrich, M., Chen, D., Chiu, G., Coteus, P., Giampapa, M., Haring, R., Heidelberger, P., Hoenicke, D., Kopcsay, G. et al.: Overview of the blue gene/l system architecture. IBM J. Res. Dev. **49**(2), 195–212 (2005)
15. Bojinov, H., Casal, A., Hogg, T.: Emergent structures in modular self-reconfigurable robots. In: Proceedings of the IEEE International Conference on Robotics and Automation (ICRA 2000), pp. 1734–1741 (2000)
16. Kosek, J., Stepanek, F., Novak, A., Grof, Z., Marek, M.: Multi-scale modelling of growing polymer particles in heterogeneous catalytic reactors. Comput. Aided Chem. Eng. **9**(2001), 177–182 (2001)
17. Terada, Y., Murata, S.: Automatic modular assembly system and its distributed control. Int. J. Robot. Res. **27**(3–4), 445–462 (2008)
18. Shen, W., Salemi, B., Will, P.: Hormone-inspired adaptive communication and distributed control for CONRO self-reconfigurable robots. IEEE Trans. Robot. Autom. **18**(5), 700–712 (2002)

Hybrid Coordinating Algorithm for Flying Robots

Kiwon Yeom(✉)

Department of Intelligent Robotics, Youngsan University, Yangsan, Korea
pragman@gmail.com

Abstract. Positioning of flying swarm robots is an interesting research area because the global behaviors must emerge from many diverse local interactions. A central issue in coordinating of swarm is enabling robots to take emergence formation by collective behavior. This paper describes a hybrid coordinating algorithm using pheromone for controlling a swarm of identical flying robots to spatially self-organize into arbitrary shapes using local communication maintaining a certain level of density. The proposed approach considers the topological structure of the organization, supports dynamic reconfiguration and self-organization.

1 Introduction

In the areas of distributed robotics systems, a lot of research effort has been developed towards controlling autonomous robots with low power requirements and simple sensor capabilities [1–3]. We aim at investigating flying robots based on minimal aerial swarm systems which have minimal capabilities for sensing and communications, which can be deployed in real-life scenarios. This research work is inspired by an application whereby several flying robots have to self-organize autonomously to establish an emergency communication network to detect multiple users located in disaster areas and relay their position information [4–7]. Flying robots can fly over difficult terrain such as flooded or debris areas [8]. Rather than relying on positioning sensors which depend on the environment and are costly, flying robots rely on proprioceptive sensors and local communication with neighbors. An approach to form an arbitrary shape in two dimensional space called the ‘ShapeBugs’ is depicted by [9]. It only requires robots to have the ability of local communication, two approximate measures for relative distance and motion, and a global compass.

We presented a modified and fast formation control approach by substitute the global compass method of [9] with continuous calculation of the error in estimation of local robot’s movement and we introduced pheromone based density control mechanism to manage and keep the overall shape as any failure in robots happens [10]. In this work, we focus on a control algorithm in which flying robots self-organize and self-sustain arbitrary 2D formations. The two-dimensional formation task is the starting point of this research, as it is simple to describe swarm applications in terms of theory or practice. For instance, given a set of flying robots and a set of points, the problem to solve is to arrange the robots on the points without being piled up on one another

D.-S. Huang et al. (Eds.): ICIC 2017, Part I, LNCS 10361, pp. 140–151, 2017.
DOI: 10.1007/978-3-319-63309-1_14

(i.e., one to one correspondence). In real world tasks, often flying robots can be involved to form a particular shape in examples such as sensor grids or sensor networks.

We propose a improved control algorithm of the previous approach of [10] by performing continuous gradient descent algorithm for searching optimal position of robot. This approach can not only accomplish arbitrary shapes by self-organization but also produces resulting the formed global shapes that are highly robust to varying numbers of robots. Briefly, our algorithm works as follows: firstly, flying robots initially wander with no information about their own coordinates or their environment. However, they have a programmed internal knowledge of the desired shape to be formed. Next, a small seed group of robots are initially located in a shape. As non-seeded robots move, they continually perform local trilaterations and gradient descent search to figure out their location by continuous local communication. At the same time, robots maintain a certain density level among themselves using pheromones and flocking-rule-based distance measurements [11]. This enables flying robots to disperse within the specific shape and fill it efficiently.

2 Related Work

There are several literatures that can be considered relevant to robotic self-assembly [9, 12–15]. These researches addressed the problem of building arbitrary shapes by self-assembling robot swarms. [16] investigated micro aerial vehicles to establish a positionless communication network. However, frequent replacement of node MAVs in the network may drift the swarm from original position. The idea behind the aerial swarm had its origins in the Beowulf Project [17]. The Beowulf Project demonstrated the possibility of creating a distributed parallel computer interconnecting through an Ethernet network using a number of cheap Linuxboxes. [18] remarked that it should be possible to configure a fleet of LinuxBots to operate across the wireless LAN as a Beowulf cluster.

Several projects are aimed at getting UAVs to fly in formation, usually under remote but high-level control. This type of project is therefore different from the biologically inspired flexibility and responsiveness of flocking pursued within a swarm. However many of the required technologies are similar. The MinuteMan project at UCLA builds a reconfigurable architecture for highly mobile multi-robot systems [19]. The intention is that the computationally capable autonomous vehicle would be able to share information across a wireless fault-tolerant network. Study of formation-flying were undertaken at MIT, within the Autonomous blimps project [20]. The University of West England developed the flying flock project slightly different from previous work [21]. The work is conceived with a minimalist approach.

Several map-based UAV applications are proposed in [22, 23]. In map-based applications, UAVs know their absolute position which can be shared locally or globally within the swarm. However, obtaining and maintaining relative or global position information is challenging for UAVs or mobile robot systems. A possible advance is to adopt a global positioning system (GPS). However, GPS is not reliable and rarely possible in cluttered areas [24]. Alternatively, wireless technologies can be used to estimate the range or angle between robots of the swarm. In this case, beacon robots can be used for a reference position to other moving robots. However, depositing beacons is generally not practical for swarm systems in unknown environments [25].

3 Aerial Robot Model

We assume a simple aerial robot that moves and turns in continuous space, which is motivated by the pieces of capabilities of real autonomous unmanned aerial robots. In addition, each robot has several simple equipment such as distance and obstacle detection sensors, a magnetic compass, wireless communication, etc.

We assume that robots are able to move in 2D continuous space, all robots execute the same program, and robots can interact only with other nearby robots by measuring distance and message exchange.

The robots' communication architecture is based on a simple IR ring architecture because we assume that robots can interact only with nearby neighbors. The robots have omnidirectional transmission, and directional reception. This means that when a robot receives a transmission, it knows roughly which direction the transmission came from (see Fig. 1). An example of such communication hardware is described in [26] (see Fig. 2).

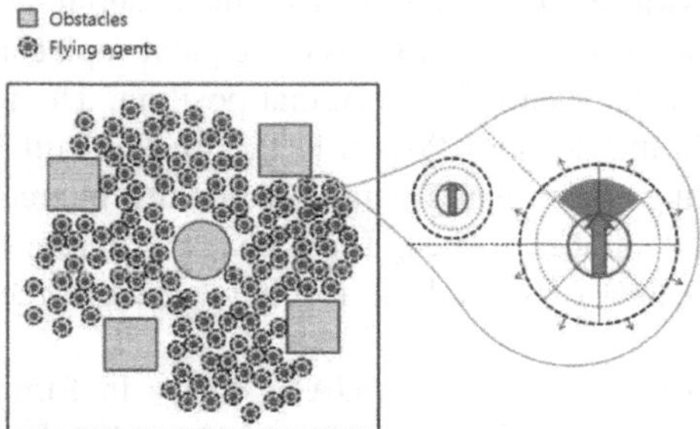

Fig. 1. Aerial robot model inspired from capabilities of real UAVs

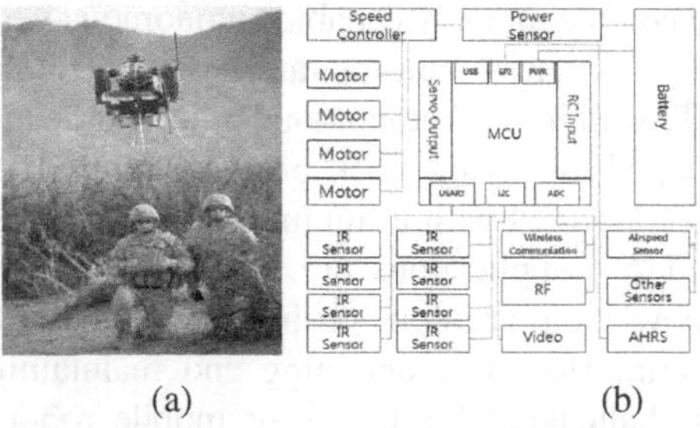

(a) (b)

Fig. 2. (a) An example of UAVs hardware. (b) UAV hardware architecture.

The robot's dynamic model is implemented using a first order flight model for simple and low-cost airframe. We assume that our robots can fly at a speed of approximately 0.5 m/s and are able to hover or make sharp turns as an example in Fig. 3. The minimum turn radius of the flying robots is assumed to be as small as possible with respect to the communication range. Later we will consider wireless communication based on the IEEE 802.11b specification (e.g., WiFi) allowing a communication range of around over 100 m. This medium might be enough for realistic scenarios because in most potential networks, ground users can use wireless communication devices which are embedded on laptops, smart phones, and other smart devices, etc.

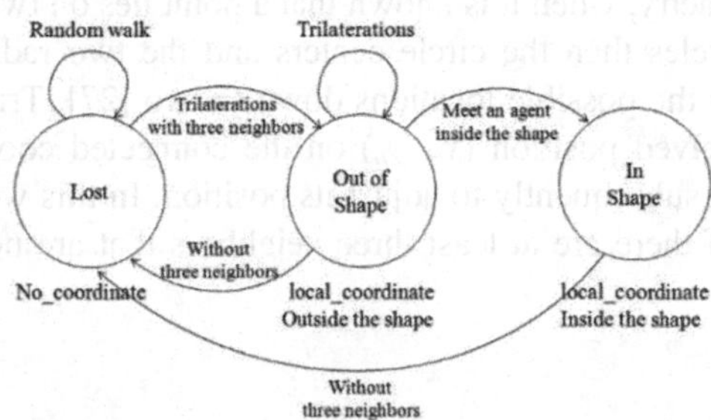

Fig. 3. An example of robot's state machine.

4 Self-organizing Formation Algorithm

In our self-organizing formation algorithm, each aerial robot has a shared map of the shape to be constructed and this should be overlaid on the robot's learned coordinate system. Initially, aerial robots are randomly scattered into the simulation world without any information about the environment. Then robots begin to execute their programmed process to decide their position using only data from their proximity sensors (i.e., distance and density) and their wireless communication link with nearby neighbors.

4.1 Robot Transition Cycle

In our model, robots have a simple transition cycle model as shown in Fig. 3. Sensing process steps is necessary because robots should compare the data before and after movement to determine distance and orientation from positioning error. After each transition of robot, the time until the next transition is set randomly from $[T_{min}, T_{max}]$, where T_{min} is 0 to the time for computation and T_{max} is approximately the time of wait or move. Robot can move only one unit or 0 unit during *Move* and *Sense* transition process. Therefore, if robot does not move, the robot's *Move* does not have any time code.

In this model, robots have three computational states such as *lost, out of shape*, and *in shape* (see Fig. 3) as describe in [9]. Although [9] is robust, it does not provide any stable state because there is no definition of simulation complete state. Unlike [9], our model uses 8 IR sensors to approximately sense the direction of the referenced robot. As mentioned earlier, this enables for robot to easily and fast approach towards inside the target shape. Once robots acquire a shared coordinate system, they begin to fill a formation shape by each robot diffuses pheromone with repulse range R_{rep} (see Fig. 6). The pheromone emission mechanism allows the formation shape to be robust against robots' death, while robots evenly spread out throughout the shape.

4.2 Hybrid Algorithm for Coordinating Flying Robot

In geometry, trilateration is the process of determining absolute or relative locations by measurement of distances using the geometry of circles or triangles. In two-dimensional geometry, when it is known that a point lies on two curves such as the boundaries of two circles then the circle centers and the two radii provide sufficient information to narrow the possible locations down to two [27]. Trilateration allows an robot to find its perceived position (x_p, y_p) on the connected coordinate system (see Fig. 4). It is also used subsequently to adjust its position. In this work, the trilateration process occurs only if there are at least three neighbors that are not in the lost state.

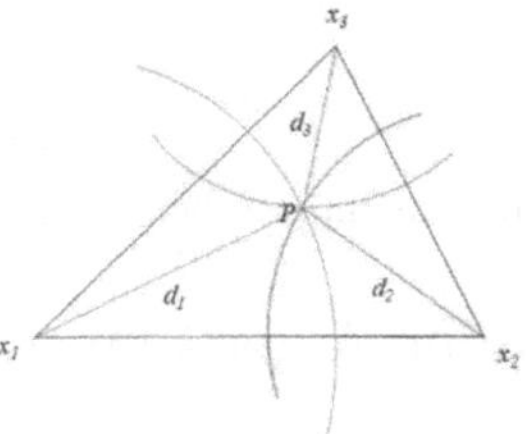

Fig. 4. Robot's trilateration

Let the positions of the three fixed anchors be defined by the vectors x_1, x_2, and $x_3 \in R^2$. Further, let $x_p \in R^2$ be the position vector to be determined. Consider three circles, centered at each anchor, having radii of d_i meters, equal to the distances between x_p and each anchor x_i. These geometric constraints can be expressed by the following system of equations:

$$\|x_p - x_1\|^2 = d_1^2 \tag{1}$$

$$\|x_p - x_2\|^2 = d_2^2 \tag{2}$$

$$\|x_p - x_3\|^2 = d_3^2 \tag{3}$$

Since $kx_p - x_i k^2 = kx_p k^2 - 2x_i \cdot x_p + kx_i k^2$, the above equations can be rewritten as follows:

$$\|x_p\|^2 - 2x_1 \cdot x_p + \|x_1\|^2 = d_1^2 \tag{4}$$

$$\left\|x_p\right\|^2 - 2x_2 \cdot x_p + \|x_2\|^2 = d_2^2 \tag{5}$$

$$\left\|x_p\right\|^2 - 2x_3 \cdot x_p + \|x_3\|^2 = d_3^2 \tag{6}$$

by subtracting the second and third equations from the first, results in the following two equations:

$$2(x_2 - x_1) \cdot x_p = d_2^2 - d_1^1 - \|x_2\|^2 + \|x_1\|^2 \tag{7}$$

$$2(x_3 - x_1) \cdot x_p = d_3^2 - d_1^1 - \|x_3\|^2 + \|x_1\|^2 \tag{8}$$

by solving the following linear system, x_p (expressed as a column vector, that is N · $x_p = \xi$) can be determined:

$$\mathcal{N} = \begin{pmatrix} x_2 - x_1 \\ x_3 - x_1 \end{pmatrix} \tag{9}$$

$$\xi = \begin{pmatrix} d_2^2 - d_1^1 - \|x_2\|^2 + \|x_1\|^2 \\ d_3^2 - d_1^1 - \|x_3\|^2 + \|x_1\|^2 \end{pmatrix} \tag{10}$$

Generally, the best fit for x_p can be regarded as the point that minimizes the difference between the estimated distance (ζ) and the calculated distance from $x_p(x_p, y_p)$ to the neighbors reported coordinate system. That is,

$$\underset{(x_p, y_p)}{\operatorname{argmin}} \sum_i \left| \sqrt{(x_i - x_p)^2} - \varsigma_i \right|. \tag{11}$$

From this information, we can learn that this problem is closely related to the *summinimization problems* that arise in least squares and maximum-likelihood estimation. Therefore we suggest simple way of search of local minima (see Eqs. 12 and 13). However, in this paper, we do not consider finding any optimal or global solution but a local minimum, because it requires a lot of computational resources and it is not suitable for a small and inexpensive device. For simplification, formula 11 can be rewritten in the form of a sum as follows:

$$Q(w) = \sum_i^n Q_i(w) \tag{12}$$

where the parameter w is to be estimated and where typically each summand function $Q_i()$ is associated with the i-th observation in the data set. We perform Eq. 13 to minimize the above function:

$$w := w - \alpha \nabla Q(w) = w - \alpha \sum_{i=1}^{n} \nabla Q_i(w) \tag{13}$$

where α is a step size. For easy understanding, we draw the pictures in Fig. 5.

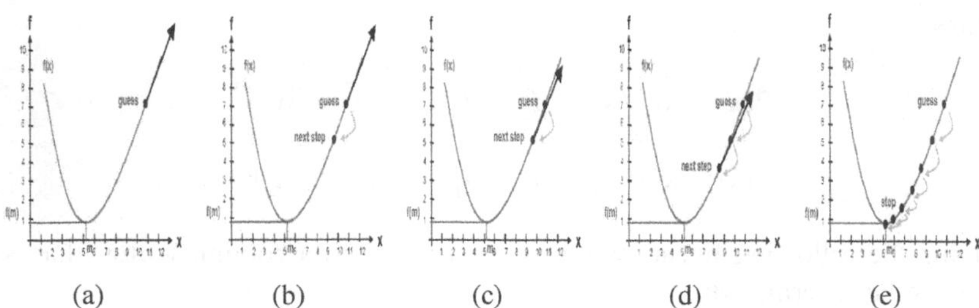

Fig. 5. (a) Initial guessing for deciding direction to find local minima. (b)–(e) stepwise searching the local minima.

4.3 Flocking Movement Control

Robots take different movement patterns according to their states. If a robot is in the lost state, it is assumed that the robot is located outside the shape or is in the initial simulation starting status. If they are outside the shape, they begin to wander randomly to find their way into the shape. When robots are inside the shape, they are considered as part of the swarm that comprises it. Once robots have acquired a coincident coordinate system in the shape, they should not take any steps so that place them outside of the shape. Then robots attempt to fill a formation shape. In this work, we achieve this control by modelling virtual pheromones in a closed container. Robots react to different densities of neighbors around them, moving away from areas of high density towards those of low density. They finally settle into an equilibrium state of constant density level throughout the shape over time. However, based on our assumption, if the number of robots is not enough to compensate for the area, the robots cannot maintain the shape any more. In other words, the swarm can respond to any loss as long as there are enough robots left to maintain a certain level of density equilibrium. Neighboring robots inside the shape with distance < Repel will repulse each other, leading to an average density of robots throughout the shape (see Fig. 6).

Fig. 6. Pheromone robot's influence ranges

This mechanism allows the shapes to be robust against robot death or addition while spreading robots evenly throughout the shape.

4.4 Pheromone Model for Density Control

Pheromone model is inspired by following factors: (1) biological discoveries about how cells self-organize into global patterns, and (2) distributed control systems for self-reconfigurable robot s [15]. Pheromone provides the common mechanism that makes it possible for robots to communicate without identifiers or addresses. The basic idea of pheromone is that a swarm is a network of robots that can dynamically communicate in the network. Robot will react to pheromone according to their local topology and state information. *Dynamic Network of Swarm Flying Robots* is specified as a network of *N* autonomous robots. Each robot has a set of connectors through which the robot can dynamically connect to other robots to forma kind of edges for communication or physical coupling. The connectors of robots are the channels it can be used to communicate with others. A channel of an robot has to be connected to the other channel of another robot to communicate. Because connectors of robots can be dynamically joined and disjoined, robots can make a sort of dynamic and reconfigurable communication network. Let $robot_i$ and $NetEdge_i$ denote the number of robots and the number of networked edges, respectively. Then the dynamic network can be mathematically written as follows:

$$DN = (robot_i, NetEdge_i) \tag{14}$$

Note that both $robot_i$ and $NetEdge_i$ can be dynamically changed because robots can autonomously join, leave, or be failed and died. The diffusion and dissipation of pheromone of a given robot is denoted by $P(x,y)$, where x and y are 2D space. We simply introduce the mechanism of diffusion and dissipation of pheromone as follows:

$$\frac{\partial P}{\partial t} = \left(\alpha \frac{\partial^2 C}{\partial x^2} + \beta \frac{\partial^2 P}{\partial y^2}\right) - \Delta E \tag{15}$$

The first term on the right is for diffusion, and α and β represent the rate of diffusion in x and y directions, respectively. The second term is for dissipation and the constant δ is the rate of dissipation. Equation 15 can be considered as a part of environment function which responsible for the implementation of the dynamic communication and other effects.

4.5 Density Control and Error Correction

The density control is based on Payton approach, but is also similar in nature to the flocking rules proposed by Reynolds. Our density control is to equalize overall density of robots at any situation. To this end, as shown Fig. 6, the robot has three different influence ranges. Each robot has a varying repellant (or repulsive power) that has a maximum value near center which is described as Collision area and a minimum value around Range zone regarding any adjacent robot. The repellant decreases at a constant ratio from center and it becomes the smallest value when it reaches a Range zone (black dotted circle area). The robot's movement vector is weighted inversely by distance. Therefore, if any two robot are close, they push away one another. This allows robots to disperse evenly at any density.

In our work, if a robot moves distance d in some direction, it may actually move $d \pm \delta$. As a result, robot's perceived coordinate systems accumulate errors over time. Thus, it should be re-calculated and re-adjusted. To this end, we accumulate previous trilaterations performed by robots in memory and average them with the recent coordinate at a certain interval of time steps (e.g., every 10 steps).

5 Experimental Results

We show that the proposed formation control algorithm can form any arbitrary shape while autonomically compensating for various errors and maintaining the shape against robot death. The system is implemented in Java 1.5.2 based on the architecture of SWARM 2.2. The user-specified shape maps are represented by bitmap images. A group of 10 robots which are in 25 × 25 pixels are seeded to trigger the first round of trilaterations. Distance sensors have a range of 10 units (around 20 m in the real world) and robots move a discrete 1 unit at every time step.

5.1 Formation of Arbitrary Shapes

Figure 9 shows several formation examples which are made by flying robots, and also shows that the same shape can be formed with different density levels that robots can accommodate. In this experiment, we basically set the initial density level of robots as 16 neighbors in target shapes. As shown in Fig. 7, at any density our virtual pheromone model causes flying robots to disperse evenly throughout user-specified shapes. As shown in Fig. 7, only connected formations are possible due to using a consensus coordinate system between robots in our formation control algorithm, and shapes have a tendency to be harder to form well (i.e., organic growth). The complete shape formation took about 1500 time steps, depending on the number of robots and the density level.

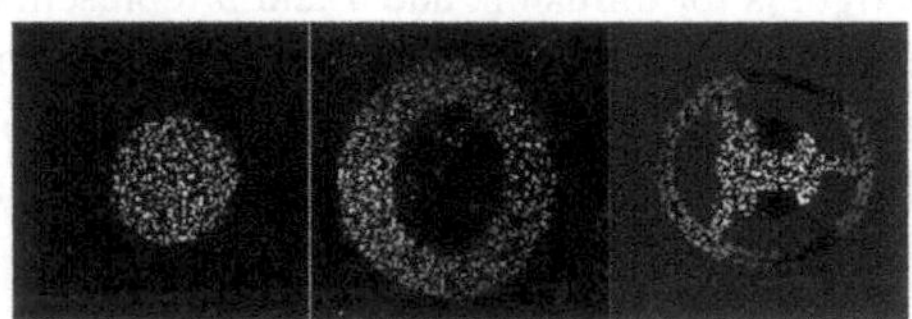

Fig. 7. Examples of formations

6 Discussion

In the proposed approach, when a robot moves, it should move to another place without negatively affecting the stability of the coordinate system for adjacent robots. To demonstrate that robot movement does not negatively affect the stability, we examined the following experiment. First, we set 1000 robots in a given 100 × 100 world. After 150 steps, these robots are to converge on a consistent coordinate system. Then, we assign each robot a probability to move randomly with a probability.

After 220 steps, the robots are no longer allowed to move (see Fig. 8(b)). For every 10 steps, the consistency is recorded. This experiment is repeated 5 times and the result is shown in Fig. 9(a). The consistency is sum of the difference of the actual distance between the two robots and the distance between their locations in the coordinate system. In other words, it is computed as $\sum_{Robot}^{all}(Distance_{i,j} - \sqrt{(x_i - x_j)^2 + (y_i - y_j)^2})$. As shown in Fig. 9, the instability of consistency jumps at time step 180. It is because when robots move, they still keep their old location information until they complete relocalization. This causes the instability during robot movement. However, after 240 steps having no more movement of robots, the consistency drops back to the level prior to movement.

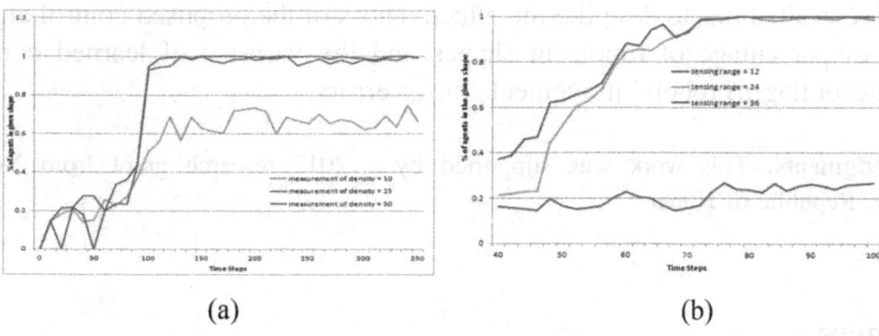

(a) (b)

Fig. 8. (a) Percentage of robots in the shape with different measurement of density. (b) Percentage of robots in the shape with different angle of sensors.

This indicates the movement does not negatively affect the coordinate system. In our experiments, the average time required to complete a stabilized shape formation is about 300 time steps, depending on the number of robots and robot density. Figure 8(a) shows the percentage of robots in the given shape in a 150 × 150 world. Most shapes are roughly formed in 100 time steps and converge after 300 time steps. In addition, rate of shape formation increases as the number of robots increases from 150. We also

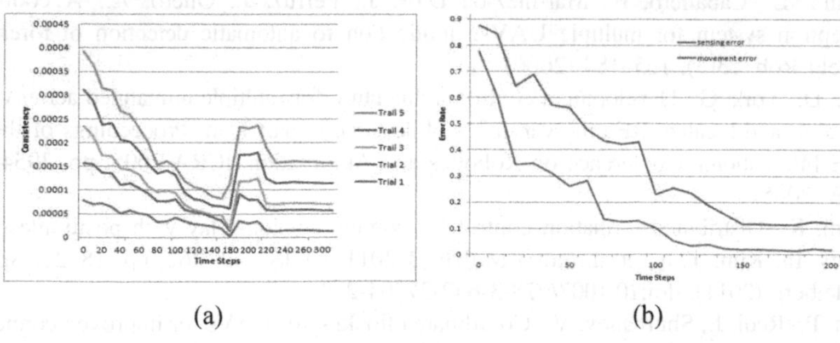

(a) (b)

Fig. 9. (a) Consistency error for 5 simulation runs. (b) Average coordinate variance under movement and sensing errors.

observe that coordinate systems very quickly propagate throughout robots when the robot density is high so that the time to stabilization is reduced. We simply tested how the robots are affected by hardware limitations. As seen in Fig. 9(b), the degree of angle of a sensor affects the performance of the robots. However, we did not evaluate robots' movement error or sensing error because those are related closely to making a consensus coordinate system among robots. Finally, we evaluated the variance of the coordinate system with respect to movement error and sensing error. Figure 9(b) shows that the variance settles to a stable value after about 300 time steps.

7 Conclusion and Future Work

We show that robots can self-organize into arbitrary user-specified shapes and maintain well the formed architecture by continuous trilateration-based on a consensus coordinate system and a virtual pheromone-based density model. We also provide several quantitative evaluations to describe the effectiveness of the proposed control algorithm in terms of percentage of robots in shapes and the variance of learned coordinate systems according to robots' movement sensor errors.

Acknowledgments. This work was supported by a 2017 research grant from Youngsan University, Republic of Korea.

References

1. Prencipe, G., Santoro, N.: Distributed algorithms for autonomous mobile robots. In: Navarro, G., Bertossi, L., Kohayakawa, Y. (eds.) TCS 2006. IIFIP, vol. 209, pp. 47–62. Springer, Boston (2006). doi:10.1007/978-0-387-34735-6_8
2. Prencipe, G.: Corda: Distributed coordination of a set of autonomous mobile robots
3. Flocchini, P., Prencipe, G., Santoro, N., Widmayer, P.: Gathering of asynchronous robots with limited visibility. Theor. Comput. Sci. **337**(1–3), 147–168 (2005)
4. Flint, M., Polycarpou, M., Fernandez-Gaucherand, E.: Cooperative control for multiple autonomous UAV's searching for targets. In: Proceedings of the 41st IEEE Conference on Decision and Control, vol. 3, pp. 2823–2828. IEEE (2002)
5. Merino, L., Caballero, F., Martínez-de Dios, J., Ferruz, J., Ollero, A.: A cooperative perception system for multiple UAVs: application to automatic detection of forest fires. J. Field Rob. **23**(3), 165–184 (2006)
6. Pack, D., York, G.: Developing a control architecture for multiple unmanned aerial vehicles to search and localize RF time-varying mobile targets: part I. In: Proceedings of the 2005 IEEE International Conference on Robotics and Automation, ICRA 2005, pp. 3954–3959. IEEE (2005)
7. Yeom, K.: Distributed formation control for communication relay with positionless flying agents. In: Kim, T.-h., et al. (eds.) MulGraB 2011. CCIS, vol. 262, pp. 18–27. Springer, Heidelberg (2011). doi:10.1007/978-3-642-27204-2_3
8. Basu, P., Redi, J., Shurbanov, V.: Coordinated flocking of UAVs for improved connectivity of mobile ground nodes. In: Military Communications Conference, MILCOM 2004, vol. 3, pp. 1628–1634. IEEE (2004)

9. Cheng, J., Cheng, W., Nagpal, R.: Robust and self-repairing formation control for swarms of mobile agents. In: Proceedings of the National Conference on Artificial Intelligence, vol. 20, p. 59. AAAI Press, MIT Press, Menlo Park, Cambridge, London (2005)
10. Yeom, K.: Pheromone inspired morphogenic distributed control for self-organization of unmanned aerial vehicle swarm. Life Sci. J. **10**(3), 979–991 (2013)
11. Elston, J., Frew, E.: Hierarchical distributed control for search and tracking by heterogeneousaerial robot networks. In: IEEE International Conference on Robotics and Automation, ICRA 2008, pp. 170–175. IEEE (2008)
12. Barnes, L., Fields, M., Valavanis, K.: Swarm formation control utilizing elliptical surfaces and limiting functions. IEEE Trans. Syst. Man Cybern. Part B Cybern. **39**(6), 1434–1445 (2009)
13. Breitenmoser, A., Schwager, M., Metzger, J., Siegwart, R., Rus, D.: Voronoi coverage of non-convex environments with a group of networked robots. In: 2010 IEEE International Conference on Robotics and Automation (ICRA), pp. 4982–4989. IEEE (2010)
14. Cortes, J., Martinez, S., Karatas, T., Bullo, F.: Coverage control for mobile sensing networks. IEEE Trans. Rob. Autom. **20**(2), 243–255 (2004)
15. Rubenstein, M., Shen, W.: A scalable and distributed approach for self-assembly and selfhealing of a differentiated shape. In: IEEE/RSJ International Conference on Intelligent Robots and Systems, IROS 2008, pp. 1397–1402. IEEE (2008)
16. Hauert, S., Winkler, L., Zufferey, J., Floreano, D.: Ant-based swarming with positionless microair vehicles for communication relay. Swarm Intell. **2**(2), 167–188 (2008)
17. Dawkins, R., Holland, O., Winfield, A., Greenway, P., Stephens, A.: An interacting multi-robot system and smart environment for studying collective behaviours. In: Proceedings of 8th International Conference on Advanced Robotics, 1997, ICAR 1997, pp. 537–542. IEEE (1997)
18. Yoxall, P.: Minuteman project, gone in a minute or here to stay-the origin, history and future of citizen activism on the United States-Mexico border. U. Miami Inter-Am. L. Rev. **37**, 517–521 (2005)
19. van de Burgt, R., Corporaal, H.: Blimp positioning in a wireless sensor network, Technische Universiteit Eindhoven (2008)
20. The flying flock. http://www.ias.uwe.ac.uk/projects.html
21. Kadrovach, B., Lamont, G.: Design and analysis of swarm-based sensor systems. In: Proceedings of the 44th IEEE 2001 Midwest Symposium on Circuitsand Systems, 2001, MWSCAS 2001, pp. 487–490 (2001)
22. Bojinov, H., Casal, A., Hogg, T.: Emergent structures in modular self-reconfigurable robots. In: Proceedings of IEEE International Conference on Robotics and Automation, ICRA 2000, pp. 1734–1741 (2000)
23. Kosek, J., Stepanek, F., Novak, A., Grof, Z., Marek, M.: Multi-scale modelling of growing polymer particles in heterogeneous catalytic reactors. Comput. Aided Chem. Eng. 9, 177–182 (2001)
24. Terada, Y., Murata, S.: Automatic modular assembly system and its distributed control. Int. J. Robot. Res. **27**(3–4), 445–462 (2008)
25. Shen, W., Salemi, B., Will, P.: Hormone-inspired adaptive communication and distributed control for CONRO self-reconfigurable robots. IEEE Trans. Robot. Autom. **18**(5), 700–712 (2002)
26. Cione, J.J., Uhlhorn, E.W., Cascella, G., Majumdar, S.J., Sisko, C., Carrasco, N., Powell, M. D., Bale, P., Holland, G., Turlington, P., Fowler, D., Landsea, C.W., Yuhas, C.L.: The first successful unmanned aerial system (uas) mission into a tropical cyclone. In: Proceedings AMS 12th Conference on IOAS-AOLS, pp. 25–30 (2008)
27. Guibin, Z., Qiuhua, L., Peng, Q., JiuZhi, Y.: A GPS-free localization scheme for wireless sensor networks. In: 12th IEEE International Conference on Communication Technology (ICCT), pp. 401–404, (2010)

Signal Processing

A Method to Detecting Ventricular Tachycardia and Ventricular Fibrillation Based on Symbol Entropy and Wavelet Analysis

Yingda Wei[1,2], Qingfang Meng[1,2(✉)], Haihong Liu[1,2], Mingmin Liu[1,2], and Hanyong Zhang[1,2]

[1] The School of Information Science and Engineering, University of Jinan, Jinan 250022, China
ise_mengqf@ujn.edu.cn

[2] Shandong Provincial Key Laboratory of Network Based Intelligent Computing, Jinan 250022, China

Abstract. Detection of ventricular tachycardia (VT) and ventricular fibrillation (VF) is crucial for the success of saving the patient's life. In this paper, we proposed a novel method for detection of VF and VT, based on the Wavelet Analysis and Symbol Entropy. The classification accuracy of symbol entropy was 80.03% with SVM, and the classification accuracy of the symbol entropy with wavelet analysis arithmetic was 99.5% with SVM. Fusion algorithm is greater than symbol entropy.

Keywords: Symbol entropy · Wavelet analysis · Ventricular Tachycardia · Ventricular Fibrillation · SVM

1 Introduction

Ventricular tachycardia (VT) and ventricular fibrillation (VF) were both life-threatening arrhythmia. Treatment protocols of VT and VF were different. The VF patients need to be electric shocked. The VT patient need drug therapy. The VF was misinterpreted ad VT, that's life-threatening. If VT is incorrectly interpreted as VF, the patients' hearts will be shocked and damaged. Therefor an efficient detecting method to distinguish VT from VF has clinical research significance [1].

In recent years, the VT and VF detecting methods mainly include time domain analysis [2], frequency-domain analysis [3] and nonlinear analysis [4]. There were overlaps of the heart rate between VT and VF, so according to the heart rate to distinguish the VT and VF while occur high error rate. To overcome this problem, scholars had proposed many quantitative analysis methods [5–9]. But these methods had certain limitations. With the rapid development of neural computing, scholars applied neural networks to distinguish VF and VT. This can distinguish two class by nonlinear [10]. But the execution time of the algorithm was high relatively.

Wang et al. [11] proposed a method based on Symbol dynamics to research VF and VT. The study showed that a sudden drop of Symbol sequence's entropy value indicated

D.-S. Huang et al. (Eds.): ICIC 2017, Part I, LNCS 10361, pp. 155–164, 2017.
DOI: 10.1007/978-3-319-63309-1_15

that the patients most likely entered the sample of ventricular tachycardia and this was a crucial sample for the clinical treatment of patients.

The origin of Wavelet happened in the late 70's created by J.Morlet, which enabling analysis signals at different scales of time and frequency. The wavelet transform (WT) decomposes a time series into components in various frequency subbands or scales with various resolutions. It is a time-scale representative for the analysis of non-stationary signals particularly that has been applied to a variety of fields including detection of the onset of epileptic seizure [12–14], analysis of VT and VF [15].

Support vector machines (SVM), has been suggested as a useful approach to improve the detection efficiency. SVM is based on structural risk minimization principle and can construct an Optimal Separating Hyperplane (OSH) in the feature space. With the minimum risk of misclassification, the OSH can classify both the training samples and the unseen samples in the test set.

In this paper, we proposed an improved symbol entropy based on wavelet analysis for detecting VT and VF. The method displayed a higher accuracy rate in the classification of VF and VT with shorter time series than original symbol entropy. Furthermore, the execution time of the algorithm was shorter than original symbol entropy.

2 The Method of Wavelet Analysis and Symbol Entropy

2.1 Wavelet Analysis

To obtain the most essential information regarding the evolution behaviors of the dynamic system: heart, decompose the ECG series into a set of components is particularly useful. The wavelet transform decomposes a time series into components in various frequency subbands or scales with various resolutions. It provides a framework for studying how frequency content changes with time. In this paper, we used the reconstruction formula of Mallat's algorithm to get five scales that decreased the frequency of ECG.

For space V_{j+1}, W_{j+1}, supposing $\phi(t)$ is the scale function and $\psi(t)$ is the wavelet function.

$$c_{j+1,k} = \sum_{m} h_0(m - 2k)c_{j,m} \quad k \in Z \tag{1}$$

$$d_{j+1,k} = \sum_{m} h_1(m - 2k)c_{j,m} \quad k \in Z \tag{2}$$

While $h_0(n)$ and $h_1(n)$ are defined by Eqs. (3) and (4)

$$h_0(n) = \langle \phi, \phi_{-1,n} \rangle \tag{3}$$

$$h_1(n) = \langle \psi, \psi_{-1,n} \rangle \tag{4}$$

$c_{j,k}$ is the scale coefficient and $d_{j,k}$ is the wavelet coefficient.

Similarly, the reconstruction formula can be defined according to the decomposition of coefficients for the same function *f(t)*.

$$c_{j-1,m} = \sum_k c_{j,k} h_0(m-2k) + \sum_k d_{j,k} h_1(m-2k) \tag{5}$$

2.2 Symbol Entropy

Symbol Entropy is a kind of time series analysis method developed from the Symbol dynamic theory, the chaotic time series analysis and the information theory. The method can react fundamental characteristics of signal. Main information of VF and VT are in low frequency. So Symbol Entropy suit for processing VF and VT.

Supposed the VF or VT sample is X, define $X = (X_1, X_2, \ldots, X_i, \ldots, X_N)$, and sample length was N.

Step 1:

We converted into Symbol sequence $S = (S_1, S_2, \ldots, S_i, \ldots, S_N)$ according a special rule. The defined rule was as followed. If the value of larger than the mean value of X, we mark as *'1'*, else we mark as *'0'*. The signifying pattern was followed.

$$S(i) = \begin{cases} '0' & x(i) \le mean(X) \\ '1' & x(i) > mean(X) \end{cases} \tag{6}$$

Here $i = (1, 2, 3, \ldots, N)$, and $x(i)$ was the value of *ith* point in X.

Step 2:

We grouped 3 adjacent symbols as a small sequence which called *"3-bit word"*. Supposed that *S = {111100101...}*, then the S was coded to be *{111,111,110,100,...}*. These were collectively called *"3-bit word"*. A 3-bit word contained symbols of *'0'* and *'1'*. The number of different combinations is 8. We defined $S' = \{s'_1, s'_2, \ldots, s'_i, \ldots, s'_M\}$. Here is 3-bit word and *M = N-3 + 1*.

Step 3:

We converted each 3-bit word of S' into decimal and marked D as the sequence of decimal. Therefor $D = \{d_1, d_2, \ldots, d_M\}$. Then we measured the histogram of D which marked as H. $H = \{h_0, h_1, \ldots, h_7\}$. And calculated probability of hi which defined as *pi = hi/M*. The Symbol Entropy was defined as Eq. (7).

$$En = -\sum\nolimits_{i=0}^{7} p(h'_i) log_2(h'_i) \tag{7}$$

2.3 Classification

In order to improve the detection precision, we used SVM classifier to discriminate VF and VT. To evaluate the generalization capability of the SVM models, data consisted of training and testing. In recent years, SVM algorithms have been successfully used in a wide number of practical classification problems, due to their good generalization capability derived from the Structural Risk Minimization principle.

Suppose the linear separable sample set is (x_i, y_i), $i = (1, 2, 3, \ldots, n)$, $x \in R_d$, and $y \in \{+1, -1\}$ denotes the class label. The idea of SVM algorithm is to locate Optimal Separating Hyperplane (OSH), while OSH can separate the samples without error and maximize the distance from either class to the separating hyperplane. The OSH must satisfy the following condition if it can classify all the samples correctly [4]:

$$y_i\left[\left(wx_i + b\right) - 1\right] + \xi_i \geq 0, i = 1, 2, \ldots, n \tag{8}$$

where ξ_i is slack variables.

Under the constraints of Eq. (8), the optimal discriminate function is

$$f(x) = sgn\left\{\sum_{i=1}^{n} a_i^* y_i K\left(x_i, x\right) + b^*\right\} \tag{9}$$

where $\mathrm{K}\left(x_i, \mathrm{x}\right)$ is a kernel function.

By minimizing following formula under the constraints of Eq. (8), the generalized OSH is determined.

$$\phi(w) = \frac{1}{2}(\omega\omega) + C\left(\sum_{i=1}^{n} \xi_i\right) \tag{10}$$

2.4 The Algorithm Based on Symbol Entropy and EMD

Our proposed algorithm is defined in the Fig. 1:

Step 1: Given a sample set X, $X = \left[X_1, X_2, \ldots, X_i, \ldots X_N\right]$. where N represents the number of samples. Defined $\mathrm{X}_i = \left[x_{i1}, x_{i2}, \ldots, x_{ij}, \ldots x_{im}\right]$, where m represents the length of X_i.

Step 2: Normalized the by the followed formula $X_i' = \dfrac{\left(X_i - \overline{x}\right)}{\sigma}$.

Where $\overline{x}$ and σ represent the mean and standard deviation of the th sample.

Step 3: Repeat *step 2* until all samples are normalized.

Step 4: Using wavelet analysis to decompose X_i by two layers. Get {ca2, cd1, cd2}. The Decompose Sketch Map is showed in Fig. 2.

Step 5: Calculate the Symbol Entropy of the {ca2, cd1, cd2}, get $En_i = \left[e_1, e_2, \ldots, e_n\right]$, where i = 1,2,3.

Step 6: Design classifier to calculate the classification accuracy for performance evaluation. We selected LIBSVM as the classifier.

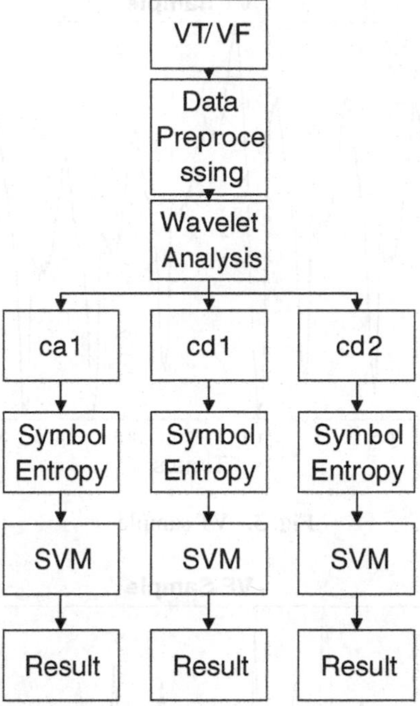

Fig. 1. Algorithm flow chart

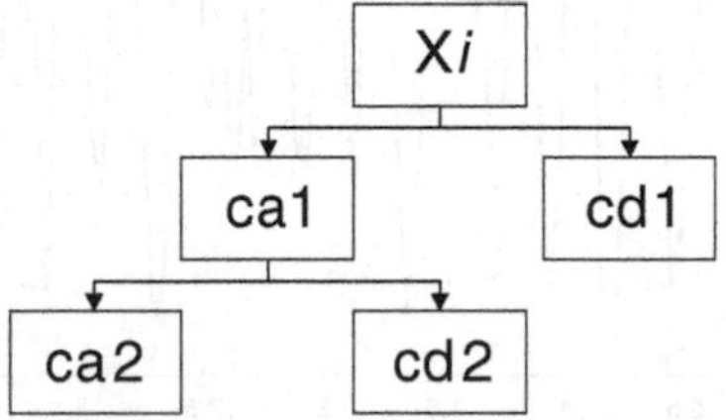

Fig. 2. Decompose sketch map

3 Experimental Results and Analysis

3.1 Data Description

The data were selected from MIT-BIH Malignant Ventricular Ectopy Database (MIT-BIH Database) and Creighton University Ventricular Tachyarrhythmia Database (CU Database). The sampling frequency of VF and VF are all 250 Hz. We extracted 100 VF samples and 100 VT samples from CU Database and MIT-BIH Database respectively. All the samples were normalized firstly.

A sample of ECG sample from CU Database and MIT-BIH Database were plotted in Figs. 3 and 4.

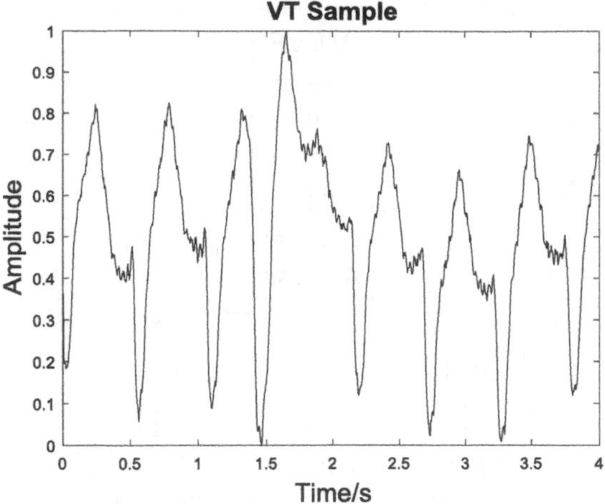

Fig. 3. VT sample

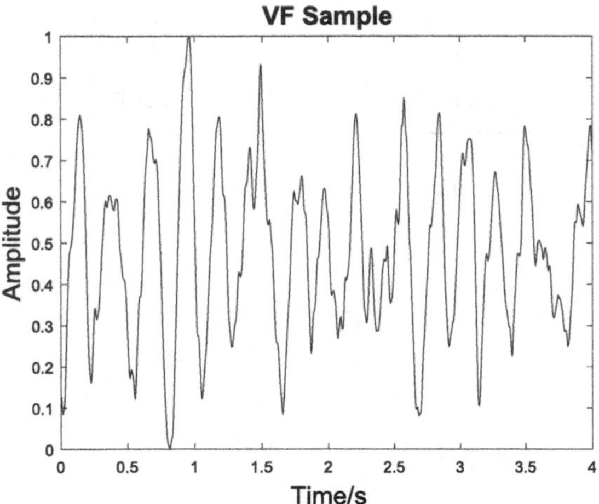

Fig. 4. VF sample

3.2 Experimental Results and Analysis

We calculated preprocessed date of symbol entropy. The result was showed in Fig. 5(a). The box-plot of symbol entropy was showed in Fig. 5(b). The Fig. 5(a) and (b) exhibited bad performance to distinguish VT and VF.

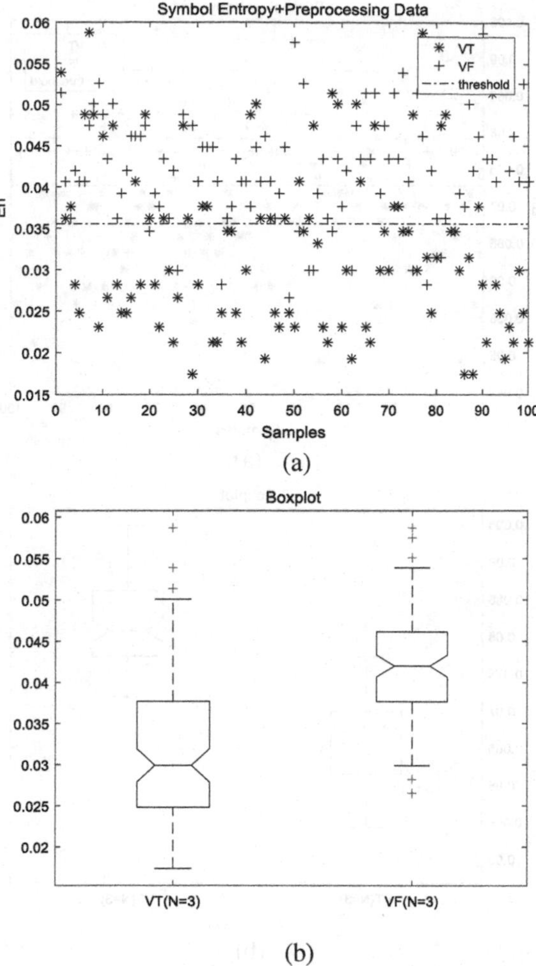

Fig. 5. (a) The symbol entropy of preprocessed date (b) The box-plot of VF and VT

We calculated {ca2, cd1, cd2} of symbol entropy respectively. The result of cd2 was showed in Fig. 6(a). The boxplot of symbol entropy was showed in Fig. 6(b). The Fig. 6(a) and (b) exhibited better performance to distinguish VT and VF than preprocessed date not be decomposed by wavelet analysis. Then the 3/4 of Ens were used to train SVM and the rest were used to test. The test result was showed in Table 1. Table 1 showed that {ca2, cd1, cd2} had a higher classification accuracy than the original. From the sensitivity and specificity, we can know that VF can be regarded as VT but VT is regarded as VF hardly.

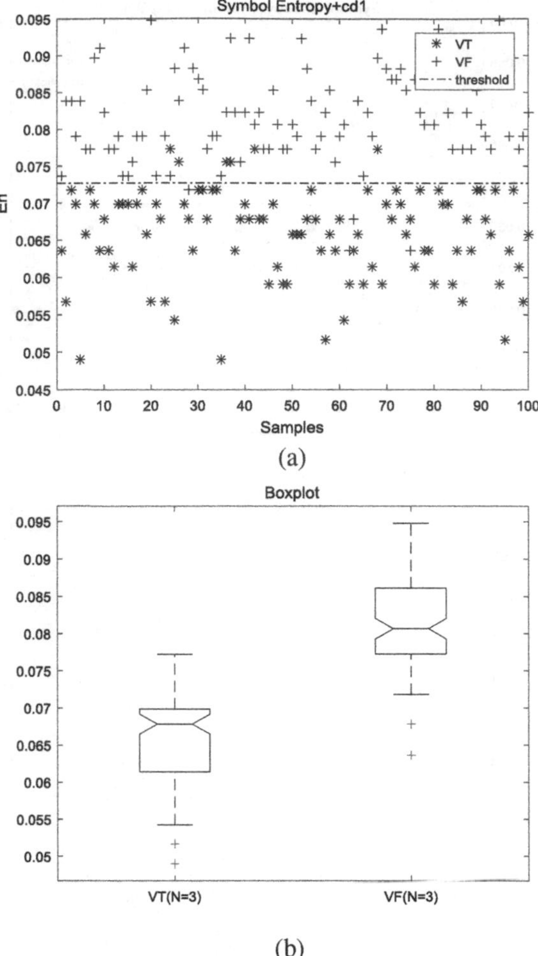

Fig. 6. (a) Symbol entropy of cd2 (b) The box-plot of VF and VT for cd2

Table 1. Classification of wavelet analysis and symbol entropy with SVM.

	SVM-g	Sensitivity	Specificity	Accuracy
Original	10^4	95.38%	91.62%	80.30%
ca2	10^{-1}	95.91%	94.11%	84.84%
cd2	10^3	99.01%	100%	99.5%
cd1	10^5	100%	96.15%	93.93%

(Regarded the VF samples as positive samples and the VT samples as negative samples.)

The SVM-c was 1000 for all models. The length of Original was 1000, but the length of cd1 was 500, the length of cd2 and ca2 was 250 because of down sampling in the wavelet analysis. Due to baseline drift in ca2, the classification accuracy of ca2 is lower than cd2 and cd1. The symbol entropy is affected by sampling rate. The sampling rate of cd2 is lower than cd1, which is suitable for symbol entropy. That's not to say more lower sampling rate is more suitable for symbol entropy. When the sampling rate is very low, called "undersampling", the information of signal will lose, whose symbol entropy is meaningless.

Xia et al. [16] proposed a method based on the Lempel-Ziv complexity and EMD to detect VF and VT. The classification accuracy was showed in Table 2. We can know that the classification accuracy of the proposed method in this paper was higher than [16].

Table 2. The comparison of different methods [16] for CU and MIT-BIH data.

	Sensitivity	Specificity	Accuracy
L-Z complexity	64.11%	62.09%	63.10%
L-Z complexity and EMD-IMF1	87.12%	100%	93.56%
L-Z complexity and EMD-IMF2	82.04%	68.16%	75.10%
L-Z complexity and EMD-IMF3	77.21%	73.18%	75.20%
L-Z complexity and EMD-IMF4	88.13%	90.28%	89.21%
L-Z complexity and EMD-IMF5	98.15%	96.01%	97.08%

4 Conclusions

An effective detection method to distinguish VF from VT has clinical research significance. In this paper, the proposed method can detect VF and VT with higher classification accuracy. The method can be improved and embedded into portable ECG monitoring wireless communication equipment to detect of VF and VT in real time. This can contribute to save the patient's life in time. There is a weakness of this paper, the proposed method should be assessed by other data-set of VF and VT.

Acknowledgment. This work was supported by the National Natural Science Foundation of China (Grant No. 61671220, 61640218, 61201428), the Shandong Distinguished Middle-aged and Young Scientist Encourage and Reward Foundation, China (Grant No. ZR2016FB14), the Project of Shandong Province Higher Educational Science and Technology Program, China (Grant No. J16LN07), the Shandong Province Key Research and Development Program, China (Grant No. 2016GGX101022).

References

1. Othman, M.A., Safri, N.M., Ghani, I.A.: A new semantic mining approach for detecting ventricular tachycardia and ventricular fibrillation. Biomed. Sig. Process. Control **8**, 222–227 (2013)
2. Kong, D.R., Xie, H.B.: Use of modified sample entropy measurement to classify ventricular tachycardia and fibrillation. Measurement **44**(3), 653–662 (2011)
3. Thakor, N.V., Zhu, Y.S., Pan, K.Y.: Ventricular tachycardia and fibrillation detection by a sequential hypothesis testing algorithm. IEEE Trans. Biomed. Eng. **37**(9), 837–843 (1990)

4. Zhang, X.S., Zhu, Y.S., Thakor, N.V., Wang, Z.Z.: Detecting ventricular tachycardia and fibrillation by complexity measure. IEEE Trans. Biomed. Eng. **46**(5), 548–555 (1999)
5. Zhang, H.X., Zhu, Y.S., Xu, Y.H.: Complexity information based analysis of pathological ECG rhythm for ventricular tachycardia and ventricular fibrillation. Int. J. Bifurc. Chaos **12**(10), 2293–2303 (2002)
6. Zhang, H.X., Zhu, Y.S.: Qualitative chaos analysis for ventricular tachycardia and fibrillation based on symbol complexity. Med. Eng. Phys. **23**(8), 523–528 (2001)
7. Owis, M.I., Abou-Zied, A.H., Youssef, A.B.M.: Study of features based on nonlinear dynamical modeling in ECG arrhythmia detection and classification. IEEE Trans. Biomed. Eng. **49**(7), 733–736 (2002)
8. Fleisher, L.A., Pincus, S.M.S., Rosenbaum, H.: Approximate entropy of heart rate as a correlate of postoperative ventricular dysfunction. Anesthesiology **78**(4), 683–692 (1993)
9. Tong, S., Bezerianos, A., Paul, J., Thakor, N.: Nonextensive entropy measure of EEG following brain injury from cardiac arrest. Phys. A **305**(3), 619–628 (2002)
10. Gamero, L.G., Plastino, A., Torres, M.E.: Wavelet analysis and nonlinear dynamics in a non extensive setting. Phys. A **246**(3), 487–509 (1997)
11. Tsallis, C.: Possible generalization of boltzmann-gibbs statistics. J. Stat. Phys. **52**, 479–487 (1988)
12. Ocak, H.: Automatic detection of epileptic seizures in EEG using discrete wavelet transform and approximate entropy. Expert Syst. Appl. **36**, 2027–2036 (2009)
13. Khan, Y.U., Gotman, J.: Wavelet based automatic seizure detection in intracer- ebral electroencephalogram. Clin. Neurophysiol. **114**(5), 898–908 (2003)
14. Burrus, C.S., Gopinath, R.A., Guo, H.: Introduction to Wavelets and Wavelet Transforms: A Primer. Prentice-Hall, Upper Saddle River (1998)
15. Sierra, G., de Jesús Gómez, M.D., Le Guyader, P., Trelles, F., Cardinal, R., Savard, P., Nadeau, R.: Discrimination between monomorphic and polymorphic ventricular tachycardia using cycle length variability measured by wavelet transform analysis. J. Electrocardiol. **31**(3), 245–255 (1998)
16. Xia, D., Meng, Q., Chen, Y., Zhang, Z.: Classification of ventricular tachycardia and fibrillation based on the lempel-ziv complexity and EMD. In: Huang, D.-S., Han, K., Gromiha, M. (eds.) ICIC 2014. LNCS, vol. 8590, pp. 322–329. Springer, Cham (2014). doi: 10.1007/978-3-319-09330-7_39

Saliency-Guided Smoothing for 3D Point Clouds

Feng Yan(✉), Fei Wang, Yu Guo, and Peilin Jiang

Institute of Artificial Intelligence and Robotics,
Xi'an Jiaotong University, 28 Xianning Road, Xi'an 710049, China
bphengyan@163.com, {wfx,pljiang}@mail.xjtu.edu.cn,
cvyuguo@gmail.com

Abstract. To efficiently process 3D unorganized point clouds with noise and significant outliers, it is important to smooth the point clouds, but still retain faithfully the original surface geometry as much as possible. In this paper, we present a simple and fast saliency-guided smoothing method of 3D point. The method consists of three stages: firstly, saliency value for each point in noisy point clouds is calculated by site entropy rate method; secondly, robust vertex normal vector is updated by neighboring noisy normal vectors with weight terms related to detected visual saliency metrics; finally, based on least-squares error criterion, vertex position is updated with the integration of vertex normal vector. Analysis and experiments show the advantages of our proposed method over similar methods in the literature on synthetic data.

Keywords: 3D unorganized point clouds · Saliency-guided smoothing · Vertex normal vector updating · Vertex position updating

1 Introduction

In recent years, 3D models have become an important and fundamental research topic, and they are used in many yields (e.g. architectural design, CAD, and et al.). With the significant development of 3D reconstruction techniques and scanning devices (e.g. depth scanners or dense multi-view stereo methods that work on RGB images of a scene), they too vague present real world scenes. However, due to the inherent physical limitations of either the scanners or the digitization processes, initial outputs of acquisition devices are often invariably corrupted with significant amounts of noise as well as outliers. Such noise can seriously affect the use of 3D models, and they often needs to be filtered at the early stage of processing. What's worse, the problem of remaining faithful to the original surface geometry is more difficult, due to the presence of outliers. It is important to create tools that adequately reposition points or remove outliers and noisy points in order to obtain pure 3D models.

This work was supported in part by Natural Science Foundation of China (No.61231018), National Science and Technology Support Program (2015BAH31F01) and Program of Introducing Talents of Discipline to University under grant B13043.

D.-S. Huang et al. (Eds.): ICIC 2017, Part I, LNCS 10361, pp. 165–174, 2017.
DOI: 10.1007/978-3-319-63309-1_16

3D models can be represented by the mesh and point cloud in computer graphics, and therefore, the smoothing algorithm of them can be divided into two categories: mesh-based and point-based. Mesh-based smoothing has been an active research topic for a long time [11]. The bidirectional filter, proposed by both Fleishman [12] and Jones [13], is used to preserve the properties of the mesh to smooth, where the bilateral filter is applied to the mesh vertex position. He et al. [14] uses L0 minimization to guide the sparseness for edge-based Laplace operator, which is effective for preserving sharp features; in this respect, this optimization is advantageous for segmented planar shapes and may not apply to non-CAD models. Yagou [15], Chen [16] and Sun [7] have also adopt the method that firstly estimates the normal of the face and then relocates the vertex position according to the normal vector. The only difference of them lies in updating face normal stage: Yagou [15] uses mean, median, and alpha-trimming filters; Chen [16] selects filters based on local sharpness; while noisy face normal vectors are filtered iteratively by weighted averaging of neighboring face normal vectors [7].

However, mesh-based smoothing needs mesh topology and maintain consistency, which makes smoothing complicated. The point-based smoothing avoids reducing memory consumption, storing point connectivity information, accelerating the speed of smoothing, and it can guarantee the same quality as mesh smoothing, or even better. So many smoothing methods of point clouds have been extensively studied recently, and some of the most popular methods are adopted [16, 18]. However they produce a smooth surface and blur the main surface detail. Subsequent approaches to attempting to solve these shortcomings include robust implicit MLS method (henceforth RIMLS) [10] in which the authors use kernel-based robust statistics and incorporate it into the implicit MLS method, thus avoid smooth and sharp features. Clarenz [19] proposed a frame of the fairing base surface by a partial differential equation based on finite element. The concept of non-local means [20] is applied to the 3D point set filtering [21].

Most of the smoothing methods are unable to preserve fine 3D features that humans are more concerned in the smoothed results due to overlooking human visual mechanism. To do so, in this paper we present a simple and fast saliency-guided smoothing method of 3D point clouds, which can remove noise effectively, while preserving geometry features which attract human attention. We firstly recommend a method that saliency value for each point is calculated by site entropy rate method; secondly, robust vertex normal vectors are updated by neighboring noisy normal vectors with a weight term related to detected visual saliency metric; subsequently, we robustly smooth the 3D points on the surface based on the integration of vertex normal vectors using a least-squares error criterion. From Fig. 1, our approach to such 3D point repositioning encourages the careful delineation and preserving sharp and fine 3D features of the surface.

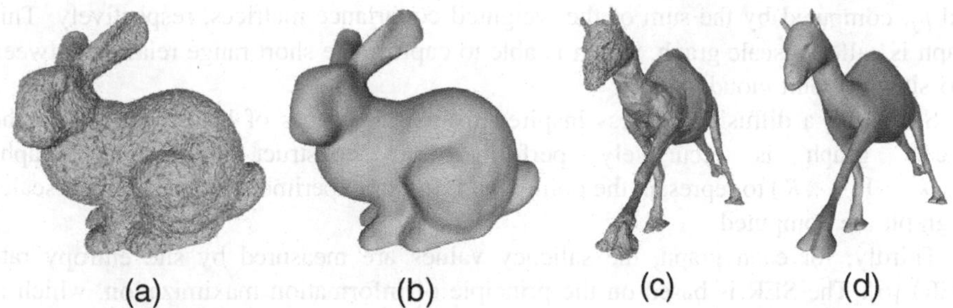

Fig. 1. Our method provides a more concise smoothing that keeps important details. (a) Bun_Zipper data with 50% Gaussian noise. (b) The smoothing result of (a). (c) Camel-gallop data with 30% Gaussian noise. (d) The smoothing result of (c).

2 Our Method

The 3D model which is the original geometry is represented by a point set $\mathbf{V}$ where $\mathbf{V} = [\mathbf{v}_1^T, \mathbf{v}_2^T, \cdots, \mathbf{v}_n^T]$, $\mathbf{v}_i^T = [v_{ix}, v_{iy}, v_{iz}]^T \in \boldsymbol{R}^3$ and $\hat{\mathbf{V}}$ denotes the displaced geometry.

In this Section, we present the details of the steps involved in our robust point cloud smoothing method. Our method consists of the following sequence of procedures:

1. Saliency value for each point is calculated by site entropy rate method (Sect. 2.1).
2. Robust vertex normal vectors are updated by neighboring noisy normal vectors with a weight term related to detected visual saliency metric (Sect. 2.2).
3. Point set repositioning (Sect. 2.3).

2.1 Saliency of Point Cloud

Visual attention plays an important role in the human visual system, which holds the post of predictor of object regions which attract human attention. In the last decade years, several algorithms of saliency detection of point cloud have been proposed [1, 2]. In our smooth framework, saliency values are computed by the method reported by Guo et al. [3] that achieves better performance on a range of raw point clouds.

For the point cloud P, the saliency value of each point is expressed by sum of site entropy rate of multi-scale graphs. Firstly, we transform the input point cloud P to an initial undirected graph G_0, where each node expresses a point p_i in P, and edges exist only in a small k neighborhood $N(p_i)$ of each point, and to eliminate the similarity measurement between two points, edge weights are computed with anisotropy metric defined as below,

$$G_0(i,j) = \exp\left(-\frac{(p_i - p_j)^T (T(p_i) + T(p_j))(p_i - p_j)}{\sigma_i + \sigma_j}\right). \quad (1)$$

Where σ_i is the average distance of point p_i to its neighbors, $\sigma_i + \sigma_j$ is taken as normalization parameter, $T(p_i)$ and $T(p_j)$ are normal voting tensors [4, 5] of vertex p_i

and p_j, computed by the sum of the weighted covariance matrices, respectively. This graph is called 0-scale graph, which is able to capture the short range relation between two sites in point cloud.

Secondly, a diffusion process inspired from the process of heat diffusion on the 0-scale graph is recursively performed to construct multi-scale graphs ($\boldsymbol{G}_k, k = 1, \cdots, K$) to represent the point cloud. In our experiments, four different scales of graph are computed.

Thirdly, for each graph, the saliency values are measured by site entropy rate (SER) [6]. The SER is based on the principle of information maximization, which is defined as,

$$SER = \pi_i \sum_j -\boldsymbol{G}_k(i,j) \log \boldsymbol{G}_k(i,j). \tag{2}$$

Where π is the value of stationary distribution of Markov chain.

SER can reflect the average total information transmitted between nodes of graph. Finally, the saliency value of each point can be obtained by a weighted sum of SERs of different scales graph.

2.2 Normal Filtering

Many works on 3D data smoothing are addressed on 3D mesh model. Our work is inspired by the idea of [7], and then we present a new method of normal vectors filtering of point clouds for the low computational cost, and even for vertexes with low errors for vertices adjacent to the ridge or angle. We perform normal updates using:

$$\hat{n}_i = \text{normalise}\left(\sum_{j \in N(i)} h_j \boldsymbol{n}_j\right). \tag{3}$$

Where n_i is the normal of the ith vertex V_i and h_j is a weight function defined as:

$$y = \begin{cases} f_n(n_i \cdot n_j - T_n) f_x(1 - s_i - T_s) & if \quad n_i \cdot n_j > T_n, 1 - s_i > T_s \\ 0 & \text{others} \end{cases}. \tag{4}$$

Here, $0 \leq T_n \leq 1$ and $0 \leq T_s \leq 1$ is a threshold that user determine, used to control averaging, and $f_n(n)$, $f_s(s)$ can be any suitably chosen monotone increasing function for $n \geq 0$ $s \geq 0$. $s_j \in [0, 1]$, which is closer to one the more significant, is the vertex saliency detected by Sect. 2.1 in our paper at the point, reflecting the saliency distribution. $f_n(n)$ is the normal weight, and $f_s(s)$ is the saliency weight. Here we use the form of $f_s(s)$, as proposed by [8]:

$$f_s(s) = r + \frac{1}{A * s^d + v}.$$

Where the upper bound of $f_s(\cdot)$ is $1/v$, and the lower bound is r, A and d reflect the grade and the start-up time of curve of $f_s(\cdot)$. The bound value of $f_s(\cdot)$ keeps a minimum distribution in $s_i = 1$ and remains unchanged in high-saliency regions. In Fig. 2, we find that $f_s(\cdot)$ is a monotonically increasing function. And experiments showed a good choice of $f_n(n)$ [10] is

$$f_n(n) = n^2.$$

The main purpose behind the weight function *Eq.* 4 is to give high weights to those vertex normal vectors close to n_i or in low-saliency region, and low weights to those far from n_i or in high-saliency region. In addition, when a vertex i is adjacent to a ridge or a corner, or the most saliency, we want to give 0 weight to the normal n_j of any vertex j which does not share the same surface with vertex i, and thus having n_j far from n_i. Note that $n_i \cdot n_j = \cos\angle(n_i, n_j)$, so $f_n(x)$ being a monotonically increasing function implies that the weight function defined by *Eq.* 2 satisfies the above requirements.

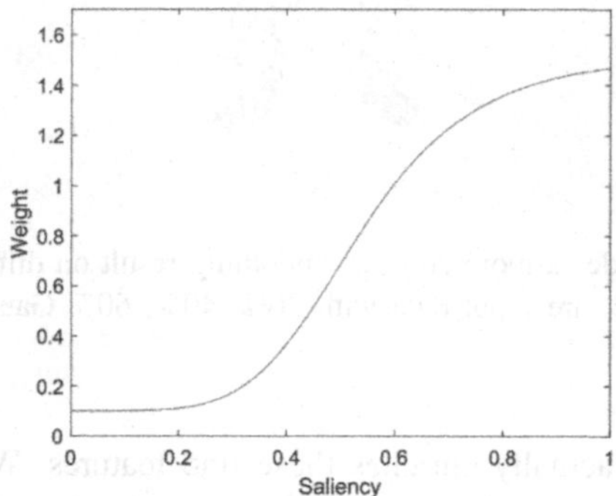

Fig. 2. Graph of the saliency weight function: r = 0.1, A = 2, d = 5 and v = 0.7.

The choice of T_n, T_s determines how many normal vectors will be used in the weighted averaging operation. $T_n = 1$ means that only normal vector with n_j equal to n_i are included in the weighted average, while $T_n = 0$ means that all normal vectors with an angle less than $\pi/2$ from n_i are used. In the same way, $T_s = 1$ means that the saliency do not affect the final results. When the point cloud has sharp edges, a larger value of T_n and t_s is appropriate, whereas when the surface is relatively smooth, T_n and T_s should be smaller (Fig. 3).

2.3 Vertex Updating

Once we have got their corrected normal vectors $\{\hat{n}_i\}_{i=1}^N$, we now reposition the vertices according to them. To do so, we use a weighted vertex update. Unlike the MLS method [10], we use a simpler fitting scheme [9] of point sets. This method is robust enough to preserve good features even in fine features such as edges and corners.

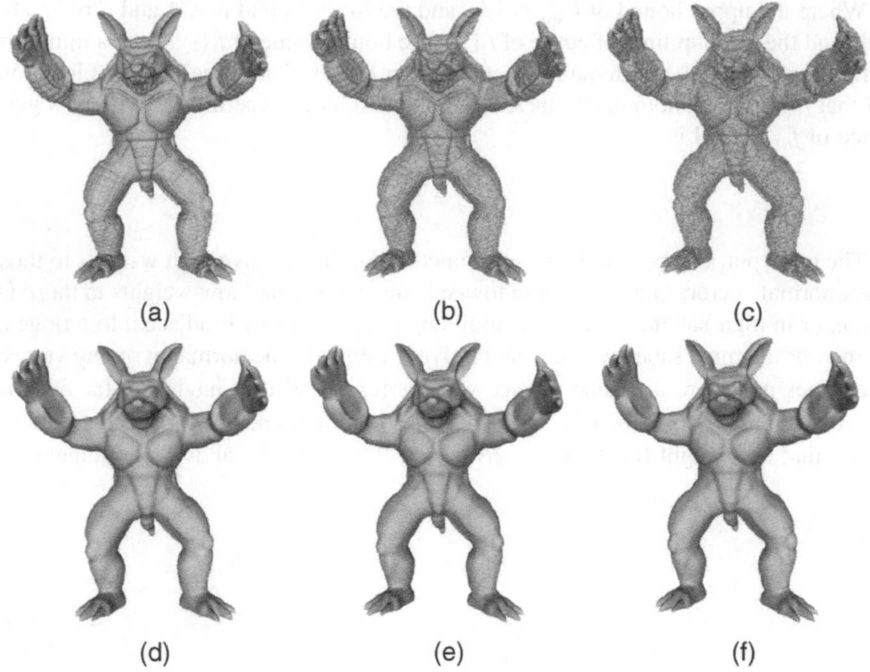

Fig. 3. Our method provides a more concise smoothing result on different levels of noise that keep important detail. (a-c) are input data with 20%, 40%, 60% Gaussian noise. (d-f) are the respective smoothing result.

Moreover, this scheme actually enriches these fine features. We solve the following minimize to estimate the new positions $\{\hat{\upsilon}_i\}_{i=1}^N$ of the 3D points,

$$\min_{\{\hat{\upsilon}_i\}_{i=1}^N} \sum_{i=1}^N \sum_{j \in N(i)} \gamma_{ij} \left\| \hat{n}_i^T (\hat{u}_i - \hat{u}_j) \right\|_2^2 + \lambda \sum_{i=1}^N \|\hat{u}_i - u_i\|_2^2. \tag{5}$$

Where:

$$\gamma_{ij} = \frac{\tau_{ij}}{\sum\limits_{j \in N(i)} \tau_{ij}}, \tau_{ij} = \exp\left(-\frac{\left\| \hat{\upsilon}_i - \hat{\upsilon}_j \right\|_2^2}{\sigma_s^2} \right)$$

are weights that adaptively set the influence of neighbors, $\lambda > 0$ is balancing term between the two cost function, $\hat{n}_i$ are the mollified normal vectors and υ_i are the noisy point positions. Since the weights γ_{ij} depend on the optimize variables $\hat{\upsilon}_i$, we use the gradient descent method to iteratively solve this minimization and it usually converges in 5-30 iterations.

We note here *Eq.* 5 that, from a qualitative point of view, if a point on one plane, moves a little along any direction, then it still is in the original plane, so the surface will not change. Since the weights γ_{ij} do not depend on the normal vectors $\hat{n}_i$, this minimization allows fine points on both sides of the structure (such as edges and corners) to move to them. This automatically defines and enhances these sharp structures. In Fig. 4(f), we show that the edges are prominently defined in our method.

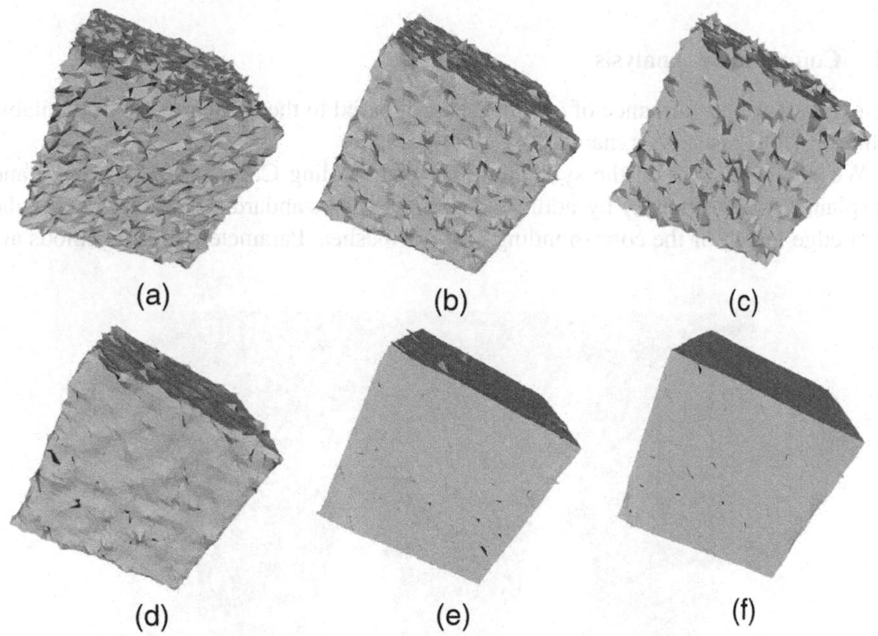

Fig. 4. Comparative smoothing result of Cube_Sf (6177 points). (a) Input data with 60% Gaussian noise. (b-f) smoothing results by [7, 12–14] and our method.

3 Results and Discussion

In this Section, we firstly devise an experiment to evaluate the performance and robustness of our filtering algorithm, and then demonstrate the efficiency of the algorithm by comparing it with several state-of-art smoothing techniques on synthetic point cloud data with noise.

Note that our method bases on point, while we use the mesh to present our noise data and the smoothed results in this paper, in order to enhance the visual effect.

3.1 Qualitative Analysis

Most of the filtering algorithms can't efficiently remove the seriously noisy points to obtain pure point cloud data including what humans pay more attention to. To demonstrate the robustness of the proposed way, we choose the typical point cloud data of Armadillo (172974 points), and then we separately add Gaussian noise standard deviation 20%, 40%, 60% of the mean edge length of the corresponding original meshes. Finally, we filter these noisy Armadillo. The top three figure of Fig. 4 show our noisy Armadillos, and the bottom is the accurately filtering results correspondingly. The results indicate that our algorithm can efficiently remove outliers and noisy points and keep the almost all important features, which attract human attention, such as eyes, nose and so on, in spite of different levels of noise.

3.2 Comparative Analysis

We evaluate the performance of our method compared to the similar methods available in the previous researches, namely [7, 12–14].

We do experiment on the synthetic data set including Cube_sf (6177 points) and Maxplanck (49132 points) by adding Gaussian noise standard deviation 60% of the mean edge length of the corresponding original meshes. Parameters of all methods are

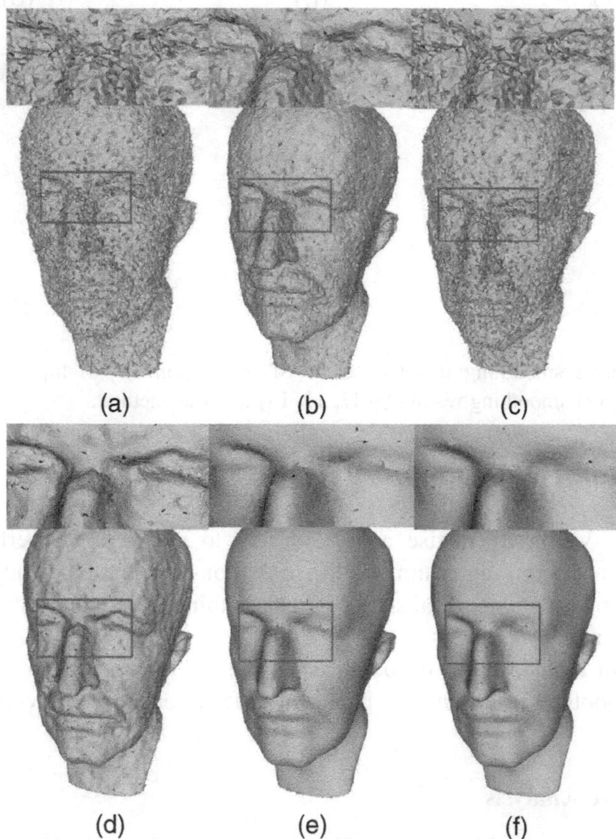

Fig. 5. Comparative smoothing result of Maxplanck (49132 points). (a) Input data with 60% Gaussian noise. (b-f) smoothing results by [7, 12–14] and our method.

adjusted to the best settings according to what corresponding papers say. Figures 4 and 5 show visual comparisons of the results obtained from the comparison method. For the Cube_sf and Maxplanck, [12–14] methods can't remove noise for high level of noise. [7] get almost the same results as our method, while [7] still leaving a slight tip of the burr on top face of cube_sf, and whose filtering result is not good on bridge of the nose of Maxplanck. Recovered important features from our method are very prominent, while removing the noise.

4 Conclusion

In this paper, a simple and fast saliency-guided smoothing method of 3D point cloud is proposed, where a new term reflecting the saliency distribution of the underlying vertex is introduced to update of vertex normal vectors. Data analysis and experiment results show that our methods can remove noise effectively, while preserving geometry features, (such as sharp edges and corners), that attract human attention, in spite of different levels of noise, and get better smoothing result compared with state-of-art smoothing technique. Our smoothing algorithms can be the basis of many point-based processing tasks.

References

1. Tasse, F.P., Kosinka, J., Dodgson, N.: Cluster-based point set saliency. In: ICCV, vol. 7, pp. 163–171 (2015)
2. Shtrom, E., Leifman, G., Tal, A.: Saliency detection in large point sets, pp. 3591–3598 (2013)
3. Guo, Yu., Wang, F., Liu, P., Xin, J., Zheng, N.: Multi-scale point set saliency detection based on site entropy rate. In: Chen, E., Gong, Y., Tie, Y. (eds.) PCM 2016. LNCS, vol. 9916, pp. 366–375. Springer, Cham (2016). doi:10.1007/978-3-319-48890-5_36
4. Medioni, G., Tang, C.K., Lee, M.S.: Tensor voting: theory and applications. Proc. Rfia **34**(8), 1482–1495 (2000)
5. Min, K.P., Lee, S.J., Lee, K.H.: Multi-scale tensor voting for feature extraction from unstructured point clouds. Graph. Model. **74**(4), 197–208 (2012)
6. Wang, W., Wang, Y., Huang, Q., Gao, W.: Measuring visual saliency by site entropy rate. In: IEEE Conference on Computer Vision and Pattern Recognition, CVPR 2010, San Francisco, CA, Usa, 13–18 June, vol. 119, pp. 2368–2375. DBLP (2010)
7. Sun, X., Rosin, P.L., Martin, R.R., Langbein, F.C.: Fast and effective feature-preserving mesh denoising. IEEE Trans. Vis. Comput. Graph. **13**(5), 925–938 (2007)
8. Meyer, M.D., Georgel, P., Whitaker, R.T.: Robust particle systems for curvature dependent sampling of implicit surfaces. In: Shape Modeling and Applications, International Conference, pp. 124–133. IEEE (2005)
9. Haque, S.M., Govindu, V.M.: Robust feature-preserving denoising of 3D point clouds. In: International Conference on 3d Vision, pp. 83–91. IEEE Computer Society (2016)
10. Ouml Ztireli, A.C., Guennebaud, G., Gross, M.: Feature preserving point set surfaces based on non - linear kernel regression. Comput. Graph. Forum **28**, 493–501 (2009)

11. Botsch, M., Pauly, M., Kobbelt, L., Alliez, P., Lévy, B., Bischoff, S., et al.: Geometric modeling based on polygonal meshes. Proc. Acm Siggraph Course Notes **29**(29), 432–441 (2007)
12. Fleishman, S., Drori, I., Cohen-Or, D.: Bilateral mesh denoising. ACM SIGGRAPH, pp. 950–953. ACM (2003)
13. Jones, T.R., Durand, F., Desbrun, M.: Non-iterative, feature-preserving mesh smoothing. ACM Trans. Graph. **22**(3), 943–949 (2003)
14. He, L., Schaefer, S.: Mesh denoising via L0 minimization. ACM Trans. Graph. **32**(4), 1–8 (2013)
15. Yagou, H., Ohtake, Y., Belyaev, A.: Mesh Smoothing via Mean and Median Filtering Applied to Face Normals. In: Proceedings of Geometric Modeling and Processing, pp. 124–131. IEEE (2002)
16. Chen, C.Y., Cheng, K.Y.: A sharpness dependent filter for mesh smoothing. Comput. Aided Geom. Des. **22**(5), 376–391 (2005)
17. Shen, C., O'Brien, J.F., Shewchuk, J.R.: Interpolating and approximating implicit surfaces from polygon soup. ACM Trans. Graph. **23**(3), 896–904 (2004)
18. Guennebaud, G., Gross, M.: Algebraic point set surfaces. ACM Trans. Graph. **26**(3), 23 (2007)
19. Clarenz, U., Rumpf, M., Telea, A.: Fairing of point based surfaces. Computer Graphics International, pp. 600–603. IEEE Computer Society (2004)
20. Buades, A., Coll, B., Morel, J.M.: A non-local algorithm for image denoising. In: IEEE Computer Society Conference on DBLP, Computer Vision and Pattern Recognition, CVPR 2005, vol. 2, pp. 60–65 (2005)
21. Jensen, R.R., Paulsen, R.R.: Second International Conference on 3D Imaging, Modeling, Processing, Visualization (2012)

Pattern Recognition

Color Two-Dimensional Principal Component Analysis for Face Recognition Based on Quaternion Model

Zhi-Gang Jia[1(✉)], Si-Tao Ling[2], and Mei-Xiang Zhao[1,3]

[1] School of Mathematics and Statistics,
Jiangsu Key Laboratory of Education Big Data Science and Engineering,
Jiangsu Normal University, Xuzhou 221116, China
zhgjia@jsnu.edu.cn
[2] School of Mathematics, China University of Mining and Technology,
Xuzhou 221116, China
[3] School of Information and Electrical Engineering,
China University of Mining and Technology, Xuzhou 221116, China

Abstract. The color two-dimensional principal component analysis (color 2DPCA) approach based on quaternion model is presented for color face recognition. Based on 2D quaternion matrices rather than 1D quaternion vectors, color 2DPCA combines the color information and the spatial characteristic for face recognition, and straightly computes the low-dimensional covariance matrix of the training color face images and determines the corresponding eigenvectors in a short CPU time. The image reconstruction theory is also built on color 2DPCA. The experiments on real face data sets are provided to validate the feasibility and effectiveness of the proposed algorithm.

Keywords: Color face recognition · Eigenface · Quaternion matrix · Color 2DPCA

1 Introduction

Color face recognition is one of the most important subjects in pattern recognition. Many classical dimensionality reduction methods such as PCA and 2DPCA have achieved success in it by transferring color to greyscale images, but their rates of face recognition still can be improved by utilizing color information directly. In this paper we present a new two-dimensional principal component analysis (2DPCA) approach based on quaternion model for face recognition and face image reconstruction, and design a fast and efficient algorithm which using only real operations for it.

The PCA-like methods for face recognition have a long history. They are based on linearly projecting the high-dimensional image space to a lower dimensional feature space. For feature extraction, the projection directions are chosen as orthonormal eigenvectors of the covariance matrix of training face images, which can maximize the total scatter of projected training samples. Sirovich and Kirby firstly applied PCA to efficiently represent the face images in [5, 14]. They made it clear that one can reconstruct any face image approximately by a facial basis (a small collection of

D.-S. Huang et al. (Eds.): ICIC 2017, Part I, LNCS 10361, pp. 177–189, 2017.
DOI: 10.1007/978-3-319-63309-1_17

images) and a mean image of face. Based on this result, the well-known eigenface method was proposed by Turk and Pentland in [16] for face recognition. Since then, many problems of the PCA method, such as the problem of dimensionality of the "face space" when eigenfaces are used, have been perfectly solved and as a result PCA has become one of the most powerful approaches in face recognition [8, 21]. As a novel technique, 2DPCA was proposed by Yang et al. [18] to improve the recognition rate of the conventional PCA in 2004. It can keep the spatial structure of a 2D face image. The main difference of 2DPCA from PCA lies in the representations of face images during the processing of face recognition.

The PCA-based face recognition technique needs to transform 2D face image matrices into 1D image vectors for preprocessing. It always has a computational difficulty at evaluating the covariance matrix and its eigenvectors due to large size and low rank. In the 2DPCA-based face recognition technique, the image covariance matrix is constructed directly using 2D face image matrices and always has a much smaller size than that of the PCA-based method. So it is relatively easier to compute the image covariance matrix and the corresponding eigenvectors with high accuracy. Including in PCA and 2DPCA, a number of technical approaches for face recognition (see also [1, 9, 10]) made achievements at using the grey-scale information associated to the face image.

Color information should be one of the most important characteristics in reflecting the structural information of the image. The brain of human relies on color cues to pinpoint identity when shape cues in images are compromised (say, by reductions in resolution), see Sinha et al. [13] for the detail. In fact, recognition performance with color images is significantly better than with grey-scale images [6]. Torres et al. [15] extended traditional PCA to color face recognition by using the R, G, B color channels, respectively. Face recognition is conducted on each color channel and the final result comes from the fusion of the three-color channels. Yang et al. [19] presented a general discriminant model for color face recognition. The model uses a set of color component combination coefficients to convert three-color channels of the face image into one channel to represent color face images in the form of one-dimensional vector for recognition. Xiang et al. [17] utilized color images matrix-representation model based on the framework of PCA and applied 2DPCA to compute the optimal projection vectors for feature extraction. All these methods have greatly improved the robust and the ratio of face recognition with applying the color information. However, the 2DPCA method is not generalized to treat color face images directly, and the mathematical theory is still in need to built color 2DPCA.

Color images require more storage space than grey-scale images with the same size, since the amount of the color image data is usually three times of the grey one. To simultaneously and mathematically deal with three channels of color image, the quaternion was applied to represent the color pixel consisting of three components [2, 3, 11]. In this way, the storage problem of color images was solved, while a new problem appeared. The existing algorithms are not efficient enough to handle the computational problems of large quaternion matrices: (1) the real or complex representation methods (see [20]) need to transform the quaternion problem into a real or complex problem with the dimension four or two times larger than the original one, which increases the storage space and the number of operations; (2) the quaternion computation can reach a high accuracy (see [12]) and it should be a good choice if the dimension of the face recognition problem were not very large and the CPU time were not an essential factor. As far as our knowledge, there are still no approaches and fast algorithms for face recognition and image reconstruction based on the framework of color 2DPCA using quaternion model.

The paper is organized as follows. In Sect. 2, we introduce color 2DPCA approach for face recognition using quaternion model. In Sect. 3, we present a computational strategy using only real operations for color face recognition. In Sect. 4, two experiments are presented for the Georgia Tech face database and the LFW face database. Finally, the conclusion is presented in Sect. 5.

2 Color Face Recognition Based on Quaternions

In this section we introduce color 2DPCA approach based on quaternion model for face recognition and the mathematical theory for the reconstruction of color face image.

2.1 Quaternions and Color Images

William Rowan Hamilton (Irish mathematician, physicist, and astronomer) invented quaternions in 1843 [4] as one of his greatest achievements. A quaternion q, including four components, is represented as:

$$q = q_0 + q_1 i + q_2 j + q_3 k$$

where q_0, q_1, q_2 and q_3 are four real numbers, i, j, k are three imaginary units and they are related to each other as follows:

$$i^2 = j^2 = k^2 = ijk = -1. \quad (1)$$

The conjugate of q is $\overline{q} = q^* = q_0 - q_1 i - q_2 j - q_3 k$ and the norm of q is $|q| = \sqrt{q^* q} = \sqrt{q_0^2 + q_1^2 + q_2^2 + q_3^2}$. A *pure quaternion* is a quaternion whose real part is zero. Let $\mathbf{H}$ denote the quaternion skew-field, which is an associative but non-commutative algebra of rank 4 over $\mathbf{R}$.

An $m \times n$ quaternion matrix is of the form $Q = Q_0 + Q_1 i + Q_2 j + Q_3 k$ with $Q_s \in \mathbf{R}^{m \times n}, s = 0, 1, 2, 3$. A *pure quaternion matrix* is a matrix whose elements are pure quaternions or zero. In RGB color space, a pixel can be represented by a pure quaternion $Ri + Gj + Bk$, where R, G and B stand for the values of Red, Green and Blue components, respectively. An $m \times n$ color image can be saved as an $m \times n$ pure quaternion matrix $Q = [q_{ab}]_{m \times n}$, in which each element $q_{ab} = R_{ab} i + G_{ab} j + B_{ab} k$. denotes one color pixel, and R_{ab}, G_{ab} and B_{ab} are nonnegative integers [7]. For example, the 4×4 color image (resized) on the left can be represented by the pure quaternion matrix on the right in Fig. 1.

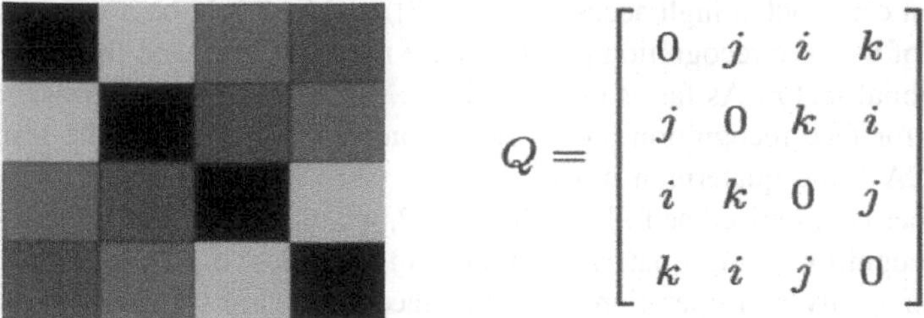

$Q = \begin{bmatrix} 0 & j & i & k \\ j & 0 & k & i \\ i & k & 0 & j \\ k & i & j & 0 \end{bmatrix}$

Fig. 1. Color image and quaternion matrix

2.2 Color 2DPCA Approach

Now we introduce a color two-dimensional principal component analysis (color 2DPCA) approach for face recognition using quaternion model, which generalizes the 2DPCA approach presented by Yang et al. [18].

Suppose that there are l training color image samples in total, denoted by $m \times n$ pure quaternion matrices $F_1, F_2, \ldots, F_l$, and the average image of all training samples is denoted by $\Psi \in \mathbf{H}^{m \times n}$. The color 2DPCA approach includes the following four steps.

(1) Compute the color image covariance (scatter) matrix of training samples by

$$G_t = \frac{1}{l}\sum\nolimits_{s=1}^{l} (F_s - \Psi)^*(F_s - \Psi) \in \mathbf{H}^{m \times n}. \tag{2}$$

Define the *generalized total scatter criterion*

$$J(v) = v^* G_t v$$

where $v \in H^n$ is a unitary column quaternion vector. Note that $J(v)$ is real and non-negative since G_t is Hermitian and positive semi-definite. The unitary quaternion vector v_{opt} that maximizes the criterion is called the optimal projection axis. Intuitively, the total scatter of projected samples is maximized once training color images $F_s(s = 1, \ldots, \ell)$ are projected onto v_{opt} as $y_s = F_s v_{opt}$. It is clear that the optimal projection axis v_{opt} is the eigenvector of G_t corresponding to the largest eigenvalue. Generally, we need to select several projection axes $v_1, \ldots, v_r$ which are orthogonal to each other and maximize the criterion $J(v)$. That is

$$\begin{cases} \{v_1, \ldots, v_r\} = \mathbf{arg\ max}\, J(v) \\ v_s^* v_t = 0, s \neq t, s, t = 1, \cdots, r \end{cases}. \tag{3}$$

These optimal projection axes are in fact the orthonormal eigenvectors of G_t corresponding to its first r largest eigenvalues.

(2) Compute the first r $(1 \leq r \leq n)$ largest eigenvalues of G_t and their corresponding eigenvectors (called eigenfaces), denoted by $(\lambda_1, v_1), \ldots, (\lambda_r, v_r)$. Define the eigenface subspace as $V = \textbf{span}\{v_1, \ldots, v_r\}$.

Next, we use the optimal projection vectors of color 2DPCA, $v_1, \ldots, v_r$, for feature extraction.

(3) Compute the projections of l training color face images in the subspace V,

$$P_s = (F_s - \Psi)V \in \mathbf{H}^{m \times r}, s = 1, \cdots, l. \tag{4}$$

The columns of the quaternion matrix $P_s, y_t = (F_s - \Psi)v_t$, $t = 1, \ldots, r$, are called the principal component (vectors) and P_s is called the feature matrix or feature image of the sample image F_s. It is worth to mention that each principle component of color 2DPCA is a vector, while the principle component of PCA is a quaternion.

With the feature matrix in hand, we use a nearest neighbor classifier for color face recognition.

(4) For a given test sample F, compute its feature matrix $P = (F_s - \Psi)V$. Find out the nearest face image F_s $(1 \leq s \leq \ell)$ whose feature matrix satisfies $\|P_s - P\| = \min$. Such F_s is output as the person who has been recognized.

The above color 2DPCA based on quaternion model can preserve the color information and the spatial information of color face images. Its computational complexity of quaternion operations is similar to the computational complexity of real operations of 2DPCA (given in [18]). Note that it is different from the C2DPCA presented in [17, Sect. 3]. C2DPCA is based on the color value representation model, where a color image is represented by a column "vector"

$$A = [c_1, c_2, \cdots, c_n]^T$$

with each element being a color vector. Based on this, the covariance matrix for C2DPCA is exactly the covariance matrix for CPCA with multiplying a constant M (see Eqs. (2) and (7) in [17]). That means the covariance matrix for C2DPCA is still high dimensional. However, the covariance matrices for color 2DPCA proposed in this paper has a much smaller dimension than that of the CPCA in [17].

Now we consider the color 2DPCA-based image reconstruction. Suppose all eigenpairs of the color image covariance matrix G_t defined by Eq. (2) are

$$(\lambda_1, v_1), \ldots, (\lambda_r, v_r), (\lambda_{r+1}, v_{r+1}), \ldots, (\lambda_n, v_n),$$

and $|\lambda_1| \geq \ldots \geq |\lambda_r| \geq |\lambda_{r+1}| \geq \ldots \geq |\lambda_n|$. Recall the definitions $V = \textbf{span}\{v_1, \ldots, v_r\}$ and $P_s = (F_s - \Psi)V$. *Define the subordinate eigenface subspace by* $\tilde{V} = \textbf{span}\{v_{r+1}, \ldots, v_n\}$ *and the subordinate feature matrix or subordinate feature image by*

$$\tilde{P}_s = (F_s - \Psi)\tilde{V} \in \mathbf{H}^{m\times(n-r)}, s = 1, \cdots, \ell. \tag{5}$$

Theorem 1. With above definitions, the original color image F_s can be reconstructed by its feature matrix P_s and subordinate feature matrix $\tilde{P}_s$ by

$$F_s = \Psi + P_s V^* + \tilde{P}_s^* \tilde{V}, s = 1, \cdots, \ell. \tag{6}$$

Proof. The color covariance matrix G_t, defined by (2), is a positive semi-definite and Hermitian quaternion matrix. That implies that its eigenvector matrix $[V, \tilde{V}]$ is unitary, and so we have

$$VV^* + \tilde{V}\tilde{V} = I_n, V^*V = I_r, \tilde{V}^*\tilde{V} = I_{n-r}, \tilde{V}^*V = 0,$$

where I_m denotes the $m \times m$ identity matrix. Defined by Eq. (4), the feature matrix P_s of the sample image F_s has the following properties:

$$P_s V^* = (F_s - \Psi)VV^*.$$

Defined by Eq. (5), the subordinate feature matrix $\tilde{P}_s$ of the sample image F_s has the following properties:

$$\tilde{P}_s V^* = (F_s - \Psi)\tilde{V}\tilde{V}^*.$$

So we have

$$P_s V^* + \tilde{P}_s \tilde{V}^* = (F_s - \Psi)\left(VV^* + \tilde{V}\tilde{V}^*\right) = F_s - \Psi,$$

which is exactly the Eq. (6).

V is generated by eigenvectors corresponding to first r largest eigenvalues of G_t, $\Psi + P_s V^*$ is always a good approximation for the color image F_s. If the number of chosen principle components $r = n$, then V is a unitary matrix and $F_s = \Psi + P_s V^*$.

From above analysis, it is easy to see that color 2DPCA has many advantages:

- easy to construct the covariance matrix accurately;
- fast to compute the eigenvectors corresponding to first r largest eigenvalues;
- convenient to reconstruct color face images from the projections.

3 Computational Problems in Color 2DPCA

Quaternion operations usually cost a longer CPU time than real and complex operations under the same conditions. In this section we present a computational strategy using only real operations for quaternion operations, which are necessary in color 2DPCA approaches.

Suppose $A = A_0 + A_1 i + A_2 j + A_3 k$, $B = B_0 + B_1 i + B_2 j + B_3 k$ and $C = C_0 + C_1 i + C_2 j + C_3 k$ are three quaternion matrices. Their addition, subtraction, multiplication and conjugate transpose can be directly defined as follows.

Addition and subtraction: $A = B \pm C \Leftrightarrow A_s = B_s \pm C_s, s = 0, 1, 2, 3.$
Multiplication: $A = B \times C \Leftrightarrow$

$$\begin{aligned} A_0 &= B_0 \times C_0 - B_2 \times C_2 - B_1 \times C_1 - B_3 \times C_3, \\ &\qquad A_2 = B_0 \times C_2 + B_2 \times C_0 - B_1 \times C_3 + B_3 \times C_1, \\ A_1 &= B_0 \times C_1 + B_2 \times C_3 + B_1 \times C_0 - B_3 \times C_2, \\ &\qquad A_3 = B_0 \times C_3 - B_2 \times C_1 + B_1 \times C_2 + B_3 \times C_0. \end{aligned}$$

Conjugate transpose: $A = B^* \Leftrightarrow A\_0 = B_0^T$ and $A_s = -B_s^T, s = 1, 2, 3.$
Norm of quaternion matrix: $||A|| = ||[A_0, A_1, A_2, A_3]||_F.$

The core work of color 2DPCA approach is evaluating eigenvectors corresponding to first r eigenvalues of the covariance matrices, which are Hermitian quaternion matrices. This work will cost a lot of CPU time if the quaternion operations are applied and will need extra storage space if the real or complex representation methods are used. Here we present a structure-preserving algorithm to solve it. Suppose that $Q = Q_0 + Q_1 i + Q_2 j + Q_3 k \in \mathbf{H}^{n \times n}$ is Hermitian, where $Q_s \in \mathbf{R}^{n \times n}, s = 0, 1, 2, 3$. There exists a unitary quaternion matrix $V \in \mathbf{H}^{n \times n}$ and a real symmetric tridiagonal matrix $D \in \mathbf{R}^{n \times n}$ such that

$$VQV^* = D$$

for a given Hermitian quaternion matrix $Q \in \mathbf{H}^{n \times n}$. As D is symmetric, there is an orthogonal matrix $X \in \mathbf{R}^{n \times n}$ such that $D = X\mathbf{diag}(\lambda_1, \cdots, \lambda_n)X^T$, where $\lambda_s \in \mathbf{R}(s = 1, 2, \cdots, n)$ are eigenvalues of D. So the Hermitian quaternion matrix Q always has a diagonalization $Q = Z\mathbf{diag}(\lambda_1, \cdots, \lambda_n)Z^*$, where $Z = V^*X$ is a unitary quaternion matrix. The algorithm is presented in the following steps.

(1) Compute $\mathrm{W_0, W_1, W_2, W_3} \in \mathbf{R}^{\mathrm{n \times n}}$ and a tridiagonal matrix $\mathrm{D} \in \mathbf{R}^{\mathrm{n \times n}}$ by the following algorithm such that $\mathrm{VQV^* = D}$ where V is a unitary quaternion matrix defined by $\mathrm{V = W_0 + W_1 i + W_2 j + W_3 k}$.

```
Function [W0, W1, W2, W3, D] = Trihermq(Q0, Q1, Q2, Q3)
for k = 1:n-1
for i = k+1:n
σ = sqrt(Q0(i,k)^2 + Q1(i,k)^2 + Q2(i,k)^2 + Q3(i,k)^2);
    if σ = 0
        G = I4;
    else
        G = (1/σ) [ Q0(i,k)  -Q2(i,k)  -Q1(i,k)  -Q3(i,k)
                    Q2(i,k)   Q0(i,k)  -Q3(i,k)   Q1(i,k)
                    Q1(i,k)   Q3(i,k)   Q0(i,k)  -Q2(i,k)
                    Q3(i,k)  -Q1(i,k)   Q2(i,k)   Q0(i,k) ];
    end
    as = Qs(i, [1:i-1, i+1:n]);
    Z = G(1,:) * [  a0   a2   a1   a3
                   -a2   a0   a3  -a1
                   -a1  -a3   a0   a2
                   -a3   a1  -a2   a0 ];
    Q0(i, [1:i-1, i+1:n]) = Z(1:n-1);
    Q2(i, [1:i-1, i+1:n]) = Z(n:2n-2);
    Q1(i, [1:i-1, i+1:n]) = Z(2n-1:3n-3);
    Q3(i, [1:i-1, i+1:n]) = Z(3n-2:4n-4);
    Q0([1:i-1, i+1:n], i) = Q0(i, [1:i-1, i+1:n])^T;
    for s = 1:3
        Qs([1:i-1, i+1:n], i) = -Qs(i, [1:i-1, i+1:n])^T;
    end
   W([i, n+i, 2n+i, 3n+i], :) = G * W([i, n+i, 2n+i, 3n+i], :);
 end
 H = house(Q0(k+1:n, k));
 for s = 1:4
    Q_{s-1}(k+1:n, :) = H * Q_{s-1}(k+1:n, :);
    Q_{s-1}(:, k+1:n) = Q_{s-1}(:, k+1:n) * H;
 end
 W = blkdiag(H, H, H, H) * W;
 W0 = W(:, 1:n); W2 = W(:, (n+1):(2*n));
 W1 = W(:, (2*n+1):(3*n)); W3 = W(:, (3*n+1):(4*n));
 D = Q0;
end
```

(2) Compute the eigenvalues $\lambda_1, \cdots, \lambda_n$ and corresponding orthonormal eigenvectors $x_1, x_2, \ldots, x_n$ of D, satisfying $D = X\mathrm{diag}(\lambda_1, \cdots, \lambda_n)X^T$ with $X = [x_1, x_2, \ldots, x_n]$. Note that $\lambda_1, \cdots, \lambda_n$ are eigenvalues of Q.

(3) Compute $Z_0 = W_0^T X$ and $Z_s = -W_s^T X$, $s = 1, 2, 3$. Then $Z = Z_0 + Z_1 i + Z_2 j + Z_3 k$ is unitary and its columns are eigenvectors of Q corresponding to the eigenvalues $\lambda_1, \cdots, \lambda_n$, respectively.

This algorithm is very fast and strongly backward stable, since in each step it uses only real operations and preserves the structures of matrices.

4 Experiments

In this section we test the color 2DPCA approach by famous color face databases. Firstly, we compare the color 2DPCA approach based on quaternion model with PCA, 2DPCA and CPCA, on their efficiencies of recognizing color faces, using the Georgia Tech face database [22]. Secondly, we test the efficiency of the color 2DPCA on image reconstruction, using the LFW face database [23].

All experiments in this section are performed on a personal computer with 2.4 GHz Intel Core i7 and 8 GB 1600 MHz DDR3 using MATLAB-R2012b.

Example 1. In this experiment, we compare color 2DPCA with PCA, 2DPCA [18] and CPCA [17] using the famous Georgia Tech face database. Georgia Tech face database contains various pose faces with different expressions on cluttered background. The samples of the cropped images are shown in Fig. 2. The first x $(= 7, 10$ or $13)$ images of each individual are chosen as the training set and the remaining as the testing set. All images in the Georgia Tech face database are manually cropped, and then resized to 33×44. pixels. The numbers of chosen eigenfaces are from 1 to 30. The face recognition rates of five PCA-like methods are shown in Fig. 3, in which the notations have the following meanings:

Fig. 2. Sample images for one individual of the **Georgia** Tech face database

- PCA-grey: PCA using greyscale of color face images,
- 2DPCA-grey: 2DPCA proposed in [18],
- CPCA: CPCA proposed in [17],
- 2DPCA-color: color 2DPCA approach proposed in Subsect. 2.2.

In Fig. 3, the figures on the top, in the middle and at the bottom respectively show the results for cases that first $x = 7, 10$ and 13 face images per individual are chosen for training.

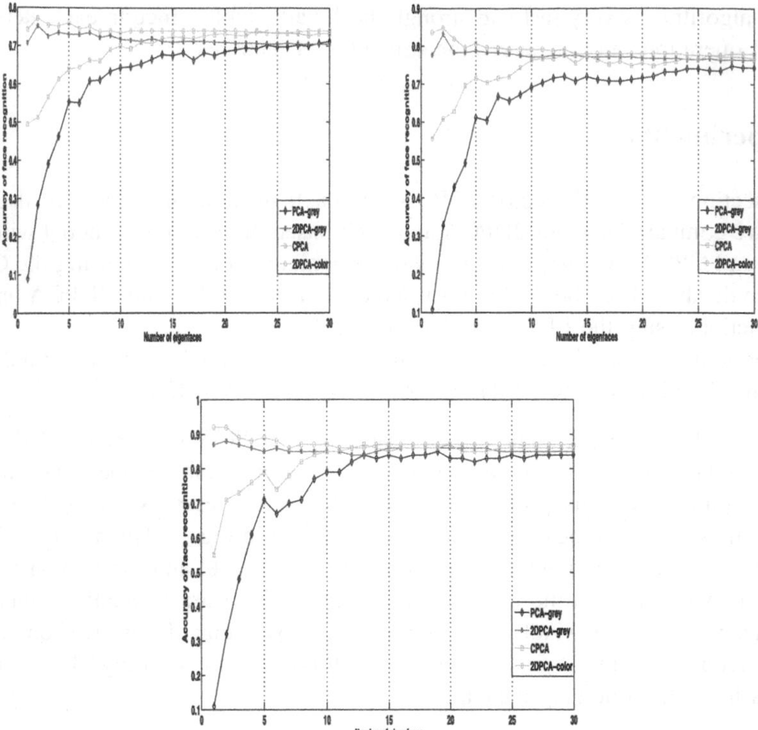

Fig. 3. Rates of face recognition by PCA-like methods based on the Georgia Tech face database

The maximal values and the CPU times of recognizing one face image in average arc presented in the Table 1, in which MR denotes the maximal face recognition rate. We can see that color 2DPCA approach in Sect. 2.2 can reach a higher recognition rate than other PCA-like methods, and it costs more CPU time than PCA and 2DPCA, but less than CPCA.

Example 2. In this experiment, we apply the color 2DPCA to reconstruct the color face images from the famous LFW face database [23]. The LFW face database contains more than 13,000 images of faces collected from the web. Each face has been labeled with the name of the person pictured. 1680 of the people pictured have two or more

Table 1. Maximal face recognition rate and average CPU time (milliseconds)

x	PCA		CPCA		2DPCA		color 2DPCA	
	MR	CPU	MR	CPU	MR	CPU	MR	CPU
7	71%	2.4	74%	19.6	75%	4.6	77%	13.9
10	75%	2.9	78%	40.7	83%	6.2	85%	19.8
13	85%	4.9	87%	123	88%	12.3	92%	44.7

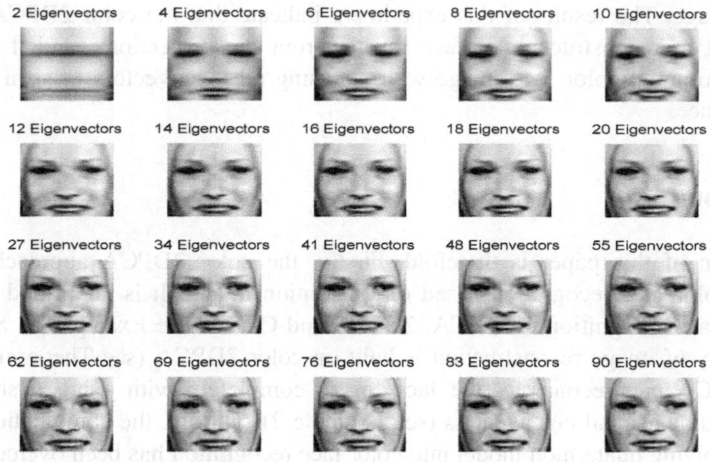

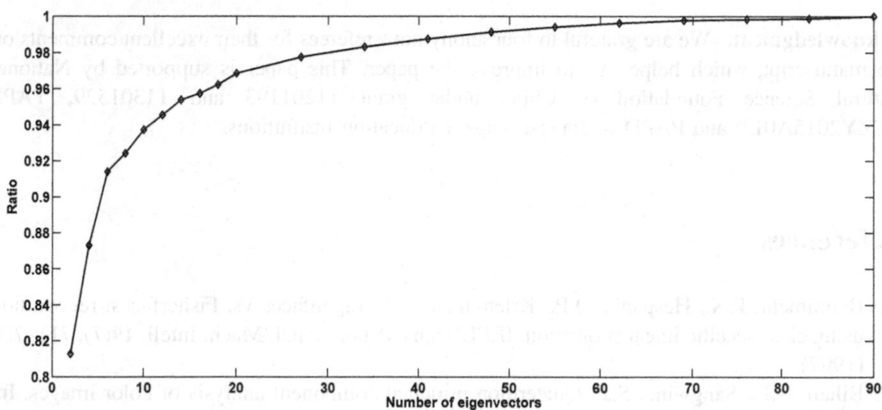

Fig. 4. Image reconstruction by color 2DPCA

distinct photos in the data set. For testing, we use 1054 color face images from the folder Frontalization/LFW3D.0.1.1-nameA.

Let F and $\tilde{F}$ respectively denote the original face image and its reconstruction. Define the ratio of image reconstruction by

$$Ratio = 1 - \frac{||\tilde{F} - F||_2}{||F||_2}.$$

In Fig. 4, we take the face image AJ-Cook-001.jpg for an example: the top picture is the reconstructions of AJ Cook's face image by color 2DPCA with different numbers of eigenvectors; the bottom picture shows the ratios of the reconstructed image over the original face image. The results of this experiment indicate that the color 2DPCA is very convenient to reconstruct color face images from the projections, and it can reconstruct the original color face image with choosing all eigenvectors to span the eigenface subspace.

5 Conclusion

The contribution of this paper is threefold. Firstly, the color 2DPCA approach is presented for color face recognition based on quaternion model. It is novel and has highest rate of face recognition over PCA, 2DPCA and CPCA (see Example 1). Secondly, the theory of image reconstruction is built on color 2DPCA (see Theorem 1). The color 2DPCA can reconstruct the face image completely with using a small number of low-dimensional eigenvectors (see Example 2). Thirdly, the computational difficulty for applying quaternion model into color face recognition has been overcome by designing structure-preserving algorithms with only real operations and without dimension expanding.

Acknowledgment. We are grateful to four anonymous referees for their excellent comments on the manuscript, which helped us to improve the paper. This paper is supported by National Natural Science Foundation of China under grant 11201193 and 11301529, TAPP (PPZY2015A013) and PAPD of Jiangsu Higher Education Institutions.

References

1. Belhumeur, P.N., Hespanha, J.P., Kriengman, D.J.: Eigenfaces vs. Fisherfaces: recognition using class specific linear projection. IEEE Trans. Pattern Anal. Mach. Intell. **19**(7), 711–720 (1997)
2. Bihan, N.L., Sangwine, S.J.: Quaternion principal component analysis of color images. In: Image Processing, pp. 809–810 (2003)
3. Denis, P., Carre, P., Fernandez-Maloigne, C.: Spatial and spectral quaternionic approaches for colour images. Comput. Vis. Image Underst. **107**, 74–87 (2007)
4. Hamilton, W.R.: Elements of Quaternions. Chelsea, New York (1969)
5. Kirby, M., Sirovich, L.: Application of the karhunenloeve procedure for the characterization of human faces. IEEE Trans. Pattern Anal. Mach. Intell. **12**(1), 103–108 (1990)
6. Luo, Y., Chen, D.: Face recognition based on color Gabor features. J. Image Graph. **13**(2), 242–243 (2006)
7. Pei, S.-C., Chang, J.-H., Ding, J.-J.: Quaternion matrix singular value decomposition and its applications for color image processing. In: Image Processing, ICIP 2003, vol. 1, pp. 805–808 (2003)
8. Pentland, A.: Looking at people: sensing for ubiquitous and wearable computing. IEEE Trans. Pattern Anal. Mach. Intell. **22**(1), 107–119 (2000)
9. Qiao, L., Chen, S., Tan, X.: Sparsity preserving discriminant analysis for single training image face recognition. Pattern Recogn. Lett. **31**, 422–429 (2010)
10. Qiao, L., Chen, S., Tan, X.: Sparsity preserving projections with applications to face recognition. Pattern Recogn. **43**, 331–341 (2010)

11. Shi, L., Funt, B.: Quaternion color texture segmentation. Comput. Vis. Image Underst. **107**, 88–96 (2007)
12. Sangwine, S., Bihan, N.L.: Quaternion toolbox for Matlab. http://qtfmsourceforge.net/
13. Sinha, P., et al.: Face recognition by humans: nineteen results all computer vision researchers should know about. In: Proceedings of the IEEE, vol. 94(11), pp. 1948–1962 (2006)
14. Sirovich, L., Kirby, M.: Low-dimensional procedure for characterization of human faces. J. Optical Soc. Am. **4**, 519–524 (1987)
15. Torres, L., Reutter, J.Y., Lorente, L.: The importance of the color information in face recognition. In: IEEE International Conference on Image Processing, vol. 3, pp. 627–631 (1999)
16. Turk, M., Pentland, A.: Eigenfaces for recognition. J. Cogn. Neurosci. **3**(1), 71–76 (1991)
17. Xiang, X., Yang, J., Chen, Q.: Color face recognition by PCA-like approach. Neurocomputing **152**, 231–235 (2015)
18. Yang, J., Zhang, D., Frangi, A.F., Yang, J.Y.: Two-dimensional PCA: a new approach to appearance-based face representation and recognition. IEEE Trans. Pattern Anal. Mach. Intell. **26**(1), 131–137 (2004)
19. Yang, J., Liu, C.: A general discriminant model for color face recognition. In: IEEE 11th International Conference on Computer Vision, pp. 1–6 (2007)
20. Zhang, F.: Quaternions and matrices of quaternions. Linear Algebra Appl. **251**, 21–57 (1997)
21. Zhao, L., Yang, Y.: Theoretical analysis of illumination in PCA-based vision systems. Pattern Recogn. **32**(4), 547–564 (1999)
22. The Georgia Tech face database. http://www.anefian.com/research/facereco.htm
23. The LFW face database. http://www.cs.umass.edu/lfw

Image Recognition Using Local Features Based NNSC Model

Li Shang[1(✉)], Yan Zhou[1], and Zhanli Sun[2]

[1] Department of Communication Technology,
College of Electronic Information Engineering,
Suzhou Vocational University, Suzhou 215104, Jiangsu, China
{sl0930, zhyan}@jssvc.edu.cn

[2] School of Electrical Engineering and Automation,
Anhui University, Hefei 230039, Anhui, China
zhlsun2006@126.com

Abstract. Non-negative sparse coding (NNSC) model can extract efficiently image features and has been used widely in image processing. However, in terms of the image feature classification, the recognition precision of NNSC is not ideal. To overcome this defect, on the basis of the basic NNSC model, considered the constraints of the sparse measure of feature basis vectors and the local features, a locality based NNSC (LNNSC) model is proposed here. In this NNSC model, the feature coefficients are learned by the optimized method that combines the gradient and multiplicative factor, and the feature basis vectors are learned by only the gradient algorithm. Selected palmprint images from PolyU database and considered different dimensions of image features, the results of feature extraction and recognition obtained by LNNSC are discussed. Furthermore, compared with other feature extraction methods, experimental results show that our NNSC model can extract image features efficiently and has quick convergence speed, as well as can model the sparse coding strategy used by the primary visual system in dealing with the nature processing. This also proves that the LNNSC model is feasibility and practicality in the theoretical research.

Keywords: Non-negative sparse coding (NNSC) · Local features · Feature extraction · Palmprint images · Image reconstruction

1 Introduction

Generally, features of one's palm are stable and remain unchanged throughout the life of a person, so in personal recognition, palmprint verification has always been a challenging topic [1–3]. In palmprint recognition, the important task is how to extract palmprint features so that one person can be easily discriminated from others. To this day, many image feature extraction methods have been developed and used widely in application, such as algorithms of non-negative matrix factorization (NMF) [4, 5], local NMF (LNMF) [5, 6], independent component analysis (ICA) [7], sparse coding (SC) [8–10], non-negative sparse coding (NNSC) [11–13] and so on. These algorithms are very fit to process high dimensional image data. In fact, algorithms of ICA and NNSC are sparse coding techniques, which are qualitatively very similar to the

D.-S. Huang et al. (Eds.): ICIC 2017, Part I, LNCS 10361, pp. 190–199, 2017.
DOI: 10.1007/978-3-319-63309-1_18

receptive fields of simple cells in the primary visual cortex (V1) in brain, and they are successfully applied to learn features of images. Although NNSC model is very efficient in implementing image extraction and recognition, the recognition rate and the convergence speed of NNSC are still not ideal in application. Therefore, to avoid defects above mentioned, considered the constraints of the sparse measure of feature basis vectors and the local features, a novel NNSC based on local features is discussed in this paper. In this NNSC model proposed in this paper, image feature coefficients are learned by the optimized method that combines the gradient and multiplicative factor, and the feature basis vectors are learned only by the gradient algorithm. Test palmprint images are selected from the PolyU database [14–16] used commonly in palmprint image processing. Considered different dimensions of feature basis vectors, palmprint features are extracted by our NNSC model. At the same time, to prove the effect of our NNSC model on feature extraction, the image reconstruction task is discussed by utilizing features extracted by our NNSC model. Further, the palmprint recognition task are implemented by classifiers of Euclidean distance, extreme learning machine (ELM) [17, 18] and radial basis function neural network (RBFNN). Otherwise, compared with other image recognition methods, such as NNSC and LNMF, in the same test condition, experimental results testify that our NNSC model can extract image features efficiently and has quicker convergence speed, as well as higher recognition rate and shorter recognition time.

2 Adding Sparseness Measure to Feature Basis

The meaning of sparse coding refers to a representational scheme where only a few units are effectively used to represent typical data vectors [4], which implies most units are close to zero while only few take significantly non-zero values. In generally, the sparsity of a random vector y is estimated by the sparseness measure function defined as follows:

$$sparseness(y) = \frac{\sqrt{n} - \left(\sum_i |y_i|\right) \Big/ \sqrt{\sum_i y_i^2}}{\sqrt{n} - 1}. \tag{1}$$

Where n is the dimensionality of y. The value of this function in Eq. (1) is selected within the range of 0 and 1. If and only if y contains only a single non-zero component, this function's value is 1. If and only if all components in y are equal, the value of this function is zero. In fact, this function is based on the relationship between the L_1 norm and L_2 norm. In basic NNSC model, the sparseness of basis vectors is unconstrained and only that of feature coefficients is constrained. To obtain useful local features from a database of images, it might make sense to require both coefficients and basis vectors to be sparse. For this reason, in our NNSC model, the feature basis vectors are added to sparseness constraint by using Eq. (1).

3 Local Features Based NNSC Model

3.1 The Cost Function

The basic NNSC model was early proposed by P.O. Hoyer in 2002 [11]. This BNNSC method adopts the idea of a parts-based representation for Sparse Coding (SC), by also putting positive constraints on the weights and the coefficients. It is similar to NMF with sparseness constraints [4], whereas NMF methods apply multiplicative update rules that don't require the definition of a learning rate. This cost function of Hoyer's NNSC model only contains two terms, the small reconstruction error and the sparse penalty function [4, 11, 12], and it doesn't consider the prior classification constraint of feature basis vectors and the constraint of feature sparseness measurement, therefore, Hoyer's NNSC model is not ideal in image feature recognition task. Considered the problems mentioned-above, on the base of Hoyer's model, the local features based NNSC is proposed and its cost function is written as in Eq. (2):

$$J(W,H) = \frac{1}{2}\sum_{i=1}^{N}\sum_{j=1}^{L}\left(V_{ij} - \sum_{i=1}^{N} W_{ik}H_{kj}\right)^2 + \lambda\sum_{k=1}^{M}\sum_{j=1}^{L} f(H_{kj}) + \gamma\sum_{i}^{N}\left(w_{ik}^T w_{ik}\right) - \beta\sum_{k}^{M}\left(h_{kj}h_{kj}^T\right). \tag{2}$$

subject to the constraints: $V \geq 0$, $\forall_i$: $w_k \geq 0$, $h_k \geq 0$, $\lambda > 0$, $\gamma > 0$, and $\|w_k\| = 1 (k = 1, 2, \cdots, m)$. Where $V = [v_1, v_2, \cdots, v_k, \cdots, v_N]^T$ denotes the non-negative training sample matrix with the size of $n \times L$, W is the feature basis matrix with the size of $n \times m$ (placed in the column), H is the weight coefficient matrix with the size of $m \times L$. And each column of matrix W contains a basis vector, while each column of H contains the weights needed to approximate the corresponding columns in V using the bases from W. The rank m of factorization usually is chosen as $m < nL/(n + L)$. The meaning of the first term and second term is the same as those of Hoyer's NNSC model. The third term is added to ensure the orthogonality between features in order to reduce data redundance, at the same time, it can also ensure the sparsity of coefficients. The four term is added to ensure the elements in W to behave the maximize representativeness. Because the information contained in basis images is represented by $\sum_{k=1}^{M}\sum_{j=1}^{L}\left(H_{kj}^2\right)$, the maximize representativeness of W can be implemented by minimizing the negative value of $\sum_{k=1}^{M}\sum_{j=1}^{L}\left(H_{kj}^2\right)$. Otherwise, to ensure the sparseness measurement of W, we use Eq. (1) to constrain each column.

3.2 The Learning Rules

Referred to the learning process of Hoyer's NNSC algorithm and LNMF, the updating of W and H is done by in turn. For coefficient matrix H, it is first updated by using the Gradient descent algorithm, and then updated by a multiplication factor. The corresponding rules are written as follows:

$$\begin{cases} H_{kl} = H_{kl} - W_{ik}^T(W_{ij}H_{kl} - V_{il}) + \lambda f'(h_k) - 2\beta H_{kl} \\ H_{kl} = \sqrt{H_{kl}\left(\frac{W_{ik}^T V_{il}}{W_{ik}H_{kl}}\right)} \end{cases}. \tag{3}$$

And fixed feature coefficient matrix H, the feature basis matrix W can be trained by the Eq. (4) deduced as follows:

$$\begin{cases} W_{ik} = \left(W_{ik}\sum_l V_{il}\frac{H_{kl}}{\sum_j W_{ik}H_{kl}}\right) \Big/ \sum_k H_{kl} \\ W_{ik} = W_{ik}/\|W_{ik}\| \end{cases}. \tag{4}$$

According to updating rules of W and H, the image features can be extracted, further, using classifiers, we can implement the image feature recognition task.

4 Experimental Results and Analysis

4.1 Data Preprocessing and Feature Basis Training

The PolyU palmprint database is used to verify the palmprint recognition method based on the local features NNSC model proposed in this paper. From the PolyU database, 600 palmprint images with the size of 128 × 128 from 100 individuals are chosen. Several images of one class of palmprint images were shown in Fig. 1. Each person has six images, and the first three images of one are used as training samples, while the remaining ones are treated as testing samples. Otherwise, to reduce the number of sources to a tractable number and provide a convenient method for calculating representations of images, each image is preprocessed to be 64 × 64 pixels. Thus the training set X_{train} and the testing set X_{test} have the same size of 300 × (64 × 64). At the same time, for the training set, considering the non-negativity, X_{train} is divided into two parts, namely, one part denoted by Y is made up of all positive elements in X_{train}, and another set denoted by Z is made up of the absolute values of all negative elements in X_{train}. Thus, Y and Z correspond to the ON channel data and OFF channel data of NNSC model. Let $X' = [Y; Z]$ be the input set of our NNSC model and consider the

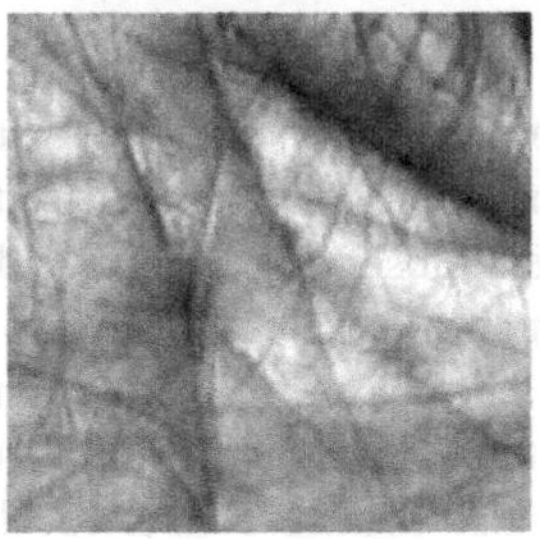
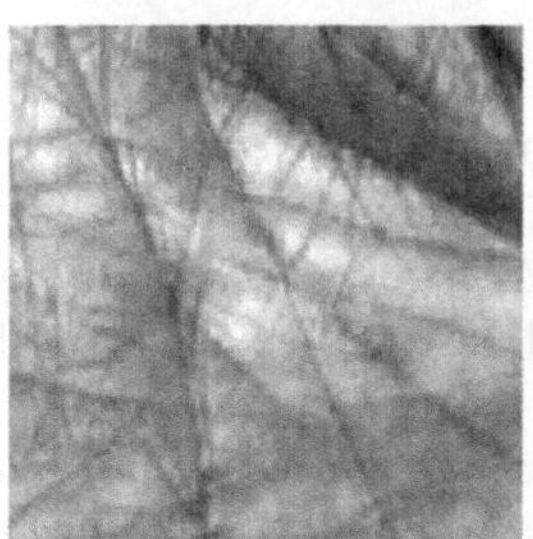
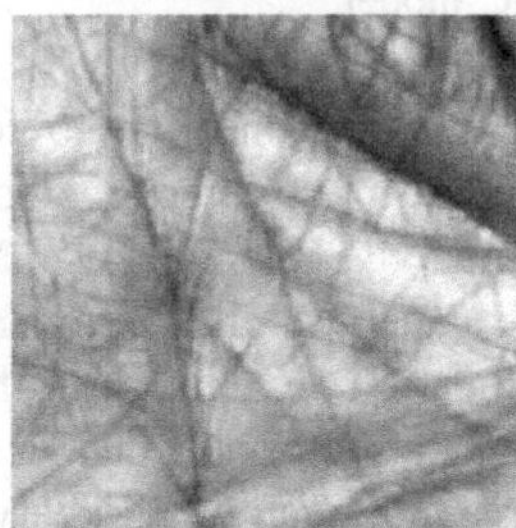

Fig. 1. One class of original palmprint images selected randomly from PolyU database.

different number of features, the feature basis images can be obtained. Limited by the length of this paper, only some feature basis images corresponding to the dimensions of 25, 36, 64 and 121 are shown in Fig. 2. Form Fig. 2, it is clear to see that feature basis images behave distinct local property.

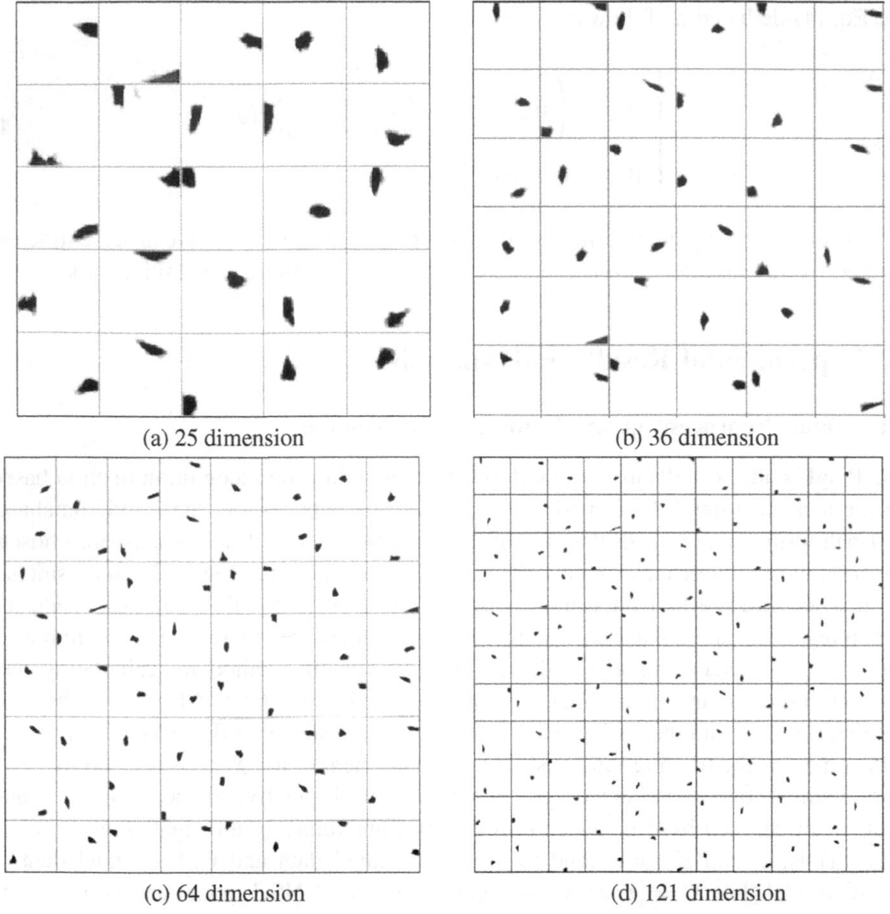

Fig. 2. Considering different feature dimensions, feature basis images obtained by our local NNSC model.

Otherwise, in order to compare the effectiveness of our NNSC model in feature extraction with other methods, feature basis images shown in Fig. 3, obtained by the original NNSC model and LNMF algorithm, were also given. Observing Figs. 2 and 3, it is clear to see that, compared the basic NNSC algorithm and LNMF algorithm with our local NNSC model, in the same feature dimension, feature basis images obtained by our model behave clearer locality, moreover, the larger the feature dimension is, the clearer the locality is. Otherwise, the LNMF basis images also behave clearer locality and sparsity when the feature dimension is the same.

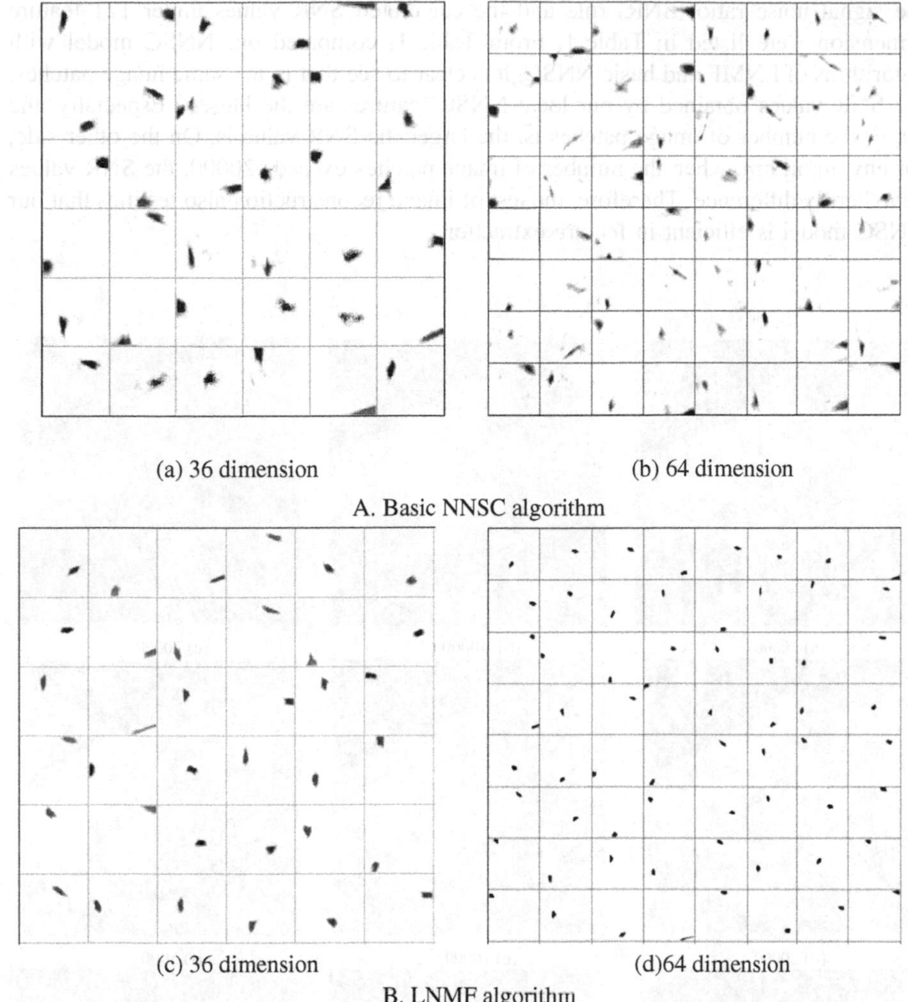

Fig. 3. Considering different feature dimensions, feature basis images obtained by the basic NNSC algorithm and LNMF algorithm.

4.2 Testifying Features

To test feature basis images obtained by our local NNSC model, the image reconstruction task is discussed for a palm image randomly selected from PolyU palm database by considering different feature dimensions. Limited by the length of this paper, and considered different image patches, only some reconstruction results of 121 palmprint features obtained by our NNSC model were given here, as shown in Fig. 4. And in the same test condition, image reconstruction task by using features of LNMF and basic NNSC were also implemented. And to prove the validity of our local NNSC model in image feature extraction, the reconstruction images were measured by using the signal noise ratio (SNR) rule and the calculated SNR values under 121 feature dimension were listed in Table 1. From Table 1, compared our NNSC model with algorithms of LNMF and basic NNSC, it is clear to see that in the same image patches, the SNR values obtained by our local NNSC features are the largest. Especially, the larger the number of image patches is, the larger the SNR value is. On the other side, for any algorithm, when the number of image patches exceeds 20000, the SNR values have hardly difference. Therefore, the test of image reconstruction also testifies that our NNSC model is efficient in feature extraction.

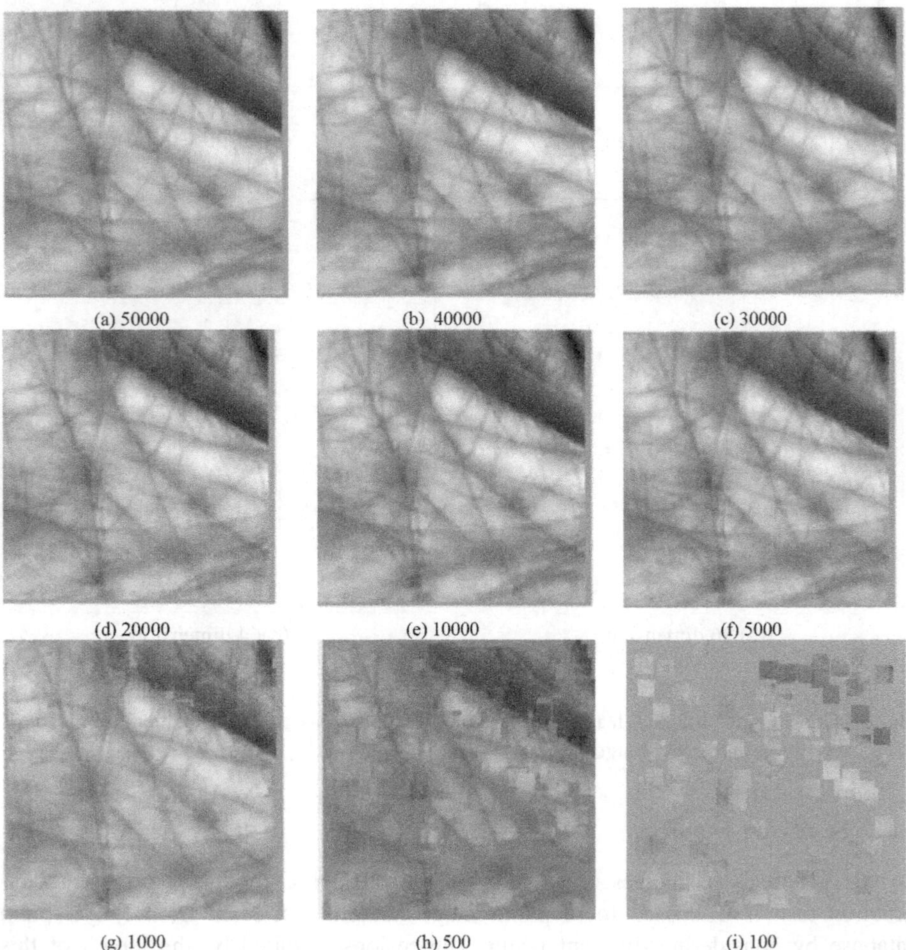

Fig. 4. Considering different image patches, the image reconstruction results obtained by our NNSC model with 121 feature dimension.

Table 1. SNR values of different methods (121 feature basis images).

Algorithms	Local NNSC	LNMF	Basic NNSC
100	16.5265	16.5246	16.4816
300	18.7271	19.0859	18.3970
500	20.7213	20.0620	19.5056
1000	23.7020	23.0440	22.0686
3000	27.4637	26.3063	25.8006
5000	28.1608	26.5222	26.0166
10000	29.0375	27.5039	26.4667
20000	29.5869	27.6108	26.6850
30000	29.6353	27.6412	26.7563
40000	29.6497	27.6580	26.7771
50000	29.6516	27.6629	26.8080

4.3 Feature Recognition

In performing the task of palmprint recognition, two classifiers were used, i.e., the Euclidean distance classifier and extreme learning machine (ELM) classifier [17, 18]. Euclidean distance classifier is the simplest distance-matching algorithm among classifiers used widely. And ELM learning has been developed to work at a much faster learning speed with the higher generalization performance, especially in the pattern recognition, so ELM is very suitable for classification problems. Moreover, many documents published have been proved to be the high efficiency of ELM classifier now [17]. In palmprint test, each image is converted a column vector in the training set and test set. And the size of the two sets are both changed to be 4096 $\times$ 300 pixels. Referring to the document [14], the palmprint recognition framework is selected to be the independent component analysis (ICA) frameworkI. In this ICA framework, the palm images are variables and the pixel values provide observation for the variables [14], namely, in the training set, the number of rows is that of the features and the number of columns is that of the image patches. Moreover, this framework is sparse within images across pixels and it produces localized features that are only influenced by small parts of the image. Here, let the number of feature dimensions to be 25, 36, 49, 64, 81 and 121, the corresponding feature recognition results by using Euclidean distance classifier and ELM classifier were respectively listed in Table 2. From Table 2, it is easy to see that when fixed the classifier, our NNSC model clearly outperform methods of LNMF and basic NNSC. But there is little difference between the recognition results of LNMF and basic NNSC. While when fixed an algorithm, it is also distinct to see that the recognition rate of ELM classifier is the best. Therefore, according to test results, it can be concluded that our NNSC method is efficient in image feature recognition task.

Table 2. Feature recognition results by using different classification under different feature dimensions.

Algorithms		25	36	49	64	81	121
Euclidean distance classifier	Local NNSC	75.03	86.17	89.13	91.651	92.34	93.33
	LNMF	73.54	81.67	85.17	90.81	92.05	93.33
	Basic NNSC	72.33	81.43	85.33	88.83	90.83	92.00
ELM classifier	Local NNSC	79.67	90.83	96.50	97.35	98.33	98.33
	LNMF	78.83	83.83	89.13	89.53	91.17	92.33
	Basic NNSC	73.24	83.23	87.33	88.52	90.62	91.67

5 Conclusions

A new local features based NNSC model and its application in image feature recognition task are discussed in this paper. In this NNSC model, the constraints of the sparse measure of feature vectors and their locality are considered. And in the process of learning this model, the feature coefficients are trained by the combination process of the gradient algorithm and the multiplicative factor, but the feature vectors are only learned by the usual gradient descent algorithm. Therefore, the learning speed is faster than the basic NNSC model. In test, the PolyU database used widely in image classification is selected, and the palmprint features extracted are proved to be efficient by the image reconstruction task. Furthermore, using classifiers of ELM and Euclidean distance to classify palmprint features obtained by algorithms of our NNSC, LNMF and the basic NNSC, experimental results prove that our NNSC model proposed here behaves the best recognition rate. Especially, when use the ELM classifier, the recognition result is satisfied.

Acknowledge. This work was supported by the grants from National Nature Science Foundation of China (Grant No. 61373098 and 61370109), the youth found of Natural Science Foundation of Jiangsu Province of China (Grant No. BK20160361), and the grant from Natural Science Foundation of Anhui Province (No. 1308085MF85).

References

1. Zhang, D., Hong, W.K., You, J.: Online palmprint identification. IEEE Trans. Pattern Anal. Mach. Intell. **25**(9), 1041–1050 (2003)
2. Lu, G., David, Z., Wang, K.: Palmprint recognition using eigenpalm features. Pattern Recogn. Lett. **24**(9), 1473–1477 (2003)
3. Li, W., David, Z., Xu, Z.: Palmprint identification by fourier tranform. Int. J. Pattern Recogn. Artif. Intell. **16**(4), 417–432 (2002)
4. Hoyer, P.O.: Non-negative matrix factorization with sparseness constraints. J. Mach. Learn. Resear. **5**, 1427–1469 (2004). MIT Press, New Work
5. Lee, D.D., Seung, H.S.: Learing the parts of objects by non-negative matrix factorization. Nature **401**, 788–791 (1999)

6. Li, H., Yang, J., Hao, Ch.P: Non-negative matrix factorization of mixed speech signals based on quantum-behaved particle swarm optimization. J. Comput. Inf. Syst. **9**(2), 667–673 (2013)
7. Hoyer, P.O., Hyvärinen, A.: Independent component analysis applied to feature extraction from colour and stereo images. Netw. Comput. Neural Syst. **11**(3), 191–210 (2000)
8. Olshausen, B.A., Field, D.J.: Emergence of simple-cell receptive field properties by learning a sparse code for natural images. Nature **381**, 607–609 (1996)
9. Thiagarajan, J.J., Ramamurthy, K.N., Spanis, A.: Multiple kernel sparse representations for supervised and unsupervised learning. IEEE Trans. Image Process. **23**(7), 2905–2915 (2014)
10. Lee, H., Battle, A., Raina, R.: Efficient sparse coding algorithms. In: Proceedings of Neural Information Processing Systems (NIPS 2007), Vancouver, B.C., Canada, pp. 801–808 (2007)
11. Hoyer, P.O.: Non-negative sparse coding. In: Proceedings of Neural Networks for Signal Processing XII, Martigny, Switzerland, pp. 557–565 (2002)
12. Hoyer, P.O.: Modeling receptive fields with non-negative sparse coding. Neurocomputing **52**(54), 547–552 (2003)
13. Shrivastava, A., Patel, V., Chellappa, M.: Multiple kernel learning for sparse representation-based classification. IEEE Trans. Image Process. **23**(7), 3013–3024 (2014)
14. Connie, T., Teoh, A., Goh, M., et al.: Palmprint recognition with PCA and ICA. Image Vis. Comput. **3**, 227–232 (2003). NZ
15. Shang, L., Su, P.G., Dai, G.P., Gu, Y.N., Zhao, Zh.Q: Palmprint recognition method using WTA-ICA based on 2DPCA. Commun. Comput. Inf. Sci. **93**, 250–257 (2010)
16. Shang, L., Cui, M., Su, P.G., Zhao, Zh.Q.: Palmprint feature extraction using weight coding based non-negative sparse coding. In: 2010 3rd International Congress on Image and Signal Processing, 16–18 October, Yantai, China, vol. 4, pp. 1905–1908 (2010)
17. Hung, G.B., Zhu, Q., Siew, C.K.: Extreme Learning Machine: Theory and Applications. Neurocomputing **70**, 489–501 (2006)
18. Huang, G.B., Chen, L., Siew, C.K.: Universal approximation using incremental constructive feedforward networks with random hidden nodes. IEEE Trans. Neural Netw. **17**(4), 879–892 (2006)

A Hashing Image Retrieval Method Based on Deep Learning and Local Feature Fusion

Yi-Liang Nie, Ji-Xiang Du(✉), and Wen-Tao Fan

Department of Computer Science and Technology,
Huaqiao University, Xiamen 361021, China
{nyl, jxdu, ftw}@hqu.edu.cn

Abstract. The multimedia information such as images and videos has been growing rapidly, how to efficiently retrieve large-scale image dataset to meet user needs is an urgent problem. The traditional method has the problem of slow retrieval and low accuracy on large-scale datasets, we propose an effective deep learning framework to generate binary hash codes for fast image retrieval, our idea is to fuse local features maps of different layers in convolutional neural networks (CNNs), and the binary hash codes can be learned by employing a hidden layer. Additionally, we train the network by combining cross entropy loss function with the triplet loss function to get better features. The approximate nearest neighbor search strategy is used to improve the quality and speed of retrieval. Experimental results show that our method outperforms several state-of-the-art hashing image retrieval algorithms on the MNIST and CIFAR-10 datasets. At last, we further demonstrate its scalability and efficacy on the CUB200-2011 and Stanford Dogs fine-grained classification datasets.

Keywords: Deep learning · Fuse local features · Approximate nearest neighbor · Hashing image retrieval

1 Introduction

In recent years, deep learning has made significant breakthroughs in various computer vision tasks, including content-based image retrieval (CBIR) [1], whose purpose is to retrieve images by analyzing the image content. With the development of CBIR, a big challenging is to extract semantics from the associated pixel-level information. Despite several hand-crafted features have been suggested to represent the images [2–5], the performance of these visual descriptors is still limited. Recent studies [6–9] have shown that the deep CNN significantly improves the performance on diverse vision tasks, such as object detection, image classification, segmentation, and image retrieval. In general, users desire to retrieve a similar image of the local details rather than just the same semantics. Therefore, this paper focuses on fusing feature maps from the CNN to get more local detailed information of image, combined with the approximate nearest neighbor image search strategy [10] and the hash code to experiment.

D.-S. Huang et al. (Eds.): ICIC 2017, Part I, LNCS 10361, pp. 200–210, 2017.
DOI: 10.1007/978-3-319-63309-1_19

2 Relative Work

Several hashing methods [11–13] have been proposed to boost the performance of approximate nearest neighbor search due to their low time and space complexity. In the early years, researchers mainly focused on data-independent hashing methods, such as Locality Sensitive Hashing (LSH) [11], it uses random projections to produce hash codes. With the development of deep learning, we can extract the characteristics of high-level semantic information from CNN, and it gradually applied to the field of image retrieval. Recent studies have shown that the use of supervised methods can improve the learning performance of hash codes. Rongkai Xia et al. [13] propose a supervised hashing approach to learn binary hashing codes for fast image retrieval through deep learning and demonstrate state-of-the-art retrieval performance on public datasets. However, in their preprocessing stage, the matrix-decomposition algorithm is used for learning the representation codes for data. It thus requires the input of a pair-wised similarity matrix of the data and is unfavorable for the case when the data size is large, because it consumes both considerable storage and computational time. Lin Kevin et al. [10], there idea is to add a hash layer to the pre-trained AlexNet network, the semantic information of labels can be embedded in the newly added hash layer through the end-to-end fine-tuning network, and the algorithm is finally obtained good precision and retrieval speed. The disadvantage of this method is that only the high-level semantic information is taken into account, but the local details of images are ignored, resulting in the user being unable to obtain the desired images they retrievaled. The DSH method proposed by Haomiao Liu et al. [14], they use a piecewise function to constrain the output of the network. When the output of the network and the expected deviation of the value are larger, the loss is larger and the algorithm is reduced the quantization error of the hash codes generated.

3 Proposed Approach

In this section, we mainly introduce our approach. Section 3.1 describes the composition of the network structure. Section 3.2 explains how to fuse local features. Section 3.3 explains how to generate hash codes. Section 3.4 describes the process of retrieving by using the approximate nearest neighbor method.

3.1 The Composition of the Network Structure

The network structure we proposed is shown in Fig. 1, The network consists of five parts. (1) The input part contains images and the corresponding labels, and images are entered by a triad. (2) The convolutional network part uses the convolution part of the GoogLeNet network, and contains the original three loss layers. (3) The local feature fusion module is mainly composed of several convolution layers, pool layers, a merge layer and a fully connected layer. (4) The hash layer consists of a fully connected layer and threshold segmentation function. (5) The loss function is primarily composed of triplet loss function and cross entropy loss functions.

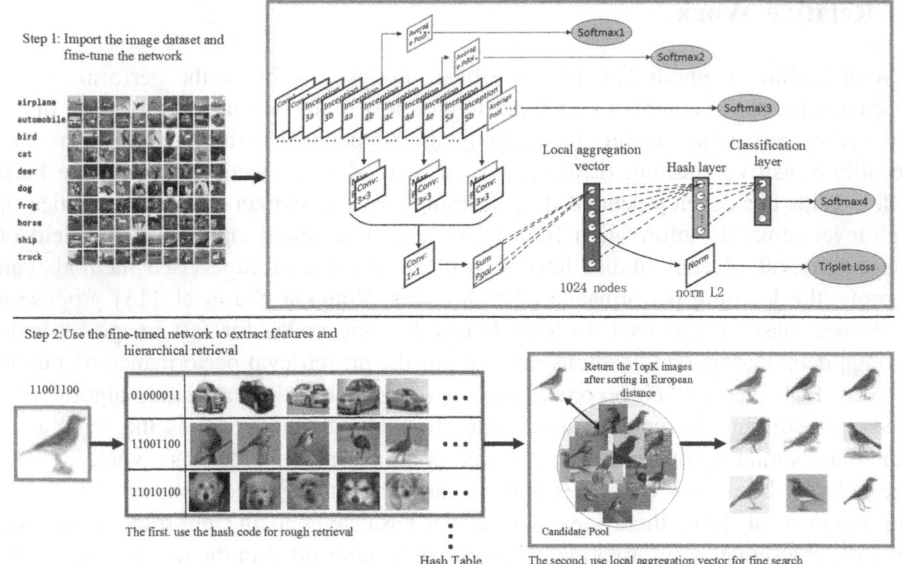

Fig. 1. The improved local feature fusion network structure based on GoogLeNet network.

3.2 Local Feature Fusion

Image retrieval based on deep learning is usually to extract feature of the last convolution layer or fully connected layer directly to calculate the similarity, but the feature of high-level has lost a lot of details of the information, and for the deeper network, the more loss of detail. For example, different clothings may be due to a certain part of the details (such as clothing texture, neckline and cuffs) to distinguish in clothing search of the electric business platform [15], so that users can not retrieve the same type of clothing, and we will encounter the same trouble in industrial precision device search: the object is similar in the overall contour, but the details of the gap is great. Therefore, the extracted features should contain as much detail as possible the local details in the case of semantic coherence.

The method proposed is to combine the feature maps of different layers in GoogLeNet network, so that we not only get the high-level semantic information of the image, but also take into account the low-level details of the texture information. After a lot of experiments, we select the three inception layers of the GoogLeNet network to make the feature fusion the best, which are inception4b, inception4e and inception5b, respectively. As shown in Fig. 2, the feature of these different scales were first sub-sampled using the maximum pooling to be converted into the same size 7×7, then we use the convolution of the size of 3×3 to compute these feature maps and activate it by ReLu of the nonlinear activation function. There are two advantages to doing this, the first is to increase the nonlinear description ability of the network, and the second is to achieve the effect of dimensionality reduction, so that each feature map of the inception layer has the same number and size ($256 \times 5 \times 5$).

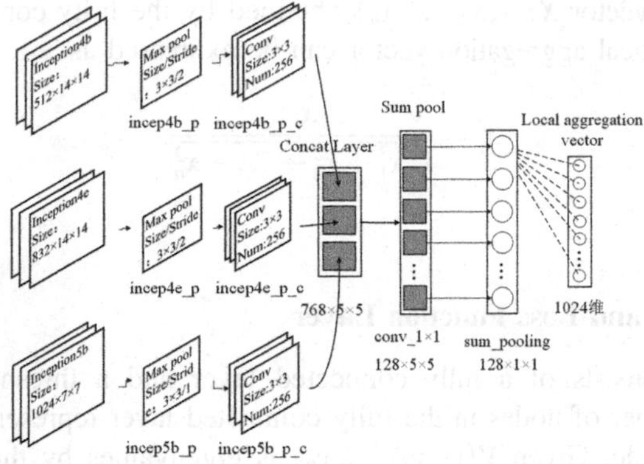

Fig. 2. The design of local feature fusion structure diagram.

Then we can combine these feature maps to generate a size of 768 × 5 × 5 features. Followed by a size of 1 × 1 linear convolution kernel to obtain a feature with a size of 128 × 5 × 5, and then create the aggregation vector by using Sum Pool [16], donate as:

$$\psi(\mathrm{I}) = \sum_{y=1}^{\mathrm{H}} \sum_{x=1}^{\mathrm{W}} \mathrm{f}(x, y) \tag{1}$$

The H and W in the formula are the height and width of the feature map, x and y are the spatial coordinates of the feature map, I is the input feature map, and $\mathrm{f}(x, y)$ is the value of the corresponding point (x, y) on the feature map. The sum pool is actually the sum of obtained for each feature map. If there are n feature maps, will eventually generate eigenvector with the size of n × 1 × 1. The H and W have a value of 5 and n is 128 to get a vector of 128 × 1 × 1 dimensions. Then we connect a 1024-node fully connected layer after sum pool and get the local aggregation vector. The parameters of local feature fusion structure are shown in Table 1.

Table 1. The Part of local convergence network configuration table.

Name	Type	Patch size/stride	Pad	Output size
incep4b_p	max pool	3 × 3/2	0	512 × 7×7
incep4e_p	max pool	3 × 3/2	0	832 × 7×7
incep5b_p	max pool	3 × 3/1	1	1024 × 7×7
incep4b_p_c	convolution	3 × 3/1	0	256 × 5×5
incep4e_p_c	convolution	3 × 3/1	0	256 × 5×5
incep5b_p_c	convolution	3 × 3/1	0	256 × 5×5
incep_concat	concat	–	–	768 × 5×5
conv_1 × 1	convolution	1 × 1/1	0	128 × 5×5
sum_pooling	sum pool	5 × 5/1	0	128 × 1×1
Local aggregation vector	inner product	1024	–	1024 × 1×1

We define the vector $X(x_1, x_2, \ldots, x_n)$ obtained by the fully connected layer. The normalization of Local aggregation vector can be expressed as:

$$x_i' = \frac{x_i}{\sqrt{x_1^2 + x_2^2 + \ldots + x_n^2}} \tag{2}$$

3.3 Hash Layer and Loss Function Layer

The hash layer consists of a fully connected layer and a threshold segmentation function. The number of nodes in the fully connected layer represents the number of bits of the hash code. Given $V(v_1, v_2, \ldots, v_n)$ is eigenvalues by the fully connected layer, and the binary hash codes $H(h_1, h_2, \ldots h_n)$ obtained by calculating the thresholds of the perceptual hash algorithm [17], expressed as:

$$H_i = \begin{cases} 1, & v_i \geq \frac{v_1 + v_2 + \cdots v_n}{n} \\ 0, & otherwise \end{cases} \tag{3}$$

The network proposed is trained by simultaneously optimizing the triplet loss function [18] and the cross entropy loss function. The triplet loss measure method can narrow the distance between two similar objects as much as possible and expand the distance between two different objects during the training process. The process of input data is first randomly selects a sample anchor from the training data set, and then randomly selects a positive sample and a negative sample, which constitutes the input triplet (anchor, positive, negative), then the loss function can be expressed as:

$$L = \sum_{i}^{N} \left[\|f(x_i^a) - f(x_i^p)\|_2^2 - \|f(x_i^a) - f(x_i^n)\|_2^2 + \alpha \right]_+ \tag{4}$$

The $f(x_i^a)$, $f(x_i^p)$ and $f(x_i^n)$ in the formula are features of the anchor image, the positive sample and the negative sample. The triplet loss function is calculated by the Euclidean distance measure, and the sign of the suffix indicates that the loss is zero when the value in the brackets is less than zero, The α is the distance between acceptable positive and negative samples.

3.4 Image Retrieval via Hierarchical Deep Search

Experiments combined with the convolution neural network for one forward operation can simultaneously extract the characteristics of multi-layer features, and we use an approximate nearest neighbor strategy to image retrieval. Figure 1 is the flow chart of the entire image retrieval system. Our method consists of two main components. The first, we import the dataset and fine-tune our network by using the pre-training model on the ImageNet. In the second, we use of the fine-tuned network to extract features and hierarchical retrieval. Hierarchical retrieval takes into account the approximate nearest neighbor strategy, the first, we use the hash code to carry on the rough retrieval, and then use the local aggregation vector to carry on the retrieval.

In the rough retrieval phase, the hamming distance is computed using the binary hash code to compute the image similarity [10]. Let $T = \{I_1, I_2, \ldots, I_n\}$ denote the dataset consisting of n images for retrieval. The corresponding binary hash codes of images are denoted as $T_h = \{H_1, H_2, \ldots, H_n\}$ with $H_i \in \{0, 1\}$. Given a query image I_r and its hash codes H_r, we can identify a pool of m candidates, $P = \{I_1, I_2, \ldots, I_m\}$, if the hamming distance between H_r and $H_i \in T_h$ is lower than a threshold.

In the fine retrieval phase, given the query image I_r and the candidate pool *P*. We use the local aggregation vector V_r to identify the top K ranked images to form the candidate pool *P*. Calculate the Euclidean distance $Dist_i$ of V_r and the query image and V_i of top K images in the candidate pool *P*, expressed as:

$$Dist_i = \|V_r - V_i^K\| \tag{5}$$

4 Experimental Results

In this section, we demonstrate the benefits of our method. We present our experimental results with the performance comparison with several state-of-the-arts on the public datasets, MNISN and CIFAR-10. Finally, we verify the scalability and efficacy of our approach on the fine-grained classification datasets, CUB200-2011 and Stanford Dogs.

4.1 Evaluation Metrics

We use a ranking based criterion [23] for evaluation. Given a query image *q* and a similarity measure, a rank can be assigned to each dataset image. We evaluate the ranking of top K images with respect to a query image *q* by a precision:

$$Precision_{top\mathrm{K}} = \frac{\sum_{i=1}^{\mathrm{K}} Rel(i)}{\mathrm{K}} \tag{6}$$

The $Rel(i)$ denotes the ground truth relevance between a query *q* and the *i*-th ranked image. Here, we consider only the category label in measuring the relevance, so $\mathrm{Re}l(i) \in \{0, 1\}$ with 1 for the query and the *i* -th image with the same label and 0 otherwise.

4.2 Comparison with the State of the Art

We use Caffe of open source deep learning framework to build the network and train it in our experiment. In order to verify the efficiency of the algorithm, we compare our method with LSH [11], SH [12], ITQ [24], BRE [25], MLH [27], ITQ-CCA [24], KSH [26], CNNH [13], CNNH+ [13], DLBHC [10].

Figure 3 shows the results of the Top10 image search on the MNIST dataset. When using the 48-bit hash codes, the MAP is 0.989. Although the accuracy is high, but the details of the handwriting are not the same, so we further perform the fine search. From the curve of Fig. 4, the method has the characteristics of high precision and stable performance compared with the existing method.

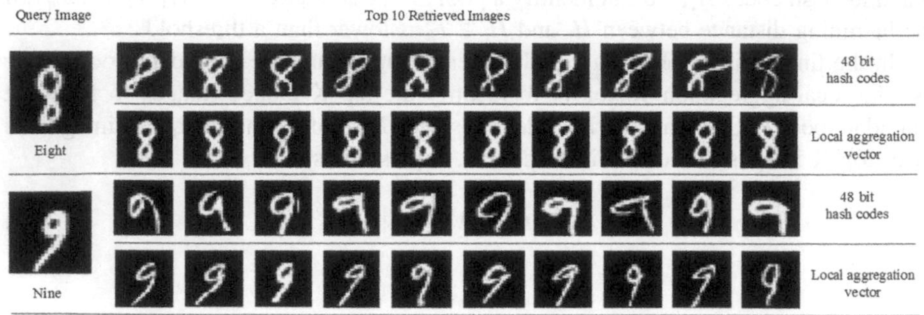

Fig. 3. Top10 image retrieval results on MNIST dataset.

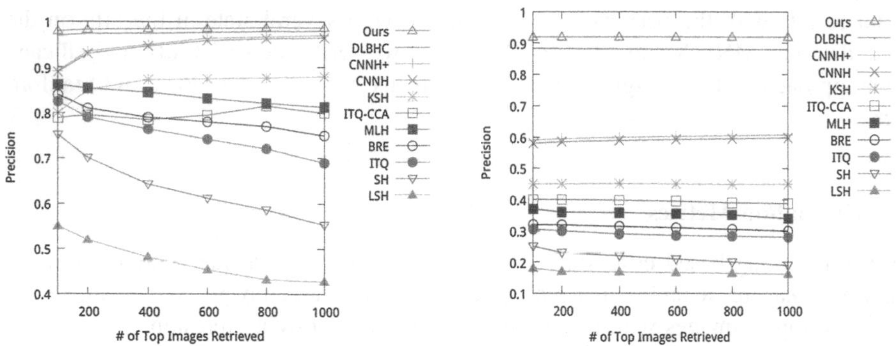

Fig. 4. Image retrieval precision with 48 bit on MNIST.

Fig. 5. Image retrieval precision with 48 bit on CIFAR-10.

Figure 5 shows the retrieval accuracy of different hash methods on the CIFAR-10 dataset. For fair comparison, our experiment uses a 48-bit hash codes and uses the hamming distance to measure the similarity between images. The retrieval is performed by randomly selecting 1000 query images from testing set for the system to retrieve relevant ones from the training set. It can be observed in the Fig. 5 that the retrieval accuracy of our method is stable at 0.92, which is 47% higher than KSH, 32% higher than CNNH+, and 3% higher than the current optimal method DLBHC.

In order to watch more intuitively the relationship between the result of the image retrieval and the length of hash bit, the experiment visualizes the search results, Fig. 6 shows the Top10 search results for two query images on the CIFAR-10 dataset, when the aircraft image is retrieved, the 48-bit and 128-bit hash code is biased to detect the

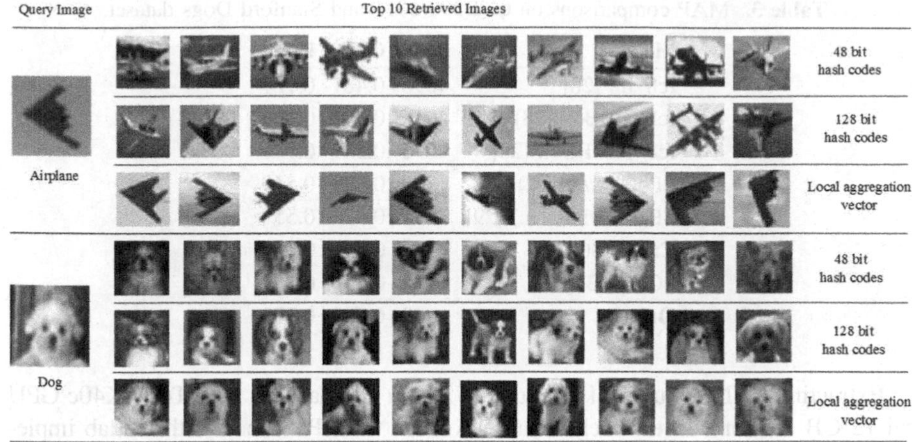

Fig. 6. Top10 image retrieval results on CIFAR-10 dataset.

associated image of the same aircraft class, including gliders, civil aircraft and fighters with different shapes and colors. The last row is a more accurate result after reordering of the 48-bit hash code using the 1024-dimensional local aggregate vector. Hash codes can retrieval the semantic similar images, but often the actual needs of users not only to retrieve the same category of images, but also need to look more similar in appearance, including the retrieval of the shape of the object, color and so on.

Table 2 shows the MAP obtained by retrieving different hash code bits on the CIFAR-10 datasets. It can be seen from Table 2 that compared with other existing methods, the algorithm has the best accuracy in retrieving CIFAR-10 datasets, and the MAP obtained in different hash bits have reached more than 0.92. At the same time, with the increase of the hash bit, the calculated MAP value is slightly increased. Table 3 shows the MAP comparisons on CUB200-2011 and Stanford Dogs dataset, our approach also achieves better performance on two fine-grained classification datasets.

Table 2. MAP comparisons on CIFAR-10 with different hash bit.

Methods	CIFAR-10			
	12bit	24bit	36bit	48bit
LSH [11]	0.134	0.145	0.153	0.162
SH [12]	0.168	0.173	0.189	0.194
ITQ [24]	0.268	0.273	0.276	0.282
BRE [25]	0.298	0.301	0.303	0.314
MLH [27]	0.310	0.321	0.335	0.340
ITQ-CCA [24]	0.367	0.372	0.376	0.388
KSH [26]	0.439	0.442	0.448	0.450
CNNH [13]	0.578	0.585	0.595	0.598
CNNH+ [13]	0.587	0.596	0.601	0.604
DLBHC [10]	0.893	0.896	0.897	0.897
Ours	**0.925**	**0.921**	**0.924**	**0.924**

Table 3. MAP comparisons on CUB200-2011 and Stanford Dogs dataset.

Methods	Dim	CUB	Dogs
fc8_im [28]	4096	0.481	0.727
fc8_gtBBox [28]	4096	0.553	0.766
fc8_predBBox [28]	4096	0.531	0.741
SPoC (w/o cen.) [29]	256	0.425	0.559
SPoC (with cen.) [29]	256	0.473	0.557
CroW [30]	256	0.597	0.684
SCDA [31]	1024	0.658	0.752
Ours	**48**	**0.682**	**0.773**

Extracting CNN features take about 0.072 s on the machine with Tesla K40c GPU and 12 GB memory. The search is carried out on the CPU mode with matlab implementation. Performing an euclidean distance measure between two 1024-dimensional vectors take about 0.614 s. In contrast, computing the hamming distance between two 48 bits hash codes take about 0.023 s in the rough retrieval phase, and fine search is about 0.009 s. The total time taken to retrieve the image is about 0.032 s. Thus, ours method is 19.2 $\times$ faster than traditional exhaustive search with 1024-dimensional features.

5 Conclusion

In this paper, we propose a hash image retrieval method based on deep learning and local feature fusion. Compared with the traditional hash algorithm, we can retrieve more images with the same semantics, and the distance of the similar sample is much smaller than the distance of the different samples by optimizing triplet loss function. The hash codes is used to approximate nearest neighbor search, so that the retrieval speed is faster and no longer linear search. At the same time, we fuse features of different layers, so that the final extracted feature not only contains high-level semantic but also has local detail. Experiments show that our method works well on four public datasets including two fine-grained classification datasets compared with other search methods, which have higher precision, faster retrieval and less memory.

Acknowledgement. This work was supported by the Grant of the National Science Foundation of China (No. 61673186, 61370006, 61502183), the Grant of the National Science Foundation of Fujian Province (No. 2013J06014, 2014J01237), the Promotion Program for Young and Middle-aged Teacher in Science and Technology Research of Huaqiao University (No. ZQN-YX108), the Scientific Research Funds of Huaqiao University (No. 600005-Z15Y0016), and Subsidized Project for Cultivating Postgraduates' Innovative Ability in Scientific Research of Huaqiao University (No. 1511314007).

References

1. Lew, M.S., Sebe, N., Djeraba, C., et al.: Content-based multimedia information retrieval: State of the art and challenges. ACM Trans. Multimed. Comput. Commun. Appl. **2**, 1–19 (2006)
2. Lowe D.G.: Object recognition from local scale-invariant features. In: Proceeding of 7th IEEE International Conference on Computer Vision, pp. 1150–1157. IEEE (1999)
3. Ojala, T., Pietikäinen, M., Mäenpää, T.: Gray scale and rotation invariant texture classification with local binary patterns. IEEE Trans. Pattern Anal. Mach. Intell. **24**(7), 971–987 (2002)
4. Dalal, N., Triggs, B.: Histograms of oriented gradients for human detection. In: IEEE Conference on Computer Vision and Pattern Recognition. pp. 886–893 (2005)
5. Oliva, A., Torralba, A.: Modeling the shape of the scene: a holistic representation of the spatial envelope. Int. J. Comput. Vis. **42**(3), 145–175 (2001)
6. Krizhevsky, A., Sutskever, I., Hinton, G.E.: ImageNet classification with deep convolutional neural networks. Adv. Neural. Inf. Process. Syst. **25**(2), 1106–1114 (2012)
7. Simonyan, K., Zisserman, A.: Very deep convolutional networks for large-scale image recognition. In: International Conference on Learning Representations (ICLR) (2014)
8. Szegedy, C., Liu, W., Jia, Y., et al.: Going deeper with convolutions. In: Proceeding of the IEEE Conference on Computer Vision and Pattern Recognition (CVPR), pp. 1–9 (2015)
9. He, K., Zhang, X., Ren, S., et al.: Deep residual learning for image recognition. In: Proceeding of the IEEE Conference on Computer Vision and Pattern Recognition (CVPR), pp. 770–778 (2016)
10. Lin, K., Yang, H.F., Hsiao, J.H., et al.: Deep learning of binary hash codes for fast image retrieval. In: Proceeding of IEEE Conference on Computer Vision and Pattern Recognition Workshops, pp. 27–35 (2015)
11. Slaney, M., Casey, M.: Locality-sensitive hashing for finding nearest neighbors. IEEE Signal Process. Mag. **25**, 128–131 (2008)
12. Weiss, Y., Torralba, A., Fergus, R.: Spectral hashing. In: Conference on Neural Information Processing Systems, Vancouver, British Columbia, Canada, pp. 1753–1760 (2008)
13. Xia, R., Pan, Y., Lai, H., et al.: Supervised hashing for image retrieval via image representation learning. In: AAAI Conference on Artificial Intelligence (2014)
14. Liu, H., Wang, R., Shan, S., Chen, X.: Deep supervised hashing for fast image retrieval. In: IEEE Conference on Computer Vision and Pattern Recognition (CVPR), pp. 2064–2072 (2016)
15. Liu, Z., Luo, P., Qiu, S., et al.: DeepFashion: powering robust clothes recognition and retrieval with rich annotations. In: IEEE Conference on Computer Vision and Pattern Recognition (CVPR), pp. 1096–1104 (2016)
16. Babenko, A., Lempitsky, V.: Aggregating deep convolutional features for image retrieval. In: IEEE International Conference on Computer Vision (ICCV) (2015)
17. Buldas, A., Kroonmaa, A., Laanoja, R.: Keyless signatures' infrastructure: how to build global distributed hash-trees. In: Riis Nielson, H., Gollmann, D. (eds.) NordSec 2013. LNCS, vol. 8208, pp. 313–320. Springer, Heidelberg (2013). doi:10.1007/978-3-642-41488-6_21
18. Schroff, F., Kalenichenko, D., Philbin, J.: FaceNet: A unified embedding for face recognition and clustering. In: Proceeding of IEEE Conference on Computer Vision and Pattern Recognition (CVPR), pp. 815–823 (2015)
19. Lecun, Y., Bottou, L., Bengio, Y., et al.: Gradient-based learning applied to document recognition. Proc. IEEE **86**(11), 2278–2324 (1998)

20. Krizhevsky, A: Learning multiple layers of features from tiny images. Computer Science Department, University of Toronto, Technical report (2009)
21. Wah, C., Branson, S., Welinder, P., Perona, P., Belongie, S.: The Caltech-UCSD Birds-200–2011 Dataset. Computation and Neural Systems Technical report, CNS-TR-2011-0012
22. Khosla, A., Jayadevaprakash, N., Yao, B., Li, F.F.: Novel dataset for fine-grained image categorization. In: IEEE Conference on Computer Vision and Pattern Recognition on First Workshop on Fine-Grained Visual Categorization (FGVC) (2011)
23. Jia, D., Berg, A.C., Li, F.F.: Hierarchical semantic indexing for large scale image retrieval. In: CVPR 2011 IEEE Conference on Computer Vision and Pattern Recognition, Colorado Springs, CO, USA, 20–25 June, pp. 785–792. IEEE Xplore (2011)
24. Gong, Y., Lazebnik, S.: Iterative quantization: A procrustean approach to learning binary codes. In: IEEE Conference on Computer Vision and Pattern Recognition, pp. 817–824. IEEE Computer Society(2011)
25. Zeiler, M.D., Fergus, R.: Stochastic pooling for regularization of deep convolutional neural networks. In: International Conference on Learning Representations (ICLR) 2013
26. Liu, W., Wang, J., Ji, R., et al.: Supervised hashing with kernels. In: IEEE Conference on Computer Vision and Pattern Recognition, vol. 157, no. 10, pp. 2074–2081. IEEE Computer Society (2011)
27. Norouzi, M.E., Fleet, D.J.: Minimal loss hashing for compact binary codes. In: International Conference on Machine Learning (ICML 2011), Bellevue, Washington, USA, June 28–July 2011, pp. 353–360 (2011)
28. Gong, Y., Wang, L., Guo, R., Lazebnik, S.: Multi-scale orderless pooling of deep convolutional activation features. In: Fleet, D., Pajdla, T., Schiele, B., Tuytelaars, T. (eds.) ECCV 2014. LNCS, vol. 8695, pp. 392–407. Springer, Cham (2014). doi:10.1007/978-3-319-10584-0_26
29. Babenko, A., Lempitsky, V.: Aggregating deep convolutional features for image retrieval. In: Proceeding of IEEE Conference on Computer Science (2015)
30. Kalantidis, Y., Mellina, C., Osindero, S.: Cross-Dimensional Weighting for Aggregated Deep Convolutional Features (2016)
31. Wei, X.S., Luo, J.H., Wu, J.: Selective Convolutional Descriptor Aggregation for Fine-Grained Image Retrieval (2016)

Anomaly Detection by Analyzing the Pedestrian Behavior and the Dynamic Changes of Behavior

Maying Shen[1,2], Xinghao Jiang[1,2(✉)], and Tanfeng Sun[1,2]

[1] School of Electronic Information and Electrical Engineering, Shanghai Jiao Tong University, Shanghai, China
{yilia,xhjiang,tfsun}@sjtu.edu.cn

[2] National Engineering Lab on Information Content Analysis Techniques (036001), Shanghai, China

Abstract. Since there is an increasing demand of security and safety assurance for public, anomaly detection has been a great focus in the field of intelligent video surveillance analysis. In this paper, a novel method is proposed for anomaly detection through the analysis of the pedestrian behavior with motion-appearance features and the dynamic changes of the behavior over time. Locality Sensitive Hashing (LSH) functions are used in the method to finally detect the abnormal behaviors. Compared to the relative works, the main novelties of this paper mainly includes: (1) the pedestrians in the image are segmented with the method of Robust Principal Component Analysis (RPCA) in the preprocessing step; (2) in order to describe the dynamic changes of behavior, the Dynamics of Pedestrian Behavior (DoPB) feature on Riemannian manifolds is proposed as the individual descriptor; (3) during the detection process, the Adaptive Anomaly Weight (AAW) with block-based optical flow tracking is used to measure the anomaly saliency. Experimental results and the comparisons with state-of-the-art methods demonstrate that the proposed method is effective in anomaly detection and localization.

Keywords: Anomaly detection · Adaptive Anomaly Weight (AAW) · Pedestrian segmentation · Dynamics of Pedestrian Behavior (DoPB) · Locality Sensitive Hashing (LSH)

1 Introduction

With the growing demand of public security, anomaly detection is an important task in the field of intelligent video surveillance analysis. Since it faces many difficulties such as defining the anomaly in different scenes and detecting anomaly successfully in crowded or complicated scenes, it is still a challenging task. In general, the existing methods for anomaly detection can be divided into global model methods, object based methods and grid based methods.

In global based methods, the input sequences or frames are often analyzed as a whole entity by exploiting the global features. Spatial temporal gradients or optical flow data [1] are commonly used in these methods. Social Force Model (SFM) [2] and other

D.-S. Huang et al. (Eds.): ICIC 2017, Part I, LNCS 10361, pp. 211–222, 2017.
DOI: 10.1007/978-3-319-63309-1_20

models that consider the interaction between pedestrians [3, 4] are often used to model the events. This type of methods often analyze the event from the overall perspective and are efficient in dealing with the crowded groups. But sometimes they are unable to locate the anomalies accurately and have limitations when some anomalies occur simultaneously.

In object based methods, authors focus their attention on the segmented objects. The typical steps mainly include object detection, target tracking and behavior analyzing. Cosar et al. [5] detected abnormal events by analyzing the objects' behaviors from the clustered trajectories. Yuan et al. [6] detected and tracked pedestrians, and used a structural context descriptor to describe each target. Although object based methods have the advantage in analyzing the individual's behavior in a relatively long period of time, they often confront the problem of detecting and tracking of people successfully in crowded scenes. Thus, they are sometimes limited to low density crowd situations.

For grid based methods, video frames are divided into grids and the motion pattern is analyzed within each block. Typical features like histogram of oriented gradient (HOG), histogram of optical flow (HOF) [5, 7] and motion boundary histogram (MBH) [5] are often used as the descriptors. Many models are applied in this type of methods to model the event, such as sparse coding [8–10], mixture of dynamic textures (MDT) model [11, 12], Gaussian model [7, 13] and hashing-based model [14]. The grid based methods usually present the events within the blocks with much lower computation cost compared to the object based methods. However, the effectiveness and the detection accuracy would decrease if an inappropriate block size is applied or the training samples are not sufficient.

In general, most of the existing methods describe the crowd behavior with motion and appearance cues, but completely ignore the dynamic changes of motion and appearance. On the other hand, the detection accuracy still needs to be improved. To tackle these problems, a novel method is proposed in this paper and the main contributions are listed as follows: (1) To deal with the crowded situation, the pedestrians are detected and segmented with the method of Robust Principal Component Analysis (RPCA) rather than the traditional detection methods. (2) In object level, the focus of our work is on the analysis of dynamic motion and appearance of each pedestrian. Thus, the Dynamics of Pedestrian Behavior (DoPB) is proposed as the individual descriptor. (3) Considering the continuity of abnormal events, we propose the Adaptive Anomaly Weight (AAW) to measure the anomaly saliency of the potential anomalous blocks. The rest of the paper is organized as follows. In Sect. 2 we give an introduction to the proposed method. In Sect. 3, the experimental results are provided and the conclusions are drawn in Sect. 4.

2 Proposed Scheme

2.1 Overview

The framework of the proposed method is illustrated in Fig. 1. For a coming input video, it is firstly divided into fixed-size blocks to extract block-based motion-appearance features with the optical flow and gray gradients. For further analysis, the pedestrians are segmented from the image with the method of RPCA.

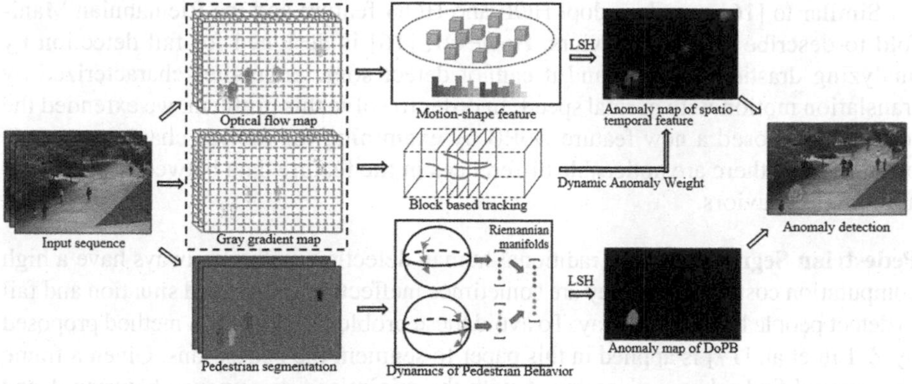

Fig. 1. Framework of the proposed anomaly detection method

Then the novel DoPB features are analyzed on the Riemannian manifolds after matching the segments between consecutive frames. Finally, Locality Sensitive Hashing (LSH) functions and AAW are used to determine the anomaly degree of each block or segment.

2.2 Motion-Appearance Feature Extraction

In our method, each video sequence is splitted into spatio-temporal blocks of fixed size $w \times w \times t$, which are overlapped between the adjacent blocks both in the temporal and spatial domain. In order to reduce the computational complexity and the impact of noise in the background, the foreground of each frame image is extracted using Visual Background Extractor (VIBE) [15]. The spatio-temporal blocks that do not contain any moving pixels are excluded. The optical flow and gray gradients of each pixel are calculated, and the motion and appearance of each block are described separately with HOF and HOG. Then HOF and HOG are connected together to form the final descriptor of the block.

2.3 Dynamics of Motion and Appearance on Riemannian Manifolds

A novel feature called Dynamics of Pedestrian Behavior (DoPB) is proposed in our method to indicate the dynamics of people's motion and appearance, based on the observations that there is a great difference between the dynamic changes of normal behaviors and abnormal activities. Specifically, the appearance and motion of a walking pedestrian change regularly over time, while the movements of wheelchairs, skateboards and cars only contain translational motion and there is a little or almost no change between consecutive frames. Additionally, most anomalies such as people jumping over the gate without payment contain drastically behaviors, resulting in appearance and motion varying significantly during a certain time interval. Inspired by the work of [16], we propose the PoPB feature.

Similar to [16], we also adopt HOF and HOG feature and the Riemannian Manifold to describe human behaviors. However, [16] is designed for fall detection by analyzing drastic behaviors and it cannot detect some anomalies characterized by translation motion with normal speed. In order to solve this problem, we extended the idea and proposed a new feature DoPB by examining the motion change rate. It is observed that there are noticeable differences in the feature data between normal and abnormal behaviors.

Pedestrian Segmentation. Traditional human detection methods always have a high computation cost. Besides, they are sometimes ineffective in crowded situation and fail to detect people located far away. To avoid these problems, the RPCA method proposed by Z. Lin et al. [17] is applied in this paper to segment the pedestrians. Given a frame D, it would find a low rank matrix A with the minimized discrepancy between A and D, following the constrained optimization:

$$\min_{A,E} \|E\|_F, \quad s.t.\ rank(A) \leq r \ and \ D = A + E \tag{1}$$

where the estimated low rank matrix A is considered as the background and the sparse matrix E represents the moving targets. Then we compute gradients of the moving targets and fill the holes to form connected regions. The connected regions are actually the final segmented pedestrians. Figure 2 shows an example of the individual segmentation result.

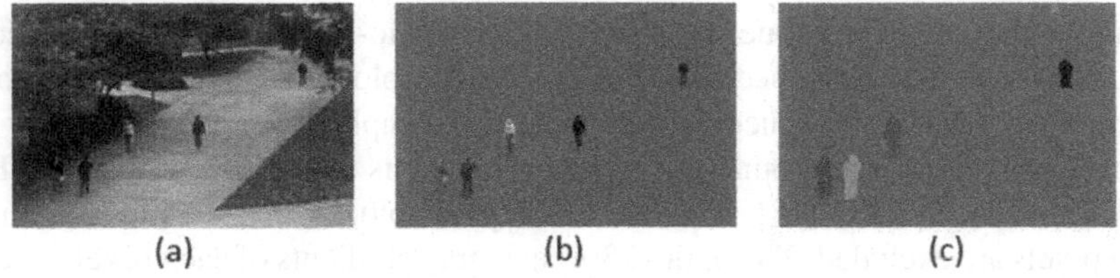

Fig. 2. Pedestrian segmentation. (a) is the original image, (b) shows the extracted moving targets and (c) is the segmented result.

Dynamics of Pedestrian Behavior (DoPB) Extraction. After pedestrian segmentation, the segments need to be matched between consecutive frames in order to analyze their dynamics. Thus, the convex hull H_{tj} is detected for the segment j in frame t and the position of the segment is indicated by the centroid (x_{tj}, y_{tj}) of H_{tj}. Since people can not move far away from their current positions in a short time, we can simply take the segment k in frame t + 1, which contains (x_{tj}, y_{tj}) inside of its corresponding convex hull $H_{(t+1)k}$, as the matched segment of segment j. To preserve the information on the contour lines of the segments, the HOF and HOG features of the pedestrians are calculated inside of convex hulls instead of segments.

As we know, the usual Euclidean calculus and conventional statistics in vector spaces may not apply in nonlinear manifolds. A Riemannian manifold is a smooth manifold and fulfills the requirement of measuring the rate of changes in behavior, since its inner product on the tangent space at each point varies smoothly from point to point. Since human's movement is coherent and smooth, Riemannian Manifold can well model human's behavior. After mapping the HOF and HOG features of the same region in successive frames onto the Riemannian Manifold, the distance between the feature points on the manifold can represent the change of motion and appearance accurately. As the simplest Riemannian manifold, the unit n-sphere is selected in our work. A unit n-sphere is defined as:

$$S^n = \{p \in R^{n+1} \mid \|p\| = 1\} \tag{2}$$

The distance between two points on the manifold is given by the geodesic distance, which in this case is the great-circle distance $d(p, q)$.

$$d(p,q) = arccos(p^T q) \tag{3}$$

To convert the HOF and HOG of the pedestrians to the manifold points, the feature vectors are normalized by the l_2 norm separately. Then the analysis of the dynamics of motion and appearance can be converted to the study of velocity statistics on the manifolds. It is worth mentioning that since the feature vectors are normalized by the l_2 norm, they are less sensitive to the perspective distortion and the size of the targets. This can lead to a simple and effective analysis. As we mentioned before, the point representing the drastic behavior moves fast, almost remains still when it is a translational motion, and moves regularly in the normal occasion. It would be easy to tell them apart.

Let S_{nf} be the manifold for the motion and S_{ng} be the manifold for the appearance. HOF and HOG descriptors are separately converted to the points on S_{nf} and S_{ng}. Here we would describe the processing steps of HOF descriptor in detail. As for HOG, it shares exactly the same steps as HOF and won't be covered again. For convenience, the motion manifold would be noted as S_n during the explanation. The normalized HOF of H_{tj} is denoted as the point p_{tj} on S_n, and its connected point that is converted from the matched convex hull $H_{(t+1)k}$ is denoted as $p_{(t+1)k}$. Then the velocity of segment j in frame t is defined as:

$$v_t^j = d\left(p_{tj}, p_{(t+1)k}\right) / \Delta t \tag{4}$$

where Δt is set to 1. To amplify the variance difference between the set of low velocities and the set of high velocities, all the velocities obtained by (4) are normalized by:

$$z_t^j = a \cdot (1 - 1/(1 + e^{-\lambda \cdot v_t^j})) \tag{5}$$

where λ is the parameter to adjust the constraint. The value of a is set to 2 to keep the projected value in the range of [0, 1]. In such a way, the variance of low velocities tends to be a large one, while it is small for the set of high velocities. Meanwhile, it still keeps the large difference on the mean value between the two sets. Thus, each segment in frame t contains a series of normalized velocities during a certain period of time T, and the constructed set of segment j is denoted as z^j:

$$z^j = \{z^j_t, z^j_{t+1}, \cdots, z^j_{t+T}\} \tag{6}$$

Then the overall dynamics during the time are described by the mean value, the variance, the maximum and the minimum value of the set. The feature vector is defined as:

$$f^j = [E[z^j], Var[z^j], max[z^j], min[z^j]] \tag{7}$$

Finally, the novel feature DoPB is constructed by concatenating the feature vectors of motion and appearance:

$$F^j = [f^j_m, f^j_a] \tag{8}$$

where f^j_m is the dynamic motion feature vector of segment *j* obtained by (7) and f^j_a is for dynamic appearance.

3 Anomaly Detection

In this part, the LSH filter model is applied to model the events. Besides, we propose AAW to measure the anomaly saliency. The key idea is presented here.

Locality Sensitive Hashing (LSH). As an important technique for fast approximate similarity search in large high-dimensional databases, LSH is wildly used in recent years. In this paper, hash functions based on p-stable distributions are selected as described in [14]. According to the key idea of LSH, the feature vectors for training are hashed into a set of buckets and each bucket collects the similar data points together. Since the feature vectors of abnormal behaviors, in ideal situation, are far away from the normal data points in the feature space, they can be projected into different buckets. The abnormal degree would be calculated with the distance between the testing bucket and its nearest training bucket as Zhang et al. did in [14]. As shown in Fig. 1, the anomaly degrees of motion-appearance feature and DoPB are computed separately in our work, and are fused by taking the sum of the values as the final anomaly degree. Then the anomaly is detected and localized by setting a threshold to the anomaly degree.

Adaptive Anomaly Weight (AAW) with Block-Based Optical Flow Tracking. Based on the fact that people who act unusually over a period of time would be anomalous in the upcoming period with a much higher probability, the AAW is proposed in this paper to focus more on the spatio-temporal blocks that are judged as abnormal during the detection. The reason why we do not apply the AAW to the anomaly detection of DoPB is that DoPB has already contained the tracked pedestrian's dynamics in a relatively long time and it is unnecessary to re-track again.

To represent the movement direction of a block B_i, the optical flow angles of the block pixels are quantified into 8 orientations. The most frequently appeared one, denoted as $O(\alpha)$, is selected as the orientation of the block. When in the testing phase, once B_i is determined as an abnormal block in successive l frames, it will be regarded as a potential target. The adjacent block B_j, which B_i is moving forward to according to

O(α), is regarded as the targeting block. It is assumed that B_i and B_j would be more likely to be abnormal in the upcoming frames. Thus, in the flowing frames, the anomaly degrees of B_i and B_j are both weighted by:

$$AAW_{i,j} = 1 + \beta \tag{9}$$

where $\beta > 0$ increases with the value of l increases, which means that the longer the abnormal behavior lasts, the higher anomaly probability it has. In order to reduce the false alarm, once the block is detected as a normal one, the AAW will be set to 1. Only when it is abnormal in another successive l frames, it is going to tracked and weighted again.

4 Experimental Analysis and Discussion

4.1 Experimental Dataset

Two commonly used datasets, which are the UCSD dataset and subway dataset, are used to measure the performance of the proposed method. The UCSD anomaly detection dataset [12] contains video clips captured in two different scenes, namely ped1 and ped2. Each clip contains around 200 frames. Ped1 contains 34 training video samples and 36 testing video samples, with the resolution of 158 × 238. Ped2 contains 16 training video samples and 12 samples for testing, with the 240 × 360 resolution. The anomalies in this dataset include bikers, skaters, small carts, wheelchairs, cars, people walking on the grass and people walking across the walkway. The subway dataset is provided by Adam et al. [18]. It has two surveillance videos with the resolution of 512 × 384, which respectively monitor the subway entrance gate and exit gate. According to [19], the anomalies can be classified as: (1) wrong direction (WD); (2) no payment (NP); (3) loitering (LT); (4) irregular interaction (II); (5) misc (MISC).

4.2 Implementation Details

Parameter Setting. The size of spatio-temporal blocks is set to 10 × 10 × 5 for UCSD ped1 clips, and it is set to 20 × 20 × 5 for all the other clips. The overlapping proportion between contiguous blocks in spatial domain is set to 50%. The value of the time duration *T* in DoPB feature is set to 10 for UCSD and 20 for subway dataset according to the fps of the video. λ is the parameter to adjust the constraint during the feature projection, and the curvature of formula (5) in the range of [0, 1] increases when the value of λ increases, leading to a larger difference between the normal and abnormal data. We set it as 8 since we found that continue increasing the value do little contribution to the detection accuracy. Only normal samples are used for training.

Evaluation Criterions. In our work, frame-level and pixel-level measurements are adopted to evaluate the performance of the method. The existence of anomaly is regarded as a "positive" and it is a "negative" on the contrary. For frame-level, it would be considered to be an anomaly if the frame contains at least one abnormal pixel. For pixel-level, it would be a detection only when the detected area covers more than 40% of the anomalous pixels in the ground truth. Then the Receiver Operating Characteristic (ROC) curve is generated by taking different value of the determinate threshold. The Area Under Curve (AUC) and Equal Error Rate (EER) are taken as the evaluation criteria.

4.3 Performance

Performance on UCSD. Figure 3 shows some examples of the detection results of the proposed method. We can see from the figures that our method is able to detect and accurately locate the anomalies occurred in this dataset, such as cars (Fig. 3(d), (h)), bikers (Fig. 3(a), (f), (g)), skateboarders (Fig. 3(b), (j)), wheelchairs with low velocity (Fig. 3(e)) and walking across the walkway (Fig. 3(c)).

Fig. 3. Examples of anomaly detection results on UCSD dataset. The first row shows examples of ped1 and the second row shows examples of ped2.

We compare the proposed method with the state-of-the-art methods on UCSD dataset in Fig. 4. The ROC curves both in frame-level and pixel-level are shown in the figure. It can be seen from the ROC curves that the proposed method significantly outperforms other methods overall. Especially the pixel-level comparison results indicate that our method locates the anomalies more accurately.

Based on the ROC curves, the AUC and EER of these methods on UCSD are calculated and listed in Table 1. The results that are not provided by the authors are marked as "-". The results of methods proposed in the early years in Fig. 4 are also not included in the table. We can see from the table that the AUC in pixel-level of our method reaches 0.80 on ped1, and the values of all the four metrics reach the best among these methods on ped2 (0.91/0.15 and 0.94/0.12 AUC/EER in pixel and frame level separately). Comparing with the state-of-the-art approaches [6, 7, 14], the proposed method has the best anomaly localization ability (in terms of the average AUC and EER) and scene universality. For [6, 14], the proposed method can achieve better AUC and EER in UCSD dataset, especially in pixel-level. The results of pixel-level on ped2 are not provided by [6], while our method achieves the best results among these methods. As for [7], it outperforms ours when performed on USCD ped1 in frame level. However, the datasets used in [7] are all captured from scenes where people walk in a vertical

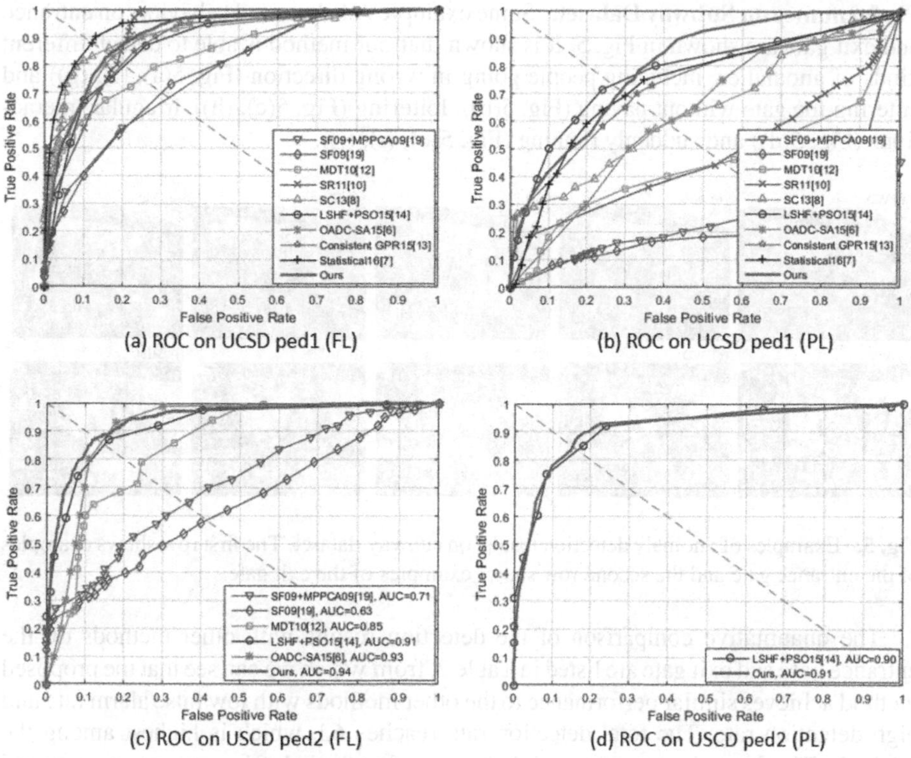

Fig. 4. The ROC performance of anomaly detection and localization on UCSD dataset on the frame-level (FL) and pixel-level (PL).

direction towards the camera. It did not provide the experimental results on scenes where people move parallel to the camera plane such as UCSD ped2. We may think that our method is more applicable to variant scenes. Besides, the detection results in pixel-level of our method are better than [7], which indicates that our method has a better ability of anomaly localization.

Table 1. Comparison of the proposed method with the other methods on the UCSD dataset on the pixel-level (PL) and the frame-level (FL).

Method	UCSD ped1				UCSD ped2			
	AUC (PL)	EER (PL)	AUC (FL)	EER (FL)	AUC (PL)	EER (PL)	AUC (FL)	EER (FL)
OADC-SA15 [6]	0.75	-	0.91	-	-		0.93	0.15
Statistical16 [7]	0.731	0.315	0.937	0.121	-	-	-	-
LSHF+PSO15 [14]	0.77	0.27	0.87	0.20	0.9	0.17	0.91	0.15
Proposed	0.80	0.29	0.91	0.16	0.91	0.15	0.94	0.12

As a result, among the four methods investigated, the proposed algorithm achieves an overall best performance in anomaly detection, anomaly localization and scene universality.

Performance on Subway Dataset. Some examples of abnormal behaviors on entrance and exit gate are shown in Fig. 5. It is shown that our method is able to detect different kinds of anomalies, including people going in wrong direction (Fig. 5(a), (f), (g)) and entering the gate without paying (Fig. 5(b)), loitering (Fig. 5(c), (h)), irregular interactions (Fig. 5(d)) and suddenly running (Fig. 5(e), (i), (j)).

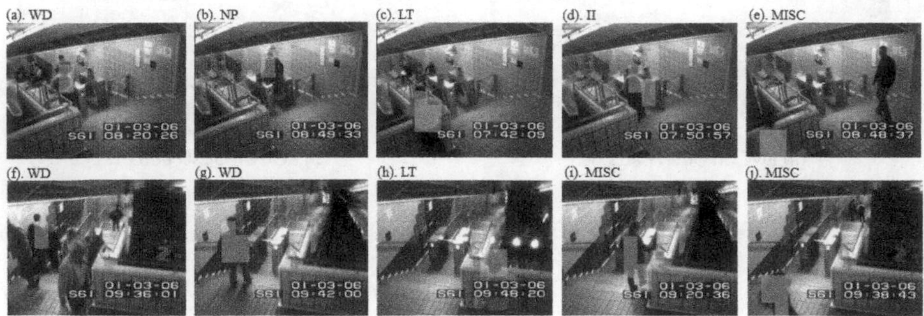

Fig. 5. Examples of anomaly detection results on subway dataset. The first row shows examples of the entrance gate and the second row shows examples of the exit gate.

The quantitative comparison of the detection results with other methods on the entrance gate and exit gate are listed in Table 2, from which we can see that the proposed method achieves similar performance to the other methods with low false alarm rate and high detection rate. The total detection rate reaches 62, which is highest among the methods. The false alarm occurs mainly because that the behavior patterns are new and they are not included in the training samples. Obviously it would be better if enough training samples are used in a practical application.

Table 2. Comparison of anomaly detection rate and false alarm rate on subway dataset

Method	Entrance gate							Exit gate				
	WD	NP	LT	II	MISC	Total	FA	WD	LT	MISC	Total	FA
Ground truth	26	13	14	4	9	66	0	9	3	7	19	0
Statistical16 [7]	24	9	13	4	8	58	4	9	3	7	19	3
LSHF+PSO15 [14]	25	9	14	4	9	61	4	9	3	7	19	2
Proposed	25	10	14	4	9	62	5	9	3	7	19	2

4.4 Discussion

When performing the proposed method on the experimental datasets, we can clearly find that some anomalies such as wheelchairs and some skateboarding behaviors have a low velocity similar to the walking speed. In addition, the low resolutions make it hard to recognize through the appearance. The proposed DoPB helps a lot since it can distinguish the translation motion and drastic motion. Moreover, there is a phenomenon that the anomaly would be failed to detect in some frames during the continuous behavior because of the relatively low anomaly degree. The AAW solves the problem by tracking the anomalous blocks and amplifying its anomaly degree in the following time. The

DoPB and AAW are indeed beneficial for detection accuracy improving. The experimental results on UCSD and subways dataset prove that the proposed method is feasible and effective.

5 Conclusion

In this paper, an anomaly detection method is proposed, which takes the motion-appearance features and the analysis of pedestrian-based dynamics into consideration. The method of pedestrian segmentation and the novel feature DoPB are proposed to represent the dynamics of motion and appearance in a period of time. Then the anomaly maps of motion-appearance feature and DoPB are obtained separately by using the method of LSH. When in the detection phase, the AWW with block-based optical flow tracking is proposed to determine the anomaly saliency of the blocks. The experimental results indicate the efficiency of DoPB and AAW on accuracy improving. The comparison with the state-of-the-art methods proves that the proposed method is able to detect and locate the anomalies accurately and it achieves a good performance.

Acknowledgement. This work was supported by the National Natural Science Foundation of China (No. 61572321, 61572320). Corresponding author is Prof. Xinghao Jiang, any comments should be addressed to xhjiang@sjtu.edu.cn.

References

1. Yi, S., Li, H., Wang, X.: Pedestrian behavior modeling from stationary crowds with applications to intelligent surveillance. IEEE Trans. Image Process. **25**(9), 4354–4368 (2016)
2. Mehran, R., Oyama, A., Shah, M.: Abnormal crowd behavior detection using social force model. In: IEEE Conference on Computer Vision and Pattern Recognition, pp. 935–942 (2009)
3. Lee, D.G., Suk, H.I., Lee, S.W.: Modeling crowd motions for abnormal activity detection. In: IEEE International Conference on Advanced Video and Signal Based Surveillance, pp. 325–330 (2014)
4. Cui, X., Liu, Q., Gao, M., Metaxas, D.N.: Abnormal detection using interaction energy potentials. In: IEEE Conference on Computer Vision and Pattern Recognition, pp. 3161–3167 (2011)
5. Cosar, S., Donatiello, G., Bogorny, V., Garate, C., Alvares, L.O., Bremond, F.: Towards abnormal trajectory and event detection in video surveillance. IEEE Trans. Circ. Syst. Video Technol. **PP**, 1 (2016)
6. Yuan, Y., Fang, J., Wang, Q.: Online anomaly detection in crowd scenes via structure analysis. IEEE Trans. Cybern. **45**(3), 562–575 (2015)
7. Yuan, Y., Feng, Y., Lu, X.: Statistical hypothesis detector for abnormal event detection in crowded scenes. IEEE Trans. Cybern. **PP**, 1–12 (2016)
8. Lu, C., Shi, J., Jia, J.: Abnormal event detection at 150 fps in matlab. In: IEEE Conference on Computer Vision and Pattern Recognition, pp. 2720–2727 (2013)
9. Wen, H., Ge, S., Chen, S., Wang, H.: Abnormal event detection via adaptive cascade dictionary learning. In: IEEE International Conference on Advanced Video and Signal Based Surveillance, pp. 847–851 (2015)

10. Cong, Y., Yuan, J., Liu, J.: Sparse reconstruction cost for abnormal event detection. In: IEEE Conference on Computer Vision and Pattern Recognition, pp. 3449–3456 (2011)
11. Li, W., Mahadevan, V., Vasconcelos, N.: Anomaly detection and localization in crowded scenes. IEEE Trans. Pattern Anal. Mach. Intell. **36**(1), 18–32 (2014)
12. Mahadevan, V., Li, W., Bhalodia, V., Vasconcelos, N.: Anomaly detection in crowded scenes. In: IEEE Conference on Computer Vision and Pattern Recognition, pp. 1975–1981 (2010)
13. Cheng, K.W., Chen, Y.T., Fang, W.H.: Gaussian process regression-based video anomaly detection and localization with hierarchical feature representation. IEEE Trans. Image Process. **24**(12), 5288–5301 (2015)
14. Zhang, Y., Lu, H., Zhang, L.: Video anomaly detection based on locality sensitive hashing filters. Pattern Recogn. **59**, 302–311 (2015)
15. Van Droogenbroeck, M., Barnich, O.: ViBe: a disruptive method for background subtraction. In: Background Modeling and Foreground Detection for Video Surveillance, pp. 7.1–7.23 (2014)
16. Yun, Y., Gu, I.Y.H.: Human fall detection in videos by fusing statistical features of shape and motion dynamics on riemannian manifolds. Neurocomputing **207**, 726–734 (2016)
17. Lin, Z., Chen, M., Ma, Y.: The augmented lagrange multiplier method for exact recovery of corrupted low-rank matrices. Eprint arXiv, vol. 9 (2009)
18. Adam, A., Rivlin, E., Shimshoni, I., Reinitz, D.: Robust real-time unusual event detection using multiple fixed-location monitors. IEEE Trans. Pattern Anal. Mach. Intell. **30**(3), 555–560 (2008)
19. Kim, J., Grauman, K.: Observe locally, infer globally: a space-time MRF for detecting abnormal activities with incremental updates. In: IEEE Conference on Computer Vision and Pattern Recognition, pp. 2921–2928 (2009)

Locality Preserving Projections with Adaptive Neighborhood Size

Wenjun Hu[1,2], Xinmin Cheng[1], Yunliang Jiang[1], Kup-Sze Choi[2], and Jungang Lou[1(✉)]

[1] School of Information Engineering, Huzhou University, Huzhou 313000, Zhejiang, China
loujungang0210@hotmail.com

[2] Centre for Smart Heath, School of Nursing, Hong Kong Polytechnic University, Hong Kong, China

Abstract. Feature extraction methods are widely employed to reduce dimensionality of data and enhance the discriminative information. Among the methods, manifold learning approaches have been developed to detect the underlying manifold structure of the data based on local invariants, which are usually guaranteed by an adjacent graph of the sampled data set. The performance of the manifold learning approaches is however affected by the locality of the data, i.e. what is the neighborhood size for suitably representing the locality? In this paper, we address this issue through proposing a method to adaptively select the neighborhood size. It is applied to the manifold learning approach Locality Preserving Projections (LPP) which is a popular linear reduction algorithm. The effectiveness of the adaptive neighborhood selection method is evaluated by performing classification and clustering experiments on the real-life data sets.

Keywords: Dimensionality reduction · Feature extraction · Neighborhood size · Locality preserving projections

1 Introduction

High dimensional data are prevalent in many application domains, such as pattern recognition, information retrieval, text categorization, computer vision and data mining. Many feature extraction techniques have been proposed in the past few decades. The most well-known techniques include Principle Component Analysis (PCA) [1, 2], Linear Discriminant Analysis (LDA) [3, 4], Local Structure Preserving Discriminant Analysis (LSPDA) [5], Local Learning Projection (LLP) [6], and Discriminative Local Learning Projection (DLLP) [7], Low-Rank Preserving Projection (LRPP) [8], Sparse Linear Embedding [9]. PCA is an unsupervised dimensionality reduction technique used to find a set of orthogonal bases by which the global information of the data in the principle component space is captured. The principle component space is obtained by solving an eigenvalue problem, which is mathematically equivalent to performing Singular Value Decomposition (SVD) in the centered data matrix. LDA is a supervised dimensionality reduction approach used to obtain the optimal projection or transformation subspace of the data, which is obtained by minimizing the within-class scatter

D.-S. Huang et al. (Eds.): ICIC 2017, Part I, LNCS 10361, pp. 223–234, 2017.
DOI: 10.1007/978-3-319-63309-1_21

matrix and maximizing the between-class scatter matrix. Since LDA only can captures global geometrical structure information of the data, Zhang et al. point three kinds of different local information (including local similarity information, local intra-class pattern variation, and local inter-class pattern variation) and incorporate them to LDA, called Complete Global-Local LDA (CGLDA) [10]. LSPDA respectively constructs the local scatter and the between-class scatter to characterize the sub-manifold and multi-manifold information. In LSPDA, the local scatter considers both the local information and the class information through different similarity between the - same-class points or between the not-same-class points, and the between-class scatter depicts the importance degrees of the not-same-class points. LLP seeks projections which can preserve the local structure of data by minimizing the local estimation error. DLLP is based on LLP and further developed for face recognition, which computes a linear transformation simultaneously through minimizing the local estimation error and maximizing the dissimilarities among all the manifolds. Furthermore, by considering the "locality" and "nonlocality", Yang et al. [11] proposed a novel method, called Unsupervised Discriminant Projection (UDP) for feature extraction, which is based on the local scatter and the nonlocal scatter. UDP is an unsupervised method and Wang [12] proposed kernel-based supervised discriminant projection (KSDP) aiming to apply it in the supervised scene.

Recently, researchers consider that the data sampled from a probability distribution may be on, or in close proximity to, a low-dimensional manifold of the ambient space [13–16]. As a result, many manifold learning algorithms have been proposed to deal with such kind of data, including ISOMAP [13], Laplacian Eigenmap (LE) [15], Locally Linear Embedding (LLE) [14], Locality Preserving Projections (LPP) [17], Rotational-Invariance-based LPP (RILPP) [18] and Two-Dimensional Maximum Embedding Difference (2DMED) [19]. The former two methods are nonlinear algorithms, while the latter two are linear algorithms. Generally, most of the manifold methods make use the notion of local invariance to detect the underlying manifold structure so that the low-dimensional representations can be obtained. Besides, the local invariance is guaranteed by usually using an adjacent graph of the sampled data set. Here, the construction of valid adjacent graph becomes an important issue which directly impacts the performance of the manifold learning approaches. That is to say, what can be considered as local? Therefore, it is noteworthy to select the neighborhood size to construct the adjacent graph for matching the local geometry of the manifold. In this paper, we focus on this problem and propose a method to adaptively select the neighborhood size, which is validated in the original LPP method.

The main contributions highlight as follows:

(1) The nearest neighbor of a sample is estimated through probability density of a given data set. Then, the neighborhood corresponding to this sample can be obtained. Specifically, the neighborhood's center is this sample and its radius is the distance between the estimated nearest neighbor and this sample. Through this neighborhood, the set of nearest neighbors can be detected. Especially, for the point on the different cluster's boundary, the obtained nearest neighbors may be more likely to come from the same cluster, which contributes to improve the performance in supervised learning and un-supervised learning.

(2) The method for constructing the set of nearest neighbors is integrated into LPP and a new LPP algorithm with adaptive neighborhood size is proposed. This improves the performance of LPP.

2 Locality Preserving Projections

LPP is one of the most popular manifold learning methods which are based on the spectral graph theory [20]. Generally speaking, the aim of manifold learning method is to find an optimal map to preserve the intrinsic geometry structure of the data manifold so that the geometry structure information of the data can be discovered. Let $\mathbf{y} = [y_1, \cdots, y_N]$ be the one-dimensional map of the data set $\mathbf{X}$. Given a k-nearest neighbor graph G with weight matrix $\mathbf{W}$, LPP attempts to obtain the optimal maps by solving the following optimization criterion:

$$\min \sum_{ij} (\mathbf{v}^T\mathbf{x}_i - \mathbf{v}^T\mathbf{x}_j)^2 \mathbf{W}_{ij}. \tag{1}$$

where $\mathbf{v} \in \Re^D$ is a basis vector (i.e. $y_i = \mathbf{v}^T\mathbf{x}_i$) and $\mathbf{W} = [\mathbf{W}_{ij}]$ is a weight matrix which reveals the neighborhood relationship between the data points. The weight matrix $\mathbf{W}$ can be defined as:

$$\mathbf{W}_{ij} = \begin{cases} \exp(-\|\mathbf{x}_i - \mathbf{x}_j\|^2 / t) & \text{if } \mathbf{x}_i \in N(\mathbf{x}_j) \text{ or } \mathbf{x}_j \in N(\mathbf{x}_i), \\ 0 & \text{otherwise.} \end{cases} \tag{2}$$

where $\exp(-\|\mathbf{x}_i - \mathbf{x}_j\|^2 / t)$ is the heat kernel function with a kernel width parameter t and $N(\mathbf{x}_j)$ denotes the set of k-nearest neighbors of $\mathbf{x}_j$. For simplicity, the weight matrix can also be defined with the so-called 0–1 weight, i.e.

$$\mathbf{W}_{ij} = \begin{cases} 1 & \text{if } \mathbf{x}_i \in N(\mathbf{x}_j) \text{ or } \mathbf{x}_j \in N(\mathbf{x}_i), \\ 0 & \text{otherwise.} \end{cases} \tag{3}$$

Clearly, the minimization criterion in (1) is an attempt to ensure that the points y_i and y_j are also close to each other if $\mathbf{x}_i$ and $\mathbf{x}_j$ are "close". That is to say, the obtained optimal map attempts to preserve the local structure of the data. In fact, the purpose of LPP is to find a group of basis vectors $\mathbf{V} = [\mathbf{v}_1, \cdots, \mathbf{v}_d] \in \Re^{D\times d}$ and obtain a subspace to preserve the local structure of the data set. With the constraint $\mathbf{v}^T\mathbf{XDX}^T\mathbf{v} = 1$ and by some algebraic steps, the objective function in (1) can be reduced to a generalized minimum eigenvalue problem $\mathbf{XLX}^T\mathbf{v} = \lambda\mathbf{XDX}^T\mathbf{v}$, where $\mathbf{L} = \mathbf{D} - \mathbf{W}$ is the Laplacian matrix [17, 19], and $\mathbf{D}$ is a diagonal matrix whose entries along the diagonal are the column sum of $\mathbf{W}$, i.e. $\mathbf{D}_{ii} = \sum \mathbf{W}_{ij}$. Details of the solution to the eigenvalue problem can be found in [17].

3 LPP with Adaptive Neighborhood Size

3.1 Estimation of Nearest Neighbor

Given a sample space $\Gamma \subset \Re^D$, let $\mathbf{X} = [\mathbf{x}_1, \cdots, \mathbf{x}_N] \subset \Gamma$ denote the data set with $\mathbf{x}_i \in \Re^D$, which is sampled from a d-dimension submanifold embedded in $\Re^D$. Suppose we have known the nearest neighbor $\mathbf{z}_i$ of each sample $\mathbf{x}_i$ in Γ, then we can determine whether $\mathbf{x}_{j \neq i}$ is the neighbor of $\mathbf{x}_i$ by comparing $\|\mathbf{x}_j - \mathbf{x}_i\|$ and $\|\mathbf{z}_i - \mathbf{x}_i\|$. Clearly, $\mathbf{x}_{j \neq i} \in N(\mathbf{x}_i)$, if $\|\mathbf{x}_j - \mathbf{x}_i\| \leq \|\mathbf{z}_i - \mathbf{x}_i\|$, and $\mathbf{x}_{j \neq i} \notin N(\mathbf{x}_i)$ otherwise, where $N(\mathbf{x}_i)$ denotes the set of nearest neighbors of $\mathbf{x}_i$. Hence, our goal is to estimate the nearest neighbor $\mathbf{z}_i$ of each sample $\mathbf{x}_i$ in Γ, which will be described in the following.

For the nearest neighbors of a given sample $\mathbf{x}_i$, we can treat them as hidden random variables. By employing the principles of the expectation-maximization algorithm [21], the nearest neighbor $\mathbf{z}_i$ can be estimated by computing the expectation of $\mathbf{x}_i$, i.e. averaging out the hidden variables, which is given by

$$\mathbf{z}_i = \mathrm{E}_{\mathbf{x}_i \sim \widetilde{\mathbf{X}}_i}(\mathbf{x}_j) = \sum_{\mathbf{x}_j \in \widetilde{\mathbf{X}}_i} p(\mathbf{x}_j)\mathbf{x}_j, \tag{4}$$

where $\widetilde{\mathbf{X}}_i = [\mathbf{x}_1, \cdots, \mathbf{x}_{i-1}, \mathbf{x}_{i+1}, \cdots, \mathbf{x}_N]$ and $\mathbf{x}_j \in \widetilde{\mathbf{X}}_i$, $\mathrm{E}_{\mathbf{x}_i \sim \widetilde{\mathbf{X}}_i}$ denotes the expectation computed with respect to $\widetilde{\mathbf{X}}_i$, and $p(\mathbf{x}_j)$ is the probability of $\mathbf{x}_j$ being the nearest neighbor of $\mathbf{x}_i$. The probability can be obtained based on standard kernel density estimation [22] as $p(\mathbf{x}_j) = k_\sigma(\mathbf{x}_i, \mathbf{x}_j) \Big/ \sum_{\mathbf{x}_l \in \widetilde{\mathbf{X}}_i} k_\sigma(\mathbf{x}_i, \mathbf{x}_l)$, where $k_\sigma(\mathbf{x}_i, \mathbf{x}_j)$ is a kernel function with a width parameter σ. All of the kernel functions described in [23] can be used for the estimation, and especially the Gaussian kernel is adopted in this study, which is formulated as

$$k_\sigma(\mathbf{x}_i, \mathbf{x}_j) = \exp(-\|\mathbf{x}_i - \mathbf{x}_j\|^2 / 2\sigma^2) \tag{5}$$

After the nearest neighbor is obtained by above estimation, the samples in $\mathbf{X}$ that belong to the nearest neighbors $N(\mathbf{x}_i)$ of the data $\mathbf{x}_i$ are then determined.

3.2 Adaptive Neighborhood Size for LPP

By using the above technique, the neighborhood size to LPP can be adaptively selected. Then, LPP algorithm is also extended to yield a new algorithm with adaptive neighborhood size. The new LPP algorithm is denoted $\mathrm{LPP_{ANS}}$ and described as follows:

Algorithm 1. LPP$_{ANS}$
Input: Data matrix $\mathbf{X}=[\mathbf{x}_1,\cdots,\mathbf{x}_N]\in\Re^{D\times N}$, kernel width parameter σ; **Output**: Basis vectors $\mathbf{V}=[\mathbf{v}_1,\cdots,\mathbf{v}_d]\in\Re^{D\times d}$.
1. Estimate the nearest neighbor $\mathbf{z}_i$ of each data $\mathbf{x}_i$ using (4); **2.** Find the neighbors $N(\mathbf{x}_i)$ of the data $\mathbf{x}_i$ in $\mathbf{X}$ by comparing $\lVert\mathbf{x}_j-\mathbf{x}_i\rVert$ and $\lVert\mathbf{z}_i-\mathbf{x}_i\rVert$, i.e. $\mathbf{x}_{j\neq i}\in N(\mathbf{x}_i)$, if $\lVert\mathbf{x}_j-\mathbf{x}_i\rVert\le\lVert\mathbf{z}_i-\mathbf{x}_i\rVert$; **3.** Compute the corresponding weight matrix $\mathbf{W}$ using (2) or (3); **4.** Compute the diagonal matrix $\mathbf{D}$ ($\mathbf{D}_{ii}=\sum\mathbf{W}_{ij}$) and the Laplacian matrix $\mathbf{L}=\mathbf{D}-\mathbf{W}$; **5.** Compute the eigenvectors and eigenvalues for the generalized eigenvector problem $\mathbf{XLX}^T\mathbf{v}=\lambda\mathbf{XDX}^T\mathbf{v}$; **6.** Output the basis vectors $\mathbf{V}=[\mathbf{v}_1,\cdots,\mathbf{v}_d]$ with the eigenvectors corresponding to the smallest d eigenvalues.

Note: In Step 2, if no neighbor of the data $\mathbf{x}_i$ has been obtained, the $k_{\min}$ points nearest to $\mathbf{x}_i$ will be considered as its neighbors $N(\mathbf{x}_i)$ for the stability of the LPP algorithm. The value of $k_{\min}$ is fixed at 2 in our experiments.

Clearly, in LPP$_{ANS}$, the size of the neighbors may be different for different point. Suppose this scene, a data set is sampled from a manifold and the manifold consists of different clusters. In this case, the k-nearest neighbors for the point $\mathbf{x}_i$ on the boundary may come from different cluster. If the set of nearest neighbors is generated through adaptive neighborhood size, the nearest neighbors may be more likely to come from the same cluster, which contributes to improve the performance in supervised learning and un-supervised learning.

4 Experiment Results

To investigate the effectiveness of LPP with adaptively selected neighborhood size (LPP$_{ANS}$), experiments are carried out to evaluate its classification and clustering performance respectively. The results are compared with that of PCA and LPP with different neighbor size.

4.1 Data Preparation and Parameter Set

Two real world data sets are employed for the experiments. The first data set is the COIL20 image library from the Columbia University. It is used in the classification experiment. The data set contains 1440 images generated from 20 objects. Each image is represented by a 1024-dimensional vector, and the size is 32 × 32 pixels with 256 grey levels per pixel. The second data set is obtained from the PIE face database, which is used in the clustering experiment. The PIE face images are created under different poses, illuminations and expressions. This database contains 41,368 images of 68 subjects. The image size is 32 × 32 pixels with 256 grey levels. In our experiment, we select 1428 images of different illuminations for clustering. In the following sections, the classification experiment is first discussed, followed by the clustering experiment.

In the experiment of LPP_{ANS}, the width parameter σ in Gaussian kernel is selected from the grids $\{2^{-7}s, 2^{-6}s, \cdots, 2^{7}s\}$, where s is the mean norm of all training data. And the weight matrix in LPP_{ANS} and LPP, is computed by Eq. (3).

4.2 Classification

In this experiment, we employ the 1-nearest neighbor (1-NN) classifier to evaluate the discriminating power of the features that are extracted in four different ways, including the proposed LPP_{ANS}, LPP with neighborhood size k = 3 ($LPP_{k=3}$), LPP with neighborhood size k = 10 ($LPP_{k=10}$) and the PCA method. Half of the images in the COIL20 library are randomly selected and used for training. The remaining images are used for testing. The test is performed with different number of the extracted features in the range of {5, 10, 20, …, 250}. For each feature number in such range, the test is executed for 10 times to evaluate the average performance. Here, the performance metric is the classification accuracy, which is computed as

$$\text{Accuracy} = \frac{1}{n}\sum_{i=1}^{n}\delta(l_o(\mathbf{x}_i), l_t(\mathbf{x}_i)),$$

where $\mathbf{x}_i$ is the test sample, $l_o(\mathbf{x}_i)$ is the true class label of $\mathbf{x}_i$, $l_t(\mathbf{x}_i)$ is the class label obtained by 1-NN, n is the size of the testing data set, and the function $\delta(l_o(\mathbf{x}_i), l_t(\mathbf{x}_i))$ equals 1 if $l_o(\mathbf{x}_i) = l_t(\mathbf{x}_i)$ and 0 otherwise. The higher the classification accuracy, the better the extracted features.

Figure 1 shows the classification results on the COIL20 data set. It can be seen that the proposed method of adaptive neighborhood size selection improves the LPP algorithm. The classification accuracy of LPP_{ANS} is higher than that of the other four methods. The performance of LPP_{ANS} also remains stable for feature number in the range of 5 to 250 features. Among the five methods, the classification accuracy of $LPP_{k=10}$ is lowest and varies considerably with the number of extracted features. When the number of extracted features is 220, the accuracy of the $LPP_{k=3}$ achieves the same classification accuracy as that of 1-NN (where all the features of the data set is used in 1-NN). The accuracy of LPP_{ANS} exceeds that of 1-NN only with the mere of 5 features, and that of PCA with the mere of 15 features.

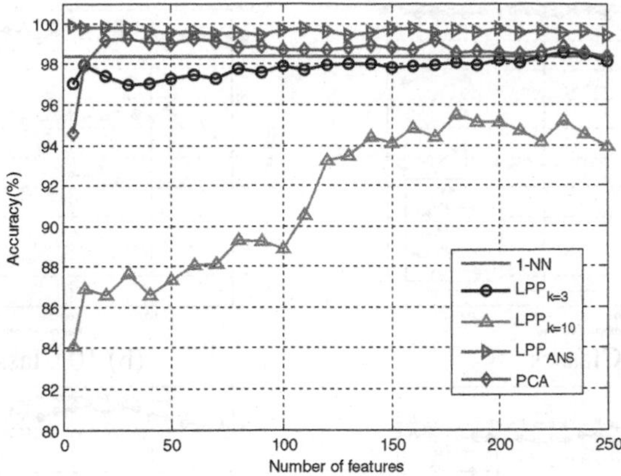

Fig. 1. Classification accuracy versus the number of extracted features on COIL20 data set

4.3 Clustering

In this experiment, we employ the K-means clustering method (KM) to evaluate the clustering performance of LPP_{ANS} with different numbers of the extracted features. The performance metrics are the clustering accuracy and normalized mutual information (NMI). Details about these two metrics can be found in [24, 25]. The experiments are conducted repeatedly with different number of clusters K in the range of $\{5, 10, 20, \ldots, 60, 68\}$ and with different feature number in the range of $\{5, 10, 20, \ldots, 250\}$. For a given value of K, the experiment is performed following the below steps:

(1) Extract the best 250 features (except for K-means clustering in original space, i.e. using all the original features);
(2) Set the feature number and Randomly select K classes from the data set;
(3) Execute the K-means clustering method for 10 times with different initialization settings and record the best results;
(4) Repeat steps (3) and (4) for 20 times (except for $K = 68$);
(5) Compute the mean and standard error of performance for the given value of K and the feature number;
(6) Change the number of clusters K and repeat steps (3) to (6) until all the values of K are selected.
(7) Change the feature number and repeat steps (3) to (7) until all the feature numbers are chosen;

Figures 2 and 3 show the clustering performance in terms of clustering accuracy and NMI respectively. The experimental results that the three LPP algorithms, including $\text{LPP}_{k=3}$, $\text{LPP}_{k=10}$ and LPP_{ANS}, achieve better performance than the PCA and KM algorithms. This demonstrates the importance of the invariant of the geometrical structure in feature extraction. Furthermore, LPP_{ANS} has the best performance among the other algorithms, as a result of its ability to select the neighborhood size in an adaptively manner in the construction of the adjacent graph for learning manifolds.

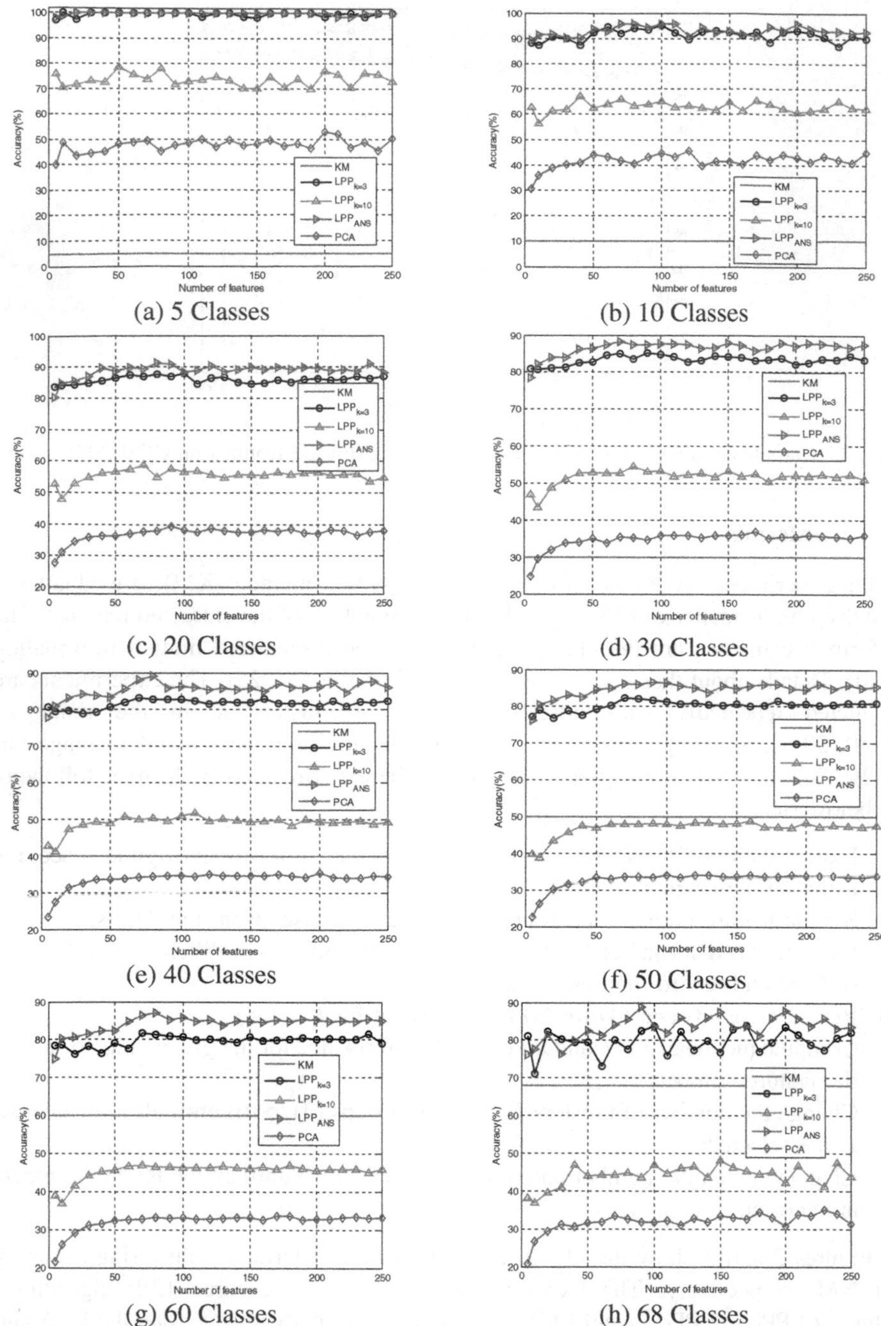

Fig. 2. Clustering accuracy versus the number of extracted features on PIE data set

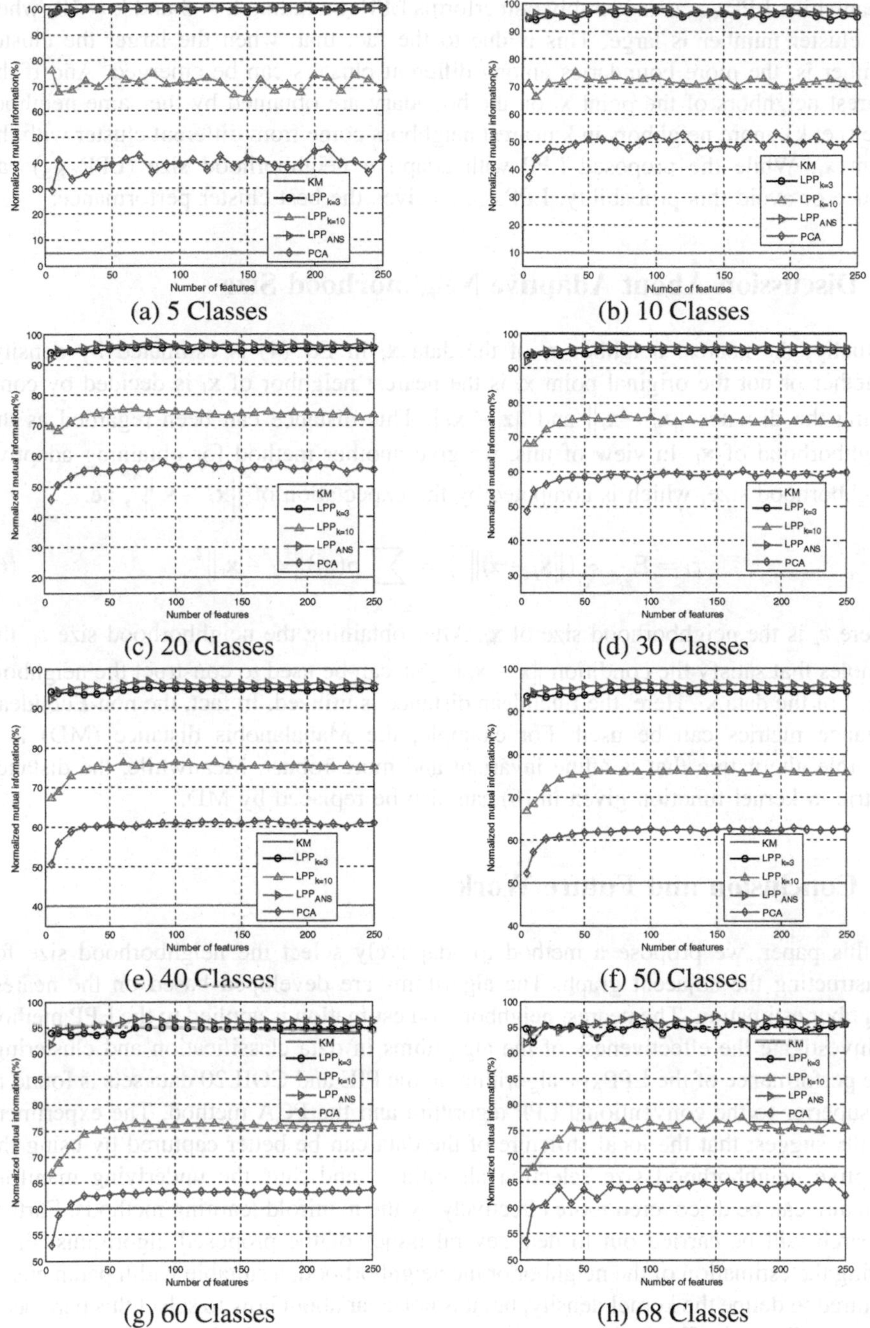

(a) 5 Classes (b) 10 Classes

(c) 20 Classes (d) 30 Classes

(e) 40 Classes (f) 50 Classes

(g) 60 Classes (h) 68 Classes

Fig. 3. Normalized mutual information versus the number of extracted features on PIE data set

Meanwhile, LPP_{ANS} considerably outperforms LPP (including k = 3 and k = 10) when the cluster number is large. This is due to the fact that, when the larger the cluster number is, the more boundaries among different clusters can be emerged. And if the nearest neighbors of the point $\mathbf{x}_i$ on the boundary are obtained by the same neighbor size (i.e. k), more neighbors in k nearest neighbors come from different cluster with the point $\mathbf{x}_i$. While the proposed LPP with adaptive neighborhood size (LPP_{ANS}) can reduce or avoid this probability, LPP_{ANS} receives the best cluster performance.

5 Discussion About Adaptive Neighborhood Size

Actually, the nearest neighbor $\mathbf{z}_i$ of the data $\mathbf{x}_i$ in Eq. (4) is estimated by density. Whether or not the original point $\mathbf{x}_j$ is the nearest neighbor of $\mathbf{x}_i$ is decided by comparing the distance $\|\mathbf{x}_j - \mathbf{x}_i\|$ and $\|\mathbf{z}_i - \mathbf{x}_i\|$. This distance can been regarded as the neighborhood of $\mathbf{x}_i$. In view of this, we give another method for obtaining adaptive neighborhood size, which is computed by the expectation of $\|\mathbf{x}_j - \mathbf{x}_i\|^2$, i.e.

$$\varepsilon_i = \mathrm{E}_{\mathbf{x}_i \sim \tilde{\mathbf{X}}_i}\left(\|\mathbf{x}_j - \mathbf{x}_i\|^2\right) = \sum_{\mathbf{x}_j \in \tilde{\mathbf{X}}_i} p(\mathbf{x}_j)\|\mathbf{x}_j - \mathbf{x}_i\|^2, \tag{6}$$

where ε_i is the neighborhood size of $\mathbf{x}_i$. After obtaining the neighborhood size ε_i, the samples that satisfy the condition $\|\mathbf{x} - \mathbf{x}_i\|^2 \leq \varepsilon_i$ can be used to construct the neighbors $N(\mathbf{x}_i)$ of the data $\mathbf{x}_i$. Here, the Euclidean distance is utilized. In fact, the non-Euclidean distance metrics can be used. For example, the Mahalanobis distance (MD) is a possible alternative that is affine invariant and more robust. Meanwhile, the distance metric in kernel function given in (6) can also be replaced by MD.

6 Conclusion and Future Work

In this paper, we propose a method to adaptively select the neighborhood size for constructing the adjacent graph. The algorithms are developed based on the nearest neighbor estimation. The nearest neighborhood estimation is applied to the LPP method to investigate the effectiveness of the algorithms in data classification and clustering. The performance of the LPP_{ANS} algorithm on the PIE and COIL20 data sets is found to be superior to the conventional LPP algorithm and the PCA method. The experiment results suggest that the local structure of the data can be better captured by using the adaptive neighborhood size selection algorithms, and thus the underlying manifold structure can be discovered more effectively by the manifold learning methods. Further research will be carried out to deal several issues of the proposed algorithms. First, during the estimation of the neighbor or the neighborhood, a suitable width parameter is required to define the kernel density, but it is not clear about how to select this parameter theoretically and effectively. Second, another method for obtaining adaptive neighborhood size or non-Euclidean distance metrics is still deep investigated. Third, the data set sampled from a low-dimensional manifold is usually noisy. It is necessary to investigate the robustness of our proposed algorithm against the noises.

Acknowledgements. This work was supported partly by the National Natural Science Foundation of China under Grant 61573137; Zhejiang Provincial Natural Science Foundation under Grants LY13F020011, LY14F010010 and LY14F020009.

References

1. Hotelling, H.: Analysis of a complex of statistical variables into principal components. J. Educ. Psychol. **24**, 417–441 (1933)
2. Jolliffe, I.T.: Principal Component Analysis. Springer, New York (1986). doi:10.1007/978-1-4757-1904-8
3. Fisher, R.A.: The use of multiple measurements in taxonomic problems. Ann. Eugen. **7**(2), 179–188 (1936)
4. Rao, C.R.: The utilization of multiple measurements in problems of biological classification. J. Roy. Stat. Soc. B **10**(2), 159–203 (1948)
5. Huang, P., Chen, C.K., Tang, Z.M., Yang, Z.J.: Feature extraction using local structure preserving discriminant analysis. Neurocomputing **140**, 104–113 (2014)
6. Wu, M., Yu, K., Yu, S., Schölkopf, B.: Local learning projections. In: Proceeding of the 24th International Conference on Machine Learning, pp. 1039–1046. ACM (2007)
7. Chen, Y., Huang, J., Xu, X.H., Lai, J.H.: Discriminative local learning projection for face recognition. Int. J. Pattern Recogn. Artif. Intell. **25**(1), 83–97 (2011)
8. Lu, Y., Lai, Z., Xu, Y., Li, X., Zhang, D., Yuan, C.: Low-rank preserving projections. IEEE Trans. Cybern. **46**(8), 1900–1913 (2015)
9. Lai, Z., Wong, W.K., Xu, Y., Yang, J.: Approximate orthogonal sparse embedding for dimensionality reduction. IEEE Trans. Neural Netw. and Learn. Syst. **27**(4), 723–735 (2016)
10. Zhang, D., He, J.Z., Zhao, Y., et al.: Global plus local: a complete framework for feature extraction and recognition. Pattern Recogn. **47**(3), 1433–1442 (2014)
11. Yang, J., Zhang, D., Yang, J.Y., et al.: Globally maximizing, locally minimizing: unsupervised discriminant projection with applications to face and palm biometrics. IEEE Trans. Pattern Anal. Mach. Intell. **29**(4), 650–664 (2007)
12. Wang, J.G.: Kernel supervised discriminant projection and its application for face recognition. Int. J. Pattern Recogn. Artif. Intell. **27**(2) (2013)
13. Tenenbaum, J., de Silva, V., Langford, J.: A global geometric framework for nonlinear dimensionality reduction. Science **290**(5500), 2319–2323 (2000)
14. Roweis, S., Saul, L.: Nonlinear dimensionality reduction by locally linear embedding. Science **290**(5500), 2323–2326 (2000)
15. Belkin, M., Niyogi, P.: Laplacian eigenmaps and spectral techniques for embedding and clustering. In: Advances in Neural Information Processing Systems, vol. 14, pp. 585–591, MIT Press (2001)
16. Seung, H.S., Lee, D.D.: The manifold ways of perception. Science **290**(12), 2268–2269 (2000)
17. He, X., Niyogi, P.: Locality preserving projections. In: Proceeding of the Conference Advances in Neural Information Processing Systems (NIPS) (2003)
18. Lai, Z., Xu, Y., Yang, J., Shen, L., Zhang, D.: Rotational invariant dimensionality reduction algorithms. IEEE Trans. Cybern. **99**, 1–14 (2016)
19. Wan, M.H., Li, M., Yang, G.W., Gai, S., Jin, Z.: Feature extraction using two-dimensional maximum embedding difference. Inf. Sci. **274**, 55–69 (2014)
20. Chung, F.R.K.: Spectral graph theory. In: Regional Conference Series in Mathematics, vol. 92 (1997)

21. Dempster, A., Laird, N., Rubin, D.: Maximum likelihood from incomplete data via the EM algorithm. J. Roy. Stat. Soc. B **39**(1), 1–38 (1977)
22. Sun, Y.J., Todorovic, S., Goodison, S.: Local-learning-based feature selection for high-dimensional data analysis. IEEE Trans. Pattern Anal. Mach. Intell. **32**(9), 1610–1626 (2010)
23. Atkeson, C., Moore, A., Schaal, S.: Locally weighted learning. Artif. Intell. Rev. **11**(15), 11–73 (1997)
24. He, X., Cai, D., Niyogi, P.: Laplacian score for feature selection. In: Advances in Neural Information Processing Systems, vol. 18. MIT Press (2005)
25. Cai, D., He, X., Han, J.: Document clustering using locality preserving indexing. IEEE Trans. Pattern Anal. Mach. Intell. **17**(12), 1624–1637 (2005)

Research on Feature Selection and Predicting ALS Disease Progression

Jin Li, Shu-Lin Wang(✉), and JingJing Wang

College of Computer Science and Electronics Engineering, Hunan University, Hunan, 410082 Changsha, China
goldliclass@163.com, smartforesting@gmail.com

Abstract. Amyotrophic lateral sclerosis (ALS) is a rapidly progressive, invariably fatal neurological disease that attacks the nerve cells responsible for controlling voluntary muscles. The disease belongs to a group of disorders known as motor neuron diseases, which are characterized by the gradual degeneration and death of motor neurons. Although ALS is incurable and fatal, with median survival of 3–5 years, treatment can extend the length and meaningful quality of life for patients. Here, to be useful clinically, we tried several feature selection methods to choose predictive features identified using ALS clinical trials dataset. The feature selection method of random frog coupled with partial least square is an exact way that can be helpful for predictive feature selection. We further apply the proposed regression method partial least square regression to predict 3–12 month ALS progression slope, as measured using the ALS functional rating scale (ALSFRS). The experiment results show that the proposed selector and predictor has shown itself to be robust to extreme outliers. It is of great benefit to accelerate ALS research and development, identify new disease predictors and potentially significantly reduce the costs of future ALS clinical trials.

Keywords: Amyotrophic lateral sclerosis · Feature selection · Random frog · Partial least square regression

1 Introduction

Amyotrophic lateral sclerosis (ALS) is a neurodegenerative disorder of upper and lower motor neurons with a progressive limbic or bulbar muscular weakness and wasting [1], and its symptoms include muscle weakness, paralysis and eventually death, usually within 3 to 5 years from disease onset. Approximately only 25% of patients survive for more than 5 years after diagnosis and 10% will survive over 10 years [2]. The fundamental cause of the disease is unknown so that the modern medicine faces with a major challenge in finding an effective treatment. At present Rilutek is the only approved medication for ALS but a limited effect on survival [3].

There are substantial obstacles to developing an effective treatment method for ALS disease. One of them is due to the heterogeneity of the disease course ranging from under a year to over ten years. In addition, the uncertainty disease evolution and prognosis is an enormous burden for patients and their families. Fortunately, the ALSFRS-R (ALS

D.-S. Huang et al. (Eds.): ICIC 2017, Part I, LNCS 10361, pp. 235–246, 2017.
DOI: 10.1007/978-3-319-63309-1_22

functional rating scale revised) score [4] is an instrument for evaluating the functional status of patients with ALS, it is used to monitor functional change in a patient over time, can therefore lead to great significance in clinical practice and clinical trial management.

The Pooled Resource Open-Access ALS Clinical Trials (PRO-ACT) platform houses the largest ALS clinical trials dataset ever created. PRO-ACT contains over 8500 ALS patient records from multiple completed clinical trials. The PRO-ACT initiative merges data from ALS clinical trials to generate an invaluable resource for accelerating discovery in the field of ALS [5]. Nevertheless, we still face the challenge that the structure of these trial data existence unknown causality of ALS and a dirty data for researchers. So, we need to exploratory analysis for the data helped identify several interesting predictors or relationship. That it, for these massive, inaccurate, noisy data, we aim to predict a patient's future ALSFRS slope as well as survival by given only his or her vector of minority numeric features.

To better address this issue, feature selection and regression prediction models are applied to solve the problem. Feature selection has the advantage of improving the performance of a predictive model. Especially for these challenging matters involving many redundant features and explore domains with hundreds to tens of thousands of features [6].

In this article, we apply a variety of methods for feature selection and using multiple predictor use 3 months of individual patient level clinical trial information to predict that patient's disease progression over the subsequent 9 months. Here, several feature selection algorithms, such as Random Frog Algorithm coupled with the Partial Least Squares (RFA-PLS) [7], Boost Decision Tree (BDT), Least Absolute Shrinkage and Selection Operator (LASSO) et al. to select numbers of finite features. Then, through the using of different prediction methods to predict clinical outcomes using only a small subset of features, such as Partial Least Squares Regression (PLSR), Kernel Ridge Regression (KRR) and Random Forest Regression (RFR) et al.

2 Materials and Methods

2.1 Problem Description

ALS clinical trials data presented in this article was accumulated consist of ALS patients that available open access on the PRO-ACT database (www.ALSdatabase.org). We transformed the original descriptive text of ALS clinical data to the quantitative data, the method of data conversion is referenced for our previous research [8]. The goal of analyzing these data will focus on predicting clinical outcome targets, the ALSFRS slope or disease progression. Specifically, by providing with ALSFRS measured between 0–3 months and predict the slope of ALSFRS changes between 3–12 months.

As in the areas of assessment, to determine the actual slope contrast with the predicted one, the first visit after month three of participation in the clinical trial is assigned as m_1. If there were visits through month 12, the first such visit after month 12 is assigned as m_2. Then, the ALSFRS slope of the training set can be calculated as

$$y_{slope} = \frac{ALSFRS(m_2) - ALSFRS(m_1)}{m_2 - m_1}, \quad (1)$$

thus, we can describe the dataset as the matrix $A_{m\times(n+1)}$, where m denotes the number of patients, and n denotes the number of features. The i-th patient can also be described as a vector $a_{i.} = (a_{i,1}, a_{i,2}, \cdots a_{i,n+1})$, where $a_{i,j}$ $(j = 1, \dot{2}, \cdots, n)$ represent the extracted features, respectively, and the $a_{i,n+1}$ represents the corresponding slope value.

2.2 Flowchart of Overview

The flowchart of our proposed models that mainly include four steps illustrated by Fig. 1 blow. (1) data cleaning and the construction of a feature vector describing each patient based on the clinical data. (2) feature selection - select a limited number of features. (3) prediction - based on the selected features to predict ALSFRS slope. (4) the evaluation of results. Specifically, the data from a given patient is firstly fed into the feature selection algorithm ("selector" in short). Then the selector selects a subset of features. Next, the prediction program ("predictor" in short) reads selected features in order to predict the disease progression as ALSFRS slope. Finally, our prediction model is evaluated by an independent validation dataset.

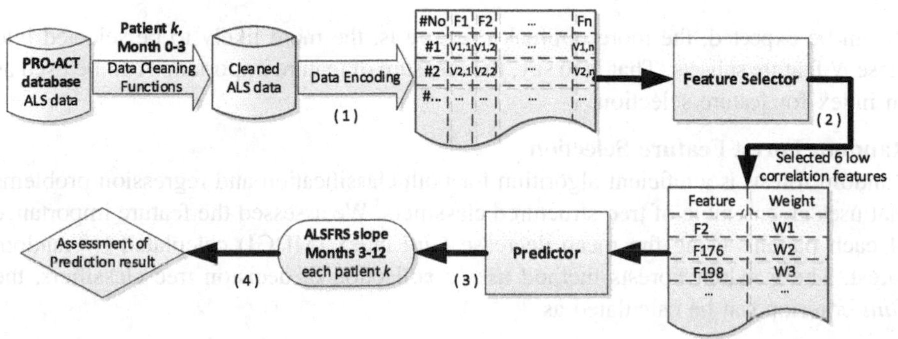

Fig. 1. The flowchart of constructing the prediction model.

2.3 Feature Selection Algorithms

We can search a large space of possible feature that extracted from the provided clinic data. In the following, we applied various feature selection methods to select effective features from the set of candidate features.

Random Frog Coupled with PLS
Random frog is a mathematically simple and computationally efficient method. Just like the framework of reversible jump Markov Chain Monte Carlo (RJMCMC) [7]. Random frog coupled with PLS means that PLS is used as a modeling method in

random frog. Before running the method, there are several parameters should be initialized [9].

(1) N: the number of iterations. In order to achieve convergence, this value needs to be sufficiently large.
(2) Q: the number between 1 and p. this variable had an impact on the iterative process at the first time but not on the overall performance of random frog.
(3) θ: a factor controlling the variance of a normal distribution from which the number of features is sampled to enter a candidate feature subset.
(4) η: this parameter is range from 0 to 1. This is the upper bound of the probability for accepting a candidate feature subset F^* whose performance is not better than F_0.

Briefly, random frog works in three steps mainly: (1) Initialized feature subset F_0 containing Q features randomly; (2) select a candidate feature subset F^* containing Q^* features from F_0, accept F^* with a certain probability as F_1, then replace F_0 with F_1, and loop this step until N iterations; (3) compute the selection probability of each feature whose value can be used as a measure of feature importance.

After N iterations, the selection probability of each feature after N iterations is computed. The frequency of the $j-$ th feature, $j = 1, 2, \ldots, p$, that has been selected in these N feature subsets is denoted as N_j. For each feature can be ranked in terms of selection probability, its selection probability can be calculated as

$$P_j = \frac{N_j}{N}, \; j = 1, 2, \ldots, p, \tag{2}$$

as can be expected, the more optimal a feature is, the more likely to be selected into these N feature subsets. That is to say, the measure of feature importance can be used as an index for feature selection.

Random Forest Feature Selection

Random forests is a efficient algorithm for both classification and regression problems that uses an ensemble of tree-structured classifiers. We assessed the feature importance of each patients using the mean decrease *Gini* index (MDGI) calculated by random forest. The Random Forests method uses a collection of decision tree classifiers, the *Gini* criterion can be calculated as

$$Gini = n_L \sum\nolimits_{k=1,\ldots,K} p_{kL}(1 - p_{kL}) + n_R \sum\nolimits_{k=1,\ldots,K} p_{kR}(1 - p_{kR}), \tag{3}$$

where p_{kL} is the proportion of class k in left node, p_{kR} is the proportion of class k in right node.

Each tree in the forest have been trained on bootstrap sample of individuals from the ALS clinic patients. The corresponding feature that with top 6 *Gini* value were selected as optimal feature candidate.

Least Absolute Shrinkage and Selection Operator

LASSO is a useful tool to achieve the shrinkage and feature selection at the same time. It achieves better prediction accuracy by shrinkage as the ridge regression.

We first explain the general LASSO setting. Let $(y_1, x_1), \ldots, (y_n, x_n)$ be n output and input pairs where $y_i \in Y$ and $x_i \in X$, assume that $X = X_1 \otimes X_2 \otimes \cdots \otimes X_d$ and X_l are subsets of $\mathcal{R}^{pl}, l = 1, \ldots, d$. Similarly, we let $x_i = (x_{i,1}, \ldots, x_{i,d})$ where $x_{i,l} \in X_l$ and $\beta = (\beta_1, \ldots, \beta_d)$ be the corresponding coefficients, where $\beta_l \in \mathcal{R}^{pl}$. Assume a given loss function F, the objective of LASSO is to minimizes the risk

$$\mathcal{R}(\beta_0, \beta) = \sum\nolimits_{i=1}^{n} F\left(y_i, \beta_0 + \sum\nolimits_{l=1}^{d} x_{il}\beta_l'\right), \tag{4}$$

subject to $|\beta_l|_1 \leq \lambda_l$ for $l = 1, \ldots, d$. Here $|\beta_l|_1 = \sum_{k=1}^{p_l} |\beta_{lk}|$.

The dataset follow a statistical model, the effects of the explanatory features $X = \{X_1, \ldots, X_d\}$ on the response ALSFRS slope Y are then estimated by the corresponding coefficients. Next, assume the ALS clinical trials data follows a linear regression model

$$Y = \beta_0 + \beta_1 X_1 + \cdots + \beta_d X_d + \varepsilon, \tag{5}$$

for instance, the LASSO estimate $\widehat{\beta_0}$, …, $\widehat{\beta_d}$ to minimize the risk mentioned above.

2.4 ALSFRS Slope Prediction

Our ultimate goal is using such a dataset to make the clinic trials more sophisticated and expensive assessments of biological information by the predictive algorithms and predictive features.

Partial Least Squares Regression

Partial least squares (PLS) have gained wide applications in the fields such as bioinformatics. It's the method that predict or analyze a set of dependent features from a set of independent predictive features.

The PLS regression model is developed from a dataset of N features with K X-variables denoted by X_{N*K} $(k = 1, \ldots, K)$. Later, the linear PLS regression model finds a few variables, which are called X-scores and denoted by s_a $(a = 1, 2 \ldots, A)$, the X-scores are good predictor of Y, formulas can be written below

$$y_{im} = \sum\nolimits_a s_{ia} w_{ma}^{'} + f_{im} \quad (\boldsymbol{Y} = \boldsymbol{T}\boldsymbol{W}' + \boldsymbol{F}), \tag{6}$$

where $\boldsymbol{W}$ denote the weights, f_{im} express the deviations between the observed and modeled responses, and comprise the elements of the Y-residual matrix $\boldsymbol{F}$.

Kernel Ridge Regression

Kernel Ridge Regression (KRR) is a simple, yet powerful method for non-parametric regression to solving a linear system, especially for dense and highly ill-conditioned system.

Assume a kernel function $k : \chi \times \chi \rightarrow \mathbb{R}$ is defined on the input domain $\chi \subseteq \mathbb{R}^d$, $k(x, z) = \langle \Psi(x), \Psi(z) \rangle$. Given training data $(x_1, y_1), \ldots, (x_n, y_n) \in \chi \times Y$ and ridge parameter λ as regulation term, then we perform linear ridge regression on $(\Psi(x_1), y_1), \ldots, (\Psi(x_n), y_n)$. Finally, the model has the form

$$f(x) = \sum\nolimits_{i=1}^{n} \alpha_i k(x_i, x), \tag{7}$$

where $\alpha_1, \ldots, \alpha_n$ can be calculated by solving the linear equation

$$(K + \lambda I_n)\alpha = y, \tag{8}$$

where $K \in \mathbb{R}^{n \times n}$ is the kernel matrix.

2.5 Performance Evaluation

The root mean square deviation (*RMSD*) and Pearson correlation coefficient (PCC, ρ) are used to assess the prediction performance. Assume $S = (s_1, s_2, \ldots, s_\mathrm{i}), i = 1, \ldots, N$ is the actual ALSFRS slope and $P = (p_1, p_2, \ldots, p_\mathrm{i}), i = 1, \ldots, N$ is the predicted one. The *RMSD* measures absolute deviations between N corresponding slope pairs. The measurement formula can be denoted as

$$RMSD = \sqrt{\frac{1}{N}\sum\nolimits_{\mathrm{i}=1}^{N} |s_\mathrm{i} - p_i|^2}, \tag{9}$$

In addition, ρ evaluates how well a prediction model is able to reveal ALSFRS trends. ρ can be expressed in

$$\rho_{S,P} = \frac{cov(S, P)}{\sigma_S \sigma_P}, \tag{10}$$

Usually, the smaller the value of *RMSD* is, the better the method performs, while better predictions lead to a higher value of the ρ, up to 1.0 for the perfect prediction.

3 Experiments

3.1 Data Collection

The experimental data from ALS clinical trials is from the PRO-ACT database that was collected between 1990 and 2015. The data was de-identified following the HIPAA de-identification conventions for personal health information to protect patient privacy.

From the raw data associated with a given patient, we first disregard those features which 60% of values are missing. After that, we divided the dataset into 10 classes according to the ALSFRS slope manually and then replace any missing value by applying nearest neighbor interpolation method. Those missing value imputation has the benefit of not changing the sample mean for that variable.

Overall, from $m = 2187$ patients of the clinical data we extract $n = 222$ features initially, including the ALSFRS scores, personal assessments as well as laboratory measurements, *etc*. All 2187 samples are divided into two groups: training set and validation set. The training set includes approximately 75% samples using Kennard-Stone algorithm partition by each different classes that mentioned above, and the remaining samples are the testing dataset.

3.2 Experimental Results

The features selected by the algorithms should be able to maintain predictability. In our experiments three kinds of feature selection methods RFA-PLS, RF and LASSO are adopted to select ALS-predictive features, we constrain that only six features in each subset of features are selected. For those selected features, we can further develop the regression model to predict the ALSFRS slope, *e.g.*, PLSR, KRR and RFR was used in our experiment.

Feature Selection Results

RFA-PLS method has several parameters affecting the performance of RFA. These parameters as well as their settings are given as follows. The number of iterations N is set to $N = 10000$. Q represents the number of features contained in the initialized feature vector. Here, for determining the optimal parameter Q, we adopt 10-Fold CV to optimize parameter Q on training set and determined to be 7. The experimental results including optimal feature subset shown in Table 1. The digital numbers in the end of column represent the series number corresponding to the feature encoding vector. As for the selection probability of each feature can be shown in Fig. 2.

Table 1. The selected six features including the value of variable importance by RFA-PLS.

Rank	Feature name	Feature sub-name	Probability	Feature index
1	onset_delta	onset_delta_log	**0.94945**	No. 222
2	onset_site	onset_site	**0.86025**	No. 3
3	Q3_Swallowing	Q3_Swallowing - three_m_slope	**0.84975**	No. 178
		Q3_Swallowing - min_scoreOfPos	0.27110	No. 175
4	Q5_Cutting	Q5_Cutting - min_scoreOfPos	**0.57615**	No. 185
5	mouth	mouth - lastTrail	**0.53480**	No. 151
		mouth - max_slope	0.44550	No. 147
6	Q1_Speech	Q1_Speech - max_scoreOfPos	**0.47290**	No. 164
		Q1_Speech - mean_scoreOfPos	0.34425	No. 166

RF method has two parameters to determine in this experiment: one is the number of features selected in bootstrap sample called as *mtry*, and another one is the number of total decision trees in the ensemble called as *ntree*. The number of trees could affect the mean squared error (*MSE*) used to calculate the percent variance. In the experiment, once the number of trees reaches 1200, *MSE* will become stable (shown in Fig. 3). Therefore, we set $mtry = 14$ (sqrt the number of features) and $ntree = 1200$ to construct the predictor model with 10-fold CV. The experimental results shown in Table 2

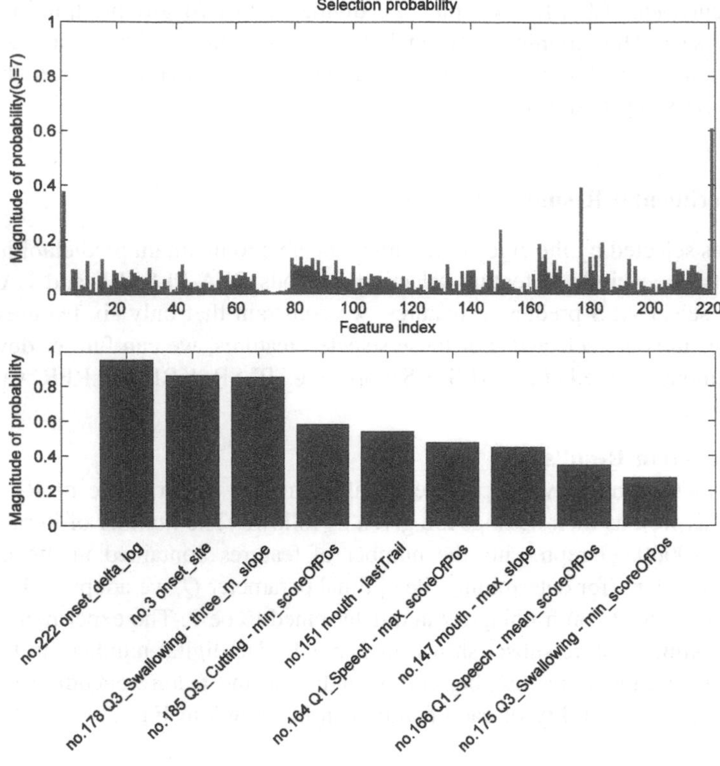

Fig. 2. The upper subplot describes the selection probability of each feature and the lower one is the magnitude of probability of the selected features.

Table 2. The selected six features including *Gini* value by RF method.

Rank	Feature name	Feature sub-name	*Gini* value	Feature index
1	Phosphorus	Phosphorus - three_m_slope	**19.261048**	No. 218
		Phosphorus - min_scoreOfPos	12.095680	No. 215
		Phosphorus - mean_scoreOfPos	11.952816	No. 216
		Phosphorus - lasttrail_first_m	11.589059	No. 217
2	onset_delta	onset_delta_log	**12.448359**	No. 222
		onset_delta	12.318448	No. 2
3	bp_systolic	bp_systolic - max_slope	**10.931289**	No. 111
		bp_systolic - mean	10.649709	No. 106
		bp_systolic - last_slope	10.469397	No. 113
4	onset_age	onset_age	**10.660689**	No. 221
5	pulse	pulse - last_slope	**10.576168**	No. 29
6	respiratory_rate	respiratory_rate - fittingSlope	**10.444450**	No. 84
		respiratory_rate - mean_slope	10.288468	No. 88

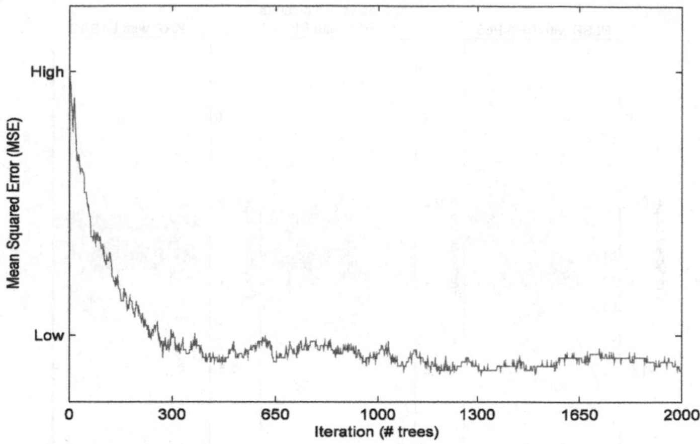

Fig. 3. Relationship between MSE and the number of trees.

lists the *MDGI* result of each feature, the top-ranked six features with the maximum *Gini* value are selected.

As for LASSO, we set the regulation term $\lambda = 100$ which is selected by experience. The sigmoid kernel with the scale parameter to the dimension of inputs is used. Table 3 presents the selected features.

Prediction Results

In our experiments, we adopt three regression methods (PLSR, KRR and RFR) to predict the ALSFRS slope. For PLSR, we apply 10-fold CV to determine the number of components. In addition, PLSR necessitates sophisticated computations and therefore its application depends on the availability of software [10]. The source codes available at www.libpls.net. For KRR, we use the polynomial kernel $k(x,z) = (\gamma x^{\mathsf{T}} z + b)^d$, and the bias in the polynomial kernel set to $b = 0.05$ and the degree set to $d = 2$, the regulation term γ is determine imperially using cross validation. An open-source implementation of the KRR algorithm is available from Joseph Santarcangelo.

Table 3. The selected six features including the weight value by LASSO.

Rank	Feature name	Feature sub-name	Weight value	Feature index
1	onset_delta	onset_delta_log	**0.1514309**	No. 222
		onset_delta	0.0092392	No. 2
2	Phosphorus	Phosphorus - three_m_slope	**0.1182800**	No. 218
3	respiratory_rate	respiratory_rate - slope	**0.0198943**	No. 83
4	mouth	mouth - lastTrail	**0.0188541**	No. 151
5	Q1_Speech	Q1_Speech - max_scoreOfPos	**0.0155864**	No. 164
		Q1_Speech - min_scoreOfPos	0.0131080	No. 165
6	Q3_Swallowing	Q3_Swallowing - three_m_slope	**0.0149737**	No. 178

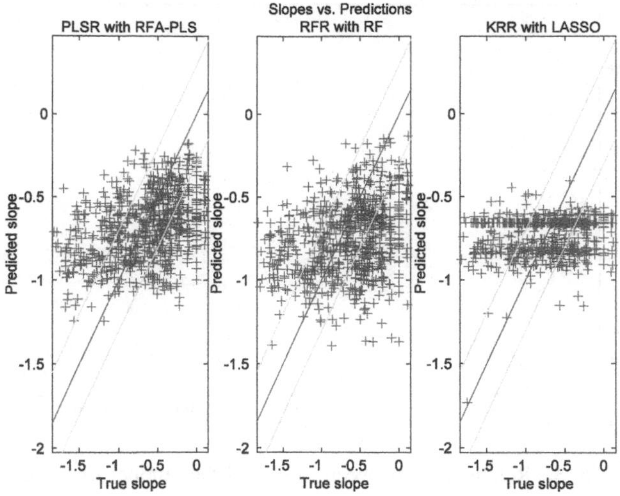

Fig. 4. The scatter diagram of prediction results based on three different models (PLSR with RFA-PLS, RFR with RF, KRR with LASSO). (Color figure online)

Table 4. The prediction results with different predictors.

Selector	Predictor	*RMSD*	*PCC*, ρ
RFA-PLS	PLSR	**0. 5184**	**0. 4538**
	KRR	0.5190	0.4419
	RFR	0.5438	0.3848
RF	PLSR	0.5417	0.3751
	KRR	0.5422	0.3569
	RFR	0.5447	0.3503
LASSO	PLSR	0.5676	0.2041
	KRR	0.5657	0.2118
	RFR	0.5947	0.1305

For the RFR, the parameters are set as follows. (1) The number of trees is set to $ntree = 1200$, owing to *MSE* against the number of trees no longer fluctuates. (2) The number of candidate predictors at each split is set to $mtry = 2$. Figure 4 intuitively illustrates the scatter diagram of the actual slope and predicted one of each testing sample. As can be seen from Fig. 4, the best performance results are obtained by PLSR with RFA-PLS while the results of the ALSFRS slope predicted by KRR with LASSO selection model is relatively bad. The red line represents the predicted value of ALSFRS slope fits the real one, the yellow one means $|s_i - p_i| \leq 0.3$. The ratio of the three plots between the two yellow lines is 0.4498, 0.4032, 0.3945 respectively.

Furthermore, we adopt *RMSD* and ρ to evaluate the performance of different models. As summarized in Table 4, it is obvious that our method combining RFA-PLS with PLSR performs the best on the validation set by $RMSD = 0.5184$ and $\rho = 0.4538$, and we could draw a conclusion that feature selection is the most important step and the appropriate features dominate the performance of the model.

4 Conclusion and Discussion

This paper aims to estimate the future disease progression of ALS patients using only a small subset of features, which is very important for patient care and making decisions regarding clinical interventions and assistive technology. Notably, the ALS prediction was to validate features that had previously been suggested to identify novel predictive features. The establishment of prediction model includes two steps. The first step is to construct the feature vector for all patients using the clinical data. The second one is to adopt different combinations of various feature selection and prediction methods. The experimental results indicate that the random frog algorithm coupled with partial least square can select candidate features effectively. Comparing with recent regression methods such as Support Vector Regression (SVR) to predict the clinical outcomes [8], however, an advantage of PLSR model is easy to determine the optimal parameters and more robustly than SVR. More importantly, PLSR is competitive in two evaluation items including *RMSD* and ρ.

In summary, our method have enabled the identification of several nonstandard, potential features of disease progression that might shed new light on disease pathways, and can be used to aid clinical care, identify new disease features and potentially significantly reduce the costs of future ALS clinical trials.

Acknowledgement. This work was supported by the grants of the National Science Foundation of China (Grant Nos. 61472467, 61672011, and 61471169) and the Collaboration and Innovation Center for Digital Chinese Medicine of 2011 Project of Colleges and Universities in Hunan Province. We also specially thank both Prize4Life and Sage Bionetworks-DREAM for providing the ALS clinical data.

References

1. Kiernan, M.C., Vucic, S., Cheah, B.C., Turner, M.R., Eisen, A., Hardiman, O., Burrell, J.R., Zoing, M.C.: Amyotrophic lateral sclerosis. Lancet **377**(9769), 942–955 (2011)
2. Drigo, D., Verriello, L., Clagnan, E., Eleopra, R., Pizzolato, G., Bratina, A., D'Amico, D., Sartori, A., Mase, G., Simonetto, M., de Lorenzo, L.L., Cecotti, L., Zanier, L., Pisa, F., Barbone, F.: The incidence of amyotrophic lateral sclerosis in Friuli Venezia Giulia, Italy, from 2002 to 2009: a retrospective population-based study. Neuroepidemiology **41**(1), 54–61 (2013)
3. Miller, R.G., Mitchell, J.D., Moore, D.H.: Riluzole for amyotrophic lateral sclerosis (ALS)/motor neuron disease (MND). Cochrane Database Syst. Rev. (3) (2012)
4. Kollewe, K., Mauss, U., Krampfl, K., Petri, S., Dengler, R., Mohammadi, B.: ALSFRS-R score and its ratio: a useful predictor for ALS-progression. J. Neurol. Sci. **275**(1–2), 69–73 (2008)

5. Kuffner, R., Zach, N., Norel, R., Hawe, J., Schoenfeld, D., Wang, L.X., Li, G., Fang, L., Mackey, L., Hardiman, O., Cudkowicz, M., Sherman, A., Ertaylan, G., Grosse-Wentrup, M., Hothorn, T., van Ligtenberg, J., Macke, J.H., Meyer, T., Scholkopf, B., Tran, L., Vaughan, R., Stolovitzky, G., Leitner, M.L.: Crowdsourced analysis of clinical trial data to predict amyotrophic lateral sclerosis progression. Nat. Biotechnol. **33**(1), 51-U292 (2015)
6. Guyon, I., Elisseeff, A.: An introduction to variable and feature selection. J. Mach. Learn. Res. **3**(6), 1157–1182 (2002)
7. Li, H.D., Xu, Q.S., Liang, Y.Z.: Random frog: an efficient reversible jump Markov Chain Monte Carlo-like approach for variable selection with applications to gene selection and disease classification. Anal. Chim. Acta **740**, 20–26 (2012)
8. Wang, S.-L., Li, J., Fang, J.: Predicting progression of ALS disease with random frog and support vector regression method. In: Huang, D.-S., Han, K., Hussain, A. (eds.) ICIC 2016. LNCS, vol. 9773, pp. 160–170. Springer, Cham (2016). doi:10.1007/978-3-319-42297-8_16
9. Yun, Y.H., Li, H.D., Wood, L.R.E., Fan, W., Wang, J.J., Cao, D.S., Xu, Q.S., Liang, Y.Z.: An efficient method of wavelength interval selection based on random frog for multivariate spectral calibration. Spectrochim. Acta Part A Mol. Biomol. Spectrosc. **111**(7), 31 (2013)
10. Li, H., Xu, Q., Liang, Y.: LibPLS: an integrated library for partial least squares regression and discriminant analysis. PeerJ (2014)

Automatic Tongue Image Segmentation for Traditional Chinese Medicine Using Deep Neural Network

Panling Qu[1], Hui Zhang[1], Li Zhuo[1(✉)], Jing Zhang[1], and Guoying Chen[2]

[1] Signal and Information Processing Laboratory,
Beijing University of Technology, Beijing, China
zhuoli@bjut.edu.cn

[2] Office of Science and Technology Development,
Beijing University of Technology, Beijing, China

Abstract. Automatic tongue image segmentation is a key technology for the research on tongue characterization in Traditional Chinese Medicine. Due to the complexity of automatic tongue image segmentation, the automation degree and segmentation precision of the existing methods for tongue image segmentation are not satisfied. To address the above problem, a method of automatic tongue image segmentation using deep neural network is proposed in this paper. In our method, an image quality evaluation method based on brightness statistics is proposed to judge whether the input image is to be segmented, and the SegNet is employed to train on the TongueDataset1 and TongueDataset2 to obtain the deep model for automatic tongue image segmentation. TongueDataset1 and TongueDataset2 are specially constructed for tongue image segmentation. The experimental results on TongueDataset1 and TongueDataset2 show that the mean intersection over union score can reach to 95.89% and 90.72%, respectively. Compared with the traditional methods of tongue image segmentation, our method can avoid the complicated process of extracting features manually, and has obvious superiority in the segmentation performance.

Keywords: Traditional Chinese Medicine · Tongue diagnosis · Automatic tongue image segmentation · Deep neural network · SegNet · Tongue image dataset

1 Introduction

Tongue diagnosis is to know the physiological function and pathological changes by observing the changes of the tongue, which is an important part in Traditional Chinese Medicine (TCM). Since the 1980s, with the development of computer technology, the digital image processing technology has been applied to the tongue characterization. Besides the tongue body, the collected tongue images often contain the background such as the human face. In order to avoid the influence of these regions for tongue features analysis, the tongue body should be segmented firstly. Therefore, the automatic tongue image segmentation is one of the key technologies in the research of the tongue characterization, the precision of which directly affects the accuracy of tongue image

D.-S. Huang et al. (Eds.): ICIC 2017, Part I, LNCS 10361, pp. 247–259, 2017.
DOI: 10.1007/978-3-319-63309-1_23

analysis. However, the difficulty of automatic segmentation of the tongue images has been greatly increased by the objective factors, such as the differences in the size and shape of the tongue bodies and the similarity between the colors of the tongue body and the lip. The existing tongue image segmentation methods can be divided into two categories, the one based on the low level features and the one based on variable models.

Tongue image segmentation algorithms based on image features can be further divided into the region segmentation, the edge detection and the segmentation based on specific theoretical tools. In [2], the binary image of H component in the HIS color space is obtained firstly; then the areas not belonging to the tongue body were removed by the clustering algorithm, and denoising was conducted by morphological method; finally the results of tongue image segmentation can be obtained. In [3], according to the similarity of both color and space, mean shift clustering was used to classify the pixels firstly, and then a tongue body detection algorithm based on principal component analysis was employed, finally the tongue area can be obtained by using Tensor Voting. A method for color tongue image segmentation using Fuzzy Kohonen Networks and genetic algorithm was proposed by Wang et al. [4].

In the middle of the 1980s, the two-dimensional variable model known as Snakes or dynamic contour model, was proposed by Kass [5] et al. The Snakes is based on the dynamic optimization of the energy function to approach the real contour of the target. In [6], a Catmull-Rom spline Snakes model was proposed and applied to automatic tongue image segmentation. In [7], the prior information of the position and color of the tongue was taken into consideration and the tongue body area was finally obtained by the Snakes model. In [8], the initial tongue boundary was firstly extracted based on a simple initial boundary extraction procedure and a two stage k-means clustering, then the Snakes model was employed to obtain a more precise tongue boundary. The parameters of Snakes model should be reset when the external environment (e.g. the illumination) changes.

Although the research work on automatic tongue image segmentation has made some progress, being affected by a series of factors, such as differences in image quality and patient's tongue and illumination, there is still no method of automatic tongue image segmentation by which different kinds of tongue images can obtain ideal segmentation results. What's more, the degree of automation and robust and the accuracy of the existing tongue image segmentation methods are not satisfied.

More recent, deep learning has presented state-of-the-art success in the field of computer vision, such as image classification, object detection and semantic segmentation. The convolutional neural network (CNN) [9] is widely used in image semantic segmentation due to its powerful ability of feature learning and representation. For example, full convolutional neural network (FCN) was firstly proposed for image semantic segmentation by Long et al. [10], in which the fully connected layer in the network for image classification was modified to the corresponding convolution layer, and the output heatmaps were used as input to the soft-max layer to achieve pixel-wise semantic segmentation. In [11], the conditional random field (CRF) was employed to further optimize the output of FCN network, by which the accuracy of segmentation had been improved. In [12], CRF with Gaussian pairwise potentials and mean-field approximate inference is formulated as Recurrent Neural Networks (RNN) to train FCN and

CRF-RNN from end to end, by which the segmentation accuracy had been further improved.

In 2015, a deep neural network for image semantic segmentation, which is called SegNet [13], was proposed by the Computer Vision and Robotics Group in University of Cambridge. It is primarily motivated by the autonomous driving and intelligent robot applications and can divide image pixels into 12 classes. SegNet is a deep convolutional encoder and decoder network, in which the encoder is utilized to extract high-level semantic features of the image and the decoder is used to reconstruct the segmentation image with the high-level semantic features. Compared with other network semantic segmentation, the network has the advantages of less training parameters, small model and high accuracy.

Taken the problems of existing tongue image segmentation methods into consideration, a method of automatic tongue image segmentation using deep neural network is proposed in this paper. In our method, SegNet is trained on the TongueDataset1 and TongueDataset2 to achieve the model for automatic tongue image segmentation. The experimental results show that the segmentation accuracy obtained by our method is much better than the traditional methods.

In the rest of this paper, Sect. 2 introduces the framework and specific implementation of the automatic tongue image segmentation method using SegNet in detail. The experimental results are shown and analyzed in Sect. 3. The conclusions are drawn in Sect. 4.

2 The Proposed Tongue Image Segmentation Method

The framework of the proposed tongue image segmentation method using SegNet has been shown in Fig. 1. Firstly, the brightness of tongue images is discriminated and images with higher or lower brightness are eliminated. Then, the SegNet is trained by the normalized images to obtain the model for tongue image segmentation. During segmentation phrase, the brightness of tongue image to be segmented is judged firstly,

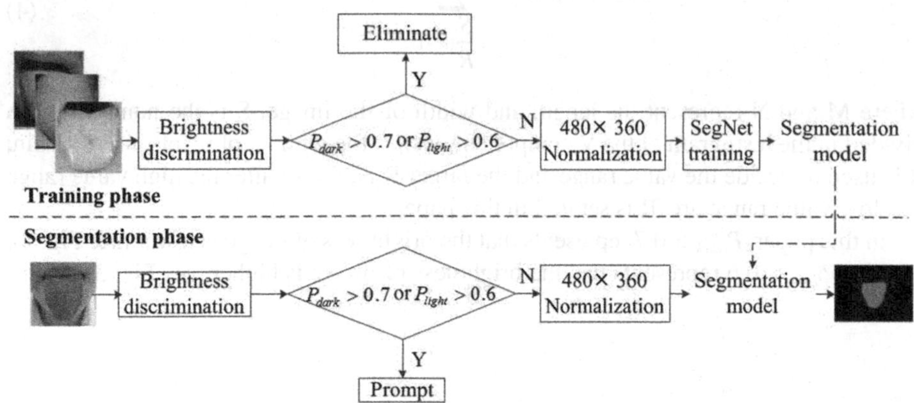

Fig. 1. The diagram of the proposed tongue image segmentation method

and the image with higher or lower brightness is not segmented. Then the normalized image is segmented by the segmentation model to achieve the final segmentation image. Next, the discrimination of brightness, the architecture of SegNet and network training are introduced in detail.

2.1 Discrimination of Brightness

Due to the influence of the acquisition environment, the brightness of the tongue image varies from a large range. Since the quality of the tongue image with higher or lower brightness is too poor to conduct the follow-up automatic analysis of tongue features (such as the tongue color, greasy etc.), these images should be judged before segmentation. Therefore, a method of image quality evaluation based on brightness statistics is proposed in this paper, by which the brightness of captured tongue image is discriminated to judge whether the image should be segmented.

First of all, the histogram statistics of V component of the tongue image in the HSV color space can be conducted, and then the proportion of the pixels with high value and low value in the whole pixels is calculated according to Eqs. (1) and (2) respectively,

$$P_{dark} = \frac{Count\_dark}{M \times N} \tag{1}$$

$$P_{light} = \frac{Count\_light}{M \times N} \tag{2}$$

in which *Count_dark* and *Count_light* can be calculated by Eqs. (3) and (4) respectively,

$$Count\_dark = \sum_{i=0}^{\frac{S}{R}} hist_i \tag{3}$$

$$Count\_light = \sum_{i=\frac{S}{R}(R-1)}^{S-1} hist_i \tag{4}$$

where M and N represent the length and width of the image, S is the number of bin divided in the histogram of the V component. $hist_i$ is the number of pixels in the ith bin, R is used to decide the value range and the larger R is, the smaller the high value range and low value range are. R is set to 3 in this paper.

In this paper, $P_{dark} > 0.7$ represents that the brightness of image is lower (see Fig. 2), and the $P_{light} > 0.6$ represents that the brightness of image is higher (see Fig. 3).

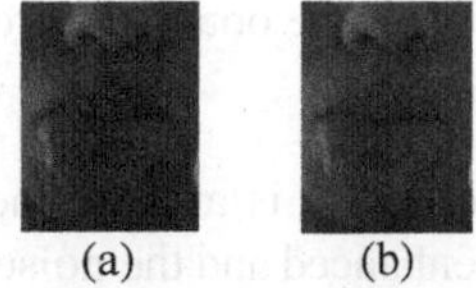

Fig. 2. The examples of tongue images with low brightness: (a) original image; (b) V channel.

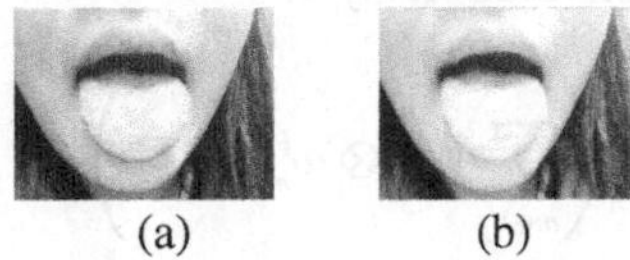

Fig. 3. The examples of tongue images with high brightness: (a) original image; (b) V channel.

2.2 The Architecture of SegNet

The architecture of SegNet is shown in Fig. 4, in which Input is the image to be segmented and Output is the segmented image with different categories being represented by different colors. SegNet is a symmetric network, including encoder network and decoder network. The encoder network is composed of the first 13 layers of VGG16 network [14], and excludes the full connection layer of VGG16 network, by which the feature maps with high resolution can be preserved and the parameters of the encoder network can be reduced. In addition, the Normalization, activation layer and pooling layer are attached to each convolutional layer. The decoder network consists of unpooling layers and convolution layers, in which the unpooling layers and pooling layers are corresponding to each other. The output of the decoder network is fed into the softmax layer to carry out multi-class classification, and then the classification probability of each pixel is obtained to realize the semantic segmentation. In this paper, the pixels of the tongue image are divided into 2 categories, namely the tongue area and the background.

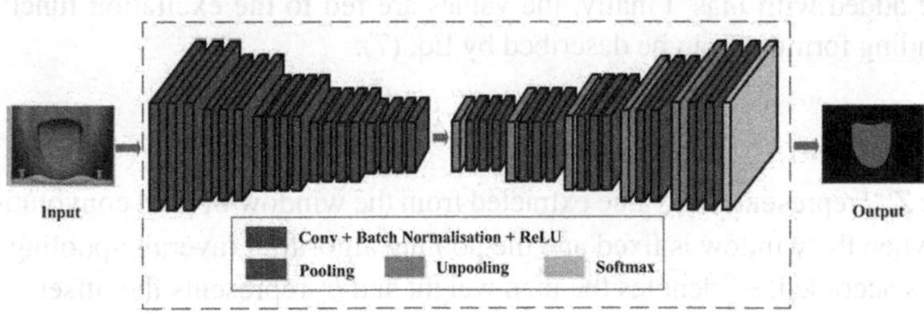

Fig. 4. The architecture of SegNet

The encoder network of SegNet is used to extract features, and the decoder network is used to enrich the image information from the unpooling layers, so that the missing information in the pooling process can be obtained through the network learning. The network is introduced in detail as follows.

Convolution Layer. Convolution layer is an extremely crucial layer in CNN. The features of original signal can be enhanced and the noise can be reduced enormously by convolution operations. The convolution layer has the property of parameter sharing, which can reduce the parameters and the computational complexity. The features maps can be gained by Eq. (5) in convolution layer.

$$y_n^l = f_l\left(\sum_{m \in V_n^l} y_m^{l-1} \otimes \omega_{m,n}^l + b_n^l\right) \tag{5}$$

Where y_n^l is nth feature map in l layer, $\omega_{m,n}^l$ denotes the convolution kernel when the feature is extracted from layer l, b_n^l is the bias. V_n^l represents the gather of characteristic pattern that linked to layer l.

Normalization and Activation Layer. The batch normalization (BN) in [15] has been used to accelerate the convergence speed and prevent the gradient from disappear. Before activation layer, each convolution layer is connected to a BN layer.

The Rectified Linear Units (ReLU) is employed as the activation function, as shown in Eq. (6). Compared to sigmoid and tanh functions, ReLU has a large accelerating effect on the convergence of the stochastic gradient descent and only requires one threshold to get the activation value, without a large number of complex computations.

$$f(x) = \max(0, x) \tag{6}$$

Pooling and Unpooling Layer. Pooling layer has the same number of feature maps with the former convolution layer. The operation of subsampling is taken on the image by pooling layer, by which the amount of data processed can be reduced while the useful information can be retained. Firstly, the input images are divided into disjoint zones and the operation of subsampling is taken on these zones. Then the values obtained from the above are added with bias. Finally, the values are fed to the excitation function. The corresponding formulas can be described by Eq. (7):

$$y_n^l = f_l\left(z_n^{l-1} \times \omega_n^l + b_n^l\right) \tag{7}$$

where Z_n^{l-1} represents the value extracted from the window of $l - 1$ convolution layer features when the window is fixed and the pooling algorithm (average-pooling or max-pooling) is accorded. ω_n^l denotes the map weight and b_n^l represents the offset.

In this paper, max-pooling is used to realize subsampling. Compared to the pooling layers in other network, the pooling layer in the SegNet has a function of index in each pooling operation. In each pooling operation, the relative positions in 2×2 blocks of the weights selected by the max operation are preserved as pooling indices, as shown in Fig. 5(a). Unpooling operation is the inverse of the pooling process, by which the input feature maps are enlarged two times, and then the feature data is laid on the corresponding position according to the pooling indices, as shown in Fig. 5(b).

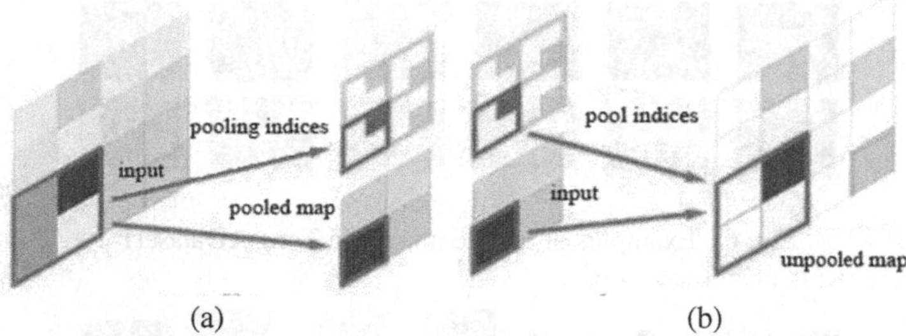

Fig. 5. Pooling and unpooling operations: (a) pooling; (b) unpooling.

2.3 Network Training

SegNet is modified based on the framework of CNN, and is also a supervised learning method. The network training process consists of two stages: forward and backward propagation. In the forward propagation stage, the input image is fed into the network and an output image is obtained through the network parameter calculation. In the back propagation stage, the error of the output image and the given label image is calculated and the network parameters are corrected by the error back propagation algorithm. Since the encoder network of SegNet is constructed by the first 13 layers of VGG16 network, it can be initialized by the classification model which is obtained by training VGG16 network on ILSVRC-2012 dataset.

3 Experimental Results and Analysis

3.1 Dataset

At present, there is no public standard dataset for evaluating the tongue image segmentation performance. To meet the requirements of practical applications, two tongue image dataset named TongueDataset1 and TongueDataset2 are constructed in this paper which are corresponding to the closed and open environment respectively. TongueDataset1 is made up of 500 tongue images, which are collected by tongue image capture machine of SIPL series [1] which is developed by the research group. The tongue image capture machine of SIPL series adopts D65 light source and has enclosed design to avoid the stray light from the outside. Therefore, the tongue images in TongueDataset1 have the characteristics of uniform illumination and relatively consistent tongue position.

TongueDataset2 is made up of 3000 tongue images, which are collected from the Internet images and captured in different environment and at different time by various image acquisition devices such as cell phones and cameras. Therefore, the tongue images in TongueDataset2 have different sizes, complex illumination, and different sizes and positions of the tongue body. The examples of tongue images inTongueDataset1 and TongueDataset2 are shown in Figs. 6 and 7.

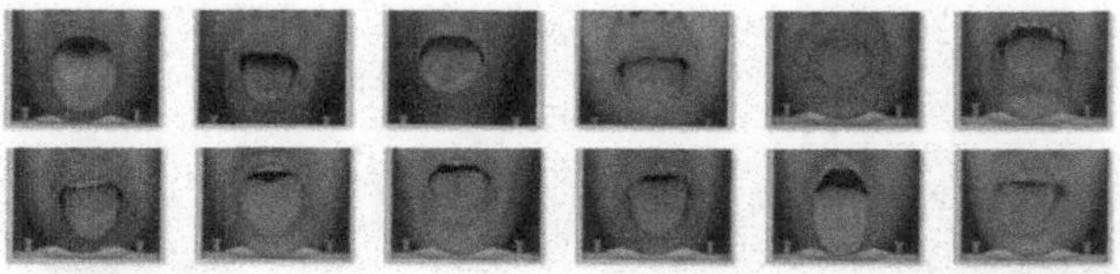

Fig. 6. Examples of tongue images in TongueDataset1

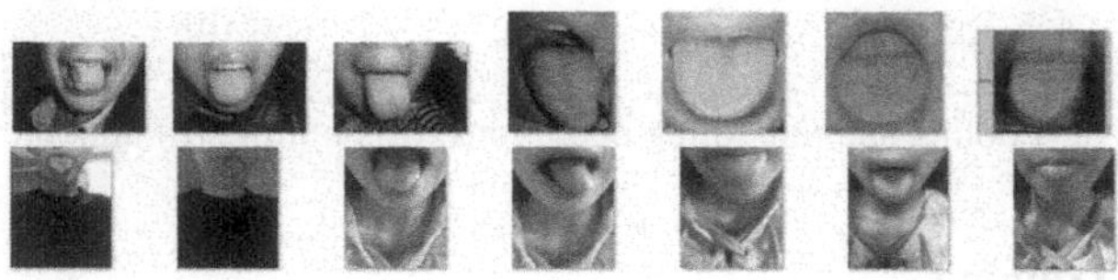

Fig. 7. Examples of tongue images in TongueDataset2

3.2 Evaluation Metrics

To evaluate the segmentation results, three commonly used performance measures in semantic segmentation have been employed, namely Global accuracy (G), class average accuracy (C) and mean intersection over Union (mIoU). G refers to the percentage of pixels correctly classified in the dataset; C refers to the mean of the predictive accuracy of all classes; mIoU is as used in the Pascal VOC2012 challenge. These three metrics can be calculated by Eqs. (8)–(10), respectively.

$$G = \frac{\sum_i n_{ii}}{\sum_i t_i} \tag{8}$$

$$C = \left(\frac{1}{n_{cl}}\right) \sum_i \frac{n_{ii}}{t_i} \tag{9}$$

$$mIoU = \left(\frac{1}{n_{cl}}\right) \sum_i \frac{n_{ii}}{t_i + \sum_j n_{ji} - n_{ii}} \tag{10}$$

where n_{ij} is the number of pixels classified as the ith class, n_{cl} is the number of pixel classes, $t_i = \sum_j n_{ij}$ is the number of pixels belonging to ith class.

3.3 Data Augmentation

The research results suggest that the number of training samples has a direct impact on the performance of the deep neural network. Due to the limited scale of the tongue image dataset TongueDataset1, the training samples are augmented by perturbing each training sample with changes in translation, rotation and horizontal flip. A training sample can produce 8 samples through rotation with 8 different degrees generated randomly between ±70° and 4 samples by translation in four directions of upper, lower, left, right. This increases the size of the tongue image dataset by a factor of 24.

3.4 Experimental Results and Analysis

The experiments are conducted by using caffe source code in Ubuntu system. The system configuration is Intel(R) Core (TM) i5-4590 M CPU 3.3 GHz with 16 G internal storage and NVIDIA GTX1080 graphics card. The stochastic gradient descent algorithm (SGD), with a fixed learning rate of 0.001 and momentum of 0.9, is chosen and the cross-entropy loss is used as the objective function to train the network in this paper.

The Influence of the Initial Methods on Segmentation Accuracy. The comparison of two initialization methods, which are the random initialization and the initialization by the classification model obtained by training the VGG16 network on ILSVRC-2012 dataset, have been carried out on TongueDataset1 and TongueDataset2 respectively with the ratio of the training set and the test set being set to 3:2. The experimental results have been shown in Table 1.

Table 1. The comparison of two initialization methods.

Initial methods	Dataset	G	C	mIoU
Random initialization	TongueDataset1	98.57%	98.25%	95.60%
	TongueDataset2	98.41%	94.79%	90.22%
Classification model (VGG16)	TongueDataset1	98.96%	98.65%	95.89%
	TongueDataset2	98.44%	95.30%	90.72%

From Table 1, mIoU of the network initialized by classification model (VGG16) on TongueDataset1 and TongueDataset2 are higher than that of random initialization by 0.39% and 0.50% respectively.

Figures 8 and 9 Show the visualization of the weights in the first convolution layer with two different initialization methods. In Figs. 8 and 9, the SegNet encoder network extracts the abstract features layer by layer, the first convolution layer extracts the image color features and the weights in the first convolution layer initialized by classification model (VGG16) on two tongue image datasets have only slight difference. The examples of segmentation results on TongueDataset1 and TongueDataset2 have been shown in Figs. 10 and 11.

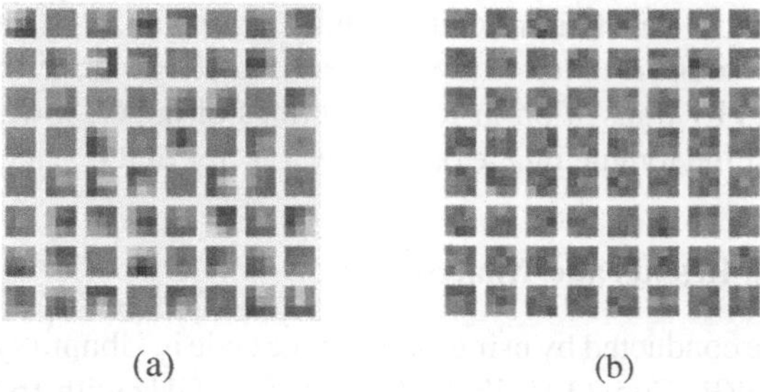

(a) (b)

Fig. 8. Visualization of the weights in the first convolution layer using two different initialization methods in TongueDataset1: (a) initialized by Classification Model (VGG16); (b) random initialization.

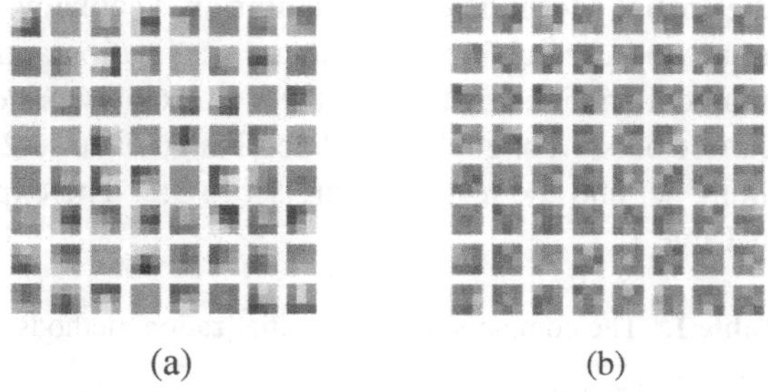

(a) (b)

Fig. 9. Visualization of the weights in the first convolution layer using two different initialization methods in TongueDataset2: (a) initialized by Classification Model (VGG16); (b) random initialization.

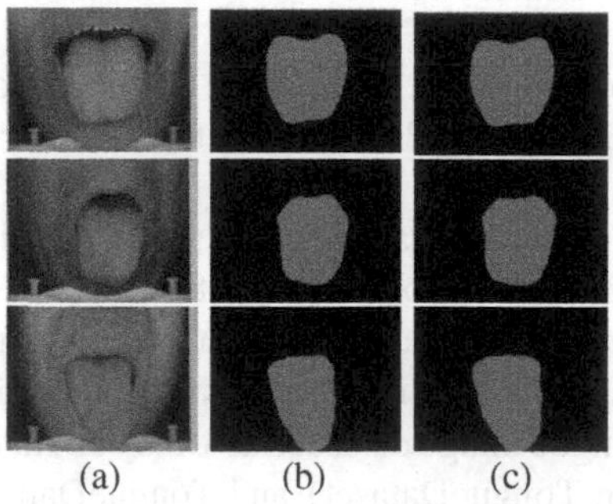

(a) (b) (c)

Fig. 10. The examples of segmentation results in TongueDataset1: (a) the images to be segmented; (b) the ground truth; (c) segmented results.

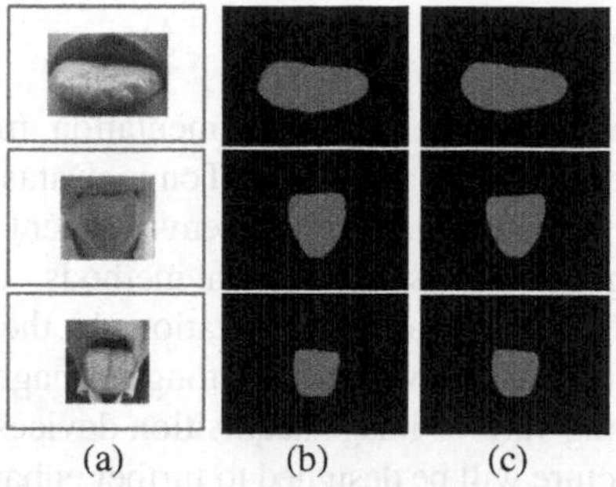

Fig. 11. The examples of segmentation results in TongueDataset2: (a) the images to be segmented; (b) the ground truth; (c) segmented results.

The Influence of Data Augmentation on Segmentation Accuracy. The experiments with data augmentation and without data augmentation on TongueDataset1 have been carried out to analyze the influence of data augmentation on Segmentation accuracy. The experimental results are shown in Table 2.

Table 2. The influence of data augmentation on segmentation accuracy.

	G	C	mIoU
With data augmentation	98.96%	98.65%	95.89%
Without data augmentation	98.34%	97.59%	95.56%

It can be seen from Table 2 that data augmentation can effectively prevent the deep models from over-fitting and promote the segmentation accuracy by 0.33%.

The Comparison between the Proposed Method and the Existing Methods. In this paper, the propose method is compared with the traditional tongue image segmentation methods proposed in [3, 8]. Both of the methods proposed in [3, 8] can't work well when the color of lip is close to the color of the tongue. Since the tongue image segmentation methods proposed in [3, 8] are applied to the closed environment, TongueDataset1 is used in this experiment. The experimental results have been shown in Table 3. It can be seen that, compared with the traditional tongue image segmentation methods, the proposed method in this paper can improve the segmentation accuracy by a large margin and effectively separate the lips and tongues. This is because the deep neural network can automatically learn effective deep features from the data without the complicated process of extracting features manually, and then the pixels can be labelled with their corresponding semantic information according to the deep features.

Table 3. The Comparison between the proposed method and the existing methods.

Method	mIoU
Cluster + PCA [3]	65.46%
Snakes [8]	81.78%
Proposed method	95.89%

4 Conclusions

In this paper, an automatic tongue image segmentation method using SegNet is proposed, and two tongue image datasets named TongueDataset1 and TongueDataset2 are specially constructed in the closed and open environment respectively. Compared with the traditional tongue image segmentation methods, our method can greatly improve the accuracy of tongue image segmentation. In the future, we will further expand the tongue image datasets, in which more tongue images will be captured under different illuminations using various image acquisition devices. Meanwhile, the neural network with new architecture will be designed to further enhance the robustness of the tongue segmentation model.

Acknowledgment. The work in this paper is supported by BJUT United Grand Scientific Research Program on Intelligent Manufacturing, the National Natural Science Foundation of China (No.61531006, No.61372149, No.61370189, No.61471013 and No.61602018), the Importation and Development of High-Caliber Talents Project of Beijing Municipal Institutions (No.CIT&TCD20150311, CIT&TCD201404043), the Beijing Natural Science Foundation (No. 4142009, No.4163071), the Science and Technology Development Program of Beijing Education Committee (No. KM201510005004), Funding Project for Academic Human Resources Development in Institutions of Higher Learning Under the Jurisdiction of Beijing Municipality.

References

1. Shen, L., Cai, Y., Zhang, X.: Collection and analysis of TCM Tongue. Beijing University of Technology Press, Beijing (2007)
2. Zhao, Z., Wang, A., Shen, L.: Color tongue image segmentation method based on mathematical morphology and HIS model. J. Beijing Univ. Technol. **25**(2), 67–71 (1999)
3. Xu, W., Kanawong, R., Xu, D., Li, S., Ma, T., Zhang, G., Duan, Y.: An automatic tongue detection and segmentation framework for computer-aided tongue image analysis. In: 13th International Conference on e-Health Networking Applications and Services, pp. 189–192. IEEE, Piscataway (2011)
4. Wang, A., Shen, L., Zhao, Z.: Color tongue image segmentation using fuzzy kohonen networks and genetic algorithm. In: Nasser, K. (ed.) Applications of Artificial Neural Networks in Image Processing, Electronic Imaging, vol. 3962, pp. 182–190. SPIE, San Jose (2000)
5. Kass, M., Witkin, A., Terzopoulos, D.: Snakes: active contour models. Int. J. Comput. Vis. **1**, 321–331 (1988)
6. Shen, S., Wang, M.: Research on the classification of image segmentation. J. Beijing Univ. Technol. **26**(3), 33–38 (2000)
7. Shi, M., Li, G., Li, F.: C^2G^2FSnake: an automatic tongue image segmentation algorithm based on prior knowledge. Sci. China Inf. Sci. **56**(9), 1–14 (2013)
8. Guo, J., Yang, Y., Wu, Q., Su, J., Ma, F.: Adaptive active contour model based automatic tongue image segmentation. In: 9th International Congress on Image and Signal Processing, BioMedical Engineering and Informatics, pp: 1386–1390 (2016)
9. Krizhevsky, A., Sutskever, I., Hinton, G.E.: Imagenet classification with deep convolutional neural networks, In: International Conference on Neural Information Processing Systems, pp. 1097–1105. Curran Associates Inc. (2012)

10. Long, S., Shelhamer, E., Darrell, T.: Fully convolutional networks for semantic segmentation. IEEE Trans. Pattern Anal. Mach. Intell. **39**(4), 640–651 (2015)
11. Chen, L.-C., Papandreou, G., Kokkinos, I., Murphy, K., Yuille, A.L.: Yuille: Semantic image segmentation with deep convolutional nets and fully connected CRFs. Comput. Sci. **4**, 357–361 (2014)
12. Zheng, Y., Romera-Paredes, V., Su, D., Huang, T.: Conditional random fields as Recurrent Neural Networks. In: IEEE International Conference on Computer Vision, pp: 1529–1537 (2015)
13. Badrinarayanan, V., Kendall, A., Cipolla, R.: SegNet: a deep convolutional encoder-decoder architecture for image segmentation. arXiv preprint arXiv:1511.00561 (2015)
14. Simonyan, Z.: Very deep convolutional networks for large-scale image recognition. Comput. Sci. (2014)
15. Ioffe, S.: Batch Normalization: accelerating deep network training by reducing internal covariate shift. Comput. Sci. (2015)

P2P Traffic Identification Method Based on Traffic Statistical Characteristics

Bing Hu and Zhixin Sun(✉)

Key Laboratory of Broadband Wireless Communication and Sensor Network Technology, Nanjing University of Posts and Telecommunications, Ministry of Education, Nanjing, China
sunzx@njupt.edu.cn

Abstract. Different from traditional P2P traffic identification methods based on keywords or well-known ports, this paper presents a new identification method of P2P traffic based on the most basic characteristics of P2P protocol. We compare and analyze P2P traffic and traditional C/S traffic using statistical analysis techniques, and obtain two statistical characteristics of P2P traffic: continuity and multi-connectivity. In addition, the mechanism of sliding window is introduced in the quantification of the statistical characteristics to establish a P2P traffic identification model. Finally, we develop a P2P traffic identification emulation system based on the proposed model, and the experiment results reveal that this new model can identify known and unknown P2P traffic effectively.

Keywords: Traffic identification · Statistical characteristic · Sliding window · Peer-to-Peer (P2P)

1 Introduction

The concept of Peer-to-Peer (P2P) network service is first proposed by Steve Crocker in 1969 [1]. Each peer in the P2P network can communicate or share files with any other peers independently and directly. In recent years, P2P technology has been paid widespread attention. The main application areas of P2P technology include information sharing, real-time communication, online games, financial service, information retrieval and network storage [2].

P2P technology is not only able to provide fast and efficient file sharing and low-cost and high-availability computing and storage resources sharing, but also holds a strong network connectivity as well as more direct and flexible information communication capability. However, the benefits of P2P technology have also brought some new problems. (1) Due to the high transmission speed of P2P, its users occupy 60%–80% of the network bandwidth, while leaving limited bandwidth for non-P2P users, which can easily cause the obstruction of enterprise and ISP bottleneck link. (2) The vast volume of P2P downloads consumes large amounts of network bandwidth, which prevents other users from accessing the network resources normally. (3) Continuous high-speed downloads of P2P users not only increase the load of network equipment, but also easily cause link congestion at peak time.

D.-S. Huang et al. (Eds.): ICIC 2017, Part I, LNCS 10361, pp. 260–272, 2017.
DOI: 10.1007/978-3-319-63309-1_24

Hence, it is urgent to identify P2P traffic and control it properly [3]. In this paper, based on the analysis of the characteristics of P2P traffic through experiments, we take the statistical characteristics of P2P traffic, continuity and multi-connectivity, as the basis of P2P traffic identification model to identify P2P traffic with known and unknown protocols. Finally, the validity of the proposed method is verified by developing a P2P traffic identification system.

The main contributions of this paper are highlighted as follows:

(1) From the perspective of the monitoring host, we analyze the characteristics of P2P traffic through experiments and the continuity and multi-connectivity of P2P traffic are selected as the basis of traffic identification model.
(2) A multi-connectivity quantization method is proposed based on an improved address aggregation method, which not only counts the number of IP address, but also saves a lot of space.
(3) According to the proposed traffic identification model, we develop a P2P traffic identification system to identify P2P traffic.

The rest of this paper is organized as below. Section 2 illustrates the related work. In Sect. 3, we describe the analysis of P2P traffic statistical characteristics. The methods for identifying P2P traffic and the strategies of quantifying its traffic characteristics are illustrated in detail in Sect. 4. In order to verify the effectiveness and practicality of the identification method, Sect. 5 presents a simulation experiment. Conclusions are provided in the last section.

2 Related Work

During the past decade, many approaches employ statistical methods to identify the P2P traffic are proposed. Reference [4] proposes a method based on the characteristics of the transmission layer communication mode of P2P application, such as the characteristics of communication protocol in transport layer and the relationship between ports and destination IP, to identify P2P traffic. More depth analysis of various kinds of traffic (such as P2P, WEB and mail) are carried out in [5]. According to the basic characteristics of various kinds of traffic, a new traffic classification and identification method, called BLINC, is proposed [5]. BLINC takes the network host as the research object, and the location analysis of network host is carried out from three aspects: (1) Social level, revealing the popularity of the monitoring host via the number of hosts that communicate with it. (2) Functional level, the role a host plays, server or client. (3) Application level, the communication characteristics of the host in transport layer. Reference [6] proposes a method that uses fundamental characteristics of P2P protocols, such as a large network diameter and the presence of many hosts acting both as servers and clients, without any application-specific information, to identify both known and unknown P2P protocols in a simple and efficient way. Reference [7] uses the statistical characteristics of P2P traffic, such as connection success rate and connection response success rate, to identify and classify P2P traffic. Based on the method proposed in [4], Ref. [8] combines all the characteristics of P2P traffic in [4] with a new characteristic, DNS query log feature, to create P2P hosts feature set. And then use flexible neural tree (FNT) model to identify P2P hosts and traffic. Reference [9] presents a novel approach which relies on counting some special traffic that are appearing frequently and steadily in the traffic generated by specific P2P applications to identify P2P traffic at a fine-grained level. Besides, it is also able to identify encrypted traffic in real-time. Reference [10] develops a framework named CUFTI (Core Users Finding and Traffic Identification) to identify and manage P2P traffic of core users, i.e. long-lived peers. The model employed the direction and payload length of the control packets at the beginning of the traffic as the characteristics to perform traffic identification. Furthermore, the model can be employed for real-time traffic identification.

According to the related works, we can find that identification method based on traffic characteristics is an effective way for identifying P2P traffic with known P2P protocols and unknown P2P protocols. In this paper, we propose a new P2P traffic identification approach based on the statistical characteristics of P2P traffic, which can effectively identify known and unknown P2P traffic.

3 Characteristics of P2P Traffic

Due to the difference between P2P network and C/S network, their traffic also shows many differences. Compared with the traditional C/S traffic, P2P traffic presents many characteristics, such as higher transfer speed, heavier traffic and occupying more bandwidth. We analyze the current popular multiple P2P application software with using Ethereal to capture P2P traffic, and with the help of the statistical function of Ethereal to study the P2P traffic deeply. Then, compared with normal applications traffic, two important statistical characteristics of P2P traffic are found, strong continuity and multi-connectivity, which will be elaborated in the next.

3.1 Continuity

Network traffic of the host computer which is running P2P applications is not necessarily all P2P traffic. In order to distinguish the normal application traffic and the P2P application traffic, the average packet length of P2P traffic is analyzed. We use the "IO Graphs" analysis method of Ethereal analysis tool to analyze the byte rate of the traffic involved in the monitoring host. The average packet length of the eMule traffic (P2P traffic) with the time interval 0.1 s and 1 s is illustrated in Fig. 1, respectively. And Fig. 2 shows the results of the Thunder analysis. Thunder is a popular P2P download software in China. As can be seen from the figures, with the increase of analysis granularity, the average packet length of P2P traffic tends to be a fixed large size.

The average packet length of web traffic (non-P2P traffic) is showed in Fig. 3. It can be seen by comparing Figs. 2 and 3. When the analysis granularity is small (such as 0.1 s), the average packet length of both Thunder and web traffics fluctuates greatly with time. However, when the analysis granularity is increased to 1 s, the average packet length of web traffic still fluctuates greatly, and sometimes tend to be 0, but the average packet length of P2P traffic is stable.

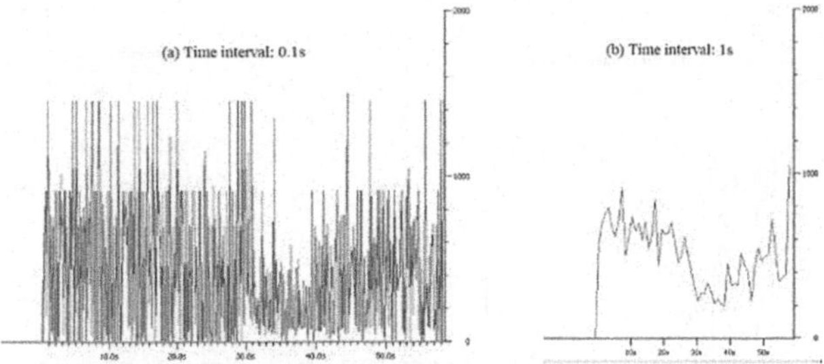

Fig. 1. Average packet length of eMule traffic.

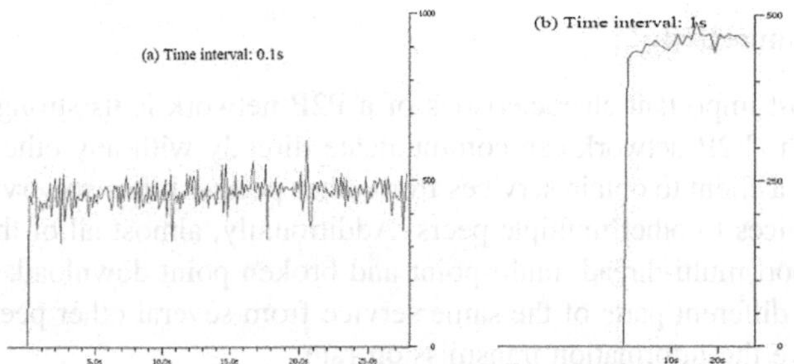

Fig. 2. Average packet length of Thunder traffic.

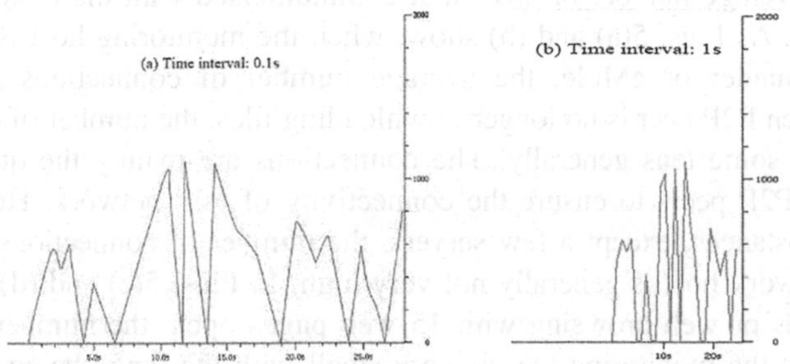

Fig. 3. Average packet length of web traffic.

The analysis results of the average packet length for FTP traffic (non-P2P traffic) are shown in Fig. 4. Compared with Fig. 1, average packet length of FTP traffic exhibits the same characteristics as that of eMule traffic. Therefore, FTP traffic (non-P2P traffic) and eMule (P2P traffic) cannot be distinguished only based on the characteristic of the average packet length.

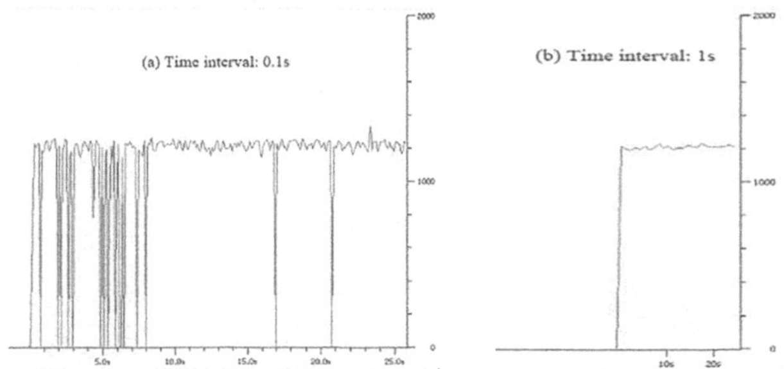

Fig. 4. Average packet length of FTP traffic.

3.2 Multi-connectivity

One of the most important characteristics of a P2P network is its strong connectivity. Each peer in the P2P network can communicate directly with any other peer, that is, any peer act as a client to obtain services from other peers while also serving as a server to provide services to other multiple peers. Additionally, almost all of the current P2P protocols support multi-thread, multi-point and broken point download. Hence, a peer can obtain the different parts of the same service from several other peers at the same time to improve the information transmission rate.

We still utilize Ethereal to capture the P2P traffic of the monitoring host and further use its "conversations" method to establish a "time-connection" table for each kind of traffic. The table records the average number of connections of various traffic, which indicates the average number of hosts that communicated with the monitoring host at different times. As Figs. 5(a) and (b) show, when the monitoring host is downloading files using Thunder or eMule, the average number of connections remains high. Moreover, when P2P peer is no longer downloading files, the number of connections is still high with some tens generally. The connections are mainly the query messages sent by other P2P peers to ensure the connectivity of P2P network. However, under normal circumstances, except a few servers, the number of connections for the traditional C/S network host is generally not very high. In Figs. 5(c) and (d), whether it is FTP downloads or web browsing with 15 web pages open, the number of hosts that connected with the monitoring host is very small with the range being from 5 to 15. Unlike P2P network, in the C/S network, client hosts only communicate with the server rather than with other client hosts.

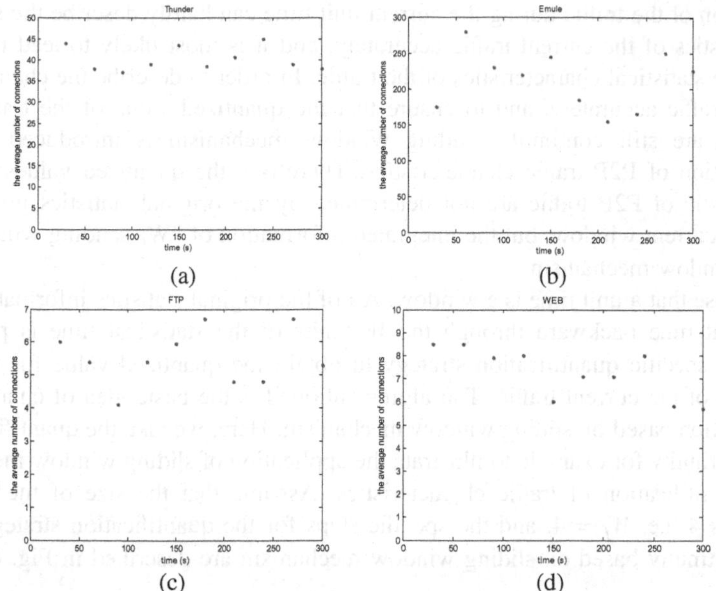

Fig. 5. Average number of connections.

4 Proposed P2P Traffic Identification Method

In this section, a P2P traffic identification method is proposed, which is based on the statistical characteristics of P2P traffic and the sliding window mechanism. The symbols used in this paper are illustrated as follows.

ti: Basic unit time for collecting the raw data of the traffic.

$f_t = (f_{1,t}, f_{2,t})$: The statistical state of the network traffic during the unit time t; $f_{1,t}$: quantized value of the traffic continuity; $f_{2,t}$: quantized value of the traffic multi-connectivity.

W_f: The size of sliding window.

TB_{ti}: The total number of bytes obtained by monitoring the host traffic during the unit time *ti*.

Statistical characteristics of P2P traffic are descriptive statistical characteristics, which need to be quantified to be used in the identification model. In addition, the quantized value can effectively reflect the strength degree of the traffic characteristics. In this paper, the quantized value of traffic characteristics is $f_{i,t} \in [0, 1],\ i = 1, 2$. The closer the value is to 1, the more obvious the corresponding characteristics. For example, the continuity of the traffic of $f_{1,t} = 0.8$ is stronger than that of $f_{1,t} = 0.6$.

4.1 Application of Sliding Window in Quantification of Traffic Characteristics

Quantifying characteristics of P2P traffic are quantifying the statistical information of specific characteristic within a unit time. However, use only the original statistics information of the traffic during the current unit time can hardly describe the statistical characteristics of the current traffic accurately, and it is most likely to lead to a great leap in the statistical characteristics of the traffic. In order to describe the change of the network traffic accurately, and to ensure that the quantized value of the traffic characteristics are still continuity, sliding window mechanism is introduced into the quantification of P2P traffic characteristics. Therefore, the quantized values for each characteristic of P2P traffic are not determined by the original statistics information about the current window, but the integrated information of "W_f sending windows" in sliding window mechanism.

Suppose that a unit time is a window. All of the original statistics information from the current time backward through the W_f units of the statistical time is processed through a specific quantification strategy to obtain the quantized value for the characteristics of the current traffic. The aforementioned is the basic idea of characteristic quantification based on sliding window mechanism. Here, we take the quantification of traffic continuity for example to illustrate the application of sliding window mechanism in the quantification of traffic characteristics. Assume that the size of the statistics window is 4, i.e. $W_f = 4$, and the specific steps for the quantification strategy of the traffic continuity based on sliding window mechanism are described in Fig. 6.

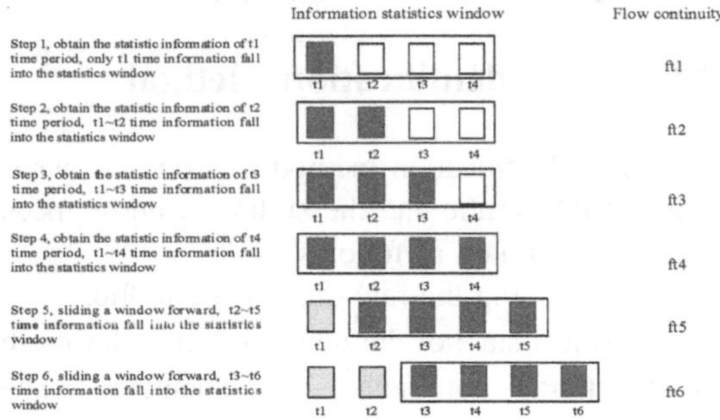

Fig. 6. Procedure of quantifying flow continuity.

(1) Initially, as step 1 shows, though the window size is 4, $t1$ is the only effective window. We can only get the statistical value TB_{t1} during the unit time $t1$, so the quantification value of the traffic continuity $f_{1,t1}$ is only decided by TB_{t1}, and the $\mathrm{TB}_{ti},\ i = 2, 3, \cdots$ is 0 by default. Then defining $f_{1,t1} = \begin{cases} 1, & TB_{t1} > 0 \\ 0, & TB_{t1} = 0 \end{cases}$. Sliding window is not moved in this step.

(2) As shown in step 2 and 3, TB_{t2} and TB_{t3} are collected during the unit time $t2$ and $t3$ respectively. The quantification value of the traffic continuity is determined by the windows that have already collected its information and the $\mathrm{TB}_{ti},\ i = 4, 5, \cdots$ is 0 by default. So the quantifying formula is

$$f_{1,ti} = \frac{\sum_{n=1}^{i} TB_n}{i},\ i = 2,3;\ TB_n = \begin{cases} 1, & TB_{tn} > 0 \\ 0, & TB_{tn} = 0 \end{cases},\ n = 1,2,3,\cdots$$

After the step 3, the window is still not fully occupied, so sliding window keep the same.

(3) As step 4 shows, the statistics of the fourth window is obtained and the four windows are all occupied, so the quantification value $f_{1,t4}$ is determined by all the statistical results of 4 time spans in the sliding window. Then defining

$$f_{1,t4} = \frac{\sum_{n=1}^{4} TB_n}{4};\ TB_n = \begin{cases} 1, & TB_{tn} > 0 \\ 0, & TB_{tn} = 0 \end{cases},\ n = 1,2,3,\cdots$$

(4) Information statistics window should move forward with a unit of time window to obtain the statistics of the fifth window, so the quantifying formula is

$$f_{1,t5} = \frac{\sum_{n=5-W_f+1}^{5} TB_n}{W_f};\ TB_n = \begin{cases} 1, & TB_{tn} > 0 \\ 0, & TB_{tn} = 0 \end{cases},\ n = 1,2,3,\cdots$$

(5) Then, continue to move forward a window to vacate a new window for obtaining the statistics of the following time span.

4.2 Quantification of Traffic Continuity

P2P traffic continuity represents the continuity of the network traffic that the P2P host participated in. If the network traffic of the monitoring host is not empty in a certain period of time, the host has a strong traffic continuity. Otherwise, if the paroxysm or return to 0 phenomenon occurs in the network traffic of the monitoring host, it means the traffic continuity of the host is weak. According to the research in Sect. 2, P2P peers have no difference from common network hosts when they do not participate in resource downloading or information sharing, in this case, the continuity of their network traffic should be very weak. In contrast, strong continuity of network traffic will appear when the hosts participate in resource downloading or information sharing.

If the total network traffic of the monitoring host is 0 in a unit time, it is considered that the host's traffic appears return to zero phenomenon in the time unit. Otherwise the host's traffic does not return to zero. Calculate the proportion of the time windows that does not return to zero from the current time back to the W_f time units. This ratio is the quantized value of the traffic continuity. As time forwards, statistical windows move forward and the statistics of those windows is always the traffic statistics of the latest W_f time units. Then, the algorithm of calculating $f_{1,t}$ is shown in Algorithm 1.

Algorithm 1

```
temp=0;
for the size of statistics window is W_f
  if TB_ti ≠ 0
    temp++;
  end if
end for
```

$$f_{1,t} = \frac{temp}{W_f}$$

4.3 Quantification of Traffic Multi-connectivity

Compared with the traditional web services clients, P2P peers have the characteristics of multi-connectivity. In a unit time, the amount of the hosts that communication with the monitoring host can effectively reflect the multi-connectivity property of P2P traffic. In this paper, we use the number of IP addresses that communicate with the monitoring host to represent the multi-connectivity of the traffic.

Since the quantification of multi-connectivity only needs to record the number of IP addresses that communicate with the monitoring host, and do not need to know the specific IP address. We propose an improved address aggregation method which can not only count the number of IP addresses, but also save a lot of space.

We divide the IP address into K segments and each segment take up a space of L bits. Set up a hash mapping table with a size of $K \times 2^L$ bits. As Fig. 7 shows, divide IPv4 address into six domains: a, b, c, d, e, f, where five domains occupy 6 bits respectively, the remaining domain only 2 bits. So we can set up a 6×64 hash mapping table, using the decimal value mapping function to map the 6 domains respectively. Since the IPv4 address is 32 bits, there are only 4 valid spaces in the last column of the hash mapping table and the remaining 60 spaces will never be mapped to. Therefore, the actual valid space of the hash mapping table is

$$hashlength * hashwidth + ((oxooo1) \ll (32\%hashwidth)) - hashlength$$

where hashwidth indicates K and hashlength is 2^L. The result that the IPv4 address 10.10.10.1 expressed in dotted decimal notation form through hash mapping is represented in Table 1.

For a packet received within a unit time, if its source IP address or destination IP address is the address of the monitoring host, the six domains of the corresponding destination IP address or source IP address are mapped into the address aggregation mapping table, at the same time, add 1 to the corresponding domain value in the address aggregation mapping table respectively. The aggregation degree of the corresponding domain of the hosts that communicate with the monitoring host is recorded by the domain value. Then, the algorithm of quantifying the multi-connectivity of network traffic by using address aggregation technology is showed in Algorithm 2.

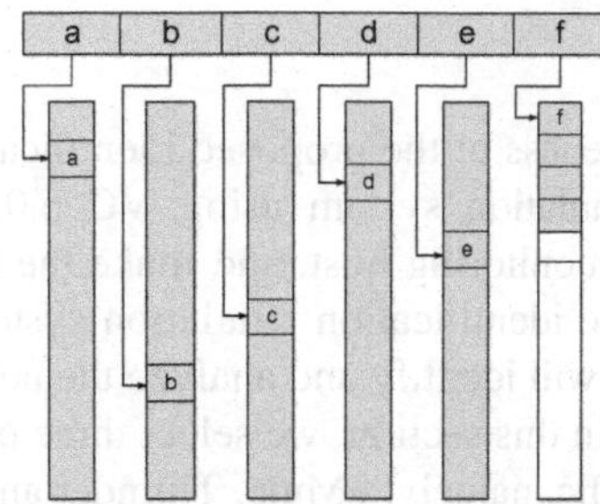

Fig. 7. hashTable: address aggregation mapping table.

Table 1. An example for address aggregation hash mapping.

IP address in dotted decimal notation form	10.10.10.1					
Binary form	00	001010	000010	100000	101000	000001
Hash value	0	10	2	32	40	1
Corresponding domain	f	e	d	c	b	a

Algorithm 2

```
temp=0;
k=0;
for the size of statistics window is W_f
    k=the number of the nonzero filed in the table hashTable during the current unit time ti ;
    temp=temp+k;
end for
```

$$f_{2,t} = \frac{temp}{hashlength * hashwidth + ((oxooo1) << (32\%hashwidth)) - hashlength}$$

P2P traffic identification and control system only record the aggregation information of the hosts that communicate with the monitoring host in the latest W_f unit time, and the statistics of the traffic multi-connectivity are carried out in accordance with Algorithm 2 during every time unit. Since the traffic multi-connectivity of the monitoring host is from the traffic statistics in the latest W_f time units, it basically reflects the real-time changes of the current traffic. Furthermore, the system only needs to keep W_f address aggregation mapping table with a fixed size for every host in the local network to achieve the statistics of traffic multi-connectivity with saving a lot of space. On the other hand, the realization of the method is very simple and effective.

5 Emulation

In order to verify the effectiveness of the proposed identification method, we develop a P2P traffic identification emulation system using VC 6.0. We install a number of different P2P software on a monitoring host, and make the host become a node in the P2P network. The P2P traffic identification emulation system is also installed on the monitoring host. The system will identify and analyze the network traffic when the host runs different P2P software. In this section, we select three popular P2P applications as the representative of P2P traffic, namely: eMule, Thunder and Vagaa. Then continue to run 24 h on the monitoring host. Similarly, all kinds of non-P2P applications are run randomly on the monitoring host, such as, FTP download and web browsing, the running time is also 24 h. The P2P traffic identification emulation system captures and stores all the network traffic generated during the application runs at the experimental network export, and sets the capture data into experimental data set. The system combines the two features mentioned above to form a feature vector, and uses the support vector machine algorithm to identify the P2P traffic.

In the system, users can set up the IP address of the monitoring host, analysis interval and the size of the analysis window. First, we reveal the influence of sliding window size on the identification accuracy by setting different size of the window. As Fig. 8 shows, when the sliding window size is set to 4, the identification accuracy is high, and with the increase of the window size, the identification accuracy is not improved obviously. Therefore, we set the sliding window size to 4 in the next experiment.

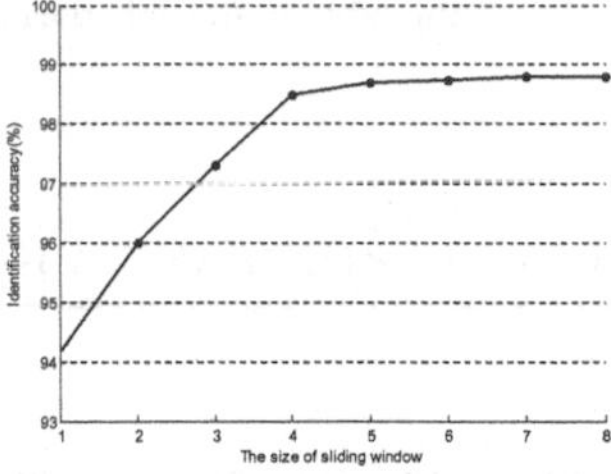

Fig. 8. The influence of sliding window size on the identification accuracy

Next, we exhibit the traffic analysis graph of P2P traffic identification emulation system for difference applications, where the red line represents the curve of traffic continuity quantization and the blue line represents the curve of traffic multi-connectivity quantization, and the identification result of the current traffic is displayed in the right box of the system. Figure 9 shows the traffic analysis graph of a P2P software with unknown protocol, i.e. Vagaa. In the test, Vagaa downloads two video files at the rate of 37 KB/s and 64 KB/s, respectively. It can be seen from the figure, the continuity quantization value of the Vagaa traffic is mainly around 1.0 and its multi-connectivity quantization value is changed between 0.4 and 0.8. The system determined the current traffic is P2P traffic.

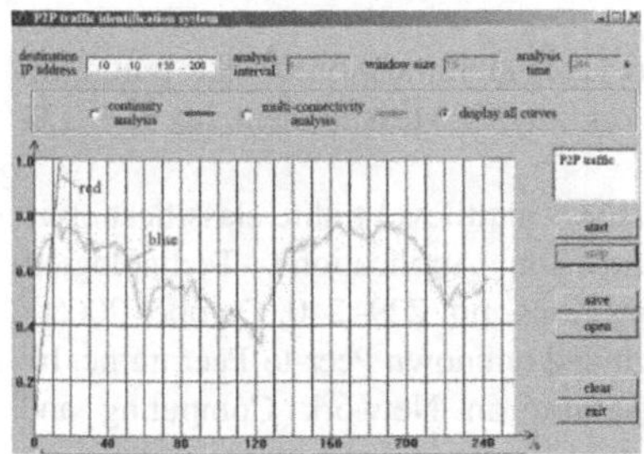

Fig. 9. Analysis graph of Vagaa traffic

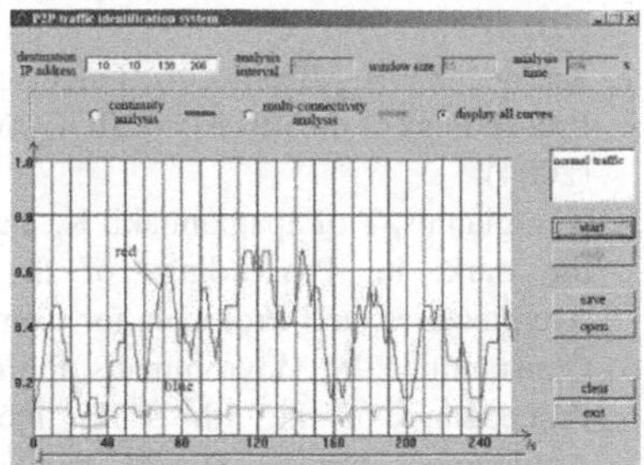

Fig. 10. Analysis graph of web traffic.

Figure 10 is the analysis graph of web traffic. As the figure shows, its continuity quantification values fluctuate greatly between 0.0 and 0.7. It indicates that the continuity of web traffic is weak. Although in the process of web traffic analysis, we open up to 20 different web pages continuously, the multi-connectivity quantification values generally stay below 0.2. The identification result shown in the right box is normal traffic, i.e. non-P2P traffic.

6 Conclusions

In this paper, we study the characteristics of P2P traffic start from the basic characteristics of a P2P network, and take the statistical characteristics of P2P traffic, continuity and multi-connectivity, as the basis of P2P traffic identification model. Compared with the traditional P2P traffic identification methods, the P2P identification method based on statistical characteristics presented in this paper has many advantages. First, the traffic identification method is easy to carry out without knowing the content of traffic itself. Secondly, the method can identify P2P traffic with known and unknown protocols. Admittedly, the P2P traffic identification method proposed in this paper cannot identify the specific traffic type that comes from which P2P software, which will serve as our future research direction.

Acknowledgments. This work is supported by the National Natural Science Foundation of China (No. 61373135, No. 61672299).

References

1. Horng, M.F., Chen, C.W., Chuang, C.S.: Identification and analysis of P2P traffic-an example of bittorrent. In: Proceedings of the First International Conference on Innovative Computing, Information and Control, pp. 266–269, 30 August–1 September 2006
2. Chen, Z.Q., Delis, A., Wei, P.: Identification and management of sessions generated by instant messaging and Peer-to-Peer systems. Int. J. Coop. Inf. Syst. **17**, 1–51 (2008)

3. Bhatia, M., Rai, M.K.: Identifying P2P traffic: a survey. Peer-to-Peer Netw. Appl. **9**, 1–22 (2016)
4. Karagiannis, T., Broido, A., Faloutsos, M.: Transport layer identification of P2P traffic. In: Proceedings of the 4th ACM SIGCOMM Conference on Internet Measurement, pp. 121–134 (2004)
5. Karagiannis, T., Papagiannaki, K., Faloutsos, M.: BLINC: multilevel traffic classification in the dark. In: Proceedings of the 2005 Conference on Applications, Technologies, Architectures, and Protocols for Computer Communications, pp. 229–240, October 2005
6. Constantinou, F., Mavrommatis, P.: Identifying known and unknown Peer-to-Peer traffic. In: Proceedings of the 5th IEEE International Symposium on Network Computing and Applications, pp. 93–102, 24–26 July 2006
7. Matsuda, T., Nakamura, F., Wakahara, Y.: Traffic features fit for P2P discrimination. In: Proceedings of the 6th Asia-Pacific Symposium on Information and Telecommunication Technologies, pp. 230–235, 9–10 November 2005
8. Chen, Z., Wang, H., Peng, L.: A novel method of P2P hosts detection based on flexible neural tree. In: Proceedings of the of 6th International Conference on Intelligent Systems Design and Applications, pp. 556–561, 16–18 October 2006
9. He, J., Yang, Y., Qiao, Y.: Fine-grained P2P traffic classification by simply counting flows. Front. Inf. Technol. Electron. Eng. **16**, 391–403 (2005)
10. Qin, T., Wang, L., Zhao, D.: CUFTI: methods for core users finding and traffic identification in P2P systems. Peer-to-Peer Netw. Appl. **9**, 424–435 (2016)

ChinFood1000: A Large Benchmark Dataset for Chinese Food Recognition

Zhihui Fu, Dan Chen, and Hongyu Li(✉)

ZhongAn Information Technology Service Co., Ltd, Shanghai, China
{fuzhihui, chendan, lihongyu}@zhongan.io

Abstract. In this paper, we introduce an 1000-category food dataset ChinFood1000 and propose a simple and effective baseline approach. To our best knowledge, the proposed ChinFood1000 dataset enjoys the largest number of food categories among all publicly available food dataset currently. The categories of the ChinFood1000 dataset are carefully selected to include the most popular Chinese dishes. The dataset includes 852 categories of Chinese dishes, together with 91 classes of drinks and snacks, 26 kinds of fruits and 31 kinds of other food. The images in the dataset present both large inter-class affinity and high intra-class variance. To illustrate the challenges presented by the dataset, a baseline based on a very deep CNN is proposed. In the experiments, the baseline approach is evaluated on three most widely used food datasets and achieves the best performance on all of them. The baseline approach is also applied to the ChinFood1000 dataset, with a promising accuracy reported.

Keywords: Chinese food recognition · Benchmark dataset · CNN · Chin-Food1000

1 Introduction

Food recognition is an attractive computer vision problem, which can be regarded as a basic building block, aiding tasks including but not limited to visual calorie estimation, recording and analyzing of food intaking and food recipe matching.

Different to typical objects like dogs and cars, food rarely has structured layout, which makes Chinese food classification more challenging. Firstly, the ingredients of Chinese food have various shapes, such as stripes, chops and blocks. Secondly, Chinese food have many cooking styles, leading to complex mixing ways of the ingredients, e.g. stewing, steaming and potting. In addition, Chinese food has a great variety of dishes. There exists at least eight series of traditional Chinese cuisines, each of which is composed of hundreds of dishes.

For now, there have been limited food datasets publicly available. PFID [1] introduced a small dataset of fast food, containing only 13 classes. Moreover, the 4545 still images and 606 stereo pairs of fast food in the dataset are captured in labs and restaurants. UEC100 [2] is mainly composed of Japanese dishes, with 100 categories. The images are provided together with bounding boxes to include both the food objects and the environment. The number of food categories is enlarged to 256 in the dataset UEC256 [5]. But it still focuses on the Japanese food rather than the Chinese dishes.

D.-S. Huang et al. (Eds.): ICIC 2017, Part I, LNCS 10361, pp. 273–281, 2017.
DOI: 10.1007/978-3-319-63309-1_25

Bossard et al. [6] introduced a western food dataset Food101, with high noise resulting in the unsatisfactory classification accuracy in the practical applications. VIREO Food-172 [7] is the only available Chinese food dataset until now. It collected 172 classes of Chinese food images labeled by both food category and food ingredients. However, the number of food categories is still small, which is hard to capture the broadness and variety of the Chinese food. As a first step to increase the number of categories for Chinese food, we introduce an 1000-category food dataset, namely ChinFood1000. The 1000 categories are selected to contain the most popular Chinese food. Figure 1 shows some examples of the images from the ChinFood1000 dataset. The images in the dataset demonstrate large inter-class similarity and intra-class variance. The first two rows illustrates the large inter-class similarities. For example, Braised tofu (row2 left) and Mapo tofu (row 2 right) are very visually similar, containing the same major ingredient tofu and share the red color. The last three rows demonstrate the large intr-class variance, e.g. Vermicelli with Minced Pork (row 5 left) has numerous cooking styles, different shapes and colors within the category.

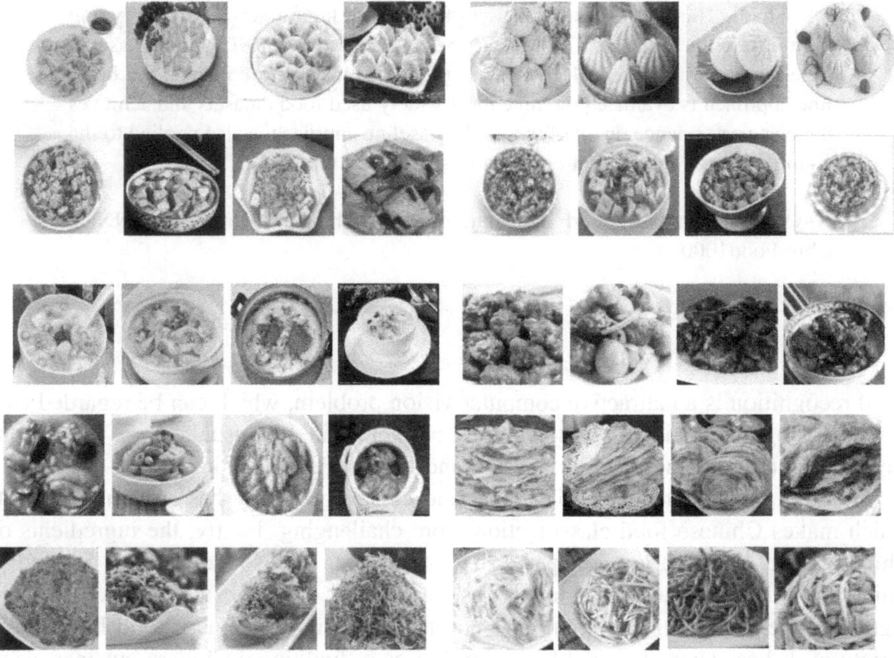

Fig. 1. Examples from ChinFood1000. Each row contains two different groups of Chinese dishes. Note the images in the dataset demonstrate large inter-class similarity and intra-class variance. The first two rows show the examples of the inter-class similarities in the dataset. For example, Braised tofu (row 2 left) and Mapo tofu (row 2 right) are very visually similar, containing the same major ingredient tofu and with close red color. The last three rows show examples of the high intra-class variances, e.g. Vermicelli with Minced Pork (row 5 left) has numerous cooking styles, different shapes and colors within the category.

Following traditional object recognition approaches, methods based on color histogram or SIFT BoW features have been widely applied to food classification [1, 3]. To further improve the robustness to the noise, Random Forest is utilized to mine discriminative parts of specific food [6]. However, the hand designed features is limited to low-level representations such as edges and textures, which restricts the classification accuracy. Liu C. et al. proposed a food recognition method [8] based on DCNN. Although the adopted network uses an efficient bottleneck structure, the accuracy is still limited.

In this paper, we propose a simple but yet effective baseline approach. Inspired by the strong modeling ability and the ease of optimization of the ResNet [9], we finetuned a deep 50-layer ResNet on ChinFood1000 as a feature learner. Upon it, a One-Vs-All logistic regressor is trained. Under the multi-stage learning framework, the number of food categories can be easily expanded and the classifiers can be trained in an online learning style. The baseline approach is evaluated on three publicly available food datasets. It is shown that the baseline approach achieves the best performances on the UEC and the Food101 datasets. It also achieves a promising accuracy on the ChinFood1000 dataset.

The rest of the paper is organized as follows: We first describe how we collect the ChinFood1000 dataset and analyse its properties in Sect. 2. The proposed baseline approach is described in Sect. 3. Experimental results are presented in Sect. 4 to show the effectiveness of the baseline approach and the challenges in the ChinFood1000 dataset.

2 The ChinFood1000 Dataset

In this paper, we introduce a 1000 categories data, the ChinFood1000 dataset, mainly composed of Chinese dishes. Some drinks, fruits and snacks are also included. To our best knowledge, for now the number of food categories is the largest among all the publicly available food datasets.

2.1 Dataset Collection

The dataset collection is divided into two steps. The names of the dishes to collect are firstly specified and selected with a naming strategy. Then images are crawled and cleaned according to the selective names.

We firstly specify the category names of the food in the dataset. The name selection strategy is shown in Fig. 2. The collected food in the dataset should be non-duplicated, and should contain the most ordinary dishes in China. Several lists of food names are obtained from Chinese cooking websites 'Xinshipu'[1], 'CNDZYS'[2] and diet website '39Jiankang'[3]. In addition to food names, we also crawled the food ingredients. As a

1 http://www.xinshipu.com/.
2 http://www.cndzys.com/.
3 http://fitness39.net/.

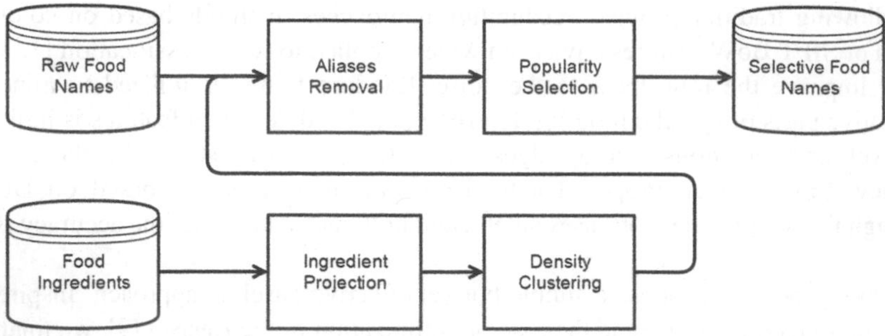

Fig. 2. The strategy for food name selection. In the projected ingredient space, density clusters are utilized to find name aliases. After aliases removal, popular food names are filtered out to get the selective food names.

result, each food in the lists corresponds a food ingredient vector. To obtain the most popular food in China, all lists are intersected to form the raw food names. However, a food name may have numerous aliases. For example, dumpling and Jiaozi are the same food. Fortunately, we observe that aliases usually share very similar major ingredients. Donate S as ingredient matrix, each row S_i is an ingredient vector contains 0s and 1s, where 1 indicates the ingredient is used in the food and vice verse. We aim to find an ingredient space that projects the ingredients to the directions with the most variations:

$$T = W\hat{S} \tag{1}$$

Where $\hat{S}$ is the centering matrix of S,W is the left singular vectors of $\hat{S}$ To find density peaks, density clustering such as DBSCAN is performed on T. After that, the density peaks are manually looped to remove the food aliases. In addition, it is assumed that the length of the names for popular food tends to be short according to the Chinese food traditions. Any name longer than a specific threshold is discarded to obtain the selective results. In practice, the threshold is configured as 7 Chinese characters. Finally, we manually add some popular food and obtain the names of food in the dataset. As a result, the food category names include 852 Chinese dishes, 91 classes of drinks and snacks, 26 kinds of fruits adn 31 kinds of other food.

With the selective food names as keyword, the relative images were down- loaded from the Baidu image search engine[4]. The search results are truncated and only the most relevant images are preserved.

In addition, since a certain image may appear in the search results of different food categories due to the close major ingredients, we perform filtering [12] on the crawled images by re- moving replicative pictures. During filtering, a table is constructed to remove the replicative images, where the keys are the MD5 hash of the image files and the value of each key is the occurrence in the dataset. Finally, we double check the images to avoid the occurrence of the irrelevant images.

[4] Baidu image search: http://image.baidu.com.

2.2 Dataset Properties

In this section, three different properties of the ChinFood101 dataset are analysed. Images Per Class and Noise shows that the images of each class are fairly balanced and the labeling is accurate. Meanwhile we define Diversity to show the intra-class variances of the dataset.

Images Per Class. We plot the distribution of the number of images for each category in Fig. 3. The amount of images of each class is fairly balanced. Interestingly, the number of images of a more popular category tends to be larger and vice verse. A possible reason is that the images of popular food is more likely to be taken and uploaded to the Internet. Thus the images are more accessible by the Internet image search engines.

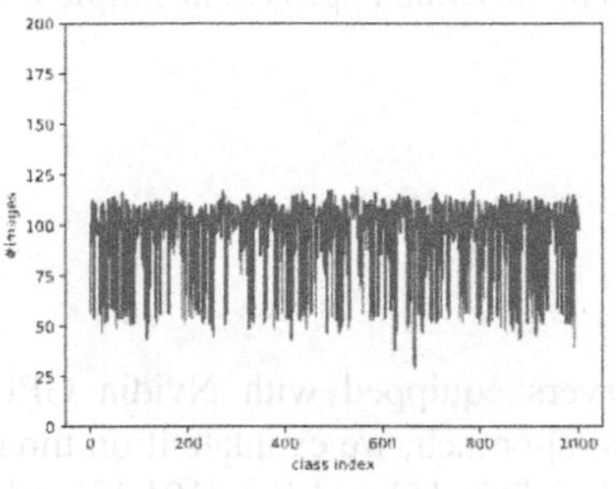

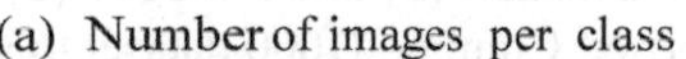

(a) Number of images per class

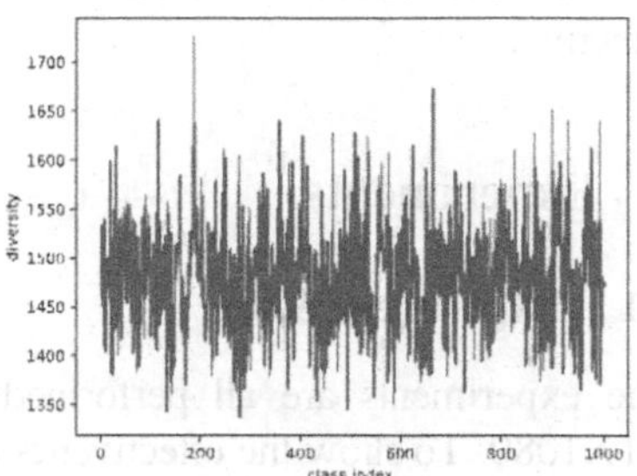

(b) Diversity of images per class.

Fig. 3. Dataset properties of ChinFood1000. Left: the number of images per class; Right: the diversity of images per class. The diversity is measured by the Lossless JPG file size of the average images in each class.

Noise. To estimate the degree of noise of the dataset, we randomly pick 5% images from the dataset and check validity of the corresponding labelings. The correctness of our labeling is about 86%, which proves that the noise is acceptable.

Diversity. Inspired by ImageNet [13], the average image of each category is computed. Lossless JPG encoding is performed on the average image. Intuitively, the average of diverse images tends to be blurrier and vice verse. We compute the average images of each category, the file sizes are depicted in Fig. 3. The classes in ChinFood1000 provides a large variance in intra-class diversity.

3 The Baseline Approach

We adopt a multi-stage learning framework for the baseline approach. There has been several papers [4, 8] that adopts DCNN as the feature learner for food recognition. However, to our best knowledge, ultra deep networks have not been used in the food classification. In this paper, a Residual Convolutional Neural Network model [9] is utilized for feature learning. The network has 3 bottleneck units with 256d output, 4 with 512d output, 6 with 1024d output and 3 with 2048d output. Every bottleneck unit has a 1 × 1 conv layer followed by a 3 × 3 conv layer and a 1 × 1 conv layer. Identity mapping is used to ease the difficulty of optimization.

The network is initialized by pretrained parameters on ImageNet. We fine- tune the network parameters via SGD to minimize the softmax log loss. The batch size is fixed as 10. The learning rate is initialized as 1e5 for the last fc layer and 1e4 for all the other layers. The learning rate is decreased by a factor of 0.1 after each epoch. The network is finetuned for 5000 iterations.

On the next stage, the output of the average pooling layer is used as the features. We apply One-Vs-All Logistic Regression on the features to classify the food. Under the multi-stage learning framework, it is relatively easy to expand the number of categories of food. Moreover for a light-weight expansion, it is very efficient since there is no need to retrain the feature learner. As a result, The baseline approach is simple but effective.

4 Experiments

4.1 Experiment Setup

The experiments are all performed on Linux servers equipped with Nvidia GPU GTX1080. To show the effectiveness of the baseline approach, we evaluate it on three publicly available food benchmarks: UEC100 [2], UEC256 [5] and Food101 [6]. We then evaluate the baseline approach upon the proposed ChinFood1000 dataset to show the challenges provided by the dataset. We split the Chin-Food1000 dataset into training/testing data. Totally 20% of images are used for testing. The training and testing splits are fixed for the convenience of bench- marking. Two measures, top-1 and top-5 accuracies, are adopted to estimate the performance of the baseline approach.

4.2 Baseline Results and Discussion

In this section, we firstly evaluate the proposed baseline approach. The base- line approach is evaluated on the following datasets: UEC100 [2], UEC256 [5] and Food101 [6]. We compare the baseline approach to both deep CNN based methods and the traditional methods. Specifically, the following approaches are utilized for comparison:

DeepFood DeepFood [8] is a state-of-the-art CNN based food classification algorithm. The authors adopt a finetuned GooLeNet [10] and report a great margin of accuracy gain over traditional methods based on hand designed features.

Deep-UEC Deep-UEC [4] is another DCNN based food classification method. OverFeat [11] is adopted as the feature learner. An One-Vs-All linear learner is used for classification.

Food101 Food101 [6] introduces a method to mine discriminative parts via Random Forests (RF), enabling mining parts simultaneously for all categories and sharing knowledge among them.

The comparison results are shown in Table 1. The proposed baseline approach achieves the best accuracies on all the three datasets. The pro- posed approach and the state-of-the-art DCNN based food classifier DeepFood both adopt a deep CNN based framework. The difference is that DeepFood utilizes a wider but shallower network while the proposed approach uses a thinner and deeper Residual Neural Network, which indicates that for food recognition, a well trained deeper network can model the food appearances better.

Table 1. Comparison with state-of-the-art methods on UEC100, UEC256 and Food101 datasets. The result of the baseline approach on the proposed Chin-Food1000 dataset is also reported.

Dataset	Method	Top-1	Top-5
UEC100	DeepFood [8]	76.3	94.6
	Deep-UEC [4]	72.2	92.0
	Proposed baseline	**80.6**	**95.9**
UEC256	DeepFood [8]	63.8	87.2
	Proposed baseline	**71.2**	**91.0**
Food101	DeepFood [8]	77.4	93.7
	Food101 [6]	50.76	NA
	Proposed baseline	**78.5**	**94.1**
ChinFood1000	Proposed baseline	44.1	68.4

Compared to the DeepFood method, the proposed baseline approach has a larger accuracy gain on the UEC datasets than the Food101 dataset, i.e. on average 5.85% top-1 accuracy gain vs. 1.1% top-1 accuracy gain. The UEC dataset is mainly composed of Japanese food, which is close to the Chinese dishes, implying that the proposed baseline is more good at modeling data like Chinese food.

In addition, we report the top-1 and top-5 accuracies of the baseline approach on the 1000-category ChinFood1000 dataset. The accuracies are promising but is much lower than the UEC and Food101 datasets, indicating that the Chin-Food1000 dataset is still challenging for the current food classification methods due to the great number of categories, the high inter-class similarities and the large intra-class variances.

We calculated the per class F1 score from the baseline approach. The best 10 categories and the worst 10 categories are respectively reported in Fig. 4. Note that for the 10 worst categories, the F1 scores are all 0, where the food is mainly composed of Chinese dishes. For instance, mushroom fried rice and rice are very similar classes. For the best 10 categories, a major composition is fruits, which enjoys more consistent visual appearances than typical Chinese dishes.

The proposed baseline approach is implemented via Caffe [14], and deployed on a Linux server with 1G memory, Intel Intel(R) Xeon(R) CPU E5-2650 and no CUDA acceleration. The runtime of the prediction for one image is around 500 ms.

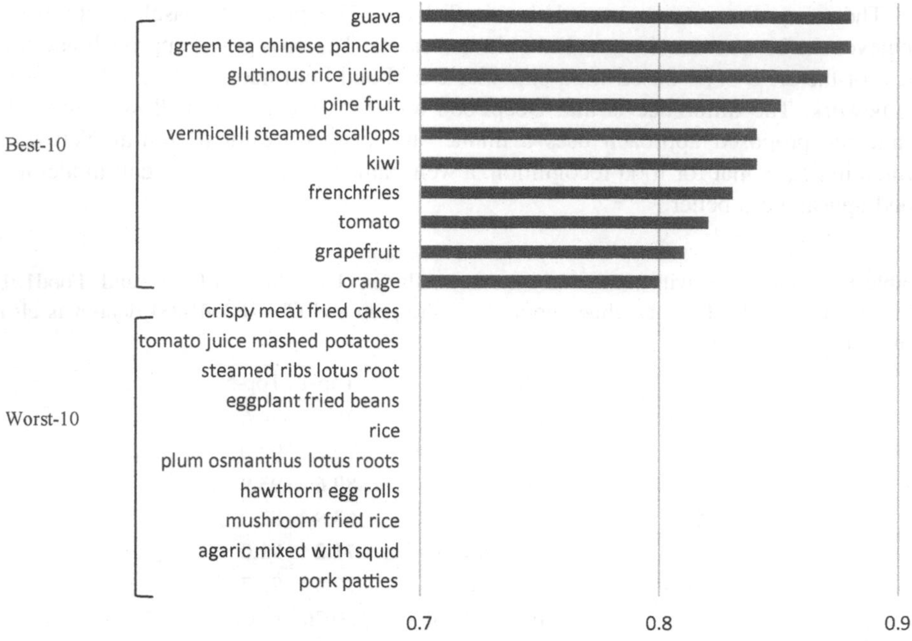

Fig. 4. The best 10 categories and the worst 10 categories by the baseline approach. The best and worst predictions are evaluated by the F1 score. For the 10 worst categories, the F1 scores are 0. The worst categories are mainly composed by Chinese dishes while the best categories are mainly fruits, which enjoy more consistent visual appearances than typical Chinese dishes.

5 Conclusion and Future Work

In this paper, we introduce the 1000-category food dataset ChinFood1000. The dataset is very challenging, with large inter-class similarities and intra-class variance. Since the categories in the dataset are carefully designed to include the most popular Chinese dishes, fast food, snacks, drinks and fruits, the dataset is meaningful and practical for real applications. We further propose a simple and effective baseline. With the power of a very deep network for food images, the baseline achieves the currently best performance on UEC100, UEC256 and Food101 datasets. We study the ChinFood1000 dataset with the baseline approach. The promising results indicate that the ChinFood1000 dataset is still challenging for the current food classification methods. Our future work will improve the classification accuracies with the help of the hierarchical structures.

References

1. Chen, M., Dhingra, K., Wu, W., Yang, L., Sukthankar, R., Yang, J.: PFID: Pittsburgh fast-food image dataset. In: ICIP (2009)
2. Matsuda, Y., Hoashi, H., Yanai, K.: Recognition of multiple-food images by detecting candidate regions. In: ICME (2012)
3. Kawano, Y., Yanai, K.: FoodCam: a real-time food recognition system on a smartphone. Multimedia Tools Appl. **74**, 5263–5287 (2014)
4. Kawano, Y., Yanai, K.: Food image recognition with deep convolutional features. In: CEA (2014)
5. Kawano, Y., Yanai, K.: Automatic expansion of a food image dataset leveraging existing categories with domain adaptation. In: Agapito, L., Bronstein, M.M., Rother, C. (eds.) ECCV 2014. LNCS, vol. 8927, pp. 3–17. Springer, Cham (2015). doi:10.1007/978-3-319-16199-0_1
6. Bossard, L., Guillaumin, M., Van Gool, L.: Food-101 – mining discriminative components with random forests. In: Fleet, D., Pajdla, T., Schiele, B., Tuytelaars, T. (eds.) ECCV 2014. LNCS, vol. 8694, pp. 446–461. Springer, Cham (2014). doi:10.1007/978-3-319-10599-4_29
7. Chen, J-J., Ngo, C-W.: Deep-based ingredient recognition for cooking recipe retrival. In: ACM Multimedia (2016)
8. Liu, C., Cao, Y., Luo, Y., Chen, G., Vokkarane, V., Ma, Y.: DeepFood: deep learning-based food image recognition for computer-aided dietary assessment. In: Chang, C.K., Chiari, L., Cao, Y., Jin, H., Mokhtari, M., Aloulou, H. (eds.) ICOST 2016. LNCS, vol. 9677, pp. 37–48. Springer, Cham (2016). doi:10.1007/978-3-319-39601-9_4
9. He, K., Zhang, X., Ren, S., Sun, J.: Deep residual learning for image recognition. In: CVPR (2016)
10. Szegedy, C., Liu, W., Jia, Y., Sermanet, P., Reed, S., Anguelov, D., Erhan, D., Vanhoucke, V., Rabinovich, A.: Going deeper with convolutions. In: CVPR (2015)
11. Sermanet, P., Eigen, D., Zhang, X., Mathieu, M., Fergus, R., LeCun, Y.: OverFeat: integrated recognition, localization and detection using convolutional networks. In: CVPR (2015)
12. Krause, J., Sapp, B., Howard, A., Zhou, H., Toshev, A., Duerig, T., Philbin, J., Fei-Fei, L.: The unreasonable effectiveness of noisy data for fine-grained recognition. In: Leibe, B., Matas, J., Sebe, N., Welling, M. (eds.) ECCV 2016. LNCS, vol. 9907, pp. 301–320. Springer, Cham (2016). doi:10.1007/978-3-319-46487-9_19
13. Deng, J., Dong, W., Socher, R., Li, L.-J., Li, K., Fei-Fei, L.: ImageNet: a large-scale hierarchical image database. In: CVPR (2009)
14. Jia, Y., Shelhamer, E., Donahue, J., Karayev, S., Long, J., Girshick, R., Guadarrama, S., Darrell, T.: Caffe: convolutional architecture for fast feature embedding. arXiv preprint arXiv:1408.5093 (2014)

Plant Species Recognition Based on Deep Convolutional Neural Networks

Shanwen Zhang[1] and Chuanlei Zhang[2(✉)]

[1] Department of Information Engineering, Xijing University, Xi'an 710123, China
wjdw716@163.com
[2] Tianjin University of Science and Technology, Tianjin 300222, China
a17647@gmail.com

Abstract. A number of the existing leaf based plan leaf recognition methods rely on the hand-crafted features of color, texture and shape, and other various features. One drawback of these methods is poor convergence and generalization. To overcome this problem, a deep convolutional neural network (DCNN) is applied to plant species recognition. The proposed method is different from the existing feature extraction based recognition approaches. The high-level features can be extracted by DCNN. The experimental results clearly demonstrate the effectiveness and efficiency of the proposed model in leaf identification in comparison with current state-of-the-art.

Keywords: Automatic plant species identification · Deep convolutional neural networks (DCNN) · Feature extraction · Deep learning (DL)

1 Introduction

Automatic plant species identification is an important and challenging research area in computer vision. Plant can be identified by leaves, flowers and fruits, etc. Among of them, leaves are often the basis for identifying plant species since they are easily observed, acquired, photographed, processed and analyzed, and leaves are present most of the year and contain much information about the taxonomic identity of the plant, while flowers and fruits for several weeks. Leaf images are collected usually using either a flat-bed scanner (scan images) or a digital camera (scan-like images). Leaves can be characterized by their color, texture and shape. The leaf color is not sufficiently discriminant to be used alone in a plant identification task, because this feature may vary with seasons and geographical locations. Thus, shape and texture are the most relevant features of the leaf. Shape analysis and recognition play an important role in the design of robust and reliable computer vision systems. Such systems rely on feature extraction to provide meaningful information and representation of shapes and images [1]. The leaf shape edge detector algorithm has been widely applied to plant species identification due to its superior performance [2]. Zhao et al. [3] proposed a new method for plant identification using shapes of their leaves. Different from existing studies which target at simple leaves, the method can accurately recognize both simple and compound leaves by capturing global and local shape information independently so that they can be

D.-S. Huang et al. (Eds.): ICIC 2017, Part I, LNCS 10361, pp. 282–289, 2017.
DOI: 10.1007/978-3-319-63309-1_26

examined individually during classification. The proposed counting-based shape descriptor is not only discriminative for classification but also computationally fast and storage cheap. Hu et al. [4] proposed a contour-based shape descriptor, called the multi-scale distance matrix, to capture the shape geometry. The descriptor is invariant to translation, rotation, scaling, and bilateral symmetry, and is further combined with a dimensionality reduction to improve its discriminative power. Chaki et al. [5] proposed an automated system for recognizing plant species based on leaf images. Plant leaf images corresponding to three plant types, are analyzed using two different shape modeling techniques, the first based on the Moments-Invariant (M-I) model and the second on the Centroid-Radii (C-R) model. Hasim et al. [6] proposed a leaf shape recognition method using centroid contour distance (CCD) as shape descriptor. CCD is an algorithm of shape representation contour-based approach which only exploits boundary information. CCD calculates the distance between the midpoint and the points on the edge corresponding to interval angle. Many existing plant species recognition methods did not capture color information, because color was not recognized as an important aspect to the identification. Kadir et al. [7] made use of the shape, vein, color, and texture features to classify plant leaves. Jin et al. [8] proposed an automatic species identification method using sparse representation of leaf tooth features.

All the above methods demonstrate that automatic species identification is suitable for some species. However, the key issue to solve the problem of robust automatic species identification is how to extract the robust classifying features, and how to deal with the different deformation of leaf character, and the large and small inter-class variations that are typical of botanical samples. Although hundreds of kinds of features can be extracted from a leaf image, it is difficult to know which features are the more discriminant features. Thus, the existing methods are insufficient to identify the complex species; various additional features are often extracted to be incorporated into the current automatic species identification method. Due to the diversity of natural leaf, traditional image feature extraction method has no fixed extraction principle, which relies on manual design feature extraction method, and leads to the performance of all these approaches depending heavily on the underlying predefined features. Feature extraction itself is a complex and tedious process which needed to be revisited every time the problem at hand or the associated dataset changed considerably. Deep learning (DL) is a representation learning approach ideally suited for image analysis challenges in pattern recognition [9–11]. Deep convolutional Neural Networks (CNNs) are multi-layer supervised networks which can learn features automatically from datasets, and has been used more widely in the field of image processing. DCNN based on automatic feature extraction method and DL based on artificial neural network, through multiple linear or nonlinear transformation of the input data, the data is mapped to the expression of the new space, can effectively extract the intrinsic character of the image [12–14]. In the plant species recognition field, we can extract the high level features of the image by learning the whole leaf image.

The rest of the paper is organized as follows. Section 2 introduces the general DCNN model. Experiments are illustrated in Sect. 3. Section 4 gives a brief summary of this paper.

2 Deep Convolutional Neural Network

DCNN is mainly composed of three parts: convolution layer, pool layer, and fully connected layer [15, 16]. Convolution layer is the upper layer of the input convolution filter; the pool is the upper layer of the input sampling. Each neuron of the convolution layer is connected with a small piece of adjacent input area (i.e. receptive field area), and the local features of the image are represented. The pool is used to extract the statistical features of the image, so that the network structure is not sensitive to small translation and deformation. Convolution layer and pool layer alternately connected and is composed of a plurality of 2-D plane features, each feature represents a plane feature mapping, fully connected layer is 1-D network layer is composed of vector, expression characteristics of the input data. Figure 1 is a DCNN basic structure used in this paper, which has a total of 9 layers: input and output layers, convolutional layer, max-pooling layer, fully-connected layer. Each layer contains trainable parameters (weights) [17–19].

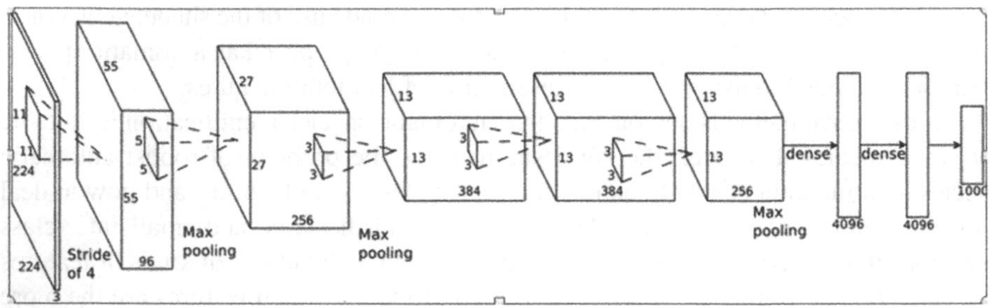

Fig. 1. DCNN basic structure

In Fig. 1, the basic parameters of the model are listed as follows: in the first convolution layer, the original 224 × 224 × 3 leaf image is convoluted by 96 convolutional kernels with size of 11 × 11 × 3 and the stride length of 4; in the second convolution layer, the output of the former layer is reduced by Stochastic-pooling as the input, and conduct convolution operate of the 256 convolution kernels with the size of 5 × 5 × 48; there are 384 convolution kernels with the size of 3 × 3 × 256 in the third convolution layer, 256 convolution kernels with the size of 3 × 3 × 192 in the fourth convolution layer, 256 convolution kernels with the size of 3 × 3 × 192 in the fifth convolution layer; 4096 neurons in the fully-connected layer; and the last layer is the output decision layer composed of softmax (the number of output nodes is equal to the number of categories), and each dimension of the output is the probability that the leaf image belongs to the category. The second, fourth and fifth convolutional layer are connected only with the feature map of the upper layer on the same GPU, and all of the features of the third layer are connected with all the features of the second layers. The neurons of the fully-connected layer are connected with all neurons in the upper layer. The standard response layer is behind the first and second convolutional layer; the Stochastic-pooling layer is behind the standard response layer and the fifth convolutional layers [15, 17]. The modified ReLUs (rectified-linear-units) function is used as the nonlinear activation function, and is applied in each convolutional layer and the fully-connected layer.

The general architecture of the DCNN based plant recognition framework is illustrated in Fig. 2, which consists of two main components: image preprocessing and DCNN based classification.

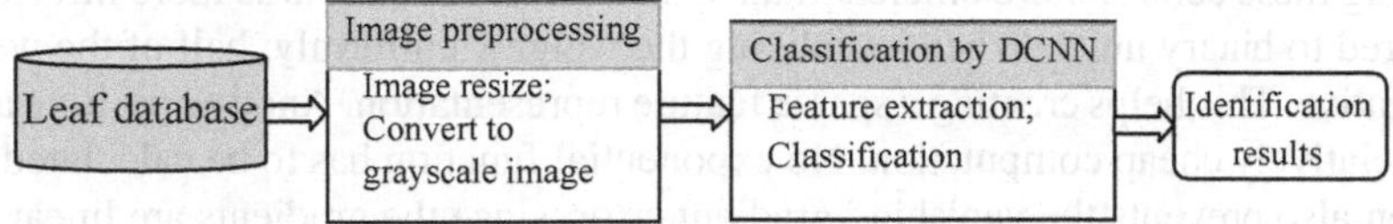

Fig. 2. Plant species recognition framework architecture based DCNN

In DCNN, the feature extraction model is the part where the network learns to detect different high-level features from the input images. It consists of a sequence of convolution and pooling layers discussed as follows.

Convolutional layer takes a square kernel of size $k \times k$, which is smaller than the input w, and perform a convolution over the input image to obtain network activations. A number of these kernels are learned such that they minimize the training error function. Let f_k be a filter with a kernel size $k \times k$ applied to the input image x. $n \times m$ is the number of input connections each CNN neuron has. The resulting output of the layer calculates as follows:

$$M_i = b_i + \sum_k W_{ik} * x_k \tag{1}$$

where $*$ is the convolution operator, x_k is the kth input channel, W_{ik} is the sub kernel of that channel and b_i is a bias term.

The convolution operation being performed for each feature map is the sum of the application of k different 2D squared convolution features plus a bias term.

To calculate a more rich and diverse representation of the input, multiple filters f_k can be applied on the input. The filters f_k are realized by sharing weights of neighboring neurons. This has the positive effect that lesser weights have to be trained in contrast to standard multilayer perceptions, since multiple weights are bound together.

Pooling layers are used to summary the information created from the layer above. Two max and average pooling operators are often used, which summarize an area of $k \times k$ into either the maximal value or the mean value. The max-pooling and average-pooling are computed as follows, respectively,

$$h_j^n(x, y) = \max\left(0, h_j^n * W_j^n\right) \tag{2}$$

$$h_j^n(x, y) = \frac{1}{K} \sum_{x \in N(x), y \in N(y)} h_j^{n-1}(x, y) \tag{3}$$

where h_j^n is the out feature mp, h_j^{n-1} is the input feature map, W_j^n is kernel.

The output size is computed in a manner similar to the convolutional layer.

A Rectified Linear Unit (ReLU) is a cell of a neural network which uses the following activation function to calculate its output given x:

$$R(x) = \max(0, x) \tag{4}$$

Using these cells is more efficient than sigmoid and still forwards more information compared to binary units. When initializing the weights uniformly, half of the weights are negative. This helps creating a sparse feature representation. Another positive aspect is the relatively cheap computation. No exponential function has to be calculated. This function also prevents the vanishing gradient error, since the gradients are linear functions or zero but in no case non-linear functions.

3 Experiments and Analysis

To validate the performance of the DCNN based plant species recognition method, we apply DCNN to leaf classification task and conduct a set of experiments using the Swedish leaf dataset, and compare with multiscale distance matrix (MDM) [4], shape based features and neural network classifiers (SFNN) [5], centroid contour distance (CCD) [6] and shape, color, and texture features (SCTF) [7]. The database contains isolated leaves from 15 different Swedish tree species, with 75 leaves per species. Figure 3A shows some representative examples. The footstalks maybe affect the recognition results, so we cut them off to construct a clean dataset, as shown in Fig. 3B.

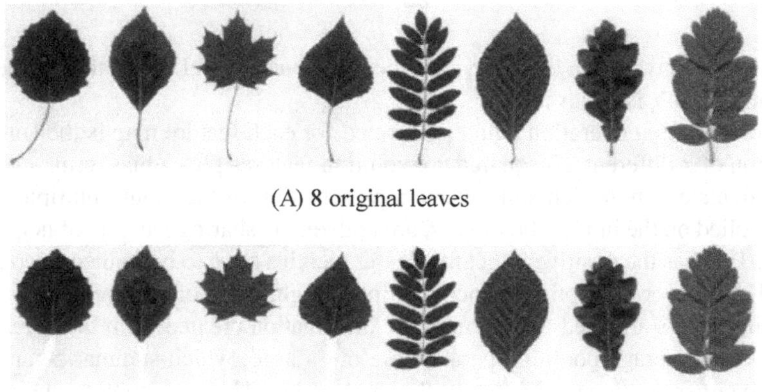

(A) 8 original leaves

(B) 8 leaves without footstalks

Fig. 3. Leaf samples from Swedish leaf dataset

Because DCNN in Fig. 1 only accepts the size of 224 × 224 gray image, and the images of the Swedish blade data set are high-definition color images. So, we firstly covert each color image into grayscale image, and then resize it to same size of 224 × 224, and then input to the DCNN model. The green channel of colorful leaf images is selected as grayscale version of them. The effect of horizontal reflection augmentation of datasets are presented but further augmentations such as shift and rotation was not applied because of the already good performance of the model but in object recognition literature the use of multiple augmentation (including shift and rotation) is also common to further improve the results. The substitution of well-known ReLU activation functions with newly suggested ELUs was also experimented that showed that although ELUs, slightly improved the quality of classification, but increased the learning time thus the use of ReLU in deep architectures seems more effective than ELUs.

The parameters in DCNN of the experiments are set as follows: as for the color leaf images, the initial learning rate is 10^{-5}, reset as 10^{-6} after 20000 iterations; momentum coefficient is 0.9; weighted decay coefficient is 0.0005. As for the depth data, the parameters are set as for the color leaf data. The configuration of the computer is inteli5 processor, 16 GB memory, GTX7502 GB graphics card, training time is about 15 h. Because the SVM classifier is more suitable for the classification of high dimensional feature data, and the existing experimental results showed that SVM is better than Softmax classifier in classification, so in the following experiments, the linear SVM classifier (LinSVM) is used to classify the features extracted from the model. The 5-fold cross validation scheme is used to test the performance of DCNN. As for each 5-fold cross validation data split, we repeat recognition experiment 50 times.

From each category, 15 leaf images are randomly selected as test set, the rest as training set. The training samples used to train DCNN, while the test samples are used to test the performance of DCNN. The process is repeated 5 times, that is, the random selection of the training set and the test set, and the final result is the average of the accuracy of the 5 sets of test sets. After choosing the training set and test set, AlexNet model is trained, and the obtained features are used to train the LinSVM classifier.

The recognition results are shown in Table 1. In Table 1, four different results from four existing approaches are listed as comparison. From Table 1, it is found that the proposed method is very effective on this dataset with the highest average accuracy of 99.26%.

Table 1. The recognition results of and the proposed method

Methods	MDM	SFNN	CCD	SCTF	Our method
Results	92.62 ± 2.38	91.62 ± 3.41	91.15 ± 3.14	92.34 ± 2.70	98.66 ± 1.14

4 Conclusion

There are many methods in automated or computer vision plant classification process, but plant species recognition based on the leaves methods are still important research area because of the complexity of the plant leaf. In this paper, an approach of using deep learning method is explored in order to automatically classify and detect plant diseases from leaf images. The developed model aims at dealing with the high resolution of the images generally used for plant species. Adapting the existing deep neural network models for larger image database can result in more complex architectures, with larger sets of parameters (more and larger layers), which can substantially increase the complexity of the model. The results showed that the proposed architecture for CNN-based leaf classification is closely competed with the latest extensive approaches on devising leaf features and classifiers. The experimental results demonstrate that the proposed deep learning based method can achieve state-of-the-art performance on plant species dataset.

Acknowledgments. This work is partially supported by the China National Natural Science Foundation under grant No. 61473237. It is also supported by Tianjin Research Program of Application Foundation and Advanced Technology under grant No. 14JCYBJC42500, and Tianjin science and technology correspondent project 16JCTPJC47300. The authors would like to thank all the editors and anonymous reviewers for their constructive advices

References

1. Souza, M.M.S.D., Medeiros, F.N.S., Ramalho, G.L.B., et al.: Evolutionary optimization of a multiscale descriptor for leaf shape analysis. Expert Syst. Appl. **63**, 375–385 (2016)
2. Xu, Q., Varadarajan, S., Chakrabarti, C., Karam, L.J.: A distributed canny edge detector: algorithm and FPGA implementation. IEEE Trans. Image Process. **23**(7), 2944–2960 (2014)
3. Zhao, C., Chan, S.S.F., Cham, W.K., et al.: Plant identification using leaf shapes-A pattern counting approach. Pattern Recogn. **48**(10), 3203–3215 (2015)
4. Hu, R., Jia, W., Ling, H., et al.: Multiscale distance matrix for fast plant leaf recognition. IEEE Trans. Image Process. Publ. IEEE Signal Process. Soc. **21**(11), 4667 (2012)
5. Chaki, J., Parekh, R.: Plant Leaf recognition using shape based features and neural network classifiers. Int. J. Adv. Comput. Sci. Appl. **2**(10), 41–47 (2011)
6. Hasim, A., Herdiyeni, Y., Douady, S.: Leaf shape recognition using centroid contour distance. IOP Conf. Ser. Earth Environ. Sci. **31**, 1–8 (2016)
7. Kadir, A., Nugroho, L.E., Susanto, A., et al.: Leaf classification using shape, color, and texture features. Int. J. Comput. Trends Technol. **1**(3), 225–230 (2013)
8. Jin, T., Hou, X., Li, P., et al.: A novel method of automatic plant species identification using sparse representation of leaf tooth features. Plos One **10**(10), e0139482 (2015)
9. Atabay, H.A.: A convolutional neural network with a new architecture applied on leaf classification. IIOABJ **7**(5), 326–331 (2016)
10. Wu, Y.H., Shang, L., Huang, Z.K., Wang, G., Zhang, Xiao-Ping: Convolutional neural network application on leaf classification. In: Huang, D.S., Bevilacqua, V., Premaratne, P. (eds.) ICIC 2016. LNCS, vol. 9771, pp. 12–17. Springer, Cham (2016). doi:10.1007/978-3-319-42291-6_2
11. Krizhevsky, A., Sutskever, I., Hinton, G.E.: ImageNet classification with deep convolutional neural networks. In: International Conference on Neural Information Processing Systems, pp. 1097–1105. Curran Associates Inc. (2012)
12. Krizhevsky, A., Sutskever, I., Hinton, G.E.: Imagenet classification with deep convolutional neural networks. In: Advances in Neural Information Processing Systems, pp. 1097–1105 (2012)
13. Zeiler, M.D., Fergus, R.: Visualizing and understanding convolutional networks. In: Fleet, D., Pajdla, T., Schiele, B., Tuytelaars, T. (eds.) ECCV 2014. LNCS, vol. 8689, pp. 818–833. Springer, Cham (2014). doi:10.1007/978-3-319-10590-1_53
14. Schmidhuber, J.: Deep learning in neural networks: An overview. Neural Netw. **61**, 85–117 (2015)

15. Sladojevic, S., Arsenovic, M., Anderla, A., et al.: Deep neural networks based recognition of plant diseases by leaf image classification. Comput. Intell. Neurosci. **2016**(6), 1–11 (2016)
16. The experimental result shows that the method for classification gives average accuracy of 93.75% when it was tested on Flavia dataset, that contains 32 kinds of plant leaves. It means that the method gives better performance compared to the original work
17. Zhang, D., Lu, G.: Review of shape representation and description techniques. Pattern Recogn. **37**(1), 1–19 (2004)
18. Simonyan, K., Zisserman, A.: Very deep convolutional networks for large-scale image recognition. arXiv preprint arXiv:1409.1556(2014)
19. Szegedy, C. et al.: Going deeper with convolutions. In: Proceedings of the IEEE Conference on Computer Vision and Pattern Recognition, pp. 1–9 (2015)

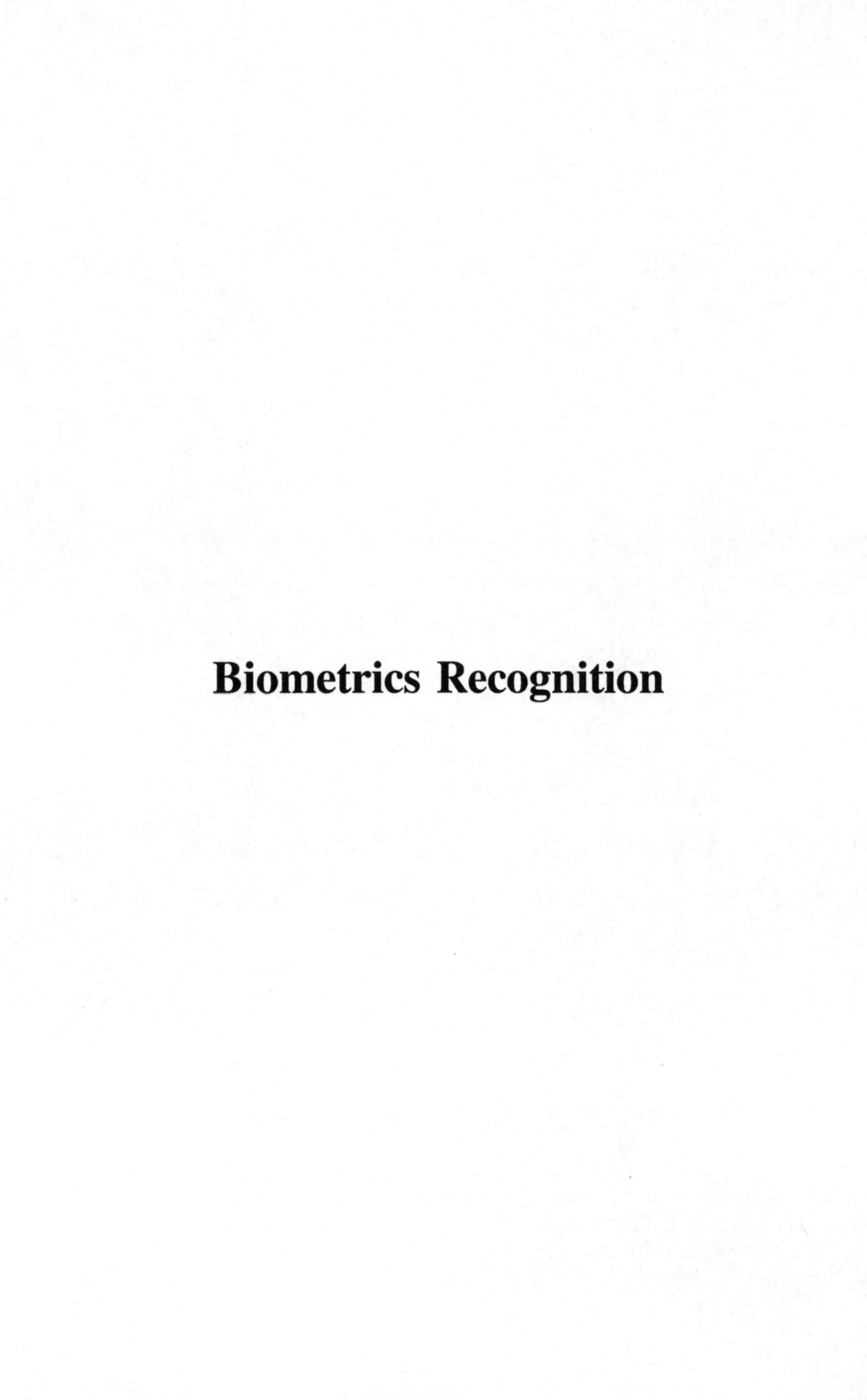

Biometrics Recognition

Effective Iris Recognition for Distant Images Using Log-Gabor Wavelet Based Contourlet Transform Features

Lasker Ershad Ali, Junfeng Luo, and Jinwen Ma(✉)

Department of Information Science,
School of Mathematical Sciences and LMAM,
Peking University, Beijing 100871, China
jwma@math.pku.edu.cn

Abstract. Distant iris recognition has become an active research topic in biometric as well as computer vision, but it is still a very challenging problem. In order to solve it effectively, we propose a novel framework by utilizing Log-Gabor wavelet based Contourlet transform (LGCT) feature descriptor with an effective kernel based extreme learning machine (KELM) classifier. The experiments are conducted on CASIA-v4 which is a typical database of distant iris images. It is demonstrated by the experimental results that our proposed LGCT features are quite effective for distant iris recognition and the highest accuracy can arrive at **95.93%** when they are fused with the convolutional neural networks (CNN) and gradient local auto-correlations (GLAC) features together.

Keywords: Iris recognition · Distant iris image · Log-Gabor wavelet · Contourlet transform · CNN · KELM

1 Introduction

Iris recognition is a process of identification and authentication of human from facial images. It has drawn much attention to researchers due to the growing demands from vast applications in different areas relevant to real life such as high security surveillance, border-crossing control, law enforcement, office management, credit-card authentication, cell phone and other wireless-device-based authentication. The earlier attempt at iris recognition system had been focused on the iris images acquired using near infrared (NIR) imaging from controlled environment [1–7]. These iris recognition systems were easy to apply for border-crossing control system, but very difficult for high security surveillance. Iris recognition from distantly acquired images under less controlled environment, i.e., distant iris recognition is to break through the limitations of those recognition systems. But distant iris recognition is more challenging because these distant images have certain noises which can degrade the image quality and cannot illustrate clear textures of iris. Some sample distant eye images of CASIA-v4 database are shown in Fig. 1.

In fact, a large number of researchers have been devoted to develop more accurate iris recognition approaches for noisy eye images. However, the existing research works

D.-S. Huang et al. (Eds.): ICIC 2017, Part I, LNCS 10361, pp. 293–303, 2017.
DOI: 10.1007/978-3-319-63309-1_27

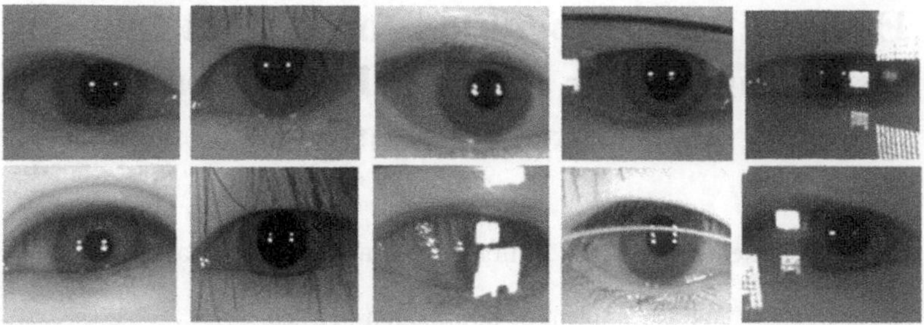

Fig. 1. Sample images of CASIA-v4 distance image under less controlled environment.

have not provided effective strategies for iris recognition from distant images and less controlled environments. A few systems have been established for distant images, but the performances of these methods are not good enough [8–13]. So, it is urgent to deeply investigate the problem of distant iris recognition and establish the effective method for it. Actually, the motivation of this research is to propose the effective feature extraction method for distant iris recognition. For this purpose, we propose LGCT feature descriptor based recognition framework. In this framework, we first apply the Log-Gabor wavelet decomposition on the segmented normalized iris image and then implement the Contourlet transform on the Log-Gabor decomposed images to obtain the smooth contour and texture features in different directions. For convenience, we refer to these features as the LGCT features. We finally adopt the KELM classifier to complete the distant iris recognition.

The rest of the paper is organized as follows. We review some recent related work in Sect. 2. Section 3 proposes the LGCT features as well as the framework of the distant iris recognition. The experimental results and comparisons are summarized in Sect. 4. Finally, we conclude briefly in Sect. 5.

2 Related Work

In this section, we review some recent related works for iris recognition. Almost all the existing iris recognition algorithms have been used binary iris codes based on Gabor filter. Fancourt et al. [8] firstly proposed the concept of distance iris recognition in 2005. Tan and Kumar [9] established a unified framework for automatic iris segmentation using distantly acquired face images, which employed the multiple higher order local pixel dependencies to robustly classify the eye region pixels into iris or non-iris regions. In addition, they utilized the Grow-Cut segmentation algorithm for remotely acquired iris images [10]. Moreover, the random walker algorithm was applied for coarse iris segmentation and then the features from periocular region (around the eye region) and segmented iris region were extracted by implementing Leung-Malik Filter (LMK) and the Log-Gabor filter respectively [11]. According to the above research results, the segmentation performance of the random walker algorithm is better than the others.

In 2012, Kumar and Chen [12] developed and formulated an approach for the iris recognition using hypercomplex (quaternionic) and sparse representation of unwrapped iris images. Recently, Tan and Kumar [13] proposed the distant iris recognition system to build the global model for Log-Gabor features and the local model for Zernike moments phase features. They combined the global and local matching scores by using weighted sum method and the recognition accuracy of their method was **95%** on CASIA-v4. They also merged the fragile bit method [14] and the personalized weight map method [15] via the weighted sum and achieved **93.8%** recognition accuracy. Kobayashi and Otsu [16] further proposed the GLAC algorithm to extract the features from images in 2008, which was very effective for human detection. Based on the idea of GLAC and the rapid development of deep learning framework in computer vision, we fused the GLAC and CNN features together for distant iris recognition in our earlier work [17]. In that work, we introduced the concept of contextual eye images and combined the features from both contextual eye images and iris images, which could lead to **98.60%** iris recognition accuracy. However, since the contextual eyes can be changed in different times, we do not consider this kind of information in this paper. Actually, it was shown by the experimental results in our previous paper that the best iris recognition accuracy was only **93.96%** when the CNN and Contourlet transform iris image features were used together. Here, we only consider the features extracting from the segmented iris image and try to improve the recognition accuracy by proposing a novel feature descriptor using the Log-Gabor wavelet based Contourlet transform.

3 Methodology

Here we present the main steps of our new iris recognition system or framework. For iris recognition, our main steps are respectively image preprocessing, iris segmentation, iris normalization, feature extraction, iris classification and finally recognition. The block diagram of our iris recognition system is shown in Fig. 2.

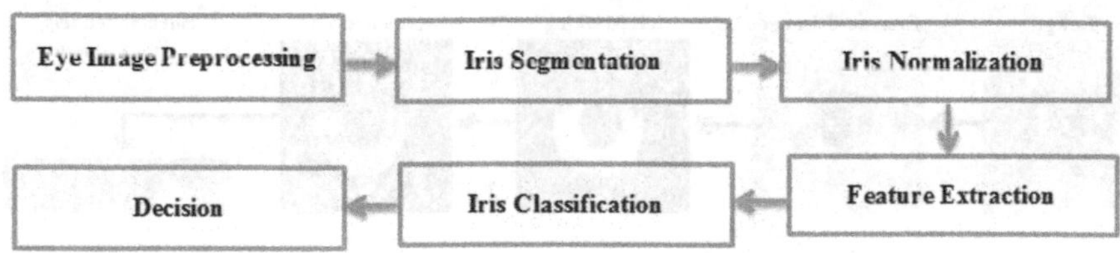

Fig. 2. The block diagram of iris recognition system.

3.1 Image Preprocessing

Distantly acquired eye images are influenced by the variation of illumination and multiple sources of noises. Illumination variation poses the difficulties for iris segmentation and recognition. So image preprocessing step is needed to remove such noises. To account the reason, the single scale retinex (SSR) algorithm is applied on the eye image to improve the overall image quality [11]. The mathematical formulation of SSR is as follows:

$$R(x,y) = log \frac{I(x,y)}{G_\sigma * I(x,y)} \quad (1)$$

where $I(x,y)$ denotes the input gray scale image, G_σ is a low pass filter with standard deviation σ and $*$ denotes the convolutional operator.

3.2 Iris Segmentation

It is necessary to separate iris from the whole eye image for iris recognition. The recognition algorithms fail to recognize correctly due to weakness of the segmentation. In our segmentation process, we have used the random walker (RW) algorithm for coarse iris segmentation and obtained the iris image and corresponding binary iris mask. The coarse segmented iris mask cannot provide detail information about limbic and pupillary boundaries. The localization of limbic and pupillary boundaries is approximated from the segmented iris region using circular model [11]. After removing the occlusion noise such as eyelashes, shadow and eyelids, the segmentation stage is completed.

3.3 Iris Normalization

Iris size varies due to the imaging distance between the eye and the acquisition device, as well as the pupillary response to ambient light. It is more convenient to normalize the segmented iris image for suitable feature extraction. Daugman's rubber sheet model is employed to normalize the segmented iris image by performing a mapping from thc cartesian coordinates to the polar coordinates [3]. Eye image preprocessing to iris normalization steps are shown in Fig. 3.

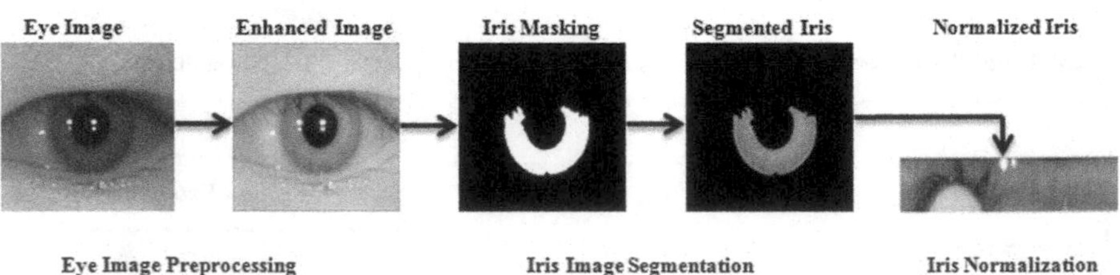

Fig. 3. The schema of the image preprocessing for iris normalization.

3.4 Feature Extraction

An appropriate feature descriptor can enhance the recognition algorithm significantly. Also the real-time consistency of a recognition framework depends on the feature extraction strategy. Moreover, the extracted features overcome the inter-class ambiguities if the iris patterns are represented through the features considerably. It is important to extract features properly from the segmented iris images for reliable performance of iris recognition system. The proposed LGCT feature extraction technique has been explained sequentially.

The wavelet transform has many applications of image processing such as image compression, noise removal, image edge enhancement and feature extraction. Log-Gabor filter of wavelet has shown its ability to extract features from iris images. But iris images consist of edges that are smooth curves and also contain two-dimensional singularities, which cannot be captured efficiently by the Log-Gabor wavelet transform. Beside, Contourlet transform is a geometrical image transform, which can represent images containing smooth contours and textures. It cannot properly capture accurate iris features from distant images. Only wavelet or Contourlet transform is not optimal choice for iris images. Thus we propose a Log-Gabor wavelet based Contourlet transform features which can capture iris information more accurately than either Log-Gabor or Contourlet transform (LGCT). LGCT consists of two stages: the first stage provides Log-Gabor wavelet decomposition for the multiscale and multidirectional spatial frequency response [17, 18]. The frequency response of a Log-Gabor filter is given as:

$$G(\eta) = \exp\left(-\frac{\left(\log\frac{\eta}{\eta_0}\right)^2}{2\left(\log\frac{\sigma}{\eta_0}\right)^2}\right) \tag{2}$$

where $\eta_0 = \frac{1}{\lambda_0}$, λ_0 is the minimum wavelength of the basis filter and $\frac{\sigma}{\eta_0}$ is the ratio of standard deviation of the Gaussian describing the Log-Gabor filter transfer function in the frequency domain. We convolve the input iris image by using First Fourier Transform (FFT) and the frequency response of Log-Gabor filter. Then we apply inverse FFT to obtain the Log-Gabor decomposed image. The decomposed image has real and imaginary parts. If $GI(x,y)$ denotes the Log-Gabor decomposed image of an original image $I(x,y)$ then it can be written as follows.

$$GI(x,y) = \mathrm{Re}GI(x,y) + \mathrm{Im}GI(x,y) \tag{3}$$

where $\mathrm{Re}GI(x,y)$ and $\mathrm{Im}GI(x,y)$ represent the Log-Gabor wavelet decomposed real and imaginary images respectively.

The second stage, we apply Contourlet transform [19] to represent directional multiresolution image for both decomposed real and imaginary images. This method utilizes a double filter bank in order to obtain a sparse expansion of typical images containing smooth contours. The Laplacian pyramid is used to detect the point discontinuities of the image. Each decomposition level creates a down sample low pass version of the Log-Gabor wavelet decomposed image and a bandpass image. The directional filter bank provides angular decomposition by using simple quincunx filter bank for fan filter. It links the point discontinuities into linear structures and can achieve perfect reconstruction.

In this stage, we use 2nd level decomposition of Contourlet transform for both $\mathrm{Re}GI(x, y)$ and $\mathrm{Im}GI(x, y)$ images. We obtain five sub-band images for $\mathrm{Re}GI(x, y)$ and combine them to construct a real feature vector $\mathrm{LGCT_{Re}}$. In the same procedure, we build an imaginary feature vector $\mathrm{LGCT_{Im}}$. Finally, the concatenation of $\mathrm{LGCT_{Re}}$ and $\mathrm{LGCT_{Im}}$ feature vectors provides LGCT feature descriptor. These LGCT features can describe the texture and edge information compactly for an input normalized iris image. Proposed LGCT feature extraction technique is illustrated in Fig. 4.

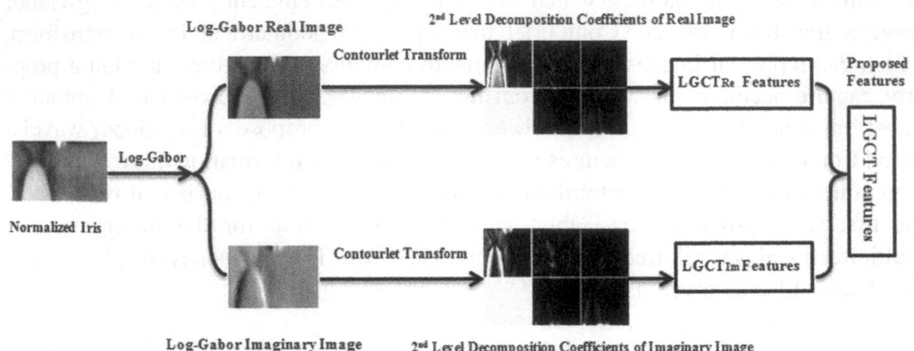

Fig. 4. The diagram of our proposed **LGCT** features.

3.5 Classifier

In our recent work, we have used K-nearest neighbor (KNN), support vector machine (SVM) and kernel extreme learning machine (KELM) algorithms [17] for iris classification. The performance of KELM is better than other two classifiers. So we have used the KELM for this recognition framework. A brief explanation about KELM classification technique is given below.

Extreme learning machine (ELM) was developed for single-hidden-layer feed-forward neural networks (SLFNs) [20]. For n training samples $\{x_i, y_i\}_{i=1}^{n}$, where $x_i \in \mathbb{R}^M$, $y_i \in \mathbb{R}^C$ and C is the class label, the single hidden layer neural network having L hidden nodes can be expressed as:

$$\sum\nolimits_{j=1}^{L} \alpha_j f\left(w_j \cdot x_i + e_j\right) = y_j, i = 1, 2, \ldots, n \tag{4}$$

where $f(\cdot)$ is a nonlinear activation function (sigmoid), $\alpha_j \in \mathbb{R}^C$ and $w_j \in \mathbb{R}^M$ are weight vectors and e_j is the bias of the j hidden node. The matrix form of n equations can be written as:

$$F\alpha = Y \tag{5}$$

where $\alpha = \left[\alpha_1^T \ldots\ldots\ldots \alpha_n^T\right]^T \in \mathbb{R}^{L\times C}$, $Y = \left[Y_1^T \ldots\ldots\ldots Y_n^T\right]^T \in \mathbb{R}^{n\times C}$ and

$$F = \begin{bmatrix} f(x_1) \\ \ldots \\ f(x_n) \end{bmatrix} = \begin{bmatrix} f(w_1 \cdot x_1 + e_1) \ldots\ldots f(w_L \cdot x_1 + e_L) \\ \ldots\ldots\ldots\ldots\ldots\ldots\ldots\ldots\ldots\ldots\ldots \\ f(w_1 \cdot x_n + e_1) \ldots\ldots f(w_L \cdot x_n + e_L) \end{bmatrix} \tag{6}$$

In the most cases $L \ll n$, the smallest norm least-square solution of Eq. (5) is

$$\alpha = F^{\dagger} Y \tag{7}$$

where $F^{\dagger} = F^T\left(FF^T\right)^{-1}$ is the Moore-Penrose generalized inverse of F. To gain a better stability and generalization $\frac{1}{\rho}(\rho > 0)$ is simply added to the diagonal of FF^T. Then the output function of the ELM can be written as:

$$h_L(x_i) = f(x_i)\alpha = f(x_i)F^T\left(\frac{I}{\rho} + FF^T\right)^{-1} Y \tag{8}$$

If the feature mapping $f(x_i)$ is unknown, a kernel matrix for ELM can be considered as follows:

$$\Psi = FF^T : \Psi_{ij} = f(x_i) \cdot f(x_j) = K\left(x_i, x_j\right) \tag{9}$$

Then the output function of KELM is given by:

$$h_L(x_i) = \begin{bmatrix} K(x_i, x_1) \\ \ldots\ldots \\ K\left(x_i, x_j\right) \end{bmatrix}^T \left(\frac{I}{\rho} + \Psi\right)^{-1} Y \tag{10}$$

Finally, the label of a test sample x_l is assigned to the index of the output node with the largest value, which is written as:

$$y_l = \operatorname*{arg\,max}_{k=1,..,C} h_L(x_l)_k \tag{11}$$

where $h_L(x_l)_k$ denotes the kth output of $h_L(x_l) = \left[h_L(x_l)_1, \ldots, h_L(x_l)_C\right]^T$.

4 Experimental Results

In this section, we have tried to evaluate our proposed framework for the distant iris recognition. For the experimental setup, CASIA-v4 database is used which is publicly available data source for distantly acquired facial images. It is created by Institute of Automation, Chinese Academy of Sciences, China [21]. The images are captured by using near infrared (NIR) imaging with 3 meters distance from the subject. The full database consists of total 2567 images from 142 subjects. All images of the first 14 subjects are used for tuning parameters in our experiments. Totally 3041 images (left and right eye images) are selected randomly from each of the rest 128 subjects to build the training databases and the other 712 images are adopted for testing. All the images except the test data are utilized for training the CNN. To compute CNN features, we have used pre-trained GoogLeNet networks by using some customize parameters in Caffe model [22] and extracted the features from the fully connected layer of the networks. The experiments are conducted by using MATLAB on 4 core GPU system.

In our experiments, all the feature vectors are normalized by using min-max normalization technique and principle component analysis (PCA) is utilized to reduce the dimension of LGCT features. The PCA transform matrix is calculated using training feature set and then applied to the test set. The RBF kernel is used to build KELM classification model. We have compared the discriminatory power of LGCT features to Haar wavelet, Log-Gabor wavelet, Contourlet transform, GLAC, CNN and space time auto correlation of gradients (STACOG) [23] by using same classification technique. Many investigations show that the feature level fusion is better than the single feature descriptor. We have fused two complementary features end by end to represent a suitable feature space for KELM. We also have integrated with CNN, GLAC and LGCT features to learn the classifier. The performances of different feature descriptors based KELM algorithms are compared by using precision, recall and F1-measure values.

The average precision, recall and F1-measure values of the proposed algorithm as well as other methods evaluated on CASIA-v4 database are shown in Table 1. Higher precision, recall and F1-measure values indicate the reliable performance of the classifier. The proposed integrated LGCT-CNN-GLAC based method achieves the highest average precision **0.9612** and average recall **0.9560**. Specifically, the F1-measure of the LGCT-CNN-GLAC is significantly better than other methods, even though it achieves **5%** improvement than proposed LGCT feature descriptor.

Table 1. Performance measures of KELM classifier.

Feature descriptors	Average recall	Average precision	F1-measure
Haar wavelet	0.8878	0.8984	0.8812
Log-Gabor wavelet	0.8615	0.8584	0.8474
Contourlet transform	0.8620	0.8734	0.8561
GLAC	0.8568	0.8774	0.8534
STACOG	0.7784	0.7786	0.7595
CNN	0.9198	0.9252	0.9142
LGCT	**0.9055**	**0.9135**	**0.9012**
LGCT-CNN	**0.9530**	**0.9571**	**0.9499**
LGCT-CNN-GLAC	**0.9560**	**0.9612**	**0.9550**

The overall recognition rates of different methods on CASIA-v4 distance database are depicted in Table 2. As can be seen, among the nine feature descriptors, our proposed LGCT-CNN-GLAC features fusion based algorithm achieves the highest overall recognition rate **95.93%**. It can be also observed that LGCT base algorithm is better than Haar wavelet, Log-Gabor, Contourlet transform, GLAC and STACOG

Table 2. The KELM classification results for CASIA-v4 image database.

Feature descriptors	Classification accuracy (%)
Haar wavelet	89.75
Log-Gabor wavelet	86.24
Contourlet transform	86.52
GLAC	86.10
STACOG	77.95
CNN	92.00
LGCT	**91.00**
LGCT-CNN	**95.51**
LGCT-CNN-GLAC	**95.93**

feature descriptors based algorithm. The experimental results demonstrate that LGCT features are comparable to CNN features. Also the integrated with CNN, GLAC and proposed LGCT features based classification algorithm provides the competitive result for this image database.

The iris recognition performances of proposed method as well as other competing methods on CASIA-v4 are shown in Table 3. It is not possible to directly compare our method with other results as well as contextual eye based iris recognition. Other methods were conducted different number of images and subjects, for example in [13], they only used 935 images from CASIA-v4 database and their recognition accuracy was **95%**. As mentioned earlier, we have used 3753 images for our experiment, it can be said that the experimental data are more complicated than others. From Table 3, we can observe that the recognition accuracy of KELM in LGCT-CNN-GLAC features fusion case is better than the state-of-the-art performance.

Table 3. Iris recognition rates from different competitive methods.

Methods	Recognition rate (%)
Fragile bits [14] and PWMap [15]	93.80
Log-Gabor and LMK [11]	93.90
Contourlet transform and CNN [17]	93.96
Log-Gabor and Zernike Moments [13]	95.00
LGCT-CNN and KELM	**95.51**
LGCT-CNN-GLAC and KELM	**95.93**

5 Conclusion

We have established an effective iris recognition framework for the distant iris images under less constrained environment which is based on KELM model with LGCT features. Our proposed LGCT feature is fundamentally different from majority of the previous ones and the fusion of the LGCT-CNN-GLAC features can improve the accuracy of the distant iris recognition considerably. The experimental results on

CASIA-v4 database demonstrate that our proposed system can achieve the better performance of the distant iris recognition than the other competitive systems.

Acknowledgment. This work was supported by the Natural Science Foundation of China for Grant 61171138. We also acknowledge the Institute of Automation (Chinese Academy of Science, China) for the contributions of the database employed in this work.

References

1. Daugman, J.G.: Biometric personal identification system based on iris analysis. United States Patent, Patent Number: 5291560 (1994)
2. Boles, W., Boashash, B.: A human identification techniques using images of the iris and wavelet transform. IEEE Trans. Sig. Process. **46**(4), 1185–1188 (1998)
3. Daugman, J.G.: How iris recognition works. IEEE Trans. Circ. Syst. Video Technol. **14**(1), 21–30 (2004)
4. Seung-In, N., Bae, K., Park, Y., Kim, J.: A novel method to extract features for iris recognition system. In: Kittler, J., Nixon, M.S. (eds.) AVBPA 2003. LNCS, vol. 2688, pp. 862–868. Springer, Heidelberg (2003). doi:10.1007/3-540-44887-X_100
5. Monro, D.M., Rakshit, S., Zhang, D.: DCT-based iris recognition. IEEE Trans. PAMI **29**(4), 586–595 (2007)
6. He, Z., Tan, T., Sun, Z., Qiu, X.: Toward accurate and fast iris segmentation for iris biometrics. IEEE Trans. Pattern Anal. Mach. Intell. **31**(9), 1670–1684 (2009)
7. Kumar, A., Passi, A.: Comparison and combination of iris matchers for reliable personal authentication. Pattern Recogn. **43**(3), 1016–1026 (2010)
8. Fancourt, C., Bogoni, L., Hanna, K., Guo, Y., Wildes, R., Takahashi, N., Jain, U.: Iris recognition at a distance. In: Kanade, T., Jain, A., Ratha, N.K. (eds.) AVBPA 2005. LNCS, vol. 3546, pp. 1–13. Springer, Heidelberg (2005). doi:10.1007/11527923_1
9. Tan, C.-W., Kumar, A.: Unified frame work for automated iris segmentation using distantly acquired face image. IEEE Trans. Image Process. **21**(9), 4068–4078 (2012)
10. Tan, C.-W., Kumar, A.: Efficient iris segmentation using grow-cut algorithm for remotely acquired iris images. In: 15th International Conference on BTAS, pp. 99–104. IEEE (2012)
11. Tan, C.-W., Kumar, A.: Towards online iris and periocular recognition under relaxed imaging constraints. IEEE Trans. Image Process. **22**(10), 3751–3765 (2013)
12. Kumar, A., Chan, T.-S.: Iris recognition using quaternionic sparse orientation code (QSOC). In: Proceedings of Computer Vision and Pattern Recognition Workshops (CVPRW), pp. 56–64 (2012)
13. Tan, C.-W., Kumar, A.: Accurate iris recognition at a distance using stabilized iris encoding and zernike moments phase features. IEEE Trans. Image Process. **23**(9), 3962–3974 (2014)
14. Hollingsworth, K.P., Bowyer, K.W., Flynn, P.J.: The best bits in an iris code. IEEE Trans. Pattern Anal. Mach. Intell. **31**(6), 964–973 (2009)
15. Wenbo, D., Zhenan, S., Tieniu, T.: Iris matching based on personalized weight map. IEEE Trans. Pattern Anal. Mach. Intell. **33**(9), 1744–1757 (2011)
16. Kobayashi, T., Otsu, N.: Image feature extraction using gradient local auto-correlations. In: Forsyth, D., Torr, P., Zisserman, A. (eds.) ECCV 2008. LNCS, vol. 5302, pp. 346–358. Springer, Heidelberg (2008). doi:10.1007/978-3-540-88682-2_27

17. Ali, L.E., Luo, J., Ma, J.: Iris recognition from distant images based on multiple feature descriptors and classifiers. In: Proceedings of IEEE 13th International Conference on Signal Processing (ICSP), pp. 1357–1362 (2016)
18. Field, D.: Relations between the statistics of natural images and the response properties of cortical cells. J. Opt. Soc. Am. **4**(12), 2379–2394 (1987)
19. Do, M.N., Vetterli, M.: The Contourlet transform: an efficient directional multiresolution image representation. IEEE Trans. Image Process. **14**(12), 2091–2106 (2005)
20. Huang, G.-B., Zhou, H., Ding, X., Zhang, R.: Extreme learning machine for regression and multiclass classification. IEEE Trans. Syst. Man Cybern. **42**(2), 513–529 (2012)
21. CASIA-v4 database: http://biometrics.idealtest.Org/dbDetailForUser.do?Id=4
22. Jia, Y., Shelhamer, E., Donahue, J., Karayev, S., Long, J., Girshick, R., Guadarrama, S., Darrell, T.: Caffe: Convolutional architecture for fast feature embedding. arXiv preprint arXiv:14085093 (2014)
23. Kobayashi, T., Otsu, N.: Motion recognition using local auto-correlation of space-time gradient. Pattern Recogn. Lett. **33**, 1188–1195 (2012). Elsevier

Facial Expression Recognition Using Double-Stage Sample-Selected SVM

Ting Yu and Xiaodong Gu(✉)

Department of Electronic Engineering,
Fudan University, Shanghai 200433, China
xdgu@fudan.edu.cn

Abstract. This paper proposes a double-stage classification model for the classification of six basic facial expressions. Inspired for the fact that an increase in the number of classes brings a drop on the accuracy for facial expression recognition, we use classifiers with fewer classes to improve the performance of a six-class expression recognition classifier. Support Vector Machine (SVM) is adopted as the classifiers due to its excellent performance in small databases. To make SVMs classify samples more precisely, selecting more support vectors trains the model. Active Shape Model (ASM) is used to locate shape points. The shape points are used as features to train the double-stage SVM, which includes a six-class SVM and a following few-class SVM with the classes corresponding to the largest classification probabilities of the former. The approach in this paper achieves an accuracy of 98.25% on the Japanese Female Facial Expression (JAFFE) database, 3.08% and 5.53% higher than those of Local Curvelet Transform method Facial Movement Features method respectively, and besides far better than six other methods.

Keywords: Facial expression recognition · Double-stage SVM · Sample-selected SVM · Active Shape Model (ASM)

1 Introduction

During the previous three decades, facial expression research has attracted more and more attention from researchers for its many applications, such as human-machine interaction, image understanding, and synthetic face animation. Considerable research in social psychology has shown that facial expressions help communication between two speakers. Regardless of country, gender, race, religion, age and culture, the facial expressions were defined as six basic facial expressions [1], including anger, disgust, fear, happiness, sadness, and surprise.

Many efforts have been made on facial expression recognition. A survey on the facial expression recognition can be found in [2, 3].The methods about facial expression recognition can be classified as the feature-based ones and the template-based ones by the way of feature extraction. And according to the data used for facial expression recognition, the methods can be divided into static image based facial expression recognition and dynamic video based facial expression recognition.

D.-S. Huang et al. (Eds.): ICIC 2017, Part I, LNCS 10361, pp. 304–315, 2017.
DOI: 10.1007/978-3-319-63309-1_28

Tu et al. adopt a dual subspace nonnegative matrix factorization to decompose facial images into identity part and expression part, removing the person-dependent influence from expression images [4]. Researchers from Queensland university of Technology solve the limitation of extracted features not fully considering facial element and muscle movements by using facial movement features [5]. Curvelet based local binary patterns are applied to the image of a face in [6] to form the descriptive feature set of the expressions.

Even though much work has been done, the facial expression recognition with a high accuracy is still difficult due to the non-rigidity and complexity of facial expression. Among the different stages of recognition, the classification is the key to the recognition performance. Bayes, SVMs, Adaboost, Neural Networks and linear programming classifiers are the major classifiers used by researchers. When used to classify samples into two categories, most of regular methods in pattern recognition give a satisfactory result. However, performances turn to be poor when the number of classes increases, especially for methods designed for two-class problems originally. The accuracy reduction results from similarities between some expressions. That is to say, the addition of more classes decreases the differences among them. So we propose the solution of conducting a second recognition for unsuccessful recognition samples using a classifier with fewer classes than the first classification. In this paper, SVM is used as the classifier.

On the other hand, the inter-class similarity produces more support vectors, which are the key of a right prediction. The too many support vectors make these key samples more possible to be divided into testing set under a random division of training and testing set, resulting in more failed recognition samples. In this paper more support vectors are selected in the training set to improve performance of SVM, serving as another solution to the accuracy reduction.

In short, the contributions of this paper are as follows. Firstly, the double-stage classification model for facial expression recognition is proposed, inspired by the drop of the accuracy resulted from added classes. Secondly, sample-selected SVM is used as the classifier of the double-stage classification model. Finally, the two solutions are combined as the double-stage sample-selected SVM to do the facial expression recognition and a recognition rate of 98.25% is obtained.

The rest of the paper is organized as follows. Section 2 is the feature extraction using Active Shape Model (ASM). Section 3 describes the double-stage classification model and the sample-selected SVM in detail. Using the Japanese Female Facial Expression (JAFFE) [7] Database as the test data, experimental results and analysis are shown in Sect. 4. Conclusions are drawn in Sect. 5.

2 Feature Extraction Based on Active Shape Model

A distinguishable representation of the expressions is essential to the recognition accuracy of the algorithm. For a certain person, the most identifiable representation of different expressions is the geometry variations of the five facial features. For example, when a person feels happy, the eyes and eyebrows become curve, the eyes narrow, and the corners of the mouth get upward compared to the neural state. However, when the

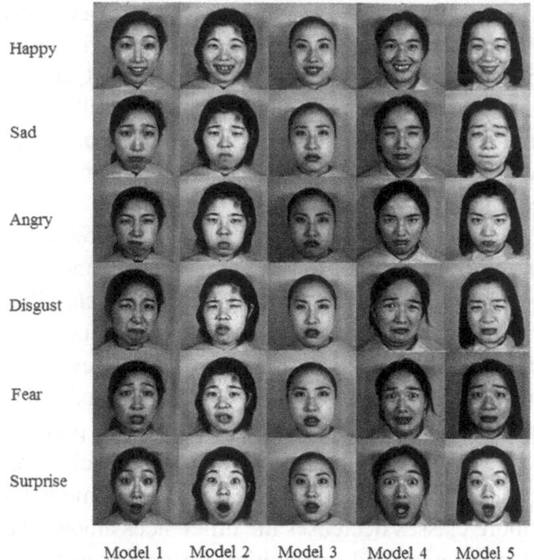

Fig. 1. Six expressions of five models in JAFFE database.

person turns to be surprise, the eyebrows get raised, the eyes widen, and the mouth open. The variations of the facial features are shown in Fig. 1.

The Active Shape Model (ASM) [8] method is adopted to extract landmarks, that is, the points on the facial features. The coordinates of the points are arranged as Eq. (1).

$$\mathbf{x}= [x_1, x_2, \ldots, x_j, \ldots, x_n, y_1, y_2, \ldots, y_j, \ldots, y_n]^{\mathrm{T}} \tag{1}$$

In the equation above, (x_j, y_j) represents the coordinate of the jth landmark point, n is the total number of the points representing the shape of five organs.

The shapes of the training samples are acquired manually and the shape model is built as

$$\mathbf{y}_s = \bar{\mathbf{x}} + \mathbf{\Phi}_s \mathbf{b}_s, \tag{2}$$

where $\bar{\mathbf{x}}$ is the mean shape of the training set, $\mathbf{\Phi}_s$ and $\mathbf{b}_s$ are relative parameters of the model.

After the model building, the ASM matching is conducted to acquire the shapes of the testing samples by comparison between the testing image and the built model. The error function to be minimized is defined in Eq. (4).

$$E = (\mathbf{x}_i - \mathbf{M}_i\mathbf{y}_i)^{\mathrm{T}}\mathbf{W}(\mathbf{x}_i - \mathbf{M}_i\mathbf{y}_i) \tag{4}$$

Here, $\mathbf{M}_i$ is the geometric transformation matrix, $\mathbf{W}$ represents the weight of each points. $\mathbf{x}_i$ and $\mathbf{y}_i$ are the shape of the testing image and the shape model in the ith iteration respectively.

$\mathbf{y}_i, \mathbf{b}_{i+1}, \mathbf{x}_i$ are computed iteratively using Eqs. (5), (6) and (7) until the minimum of Eq. (4) is reached.

$$\mathbf{y}_{i+1} = \mathbf{M}_i^{-1}\mathbf{x}_i \tag{5}$$

$$\mathbf{b}_{i+1} = \mathbf{\Phi}_{\mathbf{s}}^T(\mathbf{y}_{i+1} - \bar{\mathbf{x}}) \tag{6}$$

$$\mathbf{x}_{i+1} = \mathbf{M}_i(\bar{\mathbf{x}} + \mathbf{\Phi}_s\mathbf{b}_{i+1}) \tag{7}$$

After modeling and matching, the set of shape coordinates is used as the feature of each sample in our work.

3 Double-Stage Classification Model with Sample-Selected SVM

Different from the conventional classification classifying the sample as the "client" or the "poster", the facial expression is a multi-classification problem. The sample is determined as a one of the six possible classes. The accuracy is remarkably lower in a multi-class classification than that in a two-class classification. In other words, to seek out the correct class from more classes is more difficult than from two classes, because the feature used for discrimination is not that identifiable any more under the addition of more classes. In other words, it is likely that samples of more than one classes share some similar characteristics, followed by the similar feature values.

Taking the facial expressions as the example, "sad" is easy to distinguish from "happy" because of the obvious difference on the shape of mouth on the two expressions, which is shown on the top two rows in Fig. 1. When a person is in a "happy" mood, the mouth is open and the corners of the mouth are lifted, resulting in an angle between the lips. However, not much action happens on the mouth, or two lips are pushed together when the person is "sad". That is to say, the mouth shape is a distinguishable local characteristic for the classification between the "happy" and "sad" expressions. When it comes to the classification among six expressions, adding the "angry", "disgust', "fear", and "surprise" into classification, "angry" and "disgust" also has the property of pushing the lips together, similar as "sad". The similarity lies in the fact that the three expressions all happen when a person is in a negative mood like disliking or unsatisfied with something. In the condition, it is possible that the classifier gets stuck when telling apart "sad" from "angry" or "disgust", and the "sad" samples are likely to be regarded as the two similar classes wrongly. The local similarity also can be seen between "fear" and "surprise" with lifted eyebrows, widen eyes and open mouth. The similarity between the two expressions is extremely obvious on model 3. The shared characteristics are called inter-class expression similarity in this paper.

Based on the discussions above, the conclusion is reached that the addition of classes results in the reduction of the recognition accuracy, not only of a single class but also of the whole datasets. Based on the fact, the solution of reducing the number of classes to alleviate the poor performance of the multi-classification is proposed.

On the other hand, we select more support vectors to train the classifier, which is the other solution to the weaken discrimination of the feature caused by the shared actions among some expressions.

3.1 Double-Stage SVM Classification System

Compared with other pattern recognition methods, SVM has significant advantages on small databases over other classifiers. Another reason of us choosing SVM is that it takes less time for both training and testing.

In this paper SVM with fewer classes polishes the result of the less effective multi-class SVM. A problem following is how to reduce the classes to be classified. A natural idea is according to the probabilities of a sample belonging to the corresponding classes, which are produced by the multi-classification. In that case, two recognitions needed to be conducted to each sample, of which the first is to get the probabilities and the second get the final class. However, the process is too time-consuming. In fact, not all the single samples get unreliable results from multi-class classifier. That is, the decline in accuracy introduced by addition of more classes is caused by a part of samples, not all of them. Therefore, the samples more likely to be identified unsuccessfully by the multi-class classifier are selected out and get a second classification by a few-class classifier (compared to the multi-class classifier). Figure 2 shows the structure of the whole system, which consists of s-SVM, DM, MAX^r, and r-SVM.

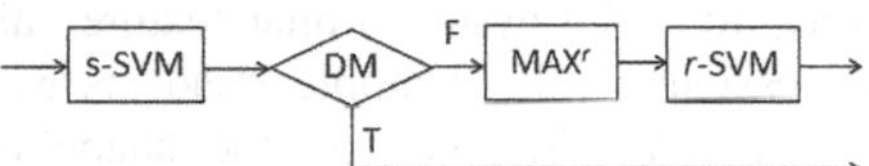

Fig. 2. The SVM-based double-stage classification system proposed in this paper.

s-SVM. s-SVM is the first stage of the recognition system. In this paper it is a six-class SVM which conducts six-class classification on the sample. The input of of the module is the feature points shown in Eq. (1). Assuming that y_l represents the recognition result and $P_l(i)$ is the probability in the corresponding situation. The relationship of the two is shown in Eq. (8).

$$P_l(i) = P(y_l = i|\mathbf{X}_l), i = 1, 2, \cdots, K. \quad (8)$$

Then K probabilities are combined into $\mathbf{P}_l$ as

$$\mathbf{P}_l = [P_l(1), P_l(2), \ldots, P_l(K)] \quad (9)$$

In order to select out the classes for the sample to be recognized the second time more easily, the elements in $\mathbf{P}_l$ are arranged in deceasing order. That is to say,

$$P_l(1) >, P_l(2) >, \ldots, > P_l(K) \tag{10}$$

Basic principles of multi-class SVM and its probability outputs. The multi-class SVM in this paper is obtained by combining binary SVMs. One-against-One combination method is adopted in our work. In [9], Platt gives the pairwise class probability in one-against-one approach as

$$r_{ij} \approx \frac{1}{1+e^{Af+B}}. \tag{11}$$

As for multi-class SVM, the method in [10] is adopted to calculate the probabilities of each situations. The decision is in Eq. (12) with more details presented in [10].

$$p_t = \frac{1}{Q_{tt}}[-\sum_{j:j\neq t} Q_{tj}p_j + \mathbf{P}^T\mathbf{QP}], \tag{12}$$

where

$$Q_{ij} = \begin{cases} \sum_{s::s\neq i} r_{si}^2 & \text{if } i = j \\ -r_{ji}r_{ij} & \text{else} \end{cases} \tag{13}$$

Decision Module (DM). As is mentioned above, the approach in this paper aims at doing a further identification on the samples whose results given by the s-SVM are unreliable, to alleviate the accuracy reduction situation caused by the too many classes. Obviously, the key is to define the samples needing the second recognition. The probabilities produced by the s-SVM are good references. They tell the similarities between a testing sample and those in the training set of each class. If the largest probability is not great enough, in other words, if there's still considerable likelihood that the sample belongs to other classes, the corresponding multi-classification result is regarded to be unsuccessful. In this condition, there maybe not much difference between the largest two, three, or even more probabilities, resulting from the similar probabilities in certain binary SVMs caused by inter-class expression similarity.

DM makes judgement to each output of s-SVM, with rules expressed as follows.

$$\text{DM} = \begin{cases} \text{F if } P_l(1) < \delta,\ 0 < \delta < 1 \\ \text{T else} \end{cases}, \tag{14}$$

where F means the classification result of a certain sample can't be trusted and the corresponding sample is sent to the following two modules, while T means the output of MM is reliable and regarded as the final result.

MAX^r Module (MAX^r). After being judged as a failure of the s-SVM, a sample needs a second classification among fewer classes, to obtain a more trustable result. However, how to select the classes and train a SVM of these classes is an obstacle, affecting the accuracy of the final result directly. No regular pattern in the probabilities produced by s-SVM is found, nor do any statistical properties of them. In this condition, a unified rule for all the samples is not reasonable.

On the other hand, all the classes which are attached a considerable probability should be included in the second recognition. Some samples are among the similar two classes, such as the "fear" and the "surprise". In this condition, the largest two probabilities may be close to each other. However, when the sample is of the class similar to two other classes, like the "sad" mentioned in the beginning of Sect. 3, then the difference among the largest three probabilities is negligible. In some extreme cases, some model makes the expressions quite slightly, and the probabilities of the six expressions distribute evenly. Model 3 is a typical example of this kind in Fig. 1. The variation among different expressions of the model is not as obvious as that of others.

To deal with the different situations in the unsuccessful results from s-SVM, a threshold is set on the sum of the biggest probabilities, of which the corresponding classes are reserved to be classified among by the following r-SVM. In other words, MAX^r is used to choose the biggest r probabilities. The strategy is expressed as

$$\mathbf{P}_l^r = [P_l(1), P_l(2), \ldots, P_l(r)],\ r = \min\left\{n, \sum_{i=1}^{n}(P_l(i)) > p\right\}, \tag{15}$$

where r is the classes selected for the second classification. The rest $l - r$ ones are removed and given no consideration any more. Obviously, r is different for different samples, it is the minimum value where the sum of the largest r probabilities is greater than p. p is a constant in the module.

***r*-SVM.** The function of the module is to go further to classify the objects among the fewer classes selected by MAX^r. It is an r-class SVM.

For each sample there is a special r-SVM. It is an adaptive module. The testing data $\mathbf{X}_l$ can only be recognized as one of the r labels. For example, if the output of MAX^r is the expression "fear" and "surprise", then the r-SVM is a binary SVM trained by the "fear" and "surprise" samples in the whole training set. The output of r-SVM is the final recognition result of the input sample.

After the processing by the whole system, the results of some samples are gained from s-SVM, and those of the others from r-SVM.

3.2 Sample-Selected SVM

SVM projects feature vectors into a high-dimensional space using kernel functions, which is illustrated in Fig. 3(a). The circles and triangles represent two different classes separately. Obviously, the optimal hyperplane is decided by the solid symbols on the edges of corresponding classes and closest to the other class. In other words, these

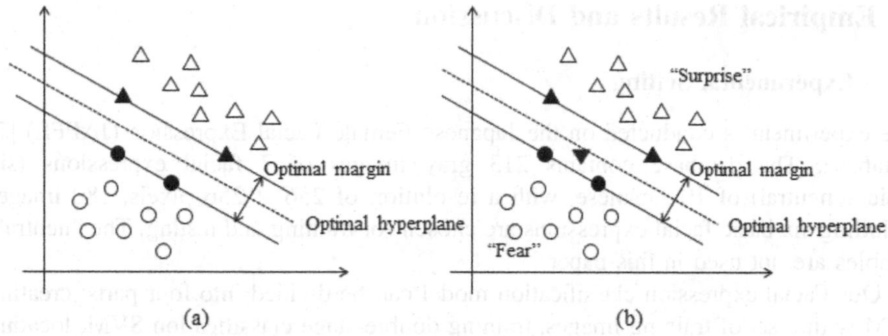

Fig. 3. (a) The illustration of SVM classification. (b) The SVM classification between "Surprise" and "Fear".

samples are the ones of the greatest similarity to the samples in the opposite class. These samples are called support vectors of the SVM.

Considering the SVM in Fig. 3(b) as the one used to classify "fear" and "surprise" samples, we get the optimal hyperplane represented by dotted line from the training process. However, a testing "surprise" sample marked by the inverted solid triangle crosses the edge and locates inside the optimal margin. It is even closer to the "fear" class than to the "surprise". In this condition, a failed recognition occurs. In other words, the real optimal hyperplane of the whole dataset locates far from the one obtained from training process due to the too many support vectors, which are difficult to be all included in the training set.

On one hand, the too many support vectors result from the inter-class similarity caused by the addition of the classes, which has been discussed in the beginning of Sect. 3, for they are the ones nearest to the opposite class in the high-dimensional space.

On the other, the phenomenon is more likely to happen on a small database, for the proportion of support vectors are more likely to be larger than larger datasets, introducing great variations to repeated experiments under random divisions of training and testing samples. In other words, the algorithm is not stable. Our work is typical of the situation. 180 samples are used in total, with 30 samples for each expression. For most of the six classes, more than 20 of the 30 samples are support vectors. That is to say, the result relies heavily on the division of the training set and testing set. If enough support vectors are divided into the training set, a high accuracy can be reached. Otherwise the model gives a poor performance with a lowest accuracy of less than 85%.

Based on the fact mentioned above, sample-selected SVM is introduced in this paper. In sample-selected SVM, as many support vectors are selected in the training set as possible. In this way, SVM can give a steadier hyperplane for different divisions of the training set and the testing set, and the hyperplane obtained from the training reflects the distribution of the whole database better.

In experiments, support vectors are figured out before the training process. In the training stage support vector samples are selected first and other samples are not chosen until the support vector samples are used up.

4 Empirical Results and Discussion

4.1 Experimental Setting

The experiment is conducted on the Japanese Female Facial Expression (JAFFE) [7] Database. The database contains 213 gray images of 7 facial expressions (six basic + neutral) of 10 Japanese, with a resolution of 256 × 256 pixels. 180 images including six basic facial expressions are chosen for training and testing. The "neutral" samples are not used in this paper.

Our Facial expression classification model can be divided into four parts: creating ASM with a set of training images, training double-stage classification SVM, locating features of testing samples, and do recognition on them. Figure 4 shows the whole procedure of the proposed recognition system.

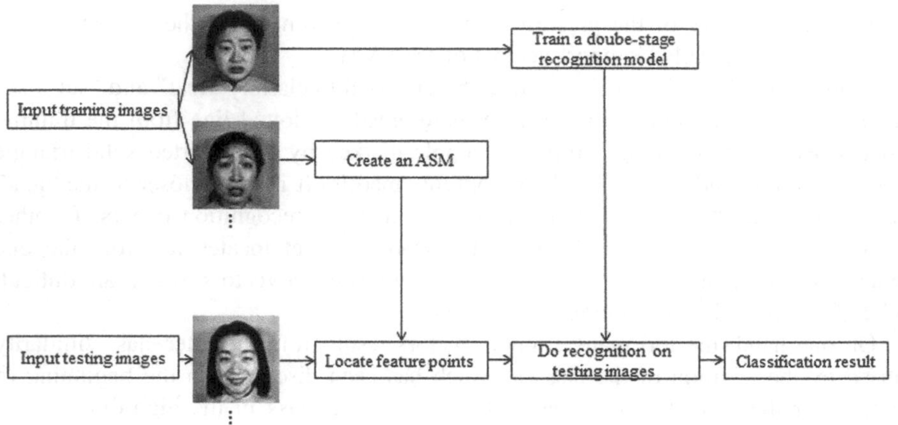

Fig. 4. The flow diagram of our facial expression classification process.

When building ASM, 66 points are labeled around the edges of the eyes, nose tip, eyebrows and mouth (20 for eyes, 20 for eyebrows, 5 for nose tip, 21 for mouth). Examples of labeling the samples are shown in Fig. 5.

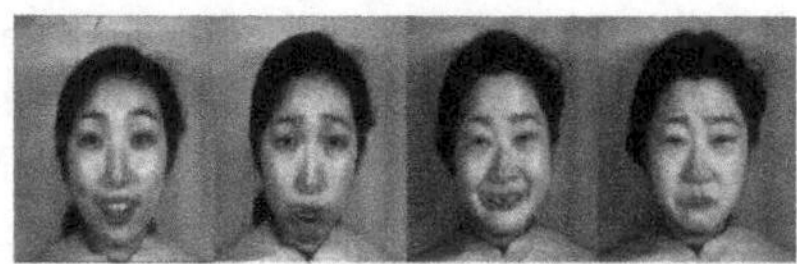

Fig. 5. Some examples of labeling the feature points

4.2 Recognition Results of the Double-Stage Sample-Selected SVM

First of all, taking the "disgust" for example, the recognition accuracies in multi class SVM and binary SVM difference are given in Table 1, to show the difference more obviously.

Table 1. The recognition accuracies of "disgust" in binary SVMs and six-class SVM

SVM	Recognition accuracy of "disgust"
Binary SVM of "disgust" and "angry"	91.6%
Binary SVM of "disgust" and "happy"	100%
Binary SVM of "disgust" and "surprise"	97.8%
Binary SVM of "disgust" and "sad"	93.8%
Binary SVM of "disgust" and "fear"	96.4%
Six-class SVM among all the expressions	86.6%

The table shows remarkable accuracy decrease with the addition of more classes. Besides, the accuracies between "disgust" and "angry" or "sad" are the lowest among all the binary classification, validating the claim that the three negative expression share similar characteristics. Moreover, the classification between "disgust" and "happy" is the only one achieving 100%, for they are of totally different moods.

Then the effectiveness of our proposed double-stage sample-selected SVM is tested. The LIBSVM software is used in this paper [11], and the RBF kernel is adopted. In fact, there is no need to train new SVMs in the second step, for the binary SVMs in the construction of six-class SVM can be reused.

Before training, we use all the 180 samples as the input of LIBSVM and get the support vectors of each class. Then the training sets are constructed as is illustrated in Subsect. 3.2 with number of the training samples varying from 1 to 29 for each expression. The rest are used for testing. The parameters δ in (14) and p in Eq. (15) are set to 0.38 and 0.6 respectively by cross-validation. The procedure is repeated 100 times for the different sizes of the training set. When 20 samples are chosen into the training set for each class, the average recognition accuracy is 98.25%. To show the effectiveness of the proposed two solutions to inter-class similarity, the recognition accuracies in four different situations are presented in Table 2. The table illustrates the effectiveness of both the double-stage SVM, in which the accuracy improvement is produced by the introduction of r-SVM, and the sample-selected SVM. When the two solutions are combined, the performance is improved further.

Table 2. Performances of the two proposed solutions

Classifier	Recognition accuracy
Six-class SVM	91.22%
Double-stage SVM	92.14%
Sample-selected six-class SVM	96.67%
Sample-selected double-stage SVM	98.25%

We compare our approach with some other facial expression recognition methods conducted on JAFFE database in Table 3. Our recognition system shows superiority over other work.

Table 3. Performance comparison of work on the JAFFE database.

Reference	Method	Year	Recognition accuracy
[12]	Combination of the Gabor transform and ASM	2009	87.5%
[13]	Gabor wavelet transform and reduce dimensionality by Principal Component Analysis (PCA) and *Local Binary Pattern* (*LBP*)	2014	90%
[14]	*Extract LBP as feature and use SVM to learn the facial expression model*	2010	91%
[5]	Facial movement features	2011	92.93%
[6]	*Curvelet based LBP*	2010	93.69%
[15]	Local curvelet transform	2016	95.17%
Proposed method	Double-stage sample-selected SVM	2017	98.25%

5 Conclusion

The accuracy decrease produced by addition of classes in the facial expression recognition is alleviated by the two solutions proposed in this paper. Compared to the original SVM, the preference to support vectors when choosing training samples leads to an increase of 5.45% in recognition accuracy. Besides, the introduction of the decision module and *r*-SVM in the double-stage SVM corrects the unsuccessful recognition results of the first six-class classification. The combination of the two solutions gives a recognition result of 98.25% in the Japanese Female Facial Expression Database, higher than eight other approaches. The empirical results demonstrate the effectiveness of the proposed solutions.

Acknowledgments. This work was supported in part by National Natural Science Foundation of China under grant 61371148. The authors appreciate the suggestion of using SVM as classifiers in our research from both Linlu Wang and Zhaohui Meng.

References

1. Ekman, P., Friesen, W.V., Ellsworth, P.: Emotion in the Human Face: Guidelines for Research and an Integration of Findings. Pergamon Press, New York (1972)
2. Fasel, B., Luettin, J.: Automatic facial expression analysis: a survey. Pattern Recogn. **36**, 259–275 (2003). doi:10.1016/S0031-3203(02)00052-3
3. Pantic, M., Rothkrantz, L.J.M.: Facial action recognition for facial expression analysis from static face images. IEEE Trans. Syst. Man Cybern. Part B (Cybern.) **34**, 1449–1461 (2004). doi:10.1109/TSMCB.2004.825931

4. Tu, Y.H., Hsu, C.T.: Dual subspace nonnegative matrix factorization for person-invariant facial expression recognition. In: Proceedings of the 21st International Conference on Pattern Recognition (ICPR2012), pp. 2391–2394 (2012)
5. Zhang, L., Tjondronegoro, D.: Facial expression recognition using facial movement features. IEEE Trans. Affect. Comput. **2**, 219–229 (2011). doi:10.1109/T-AFFC.2011.13
6. Saha, A., Wu, Q.M.J.: Facial expression recognition using curvelet based local binary patterns. In: 2010 IEEE International Conference on Acoustics, Speech and Signal Processing, pp. 2470–2473 (2010). doi:10.1109/ICASSP.2010.5494892
7. Lyons, M., Akamatsu, S., Kamachi, M., Gyoba, J.: Coding facial expressions with Gabor wavelets. In: Proceedings of Third IEEE International Conference on Automatic Face and Gesture Recognition, pp. 200–205 (1998). doi:10.1109/AFGR.1998.670949
8. Cootes, T.F., Taylor, C.J., Cooper, D.H., Graham, J.: Active shape models-their training and application. Comput. Vis. Image Underst. **61**, 38–59 (1995). doi:10.1006/cviu.1995.1004
9. Lin, H.-T., Lin, C.-J., Weng, R.C.: A note on Platt's probabilistic outputs for support vector machines. Mach. Learn. **68**, 267–276 (2007). doi:10.1007/s10994-007-5018-6
10. Wu, T.F., Lin, C.J., Weng, R.C.: Probability estimates for multi-class classification by pairwise coupling. J. Mach. Learn. Res. **5**, 975–1005 (2004)
11. Chang, C.-C., Lin, C.-J.: LIBSVM: a library for support vector machines. ACM Trans. Intell. Syst. Technol. **2**, 1–27 (2011). doi:10.1145/1961189.1961199
12. Wu, P., Li, X.H., Zhou, J.L., Lei, G.: Face expression recognition based on feature fusion. In: 2009 International Workshop on Intelligent Systems and Applications, pp. 1–4 (2009). doi:10.1109/IWISA.2009.5072861
13. Abdulrahman, M., Gwadabe, T.R., Abdu, F.J., Eleyan, A.: Gabor wavelet transform based facial expression recognition using PCA and LBP. In: 2014 22nd Signal Processing and Communications Applications Conference (SIU), pp. 2265–2268 (2014). doi:10.1109/SIU.2014.6830717
14. Yu, K., Wang, Z., Zhuo, L., Feng, D.: Harvesting web images for realistic facial expression recognition. In: 2010 International Conference on Digital Image Computing: Techniques and Applications, pp. 516–521 (2010). doi:10.1109/DICTA.2010.93
15. Uçar, A., Demir, Y., Güzeliş, C.: A new facial expression recognition based on curvelet transform and online sequential extreme learning machine initialized with spherical clustering. Neural Comput. Appl. **27**, 131–142 (2016). doi:10.1007/s00521-014-1569-1

Image Processing

MMW Image Restoration Using the Combination Method of Modified Fuzzy RBFNN and Sparse Representation

Li Shang[1(✉)], Yan Zhou[1], and Zhanli Sun[2]

[1] Department of Communication Technology,
College of Electronic Information Engineering, Suzhou Vocational University,
Suzhou 215104, Jiangsu, China
{sl0930,zhyan}@jssvc.edu.cn
[2] School of Electrical Engineering and Automation, Anhui University,
Hefei 230039, Anhui, China
zhlsun2006@126.com

Abstract. Common fuzzy radial basis function neural network (F-RBFNN) behaves the capability of the approximation of nonlinear noise signal and can denoise images well. In F-RBFNN, the knowledge expression of fuzzy logic and the reasoning ability are combined with the RBFNN's capabilities of fast learning and generalization. And the F-RBFNN's structure and parameters can be adjusted according to the real problem. But this model can not overcome the detect of redundant fuzzy rule so that the optimized learning speed is too slow. To avoid this problem, considering the modified structure and learning algorithm of the antecedent and subsequent network of F-RBFNN, an improved F-RBFNN is proposed and used to denoise millimeter wave (MMW) images. At the same time, to obtain high-quality restored image, the method of sparse representation with the self-adaptive denoising property is used again to denoise results obtained by F-RBFNN. Using the relative single noise ratio (RSNR) criterion to measure denoised images, simulation experimental results show that, compared with other denoising methods such as F-RBFNN, RBFNN and K-SVD and so on, this combination denoising method can obtain better restored images.

Keywords: Fuzzy RBFNN · Sparse representation · K-SVD algorithm · Millimeter wave (MMW) images · Non-linear denoising · Image reconstruction

1 Introduction

At present, fuzzy systems (FSs) and neural networks (NNs) are main two types of artificial intelligence technology [1–3]. The both can simulate man's intelligent action without accurate mathematical models, and can solve many complex, uncertain and non-linear problems, which can not be solved by the traditional techniques [4–6]. Otherwise, FS are good at utilizing man's experiential knowledge and are usually used to express direct and senior logical information [5, 6]. However, in FS, some problems, such as atutomatical extraction of fuzzy rules, fuzzy membership functions' automatic generation and optimization, are difficult to solve and they hinder the development of

D.-S. Huang et al. (Eds.): ICIC 2017, Part I, LNCS 10361, pp. 319–328, 2017.
DOI: 10.1007/978-3-319-63309-1_29

the fuzzy information processing technology [6, 7]. On the other hand, NN can well obtain knowledge of data representation by learning, but this knowledge is implicitly represented by using weight or threshold form, so, it is difficult in understanding [7]. Fuzzy neural networks (FNNs) consider the complementarity of FSs and NNs, and behave the self-adapting denoising capability, the property of self-organizing and self-learning, as well as the strong parallel computing power [7]. Namely, FNNs not only behave NNs' optimization capability and connection structure, but also behave FSs' if-then rules, which are similar to man's form of thinking. Because of the advantages of radial basis function (RBF), such as RBF's strong parallel computation, self-learning, fault tolerance, basis function's radial symmetry and smooth, etc., RBF network is selected as the neural network. Thus, combined fuzzy technology, the fuzzy RBF neural network (F-RBFNN) is used here to process low resolution (LR) images. RBFNN has been proved to be a universal global function approximator [8] and it can approximate any non-linear function with any precision. To reduce the learning time of RBFNN and the number of fuzzy rules, the common RBFNN is modified. This modified RBFNN behaves better global approximation capability and quicker convergence speed, and can well denoise non-linear noise. Otherwise, to denoise linear noise and obtain clearer restored images, sparse representation is used to again locally denoise images, which have been denoised by RBFNN in this paper [9, 10]. Sparse representation is a dictionary learning method, and it can efficiently represent images' structure with few atoms [9, 10]. Typical sparse representation is the K-singular value decomposition (K-SVD) model. It can synchronously denoise noise and learn the dictionary [11, 12]. Now K-SVD has been used successfully in restoring images [12–15]. Therefore, combined the advantages of F-RBFNN and K-SVD, a new image restoration method is discussed here. Using this method, the non-linear noise and linear noise both can be denoised efficiently. In test, the real LR images, i.e., the millimeter wave (MMW) images, are used to testify our method proposed. Further, using the relative signal noise ratio (RSNR) as the measurement criterion of restored images, the validity of our method is also proved by comparing it with algorithms of the wavelet, RBFNN and K-SVD.

2 The Theory of RBFNN

2.1 The Structure of Basic RBFNN

RBFNN is equivalent to the function used in fuzzy reasoning process. According to this equivalence, the system with two different structures can be united in function. Thus, neural networks are not unseen, and all nodes and parameters exhibit certain meanings [9, 10]. The basic F-RBFNN structure is shown in Fig. 1. It is clear to see that the F-RBFNN contains two parts, the premise part network and the consequent part network. The former comprises four layers: the input layer, the fuzzy layer, the rule layer and the normalized layer [12–14]. And the latter comprises three layers: the input layer, the hidden layer and the output layer [12–14], and the relation between each layer is linear. The premise of fuzzy implication indicates a fuzzy subspace of the input space, and each consequent expresses a local input-output relation in the subspace corresponding to the premise part network.

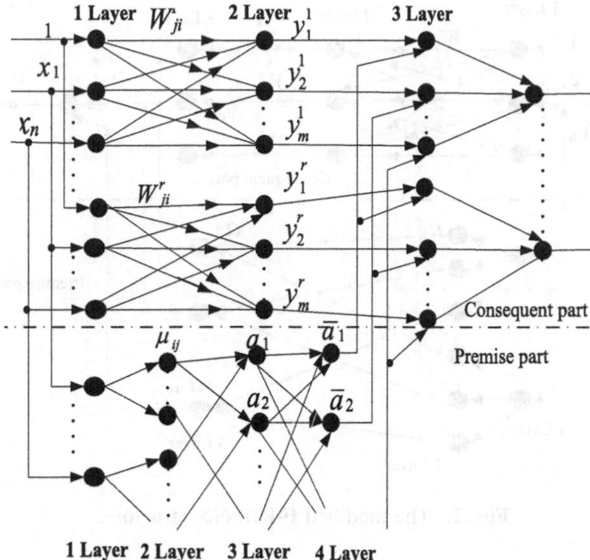

Fig. 1. The basic F-RBFNN structure.

2.2 The Structure of Modified RBFNN

Although the basic F-RBFNN exhibits some advantages, but this network also have unavoidable drawbacks, for example, the network's structure and parameters' learning algorithm are very complex, the learning speed is slow, the number of fuzzy rules in the premise part and the number of neurons can not be defined certainly and so on. So, to avoid these drawbacks, a modified F-RBFNN is obtained, shown in Fig. 2. Clearly, in Fig. 2, the premise part become three layers, and the number of nodes of the fuzzy rule layer also reduces greatly. Otherwise, to enhance noisy robust, a constrain is also considered in the membership function.

In Fig. 2, the premise part contains the input layer, the membership computation layer and the fuzzy reasoning layer. In the input layer, each node connects directly the input vector x_1. This layer transfers input matrix $X = [x_1, x_2, \cdots, x_n]^T$ to the next layer, namely $f_i^{(1)} = x_i (i = 1, 2, \cdots, n)$. And the number of nodes is n. In the second layer, each node denotes a linguistic variable value. This layer's function is to compute the membership function μ_{ij}, which is the membership degree of input components belonging to the fuzzy set of linguistic variable values. The membership function is computed and constrained by the following forms:

$$\begin{cases} f_{ij}^{(2)} = \mu_{ij} = \exp\left[-\left(f_i^{(1)} - c_{ij}\right) \middle/ \sigma_{ij}^2\right] (j = 1, 2, \cdots, m_i) \\ g(\mu_{ij}) = -\sum\limits_{i=1}^{M} \left(\mu_{ij} - \frac{1}{2}\right)^2 \end{cases} . \tag{1}$$

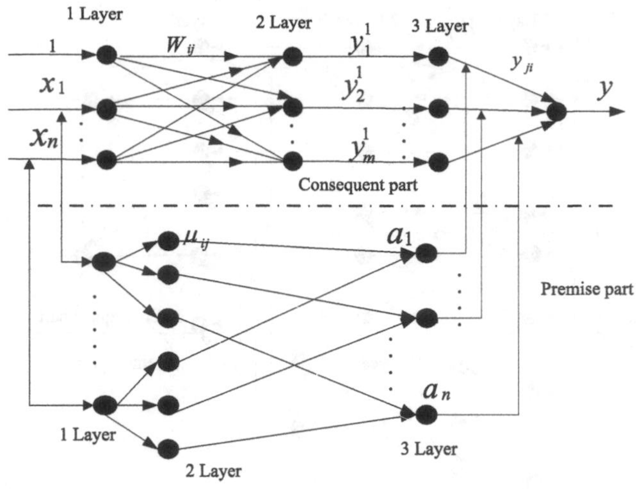

Fig. 2. The modified F-RBFNN structure.

where c_{ij} and σ_{ij} are respectively the membership function's center and width, and m_i is the number of fuzzy segmentation of x_i. The number of nodes in the second is $\sum_{i=1}^{m} n_i$. The third layer is used to compute the conformity of each rule. Each node denotes a fuzzy rule and its function is used to match the premise part of fuzzy rules, and the third layer's expression is defined as follows

$$f_l^{(3)} = \mu_{11}\,\mu_{22} \cdots \mu_{ln} \quad (l = 1, 2, \cdots, m). \tag{2}$$

Where the number of nodes is m and $m = \prod_{i=1}^{n} m_i$. Clearly, in this layer, the number of nodes is reduced greatly and the redundancy of fuzzy rules is also decreased.

In the consequent part network in Fig. 2, the first layer is the input layer and $y^{(1)} = x_i$. The second layer is the middle layer. This layer has m nodes and computes the consequent part of each rule, namely, $y_l^{(2)} = \sum_{i=1}^{n} (w_{li})^2 x_i$. The last layer is the output layer and the output components are computed by $y_l^{(3)} = \sum_{l=1}^{m} f_l^{(3)} y_l^{(2)}$.

Assumed that the error function is denoted by the Eq. of $E = \frac{1}{2}\sum_{l=1}^{m} (\hat{y}_l - y_l)^2$, where $\hat{y}_l$ and y_l represent respectively the expected output and the actual output. Utilizing gradient optimization algorithm to train w_{li}, σ_{ij} and c_{ij}, the responding computation forms are deduced as follows:

$$\begin{cases} w_{li}(t+1) = w_{li}(t) + \beta(\hat{y}_l - y_l)\, x_i\, w_{li} \\ \sigma_{ij}(t+1) = \sigma_{ij}(t) - \beta(\hat{y}_l - y_l)\, y_l^{(2)} \left[2(x_i - c_{ij})^2 \middle/ \sigma_{ij}^2 \right] \\ c_{ij}(t+1) = c_{ij}(t) - \beta(\hat{y}_l - y_l)\, y_l^{(2)} \left[2(x_i - c_{ij}) \middle/ \sigma_{ij}^2 \right] \end{cases}. \quad (3)$$

3 Sparse Representation Based Image Denoising

3.1 Sparse Representation Idea

Supposed that $x \in \Re^N$ is the observed signal, then utilized the sparse representation's idea, x can be approximately represented as $x \approx Ds$, satisfying $\|x - Ds\|_p \leq \varepsilon$, where $s \in \Re^K$ is a vector with very few ($\ll K$) nonzero entries [12]. In common application, this sparsest representation is the solution of the following formula:

$$(P_0)\ \min_s \|s\|_0 \ \textit{subject to}\ \|x - Ds\|_2^2 \leq \varepsilon. \quad (4)$$

Where the symbol $\|\cdot\|_0$ and $\|\cdot\|_2$ are respectively the l^0 and l^2 norm, counting the nonzero entries of a vector. Generally, for an image, to improve its reconstructed version's quality, it is randomly sampled L times with $p \times p$ image patch to obtain the image patch set $X = \{x_1, x_2, \cdots, x_L\} \in \Re^{N \times L}$, thus, the dictionary D can be learned from X, and the optimized problem is described as follows:

$$\{D, S\} = \arg \min_{D,S} \|X - DS\|_F^2 + \lambda \|S\|_1 . \quad (5)$$

Subject to $\|d_k\|_2^2 \leq 1 (k = 1, 2, 3, \cdots, K)$, and d_k is the *kth* column atoms of the dictionary D, and S denotes the sparse coefficient matrix.

3.2 Modified K-SVD Denoising Model

Let Y denote a clear image patch sct, $\tilde{Y}$ denote the noise version of Y. Assumed that the over-complete dictionary D is known and R_{ij} is the extraction mark of lapped image patches, where each image patch $Y_{ij} = R_{ij}\, Y$ with $p \times p$ pixels in every location has a sparse representation with bounded error. In other words, R is the extraction matrix with the size of $p \times N$ pixels, which extracts the (i,j) block from an image with $N \times N$ pixels. Then, the common K-SVD denoising model is written as

$$J(\hat{D}_{ij}, \hat{Y}) = \arg \min_{s_{ij}, U} \left[\lambda \left\| Y - \tilde{Y} \right\|_2^2 + \sum_{i,j} \mu_{ij} \left\| S_{ij} \right\|_1 + \sum_{i,j} \left\| D\, S_{ij} - R_{ij}\, Y \right\|_2^2 \right]. \quad (6)$$

To assure the maximum sparseness, the constraint term of $\sum_{i,j} \left(D_{ij}^T D_{ij} \right)$ is considered in Eq. (6). And then, the denoising model is written as follows

$$J(\hat{D}_{ij},\hat{Y}) = \arg\min_{s_{ij},U}\left[\lambda\left\|Y-\tilde{Y}\right\|_2^2 + \sum_{i,j}\mu_{ij}\left\|S_{ij}\right\|_1 + \gamma\sum_{i,j}\left(D_{ij}^T D_{ij}\right) + \sum_{i,j}\left\|D S_{ij} - R_{ij} Y\right\|_2^2\right]. \quad (7)$$

In Eq. (7), the first term controls the degree of the approximation of Y and $\tilde{Y}$ by controlling the relational expression of $\left\|Y-\tilde{Y}\right\|_2^2 \le Const \cdot \sigma^2$. And the larger the parameter σ is, the smaller the parameter λ is. The second and the third terms are parts of the image priors that makes sure that, in the constructed image patch set Y, each patch Y_{ij} with the size $p \times p$ pixels in every location has a sparse representation with bounded error, where Y_{ij} is calculated by using the Equation of $Y_{ij} = R_{ij} Y$.

4 Our Image Restoration Method

Utilizing F-RBFNN's capability of non-linear function approximator and the self-adapting denoising of K-SVD model, a combined image denoising method is proposed here. At first, a noise image $\tilde{Y}$ is non-linearly filtered by the modified F-RBFNN. Let $y(t)$ be the clear signal, $n(t)$ be the noise signal, and $\tilde{y}(t)$ be the noise version of $y(t)$, where $\tilde{y}(t)$ is computed by $\tilde{y}(t) = y(t) + F(n(t), n(t-1))$. The function $F(\cdot)$ is the non-linear matching function. The non-linear function $F(n(t), n(t-1))$ can be modeled by using F-RBFNN, and the noise can be estimated. Let $\hat{F}(n(t), n(t-1))$ denote the approximated noise, and then, the denoised signal $\hat{y}(t)$ can be obtained according to the bias between $\tilde{y}(t)$ and $\hat{F}(n(t), n(t-1))$. Here, the mathematical modeling of non-linear function $F(\cdot)$ is defined as the following form

$$F(t) = \gamma \cdot \frac{n^3(t) * \cos(n(t-1))}{1 + n^2(t-1)}. \quad (8)$$

Further, to get clearer restored images, for the denoised results obtained by F-RBFNN, we use the modified K-SVD denoising model to denoise locally again. The denoising process of K-SVD algorithm is briefly summarized as follows: Each LR image is sampled randomly image patches of size $p \times p$ pixels, and each image patch is converted into a column vector. Using K-SVD denoising model to train these image patches, the dictionary D and sparse coefficients S can be learned. Based on D and S, the restored image can be estimated by $\hat{Y} = DS$.

5 Experimental Results and Analysis

In test, the MMW image was generated by the State Key Lab. of Millimeter Waves of Southeast University, which is our cooperation group. A MMW image with 41 × 41 pixels and its imaging object, i.e., the original toy gun, were shown in Fig. 3(a) and (b). It was also noted that the MMW image's size was too small to be used directly in application, so, its size was extended into 128 × 128 pixels in test. In this MMW image, the odd columns and even columns are divided into two matrices with the size of 64 × 64 pixels. One was used as training samples, denoted by X_1, and another was used as checking samples, denoted by X_2. Utilizing F-RBFNN to train X_1 to approximate non-linear noise, the noise image and the denoised result of F-RBFNN were respectively obtained, shown in Fig. 4(a) and (b). Compared Fig. 3(b) and Fig. 4(b), the noise in Fig. 3(b) was reduced greatly.

And then, the modified K-SVD denoising model shown in Eq. (7) was again used to denoise locally the filter result of F-RBFNN. To reduce computation work, the F-RBFNN's filter image was sampled randomly image patches of the size of 8 × 8 pixels 50000 times, thus, the training input set of K-SVD is 64 × 50000 pixels. The original dictionary was selected as discrete cosine transform (DCT) dictionary and the redundancy rate is set to be 4. The K-SVD dictionary and the denoised image were respectively shown in Fig. 5(a) and (b), at the same time, to prove the efficiency of our method denoted by F-RBFNN + K-SVD, the denoised results of RBFNN and K-SVD also shown in Fig. 5(c) and (d).

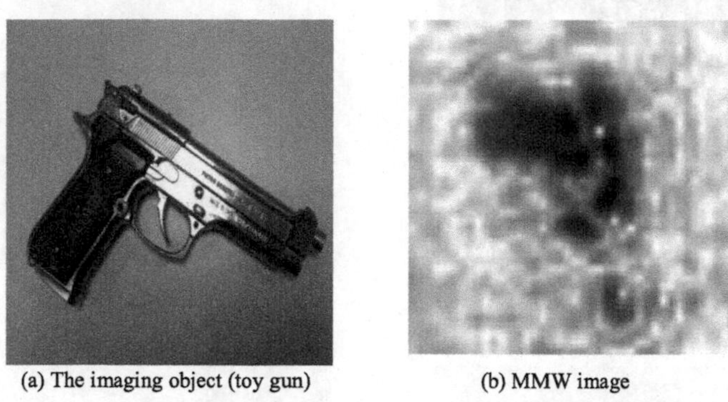

(a) The imaging object (toy gun) (b) MMW image

Fig. 3. The original image and MMW image.

However, it is noted that the original imaging object of the MMW image is unknown and there is much noise existed in the MMW image, so the quality of restored MMW image can not be measured by SNR criterion. Here, the Relative SNR (RSNR) criterion is used to estimate the MMW image. And the RSNR values are calculated by using the following formulas:

$$\mathrm{RSNR} = \frac{1}{\sqrt{NM}}\left[\sum_{i=1}^{N}\sum_{j=1}^{M}\hat{I}(i,j)\right] / \sqrt{\sum_{i=1}^{N}\sum_{j=1}^{M}\left[\hat{I}(i,j) - \tilde{I}(i,j)\right]^2}. \tag{9}$$

Where $I(i,j)$ denotes the input image with the size of $N \times M$, $\hat{I}(i,j)$ denotes the restored image and $\tilde{I}(i,j)$ denotes the mean of $\hat{I}(i,j)$. The values of RSNR the MMW image were listed in Table 1. Compared the original MMW image and denoised results obtained by different denoising algorithms, as shown in Figs. 4(b), 5(b) to (d), it is clearly to see that the noise in the background of MMW image has been reduced at various degrees. Moreover, among these denoised results, that image obtained by

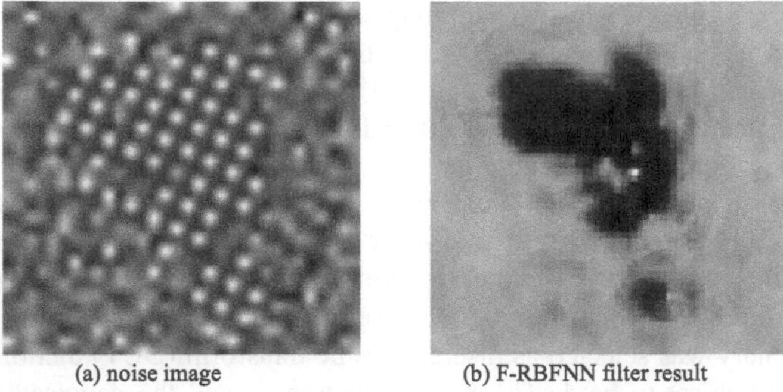

Fig. 4. The noise image and MMW image filtered by the F-RBFNN.

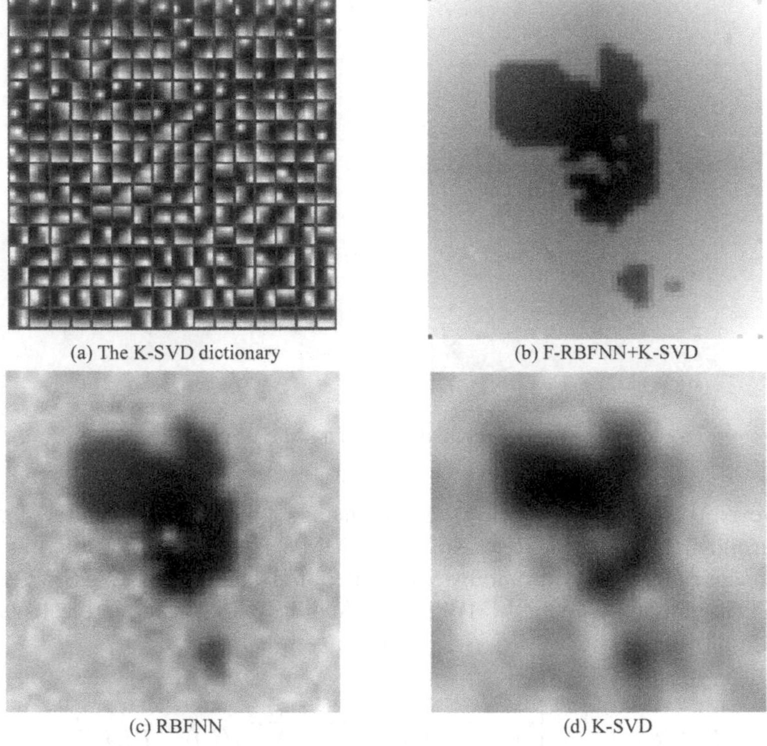

Fig. 5. K-SVD dictionary and restored images by different algorithms. (a) The K-SVD dictionary. (b) F-RBFNN + K-SVD algorithm. (c) RBFNN algorithm. (d) K-SVD algorithm.

F-RBFNN + KSVD method exhibits the clearest contour. And that obtained by RBFNN has the worst visual effect. Otherwise, form Table 1, we also see that the RSNR value of our method is the largest. Therefore, experimental data prove that the efficiency of our method in image restoration.

Table 1. RSNR values of different denoising methods.

Algorithms	Restored image	The original MMW image
	RSNR	RSNR
F-RBFNN + K-SVD	22.377	12.368
F-RBFNN	17.424	
K-SVD	14.629	
RBFNN	13.236	

6 Conclusions

A new MMW image restoration method is discussed in this paper by using the combination algorithm of the modified F-RBFNN and K-SVD denoising model, denoted by F-RBFNN + K-SVD in this paper. In the modified F-RBFNN, the structure and parameters can be adjusted according to the real problem and avoid the redundancy problem. The non-linear noise can be approximated by using F-RBFNN. Otherwise, in modified K-SVD denoising model, the sparse maximization is considered, so, this K-SVD denoising model's convergence speed is fast, at the same time, it behaves local self-denoising property. Therefore, our method can denoise non-linear and Gaussian noise existed in MMW images, meanwhile, it behaves local denoising property. Utilized the RSNR criterion as the measurement of restored images, and compared with denoising methods of RBFNN, F-RBFNN and K-SVD, experimental results prove that our method denoted by F-RBFNN + K-SVD is in deed efficient. And this method can be popularized in low resolution image restoration.

Acknowledgement. This work was supported by the grants from National Nature Science Foundation of China (Grant No. 61373098 and 61370109), the youth found of Natural Science Foundation of Jiangsu Province of China (Grant No. BK20160361), and the grant from Natural Science Foundation of Anhui Province (No. 1308085MF85).

References

1. Takagi, T., Sugeno, M.: Fuzzy identification of systems and its applications to modeling and control. IEEE Trans. Syst. Man Cybern. **1**, 116–132 (1985)
2. Sugeno, M., Kang, G.T.: Structure identification of fuzzy model. Fuzzy Sets Syst. (SO165-0114) **28**(1), 15–33 (1988)
3. Liu, J.Z., Hu, Y., Liu, Z.W.: Effective approach for t-s fuzzy modeling and its application. J. Syst. Simul. **25**(12), 2828–2832 (2013)
4. Wei, F.: Study on filtering image impulse noise based on fuzzy neural network method. Laser J. **37**(1), 131–134 (2016)
5. Guillén, A.N., Rojas, I., González, J., Pomares, H., Herrera, LJ., Prieto, A.: A fuzzy-possibilistic fuzzy ruled clustering algorithm for RBFNNs design. In: Greco, S., Hata, Y., Hirano, S., Inuiguchi, M., Miyamoto, S., Nguyen, H.S., Słowiński, R. (eds.) RSCTC 2006. LNCS(LNAI), vol. 4259, pp. 647–656. Springer, Heidelberg (2006). doi:10.1007/11908029_67

6. Yang, N., Chen, H.J., Li, Y.F.: Coupled parameter optimization of PCNN model and vehicle image segmentation. J. Transp. Syst. Eng. Inf. Technol. **12**(1), 48–54 (2012)
7. Lu, J.Z., Zhang, Q.H., Xu, Z.Y., Peng, Z.M.: Image super-resolution reconstruction algorithm using over-complete sparse representation. Syst. Eng. Electron. **34**(2), 403–408 (2012)
8. Sun, J., Xu, Z.B., Shum, H.Y.: Gradient profile prior and its applications in image super-resolution and enhancement. IEEE Trans. Image Process. **20**(6), 1529–1542 (2011)
9. Luo, J.H., Son L.X., Li L.M., Luo S.X.: Nonlinear filtering based fuzzy RBF neural network. Comput. Simul., 133–136 (2006)
10. Luo, J.H., Li, L.M., Ye, D.X., Zhou, H.L.: Noise cancellation based on improved RBF fuzzy neural network filtering. J. Syst. Simul. **19**(21), 4918–4921 (2007)
11. Liu, J., Tai, X.C., Huang, H.: A weighted dictionary learning model for denoising images corrupted by mixed noise. IEEE Trans. Image Process. **22**(3), 1108–1120 (2013)
12. Aharon, M., Elad, M., Bruckstein, A.: K-SVD: an algorithm for designing overcomplete dictionaries for sparse representation. IEEE Trans. Signal Process. **54**(11), 4311–4322 (2006)
13. Yang, J.C., Wright, J., Huang, T., Ma, Y.: Image super-resolution via sparse representation. IEEE Trans. Image Process. **19**(11), 2861–2873 (2010)
14. Elad, M., Datsenko, D.: Example-based regularization deployed to super-resolution reconstruction of a single image. Comput. J. **52**(1), 15–30 (2008)
15. Yang, J.C., Wright, J., Huang, T., Ma, Y.: Image super-resolution via sparse representation of raw image patches. In: IEEE Computer Society Conference on Computer Vision and Pattern Recognition (CVPR). LNCS, vol. 322, pp. 1–8. Springer, Heidelberg (2008)

A Novel Detection of Ventricular Tachycardia and Fibrillation Based on Degree Centrality of Complex Network

Haihong Liu[1,2], Qingfang Meng[1,2(✉)], Qiang Zhang[3], Yingda Wei[1,2], Mingmin Liu[1,2], and Hanyong Zhang[1,2]

[1] School of Information Science and Engineering, University of Jinan, Jinan 250022, China
ise_mengqf@ujn.edu.cn

[2] Shandong Provincial Key Laboratory of Network Based Intelligent Computing, Jinan 250022, China

[3] Institute of Jinan Semoconductor Elements Experimentation, Jinan 250014, China

Abstract. With the increasing number of cardiovascular disease, some scholars studied it deeply and found that vast majority of sudden cardiac death was due to ventricular fibrillation (VF) or sustained ventricular tachycardia (VT). However, they take different treatment measures. As for patients with VF, we must take defibrillation measure; and patients with VT, we should take low-energy complex heart rate measure. If we misjudge them, the result would be horrific even taking patients' life. So in this paper, we put up with a novel detection based on degree centrality of complex network to distinguish the VT and VF signals. We utilize the characteristics of complex network to analyze the VF and VT signal. At first, we convert the time series into complex network domain by using horizontal visibility graph. Then we analyze the complex network and extract the degree centrality as the single feature to classify the VF and VT signals. Experimental results show that the classification accuracy is up to 99.5%.

Keywords: Cardiovascular disease · Ventricular fibrillation · Ventricular tachycardia · Horizontal visibility graph · Degree centrality

1 Introduction

In recent years, the number of cardiovascular disease is increasing. Researchers found that ventricular fibrillation and sustained ventricular tachycardia account for the very great proportion in the death cases. As for ventricular fibrillation, we must take defibrillation measure. But for patients with ventricular tachycardia, we should take low-energy complex heart rate measure. If we misjudge them, the therapeutic measure would be wrong and the result would be very horrific even taking the patients' life. So the detection for ventricular fibrillation and ventricular tachycardia is needed urgently. Furthermore it would improve the level of people's health. In the past, the detection for ventricular fibrillation and ventricular tachycardia mainly depended on cardiac rhythm. However this method would make mistakes when distinguishing the two

D.-S. Huang et al. (Eds.): ICIC 2017, Part I, LNCS 10361, pp. 329–337, 2017.
DOI: 10.1007/978-3-319-63309-1_30

electrocardiogram(ECG) signals. Later, many methods were proposed to complete this detection, which relied on quantitative analysis of arrhythmia signal [1–5]. But those methods also have some limitations, such as the accuracy is not high. Soon some researchers come up with detections based on traditional features such as sample entropy [6], approximate entropy [7], time-dependent entropy (TDE) [8] and multi-resolution entropy (MRE) [9]. Lempel and Ziv [10] proposed Lempel-Ziv complexity to detect the ECG signal. Through study the ECG signal deeply, Xia and Meng [11] combined the Lempel-Ziv complexity with wavelet transform or empirical mode decomposition (EMD) [12] for detection. Before analyzing the signal, the original signal was decomposed with wavelet transform or EMD firstly. The result showed this method can get better classification accuracy. Later, the theory of complex network provided a new way to analyze time series. A study showed that the complex network can study the deep nonlinear dynamical characteristic of time series. Jie zhang [13] researched pseudo-periodic time series by the theory of complex network [14]. Zhang and Small [15] put up with the algorithm, which can convert the pseudo-periodic time series into complex network domain. The scholars studied many different time series [16–19] using proximity network and researched the time series utilizing the directed weighted complex network [20]. At the same time, they analyzed the time series of multi-scale characteristic of the complex network. In 2008, Lacasa proposed a new conversion algorithm, which is the visibility graph [21, 22]. Visibility graph [23] has some progress compared to proximity network, which makes it possible to convert any time sequence into complex network domain. Soon the visibility graph was applied in many areas widely. This algorithm was applied to study the exchange rate series [24] and fractional Brownian motions [25]. Afterwards, horizontal visibility graph [26] was proposed by some researchers, which improves the visibility graph.

In this paper, we utilize the complex network theory to classify the ventricular fibrillation and ventricular tachycardia, which opens the new door to analyze the ECG signal. We convert the time series into complex network at first. Then we utilize the characteristics of complex network to analyze the ventricular fibrillation and ventricular tachycardia and get the better performance. The idea of this paper and the classification feature both provide great help for clinical research.

2 The Feature Extraction Method

This section mainly involves the conversion algorithm, which completes the conversion from time series to complex network and focuses on introducing the feature extraction method for ventricular fibrillation and ventricular tachycardia signal. In this process, we adopt horizontal visibility graph to construct complex network corresponding to original time series. This conversion algorithm has more advantages than proximity network, which does not need to select any parameter. Then we propose a novel detection based on degree centrality of complex network.

2.1 Horizontal Visibility Graph

The calculation process of horizontal visibility graph is based on the theory of visibility graph. We introduce the process of this algorithm in detail followed. At first, an ECG is denoted as $\{x_i\}$ i ∈ [1 M], where the x_i presents the i_{th} sampling point in the ECG signal, which the length of the signal is M. We treat each sampling point of original time series as a node of complex network, which construct node set. Whether there is an edge between nodes depends on local convex constraints. The local convex constraints is: as for two sampling point x_i and x_j, there is an edge if $x_k < \min(x_i, x_j)$, for all k with $t_i < t_k < t_j$ if not, there is no edge between the node x_i and x_j. t_i stands for the i_{th} time. According to this principle, we can get the edge set. As for visibility graph, the criteria is $x_k < x_j + (x_i - x_j)\frac{t_j - t_k}{t_j - t_i}$. In other words, if the connecting line between this two times samples values can't be separated by any middle time sample point values, this two points are connected. Through the calculation, we can get the adjacency matrix. In order to make comparison, we introduce the general steps of proximity network. Firstly the ECG signal is divided up into several non-overlapping cycles and each cycle is treated as a node of complex network. Then calculate the distance between two nodes and measure the similarity between them according to the distance. We can get the distance matrix from the above step. Secondly, we choose an appropriate threshold (th) to determine whether there is an edge between every pair nodes. The rule is: as for two sampling point x_i and x_j, if the distance is higher than the threshold, there is no edge between the node x_i and x_j; if the distance is lower than the threshold, there is edge between the node x_i and x_j. Therefore we can get adjacency matrix which is composed of 1 or 0.

In the process of converting the original time series into complex network, we use the horizontal visibility graph. Because, this conversion algorithm has some advantages that the other methods can't match. In the horizontal visibility graph, there is no threshold involved. So it can't cause the error to the analysis of the complex network. However, as for proximity network, it must select the better threshold to get adjacent matrix. As we all know, as long as there is threshold, it will produce errors inevitably. Compared with proximity network, the method we utilized decreases the subjectivity greatly and we can analyze the ECG signal precisely. The experimental results confirm that detection algorithm we proposed has huge potential for ventricular tachycardia and fibrillation detection.

2.2 The Degree Centrality of Complex Network

In complex network, the theory provides many statistical properties, which presents the relationship between the nodes and reflects the internal relation of the original time series. The degree centrality of complex network is defined as:

$$c_D(x) = \frac{k(x)}{n-1} \tag{1}$$

In Formula (1), $k(x)$ presents the degree of the node; in other words, the degree of one node is the number of edges which is connected with this node; n presents the total number of nodes in complex network; $n-1$ presents the largest possible number of edges of one node. We extract the degree centrality of the complex network corresponding to the original ECG signal as the single feature to detect the ventricular tachycardia and fibrillation signal. The experimental result shows the degree centrality can do well in classifying the ventricular tachycardia and fibrillation signal. So we can excavate more information of original time series from the degree centrality of complex network.

3 Result and Analysis

3.1 Data Description

All the simulations were based on a 2.60 GHz quad-core Inter Pentium processor with 4 GB memory. The code was executed in environment of MATLAB 7.0.

The data were selected from MIT-BIH Malignant Ventricular Ectopy Database (MIT-BIH Database) and Creighton University Ventricular Tachyarrhythmia Database (CU Database). The sampling frequency of VF and VF are all 250 Hz and the sample's duration is 4 s. We select VT samples from CU database and VF samples from MIT-BIH database. In this paper, experimental data contain one hundred VF samples and one hundred VT samples. Each sample has one thousand sampling points. One of VF and VT samples are plotted in Fig. 1. From the Fig. 1, we can see that the two ECG signals have different waveforms.

3.2 Classification Experiment Result

Researchers show that the statistical properties of complex network can reveal the deep dynamical structure of nonlinear time series. In the past, many people used the Lempel-Ziv complexity combined with wavelet transform and empirical mode decomposition (EMD) to detect the ventricular tachycardia and fibrillation signal. The original signal is decomposed with wavelet transform or EMD firstly. Then we analyze the decomposed signal and the results show the performance is better than the previous methods. To make further improvement on efficiency, we analyze the ECG signal based on the complex network theory for the first time. We convert the original data into complex network domain. Then we analyze the complex network and extract the degree centrality as the feature to classify the ventricular tachycardia and fibrillation signal. According to the Formula (1), we calculate the degree centrality of each node. Then we calculate the average of the feature values of each sample as the feature to classify the data. The feature we extracted reveals the connection between the nodes and reflects the internal relation of the time series. Ventricular tachycardia and fibrillation signal have different characteristics, so the feature values of ventricular tachycardia and fibrillation signal should have different values. Figures 2 and 3 plot the mean of degree centrality of complex network in line graph and boxplot graph respectively. In Fig. 2, the first one hundred data and the second hundred data belong to

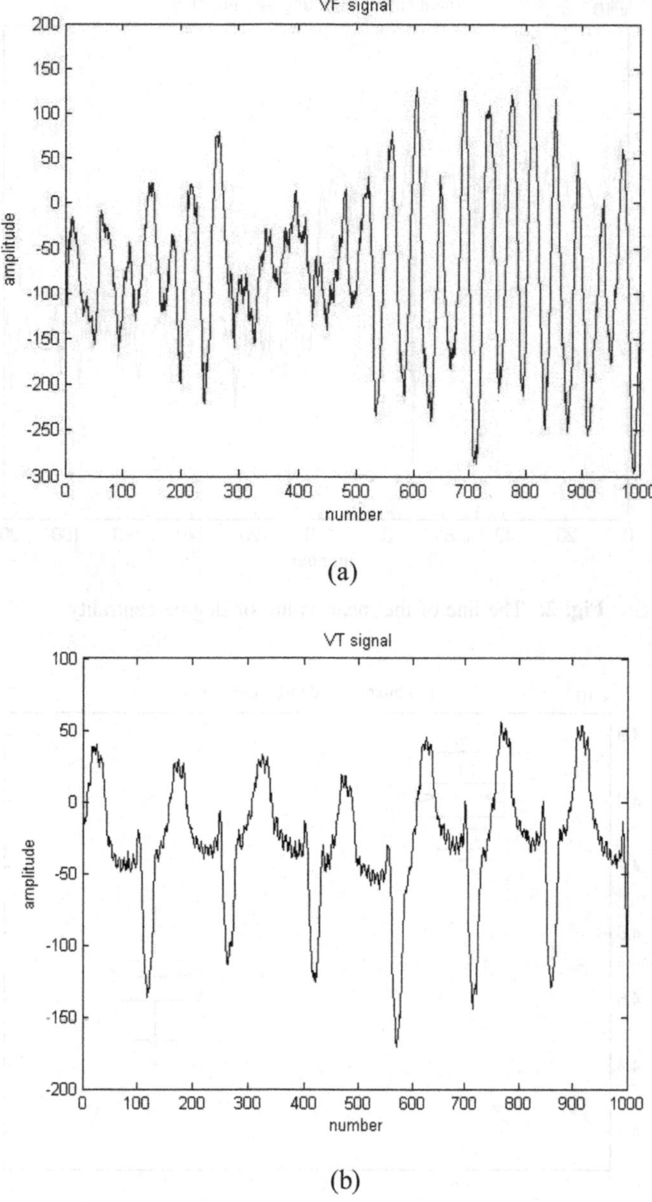

Fig. 1. (a): one of VF samples of ECG signal, (b) one of VT samples of ECG signal

the ventricular fibrillation signal and ventricular tachycardia signal respectively. We can see that only the seventieth sample value of ventricular fibrillation is lower than that of ventricular tachycardia. In Fig. 3, we can see that the values have a big difference between the two signals. From the Figs. 2 and 3, we can clearly find that the

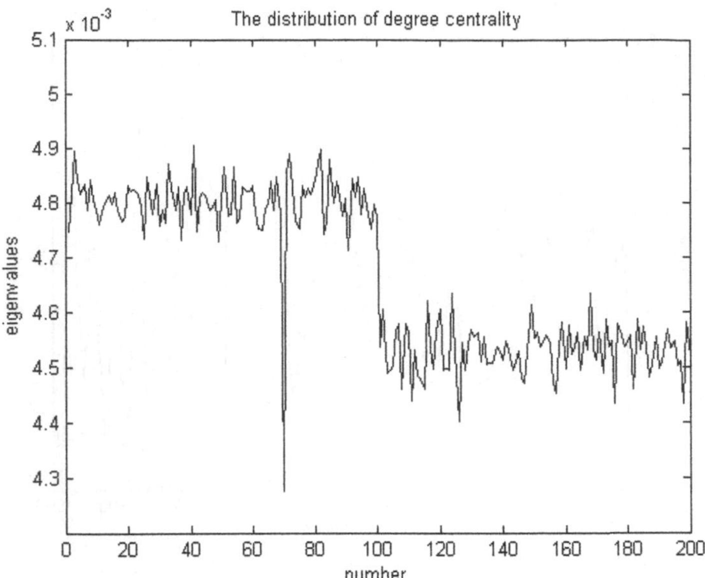

Fig. 2. The line of the mean value of degree centrality

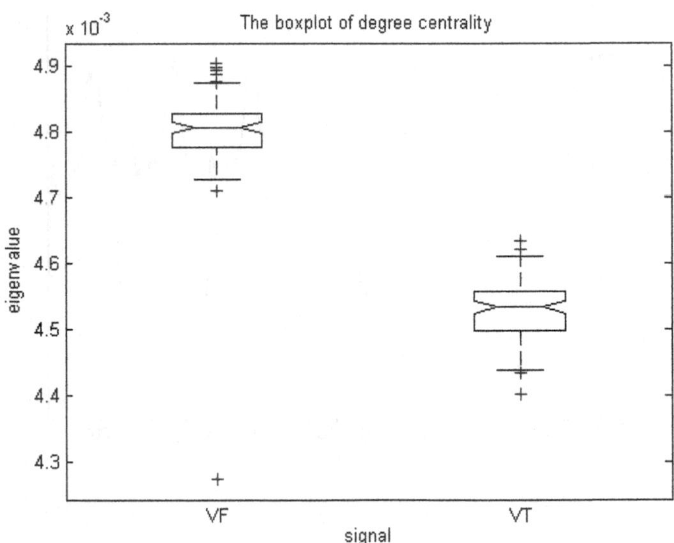

Fig. 3. The boxplot graph of the mean value of degree centrality

values of ventricular fibrillation are higher than the ventricular tachycardia generally. It means the ventricular fibrillation and ventricular tachycardia have different eigenvalues, so we can classify the experiment data by studying the different values of degree centrality of complex network.

Table 1. The classification results of the proposed method and other existing methods

Feature	Threshold	*ACC* (%)
Lempel-Ziv complexity	0.1823	63.10
Sample entropy	0.1823	90.50
Approximate entropy	0.2142	93.50
Lempel-Ziv complexity + EMD	0.1356	97.08
Degree centrality	0.0046	99.50

In the Table 1, as for existing methods, we make some contrast and analysis. Threshold is the value to classify the signal well. Accuracy (ACC) presents the total accuracy to classify the signal correctly. From the Table 1, we can clearly find our method has the higher classification accuracy than other methods. Compared to the previous methods such as Lempel-Ziv complexity, sample entropy and approximate entropy, the degree centrality of complex network has better classification accuracy. When we use the Lempel-Ziv complexity, sample entropy and approximate entropy as the feature, the classification accuracy is 63.1%, 90.5% and 93.5% respectively. However, the classification accuracy of degree centrality we selected is up to 99.50%. In order to observe the classification result visually, we plot the Fig. 4.

Figure 4 shows the best classification result when we select the degree centrality as the feature. In Fig. 4, each '+' presents the ventricular tachycardia signal and each '*' presents the ventricular fibrillation signal. From the Fig. 4, we can see the points '*' are higher than the points '+' except for the seventieth points. When the threshold is

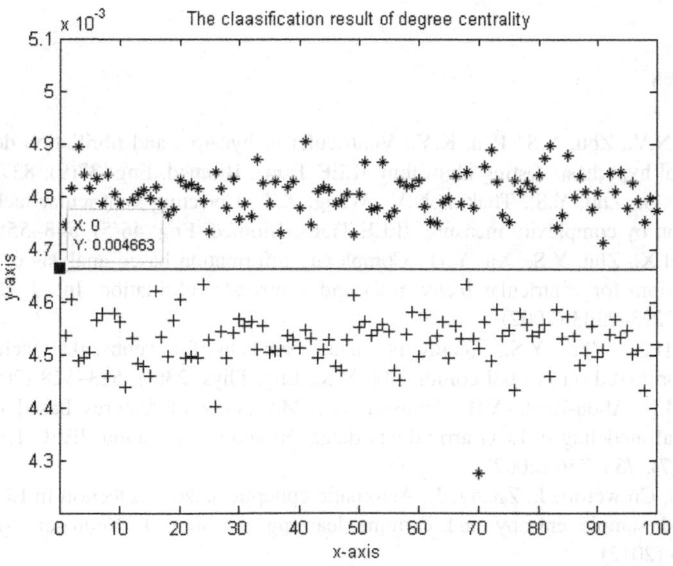

Fig. 4. The best classification result of the degree centrality

0.004663, the classification result is optimal. In conclusion, the degree centrality does well in classifying the two ECG signals.

4 Conclusions

In recent years, the number of cardiovascular disease is increasing and some researchers study the deep nonlinear dynamical characteristic of time series utilizing the topological properties of complex network. Complex network theory also gives a new perspective to analyze the nonlinear time series. Thanks to the conversion algorithms, which transform the time series into complex network, the complex network is applied to analyze the different time series. In this paper, we study the ventricular tachycardia and fibrillation signal according to the complex network. Through our analysis, we extract the degree centrality of complex network as the classification feature and it does well in classifying the ECG signal, which depicts that this feature has more information to express the original time series compared to other features. The classification accuracy of the degree centrality is higher than other traditional features, which is up to 99.5%. The research idea of this paper and the classification algorithm both provide great help for clinical research and application.

Acknowledgment. This work was supported by the National Key Research And Development Plan (No. 2016YFC0106001), the National Natural Science Foundation of China (Grant No. 61671220, 61640218, 61201428), the Shandong Distinguished Middle-aged and Young Scientist Encourage and Reward Foundation, China (Grant No. ZR2016FB14), the Project of Shandong Province Higher Educational Science and Technology Program, China (Grant No. J16LN07), the Shandong Province Key Research and Development Program, China (Grant No. 2016GGX101022).

References

1. Thakor, N.V., Zhu, Y.S., Pan, K.Y.: Ventricular tachycardia and fibrillation detection by a sequential hypothesis testing algorithm. IEEE Trans. Biomed. Eng. **37**(9), 837–843 (1990)
2. Zhang, X.S., Zhu, Y.S., Thakor, N.V., Wang, Z.Z.: Detecting ventricular tachycardia and fibrillation by complexity measure. IEEE Trans. Biomed. Eng. **46**(5), 548–555 (1999)
3. Zhang, H.X., Zhu, Y.S., Xu, Y.H.: Complexity information based analysis of pathological ECG rhythm for ventricular tachycardia and ventricular fibrillation. Int. J. Bifurc. Chaos **12**(10), 2293–2303 (2002)
4. Zhang, H.X., Zhu, Y.S.: Qualitative chaos analysis for ventricular tachycardia and fibrillation based on symbol complexity. Med. Eng. Phys. **23**(8), 523–528 (2001)
5. Owis, M.I., Abou-Zied, A.H., Youssef, A.B.M.: Study of features based on nonlinear dynamical modeling in ECG arrhythmia detection and classification. IEEE Trans. Biomed. Eng. **49**(7), 733–736 (2002)
6. Song, Y., Crowcroft, J., Zhang, J.: Automatic epileptic seizure detection in EEGs based on optimized sample entropy and extreme learning machine. J. Neurosci. Methods **210**, 132–146 (2012)

7. Fleisher, L.A., Pincus, S.M.S., Rosenbaum, H.: Approximate entropy of heart rate as a correlate of postoperative ventricular dysfunction. Anesthesiology **78**(4), 683–692 (1993)
8. Tong, S., Bezerianos, A., Paul, J., Thakor, N.: Nonextensive entropy measure of EEG following brain injury from cardiac arrest. Phys. A **305**(3), 619–628 (2002)
9. Gamero, L.G., Plastino, A., Torres, M.E.: Wavelet analysis and nonlinear dynamics in a nonextensive setting. Phys. A **246**(3), 487–509 (1997)
10. Lempel, A., Ziv, J.: On the complexity of finite sequences. IEEE Trans. Inf. Theory **22**, 75–81 (1976)
11. Xia, D., Meng, Q., Chen, Y., et al.: Classification of ventricular tachycardia and fibrillation based on the lempel-ziv complexity and EMD. Intell. Comput. Bioinform., 322–329 (2014)
12. Arafat, M.A., Sieed, J., Hasan, M.K.: Detection of ventricular fibrillation using empirical mode decomposition and bayes decision theory. Comput. Biol. Med. **39**, 1051–1057 (2009)
13. Zhang, J., Sun, J., Luo, X., Zhang, K., Nakamurad, T.: Michael Small.: Characterizing pseudoperiodic time series through the complex network approach. Physica D **237**, 2856–2865 (2008)
14. Han, S.-Y., Chen, Y.-H., Tang, G.-Y.: Sensor fault and delay tolerant control for networked control systems subject to external disturbances. Sensors **17**, 700 (2017)
15. Zhang, J., Small, M.: Complex network from pseudoperiodic time series: topology versus dynamics. Phys. Rev. Lett. **96**, 238701 (2006)
16. Milo, R., Shen-Orr, S., Itzkovitz, S., Kashtan, N., Chklovskii, D., Alon, U.: Network motifs: simple building blocks of complex networks. Science **298**, 824–827 (2002)
17. Small, M., Zhang, J., Xu, X.: Transforming time series into complex networks. In: Zhou, J. (ed.) Complex 2009. LNICST, vol. 5, pp. 2078–2089. Springer, Heidelberg (2009). doi:10.1007/978-3-642-02469-6_84
18. Xiang, R., Zhang, J., Xu, X.K., Small, M.: Multiscale characterization of recurrence-based phase space networks constructed from time series. Chaos **22**, 013107 (2012)
19. Gao, Z.: N. Jin.: Complex network from time series based on phase space reconstruction. Chaos **19**, 033137 (2009)
20. Wang, F., Meng, Q., Chen,Y.: The neoteric feature extraction method of epilepsy EEG based on the vertex strength distribution of weighted complex network. In: 2014 IEEE World Congress on Computational Intelligence, IJCNN, pp. 3234–3239 (2014)
21. Marwan, N., Donges, J.F., Zou, Y., Donner, R.V., Kurths, J.: Complex network approach for recurrence analysis of time series. Phys. Lett. A. **373**, 4264–4254 (2009)
22. Lacasa, L., Luque, B., Ballesteros, F., Luque, J., Nuno, J.C.: From time series to complex networks: the visibility graph. Proc. Natl. Acad. Sci. USA **105**, 4972–4975 (2008)
23. Lacasa, L., Toral, R.: Description of stochastic and chaotic series using visibility graphs. Phys. Rev. E **82**, 036120 (2010)
24. Yang, Y., Wang, J., Yang, H., Mang, J.: Visibility graph approach to exchange rate series. Phys. A Stat. Mech. Appl. **388**, 4431–4437 (2009)
25. Ni, X.H., Jiang, Z.Q., Zhou, W.X.: Degree distributions of the visibility graphs mapped from fractional Brownian motions and multifractal random walks. Phys. Lett. A **373**, 3822–3826 (2009)
26. Luque, B., Lacasa, L., Ballesteros, F., Liuque, J.: Horizontal visibility graphs: exact results for random time series. Phys. Rev. E **80**, 046103 (2009)

Deep Convolutional Networks-Based Image Super-Resolution

Guimin Lin[1,2], Qingxiang Wu[1(✉)], Xixian Huang[1], Lida Qiu[2], and Xiyao Chen[2]

[1] Key Laboratory of Optoelectronic Science and Technology for Medicine of Ministry of Education, College of Photonic and Electronic Engineering, Fujian Normal University, Fuzhou 350007, People's Republic of China
{gmlin, qxwu}@fjnu.edu.cn

[2] Department of Physics and Electronic Information Engineering, Minjiang University, Fuzhou 350108, People's Republic of China

Abstract. Convolutional neural networks (CNN) have been successfully applied in many fields of image processing, such as deblurring, denoising and image restoration. Estimating a high quality high-resolution image from one or a set of low-resolution images is a non-linear mapping, which can be formulated as a regression problem. According to the image formation process, a Deep Convolutional Network-based image Super-Resolution model DCNSR is proposed and is trained using end-to-end. Several key components of DCNSR, which would affect the training time and the effectiveness of reconstruction super-resolution image, are firstly demonstrated. Then, the deblurring performance is evaluated. Finally, comparisons with the results in state-of-the-arts are presented. Experimental results demonstrate that the proposed model achieves a notable improvement in terms of both quantitative and qualitative measurements.

Keywords: Image super-resolution · Deep convolutional networks · Gaussian process regression · Deblurring

1 Introduction

The objective of super-resolution (SR) is recovering the original high-resolution (HR) image from one or more low-resolution (LR) images by inferring all the missing high frequency contents, based upon reasonable assumptions or prior knowledge about the imaging process. The inverse process is generally an ill-posed problem since the solution with the reconstruction constraint is not unique. To improve the reconstruction accuracy for image SR, recent state-of-the-art methods mostly adopt the example-based [1] approaches. According to the learning strategies, example learning-based methods can be further classified into two major sub-categories: coding-based methods and regression-based methods. For regression-based methods, there are several models such as support vector regression, kernel regression, neighborhood regression, linear regression, Gaussian process regression (GPR) [2], and convolutional neural network

D.-S. Huang et al. (Eds.): ICIC 2017, Part I, LNCS 10361, pp. 338–344, 2017.
DOI: 10.1007/978-3-319-63309-1_31

(CNN) [3, 4]. GPR has been successfully applied to image SR due to its non-parametric statistical approach and nonlinear mapping capability. For example, He et al. [2] proposed a two-step GRP-based SR model called SRGPR. Despite its effectiveness, its noticeable computational cost may limit the applicability in practice.

In this paper, inspired by the great success of deep learning technique [5–7] in computer vision tasks, we translate the two-step SRGPR pipeline to a deep convolutional network (DCN) that directly learns an end-to-end mapping between low- and high- resolution images. The proposed framework is referred as DCNSR. Comparing with the SRCNN in [3, 4], our proposed method is novel in the following aspects: (1) According to the principle of CNN, the DCNSR is designed as four convolutional layers with reasonable filter sizes to improve the reconstruction accuracy. (2) Exploiting the Parametric Rectified Linear Unit (PReLU) [8] as activation function and Mean Squared Error (MSE) as loss function with a weight decay term, which speeds up the convergence during the optimization and further improve SR performance.

The proposed model achieves notable improvement over SRCNN [3, 4] in terms of both recovery accuracy and human perception. Moreover, with the correct understanding of each layer's physical meaning, the parameters of DCNSR are initialized in a more principled way, which helps to save the training-time. The remainder of this paper is organized as follows: In Sect. 2, the proposed framework and some key components are presented. The databases used for evaluation and the experiment results are demonstrated in detail in Sect. 3. Finally, the main conclusions are presented in Sect. 4.

2 Deep Convolutional Networks for SR

To achieve a high quality high-resolution image, inspired by the deblurring step in super-resolution [2], a deblurring layer is introduced after the reconstruction layer in SRCNN [4] to form a new Deep Convolutional Network for single image Super-Resolution – DCNSR. Figure 1 illustrates an overview of the proposed approach. Given an LR image, we first upsample it to the desired size using bicubic interpolation, which is the only pre-processing. The interpolated image is denoted as Y and still call Y as an "LR" image for convenient, the reconstructed image as $\tilde{\mathrm{X}}$ and the ground truth HR image as X. In the training phase, the goal is to learn a nonlinear mapping F that can produce an HR image $\tilde{\mathrm{X}}$ from the interpolated image Y. Learning is formulated as the inverse mapping $F: Y \mapsto \tilde{X}$ as a nonlinear regression problem, where Y is an interpolated LR image, $\tilde{\mathrm{X}}$ is an HR image with the same size of X, and $\tilde{\mathrm{X}}$ is as similar as possible to X. We wish to learn the inverse mapping F using a CNN model. The CNN model consists of four layers: feature extraction layer, nonlinear mapping layer, restoration layer and deblurring layers. To train the model effectively and efficiently, the important concepts are represented as follows.

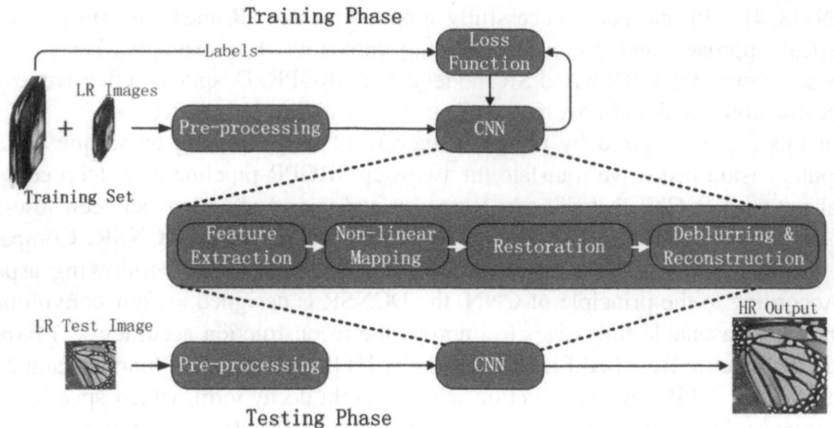

Fig. 1. Overview of the proposed method DCNSR.

2.1 Initialization

Without suitable initialization for the parameters of a deeper CNN model, back-propagated gradients could vanish or explode, thus such model would be difficult to achieve convergence. From previous works [5], it is known that the first layer of CNN mainly works as specific edge-like feature extraction, which is similar to 2D Gabor filter (formulated in Eq. (1)). So several Gabor filters are exploited with different wavelengths λ and orientations θ to initialize the feature extraction layer's parameters to shorten the training-time. The other layers' parameters are randomly initialized from zero-mean Gaussian distribution with standard deviation 0.001.

$$\begin{aligned} g(x, y; \lambda, \theta, \psi, \sigma, \gamma) &= \exp(-\frac{x'^2 + \gamma^2 y'^2}{2\sigma^2})\cos(2\pi\frac{x'}{\lambda} + \psi) \\ x' &= x \cos\theta + y\sin\theta \\ y' &= -x \sin\theta + y\cos\theta \end{aligned} \tag{1}$$

2.2 Activation Function

In CNN models, a layer generally comprises convolutional operations and an activation function, which is responsible for non-linearly transforming the input data x. Parametric Rectified Linear Unit (PReLU) [8], which introduces a small slope on the negative part of the function to avoid zero gradients, is defined as

$$f(x) = \max(0, x) + \alpha \min(0, x) \tag{2}$$

where α is the learnable parameter. PReLU improves model fitting with nearly zero extra computational cost and little over-fitting risk especially when the coefficient α is shared by all channels of one layer. The parameter α can be learned by end-to-end training, which will lead to more specialized activations and achieve better results.

2.3 Loss Function

It is the function to be minimized during training. Given a set of ground truth HR images $\{X_i\}$ and their corresponding interpolated LR images $\{Y_i\}$, a regression model can be trained and its parameters are determined by minimizing a regularized Mean Squared Error (MSE) function $\hat{\mathrm{L}}(\Theta)$ given by,

$$\mathrm{L}(\Theta) = \frac{1}{2N}\sum_{i=1}^{N}\|F(Y_i,\Theta) - X_i\|^2, \quad \hat{\mathrm{L}}(\Theta) = \mathrm{L}(\Theta) + \frac{\lambda}{2}\|W\|^2 \tag{3}$$

where $\mathrm{L}(\Theta)$ is the loss term, $\Theta = \{W_1, W_2, W_3, W_4, B_1, B_2, B_3, B_4, \alpha_1, \alpha_2, \alpha_3\}$, W_i is the i-th convolutional layer's weights, B_i is the bias term of each layer, N is the number of training samples, and α_i is the coefficient of the i-th layer's activation function PReLU. The weight decay term $\frac{\lambda}{2}\|W\|^2$ is not merely a regularizer, but also improves generalization and speeds up convergence [5].

2.4 Training

To Learn the end-to-end mapping function F, it requires the estimation of network parameters Θ by minimizing the loss function Eq. (3). The loss is minimized employing a Stochastic Gradient Descent (SGD) algorithm with momentum. The update rules for weight W are

$$v_{t+1} = \mu \cdot v_t - \eta \cdot \frac{\partial L}{\partial W_t^l} - \eta \cdot \lambda \cdot W_t^l, \quad W_{t+1}^l = W_t^l + v_{t+1} \tag{4}$$

Where $l \in \{1, 2, 3, 4\}$ is the index of layers, W_t^l and v_t are the weight of l-th layer's current weights and the previous weight update at iteration t, μ is the momentum, η is the learning rate, λ is the weight decay, and $\frac{\partial L}{\partial W_t^l}$ is the derivative. In the training phase, the momentum μ = 0.9, weight decay λ = 0.0005 and the batch size is 128. The learning rate η is always equal to 10^{-4} for all layers during training.

3 Experiments and Discussions

In this section, the 91-image dataset [9] are used as base training set, which is augmented with flip and rotations. The LR-HR image pairs for both training and evaluation are generated from original images and downsampled with bicubic interpolation method. The performance of the proposed DCNSR were evaluated using Set5 dataset [10] with a resolution enhancement factor of 3.

3.1 DCNSR Architecture

The detailed configurations of DCNSR is shown in Table 1. The number of outputs in each layer are n1 = 64, n2 = 32, n3 = 16, and n4 = 1. After convolutional operations, a non-linear activation function PReLU is applied in place to each convolutional layer except the last one. The feature maps can be padded before convolution in convolutional layers, and the padding sizes of each layer are shown in third column. The input sizes of each layer are listed in the last column.

Table 1. DCNSR Configuration –number of channels, filter size, padding size, input size for each layer and parameters.

	n	Filter size	Padding	Input
Layer 1	64	9	0	$1 \times 33 \times 33$
Layer 2	32	1	0	$64 \times 25 \times 25$
Layer 3	16	5	1	$32 \times 25 \times 25$
Layer 4	1	3	0	$16 \times 23 \times 23$
Parameters	–	20,176		–

The model is trained on a PC with one Intel i7-6700 3.4 GHz CPU and one GTX960-4G GPU. We use the Set5 dataset as the validation set during the training stage, and performance (e.g., PSNR) is evaluated only on the luminance channel. After 9×10^6 iterations, PSNR metric for our proposed DCNSR is 32.84 dB.

3.2 Effectiveness of Deblurring

To estimate the effectiveness of DCNSR in deblurring, we compare our proposed DCNSR with SRCNN, which is represented in Fig. 2. We first conduct the SR operation on interpolated LR image for both DCNSR and SRCNN with a baby image from Set5. The PSNR value of DCNSR and SRCNN is 35.08 dB, 35.01 dB, respectively. From the zoom-in eye region, we can find that the two models do well in reconstructing the eyelashes and achieve the similar PSNR. Then we perform the SR operation again

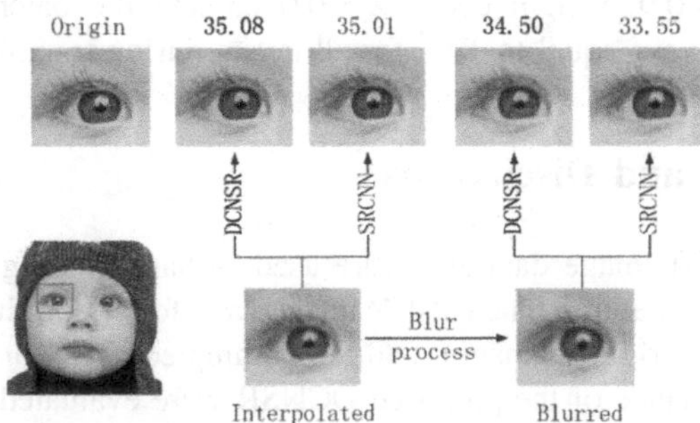

Fig. 2. Comparison with SRCNN in reconstructing HR images from the blurred LR images.

by inputting a blurred baby image. The result shows that the PSNR of SRCNN output has sharply declined to 33.55 dB, while the DCNSR still attains 34.50 dB, which is higher near 1 dB than SRCNN. The result implies that the proposed model DCNSR not only can reconstruct high quality HR images from regular LR images, but also can well estimate HR images from blurred LR images.

3.3 Comparison with State-of-the-Arts

The proposed DCNSR is compared with other recent SR methods on all the images in Set5 for upscaling factor 3. Considering that human eyes are more sensitive to the luminance component than to the chrominance components, we perform the comparison only in the luminance channel. Table 2 shows the results for sparse coding-based method ScSR [9], Gaussian process regression based method SRGPR [2], anchored neighborhood regression method ANR [11], CNN-based method SRCNN [4] and the proposed DCNSR on the Set5 dataset. The metrics showed in the table are PSNR (dB), SSIM index, and FSIM index from top to bottom, respectively. Higher values for these metrics indicate better performance.

Table 2. Objective Quality Assessment Results of PSNR (dB), SSIM and FSIM on the Set5

ScSR	SRGPR	ANR	SRCNN	**DCNSR**
31.83	30.38	31.92	32.52	**32.84**
0.8955	0.8802	0.8958	0.9048	**0.9101**
0.9266	0.9171	0.9229	0.9396	**0.9439**

Based on the results in Table 2, the proposed method performs consistently better than all previous methods in terms of three quantitative assessments. Note that our DCNSR results are based on the checkpoint of 9×10^6 back-propagations, which is less two order magnitude than the SRCNN (i.e., 8×10^8) and achieves better results.

4 Conclusions

In summary, we propose a novel four-layer CNNs-based SR method, which outperforms recent state-of-the-arts methods. These goals are achieved by careful selection of filter size and numbers of channels in each layer, and using PReLU as activation function. Quantitative assessments on the benchmark test images suggest the effectiveness of the proposed SR method.

Although our method is effective in SR reconstruction, the model needs to be trained again to attain a good performance when the SR factor is changed. It is time-consuming to train a model from scratch. It would be better to use transfer learning technique to reduce training time, or cascade several different upscaling factor in a model. And whether a deeper architecture can achieve better performance. We will leave them to future work.

Acknowledgments. The authors would like to thank supports from the National Natural Science Foundation of China under Grant 51277091 and Grant 61179011, the Program for Changjiang Scholars and Innovative Research Team in University (Grant No. IRT_15R10), the Natural Science Foundation of Fujian Province of China under Grant 2011J01219 and the Educational Research Projects for Young and Middle-aged Teachers in Fujian Province under Grant JA15433.

References

1. Yang, C.-Y., Ma, C., Yang, M.-H.: Single-image super-resolution: a benchmark. In: Fleet, D., Pajdla, T., Schiele, B., Tuytelaars, T. (eds.) ECCV 2014. LNCS, vol. 8692, pp. 372–386. Springer, Cham (2014). doi:10.1007/978-3-319-10593-2_25
2. He, H., Siu, W.C.: Single image super-resolution using Gaussian process regression. In: 2011 IEEE Conference on Computer Vision and Pattern Recognition (CVPR), pp. 449–456. IEEE (2011)
3. Dong, C., Loy, C.C., He, K., Tang, X.: Learning a deep convolutional network for image super-resolution. In: Fleet, D., Pajdla, T., Schiele, B., Tuytelaars, T. (eds.) ECCV 2014. LNCS, vol. 8692, pp. 184–199. Springer, Cham (2014). doi:10.1007/978-3-319-10593-2_13
4. Dong, C., Loy, C.C., He, K., Tang, X.: Image super-resolution using deep convolutional networks. IEEE Trans. Pattern Anal. Mach. Intell. **38**, 295–307 (2016)
5. Krizhevsky, A., Sutskever, I., Hinton, G.E.: ImageNet classification with deep convolutional neural networks. Adv. Neural. Inf. Process. Syst. **25**, 1–9 (2012)
6. Liu, W., Anguelov, D., Erhan, D., Szegedy, C., Reed, S., Fu, C.-Y., Berg, Alexander C.: SSD: single shot multibox detector. In: Leibe, B., Matas, J., Sebe, N., Welling, M. (eds.) ECCV 2016. LNCS, vol. 9905, pp. 21–37. Springer, Cham (2016). doi:10.1007/978-3-319-46448-0_2
7. Hradiš, M., Kotera, J., Zemčík, P., Šroubek., F.: Convolutional neural networks for direct text deblurring. In: Proceedings of the British Machine Vision Conference (BMVC), pp. 6.1–6.13. BMVA Press (2015)
8. He, K., Zhang, X., Ren, S., Sun, J.: Delving deep into rectifiers: surpassing human-level performance on imagenet classification. In: Proceedings of the IEEE International Conference on Computer Vision, pp. 1026–1034 (2015)
9. Yang, J., Wright, J., Huang, T.S., Ma, Y.: Image super-resolution via sparse representation. IEEE Trans. Image Process. **19**, 2861–2873 (2010)
10. Bevilacqua, M., Roumy, A., Guillemot, C., Morel, A.: Low-complexity single-image super-resolution based on nonnegative neighbor embedding. In: Proceedings of the British Machine Vision Conference, pp. 1–10 (2012)
11. Timofte, R., De, V., Van Gool, L.: Anchored neighborhood regression for fast example-based super-resolution. In: Proceedings of the IEEE International Conference on Computer Vision, pp. 1920–1927 (2013)

Traffic Road Sign Detection and Recognition in Natural Environment Using RGB Color Model

Han Huang[1(✉)] and Ling-Ying Hou[2]

[1] College of Engineering, Northeast University, Boston, MA, USA
huang.hanl@husky.neu.edu

[2] Dean's Office, Nanchang Institute of Technology, Nanchang 330099, Jiangxi, China

Abstract. Traffic sign detection and recognition play crucial roles on the Intelligent Transportation System. So far, color-based traffic sign detection and segmentation have been widely used for feature extraction and detection. This paper presents an analysis of the performance of five different color models for the color segmentation and subsequent detection of traffic signs in two-dimensional static images that obtained in real-world environment. Firstly, using color thresholding techniques to isolate relevant color region (red, blue) from the image. The regional morphology processing algorithms is applied in order to extract traffic sign's region of interesting (ROI), it could remove the noise and isolate the traffic sign. Then, a rectangle region in the original image to be selected according as its shape property. Finally, a way of quantitatively evaluate the performance of the different color space detection algorithm on the widely-used German Traffic Sign Detection Benchmark (GTSDB) has been proposed.

Keywords: Traffic sign detection · RGB color model · Thresholding segmentation · Image segmentation

1 Introduction

With the development of electronic industry, a large number of image acquisition devices and display devices are used in daily life. Typical devices, such as, color TV, video camera, image scanner, digital camera, computer and mobile phone display, video projector, multicolor LED display, etc. are based on RGB (Red-Green-Blue) color model that applied it for the sensing, representation, and display of images in electronic systems. RGB color space is the most commonly used color space in digital devices, which uses red, green and blue to create a combination of almost all possible colors. In this color space, an imported image on a computer is thus transformed into 3 matrices within which every pixel is defined by 3 values one every for the red, blue, and green elements of the picture element scalar. The advantage of the RGB color model is that its simplicity and real-time performance. If a real time process is the system aim, the best choice would be the use of RGB because there will be no transformation. The application of RGB color model for traffic signal detection has

D.-S. Huang et al. (Eds.): ICIC 2017, Part I, LNCS 10361, pp. 345–352, 2017.
DOI: 10.1007/978-3-319-63309-1_32

more practical value and significance. Hence, in this study, we focus on four representative RGB color thresholding methods which are commonly used in the traffic signs detection and presents a quantitative comparison of several segmentation methods.

The rest of the paper is organized as follows. Section 2 takes a brief look on related research works on color traffic signs segmentation. Section 3 introduces the four color thresholding algorithm had been applied for the road signs detection. Section 4 describes the template match for traffic signs classification, and Sect. 5 shows the result and future research.

2 Related Work

Road and traffic signs usually have been designed with a fixed 2-D shapes (such as, triangles, circles, square, or rectangles etc. al.) and colors (red, blue, yellow or white). So make it distinguished from the natural environment and/or man-made backgrounds and easily recognizable by drivers []. For these reason, traffic signal detection can be divided into two categories: color based and shape based. Color-based detection methods make use of color information which segment the typical colors of traffic signs, it selects a region of interest (ROI) for further processing. For example, Piccioli et al. [1] use color to extract the possible locations of signs in the image. Then a geometrical analysis of the edges has been applied to identify triangular and circular road signs from images. M. Benallal and J. Meunier [2] have analyzed the RGB components of different road signs from sunrise to sunset. They had found that the difference between R and G and the difference between R and B channels can be used for road sign detection. Escalera et al. [3] used a color thresholding technique in the normalized RGB space to select road sign regions from their backgrounds, which contains five colors. Ritter et al. [4] divided RGB components with the sum of three components within the images, and identify color combinations for traffic signs efficient detection. Xin et al. [5] set different RGB thresholds for different lighting conditions. Regions of Interest (ROI) are extracted to form the standardized traffic signs region using a simple standardized image consisting of standard color of eight categories. The performance of color based road sign detection is often faced with unfavorable situation like as dim lighting, or adverse weather conditions such as fog, rain and snow. Others color models, such as Hue–Saturation–Value (HSV) [6], HSI [7], YCbCr [8], YUV and CIELAB [9], have been used in an attempt to avoid these disadvantage. Gao et al. [10] used the CIECAM97 color model, but the author also found that still needed separate sets of thresholds for images captured under different weather conditions. In reference [11], the authors presented a quantitative comparison of several segmentation methods that it could be classified into color-space thresholding, edge detection, and chromatic/achromatic decomposition. Recently, Wali S.B et al. [12] and Anjan Gudigar [13] provided a comprehensive survey on traffic sign detection and recognition methods.

Thresholding of essential colors is used to segment ROIs for traffic signs detection. Image segmentation is usually accomplished by comparing pixel values with a set of determined thresholds.

Algorithm 1: RGB (Red-Green-Blue) is the most popular color model for processing and storing of digital image data, which describe color as a combination of red, green and blue components. In [14], the authors proposed a very simple algorithm for red traffic sign image segmentation using three thresholds Δ_{RG}, Δ_{RB} and Δ_{GB} as following expression:

$$\text{Red}(i,j) = \begin{cases} True & \begin{matrix} \textit{If } r(i,j) > g(i,j) \\ \& \, r(i,j) - g(i,j) > \Delta_{RG} \\ \& \, r(i,j) - b(i,j) > \Delta_{RB} \end{matrix} \\ False & \textit{Otherwise} \end{cases} \tag{1}$$

Segmentation of other colors such as blue or yellow signs can also be applied this algorithm by following the same principles.

Algorithm 2: Although, RGB space can be directly used for threshold processing, and can achieve good results. However, there is a strong correlation between the three color channels in RGB space. One solution named color space normalization technique which can reduce the correlation between the three channels and enhance the discriminating power of the normalized color space uses three normalized components called r, g and b. It is defined as follows:

$$r = \frac{R}{R+G+B};\; g = \frac{G}{R+G+B};\; b = \frac{B}{R+G+B} \tag{2}$$

where R, G and B are the color components of the input image. In the normalization RGB space, the mask equations for red, blue and yellow colors extraction are the following [11, 13]:

$$\begin{aligned} \text{Red}(i,j) &= \begin{cases} True & \begin{matrix} \textit{If } r(i,j) \geq ThR \\ \& \; g(i,j) \leq ThG \end{matrix} \\ False & \textit{Otherwise} \end{cases} \\ \text{Blue}(i,j) &= \begin{cases} True & \textit{If } b(i,j) \geq ThB \\ False & \textit{Otherwise} \end{cases} \\ \text{Yellow}(i,j) &= \begin{cases} True & \textit{If } r(i,j) + g(i,j) \geq ThY \\ False & \textit{Otherwise} \end{cases} \end{aligned} \tag{3}$$

Algorithm 3: It assume that each RGB pixel gray extracted values were U_r, U_g, U_b, and then calculate the percentage of each component as following:

$$\begin{aligned} K_r &= U_r \,/\, (U_r + U_g + U_b) \\ K_g &= U_g \,/\, (U_r + U_g + U_b)\,. \\ K_b &= U_b \,/\, (U_r + U_g + U_b) \end{aligned} \tag{4}$$

Then, the corresponding image is transferred into binary image according to a setting threshold by trial experience. In this algorithm, three kinds of color pixels are calculated using the same weight, which saves the calculation time, however, the feature of images are not reflected, so it is suitable for handling a large number of different types of images.

Algorithm 4: For a color image, the value of each pixel $v = [v_R, v_G, v_B]$ in the RGB space is converted as follows [15]:

$$\begin{aligned} C_{red} &= \max(0, \min((v_R - v_G)/S, (v_R - v_B)/S)) \\ C_{blue} &= \max(0, \min((v_B - v_R)/S, (v_B - v_G)/S)) \\ C_{yellow} &= \max(0, \min((v_G - v_B)/S, (v_R - v_B)/S)) \end{aligned} \tag{5}$$

This algorithm pays more attention to the level of the color image itself, and the distribution of the color pixels on the RGB dimensions determine the value of the transformation.

Therefore, this algorithm can better highlight the image changes and the edge of the distribution.

Just like algorithm 3, a binary image had been obtained according to a setting threshold by experiments.

Algorithm 5: At first, a RGB image is transformed into a grey-level image in Ohta color space just like Eq. (6).

$$\begin{bmatrix} I_1 \\ I_2 \\ I_3 \end{bmatrix} = \begin{bmatrix} \frac{1}{3} & \frac{1}{3} & \frac{1}{3} \\ 1 & 0 & -1 \\ -\frac{1}{2} & 1 & \frac{1}{2} \end{bmatrix} \begin{bmatrix} R \\ G \\ B \end{bmatrix} \tag{6}$$

The new normalized components P 1 and P 2 are given by

$$\begin{aligned} P_1 &= \frac{1}{\sqrt{2}} \frac{R - B}{R + G + B} = \frac{1}{3\sqrt{2}} \frac{I_1}{I_2} \\ P_2 &= \frac{1}{\sqrt{6}} \frac{2G - R - B}{R + G + B} = \frac{2}{3\sqrt{6}} \frac{I_3}{I_1} \end{aligned} \tag{7}$$

Using these normalized components, the colors can be classified as follows [15]:

$$\begin{aligned} \text{Red}(i,j) &= \begin{cases} True & If\ P_1(i,j) \geq ThR_1 \\ & \&\ P_2(i,j) \leq ThR_2 \\ False & Otherwise \end{cases} \\ \text{Blue}(i,j) &= \begin{cases} True & If\ P_1(i,j) \geq ThB_1 \\ & \&\ P_2(i,j) \leq ThB_2 \\ False & Otherwise \end{cases} \\ Yellow(i,j) &= \begin{cases} True & If\ P_1(i,j) \geq ThY_1 \\ & \&\ P_2(i,j) \leq ThY_2 \\ False & Otherwise \end{cases} \end{aligned} \tag{8}$$

3 Experiments Results and Discussion

One of the most widespread databases which is the German Traffic Sign Recognition Benchmark (GTSRB) has been used for evaluation of the algorithm objectively [22]. The GTSDB testing dataset consists of 600 images (161 prohibitory signs and 64 danger signs) of size 1360 × 800 pixels with PPM format. A sample image from the GTSRB database can be found in Fig. 1(a). Manual counts the numbers of traffic signs in 600 images, of which the number of traffic signs with red is 726 and with blue is 122. The threshold values of the six algorithms which been used are presented on Table 1.

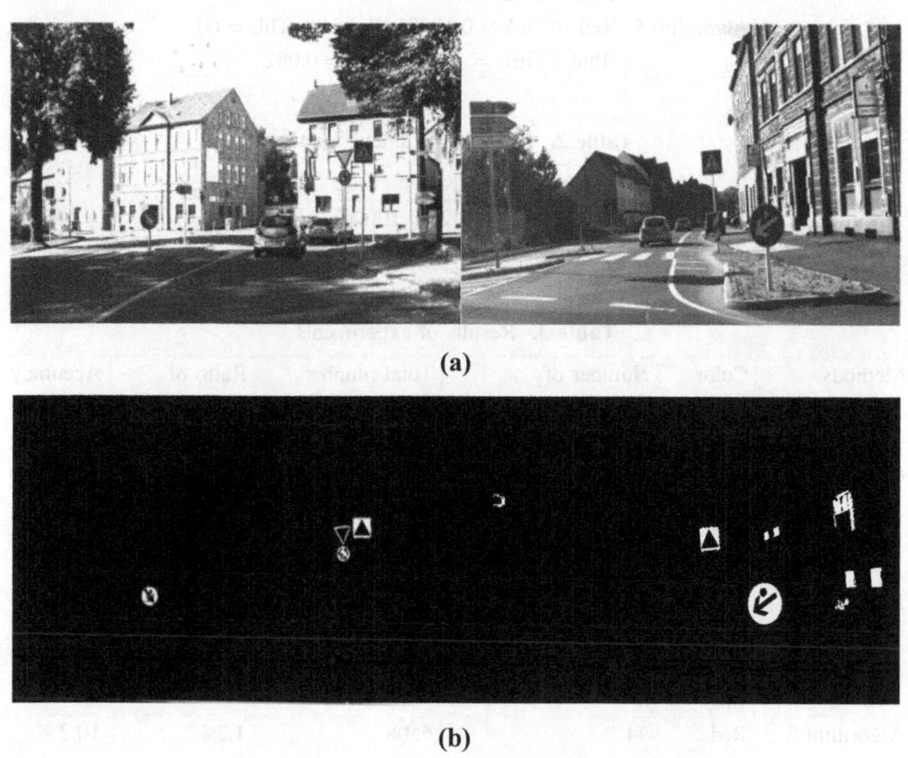

Fig. 1. (a) Original image (b) ROI locking

After color image has been thresholding, many non-traffic signal areas will be shown. To remove noises in the possible candidates, morphological operating is applied. After that, several features of the detected connected component regions are used to further reduce the number of candidates. These features are width, height, aspect ratio, region perimeter and area, and bounding-box perimeter and area. The parameters that have been used are listed in Table 2.

From Table 3, we conduct a comparison with five thresholding algorithms to evaluate the results of detection. The number of candidate regions produced by those

Table 1. Threshold values for different method.

Methods	Color	Threshold values
Algorithm 1	Red	R > 25, (R-B) > 25, and (R-G) > 25.
	Blue	B > 25, (B-R) > 25, and (B-G) > 25.
Algorithm 2	Red	ThR = 0.4
	Blue	ThB = 0.4
Algorithm 3	Red	Kr > 0.4 && Kg < 0.3
	Blue	Kr > 0.4
Algorithm 4	Red	ThR = 0.4
	Blue	ThB = 0.4
Algorithm 5	Red	ThA = 0.17, ThW = 180, ThL = 60
	Blue	$ThB_1 = -0.04$, $ThB_2 = 0.082$

Table 2. Different parameter values

Parameter	Range
Area	[260, 10000]
Aspect ratio	[0.7, 1.6]

Table 3. Results of experiments

Methods	Color ROI	Number of detection correct ROI	Total number detection ROI	Ratio of traffic signs	Accuracy
Algorithm 1	Red	100	3158	4.7%	13.8%
	Blue	49			40.2%
Algorithm 2	Red	34	4675	1.0%	4.7%
	Blue	14			11.5%
Algorithm 3	Red	135	2431	8.2%	18.6%
	Blue	65			53.3%
Algorithm 4	Red	7	91	7.78%	0.9%
	Blue	0			0
Algorithm 5	Red	74	6508	1.2%	10.2%
	Blue	8			6.6%

algorithms is greater than the number of traffic sign regions. Among which, algorithm 3 proves to be able to give the highest accuracy, it's accuracy is higher than 50%. Although, color in RGB space can provide faster for searching traffic sign areas, it's accuracy was lower, the reason is the diversity of lighting condition. At the same time, we have used a fixed threshold in our experiments, the number of candidate regions can be significant influenced by different threshold and database. Further research is how to select the thresholds dynamically.

4 Conclusions

One of the most difficult jobs in traffic sign detection online is to deal with outdoor images. In this paper, five color segmentation methods based on RGB model is presented. All methods are tested on 600 images from German Traffic Sign Recognition Benchmark (GTSRB) database under different backgrounds. From that, the algorithm which calculate the percentage of each component obtained regions with traffic signs by means of thresholding can give a satisfying result. In future, more robust and the shadow-highlight invariant method such as shape analysis, different color model should be considered.

Acknowledgement. This work was supported by the National Natural Science Foundation of China (Grant No. 61472173), Natural Science Foundation of Jiangxi Province of China, No. 20161BAB202042, the grants from the Educational Commission of Jiangxi province of China, No. GJJ151134.

References

1. Piccioli, G., De Micheli, E., Parodi, P., et al.: Robust method for road sign detection and recognition. Image Vis. Comput. **14**(3), 209–223 (1996)
2. Benallal, M., Meunier, J.: Real-time color segmentation of road signs. In: 2003 IEEE Canadian Conference on Electrical and Computer Engineering CCECE 2003, vol. 3, pp. 1823–1826. IEEE (2003)
3. De La Escalera, A., Moreno, L.E., Salichs, M.A., et al.: Road traffic sign detection and classification. IEEE Trans. Industr. Electron. **44**(6), 848–859 (1997)
4. Ritter, W., Stein, F., Janssen, R.: Traffic sign recognition using colour information. Math. Comput. Model. **22**(4–7), 149159–157161 (1995)
5. Xin, L., Shuangdong, Z., Chen, K.: Method of traffic signs segmentation based on color-standardization. Proc. Int. Conf. Intell. Human Mach. Syst. Cybern. **2**, 193–197 (2009)
6. Chen, X., Yang, J., Zhang, J., et al.: Automatic detection and recognition of signs from natural scenes. IEEE Trans. Image Process. **13**(1), 87–99 (2004)
7. Nguwi, Y.-Y., Kouzani, A.Z.: Detection and classification of road signs in natural environments. Neural Comput. Appl. **17**(3), 265–289 (2008)
8. Balali, V., Golparvar-Fard, M.: Evaluation of multiclass traffic sign detection and classification methods for US roadway asset inventory management. J. Comput. Civ. Eng. **30**(2), 04015022 (2015)
9. Fleyeh, H.: Color detection and segmentation for road and traffic signs. In: IEEE 2004 Conference on Cybernetics and Intelligent Systems, vol. 2, pp. 809–814. IEEE (2004)
10. Gao, X.W., Podladchikova, L.N., Shaposhnikov, D.G., Hong, K., Shevtsova, N.: Recognition of traffic signs based on their colour and shape features extracted using human vision models. J. Vis. Commun. Image Represent. **17**(4), 675–685 (2006)
11. Gómez-Moreno, H., Maldonado-Bascón, S., Gil-Jiménez, P., et al.: Goal evaluation of segmentation algorithms for traffic sign recognition. IEEE Trans. Intell. Transp. Syst. **11**(4), 917–930 (2010)
12. Wali, S.B., Hannan, M.A., Hussain, A., et al.: Comparative Survey on Traffic Sign Detection and Recognition: a Review. Przeglad Elektrotechniczny, ISSN, pp. 0033–2097 (2015)

13. Gudigar, A., Chokkadi, S., Raghavendra, U.: A review on automatic detection and recognition of traffic sign. Multimed. Tools Appl. **75**(1), 333–364 (2016)
14. Kuo, W.J., Lin, C.C.: Two-stage road sign detection and recognition. In: 2007 IEEE International Conference on Multimedia and Expo, pp. 1427–1430. IEEE (2007)
15. Vertan, C., Boujemaa, N.: Color texture classification by normalized color space representation. In: Proceeding of International Conference on Pattern Recogonization, Barcelona, Spain, pp. 580–583. (2000)

Active Contour Integrating Patch-Level and Pixel-Level Features

Xinyue Mao, Yufei Chen(✉), Xianhui Liu, and Weidong Zhao

CAD Research Center, Tongji University, Shanghai 201804, China
april337@163.com

Abstract. Extracting a foreground target in a complicated environment is of great interest in image processing and computer vision. This paper proposed an image segmentation method based on patch-level and pixel-level features. First, we extract patch-level texture feature of the image by building a dictionary of image patches. Second, we use texton theory to extract pixel-level texture feature. And pixel-level color feature is extracted according to the pixel color distributions. Third, a probability image is generated by combining the detection results of the texture and color features. Finally, we apply curve evolution in the probability image to segment the image into separated regions with similar features. Experimental results on various images from the Berkeley dataset show that our approach is more robust and accurate compared with four other methods.

Keywords: Color feature · Texture representation · Active contour · Curve evolution · Image segmentation

1 Introduction

Image segmentation plays an important role in image processing and image analysis. However, there are still many challenging problems because of noise and intensity inhomogeneity. The objective of image segmentation is to separate an image into several meaningful parts where inner pixels are considered as the representation of an object with respect to a certain characteristic [1]. Countless methods have been proposed over the years [2, 3]. Level set [4] methods is the most popular one which can be categorized into two groups: edge-based and region-based. Edge-based methods utilize gradient to force the curve so they can easily segment target whose boundary is defined by sharp gradient. Region-based methods use region information inside and outside the evolving contour and have better performance. There are several popular region-based methods like, Chan-Vese model [5] and Mumford-Shah functional [6].

Color and texture are essential features for image segmentation. And texture segmentation has become the most challenging problems in image segmentation [7]. The average grey level is similar doesn't mean textures are similar. Thus, texture must be dealt with differently. Rousson et al. estimate the joint probability of the elements of the structure tensor to evolve the level set [8]. The downside is when the image texture at different scale the segmentation might fail. Recently, a new segmentation technique based on sparse dictionaries is proposed [9]. Gao et al. [10] combined sparse

D.-S. Huang et al. (Eds.): ICIC 2017, Part I, LNCS 10361, pp. 353–365, 2017.
DOI: 10.1007/978-3-319-63309-1_33

dictionaries with active contour. However, once the user input is updated, the dictionary is learned again and waste lots of time.

To overcome the above problems, this paper proposed a texture and color feature based image segmentation method. This method exhibits several advantages. First, our method can deal with multiple objects and complex textures. Second, our method can produce accurate and smooth boundary after evolution. Third, it provides the user interface using user's input as the initial curve.

1.1 Related Work

1.1.1 Chan-Vese Model

Chan and Vese proposed a new model for active contours. Chan-Vese model takes a controlled closed curve and takes the difference between its internal and external energy into account. Consider image I: $\Omega \rightarrow R$ containing a target and a background and C is a close contour. The energy function is defined by:

$$\begin{aligned} F(c_1, c_2, C) &= \mu \cdot Length(C) + v \cdot Area(inside(C)) \\ &+ \lambda_1 \int_{inside(C)} |u_0(x,y) - c_1|^2 dxdy + \lambda_2 \int_{outside(C)} |u_0(x,y) - c_2|^2 dxdy \end{aligned} \tag{1}$$

where c_1 and c_2 are the average intensity level inside and outside of the contour.

In the level set method [11], the zero level set of a Lipschitz function ϕ: $\Omega \rightarrow$ R,

$$\begin{cases} C = \partial\omega = \{(x,y) \in \Omega : \phi(x,y) = 0\} \\ inside(C) = \omega = \{(x,y) \in \Omega : \phi(x,y) > 0\} \\ outside(C) = \Omega \backslash \bar{\omega} = \{(x,y) \in \Omega : \phi(x,y) < 0\} \end{cases} \tag{2}$$

The minimization problem is settled by taking the Euler-Lagrange equations and obtain the level set function using the standard gradient descent method

$$\frac{\partial \phi}{\partial t} = \delta_\varepsilon(\phi)[\mu div(\frac{\nabla \phi}{|\nabla \phi|}) - \lambda_1 (I - c_1)^2 + \lambda_2 (I - c_2)^2] \tag{3}$$

Keeping fixed ϕ and minimizing the energy function until the optimal ones that minimize the energy function are obtained or reach the predetermined maximum iteration number.

In fact, the C-V model is based on the large gray difference between object and background. It may fall into local minima very easily during the evolution process and have poor performance for images with intensity inhomogeneity. In addition, AE Rad et al. find that C-V models is faster and more accurately when the initial curve surrounds objects boundary [12].

1.1.2 Texture Representation

Texture is an important property of any surface, and along with shading, stereo, contours and motions can be used to infer depth from a 2-dimensional image [13]. [14] present prototypes textons which are a small set of prototype response vectors. Each texture is analyzed using a filters bank. Motivated by [15, 16] defines an intrinsic texture feature descriptor based on the shape operator of the texture manifold. The shape operator is a linear operator which calculates the bending of a surface in different directions [17]. The texture manifold is represented in the form of X, and two principal curvatures κ_1, κ_2 are used as an intrinsic and efficient descriptor:

$$\kappa_{1,2} = \left(-\beta \pm \sqrt{\beta^2 - 4\alpha\gamma}\right) \Big/ 2\alpha \tag{4}$$

$$\begin{cases} \alpha = (1 + I_x^2)^2 - I_x I_y \\ -\beta = \frac{1}{z}\left[I_{xx}(1 + I_y^2) + I_{yy}(1 + I_x^2) - I_{xy}(I_x I_y)\right] \\ \gamma = \frac{1}{z^2}\left[I_{xx} I_{yy} - (I_{xy})^2\right] \end{cases} \tag{5}$$

The method is fast and easy, but sensitive to noise and the initial contour.

Recently, many researchers obtained excellent results by using sparse modeling and dictionary learning [18, 19]. The idea of sparse image representation is using an over complete dictionary of image patches to reconstruct image [20]. The dictionary consist of the texture classes in the given image. R Sarkar et al. introduced DL2S (Dictionary Learning Level Set) combining sparse texture and active contour in presence of significant clutter and heterogeneous [21]. Users input the background and foreground and sparse dictionaries are built for the two classes. In [21], dictionary is calculated by solving the optimization problem:

$$D_k = \arg\min_{D, y_i} \sum_{i=1}^{N} \left\| f_i - D^T y_i \right\|_2^2 \text{ such that } \|y_i\|_0 \le \theta, \forall i \tag{6}$$

where $F = [f_1, \ldots, f_N]$ indicates the set of N discretized, coefficient vector y_i map one to one with the ith training image and θ sets the level of sparsity. Different from these methods, [22] use one dictionary to code an image and another dictionary to keep the label information. Two separate dictionaries are coupled together with the same spatial extension. The intensity dictionary D is used to model textures by minimizing the residual error

$$D = \arg\min_D \sum_{i-1}^{o} \left\| d_i^* - x_i \right\|_2 \tag{7}$$

where o is the number of training samples.

2 Active Contour Integrating Patch-Level and Pixel-Level Features

2.1 Patch-Level Features

Using patches as feature vector was first proposed for texture synthesis [23]. The use of patches can reduce the influence of noise and expression texture information in a time-saving way. Inspired by the idea of building dictionary, we aim to extract patch-level texture features by establishing a dictionary. A good dictionary needs to contain unique atoms for a kind of texture and do not miss image information. First, we randomly pick appropriate amounts of patches from the given image. It is key to choose the right number of selected image patches due to the computational complexity and representation of clustering results. Next, we use k-means algorithm to cluster the chosen image patches into K_D classes. Euclidian distance is utilized in the cluster processing. The cluster centers are the elements in the dictionary. After the above two steps, we get a texture dictionary which contains the background information, the texture object information and the border information of background and object. Figure 1 shows a dictionary computed from a small color textured image. Then, we select Euclidian distance to assign overlapping patches to the closest atom:

$$A(x,y) = \arg\min_i \|I(x,y) - d(i)\|_2^2,\ i \in \{1,\cdots,n\} \tag{8}$$

Here, *I*(x, y) is an intensity image, *d*(*i*) is the *i*th column of D_T. Using the dictionary, we obtain an assignment image consists of patch-level texture feature. So image patches pixels corresponding to same number of pixels from the texture dictionary atom. If user input a RGB or other kind of image, we execute on all color channels. The advantages of using texture dictionary extracted from image patches is reducing the cost in the comparison of all the patches. In addition, it ignores the effect of the noise and different sizes of the same texture.

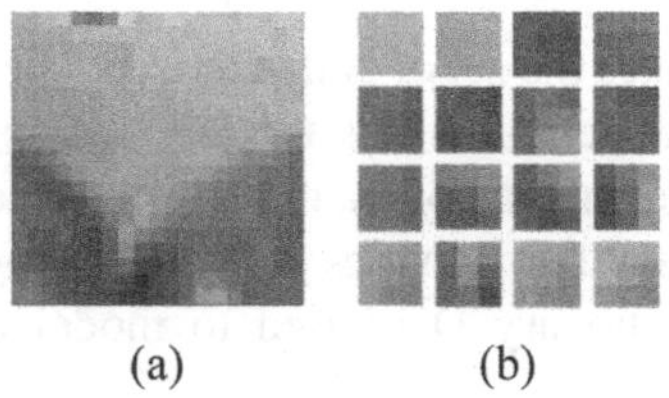

Fig. 1. (a) A small image, (b) A dictionary computed from the image: elements 1–2: background, elements 3–9: object, elements 10–16: transitions from the object to the background.

2.2 Pixel-Level Features

Although using patches has its good side, it still suffers from lots of troubles. In fact, the similarities between patches cannot describe irregular shaped objects accurately. So it may over-smooth the details around image boundary.

In image processing, Gabor filter named after Dennis Gabor [24], is a linear filter used to detect edges and it has been found to be particularly appropriate for represent and discriminate textures [25]. It is a frequency transform approach and able to model the frequency and orientation sensitivity characteristic of human being visual system. A Gaussian function multiplied by a sinusoidal wave define its impulse response:

$$g(x, y; \lambda, \theta, \psi, \sigma, \gamma) = \exp\left(-\frac{x'^2 + \gamma^2 y'^2}{2\sigma^2}\right)\exp\left(i(2\pi\frac{x'}{\lambda} + \psi)\right) \tag{9}$$

λ: the wave length of the sinusoidal factor, θ: the orientation of the normal to the parallel stripes of a Gabor function, ψ: the phase offset, σ: the standard deviation of the Gaussian envelope and γ: the spatial aspect ratio, and specifies the ellipticity of the support of the Gabor function.

The original image reconstructed from the Gabor filter can contain both texture feature and part of boundary information. We characterize pixel-level textures by their

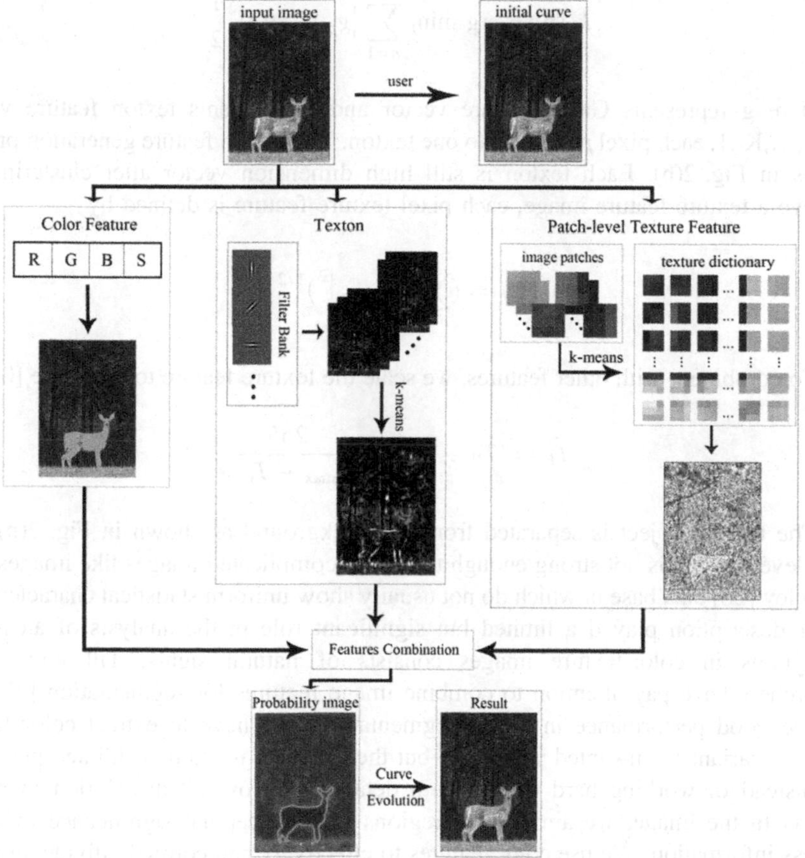

Fig. 2. General workflow of our method, (a) color features, (b) pixel-level texture features: texton, (c) patch-level texture features: texture dictionary

responses to a filter bank which composed of spatial-frequency and orientation Gabor filters. Increasing the number of orientations and scales can improve the reconstruction quality. Then every pixel are transformed into N dimensional vectors. It is hard to estimate the density of the feature data, due to such data dimension is too high. Besides, there are some irrelevant features and noise, and only select the most relevant filter responses can lead to a good reconstruction.

Actually, the filter responses are not totally different at each pixel over the texture which means several distinct filter responses can represent the others. With the idea of textures have spatial repeating characteristics, we use k-means clustering these vectors into K_G centers. Assigning data vectors to the nearest of the K_G center and then updating each of the K_G centers to the mean of the data vectors assigned to it [14]. Continuing these two steps until reaching a local minimum of the criterion and the algorithm. The centers are the textons, and the associated filter response vectors are called the appearance vectors, $c_k \in R_N$, $k = 1, \ldots, K_G$. After the clustering step, the distinct prototype vectors can be obtained.

$$k^*(p) = \arg\min_k \sum_{n=1}^{N} \left|g_{p,n} - t_{k,n}\right|^2 \frac{1}{2} \tag{10}$$

Here g represents Gabor feature vector and t represents texton feature vector, $k \in \{1, \ldots, K_G\}$, each pixel p mapped to one texton. The texton feature generation process shows in Fig. 2(b). Each texton is still high dimension vector after clustering. To achieve a texture feature image, each pixel texture feature is defined by

$$T_P = \Big(\sum_{n=1}^{N} \left|t_{k^*(p),n}\right|^2\Big)^{1/2} \tag{11}$$

To combine it with other features, we scale the texture feature to the range [0,255].

$$\widetilde{T_P} = (T_P - T_{\min}) \times \frac{255}{T_{\max} - T_{\min}} \tag{12}$$

The texture object is separated from the background as shown in Fig. 2(b). The gray level feature is not strong enough to handle complicated images like images from Berkeley [26] data base in which do not usually show uniform statistical characteristics. Color description played a limited but significant role in the analysis of an image, contiguous in color-texture images consists of natural sights. Till now, many researchers have pay attention to combine image features for segmentation [27], and receive good performance in image segmentation. We have to extract color feature which invariant to unwanted variations, but the indexing method is still an open issue.

Instead of working hard to acquire a detailed and overall description of all the objects in the image, we aim to find regions with perceptual significance to ignore useless information. We use color features to effectively and correctly divide an image into several contiguous regions. In this paper, the parameter Kc = 5, which decides the maximum number of dominant colors in clustering process. Because of a small set of color categories makes it efficient to represent an image, and offers an easier way to capture invariant properties of the object. We use k-means to represent the pixel color distributions and one color to replace similar colors to segment an image into Kc classes. And we also take spatial features S into account by calculating the spatial distance between pixels. Spatial information can make clustering result more close to human perception. The color feature vector at pixel p is:

$$C_p = (r_p, g_p, b_p, S) \tag{13}$$

where r_p, g_p, b_p corresponding to three color channels of a RGB image.

As results, each pixel is taken over by the dominant color of the class which it belongs to. The forming class-map $\widetilde{C}$ shown in Fig. 2(a). We select the class-map $\widetilde{C}$ as our color feature which contains both color and spatial information. The class-map of the given image can also be regarded as an unusual form of texture composition.

2.3 Feature Combination

After we extract texture features and color features from an image, it is a problem that how to combine multiple visual features together. We aim to transform the input image into probability image using the multiple features similar to [28].

We first band pixel-level features (color feature and texture feature acquired from filter bank) together. We use texture feature $\widetilde{\mathrm{T}}$ and three channels (Taking RGB image as an example) of the obtained class-map $\widetilde{\mathrm{C}}$ to form a new 4-D vector at pixel p, $\mathrm{CT_p} = \left(\widetilde{\mathrm{T}}_\mathrm{p}, \widetilde{\mathrm{C}}^\mathrm{r}_\mathrm{p}, \widetilde{\mathrm{C}}^\mathrm{g}_\mathrm{p}, \widetilde{\mathrm{C}}^\mathrm{b}_\mathrm{p}\right)$. Then dividing them into $\mathrm{K_{CT}}$ classes by k-means and Euclidean distance to obtain a new feature vector at each pixel, with pixel-level texture feature and color feature. This four channels image is then combined with the assignment image obtained in the previous section in order to achieve an augmented image, each pixel p is mapping to 1-D vector by

$$\widetilde{CT_p} = w_1\widetilde{T_p} + w_2\left(\frac{\widetilde{C_p}^r + \widetilde{C_p}^g + \widetilde{C_p}^b}{3}\right) + w_3A_p \tag{14}$$

where w_1, w_2, w_3 are the weight. For our experiments we have chosen $w_1 = 0.21$, $w_2 = 0.09$, $w_3 = 0.7$. The mixed feature vector contains color feature of RGB image, pixel-level and patch-level texture features. This enables us to acquire an augmented image with the color and texture characteristics to differentiate different objects.

In this work, we let users to choose a target with interest by drawing an initial curve. Using this curve we turn the original image into a label image L.

$$L(x,y) = \begin{cases} 1 & \text{if}(x,y)\text{inside the curve} \\ 0 & \text{otherwise} \end{cases} \tag{15}$$

In patch-level texture feature extraction, we get an assignment image and a binary relation between image and texture dictionary. The probability image is established from the average of overlapping label patches from dictionary labels according to augmented image. As a result, the obtained image P_{in}: $\Omega \rightarrow [0, 1]$ contains a pixel-wise probability of being part of inside. The probability image P_{in} are found to force the curve, and then evolution can be driven by the probability on the image.

2.4 Curve Evolution

We take the user's input curve as the zero level set and get a probability image after feature extraction step. This probability image defines forces evolving the curve. When the curve located in a position with large P_{in} it has to move outwards. Otherwise, it has to move inwards. The label image L is updated every step due to the evolution of the curve. Therefore the probability image is constantly changing as well. And finally the evolution should stop at where P_{in} = Pout ($P_{out} = 1 - P_{in}$) or reach the maximum permission iterative number. We define the curve evolution as [28]

$$\frac{\partial \phi}{\partial t} = \frac{1}{2} - P_{in} + b\kappa|\nabla \phi| \tag{16}$$

where $\kappa = \nabla . \left(\frac{\nabla \phi}{|\nabla \phi|}\right)$ defines curvature of the curve, b is used to minimize the length of the curve. Our method is also applicable to n multi-label situation by evolving each level set ϕ_n with it corresponding probability image. An example of two-label input image and its segmentation result shown in Fig. 3.

Fig. 3. Multi-object segmentation based on the proposed method, left is the input image, middle shows two labels mark two objects, right presents the segmentation result.

3 Experiments

In this Section, we make experiments to estimate the performance of the proposed method. The test images are from the Berkley segmentation dataset [26]. And we also show a comparison between our method and other four methods: Chan-Vese model [5], Texture-Aware Fast Global Level Set Evolution (TAFGLSE) [29], Dictionary snakes [30] and Dictionary based image segmentation (DBIS) [28] using the segmentation results. Precision is the fraction of retrieved instances that are relevant, while recall is the fraction of relevant instances that are retrieved. A measure that combines precision and recall is the harmonic mean of precision and recall, the traditional F-measure:

$$\begin{aligned} &\text{Precision} = \frac{tp}{tp + fp} \\ &\text{Recall} = \frac{tp}{tp + fn} \\ &\text{F - measure} = 2 \cdot \frac{\text{precision} \cdot \text{recall}}{\text{precision + recall}} \end{aligned} \tag{17}$$

We evaluate the performance of our method with a set of 50 color textured images of various natural scenes, which were obtained from the Berkeley image segmentation dataset. The parameters in our method are set the same for all the 50 images.

Table 1. The average values of the precision, recall and F-measure of segmentation results on the Berkeley datasets of 50 color textured images

Method	Precision (%)	Recall (%)	F-meature (%)	The variance of F-measure (%)
Chan-Vese [5]	92.24	22.67	30.29	6.01
TAFGLSE [29]	39.37	65.58	38.14	7.26
Dictionary snakes [30]	85.20	68.49	70.35	6.49
DBIS [28]	81.12	83.33	79.32	4.51
Our method	84.85	84.78	83.22	2.59

Figure 4 shows the color-texture segmentation results of six color textured images from the Berkeley datasets. Patch size decides the quality of texture dictionary. A large size can contain more texture and color information but cannot maintain border information. By conducting many experiments, we finally decide the patch size is 3×3 pixels. The Chan-Vese model cannot find the objects we are interested in. It brings out incorrect segmentation results. TAFGLSE use the geometric active contours and the image thresholding frameworks for segmentation. This combination makes their method handle complex shapes easily and ensure its convergence towards the global minimum. As a result, TAFGLSE is the fastest one in the segmenting process in our experiment. When faced with simple pictures, it shows efficient segmentation behavior. Nevertheless, it gives over segmented results in the images with complex background or texture. Dictionary snakes has better results than the previous two methods and able to segment objects roughly. Evolving a contour for pulling textures from each other based on active contour and texture characterization with dictionary. But it cannot handle the problem that segment same objects at the same time and the over segmentation is still exist. Only when the initial curve contains all the same objects, it can get multi-objects segmentation result. In the result in Dictionary based image segmentation, we can see it has solved the problems in dictionary snakes and has more accurate segmentation are not good. For the picture in the third line, it segments the tiger and its reflection in the water together because their colors are similar. For the picture in the sixth line, it seems cannot distinguish the difference between the horses and white part of background and the difference between the white horse and the black one. Background information being

Fig. 4. Segmentation results on the BSDS500. From left to right: input image, Chan-Vese model, Texture-Aware Fast Global Level Set Evolution (TAFGLSE), dictionary snakes, dictionary based image segmentation(DBIS) and our method. The green line represents the boundary of the segmentation. (Color figure online)

erroneously divided when object size is relative small. As shown in Fig. 4, our method has better performance than the others giving more accurate segmentation results. Due to the randomness of selecting image patches, the segmentation results for the same image may be different (shown in Fig. 5).

Fig. 5. Three segmentation results on the star fish. From left to right: input image, first time, second time, third time. From top to bottom: dictionary snakes, DBIS, proposed method

Figure 6 shows the F-measure of all 50 tested images of each method, in which the cyan line, green line, blue line, yellow line and red line are the results of Chan-Vese, TAFGLSE, Dictionary snakes, DBIS and our method, respectively. The average Precision, Recall, F-measure of all 50 tested images are shown in Table 1. The higher rate in this measurement shows better performance of segmentation results. It can be seen that the C-V model gets the lowest accuracy.

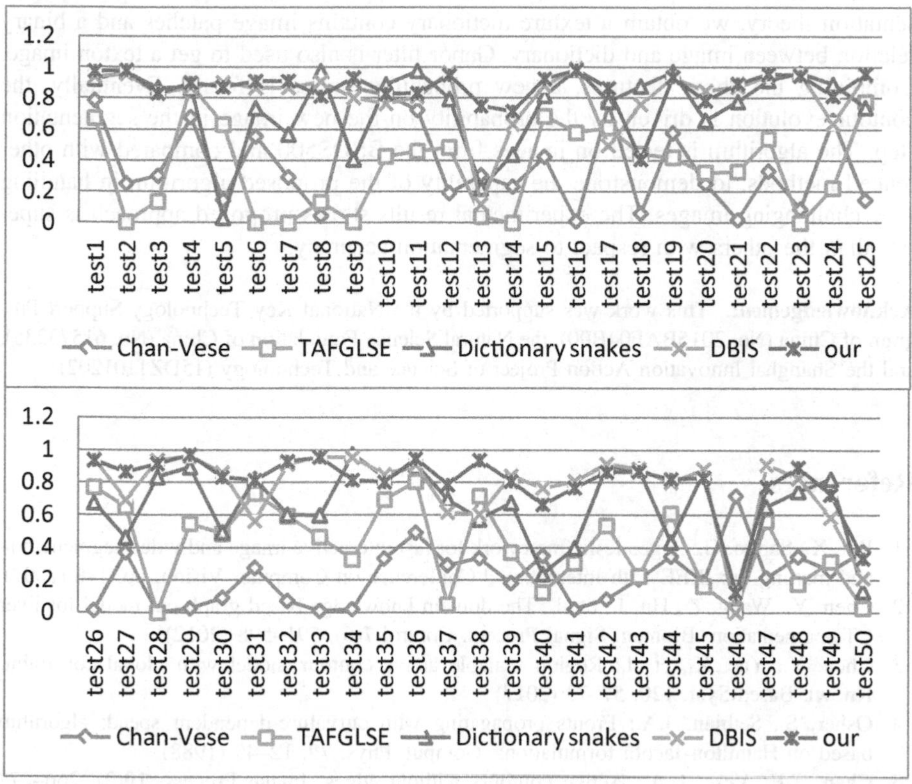

Fig. 6. The F-measure of 50 tested images selected from Berkeley data set. The cyan line, green line, blue line, yellow line and red line are the results of Chan-Vese, TAFGLSE, Dictionary snakes, DBIS and our method, respectively. (Color figure online)

As we can see in Table 1 for the F-measure, when compared with other methods, our proposed method demonstrates better quantitatively with 83.22%. The C-V model and TAFGLSE do not work well for images with complex textured objects and backgrounds. In contrast, our method is very general and works well. The performance of our method is 12.9% higher than Dictionary snakes, and 3.9% higher than DBIS. The variance of F-measure shows our method is more robust than dictionary snakes and dictionary based image segmentation. Proposed approach can not only detect objects without strong edges, but also apply to image segmentation for complex background. Besides, if there are more than one same object in the image, our method do not need to select them all. All the same objects can be separated by labeling one object.

4 Conclusions

In this work, we present a color and texture features based image segmentation method. We use different features after considering several aspects of the difficult problem. The first step is to extract and describe features. User is asked to draw an initial contour which is used as a label image. We extract the color features from the original color information and the spatial relationship between pixels. Utilizing the sparse representation theory, we obtain a texture dictionary contains image patches and a binary relation between image and dictionary. Gabor filter is also used to get a texton image. Combining the above features, a new probability image is found. Eventually, the contour evolution is driven by the probability on the new image in the segmentation step. The algorithm is tested on images from the BSDS500 and compared with other related methods, to demonstrate the capability of the proposed algorithm in handling very challenging images. The experimental results shows proposed approach is superior than the others with respect to segmentation accuracy.

Acknowledgement. This work was supported by the National Key Technology Support Program of China (No. 2015BAF04B00), the Natural Science Foundation of China (No. 61573235), and the Shanghai Innovation Action Project of Science and Technology (15DZ1101202)

References

1. Bai, X., Sapiro, G.: A geodesic framework for fast interactive image and video segmentation and matting. In: IEEE 11th International Conference on Computer Vision, pp. 1–8 (2007)
2. Chen, Y., Wang, Z., Hu, J., et al.: The domain knowledge based graph-cut model for liver CT segmentation. Biomed. Signal Process. Control **7**(6), 591–598 (2012)
3. Chen, Y., Yuc, X., et al.: Region scalable active contour model with global constraint. Knowl. Based Syst. **120**, 57–73 (2017)
4. Osher, S., Sethian, J.A.: Fronts propagating with curvature-dependent speed: algorithm based on Hamilton-Jacobi formulations. Comput. Phys. **79**, 12–49 (1988)
5. Chan, T.F., Vese, L.A.: Active contours without edges. Image Process. **10**(2), 266–277 (2001)
6. Mumford, D., Shah, J.: Optimal approximations by piecewise smooth functions and associated variational problems. Commun. Pure Appl. Math. **42**, 577–685 (1989)
7. Nawal, H., et al.: Fast texture segmentation based on semi-local region descriptor and active contour. Numer. Math. Theor. Methods Appl. **2**, 445–468 (2009)
8. Rousson, M., Broxm, T., Deriche, R.: Active unsupervised texture segmentation on a diffusion based feature space. CVPR **2**(144), 699 (2003)
9. Mairal, J., Bach, F., Ponce, J.: Task-driven dictionary learning. IEEE Trans. Pattern Anal. Mach. Intell. **2**, 791–804 (2012)
10. Gao, Y., Bouix, S., Shenton, M., Tannenbaum, A.: Sparse texture active contour. IEEE Trans. Image Process. **22**, 3866–3878 (2013)
11. Osher, S., Sethian, J.A.: Fronts propagating with curvature-dependent speed: Algorithms based on Hamilton-Jacobi formulation. J. Comput. Phys. **79**, 12–49 (1988)
12. Rad, A.E., Rahim, M.S.M., et al.: Morphological region-based initial contour algorithm for level set methods in image segmentation. Multimed. Tools Appl. **76**(2), 2185–2201 (2016)

13. Barley, A., Town, C.: Combinations of feature descriptors for texture image classification. Int. J. Comput. Vis. **02**(3), 67–76 (2014)
14. Thomas, L., Jitendra, M.: Representing and recognizing the visual appearance of materials using three-dimensional textons. Int. J. Comput. Vis. **43**, 29–44 (2001)
15. Sochen, N., Kimmel, R., Malladi, R.: A general frame-work for low level vision. IEEE Trans. Image Process. **7**(3), 310–318 (1998)
16. Houhou, N., et al.: Fast texture segmentation model based on the shape operator and active contour. In: IEEE Conference on Computer Vision and Pattern Recognition, pp. 1–8 (2008)
17. Gray, A.: Modern Differential Geometry of Curves and Surfaces with Mathematica. CRC Press, Boca Raton (1996)
18. Elad, M.: Sparse and Redundant Representations: From Theory to Applications in Signal and Image Processing. Springer, New York (2010)
19. Mairal, J., Bach, F., Ponce, J.: Task-driven dictionary learning. IEEE Trans. Pattern Anal. Mach. Intell. **34**(4), 791–804 (2012)
20. Gao, Y., Bouix, S., Shenton, M., Tannenbaum, A.: Sparse texture active contour. IEEE Trans. Image Process. **22**(10), 3866–3878 (2013)
21. Sarkar, R., Mukherjee, S.: Acton dictionary learning level set. IEEE Signal Process. Lett. **22** (11), 2034–2038 (2015)
22. Dahl, A., Larsen, R.: Learning dictionaries of discriminative image patches. In: Procedings of the British Machine Vision Conference, pp. 77.1–77.11(2011)
23. Efros, A., Leung, T.: Texture synthesis by non-parametric sampling. In: IEEE International Conference on Computer Vision, vol. 2, pp. 10–33 (1999)
24. Gabor, D.: Theory of communication. J. Inst. Electric. Eng. **93**, 429–457 (1946)
25. Rad, A.E., Rahim, M.S.M., Kolivand, H., Amin, I.B.M.: Gabor filters as texture discriminator. Multimed Tools Appl. **61**(2), 103–113 (1989)
26. Martin, D., Fowlkes, C., Tal, D., Malik, J.: A database of human segmented natural images and its application to evaluating segmentation algorithms and measuring ecological statistics. In: Computer Vision, pp. 416–423 (2001)
27. Ooi, W.S.: CP lim fusion of colour and texture features in image segmentation: an empirical study. Imag. Sci. J. **57**(1), 8–18 (2013)
28. Dahl, AB., Dahl, VA.: Dictionary based image segmentation. In: SCIA 2015, pp. 26–37 (2015)
29. Balla-Arabé, S., Gao, X., Xu, L.: Texture-aware fast global level set evolution. In: Sun, C., Fang, F., Zhou, Z.-H., Yang, W., Liu, Z.-Y. (eds.) IScIDE 2013. LNCS, vol. 8261, pp. 529–537. Springer, Heidelberg (2013). doi:10.1007/978-3-642-42057-3_67
30. Dahl, AB., Dahl, VA.: Dictionary snakes. In: International Conference on Pattern Recognition (2014)

Study on Prior-Altitude-Locating-Algorithm for Video Target Based on UAV

Wen-Bo Zhao[1(✉)] and Hai-Long Ding[2]

[1] Army Officer Academy of PLA, Hefei 230031, Anhui, China
656797226@qq.com
[2] Shenyang Artillery Academy of PLA, Shenyang 110867, Liaoning, China

Abstract. Aim at the requirement of the fast locating and positioning of the video multi target for the high speed military unmanned aerial vehicle (UAV), Using the video surveillance, UAV telemetry data and the priori elevation information of the target area to reconstructing the image of UAV video surveillance center and Establish UAV camera surveying coordinate system and also derivation positioning error statistical characteristics. This paper proposed a priori height video targeting algorithm (prior-altitude-locating-algorithm). Simulation shows that the prior-altitude-locating-algorithm can be used to solve position of ground/sea target reconnoitred by UAV video. The forward and reverse processes of prior-altitude-locating-algorithm are scientific, reasonable and fit the actual situation.

Keywords: Video-target · Prior-altitude-locating-algorithm · Photogrammetry

1 Introduction

Based on the single-point collinear location principle of UAV video location, a two-point spatial painting location method is proposed by Sun Chao [1], which is research on the location of ground/sea target by video UAV. And flight verifies the accuracy of two-point spatial painting location method (60 m) is better than single point collinear location (100 m). Based on optical flow, normalized product correlation and Gaussian tower decomposition, Chen Huawang [2] proposed pixel matching target location algorithm, in which using the UAV location, UAV attitude and camera attitude to estimate the target area, realizing precise target location based on UAV video matching. Li Dajian [3] solves the target coordinates through the aircraft coordinate system, camera screen coordinate system, ground coordinate system and its coordinate transformation modeling, with use of UAV attitude, photoelectric stability platform (photoelectric platform) attitude and UAV location parameters. Monte Carlo simulation shows that ground target location accuracy is improved from about 100 m to nearly 80 m at flying height of 2500 m when UAV attitude angle measurement accuracy has improved (yaw angle 2.0° 1.5 °, roll angle 1.0° 0.75°, pitch angle 1.0° 0.75°).

The National Nature Science Fund Project 61273001, Anhui Province Nature Science Fund Project 11040606M130.

D.-S. Huang et al. (Eds.): ICIC 2017, Part I, LNCS 10361, pp. 366–375, 2017.
DOI: 10.1007/978-3-319-63309-1_34

Wang Fengjuan [4] proposed UAV passive positioning method, which let UAV video tracking known landmarks, with the use of current and previous moment of UAV attitude, location, and photoelectric platform attitude parameters, based on triangular cosine theorem to solve the current location of the UAV, the method can also solve the location of ground/sea targets.

The above research theory is correct and feasible, but without fully consideration of technical and tactical characteristics of military UAV equipment, without close combination of battlefield using needs about UAV equipment. In this paper, for actual demand of high precision and fast target location of UAV video reconnaissance in information warfare, and for the problem that laser range finder installed on the UAV can only measure the flight height, we proposed prior-altitude-algorithm to solve the location of video visible ground/sea target, by use of UAV reconnaissance video and telemetry parameters. We given the algorithm usage flow and deduced the error statistics characteristic of this proposed algorithm.

2 The Principle of Prior-Altitude-Locating-Algorithm

Based on the knowledge of photogrammetry, let ENU coordinate system of UAV centroid be space-assisted coordinate system S-uvw (referred to as U-ray system) [5], let ENU coordinate system at the plane projection point of UAV (referred to as ground D system) be ground photogrammetric coordinate system D-XYZ [5], the pixel plane M system is the plane coordinate system P-xyz [5], also is pixel spatial coordinate system S-xyz [5].

As shown in Fig. 1, the basic principle of prior-altitude-algorithm is as follows: Using the coordinate transformation method [6], based on telemetry parameters such as attitude angle of photoelectric platform, attitude angle of UAV, and heading angle of UAV, coordinate system of pixel point is transformed from pixel plane M system to UAV W system. The elevation of ground point is known as a priori condition, and based on principle of photogrammetric center conformation [6], we deduce point coordinates of pixel point in plane M coordinate system (ground D system), so as to complete spatial resolution of ground point corresponding with video pixel point, which is called priori hcight video targeting algorithm (priori elevation algorithm).

The prior-altitude-locating-algorithm consists of 12 coordinate systems (see Fig. 1) and their mutual transformation (see Fig. 2). Coordinate systems of priori elevation algorithm mainly include ground station K coordinate system [6], ground D coordinate system [5], UAV W coordinate system [5], UAV U coordinate system [6], track S coordinate system [6], UAV N coordinate system [6], base B coordinate system [6], azimuth ring A coordinate system [6], high and low ring F coordinate system [6], roll ring R coordinate system [6], pixel plane M coordinate system [6] and pixel V coordinate system [6]. The core work of the location solution is direction cosine solution from plane M coordinate system to UAV W coordinate system: to calculate transformation relationship of space posture in pixel plane M coordinate system and UAV W coordinate system by use of telemetry parameters such as attitude angle of photovoltaic platform, attitude angle of UAV, and heading angle of UAV, mainly achieved by solving the direction cosine of the pixel plane M coordinate system to UAV W

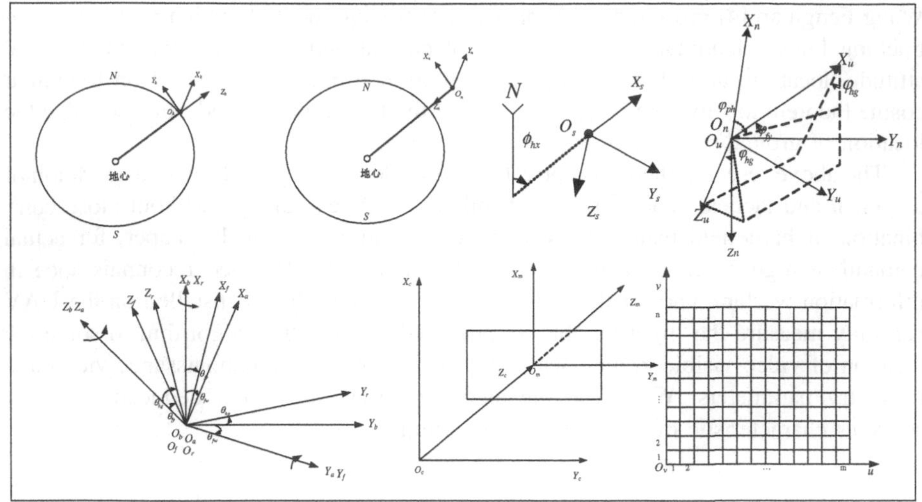

Fig. 1. Coordinate system of prior-altitude-locating-algorithm

coordinate system. Based on coordinate transformation of video target, we realize direction cosine calculation of transformation from photogrammetric pixel spatial coordinate system (pixel plane M coordinate system) to UAV W coordinate system by studing the coordinate transformation matrix of pixel point from pixel plane M coordinate system to UAV W coordinate system. The important work of location and resolution is center conformation of UAV video reconnaissance: to complete conformation from any pixel point to corresponding any object point and inverse conformation from any object point to corresponding any pixel point, relying on center projection conformation relationship of UAV W coordinate system and ground D coordinate system.

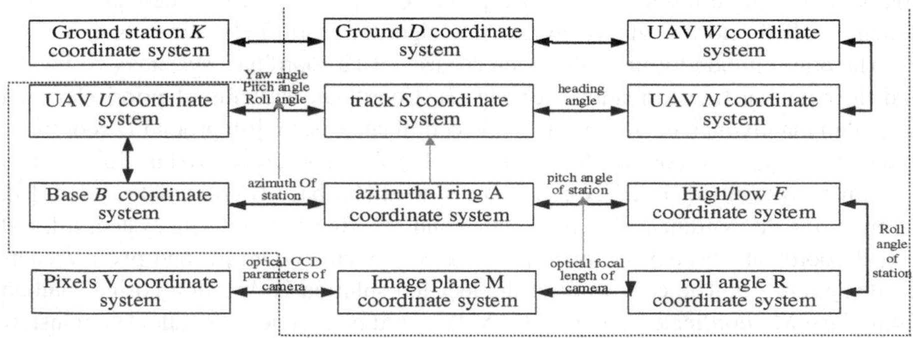

Fig. 2. Solution flow of prior-altitude-locating-algorithm

3 Steps of Prior-Altitude-Locating-Algorithm

About video visible target T, assuming that its coordinates in pixel V coordinate system are $T_v[u \quad v \quad 1]^T$, its coordinates in pixel plane M coordinate system are $T_m[x_m \quad y_m \quad f]^T$, its other coordinates are $T_i[x_i \quad y_i \quad z_i]^T$. $i = r, f, a, b, u, s, n, w, d, k$ are respectively on behalf of rolling ring R coordinate system, high and low ring F coordinate system, azimuth ring A coordinate system, base B coordinate system, UAV U coordinate system, track S coordinate system, UAV N coordinate system, UAV W coordinate system, ground D coordinate system and ground station K coordinate system of target T. Assuming elevation of the ground target point area is known, based on principle of coordinate transformation, the steps of prior-altitude-locating-algorithm are summarized as follows:

(1) transformation from pixel V coordinate system to image plane M coordinate system

$$\begin{cases} x_m = (v - v_0)s_v \\ y_m = (u - u_0)s_u \end{cases} \tag{1}$$

$$\begin{cases} u = y_m / s_u + u_0 \\ v = x_m / s_v + v_0 \end{cases} \tag{2}$$

s_v and s_u are CCD pixel physical sizes, the median of number of CCD pixels row.

(2) transformation between pixel plane M coordinate system and UAV W coordinate system

$$T_w = \boldsymbol{M}_n^e \boldsymbol{M}_s^n \boldsymbol{M}_u^s \boldsymbol{M}_a^b \boldsymbol{M}_f^a \boldsymbol{M}_r^f T_m \tag{3}$$

$$T_m = \boldsymbol{M}_f^r \boldsymbol{M}_a^f \boldsymbol{M}_b^a \boldsymbol{M}_s^u \boldsymbol{M}_n^s \boldsymbol{M}_e^n T_w \tag{4}$$

$\boldsymbol{M}_i^j$ is rotate transformation matrix from the i-th coordinate system to the j-th coordinate system, where $i, j = r, f, a, b, u, s, n, w$ and $\boldsymbol{M}_i^j = \left(\boldsymbol{M}_j^i\right)^T$. Let $\boldsymbol{M}_w^m = \boldsymbol{M}_f^r \boldsymbol{M}_a^f \boldsymbol{M}_b^a \boldsymbol{M}_s^u \boldsymbol{M}_n^s \boldsymbol{M}_n^e$, so, $\boldsymbol{M}_w^m = \begin{bmatrix} m_{11}^w & m_{12}^w & m_{13}^w \\ m_{21}^w & m_{22}^w & m_{23}^w \\ m_{31}^w & m_{32}^w & m_{33}^w \end{bmatrix}$ $\boldsymbol{M}_m^w = \begin{bmatrix} w_{11}^m & w_{12}^m & w_{13}^m \\ w_{21}^m & w_{22}^m & w_{23}^m \\ w_{31}^m & w_{32}^m & w_{33}^m \end{bmatrix}$.

(3) transformation between UAV W coordinate system to ground D coordinate system

According to three-point collinear principle [5] of projection center, pixel point and object point, we can see that the transformation of target T from UAV W coordinate system to ground D coordinate system is as follows.

$$\begin{bmatrix} x_d - X_s \\ y_d - Y_s \\ z_d - Z_s \end{bmatrix} = \lambda \begin{bmatrix} x_w \\ y_w \\ z_w \end{bmatrix} \tag{5}$$

λ is proportion factor, $[X_s \ Y_s \ Z_s]^T$ are coordinates of UAV centroid in ground station K coordinate system, when z_d is known, and origin point of ground D coordinate system is vertical projection point of UAV on the ground plane, the solution expression of object on ground is expressed as $X_s = 0$, $Y_s = 0$, $Z_s = H$ (H is flight height). x_d and y_d can be calculated according to Eq. (5), and we can get the following formulas.

$$\begin{cases} x_d = X_s + (z_d - Z_s)\frac{x_w}{z_w} = (z_d - H)\frac{x_w}{z_w} \\ y_d = Y_s + (z_d - Z_s)\frac{y_w}{z_w} = (z_d - H)\frac{y_w}{z_w} \\ z_d = z_d \end{cases} \tag{6}$$

$$\begin{bmatrix} x_w \\ y_w \\ z_w \end{bmatrix} = \frac{1}{\lambda} \begin{bmatrix} x_d - X_s \\ y_d - Y_s \\ z_d - Z_s \end{bmatrix} \tag{7}$$

Equation (7) represents transformation between ground D coordinate system and UAV W coordinate system.

(4) transformation between imaging plane M coordinate system and UAV W coordinate system

Comparing steps (3) and (6), we obtain transformation of target T from imaging plane M coordinate system to ground D coordinate system as follows.

$$\begin{aligned} x_d &= X_s + (z_d - Z_s)\frac{m_{11}^w x_m + m_{12}^w y_m + m_{13}^w f}{m_{31}^w x_m + m_{32}^w y_m + m_{33}^w f} \\ y_d &= Y_s + (z_d - Z_s)\frac{m_{21}^w x_m + m_{22}^w y_m + m_{23}^w f}{m_{31}^w x_m + m_{32}^w y_m + m_{33}^w f} \end{aligned} \tag{8}$$

$$\begin{aligned} x_m &= f\frac{w_{11}^m(x_d - X_s) + w_{12}^m(y_d - Y_s) + w_{13}^m(z_d - Z_s)}{w_{31}^m(x_d - X_s) + w_{32}^m(y_d - Y_s) + w_{33}^m(z_d - Z_s)} \\ y_m &= f\frac{w_{21}^m(x_d - X_s) + w_{22}^m(y_d - Y_s) + w_{23}^m(z_d - Z_s)}{w_{31}^m(x_d - X_s) + w_{32}^m(y_d - Y_s) + w_{33}^m(z_d - Z_s)} \end{aligned} \tag{9}$$

Equation (9) represents the transformation of the ground D system to the image plane M system.

(5) transformation between ground D coordinate system and ground station K coordinate system

$$\begin{bmatrix} x_k \\ y_k \\ z_k \end{bmatrix} = (M_k^d)^T \begin{bmatrix} x_d \\ y_d \\ z_d \end{bmatrix} + \begin{bmatrix} x_k^d \\ y_k^d \\ z_k^d \end{bmatrix} \tag{10}$$

$D_k = \begin{bmatrix} x_k^d & y_k^d & z_k^d \end{bmatrix}^T$ are coordinates of origin point of ground D coordinate system in ground station K coordinate system, M_k^d is rotation transformation matrix from ground station K coordinate system to ground D coordinate system, the expression is as follows.

$$\begin{cases} M_k^d = M_e^d (M_e^s)^T \\ M_e^d = \begin{bmatrix} -sin(L_u) & cos(L_u) & 0 \\ -cos(L_u)sin(B_u) & -sin(L_u)sin(B_u) & cos(B_u) \\ cos(L_u)cos(B_u) & sin(L_u)cos(B_u) & sin(B_u) \end{bmatrix} \\ M_e^s = \begin{bmatrix} -sin(L_s) & cos(L_s) & 0 \\ -cos(L_s)sin(B_s) & -sin(L_s)sin(B_s) & cos(B_s) \\ cos(L_s)cos(B_s) & sin(L_s)cos(B_s) & sin(B_s) \end{bmatrix} \end{cases} \tag{11}$$

$$\begin{bmatrix} x_d \\ y_d \\ z_d \end{bmatrix} = M_k^d \left(\begin{bmatrix} x_k \\ y_k \\ z_k \end{bmatrix} - \begin{bmatrix} x_k^d \\ y_k^d \\ z_k^d \end{bmatrix} \right) \tag{12}$$

Formula (12) is rotation transformation matrix from ground station K coordinate system to ground D coordinate system, L_k, B_k, L_d and B_d are latitude and longitude of ground station and origin point of ground D coordinate system.

4 Simulation Verification and Conclusion Analysis

In this paper, we use MATLAB software to do simulation analysis and theoretical verification. For ease of calculation, CCD parameters are assumed to be as follows: number of array pixels is m = 161, number of array vertical pixels is n = 91, length and width of pixels cell are both 7.4 μm. Assuming that the focal length of the camera is XX ~ XXX mm, resolution of ground target is $GSD = 5\,\text{m}$, minimum number of pixels for CCD imaging is X = 2.

We use geometry dilution of precision (GDOP) as metrics to test location performance of location algorithm in this paper. The mathematical expression of GDOP is as follows:

$$GDOP = \sqrt{\sigma_X^2 + \sigma_Y^2 + \sigma_Z^2} \tag{13}$$

σ_X^2, σ_Y^2, and σ_Z^2 are location variance of three axes of ground station K coordinate system.

In this simulation, the UAV uniform fly with planning heading angle at a certain height along the earth round. At a certain time t_k, we use prior-altitude-locating-algorithm to locate ground target and solve coordinates of target on instantaneous view field. Flying head angle of UAV 60°. The starting positions of simulation are $[118° \ 32° \ 1000]^T$ and $[119.84109° \ 32.90126° \ 1000]^T$. Flying distance is 200 km. Position of UAV ground station is $[118° \ 32° \ 40]^T$. We use WGS84 coordinate system as earth coordinate system. Long axis radius of ellipsoid earth is 6378137 m. The square of first eccentricity is 0.0818191908425523.

Simulation scene of UAV with uniform linear motion is as shown in Fig. 3. Ground shelter, that we get by use of prior-altitude-locating-algorithm, corresponding to instantaneous position of UAV, is as shown in Fig. 4. Coordinates of instantaneous view field in pixel V coordinate system are as shown in Fig. 5. Shapes of instantaneous field in ground station K coordinate system are as shown in Figs. 6, 7, 8, 9, 10, 11, 12, 13, 14, 15, 16, 17 and 18. At every moment planning heading angle of UAV is 60°, flight height is 1000 m. But attitude angles of UAV and attitude angles of photoelectric platform are different each other at each moment. Figure 6 shows instantaneous view field when all posture angles are 0. Figures 7, 8, 9, 10, 11 and 12 show instantaneous view field correspond to UAV with different attitude angles (attitude angles of UAV mean heading angle, pitch angle and roll angle). Figures 13, 14, 15 and 16 show instantaneous view field correspond to different photoelectric platform attitude angles. Figures 17 and 18 show instantaneous view field of UAV with attitude angle and non-zero photoelectric platform attitude angle. Figure 17 corresponds to camera with small focal length XX mm, Fig. 18 corresponds to camera with large focal length XXX mm.

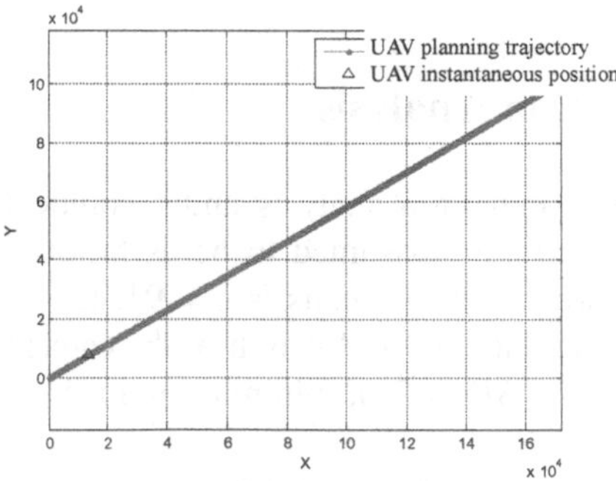

Fig. 3. Simulation scene of UAV doing uniform linear

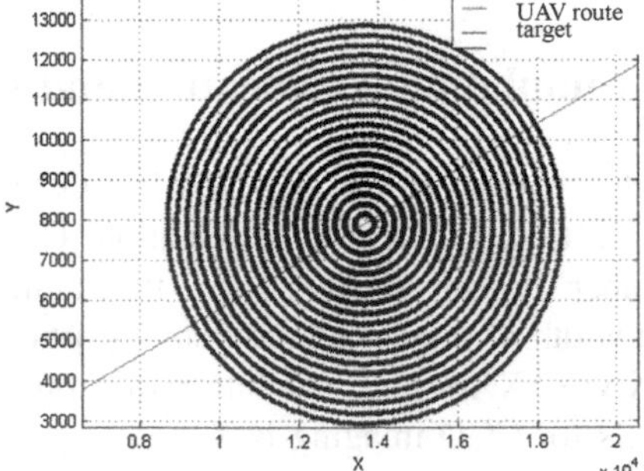

Fig. 4. Targets in accommodation scope corresponding to instantaneous position of UAV

As shown in Fig. 5, we can get coordinates in pixel V coordinate system through reverse process of prior-altitude-locating-algorithm. Horizontal width is 161 pixels, and vertical width is 91 pixels. The calculated value is very close to integer, and distribution of pixels is uniform, indicating that priori altitude algorithm is correct. As shown in

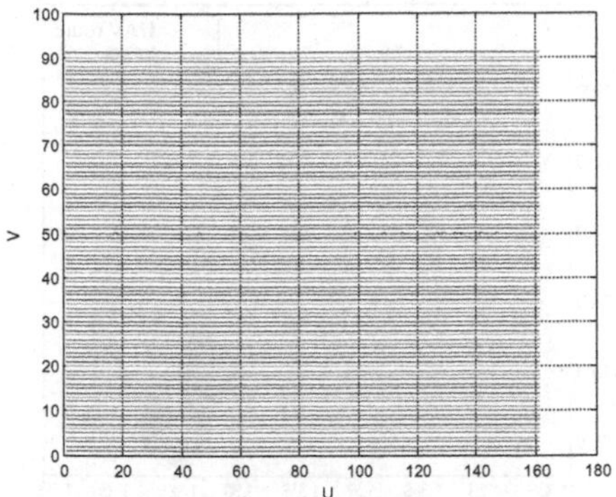

Fig. 5. Coordinates in pixel V coordinate system when attitude corresponding to instantaneous field of view platform are zero

Fig. 6. Instantaneous field of view angles of UAV and photovoltaic

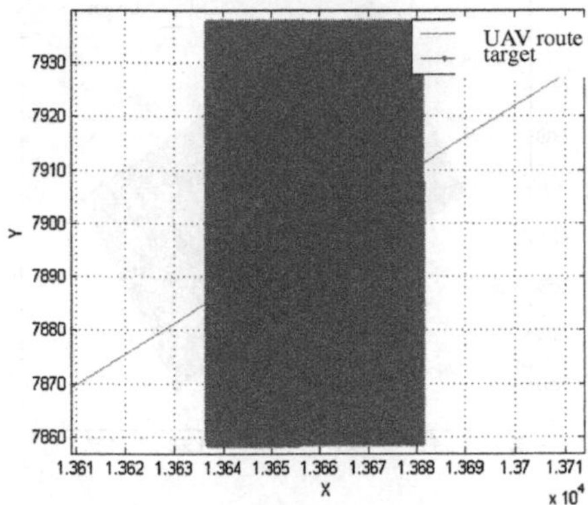

Fig. 7. Instantaneous field of view when yaw angle is 30°

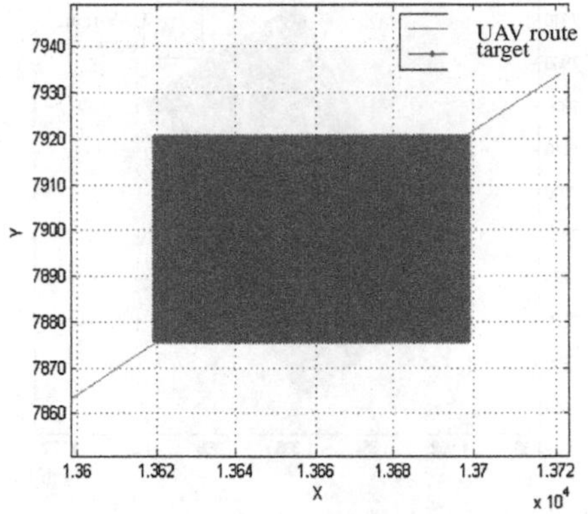

Fig. 8. Instantaneous field of view when angle is 60°

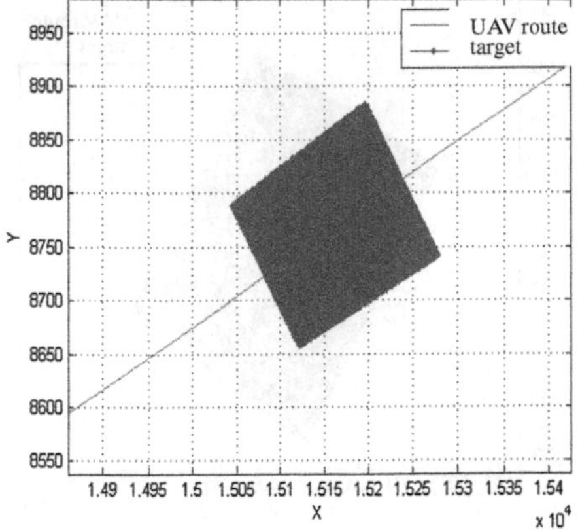

Fig. 9. Instantaneous field of view when pitch when pitc angle is 60°

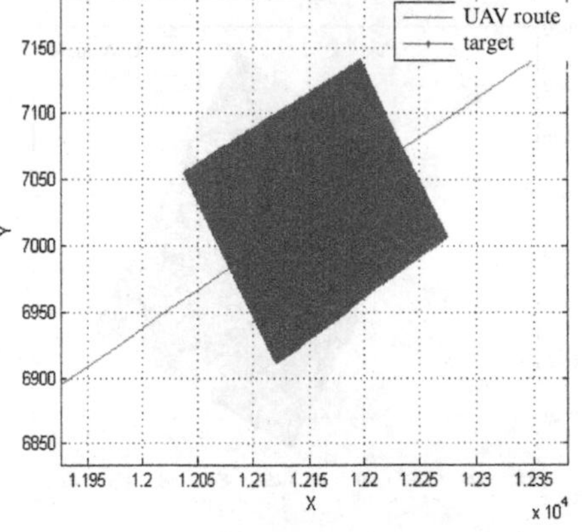

Fig. 10. Instantaneous field of view angle is −60°

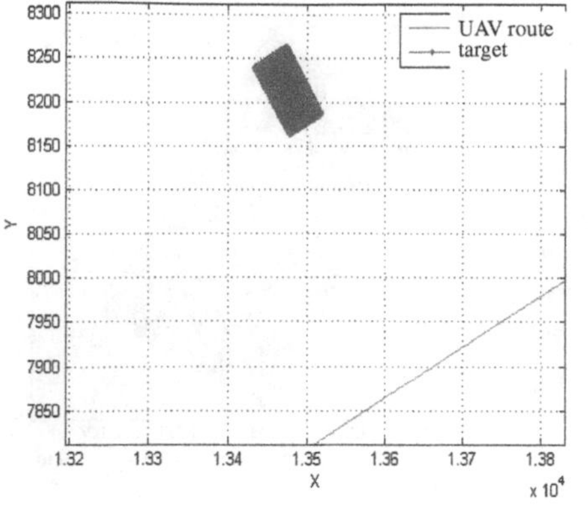

Fig. 11. Instantaneous field of view when roll angle is 20°

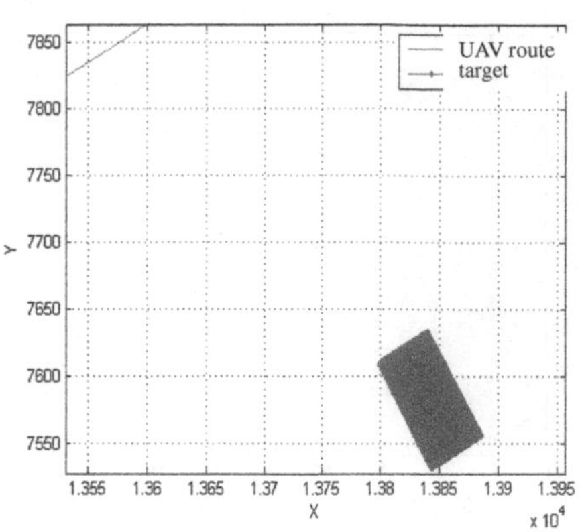

Fig. 12. Instantaneous field of view when angle is −20°

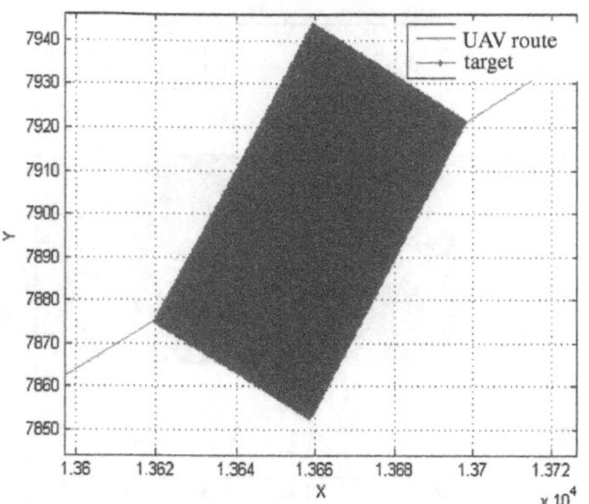

Fig. 13. Instantaneous field of view when azimuth of photovoltaic platform is 60°

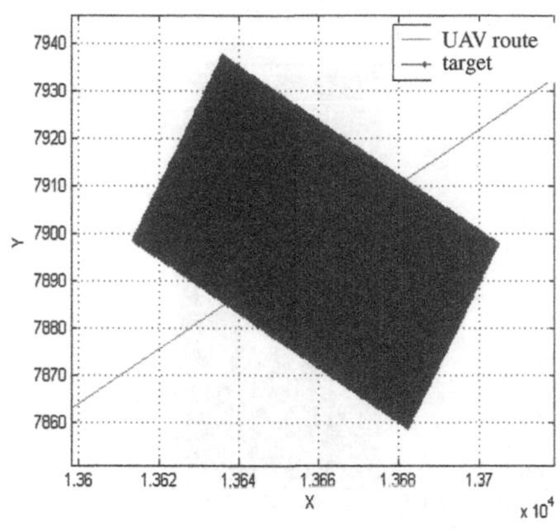

Fig. 14. Instantaneous field of view when of photovoltaic platform is −30°

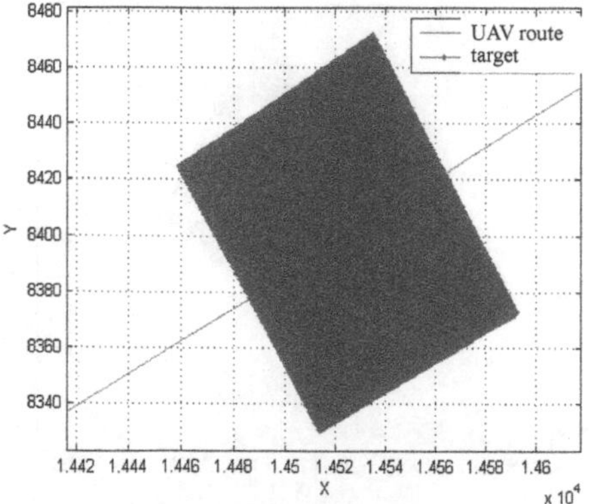

Fig. 15. Instantaneous field of view when pitch pitch angle of photovoltaic platform is 45°

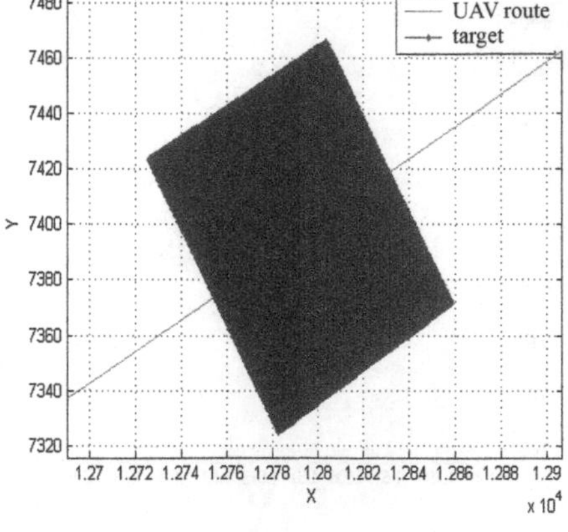

Fig. 16. Instantaneous field of view when angle of photovoltaic platform is −45°

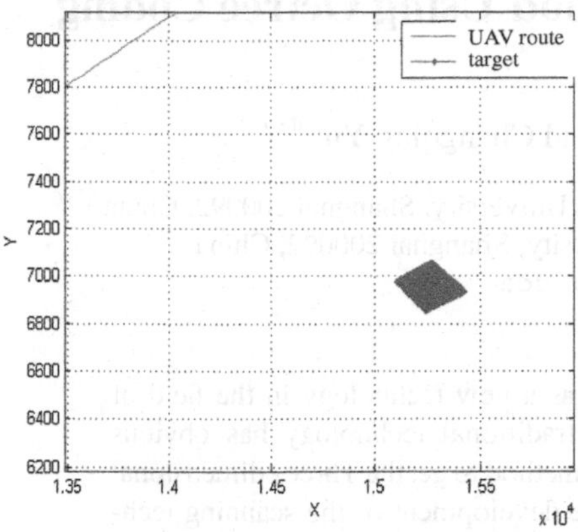

Fig. 17. Instantaneous field of view when attitude angles photoelectric platform of small focal length UAV are not zero (yaw angle −60°, pitch angle −30°, roll angle −45°, azimuth of photoelectric platform −30°, pitch angle −10°)

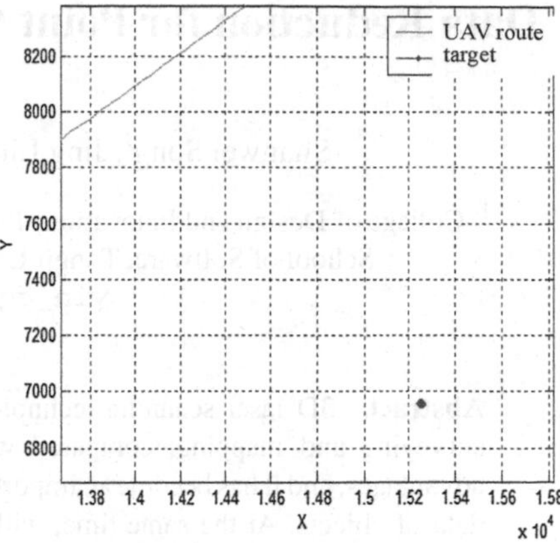

Fig. 18. Instantaneous field of view when attitude angles of photoelectric platform of large focal length UAV are not zero (yaw angle −60°, pitch angle −30° roll angle −45°, azimuth of photoelectric platform −30°, pitch angle −10°)

Figs. 6, 7, 8, 9, 10, 11, 12, 13, 14, 15, 16, 17 and 18, when attitude angles of UAV or attitude angles of photovoltaic platform are not zero, we can get different types of instantaneous quadrilateral field of view by priori altitude algorithm. Different focal lengths of camera correspond to different instantaneous field of view. The larger the focal length is, the smaller the instantaneous field of view is. On the contrary, the instantaneous fields of view get more large. From the logical analysis, we can know that forward process of priori altitude algorithm is scientific and reasonable, fitting the actual situation.

References

1. Sun Chao, D., Jiyan, D.L.: A technology of UAV positioning to target by two-point space rendezvous. Ordnance Ind. Autom. **30**(6), 35–41 (2011)
2. Chen, H.: Study on application of image registration technology in location and calibration. Northwestern Polytechnical University 3 (2004)
3. Dajian, L., Min, Q.: Monte-carlo simulation analysis for UAV ground target position accuracy. Comput. Simul. **28**(7), 75–78 (2011)
4. Wang, F.: Study on locating method of UAV based on image tracking. Xidian University 1 (2009)
5. Wang, P., Xu, Y.: Photogrammetry, vol. 6, pp. 22–45. Wuhan University Press (2012)
6. Bi, Y.: Study on control system of multi-frame photoelectric. Doctoral thesis of Graduate University of Chinese Academy of Sciences, vol. 10, pp. 10–28 (2013)

Data Reduction for Point Cloud Using Octree Coding

Shanwei Song[1], Jing Liu[2], and Changqing Yin[2(✉)]

[1] College of Design and Innovation, Tongji University, Shanghai 200092, China
[2] School of Software, Tongji University, Shanghai 200092, China
yin_cq@qq.com

Abstract. 3D laser scanning technology as a new technology in the field of surveying and mapping, compared with traditional technology has obvious advantages, and it has become an important method to get the Three - dimensional data of objects. At the same time, with the development of the scanning technology, we get more and more point cloud data. Therefore, how to streamline the cloud data, remove the invalid data and retain the necessary data has become an important research content. In this paper, we focus on the point cloud data preprocessing, analyze the shortcomings of the current point cloud data reduction method, and propose a uniform and robust algorithm based on octree coding. Apply the octree coding method to divide the point cloud neighborhood space into multiple sub-cubes with the specified side length, and keep the nearest point of each sub-cube from the center point to realize the simplification of the point cloud.

Keywords: Computer application · Data reduction · Octree coding · Streamline data

1 Introduction

In reverse engineering, the data obtained by the non-contact measurement method has the characteristics of disorder and magnanimity [1]. Disorderly refers to the characteristics of each data point with only the three-dimensional coordinate values of information, and there is no clear spatial neighborhood information, does not favor the neighborhood search data points, and neighborhood data points' search speed affect the scattered data processing and surface reconstruction efficiency is one of the main factors. Huge amounts of data will be generated in each measurement activity. Huge amounts of data often refers to the characteristics of close, this will affect the speed of surface reconstruction, and the reconstruction of surface curvature is small can also affect the surface smoothness. At the same time, too much data can make the computer processing speed slow. In prior to surface reconstruction, therefore, we need to establish a data of spatial neighborhood relation and data reduction. In this paper, we use octree encoding spatial neighborhood partition method, and based on this, we advance a new method of point cloud data evenly to streamline.

D.-S. Huang et al. (Eds.): ICIC 2017, Part I, LNCS 10361, pp. 376–383, 2017.
DOI: 10.1007/978-3-319-63309-1_35

1.1 The Refinement Methods of Point Cloud

In recent years, people do a lot of research about data compaction, mainly including streamline curvature and uniform streamline method. In the streamline curvature method. A method is that use bounding box method to construct segments, then using segmentation point cloud processing into line structure, recycling Angle, high string combined code line by line to streamline. Zhou Lv [2] using the parabolic fitting to solve local curvature, then according to the deviation of point cloud streamline curvature. In uniform streamline method, another method is that by defining the sample cube to a certain point, for the rest of the points within the cube to the distance of the point at which, according to the average distance and the user to specify to retain some percentage of cuts.in this method, there is no spatial neighborhood of point cloud data partition in advance, so in the process of retrieving sampling cube, it is necessary to determine the point cloud data, so the large amount of calculation when dealing with huge amounts of data, and cannot be based on the specified distance concise point cloud.

To solve the shortcomings of the above methods, in this paper, we puts forward a new uniform to streamline method principle.

(1) According to the specified distance, we use octree coding method to point cloud divides the spatial neighborhood space bounding box of point cloud data (minimum circumscribed cube) is divided into multiple points to specify the spacing d0 side length.

(2) Streamline respectively for each cube data, as shown in Fig. 1, if after the division of two adjacent sub cubes in 8 data points p1~p8, keep each cube apart from the center of the nearest point in the p3 and p3. Due to adjacent cube center distance of d0, so reduced the distance between the point cloud p3 and p7 is approximate to d0.

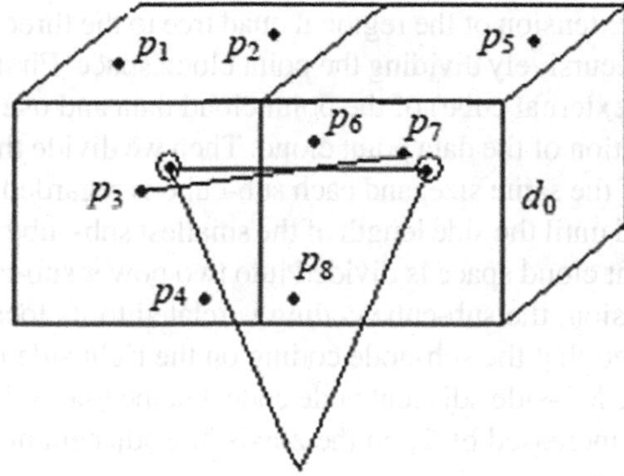

Fig. 1. The sub-cube of point cloud

1.2 The Division of Spatial Neighborhood

In the process of streamline, filtering, feature extraction, etc., We need to get the unit of normal vector, micro tangent plane and curvature value. These information requires k neighbor search data points, the data point to find k with that point Euclidean distance nearest point. The general search method is the exhaustion. Calculate a point and the Euclidean distance rest of the points set and according to the smallest, select a row in front of the K point and K nearest neighbor points for that point. This method takes a lot of time to process data. Therefore, in order to establish the good spatial point cloud neighborhood structure is the key to improve the data point k nearest neighbor search speed. In next, we will introduce the octree method.

2 The Spatial Neighborhood Division

In the process of streamlining, filtering and feature extraction of scattered data, we need to obtain the information such as the micro-cut plane and the curvature value of the data point at the corresponding point of its profile. This requires searching the k neighborhood of the data point, The data points are concentrated to find k points closest to the Euclidean distance. The general search method is the exhaustive method: calculate the distance between the point and the other point of the Euclidean distance, and sorted them, select the front of the K points for the point of the k nearest neighbors. This approach is time-consuming and inefficient for massive amounts of data. So, we need to improve it.

Next, we will introduce an efficient method of spatial neighborhood partitioning – octree method.

2.1 The Principles of Octree Method

The octree tree is the extension of the regional quad tree to the three-dimensional space, which is realized by recursively dividing the point cloud space. Firstly, we construct the space bounding box (external cube) of the point cloud data and use it as the root model of the topological relation of the data point cloud. Then we divide the external cube into eight subcategories of the same size, and each sub-cube is regarded as the root node. so recursively segmented until the side length of the smallest sub-cube is equal to the given dot pitch, and the point cloud space is divided into two power sub-cubes. In the process of octagonal tree division, the sub-cube coding is related to its location [3]. As shown in Fig. 2, it is stipulated that the sub-node coding on the right side of the x-plane in the x-axis is more than the left-side adjacent node code. On the y-axis, he sub- The adjacent node position code is increased by 2; on the z-axis, the adjacent node position code on the upper side of the z-s.

The octree space model can be represented by an n-layer octree, and each cube in the octree space model and the octree The location of the nodes in the octree space model can be represented by the octet coding Q of the corresponding node

$$Q = q_{n-1} \cdots q_m \cdots q_1 q_0 \tag{1}$$

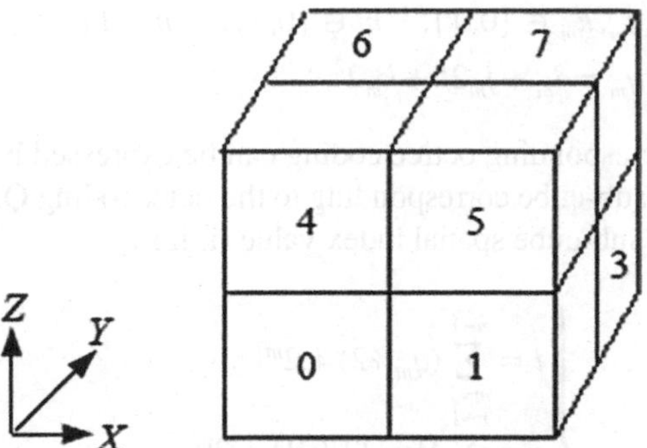

Fig. 2. Octagonal tree spatial division model

Where qm is an octal number, m ∈ {0,1,.., n. 1}, qm denotes the number of the node between its sibling nodes, and qm + 1 denotes the parent node of qm node in its sibling node Between the serial number. Thus, from q0 to qn-1, the path of each leaf node in the octree to the root is exactly represented.

2.2 The Data Encoding of Point Cloud

Point cloud data encoding steps are:

Determine the point cloud octree Hierarchical n, satisfy

$$d_0 * 2^n \geq d_{\max} \tag{2}$$

Where d0 is the reduced point of the specified point, dmax is the maximum edge of the point cloud bounding box. Determine the sub-cubes where the point cloud data points are located Code, assuming the data point P (x, y, z), where the sub-cube is empty. The index value is (i, j, k), and Q corresponds to the sub-cube. Node of the octree tree. The relationship between the three can be expressed by the formula (3) ~ (5) expressed in [6–8].

$$\begin{cases} i = \lfloor (x - x_{\min})/d_0 \rfloor \\ j = \lfloor (y - y_{\min})/d_0 \rfloor \\ k = \lfloor (z - z_{\min})/d_0 \rfloor \end{cases} \tag{3}$$

Where xmin, ymin, zmin represents the minimum vertex coordinate value of the bounding box corresponding to the root node, and [...] is the rounding operator. Converts the index value (i, j, k) to a binary representation.

$$\begin{cases} i = i_0 2^0 + i_1 2^1 + \ldots + i_m 2^m + \ldots + i_{n-1} 2^{n-1} \\ j = j_0 2^0 + j_1 2^1 + \ldots + j_m 2^m + \ldots + j_{n-1} 2^{n-1} \\ k = k_0 2^0 + k_1 2^1 + \ldots + k_m 2^m + \ldots + k_{n-1} 2^{n-1} \end{cases} \tag{4}$$

$$i_m, j_m, k_m \in \{0,1\}, \quad m \in \{0,1,\ldots,n-1\}$$
$$q_m = i_m + j_m 2^1 + k_m 2^2 \tag{5}$$

The sub-cube corresponding octree coding can be expressed by the formula (1). At the same time, if the sub-cube corresponding to the octet coding Q, can also reverse the data point where the sub-cube spatial index value (i, j, k).

$$\begin{cases} i = \sum_{m=1}^{n-1} (q_m \% 2) * 2^m \\ j = \sum_{m=1}^{n-1} \left([q_m/2]\%2\right) * 2^m \\ k = \sum_{m=1}^{n-1} \left([q_m/4]\%2\right) * 2^m \end{cases} \tag{6}$$

Where (qm% 2) represents the remainder of m q divided by 2, and qm is the result of dividing qm by 2.

When searching for the k nearest neighbor of a data point, it can be searched for the sub-cube where the data point is located and its surrounding 26 subcategories. If the index value of a sub-cube in space is (i, j, k), the index value of 26 sub-cubes around it can be represented by (i ± δ, j ± δ, k ± δ), δ ∈ {0,1}. Therefore, after the octagonal tree after the point cloud, only in the local k proximity search, significantly improved the search speed.

3 The Case of Applications

Based on the octagonal tree coding of the uniform stream of the algorithm shown in Fig. 3, in the UG software platform using C++ language secondary development of the

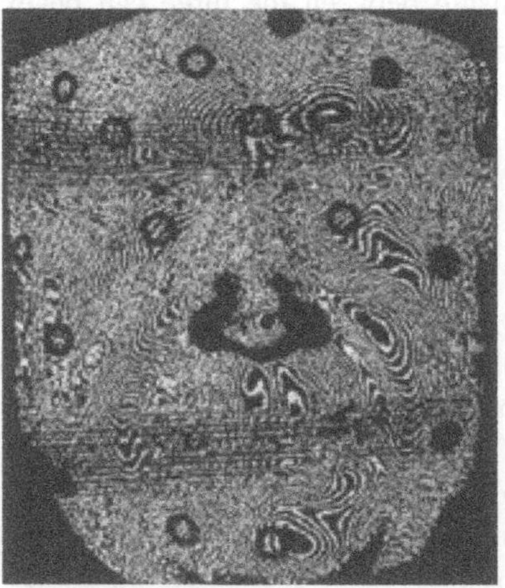

Fig. 3. Original image

point cloud data preprocessing module. And we process the point cloud data of human face. Figure 3 shows the original point cloud data, containing 24560 point cloud data. Figure 4 is the graph after streamlined. Figure 5 is the graph after Data refactoring, including the 19880 points cloud data and Fig. 6 is the reconstructed image with 22120 points cloud data.

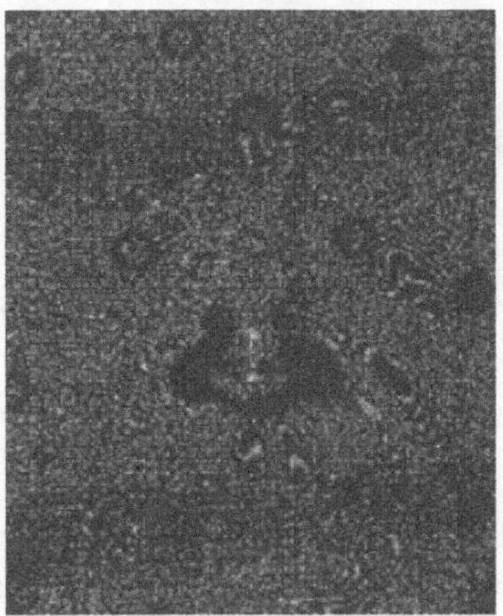

Fig. 4. Streamlined image

Fig. 5. Reconstructed image

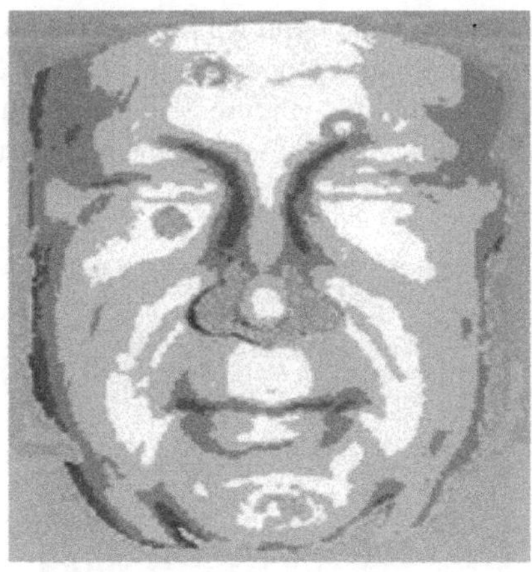

Fig. 6. Reconstructed image

4 Conclusions

The algorithm of point cloud proposed in this paper can make a quick and effective reduction of the massive data of reverse engineering measurement. The algorithm has the following characteristics:

(1) From the perspective of the overall space point of view of the uniform cloud of light, the effect is better.
(2) due to the reduction of the cloud before the cloud division, so the algorithm is simple and efficient.
(3) users through the set of simple dot pitch, you can easily achieve point cloud streamlining.

References

1. Zhang, Z., Gong, J., Zhang, Y.: An efficient 3D R-tree spatial index method for virtual geographic environments. ISPRS J. Photogram. Remote Sens. **62**, 217–224 (2002)
2. Zhang, Y.J., Liang, Y., Xu, G.: A robust 2-refinement algorithm in octree or rhombic dodecahedral tree based all-hexahedral mesh generation. Comput. Methods Appl. Mech. Eng. **256**, 88–100 (2010)
3. Xie, F., Zhao, J., Ju, F.: The point cloud collection of the incisor teeth of beaver and reconstruction of its curved surface. In: Materials Research, pp. 3–390 (2010)
4. Jia, A.K.: Data clustering: 50 years beyond Kmeans. Pattern Recogn. Lett. **31**(8), 651–666 (2009)
5. West, G., Bei, T., Niu, D.: ()utlier detection and robust normal curvature estimation in mobile laser scanning 3D point cloud data. Pattern Recogn. **48**(4), 140–1419 (2012)

6. Wang, Y., Hao, W., Ning, X.: Automatic segmentation of urban point clouds based OH the Gaussian map. T P Photogram. Rec. **28**(144), 342–361 (2013)
7. Mittle, S., Tu, Z.: Semi—supervised kernel mean shift clustering. IEEE Trans. Pattern Anal. Mach. Intell. **36**(6), 120–1215 (2012)
8. Shi, Q., Liang, J., Ao, Q.: Adaptive simplification of point cloud using k-means clustering. Comput. Aided Des. **43**(8), 910–922 (2012)

Seat Belt Detection Using Convolutional Neural Network BN-AlexNet

Bin Zhou[1,2(✉)], Dongfang Chen[1,2], and Xiaofeng Wang[1,2]

[1] College of Computer Science and Technology, Wuhan University of Science and Technology, Wuhan, China
johnbeanzb@163.com

[2] Hubei Province Key Laboratory of Intelligent Information Processing and Real-Time Industrial, Wuhan, China

Abstract. To tackle the problems of the dependence on source image clarity, the underutilization of source image information and the dependence on human-designed features in existing seat belt detection methods, a seat belt detection method using convolutional neural network (CNN) is proposed. In this paper, an improved convolutional neural network (called the BN-AlexNet) which adds the Batch Normalization (BN) module to the traditional convolutional neural network AlexNet is built to further enhance the classification ability of the convolutional neural network and greatly reduce the training difficulty. Later the confidence of detection results is analyzed, and the 95% confidence interval is used to set the rejection area. The result shows that the method achieves 92.51% correct detection rate by rejecting 6.50% test samples. Compared with the traditional methods based on image processing, the proposed method has higher correct detection rate.

Keywords: Seat belt detection · Convolutional Neural Network (CNN) · Batch Normalization (BN) · Confidence · Rejection area

1 Introduction

According to statistics, more than 1.2 million people die each year in vehicle accidents, including 40–50 percent people who are not wearing seat belts [1]. Traffic laws at home and abroad clearly stipulate that drivers and passengers in the moving vehicles should be required to wear the seat belt, or else they will be punished. This is an important measure to ensure driving safety and reduce the death rate of vehicle accidents. Recently, the camera on the road has taken the clear pictures in which people can see the clear view in the front row of the interior of the moving vehicles, and judge whether drivers and passengers are wearing seat belts. In the case of the daily generation of hundreds of thousands or even millions of pictures, the method of manual detection is inefficient, and the workload is heavy.

At present, the intelligent transportation systems at home and abroad have developed, and the module of the seat belt detection has been studied by many scholars. The main process is divided into two steps: the first step is to locate the position of the driver, in order to narrow the detection range. The driver's location is mainly based on

D.-S. Huang et al. (Eds.): ICIC 2017, Part I, LNCS 10361, pp. 384–395, 2017.
DOI: 10.1007/978-3-319-63309-1_36

the localization of license plate and windscreen. The license plate location algorithm has been very mature [2–4], and the windscreen detection also has related research [5]. The second step is to detect the seat belt. Most efforts are devoted to the study of edge detection [6, 7]. On this basis, some methods are to add the eigenvector extracted from the edge detection results to the machine learning (such as Support Vector Machine (SVM)) [8]. The above methods are based on the features extracted manually, which are easy to understand. But the common problems are lack of the overall consideration, and largely limited by the source image clarity, man-made rules and effective features, which result in the decrease of detection effect. Therefore, it's urgent to study new ideas.

In recent years, Convolutional Neural Network (CNN), as an efficient method in Deep Learning (DL), has been improving due to the rapid improvement of computer performance. Especially in the field of pattern recognition, CNN is widely used, because the input data is source images, and avoid the complex operation of image preprocessing.

However, the information of input image is limited to the driver being driving the vehicle. The first step is to locate the driver's position, and the positioning may be inaccurate. If force to input the inaccurate images to the model to test, the results are likely to be wrong. And because the training sample is less affected by light, occlusion, clothing stripes and so on than the actual test images, the actual detection accuracy will be much lower than the test value.

To solve the above problems, reduce the training difficulty and further improve the accuracy of seat belt detection, a combined algorithm of improved BN-AlexNet and mechanism with confidence levels is proposed. Firstly, the BN-AlexNet is constructed by adding BN module into the traditional CNN AlexNet, and then input the labeled images to train out a model. Secondly, the confidence mechanism is designed, according to the output distributions of input samples collected by using the Bootstrap method. Finally, the test samples are estimated by softmax function, of which the outputs are outside the rejection area. The experimental results show that compared with the traditional image processing method [9], the proposed method has higher accuracy.

2 A CNN BN-AlexNet for Seat Belt Detection

The CNN BN-AlexNet is developed by AlexNet. AlexNet is a network structure, designed by Alex and others [10]. The overall structure of AlexNet consists of eight weighted layers: the first five are convolutional layers, the last three are fully-connected layers, and the last one is connected to the softmax layer, resulting in a classification result. Among them are the activation layers control the range of parameter values, the pooling layers compress parameter quantity and the local response normalization layers prevent a large number of parameters from getting into the saturated interval. BN-AlexNet adds the Batch Normalization (BN) module to AlexNet. This module was first used in GoogLeNet [11], and its top-5 error rate was 6.67% on ImageNet, which achieved human image recognition. In this paper, we don't use GoogLeNet and other complex network structures to train, because they are hard to train.

2.1 Introduction to BN Algorithm

In the training of deep network, in order to overcome the problem of internal covariate shift and the saturated interval of Sigmoid activation function, Google engineers normalize the data $x = (x_1, x_2, \ldots, x_n)$ entered into the activation layer, that is, the mean is 0 and the standard deviation is 1. The process is as follows.

$$\mu_B \leftarrow \frac{1}{m}\sum_{i=1}^{m} x_i \tag{1}$$

$$\sigma_B^2 \leftarrow \frac{1}{m}\sum_{i=1}^{m} (x_i - \mu_B)^2 \tag{2}$$

$$\hat{x}_i \leftarrow \frac{x_i - \mu_B}{\sqrt{\sigma_B^2 + \varepsilon}} \tag{3}$$

where m is the number of a batch of data, μ_B is the mean of the data in same dimension, σ_B is the standard deviation of the data in same dimension, $\widehat{x}_i$ is the processed data in same dimension, and ε is added to prevent divisor of third formulas from being invalid.

However, this processing destroys the distribution of original data, so they come up with the learning parameters γ and β, and use the following formula to get the transformed data $y = (y_1, y_2, \ldots, y_n)$.

$$y_i = \gamma \hat{x}_i + \beta \tag{4}$$

The learning parameters γ and β are obtained by using the stochastic gradient descent algorithm, which makes the processed vector data y close to the input vector data x.

2.2 The BN Module in BN-AlexNet

The BN module in BN-AlexNet is implemented by BatchNorm layer and Scale layer. The BatchNorm layer is responsible for computing and the Scale layer is responsible for adding bias.

In the training phase of CNN, the data (type is BLOB) passing through the BN module is expressed as four-dimensional matrix (m, f, h, w), where m is the number of a batch of data, f is the number of the feature map of one convolutional layer, h is the height of feature map, w is the width of feature map. The way of learning parameters γ and β in the BN module being used is similar to the mechanism of sharing weight, that is, treating each feature map in one convolutional layer as a neuron. Therefore, each feature map in one convolutional layer has their own learning parameters γ and β. BN module is to calculate the mean and variance of all pixels in one feature map corresponding to a batch sample, and then normalize this feature map.

In the test phase of CNN, the mean E[x] and the standard deviation Var[x] in BN module are derived from the training μ_B and σ_B. Calculated as follows.

$$E[x] \leftarrow E_B[\mu_B] \tag{5}$$

$$Var[x] \leftarrow \frac{m}{m-1} E_B[\sigma_B^2] \tag{6}$$

where E[x] is the mean of the μ_B of all batches of training samples, and Var[x] is an unbiased estimator of the σ_B of all batches of training samples.

Finally, the original data x is converted to y with the corresponding learning parameters γ and β.

$$y = \frac{\gamma}{\sqrt{Var[x] + \varepsilon}} \cdot x + (\beta - \frac{\gamma E[x]}{\sqrt{Var[x] + \varepsilon}}) \tag{7}$$

In the paper [12], BN module is proposed to put in front of the activation layer, but that is just for the drawbacks of Sigmoid activation function. In the deep network with Sigmoid activation function, the training parameters enter into the saturated interval, causing the gradient to vanish, and there is almost no signal via the neurons to the weight and then to the data. In order to make the input data of the activation layer more concentrated in the unsaturated interval, the BN module is used to process the data. However, the ReLU activation function [13] doesn't have the problem of saturated interval, because the input of x < o is changed to 0 and the input of $x \geq 0$ doesn't change, it destroys the data distribution, resulting in the data distributed to one side. So, the BN module is contrary to the purpose of data normalization. Therefore, it's more suitable to apply the BN module behind the ReLU activation layer. In order to verify the correctness of this conclusion, a set of comparative experiments were carried out to compare the BN-AlexNet1 of the BN module in front of the activation layer with the BN-AlexNet2 of the BN module behind the activation layer.

The data set is ImageNet-2012, which contains about 1.2 million training images, 20 thousand validation images and 150 thousand test images, divided into 1000 different categories. In order to reduce the training time, the size of the input data set is reduced from 227 × 227 ruled by original AlexNet to 131 × 131, and fc6 and fc7 layers have 2048 neurons instead of 4096. The processed data set is input into two neural network structures for training, and the training results are shown in Fig. 1.

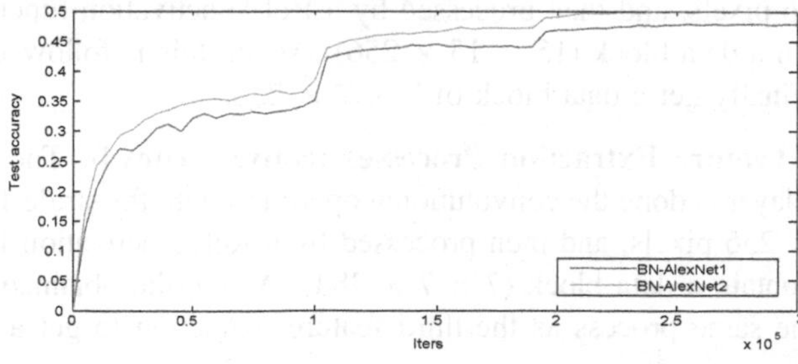

Fig. 1. Recognition accuracy of BN-AlexNet1 and BN-AlexNet2 during training

Figure 1 shows that the final recognition accuracy of BN-AlexNet2 is about 2% higher than that of BN-AlexNet1. So BN-AlexNet uses network structure BN-AlexNet2. The network structure is shown in Fig. 2. See the following training process for more details.

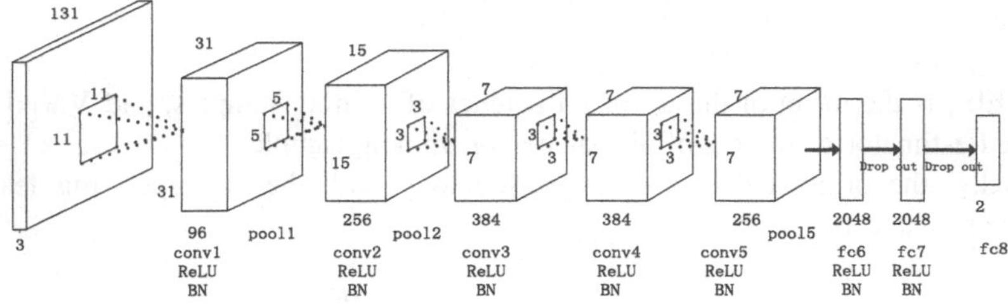

Fig. 2. The structure of BN-AlexNet. Conv1 ~ conv5 are convolutional layers, pool1 and pool2 are max-pooling layers, fc6 ~ fc8 are fully-connected layers.

2.3 The Construction and Configuration of BN-AlexNet

The neural network BN-AlexNet is constructed and configured by using Xavier [14] to initialize the network parameters, and set up the network structure as shown in Fig. 2. Each cuboid in the Fig. 2 represents a subunit that stores data. The configuration parameters of this network are as follows: the weight learning rate is 0.01, each training 1000 times the weight learning rate is reduced to 10 percent of original rate, the decay degree of weight learning rate is 0.005. Then put the training images with size of 131 × 131 pixels into the neural network in shuffled way.

2.4 The Forward Propagation Phase

The First Feature Extraction Process (conv1 ~ pool1). The input training image (131 × 131 × 3) is done the convolutional operation with the stride 4 by 96 filters of size 11 × 11 × 3 pixels and then processed by a ReLU activation function and a BN module to obtain a data block (31 × 31 × 96). And the data block is compressed into 15 × 15 × 96 by the pooling layer (kernel size 3 × 3, type Max, stride 2).

The Second Feature Extraction Process (conv2 ~ pool2). The 15 × 15 × 96 output of pool1 layer is done the convolutional operation with the stride 1 by 256 filters of size 5 × 5 × 96 pixels, and then processed by a ReLU activation function and a BN module to obtain a data block (15 × 15 × 256). Again, this is followed by the same pooling layer. Finally get a data block of 7 × 7 × 256.

The Next two Feature Extraction Processes (conv3 ~ conv4). The 7 × 7 × 256 output of pool2 layer is done the convolutional operation with the stride 1 by 384 filters of size 3 × 3 × 256 pixels, and then processed by a ReLU activation function and a BN module to obtain a data block (7 × 7 × 384). Again, the obtained data block is processed by the same process as the third feature extraction to get a data block of 7 × 7 × 384.

The Fifth Feature Extraction Process (conv5 ~ pool5). The 7 × 7 × 384 output of conv4 layer is done the convolutional operation with the stride 1 by 256 filters of size 3 × 3 × 384 pixels, and then processed by a ReLU activation function and a BN module to obtain a data block (7 × 7 × 256). Again, this is followed by the same pooling layer. Finally get a data block of 3 × 3 × 256.

The First And Second Fully-Connected Layers (fc6 ~ drop7). The 3 × 3 × 256 output of conv5 layer is done the fully- connected operation by 2048 neurons, and then processed by a ReLU activation function and a BN module to obtain a data block (1 × 1 × 2048). The following drop6 layer is used to reduce the number of training parameters and prevent over-fitting, so that some neurons in the fc6 layer are not involved in the connection, and in which the threshold is set to 0.3. The second fully-connected layer process is the same as the first fully-connected layer. After the process of second fully-connected layer, the data block of 1 × 1 × 2048 is obtained.

The Third Fully-Connected Layer (fc8). The 1 × 1 × 2048 output of drop7 layer is done the fully- connected operation by 2 neurons to obtain a data block (1 × 1 × 2).

The Classifier (loss). This layer consists of the Softmax layer and MultinomialLogisticLoss layer. The Softmax layer is used to calculate the probability of each category, and the MultinomialLogisticLoss layer is used to calculate the loss of estimated and expected. During the testing phase, only the Softmax layer works. For data block of 1 × 1 × 2, the probability of two categories and the loss of estimated and expected are calculated.

2.5 The Back Propagation Phase

Using the common back propagation (BP) algorithm [15] to update the weight.

2.6 The Testing Phase

With repeating the previous two phases and constantly updating the weight, the trained network model is obtained. The input test image is processed by the forward propagation phase, and is eventually determined which category it belongs to according to the maximum value of output probability.

3 The Confidence Mechanism of CNN

In order to improve the detection accuracy and reduce the false accept rate, the method of rejecting some samples to detect is proposed. In the actual test, the classification effect of CNN is ineffective for the images outside the training sample. Therefore, it's necessary to estimate the data distribution outside the training sample by analyzing the confidence level of the CNN output with statistical inference method, and set the optimal rejection area of the sample.

The analysis of confidence measures the reliability of the whole algorithm. In order to obtain the data distribution outside the training sample, this paper adopts the Bootstrap method proposed by statistician Efron [16]. As for the reliability of this method, the statistician Hall gives a sound proof [17]. The Bootstrap method is an ingenious method of estimating confidence interval using the single data, which improves the efficiency of statistical inference in scientific research.

The basic method: $X = (X_1, X_2, \ldots, X_n)$ is a sample from the total sample, of which the capacity is n. The Bootstrap sample $X^* = (X_1^*, X_2^*, \ldots, X_n^*)$ is obtained by resampling n sample data from the sample X. Successively and independently take B (normally 1000) Bootstrap samples of capacity n from the sample data X, and calculate the Bootstrap estimation $\left[\hat{\alpha}_i^*, \hat{\beta}_i^*\right]$ of 95% confidence interval $[\alpha, \beta]$ in the ith (i = 1,2,..., B) Bootstrap sample. See below for more details.

Firstly, the Bootstrap sample $X_i^* = \left(X_{1^i}^*, X_{2^i}^*, \ldots, X_{n^i}^*\right)$ are arranged from small to large, get

$$X_{(1)^i}^* \le X_{(2)^i}^* \le \cdots \le X_{(n)^i}^* \tag{8}$$

Then to compute the approximate quantile $X_{5\%/2^i}^*$ and $X_{1-5\%/2^i}^*$ of X_i^* according to

$$P\{X_{5\%/2^i}^* \le X_i^* \le X_{1-5\%/2^i}^*\} = 1 - 5\% \tag{9}$$

Designate $k_1 = \left[n \times \frac{5\%}{2}\right]$, $k_2 = \left[n \times \left(1 - \frac{5\%}{2}\right)\right]$. In the formula 9, $X_{(k_1)^i}^*$ and $X_{(k_2)^i}^*$ are respectively used as the estimations of quantile $X_{5\%/2^i}^*$ and $X_{1-5\%/2^i}^*$, and get the following approximate equation.

$$P\{X_{(k_1)^i}^* \le X_i^* \le X_{(k_2)^i}^*\} = 1 - 5\% \tag{10}$$

So get $\hat{\alpha}_i^* = X_{(k_1)^i}^*, \hat{\beta}_i^* X_{(k_2)^i}^*$. Assume that the 95% confidence interval of the total sample F is $\left[\hat{\alpha}, \hat{\beta}\right]$, and is approximated by the average confidence interval $\left[\hat{\alpha}^*, \hat{\beta}^*\right]$ of B Bootstrap samples, as the formulas below.

$$\hat{\alpha} = \hat{\alpha}^* = \frac{\hat{\alpha}_1^* + \hat{\alpha}_1^* + \cdots + \hat{\alpha}_B^*}{B} \tag{11}$$

$$\hat{\beta} = \hat{\beta}^* = \frac{\hat{\beta}_1^* + \hat{\beta}_2^* + \cdots + \hat{\beta}_B^*}{B} \tag{12}$$

The rejection area is set outside the 95% confidence interval $\left[\hat{\alpha}, \hat{\beta}\right]$.

After the test image is processed by the trained network, each category outputs a value P, the sum of P is equal to 1. Which category has the highest value, indicating that the test image belongs to this category. The following is the sampling process for training samples.

After the training sample of wearing seat belts is resampled with replacement method, a Bootstrap sample with capacity 1000 is obtained. And repeat 1000 times to get 1000 Bootstrap samples. The 1000 Bootstrap samples are entered into the convolutional neural network. The P-value distribution of one Bootstrap samples is shown in Fig. 3. According to the method mentioned above, and use the P-value distribution to get the 95% confidence interval [0.5891, 0.9423], that is, the rejection area is [0.5, 0.5891) and (0.9423, 1).

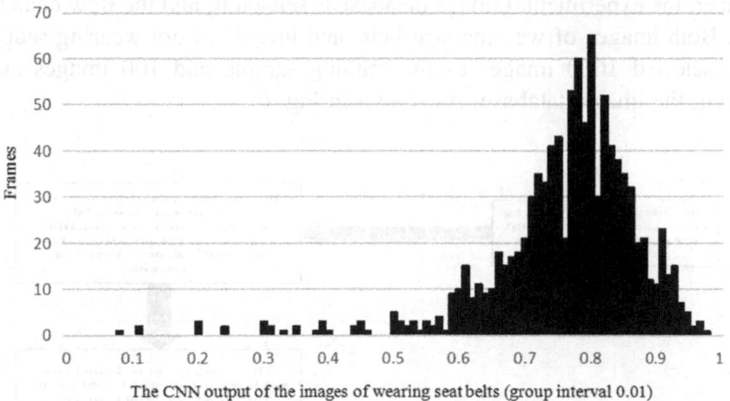

Fig. 3. P-value of one Bootstrap samples (wearing seat belts)

As for the training sample of not wearing seat belts, do the same operation as training sample of wearing seat belts. The P-value distribution of one Bootstrap samples is shown in Fig. 4. The 95% confidence interval is [0.6051, 0.9532] and the rejection area is [0.5, 0.6051) and (0.9532, 1).

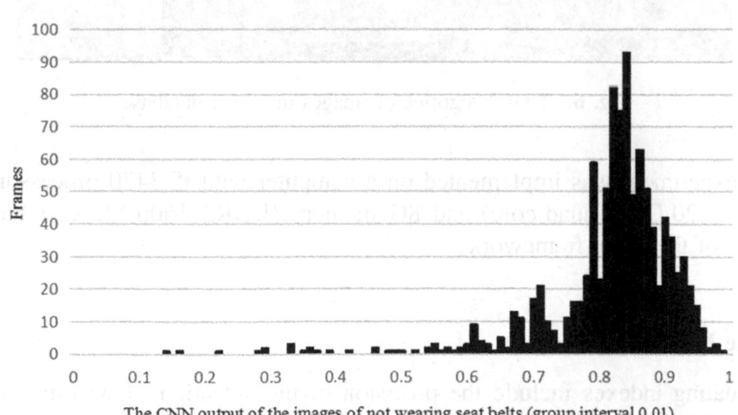

Fig. 4. P-value of one Bootstrap samples (not wearing seat belts)

For a test image, let the value P of wearing seat belts be P1, and the value P of wearing seat belts be P2, as long as one of P1 an P2 is in the rejection area, it is determined to be the rejection image.

4 Experiments and Analysis

4.1 The Dataset

In this paper, the experimental image database is self-built, and the flow chart is shown in Fig. 5. Both images of wearing seat belts and images of not wearing seat belts are randomly selected 1000 images as the training sample and 100 images as the test sample from the image database. As shown in Fig. 6.

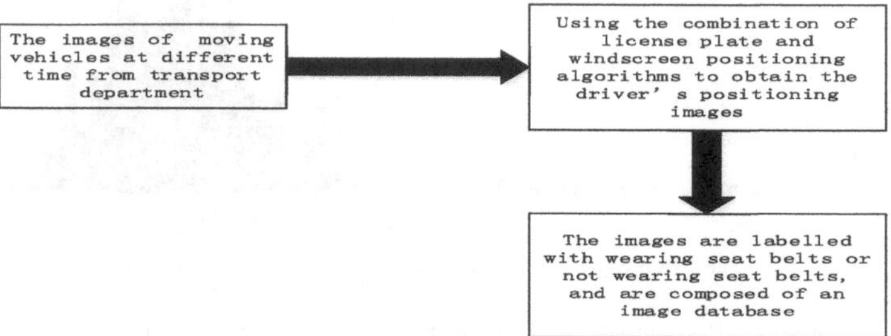

Fig. 5. The flow chart of data collection

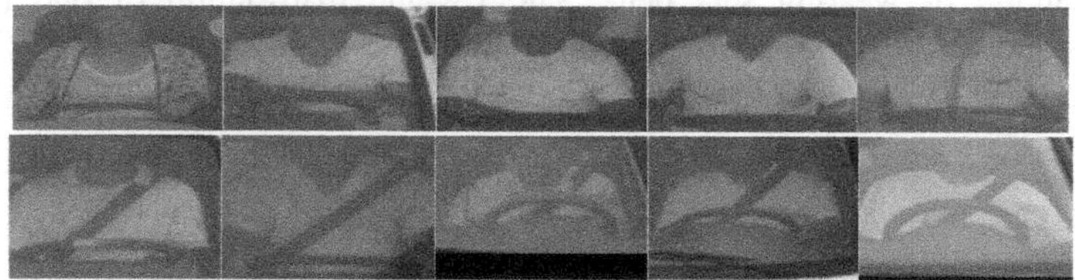

Fig. 6. Two categories of images in image database

This experiment was implemented on a computer with i5-3470 processors (main frequency 3.20 GHz, quad-core) and 8G memory (DDR3 1600 MHz), which takes advantage of the Caffe framework.

4.2 The Evaluating Indexes

The evaluating indexes include the precision of the detection of wearing seat belts (Prec_Y), the precision of the detection of not wearing seat belts (Prec_N), the precision of all samples (Prec_T) and the rejection rate (RR):

$$\Pr ec\_Y = \frac{TY}{FN + TY} \tag{13}$$

$$\Pr ec\_N = \frac{TN}{TN + FY} \tag{14}$$

$$\Pr ec\_T = \frac{TY + TN}{FN + TY + TN + FY} \tag{15}$$

$$RR = \frac{RY + RN}{FN + TY + RY + TN + FY + RN} \tag{16}$$

where TY is the number of the images of wearing seat belts predicted to be the image of wearing seat belts by network, FN is the number of the images of wearing seat belts predicted to be the image of not wearing seat belts by network, TN is the number of the images of not wearing seat belts predicted to be the image of not wearing seat belts by network, FY is the number of the images of not wearing seat belts predicted to be the image of wearing seat belts by network, RY is the number of the images of wearing seat belts in the rejection area, RY is the number of the images of not wearing seat belts in the rejection area.

4.3 The Experimental Result

In order to verify the effectiveness of the network structure designed in this paper, four kinds of network structures are selected to do the seat belt detection: (1) Select AlexNet. (2) Select VGGNet-16 [18]. (3) Select GoogLeNet (4) Select BN-AlexNet. In order to verify the effectiveness of the confidence mechanism designed in this paper, the confidence mechanism is added to the four network structures. Finally, the method in this paper compares with the traditional method of extracting feature manually. The results are shown in Tables 1, 2 and 3.

Table 1. The influence of network structure

Network structure	Prec_Y/%	Prec_N/%	Prec_T/%	Training time/h
AlexNet	84.00	83.00	83.50	8.67
VGGNet-16	92.00	88.00	90.00	14.67
GoogLeNet	91.00	88.00	89.50	6.16
BN-AlexNet	90.00	86.00	88.00	3.03

Table 2. The influence of confidence mechanism

Method	Prec_Y/%	Prec_N/%	Prec_T/%	RR/%
AlexNet + Confidence mechanism	90.11	86.02	88.04	8.00
VGGNet-16 + Confidence mechanism	93.62	90.43	92.03	6.00
GoogLeNe + Confidence mechanism	92.71	91.40	92.06	5.50
BN-AlexNet + confidence mechanism	95.70	89.36	92.51	6.50

Table 3. The comparison of different methods

Method	Prec_Y/%	Prec_N/%	Prec_T/%	RR/%
The traditional image processing method [9]	81.00	76.00	78.50	
Based on Hough transform + DL(SVM) [8]	83.00	81.00	82.00	
The proposed method	95.70	89.36	92.51	6.50

As shown in Table 1, the accuracy of the improved network structure BN-AlexNet is about 5 per cent higher than that of original AlexNet, and the time spent on training is greatly reduced. And two other network structures are a little higher than BN-AlexNet, because they have deeper layers. Of course, they need more time to train. Compare GoogLeNet with VGGNet-16, the training time of GoogLeNet is much shorter than that of VGGNet-16, because the BN module in GoogLeNet plays an important role.

According to Table 2, by rejecting to detect a number of special images, the accuracy of detection is increased, and the improved BN-AlexNet with confidence mechanism reached 92.51% detection rate by rejecting the 6.50% of test sample. The result of other two network structures with Confidence mechanism are similar to that of BN-AlexNet.

In Table 3, compared with the traditional method, the accuracy of this method is improved, but the rejection rate is added. If not considered the confidence mechanism, and combined with Table 1, the accuracy of seat belt detection using CNN is higher than that of traditional methods.

Overall, BN-AleNet with Confidence mechanism is more suitable for seat belt detection, because it is easy to train and has high detection rate.

5 Conclusion and Future Work

In this paper, based on the traditional network structure AlexNet, the network structure BN-AlexNet for seat belt detection is proposed. By adding the BN module, the network makes the training parameter distribution more concentrated, and gets better classification ability. Due to the significant improvement of BN-AlexNet training speed, the network structure can be used to train on the computer without GPU. At the same time, by studying the confidence of pattern recognition, the Bootstrap is applied to the convolutional neural network BN-AlexNet to evaluate the confidence of the network output and set the rejection area. By rejecting a part of sample, the false accept rate of the model is reduced and the classification ability is enhanced. Because the performance of this method depends on the training data and the convolutional neural network structure, in the future work, we will improve the detection ability by collecting more available data sets and improving the network structure on the basis of this method.

Acknowledgement. This work is supported by National Natural Science Foundation of China, NO. 61273225 and NO. 61572381.

References

1. Toroyan, T.: Global status report on road safety. Inj. Prev. **15**(4), 286 (2009)
2. Zhang, T., Luo, X., Zhu, X.: License plate location based on singular value feature. In: IEEE International Conference on Computer Science and Information Technology, pp. 283–287 (2010)
3. Su, J., Ma, Z.: Car license plate location based on the density and projection. In: International Conference on Computational Intelligence and Natural Computing (CINC), pp. 409–412 (2009)
4. Chen, B., Cao, W., Zhang, H.: An efficient algorithm on vehicle license plate location. In: IEEE International Conference on Automation and Logistics (ICAL), pp. 1386–1389 (2008)
5. Hou, D.: Study on Vehicle Window Detection Technology. Beijing Jiaotong University (2011)
6. Agaian, S., Almuntashri, A.: Noise-resilient edge detection algorithm for brain MRI images. In: 31st Annual International Conference of the IEEE EMBS, Minneapolis, pp. 3689–3692 (2009)
7. Srivastava, G., Verma, R., Mahrishi, R., Rajesh, S.: A novel wavelet edge detection algorithm for noisy images. In: International Conference on Ultra Modern Telecommunications & Workshops ICUMT 2009, pp. 1–8 (2009)
8. Xie, T., Ping-An, M., Dai, S., et al.: Application of an improved method for extracting safety belt image feature. Inf. Technol. (2015)
9. Guo, H., Lin, H., Zhang, S., et al.: Image-based seat belt detection. In: IEEE International Conference on Vehicular Electronics and Safety, pp. 161–164 (2011)
10. Krizhevsky, A., Sutskever, I., Hinton, G.: ImageNet classification with deep convolutional neural networks. In: International Conference on Neural Information Processing Systems, pp. 1097–1105. Curran Associates Inc. (2012)
11. Szegedy, C., Liu, W., Jia, Y., et al.: Going deeper with convolutions. pp. 1–9 (2014)
12. Ioffe, S., Szegedy, C.: Batch Normalization: Accelerating Deep Network Training by Reducing Internal Covariate Shift. Computer Science (2015)
13. Nair, V., Hinton, G.: Rectified linear units improve restricted Boltzmann machines. In: International Conference on Machine Learning, pp. 807–814 (2010)
14. Glorot, X., Bengio, Y.: Understanding the difficulty of training deep feedforward neural networks. J. Mach. Learn. Res. **9**, 249–256 (2010)
15. Bouvrie, J.: Notes on convolutional neural networks. Neural Nets (2006)
16. Efron, B.: The Jackknife, the bootstrap and other resampling plans. J. Am. Stat. Assoc. **78** (384), 316–331 (1982)
17. Hall. P.: The bootstrap and edgeworth expansion. Math. Gaz. **88**(421) (1992)
18. Simonyan, K., Zisserman, A.: Very Deep Convolutional Networks for Large-Scale Image Recognition. Computer Science (2014)

A Study on Lung Image Retrieval Based on the Vocabulary Tree

Kun Liu[1,2(✉)], Qing Chen[2], and Kun Ma[1,2]

[1] Shandong Provincial Key Laboratory of Network Based Intelligent Computing, University of Jinan, Jinan, People's Republic of China
ise_liuk@ujn.edu.cn

[2] School of Information Science and Engineering, University of Jinan, Jinan, People's Republic of China

Abstract. Nowadays, the image retrieval technology has aroused the wide concern and achieved the significant effect in the area of medical image retrieval. However, the traditional image retrieval method based on the image bottom-layer features is inefficient. With the increase in the volume of retrieval data, its shortcoming has become obvious. To promote the accuracy of retrieval, Scale-invariant feature transform (SIFT) lung feature extraction algorithm was chosen as the descriptor in this study. For such case, the method of vocabulary tree was introduced to extract the recognition features of lung image, as well as the medical retrieval of lung images using the obtained features. In the development environment of MATLAB, the reading, storage and retrieval of medical images were performed to create a full set of algorithmic retrieval system finally. The vocabulary tree-based retrieval method and the medical image processing were employed to denoise the images, SIFT extraction to extract the image features, K-means clustering that mapped to the feature space to turn the extracted features into the vocabulary tree. According to the experimental results, the algorithm proposed in this study is able to achieve good results.

Keywords: Image processing · Vocabulary tree · SIFT extraction · K-means clustering · Lung image retrieval

1 Introduction

With the rapid development of computer technology, it is required to achieve the automatic retrieval of images in the database in the medical field and the image retrieval based on the computer can thus be used to promote the efficiency and accuracy. Presently, there are mainly two methods for the medical image retrieval: one is the language of image and the other one is the visual characteristics of image. The image retrieval belongs to the matching of same contents. It is matched if the similarity is high, otherwise the matching is failed. Such retrieval can reduce the human intervention. Meanwhile, the retrieval technology is also constantly improved to meet the demands of customers.

Accordingly, in the process of study, the general and specific characteristics of different lung images should be considered comprehensively. Many scholars at home and abroad have proposed the new algorithm for the medical retrieval of lung images.

D.-S. Huang et al. (Eds.): ICIC 2017, Part I, LNCS 10361, pp. 396–407, 2017.
DOI: 10.1007/978-3-319-63309-1_37

For instance, CART decision tree was employed to train the feature vectors of lung image and ADABOOST learning algorithm was used for the retrieval of medical image. Besides, a method that integrated the one-dimensional Otsu method and two-dimensional variance method could be adopted for the processing of pulmonary parenchyma to obtain the region of interest for the lung CT images. Such method used the classifier for the classification of features.

In this study, the method of vocabulary tree was employed in the retrieval of medical images on windows platform, on which MATLAB was involved to realize the retrieval. The main works were as follows: first of all, different methods of image processing, as well as SITF algorithm, were chosen to extract the features of lung images and the obtained feature images were displayed in MATLAB; then K-means clustering was proceeded for the obtained feature vectors of lung images to generate the vocabulary tree and analyze and study the features of obtained lung image using SIFT. Besides, it was clustered in the form of vocabulary tree based on K-means and the distance was used to deal with the similarity measure of vocabulary tree. Finally, the experimental data of lung image retrieval was summarized and improved. Section 3 referred to the theoretical study on the medical image retrieval based on the vocabulary tree.

2 Related Work

2.1 Vocabulary Tree

Introduction to the Vocabulary Tree. The vocabulary tree was firstly proposed by John J. Lee etc. of Massachusetts Institute of Technology (MIT) [1]. The data structure such as the vocabulary tree has the good adaptability and expandability in the field of image retrieval [2]. Presently, the tree algorithm of large-scale image retrieval has been widely applied. On the basis of feature extraction, the step of image clustering is added. In the process of clustering, many layers of cluster are generated. Accordingly, the index structure is then employed to produce the data structure that is similar to the tree structure. In this way, it can not only promote the time efficiency of lung image retrieval, but also guarantee the quality of retrieval.

In the practical application, the visual vocabulary tree is some kind of clustering algorithm through the unsupervised learning to achieve the set of local features. K-means clustering at the different levels corresponds to the different visual vocabulary tree. The vocabulary tree that consists of visual words is an efficient data structure for the image retrieval. In the situation with plenty of words, we do not make use of all visual words to search the matching images, but allow the searching of visual words in the linear time. Similarly, with the constant addition of new images in the database, the scale of database will be increased, so will the vocabulary tree, in order to adapt to the image retrieval under the new circumstances. The vocabulary tree is thus adaptive. After the creation of visual words, they will be regarded as the leaf nodes of vocabulary tree, which can greatly promote the efficiency of retrieval.

K-means Clustering. K-means is the function to optimize the target, which means the distance from the image feature point to the center of circle. K-means clustering is the algorithm to minimize the function under the constraint condition, as the algorithm with the continuous iterative operation [3]. Such algorithm is measured by the Euclidean distance. The clustering corresponds to the vector of some initial cluster center to get such optimal results.

K-means algorithm has two disadvantages that are related to the setting of initial value. Where, K value is preset, but its selection is hard to measure. In most cases, it is unable to know how many centers are suitable for the given sample set. K-means algorithm uses the random points; while the different random points will lead to the different results, which may easily result in the local optimal solution. But K-means clustering is really fast.

Construction of Vocabulary Tree. The visual vocabulary tree belongs to the data architecture of image query by words. The larger quantity of visual vocabulary tree, the bigger advantage of such tree structure than other methods [4]. In the process of query by the vocabulary tree, there is no need to query all images, which can thus improve the efficiency [5].

For SIFT extraction of training sample images, all images can be extracted from the database to finish K-means clustering in the hierarchical form. The main steps are as follows:

(1) Put the library of original features at the first layer to finish K-means clustering, which can obtain the clustering center of k clusters;
(2) Put those k centers at the next layer of vocabulary tree to finish K-means clustering;
(3) Repeat step 2 till meeting the demand of the Lth layer.

Two parameters can be confirmed in the vocabulary tree: one is the length of vocabulary tree L; the other one is the degree of tree k. In K-means clustering, k means the number of clusters.

To form the visual vocabulary tree, it is required to extract SIFT feature points of image to create SIFT feature vector and then K-means clustering is implemented to obtain the clustering center of each cluster [6]. One visual word is one clustering center. After L times of clustering, the visual vocabulary tree can be achieved. The flow chart of vocabulary tree is shown in Fig. 1:

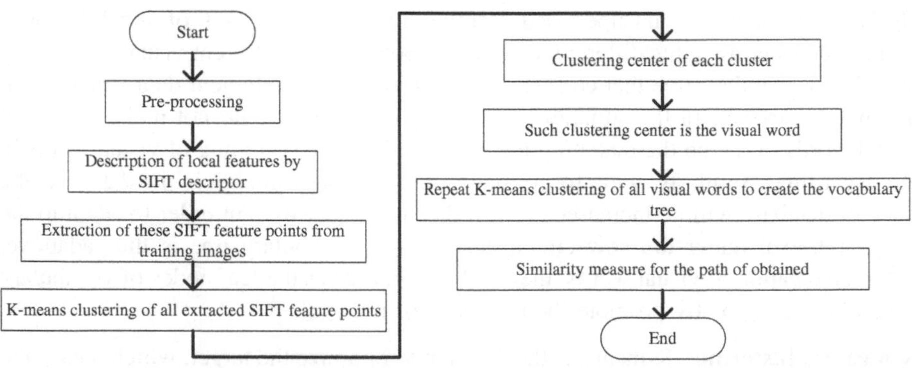

Fig. 1. Flow chart of vocabulary tree

2.2 Theoretical Studies on Similarity Measure

A visual word is a node in the tree structure. The degree of correlation between the image and vocabulary tree is expressed by defining a weight. The similarity measure module is employed to calculate the similarity between the feature vector of training sample images and the one of testing sample images. The experimental results were thus obtained according to the degree of similarity. By comparing the similarity with images, the images that meet the user's requirements can be acquired.

After the construction of vocabulary tree, the similarity measure is then executed for SIFT feature vector that is based on the visual vocabulary tree. SIFT features extracted from the training samples are quantified into the vocabulary tree. At the first layer of vocabulary tree, the dot product operation is implemented on K clustering centers. The clustering center that is closest to these points is found and iterated layer by layer till the leaf nodes of vocabulary tree. The length of vocabulary tree is L. For each piece of image, KL times of dot product operation are required.

By comparing SIFT feature vector of the image and the feature vector of testing sample in the database and searching the degree of similarity in the vocabulary tree from the top to the bottom, the degree of similarity between two images is thus obtained.

According to TD-TDF method, the feature vector of query image A and database image B is expressed by the weight vector to get a normalized correlation coefficient, as shown in Eq. (1):

$$s(A, B) = \left\| \frac{A}{\|A\|} - \frac{B}{\|B\|} \right\| \tag{1}$$

Generally, it is required to conduct the initial processing of lung images, such as the assignment operation on the nodes of vocabulary tree and normalization of images from the lung image sample library. With the norm of LP, the distance between the sample images is defined as Eq. (2):

$$\begin{aligned} s(A, B) &= \|A - B\|_p^p = \sum_i |A^i - B^i|^p \\ &= \sum_{i \mid B^i = 0} |A^i|^p + \sum_{i \mid A^i = 0} |B^i|^p + \sum_{i \mid A^i != 0, B^i != 0} |A^i - B^i|^p \\ &= \|A\|_p^p + \|B\|_p^p + \sum_{i \mid A^i != 0, B^i != 0} \left(|A^i - B^i|^p - |A^i|^p - |B^i|^p \right) \\ &= 2 + \sum\nolimits_{i \mid A^i != 0, B^i != 0} \left(|A^i - B^i|^p - |A^i|^p - |B^i|^p \right) \end{aligned} \tag{2}$$

In general, L2-norm can be used to convert Eq. (1) into Eq. (2):

$$s(A, B) = \|A - B\|_p^p = \|A - B\|_2^2 = 2 + 2 \sum\nolimits_{i \mid A^i != 0, B^i != 0} A^i * B^i \tag{3}$$

After the calculation of similarity measure of lung images, the retrieval efficiency can be greatly improved.

3 Lung Image Retrieval Based on Vocabulary Tree

The lung CT images are obtained after the study on the image retrieval algorithm based on the vocabulary tree.

3.1 System Development Environment

Development Tool: MATLAB R2014b

Web Server: 202.194.64.24

Hardware Requirements: recommended Xeon processor, 16G and above memory and 500G and above hard disc.

3.2 System Design

System Flow. It is the lung CT retrieval system based on the vocabulary tree, which integrates the global feature retrieval and single feature retrieval. The flow chart of system is shown in Fig. 2:

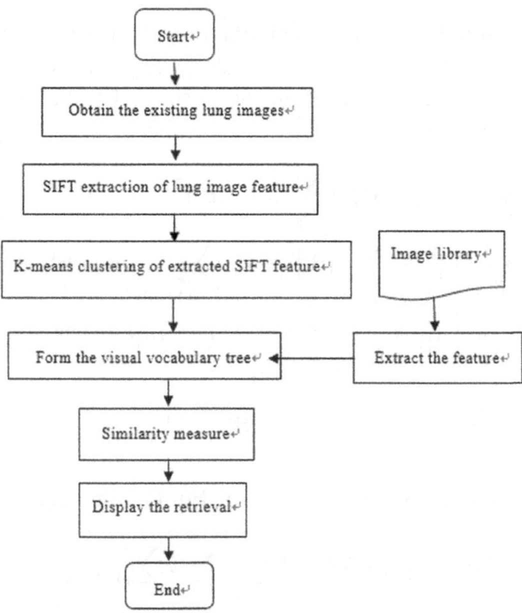

Fig. 2. System flow chart

Total Design of the System. The design is to achieve the retrieval of lung CT images. The extraction of feature points can be fulfilled using SIFT or SURF algorithm. Afterwards, the special relationship between the image feature points and the partition edge of images is used for the clustering operation. By comparing the feature vector between two images (the lung image and the testing one), the similarity can thus be obtained. The flow chart of system is shown in Fig. 3:

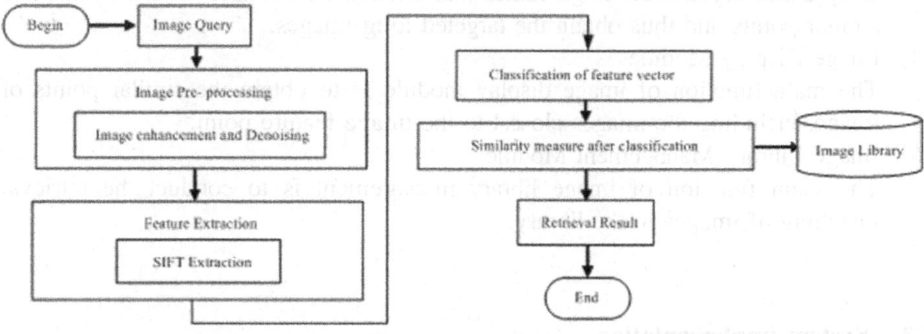

Fig. 3. System flow chart

System Module Design. This system is designed to build the application environment of lung image retrieval, mainly includes the modules of lung image query, feature extraction, similarity matching, image display and image library management. The system module design is shown in Fig. 4:

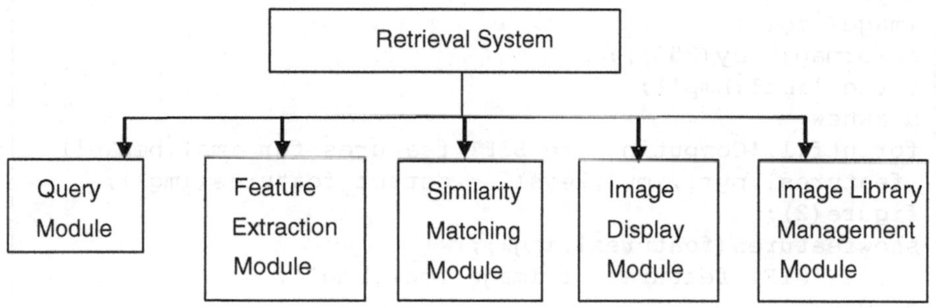

Fig. 4. System modules

(1) Query Module
The main function of query module is to provide the user with the all-round way of retrieval, namely implementing the retrieval of lung images and meeting various demands of the user.

(2) Feature Extraction Module
It mainly consists of SIFT extraction and description algorithm. The feature vector of SIFT feature points after the extraction of feature contributes to the further operation.
(3) Similarity Matching Module
The main function of similarity matching module is to complete the similarity matching between the description of training sample images and the one of testing sample images, in order to guarantee that the matched images have the same and similar points and thus obtain the targeted lung images.
(4) Image Display Module
The main function of image display module is to obtain the similar points of image, including the images closest to the image feature point.
(5) Image Library Management Module
The main function of image library management is to conduct the retrieval matching of images in the library.

3.3 System Implementation

Implementation of SIFT Extraction. The system is developed in MATLAB programming environment. the example contains many pieces of lung image; SIFT feature is extracted from many lung images; and script_eample.m is an executable file, taking 3 pieces for an example. The core code is as follows (Figs. 5, 6, 7, 9 and Table 1):

```
img1 = imread('ima1.bmp');
figure(1);
image(img1);
colormap(gray(256));
title('ima1.bmp');
drawnow;
fprintf(1,'Computing the SIFT features for ima1.bmp\n')
[features1,pyr1,imp1,keys1] = detect_features(img1);
figure(2);
showfeatures(features1,img1);
title('SIFT features of image ima1.bmp');
drawnow;
```

The demonstration result is shown as follows:

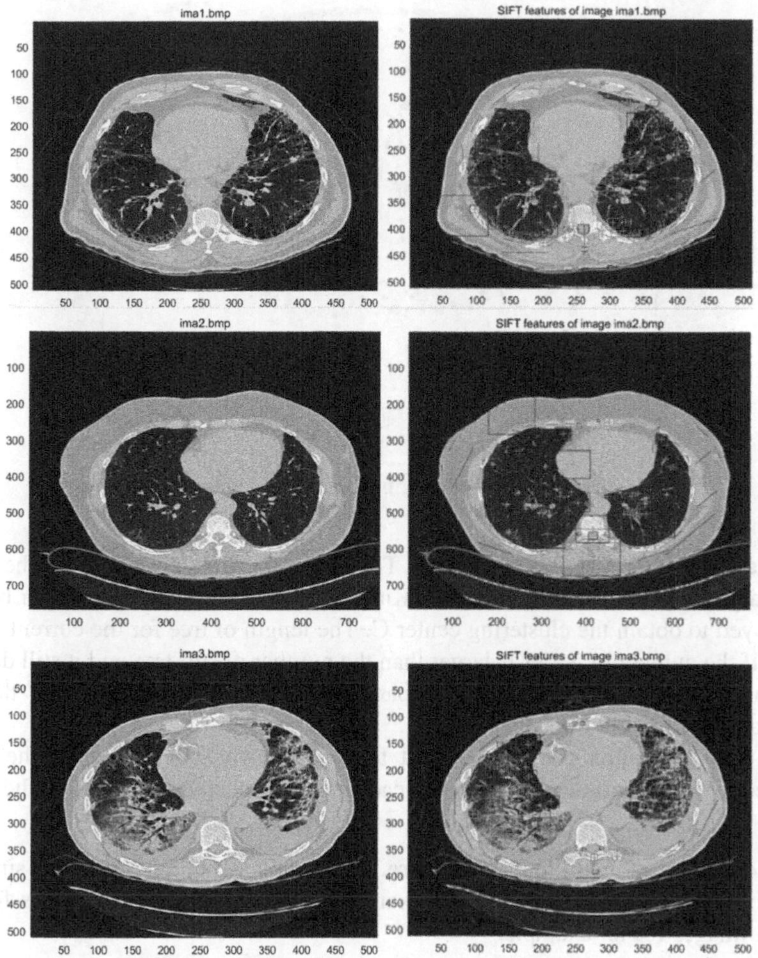

Fig. 5. SIFT extraction of lung image feature

Implementation of K-means Clustering. The implementation code of K-means clustering is written in the file of script_eample.m. The packed K-means clustering function is recalled from matlab tools and the Euclidean distance is employed. After 5 repeats of operation, the graph of relation between A(idx == 1,1) and A(idx == 1,2) is drawn, where 'r.' means the line type that uses the red '.' and the scale of 12 as the function curve; the graph of relation between A(idx == 2,1) and A(idx == 2,2) is drawn, where 'b.' means the line type that uses the blue '.' and the scale of 12 as the function curve; the graph of relation between two clustering centers is drawn, where 'kx' means the line type that uses the black 'x' and the scale of 12 as the function curve. The line width is 2. Finally, the annotation for the function curve is added on the top left of image, with the demonstration result shown as follows:

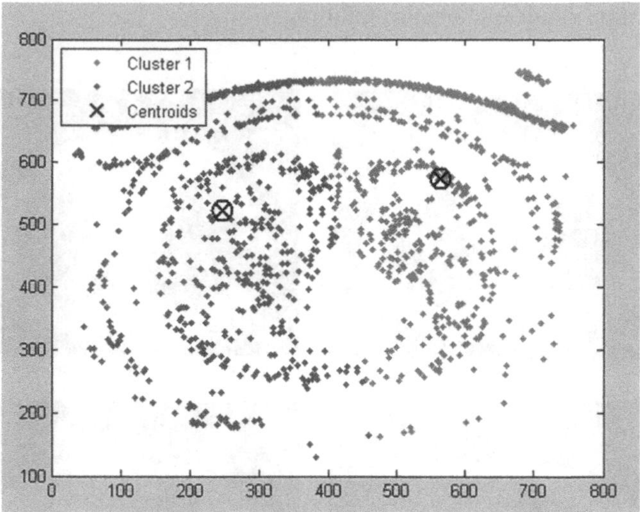

Fig. 6. Feature vector of lung image using K-means Clustering

Implementation of Vocabulary Tree. The code for the clustering in the visual vocabulary tree is written in the file of HiKmeans.m. K-means clustering layer by layer is employed to obtain the clustering center C. The length of tree for the current node is judged. If the number of nodes is larger than the number of sub-tree and it still does not reach the maximal depth of tree, the partition of tree is continued and then the vocabulary tree is built after the iteration.

The implementation codes to weight the path of vocabulary tree in the file of path2Vec.m, which gets the path vector of each image respectively and assign the weight as well.

Implementation of Similarity Measure. The implementation code of similarity measure is written in the file of script_example.m. The details are shown as follows, taking 3 images for an example:

```
imv1=sum(v1,2)/sum(v1(:));
imv2=sum(v2,2)/sum(v2(:));
imv3=sum(v3,2)/sum(v3(:));
figure(41);
bar(imv1);
figure(42);
bar(imv2);
figure(43);
bar(imv3);
```

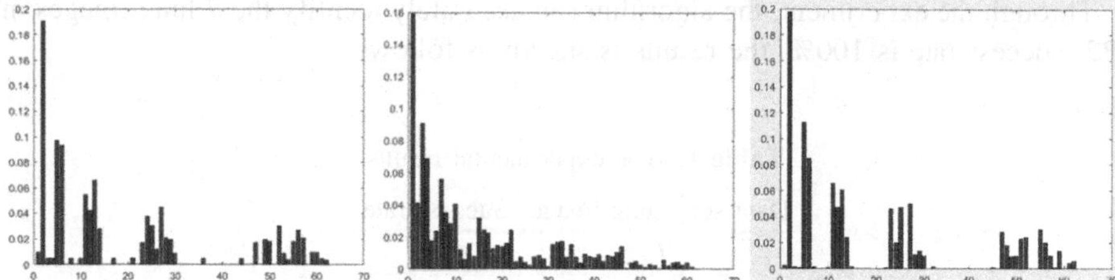

Fig. 7. Histogram of feature vector

The demonstration results are shown as follows:

The results of lung image retrieval are shown in Fig. 8:

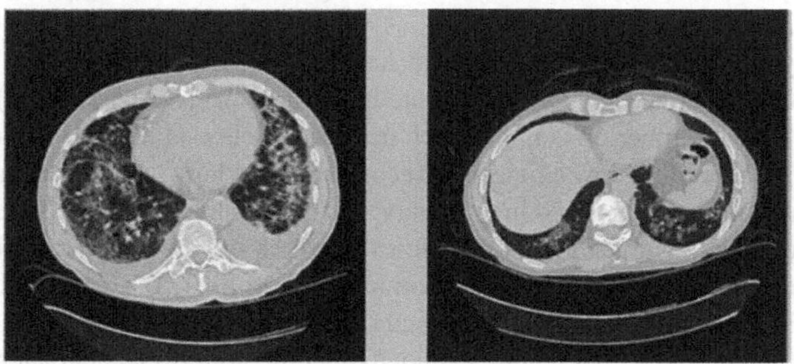

Fig. 8. Retrieval results

Experiment. Two data sets were designed for this experiment, one of which is a training set-DB1(10 images, image resolution is 1074 * 768), another is a test set-DB2 (11 images, image resolution is 1024 * 768, 7 lung images and 4 unlung images), lung images include the following categories:

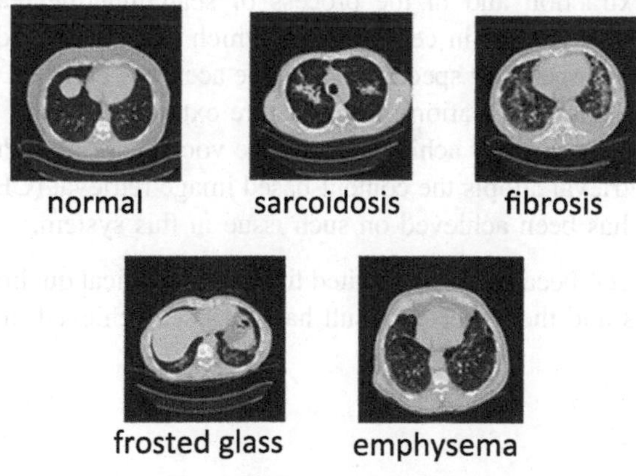

Fig. 9. LungImages categories

Through the experiment, the algorithm can accurately identify the 7 lung images in DB2, success rate is 100%, the results is shown as follows:

Table 1. The experimental results

Data set	Lung image	Success rate
11	7	100%

4 Conclusion

The study and implementation of lung CT image retrieval based on the vocabulary tree were discussed in this paper in details from the aspects of image pre-processing, feature extraction by SIFT algorithm, K-means clustering into the vocabulary tree and similarity measure.

Presently, the medical image retrieval has been widely applied. The traditional method of lung image retrieval was the content-based image retrieval (CBIR). But with the increase in the number of samples, such traditional retrieval method could not meet the demand because of the low efficiency. Accordingly, the lung image retrieval based on the vocabulary tree was proposed in this study. The vocabulary tree adopts the tree-form data structure with the advantage of fast retrieval, which has aroused the wide concern.

SIFT feature extraction is one of common algorithms to extract the features, because SIFT extraction is the method based on the local image feature and the image extraction based on the global feature cannot extract the non-global characters of image well. Besides, in some cases, the global feature cannot reflect the real characters of the object. K-means clustering is the algorithm to use the minimal solution of convex optimal problem and conduct the continuous iteration. Such algorithm chooses Euclidean distance as the similarity measure and the classification represents the vector of one clustering center, which is method to find such optimal category.

Meanwhile, the system has been improved in many aspects.

For instance:

(1) After SIFT extraction and in the process of searching the feature vector, the searching was carried out in certain areas, which can greatly reduce the calculation, improve the operating speed, promote the accuracy and also avoid the impact of the complex global situation on the feature extraction.
(2) The lung image retrieval is achieved using the vocabulary tree. Presently, most of the medical retrieval adopts the content-based image retrieval (CBIR). The certain improvement has been achieved on such issue in this system.

As a matter of fact, because of the limited time and technical quality, there has been many shortcomings and the expected result has not been achieved in some place.

For instance:

(1) In the process of K-means clustering, the selection of K-means was random according to the extraction results of lung features, which did not produce the desired effect. Accordingly, in the retrieval using the vocabulary tree later, the lung image retrieval was not fastest.
(2) SIFT algorithm might not be the best way to extract the lung CT image features. Because of the limited time, it was not compared with LBP extraction algorithm or SURF algorithm that was used to extract the lung features and thus the experimental results might not be quite accurate.

In conclusion, the lung image retrieval is a challenging and interesting research topic. Despite the development of vocabulary tree method by the scholars, it has been seldom applied to the lung images. It is expected that the greater breaking progress can be made on such topic.

References

Jian, Z., Yu, X., Chen, X., et al.: Research of image retrieval based on vocabulary tree in virtual university. In: International Conference on E-Business and E-Government, pp. 4313–4316 (2010)

Seo, H., Lee, J.: Object recognition algorithm using a vocabulary tree and a pre-matching array. Artif. Life Robot. **15**(1), 93–96 (2010)

Wu, Z., Yang, Z., Sun, H., et al.: Hybrid radar emitter recognition based on rough k -means classifier and SVM. EURASIP J. Adv. Sig. Process. **2012**(1), 198 (2012)

Masami, A., Jitendra, A.: Decision tree-based acoustic models for speech recognition. EURASIP J. Audio Speech Music Process. **2012**(1), 10 (2012)

Jason, K.: Content-based image retrieval using deep belief networks. University of Kansas (2010)

Carlos, M., Travieso, M., Jesús, B.: Fused intra-bimodal face verification approach based on scale-invariant feature transform and a vocabulary tree. J. Pattern Recogn. Lett. **36**, 254–260 (2014)

A More Robust Active Contour Model with Group Similarity

Shaozhu Chen, Xiaodong Zhao, Xianhui Liu(✉), and Yufei Chen

CAD Research Center, Tongji University,
Shanghai 201804, People's Republic of China
lxh@tongji.edu.cn

Abstract. Image segmentation based on active contour model has been widely used in recent years. However, in real application, image segmentation results are always leaded to corrupt because of partial missing of target boundaries or some misleading factors. In order to solve the issue, Zhou proposed a model called active contour with group similarity. But Zhou's model is purely based on global information of image and cannot deal with image segmentation with intensity inhomogeneity. In order to optimize Zhou's model and make it more robust, we construct a new energy function which combines global feature with local feature of image. The local feature can solve the issue of image segmentation with intensity inhomogeneity. And the global feature can make our model less sensitive to the initial position of the curve or noise. The experiment results have proved that, our method can achieve satisfying segmentation results on image segmentation with intensity inhomogeneity and show less sensitive to the initial position of the curve.

Keywords: CV model · LBF model · Group similarity · Robust · Global information · Local information

1 Introduction

Image segmentation is one of the basic technologies in image primarily processing, and it aims to segment target from the whole image. Now, there are many kinds of active contour models, and they can be roughly categorized into two classes: parametric active contour model [1] and geometric active contour model [2].

Level set model [3, 20] proposed by Osher and Sethina in 1988 belongs to geometric active contour model, which is a widely used recently.

Chan-Vese model [4] proposed by Chan and Vese in 2001 is a typical level set model. It can get satisfying segmentation results on image with weak boundary, but it is not capable of segmenting image with intensity inhomogeneity. In order to solve the defect of CV model, Li et al. proposed LBF model [5, 6] which is the improvement of CV model. LBF model can segment image with intensity inhomogeneity by using the local feature of the image. But it's sensitive to the initial position of the curve, which could lead to segmentation results corrupted.

Besides, there are other problems of image segmentation in real application, like missing or misleading features in images. Image segmentation based on shape prior

D.-S. Huang et al. (Eds.): ICIC 2017, Part I, LNCS 10361, pp. 408–419, 2017.
DOI: 10.1007/978-3-319-63309-1_38

model [7, 14–17, 19] is used to solve this issue. But it needs a large set of annotated data to learn, which may be lack of practicability. So Zhou proposed a model called active contour with group similarity [8, 20] to solve the problem.

But Zhou's model is based on CV model, and it's only suitable for image segmentation with intensity homogeneity. Thus, in this paper, we introduce a new model, which synthesize CV model and LBF model to make it more robust to fit image segmentation with intensity inhomogeneity. Besides, our method will overcome LBF model's defect [18, 25] that it is sensitive to initial position of the curve and noise. Our model can characterize a certain point in grayscale distribution. For the uneven distribution of intensity in local area, the evolving process is primarily controlled by the local fitting items of our model at that point. And if intensity distribution is even, it's primarily guided by global fitting item to control curve moving toward the target boundary and avoid being trapped in local optimum of the energy function. Then we use the energy function to optimize Zhou's model to make it more robust.

2 Related Works

2.1 Chan-Vese Model

Chan and Vese proposed Chan-Vese model in 2001, which is based on the assumption that the image is piecewise constant. Assume that the whole image contains two piecewise constant region, the energy function can be described as

$$\mathrm{E}^{cv} = \lambda_1 \int_\Omega |I - \mathrm{c}_1|^2 H(\varphi)dx + \lambda_2 \int_\Omega |I - \mathrm{c}_2|^2 (1 - H(\varphi))dx + v \int_\Omega \delta(\varphi)|H(\varphi)|dx \quad (1)$$

Inside the energy function (1), φ is the zero level set, c_1 and c_2 are the average gray-value inside and outside of the evolving curve respectively. $\delta(\varphi)$ is Dirac function and $H(\varphi)$ is Heaviside function.

CV model is less sensitive to the initial position of evolving curve and more robust to noise. But it cannot handle image segmentation with intensity inhomogeneity well.

2.2 LBF Model

In order to overcome the defect of CV model, Li proposed LBF model. The energy function of LBF model can be described as:

$$\begin{aligned}\varepsilon^{Fit}(\varphi, f_1(x), f_2(x)) = \lambda_1 \int_\Omega K_\sigma(x-y)|I(y) - f_1(x)|^2 M_1(\varphi)dy \\ + \lambda_2 \int_\Omega K_\sigma(x-y)|I(y) - f_2(x)|^2 M_2(\varphi)dy\end{aligned} \quad (2)$$

Inside the energy function (2) of LBF model, $M_1(\varphi) = H(\varphi)$, $M_2(\varphi) = 1 - H(\varphi)$, λ_1 and λ_2 are both positive integer, $f_i(x)$ is local fitting parameter on point of the image. K_σ is Gaussian kernel function with $\sigma > 0$.

Considering keeping stability of LBF model, Li add a distance regularization item [9] to restrain the evolving process of LBF model. So, the whole energy function of LBF model is:

$$\begin{aligned} E^{RSF} &= \int_{\Omega} \varepsilon^{Fit} dx + v \int_{\Omega} \delta(\varphi)|H(\varphi)| dx + \frac{1}{2}\mu \int_{\Omega} (|\nabla \varphi| - 1)^2 dx \\ &= E^{LIF} + vL(\varphi) + \mu P(\varphi) \end{aligned} \tag{3}$$

Inside the energy function (3) above, $L(\varphi)$ is the term of curve length, $P(\varphi)$ is the regularization term of distance, and E^{LIF} is the term of local energy. v and μ are both positive.

LBF model takes advantage of the local image information, thus it could have good segmentation results on images with intensity inhomogeneity. But it is sensitive to the initial position of the curve.

2.3 Image Segmentation Based on Group Similarity

In the real image segmentation, image segmentation is a difficult work, especially when the target boundaries are partially missing or covered by other misleading features. To solve the problem, image segmentation based on shape prior is proposed. However, it must learn from a large set of annotated data [21–23], and get a prior shape as a constraint term of the model, which is not available in real application. Thus, Zhou proposed a model to solve the issue. Zhou gets a low rank matrix [10, 24] from the affine transformation of target in each frame. He transforms the low rank matrix to a constraint term of nuclear norm because of the convex character of nuclear norm. So the whole energy function of Zhou's model can be described as:

$$E^{GS} = \min_{X} \sum_{i=1}^{n} f_i(C_i) + \lambda \|X\|_* \tag{4}$$

$$f_i(C_i) = \int_{\Omega_1} (I_i(x) - \mu_1)^2 dx + \int_{\Omega_2} (I_i(x) - \mu_1)^2 dx + \beta length(C_i) \tag{5}$$

Inside the energy function (5), Ω_1 and Ω_2 represent the region inside and outside the curve respectively. μ_1 and μ_2 denote the mean intensity of Ω_1, Ω_2, $length(C_i)$ is the term of curve length.

3 Our Model

As we discussed above, CV model is suitable to segment images with intensity homogeneity. It's not sensitive to the initial position of curve, and it is robust to noise. But it cannot deal with images with intensity inhomogeneity. On the contrary, LBF model takes advantage of local information of image. It can handle image segmentation with intensity inhomogeneity. But it's sensitive to noise and initial position of curve.

So we proposed a new energy function, which combines local information with global information of image. The whole energy function can be described as:

$$\begin{aligned} E^{new}(\varphi, f_1(x), f_2(x), c_1, c_2, \omega) &= \int g(x,\omega)E^{RSF} + [1 - g(x,\omega)]E^{CV}dx \\ &= \lambda_1 \int [1 - g(x,\omega)] \int |I(x) - c_1|^2 H(\varphi(x))dx \\ &+ \lambda_2 \int [1 - g(x,\omega)] \int |I(x) - c_2|^2 [1 - H(\varphi(x))]dx \\ &+ \lambda_2 \int g(x,\omega) \int K_\sigma(x-y)|I(y) - f_2(x)|^2 [1 - H(\varphi(y))]dxdy \\ &+ \lambda_1 \int g(x,\omega) \int K_\sigma(x-y)|I(y) - f_1(x)|^2 H(\varphi(y))dxdy \end{aligned} \tag{6}$$

Inside the energy function (6) above, we construct a character function $g(x,\omega)$ to control the evolving process of curve based on whether the intensity distribution is homogeneity or not. We construct the $g(x,\omega)$ as:

$$g(x,\omega) = \left| \frac{2}{1 + e^{-(f_1(x) - f_2(x))^* \omega}} - 1 \right| \tag{7}$$

Inside the function (7), ω is a positive number which controls the scope of intensity distribution homogeneity. When the absolute number of $f_1(x) - f_2(x)$ is close to 0, it means that the intensity distribution is homogeneity around the point. So the global fitting term will be allocated more weight to control the evolving process to avoid being trapped in local optimum by purely using local fitting term. On the contrary, it means the intensity distribution is inhomogeneity around the point, so we allocate more weight on the local fitting term.

In order to prevent energy function re-initialization, we add a regularization item and a term of the length of curve to restrain the evolving process. Thus, we get the final energy function:

$$E^{our}(\varphi, f_1(x), f_2(x), c_1, c_2) = E^{new}(\varphi, f_1(x), f_2(x), c_1, c_2) + \nu L(\varphi) + \mu P(\varphi) \tag{8}$$

Then we use our energy function (8) to optimize Zhou's model and make it capable to segment images with intensity inhomogeneity. The whole energy function with nuclear norm can be described as below:

$$E^{GS} = \min_X \sum_{i=1}^{n} f_i^{our}(C_i) + \lambda \|X\|_* \tag{9}$$

$$f_i(C_i) = E^{our}(\varphi, f_1(x), f_2(x), c_1, c_2) \tag{10}$$

In our model (9), we use a method called proximal gradient (PG) [11, 12] to solve the following category of problems:

$$\min_X F(X) + \lambda R(X) \tag{11}$$

In function (11), $F(X)$ is a differentiable function. $R(X)$ is a convex penalty which may not be smooth. In our energy function, $F(X) = \sum_{i=1}^{n} f_i(C_i)$ and $R(X) = \|X\|_*$. The main process in PG method is to make the following quadratic approximate to $F(X)$, which is based on the previous X' in each iteration. It can be described as:

$$\begin{aligned} Q_u\left(X, X'\right) &= F\left(X'\right) + \left\langle \nabla F\left(X'\right), X - X' \right\rangle + \frac{\mu}{2}\|X - X'\|_F^2 + \lambda R(X) \\ &= \frac{\mu}{2}\left\| X - \left[X' - \frac{1}{\mu}\nabla F\left(X'\right)\right]\right\|_F^2 + \lambda R(X) + const \end{aligned} \tag{12}$$

Inside the quadratic approximation (12), $\langle .,. \rangle$ denotes the inner product, $\|.\|_F$ means the Frobenius norm, and μ is a constant parameter. And if $F(X)$ is differentiable with Lipschitz continuous gradient, the function will converge to a stationary point in (12) with a convergence rate of $o\left(\frac{1}{k}\right)$. So, we can get the following function:

$$\begin{aligned} X^{k+1} &= \arg\min_X Q_\mu\left(X, X^k\right) \\ &= \frac{1}{2}\left\| X - \left[X^k - \frac{1}{\mu}\nabla F\left(X^k\right)\right]\right\|_F^2 + \frac{\lambda}{\mu}R(X) \end{aligned} \tag{13}$$

Then, to solve the problem of the updating step in function (13), we need to use the following lemma, which has been proved in [13].

Lemma 1. Given $X \in R^{m \times n}$, the solution to the problem $\min_X \frac{1}{2}\|X - Z\|_F^2 + \alpha\|X\|_*$ is $X^* = D_\alpha(Z)$, in which $D_\alpha(Z) = \sum_{i=1}^{\min(m,n)} (\sigma_1 - \sigma_2)_+ \mu_i v_i^T$. μ_i and v_i^T are the left and right singular vectors of Z respectively, σ_i is the singular value, and $(\cdot)_+ = max(\cdot, 0)$. Thus, the updating step of our algorithm is $X^{k+1} = D_{\frac{\lambda}{\mu}}\left(X^k - \frac{1}{\mu}\nabla F\left(X^k\right)\right)$. And now, we need to find $\nabla F\left(X^k\right)$. $F(X)$ can be thought as a decomposable part which is a sum of quantities of each frame. Hence, $\nabla F\left(X^k\right) = \left[\nabla f_1\left(C_1^k\right), \cdots, \nabla f_n\left(C_n^k\right)\right]$. In our method, for each $\nabla f_i\left(C_i^k\right)$,

$$\nabla f_i\left(C_i^k\right) = \delta_\varepsilon(\phi)\left(F^{CV} + F^{RSF}\right) + v\delta_\varepsilon(\phi) \cdot div\left(\frac{\nabla\phi}{|\nabla\phi|}\right) + \mu\left(\nabla^2\varphi - div\left(\frac{\nabla\phi}{|\nabla\phi|}\right)\right) \tag{14}$$

$$F^{CV} = -\lambda_1[1 - g(x)](I(x) - c_1)^2 + \lambda_2[1 - g(x)](I(x) - c_2)^2$$

$$F^{RSF} = -\lambda_1 e_1 + \lambda_2 e_2, e_i = \int_{\Omega} g(y) k_\sigma(y - x) |I(x) - f_i(y)|^2 dy$$

In order to make the convergence of the PG method faster, we use the Nesterov method [12] to accelerate the evolving process. The Nesterov method uses an intermediate point Y^k to replace the updating X, which is a linear extrapolation from X^{k-1} and X^k.Thus, the updating process can be described as:

$$Y^k = X^k + \frac{t^{k+1} - 1}{t^k} \left(X^k - X^{k-1} \right) \; (t^k \text{ is determined online})$$

4 Experiments Results

In this section, we discuss the implementation of our method and the segmentation results between different models. In the implementation of our method, we firstly make $X^0 = [C_0, \cdots, C_0]$ as the initial condition of the active contour, in which C_0 is the manually initialized curve.

Firstly, we start contrast experiments with Zhou's model on images with intensity inhomogeneity:

4.1 Image Segmentation Results on Intensity Inhomogeneity

The first row in Figs. 1, 2 and 4 is origin images which are processed to be intensity inhomogeneity. The second row and forth row are the segmentation results of Zhou's model, while the third and the fifth row are the segmentation results of our method. It's obviously that our model can handle image segmentation with intensity inhomogeneity well, and it can achieve integral contours even though some boundaries are partial missing or occlusion.

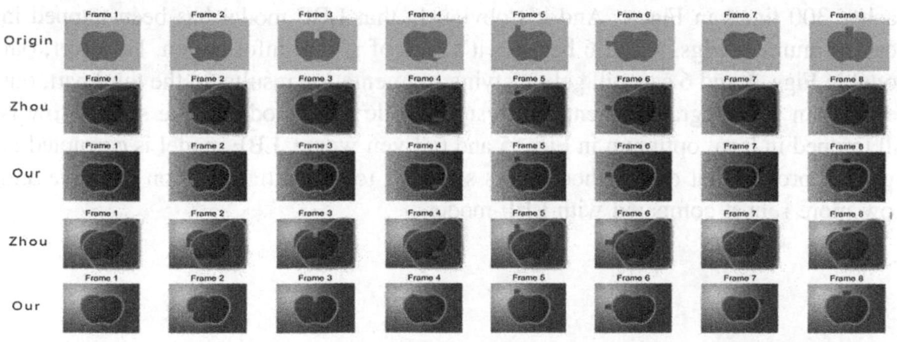

Fig. 1. Comparison on Zhou's model and our model on intensity inhomogeneity on case two

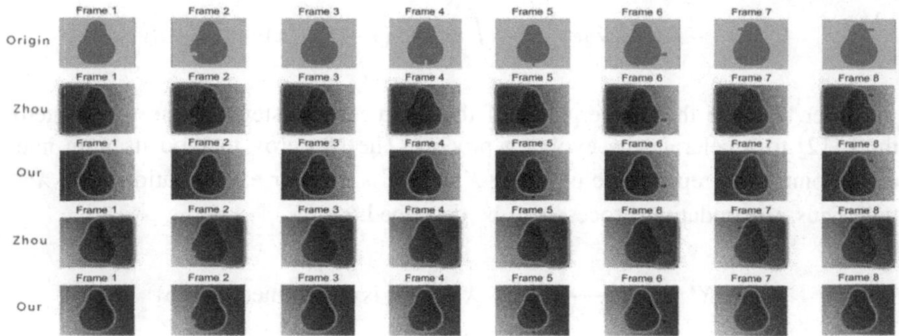

Fig. 2. Comparison on Zhou's model and our model on intensity inhomogeneity on case three

As the experiment results showed, it's obviously that Zhou's model cannot deal with intensity inhomogeneity well. However, in our method, we combine global information with local information, and we also construct a character function $g(x, \omega)$ to control the evolving process of curve. If the intensity distribution is homogeneity in local area, the global information term will primarily lead the process of the evolving. On the contrary, the local information term will primarily guide the processing. Thus, it can not only handle image segmentation with intensity inhomogeneity well, but also be less sensitive to the initial position of curve.

Then, we take global information into account to deal with the issue. The contrast experiments between our method and LBF model have proved that our method is less sensitive to the initial position of the curve.

4.2 Contrast Experiments Between Our Model and LBF Model

These three experiments in Figs. 4, 5 and 6 are all showed in three parts. The first part is the initial position of curve in our model and LBF model. Firstly, we choose the same initial position of the curve for two models. In the second part, our method in first row, gets better segmentation results than the second row (LBF model) when the iterator reaches 300 times in Fig. 4. And it's obviously that LBF model has been trapped in local optimum in Figs. 5 and 6 because it's lack of global information. However, our model in Figs. 5 and 6 can still get satisfying segmentation results. In the third part, our method can get integral segmentation results, while LBF model in the second row is still trapped in local optimum in Figs. 5 and 6. Even worse, LBF model is corrupted in Fig. 4. It proved that our method is less sensitive to the initial position of curve and show more robust compared with LBF model.

4.3 Quantitative Evaluation of Segmentation Results

In this section, we make a quantitative evaluation on these contrast experiments above. And it could help readers easier to understand our method. Here, we will use precision, recall and Dice index to evaluate segmentation results for all contrast experiments.

$$D = \frac{2|s_t \cap s_g|}{|s_t| + |s_g|} \tag{15}$$

In our model, s_g denotes the integral boundary of target which is achieved by manual operation. s_t is the real segmentation results achieved by segmentation models. Now, we assess the performance of our model based on several contrast experiments above.

Each contrast experiment is implemented on a test set which contains 5 group sequence images with 8 frames, and each group is in different intensity inhomogeneity.

Firstly, we assess the performance between our model and Zhou's model on image segmentation with intensity inhomogeneity.

The quantitative evaluation results of image segmentation on intensity inhomogeneity between our model and Zhou's model are showed in Table 1. There are three quality indexes in Table 1, such as precision, recall and Dice. For each index, the number denotes the goodness of fit between s_g and $s_t n$ in function (15). It's obviously that our model gets a higher number on each index for all four test cases compared with Zhou's model. And the segmentation results in Figs. 1, 2 and 3 has proved that, our model is more capable to handle image segmentation with intensity inhomogeneity.

Table 1. The average values of the precision, recall and Dice index of segmentation results between our model and Zhou's model on intensity inhomogeneity

Test case	Precision (%)	Recall (%)	Dice (%)	Model
Case one	97.15	91.80	94.34	Our
	80.35	35.45	47.38	Zhou
Case two	99.33	95.51	97.37	Our
	95.67	42.20	58.50	Zhou
Case three	99.08	96.69	97.87	Our
	99.66	46.31	63.17	Zhou
Case four	98.87	94.73	96.75	Our
	92.48	45.03	60.44	Zhou

Then, we start to assess the performance between our model and LBF model on sensibility on initial position of curve.

Considering the segmentation results showed in Figs. 4, 5 and 6, we can obviously notice that LBF model has been trapped in local optimum in Figs. 5 and 6. Even worse, it is corrupted in Fig. 4. Thus, it's no doubt that three indexes of LBF model are much lower than ours on each test case. And it has proved that our model is less sensitive to the initial position of curve compared with LBF model (Fig. 8).

Table 2. The average values of the precision, recall and Dice index of segmentation results between our model and LBF model on sensibility to initial position

Test case	Precision (%)	Recall (%)	Dice (%)	Model
Case one	97.78	91.93	94.74	Our
	39.82	22.18	28.21	LBF
Case two	99.30	95.03	97.12	Our
	63.98	96.33	76.88	LBF
Case three	99.18	96.91	98.03	Our
	14.01	95.56	24.03	LBF

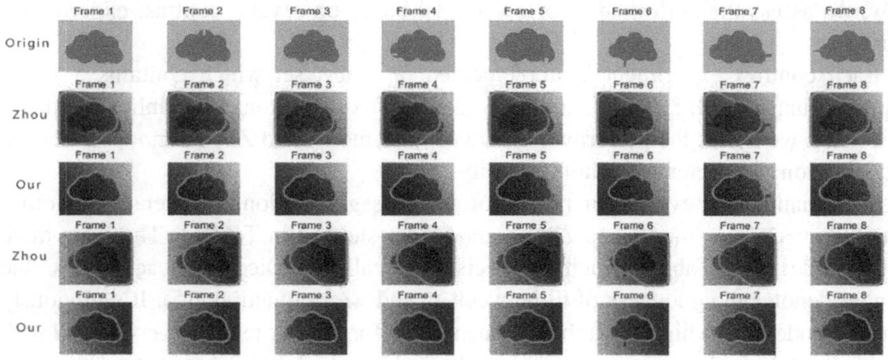

Fig. 3. Comparison on Zhou's model and our model on intensity inhomogeneity on case four

Fig. 4. Comparison on iterations of our model and LBF model on the sensitivity to initial position on case one

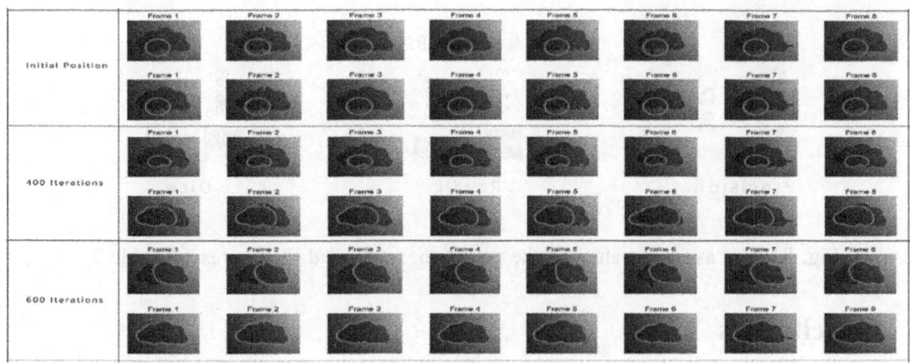

Fig. 5. Comparison on iterations of our model and LBF model on the sensitivity to initial position on case two

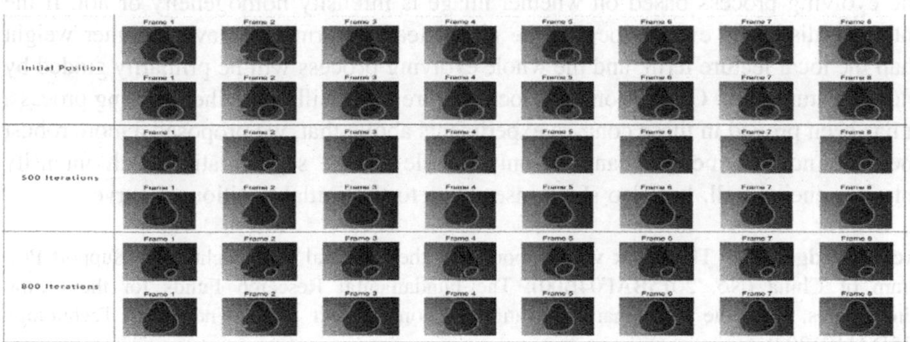

Fig. 6. Comparison on iterations of our model and LBF model on the sensitivity to initial position on case three

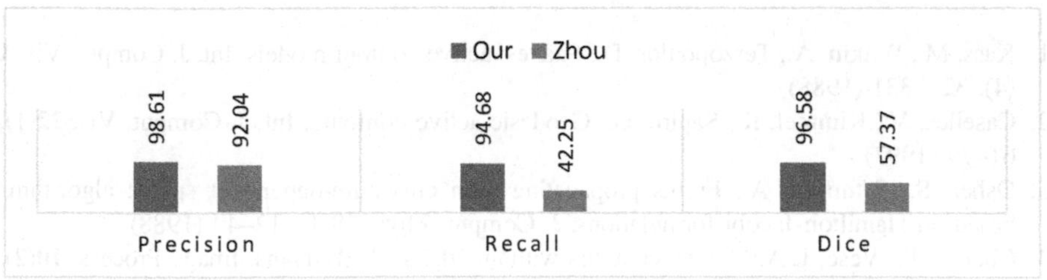

Fig. 7. The average values of the precision, recall and dice index of segmentation results of Table 1

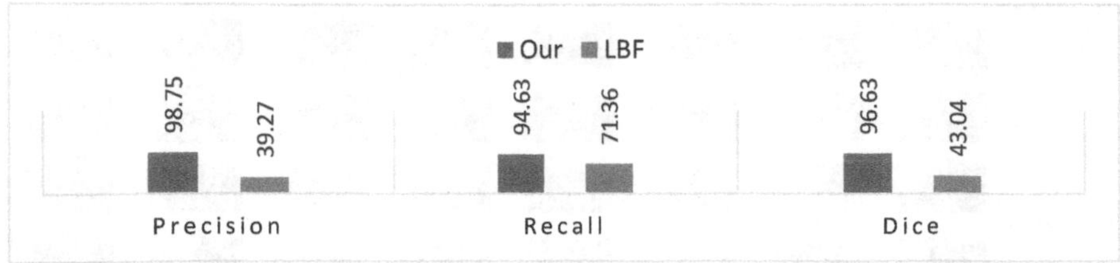

Fig. 8. The average values of the precision, recall and dice index of Table 2

5 Conclusions

In this paper, we proposed a new energy function, which combine global feature with local feature together, to optimize Zhou's model and make Zhou's model more robust. The main principle of our method is that, we construct a character function to control the evolving process based on whether image is intensity homogeneity or not. If the intensity distribute evenly locally, the global feather term will have a higher weight than the local feature term, and the whole evolving process will be primarily guided by global feature term. On the contrary, local feature term will guide the evolving process. It has been proved in these contrast experiments above that, we proposed a more robust model. And our method can not only handle image segmentation with intensity inhomogeneity well, but also show insensitive to the initial position of curve.

Acknowledgement. This work was supported by the National Key Technology Support Program of China (No. 2015BAF04B00), The Fundamental Research Funds for the central Universities, and the Shanghai Innovation Action Project of Science and Technology (15DZ1101202).

References

1. Kass, M., Witkin, A., Terzopoulos, D.: Snakes: active contour models. Int. J. Comput. Vis. **1**(4), 321–331 (1988)
2. Caselles, V., Kimmel, R., Sapiro, G.: Geodesic active contours. Int. J. Comput. Vis. **22**(1), 61–79 (1997)
3. Osher, S., Sethian, J.A.: Fronts propagating with curvature-dependent speed: algorithms based on Hamilton-Jacobi formulations. J. Comput. Phys. **79**(1), 12–49 (1988)
4. Chan, T.F., Vese, L.A.: Active contours without edges. IEEE Trans. Image Process. **10**(2), 266–277 (2001)
5. Li, C., Kao, C.Y., Gore, J.C., et al.: Implicit active contours driven by local binary fitting energy. In: IEEE Conference on Computer Vision and Pattern Recognition, CVPR 2007, pp. 1–7. DBLP (2007)
6. Li, C., Kao, C.Y., Gore, J.C., et al.: Minimization of region-scalable fitting energy for image segmentation. IEEE Trans. Image Process. **17**(10), 1940–1949 (2008)
7. Rousson, M., Paragios, N.: Shape priors for level set representations. In: Heyden, A., Sparr, G., Nielsen, M., Johansen, P. (eds.) ECCV 2002. LNCS, vol. 2351, pp. 78–92. Springer, Heidelberg (2002). doi:10.1007/3-540-47967-8_6

8. Zhou, X., Huang, X., Duncan, J.S., et al.: Active contours with group similarity. pp. 2969–2976 (2013)
9. Li, C., Xu, C., Gui, C., et al.: Level set evolution without re-initialization: a new variational formulation. In: IEEE Computer Society Conference on Computer Vision and Pattern Recognition, CVPR 2005, vol. 1, pp. 430–436. IEEE Xplore (2005)
10. Candes, E., Recht, B.: Exact matrix completion via convex optimization. Commun. ACM **55** (6), 111–119 (2012)
11. Beck, A., Teboulle, M.: A fast iterative shrinkage-thresholding algorithm for linear inverse problems. SIAM J. Imaging Sci. **2**(1), 183–202 (2009)
12. Nesterov, Y.: Gradient methods for minimizing composite objective function (2007)
13. Cai, J.F., Candès, E.J., Shen, Z.: A singular value thresholding algorithm for matrix completion. SIAM J. Optim. **20**(4), 1956–1982 (2010)
14. Leventon, M.E., Grimson, W.E.L., Faugeras, O.: Statistical shape influence in geodesic active contours. In: IEEE Conference on Computer Vision and Pattern Recognition, 2000 Proceedings, vol. 1, pp. 316–323. IEEE (2000)
15. Cootes, T.F., Taylor, C.J., Cooper, D.H., et al.: Active shape models-their training and application. Comput. Vis. Image Underst. **61**(1), 38–59 (1995)
16. Tsai, A., Yezzi, A., Wells, W., et al.: Model-based curve evolution technique for image segmentation. In: Proceedings of the 2001 IEEE Computer Society Conference on Computer Vision and Pattern Recognition, CVPR 2001, vol. 1, pp. I-I. IEEE (2001)
17. Etyngier, P., Segonne, F., Keriven, R.: Shape priors using manifold learning techniques. In: IEEE 11th International Conference on Computer Vision, ICCV 2007, pp. 1–8. IEEE (2007)
18. He, C., Wang, Y., Chen, Q.: Active contours driven by weighted region-scalable fitting energy based on local entropy. Sig. Process. **92**(2), 587–600 (2012)
19. Riklin-Raviv, T., Kiryati, N., Sochen, N.: Prior-based segmentation by projective registration and level sets. In: Tenth IEEE International Conference on Computer Vision, ICCV 2005, vol. 1, pp. 204–211. IEEE (2005)
20. Lv, P., Zhao, Q., Gu, D.: Segmenting similar shapes via weighted group-similarity active contours. In: IEEE International Conference on Image Processing, pp. 4032–4036. IEEE (2015)
21. Sidi, O., Kaick, O.V., Kleiman, Y., et al.: Unsupervised co-segmentation of a set of shapes via descriptor-space spectral clustering. ACM Trans. Graph. **30**(6), 126:1–126:10 (2011)
22. Srivastava, A., Joshi, S.H., Mio, W., et al.: Statistical shape analysis: clustering, learning, and testing. IEEE Trans. Pattern Anal. Mach. Intell. **27**(4), 590–602 (2005)
23. Malladi, R., Sethian, J.A., Vemuri, B.C.: Shape modeling with front propagation: a level set approach. IEEE Trans. Pattern Anal. Mach. Intell. **17**(2), 158–175 (1995)
24. Fazel, M.: Matrix rank minimization with applications. Ph. D. thesis, Stanford University (2002)
25. Chen, Y., Yue, X., Da Xu, R.Y., Fujita, H.: Region scalable active contour model with global constraint. Knowl. Based Syst. **120**, 57–73 (2017)

A New Infrared and Visible Image Fusion Algorithm in NSCT Domain

Xiaochun Wang[1], Lijun Yao[2], Ruixia Song[2], and Huiyang Xie[1(✉)]

[1] College of Sciences, Beijing Forestry University, Beijing 100083, China
xhyang@bjfu.edu.cn
[2] College of Sciences, North China University of Technology, Beijing 100144, China

Abstract. Infrared and visible image fusion can produce a composite image which has high contrast and rich background details of the scene. In view of the defects of some existing infrared and visible fusion method, such as low contrast and unclear background details, we propose a novel multi-scale fusion method based on the combination of non-sampled contourlet transform (NSCT), sparse representation and pulse coupled neural network. In our method, the source images are firstly decomposed into one low frequency sub-band and high frequency sub-bands at different scales and directions using NSCT. Fusion rules based on the sparse representation and modified PCNN are developed, and then used for fusion of the low sub-band and high frequency sub-bands, respectively. In the modified PCNN developed in this paper, we use Sum-Modified-Laplacian and Log-Gabor energy as values of the linking strength instead of setting it a constant. Each of the linking strength corresponds to an ignition map, the average of the two results is taken as the final PCNN output. The fused image are finally obtained by performing the inverse NSCT. Comparison experiment results show that the fused image produced by the proposed method has high contrast and rich details, as well as the greatly improved objective evaluation indexes values.

Keywords: Infrared image · Image fusion · Non-subsampled contourlet transform · Sparse representation · Pulse coupled neural network

1 Introduction

Infrared image and visible image acquired by different sensors have different ways of detecting the target. Infrared image is produced by sensing the emitted radiations, while visible image are generally produced from reflected radiation. Infrared images are nearly invariant to changes in ambient illumination, so they can discriminate targets from backgrounds and are capable of showing the important and obscure objects under all lighting conditions. However, they have the disadvantages of low contrast, little textures, and always lack detailed information. On the contrary, visible images have high spatial resolution, contain plentiful texture details, and can clearly reveal the scene information. But they cannot detect the obscured and disguised targets, moreover visible image's quality will dramatically degrade in low illumination. Image fusion is process of integrating information from two or more images into a single composite image.

D.-S. Huang et al. (Eds.): ICIC 2017, Part I, LNCS 10361, pp. 420–431, 2017.
DOI: 10.1007/978-3-319-63309-1_39

Using image fusion, we can integrate complementary information from infrared and visible images into a single image which not only retains textures but also maximizes target-background discriminability. Infrared and visible image fusion is widely used in computer vision, astronautics, target identification and military affairs [1], especially in surveillance applications to enhance image quality of ambient scenes.

Currently, infrared and visible image fusion is mainly performed in the frequency domain. Among frequency-based fusion algorithms, multi-scale transform methods including pyramid transform, wavelet transform, ridgelet transform, curvelet transform, contourlet transform [2] and nonsubsampled contourlet transform (NSCT) have become a research hotspot owing to their effectiveness in implementation. Compared with pyramid transform and wavelet transform, NSCT [3] is more effective in representing smooth contours owing to its direction selectivity, multiresolution characteristics and sift-invariance. It is widely used in image fusion [4, 5]. NSCT consists of two parts: Nonsubsampled Pyramid filter bank (NSPFB) and Nonsubsampled directional filter bank (NSDFB). NSPFB is used for multiscale decomposition, and NSDFB processes the multi-direction decomposition of the bandpass. In NSCT domain, average-selection rule is usually used for low frequency coefficients fusion and maximum-selection rule for high frequency sub-bands. The convention NSCT based fusion rule is simple and can eliminate false edge and spectral distortion, but has the disadvantage of reducing contrast and being sensitive to noise. For this reason, many fusion rules have been successively proposed. Zhang L. et al. presented an improved fusion rule that uses mean of the weighted averaged coefficients to select low frequency coefficients, a combination of maximum and local mean-square deviation fusion rules to select high frequency sub-band coefficients [6]. Yang G. et al. proposes a fusion rule based on NSCT and PCNN, in which the sub-band coefficients filtering is first performed and then the obtained results are fused based on PCNN [7].

Recently, spare representation (SR) and pulse coupled neural network (PCNN) have been widely applied in the field of image fusion [8, 9]. Sparse representation adopts atomic structure as the basic unit, so it can capture internal structure between pixels on a deeper level. PCNN is a two-dimensional feedback neural network [7], and has synchronous excitation and global characteristics. Compared with conventional image processing methods, PCNN has advantages of robustness against noise, independence of geometric variations, capability of bridging minor intensity variations in input patterns, etc. It is widely used in the field of image fusion. In NSCT domain, PCNN is often used in the high-pass sub-bands to get better fusion effect, but the traditional PCNN model sets a fixed value as the linking strength which might result in loss of source image details and unsatisfied fusion effect [10].

Based on above observation, this paper proposes a new image fusion algorithm, in which sparse representation and modified PCNN model are used for fusion of the low and high frequency coefficients in NSCT domain, respectively. In the modified PCNN model, we use Sum-Modified-Laplacian (SML) and Log-Gabor energy as linking strengths. Each of the linking strength corresponds to an ignition map, the average of the two results is taken as the final PCNN output. This method combines the advantages of NSCT, sparse representation and PCNN to get more satisfied fusion result.

2 Related Theory and Fusion Rule

2.1 Fusion Rule Based on Spare Representation

In recent years, sparse representation has become a hot research field of image processing. It is a new type of image information representation theory. The underlying motivation for sparse decomposition problems is that even though the signal is in high-dimensional space, it can be obtained in some lower-dimensional subspace due to it being sparse. Process of signal sparse representation is to represent the signal as a combination of atoms in an overcomplete dictionary with the fewest number of nonzero coefficients [2]. Let x be the input signal which is going to be sparsely represented in the overcomplete dictionary $D \in R^{n \times m}$ with $n \gg m$. Each column in D is called an atom. The sparse representation amounts to solving the following mathematic problem:

$$\min_{\alpha} \|\alpha\|_0 \quad s.t. \ \|x - D\alpha\|_2^2 < \varepsilon \tag{1}$$

where $\|\alpha\|_0$ counts the nonzero elements of the vector α, $\varepsilon\,(\varepsilon > 0)$ is error tolerance. The algorithm to solve the problem (1) of finding the sparsest representation of a signal in a given overcomplete dictionary is called the sparse decomposition algorithm. Many sparse decomposition algorithms have been introduced so far. In this paper, the orthogonal matching pursuit (OMP) algorithm is employed.

A fundamental consideration in employing the above model is the choice of the dictionary D. To realize sparse representation, atoms of the dictionary should capture the most salient features of the signals. Broadly speaking, there are two ways to construct a dictionary. In some applications, a dictionary is adaptively chosen from some predefined dictionaries, such as DCT, wavelet and curvelet transforms to efficiently represent the given signal. Although there do exist predefined dictionaries which can be used to sparsely approximate or represent a given signal, a more effective way to construct a dictionary is learning the dictionary from a set of training signals of a given class. In this case, dictionary is typically represented as an explicit matrix D, and the atoms which are the columns of D are chosen from a set of training signals by training algorithms [8, 9]. Predefined dictionaries have simple structure, but they cannot adapt to different image data, while learned dictionaries have strong adaptability and capability of better matching all kinds of signals. Usually, learned dictionaries which are formed using machine learning techniques are capable to give more sparse solution as compared to predefined transform matrices. In this paper, sparse representation with learned dictionary is employed.

Note that the low frequency coefficients in NSCT are often not sparse enough, in this paper, their sparse representation is first calculated, then the obtained coefficient matrices of the source images are merged into one using the absolute-maximum-selection rule. The fusion process is as follows [11]:

(1) Using sliding window with step 2, divide the low frequency sub-bands L_A, L_B of infrared image and visible image, from left-top to right-bottom, into T patches of size 8×8, which are denoted as $\{P_A^i\}_{i=1}^T$ and $\{P_B^i\}_{i=1}^T$, respectively. Convert

each patch into a $n \times 1$ column vector denoted by $\{V_A^i, V_B^i\}$, and a new vector $\{V_A'^i, V_B'^i\}$ is obtained by removing the average of $\{V_A^i, V_B^i\}$ from $\{V_A^i, V_B^i\}$, that is

$$V_A'^i = V_A^i - \overline{V_A^l}, \quad V_B'^i = V_B^i - \overline{V_B^l} \tag{2}$$

(2) The sparse coefficients $\{\alpha_A^i, \alpha_B^i\}$ of $\{V_A'^i, V_B'^i\}$ are calculated using OMP method and Eq. (1), the obtained sparse coefficients are merged using the maximum L_1 selection rule.

$$\alpha_F^i = \begin{cases} \alpha_A^i, & \|\alpha_A^i\|_1 > \|\alpha_B^i\|_1 \\ \alpha_B^i, & \text{otherwise} \end{cases} \tag{3}$$

For column vector $\{V_A^i, V_B^i\}$ formed by the ith sub-band patch, we have following fusion result:

$$V_F'^i = D\alpha_F^i \quad V_F^i = \begin{cases} V_F'^i + \overline{V_A^l}, & \|\alpha_A^i\|_1 > \|\alpha_B^i\|_1 \\ V_F'^i + \overline{V_B^l}, & \textit{otherwise} \end{cases} \tag{4}$$

Finally, we have fused vectors $\{V_F^i\}_{i=1}^T$.

(3) Reshape vector V_F^i into a sub-band block P_F^i with size 8×8, add the block to L_F at its responding position, the fused low frequency coefficients L_F are obtained. The fusion process is depicted in Fig. 1.

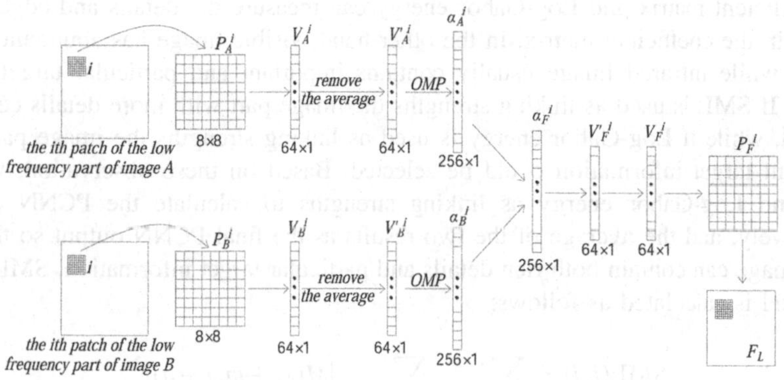

Fig. 1. Schematic diagram of the fusion process of the low frequency coefficients of infrared and visible images based on SR.

2.2 Pulse Coupled Neural Network (PCNN)

PCNN model is a feedback network with each neuron interconnected and characterized by the global coupling and pulse synchronization of neurons. It is widely used in image fusion. The simplified PCNN model is described as iteration by the following equation:

$$F_{ij}(n) = I_{ij}$$

$$L_{ij}(n) = \exp(-\alpha_L)L_{ij}(n-1) + V_L \sum W_{ijpq} Y_{pq}(n-1)$$

$$U_{ij}(n) = F_{ij}(n)(1 + \beta L_{ij}(n))$$

$$\theta_{ij}(n) = \exp(-\alpha_\theta)\theta_{ij}(n-1) + V_\theta Y_{ij}(n-1)$$

$$Y_{ij} = \begin{cases} 1, & U_{ij}(n) \geq \theta_{ij}(n) \\ 0, & U_{ij}(n) < \theta_{ij}(n) \end{cases} \tag{5}$$

In above equations, n denotes the iterative number and ij refers to a location (i, j) of a neuron, F_{ij} is the feeding input, I_{ij} represents the external sources, $L_{ij}(n)$ is linking input, $W_{ijpq}(n)$ is linking weight from neuron (i, j) to neuron (p, q), $U_{ij}(n)$ is the total internal activity, β is the linking strength, $\theta_{ij}(n)$ is the dynamic threshold value, V_L and V_θ are amplification factors of linking input and threshold, respectively. α_L and α_θ are attenuation time constants of $L_{ij}(n)$ and $\theta_{ij}(n)$, respectively. $Y_{ij}(n)$ stands for the pulse output of neuron and it gets either the value 0 or 1. $Y_{ij}(n) = 1$ means that ignition takes place one time. In each iteration, each neuron would go through three steps of receiving, modulating, and pulse generating, and finally generates a pulse output.

In PCNN model, the linking strength β is an important parameter, which reflects the correlation between a pixel and its neighbors. It is well known that human eyes can perceive area with obvious characteristics more easily. So, it is not so reasonable that all neurons have the same value of β, i.e. let β be a constant as in convention PCNN model. Note that Sum-Modified-Laplacian (SML) can describe magnitude of energy of the coefficient matrix and Log-Gabor energy can measure the details and edge information in the coefficient matrix. In the other hand, visible image has abundant object details, while infrared image usually contains important and particular target information. If SML is used as linking strengths the image part with more details could be selected, while if Log-Gabor energy is used as linking strengths the image part with important target information could be selected. Based on these observation, we use SML and Log-Gabor energy as linking strengths to calculate the PCNN output, respectively, and the average of the two results as the final PCNN output so that the fused image can contain both rich details and particular target information. SML value of a pixel is calculated as follows:

$$\mathrm{SML}(i,j) = \sum\nolimits_{m=-1}^{1} \sum\nolimits_{n=-1}^{1} [ML(i+m, j+n)]^2 \tag{6}$$

where ML represents the Modified-Laplacian of a pixel which is calculated as:

$$\mathrm{ML}(i,j) = |2\mathrm{C}(i,j) - \mathrm{C}(i-1,j) - \mathrm{C}(i+1,j)| + |2\mathrm{C}(i,j) - \mathrm{C}(i,j-1) - \mathrm{C}(i,j+1)|. \tag{7}$$

in Eq. (7), $\mathrm{C}(i, j)$ denotes the intensity value of gray scale image.

Let $LG^x_{kl}(i, j)$ be the Log-Gabor energy of the high-frequency sub-band (at the kth scale, lth direction) of image x in NSCT domain at the pixel (i, j), it is given as:

$$LG_{kl}^{x}(i,j) = \frac{1}{9}\sum\nolimits_{m=-1}^{1}\sum\nolimits_{n=-1}^{1} T_{kl}^{x}(i+m, j+n) \quad (8)$$

$$T_{kl}^{x}(i,j) = \sum_{u=1}^{U}\sum_{v=1}^{V}\sqrt{real(H_{kl}(i,j) * g_{kl}^{uv}(i,j))^2 + imag(H_{kl}(i,j) * g_{kl}^{uv}(i,j))^2} \quad (9)$$

where $H_{kl}(i,j)$ is the coefficient located at (i,j) in high-frequency sub-images of the source image x at kth scale, lth direction, $*$ denotes convolution operation, $g_{kl}^{uv}(i,j)$ corresponds to Log-Gabor wavelets in scale u and direction v, real is the real part, and imag is the imaginary part (For more information, please see Ref. [13]).

PCNN is a single layer, two-dimensional, laterally connected network of integrate and fire neurons. When PCNN is used for image processing, each neuron in the network corresponds to one pixel in an input image, and gray value of the pixel is taken as an external stimulus of the neuron in channel. The number of neurons in the network is equal to the pixel number of the input image. A neuron outputs results in two states, pulse and nonpulse. The number of pulse is proportional to external stimulus, namely, the neuron with strong stimulus has large pulse number, while the neuron with weak stimulus has small pulse number. Thus, the infrared target in an infrared image and detailed part in a visible light image are corresponding to the bright part of the ignition map. Using Log-Gabor energy as linking strength can improve the fused image's clarity. For detailed part, we have the following fusion rule:

$$C_p^F(i,j) = \begin{cases} C_p^A(i,j) & \text{if } Y_p^A(i,j) > Y_p^B(i,j) \\ C_p^B(i,j) & \text{otherwise} \end{cases} \quad (10)$$

where $C_p^X(i,j)$ is value of the element at the ith row, jth column of the pth detailed information matrix of image X, capital letter F, A and B represent the fused image and two source images. $Y_p^A(i,j)$ and $Y_p^B(i,j)$ are the PCNN output of the pth detailed information matrices of image A and B obtained by Eq. (5), respectively.

3 Proposed Fusion Algorithm

In this paper, we propose an infrared and visible image fusion scheme in NSCT domain, in which NSCT is firstly employed to decompose each of the source images into the low frequency sub-band and high frequency sub-bands at different scales and directions. Fusion rules are applied on the low-pass sub-band and high-pass sub-bands respectively. According to the fact that the low frequency NSCT coefficients are not very sparse and the high-frequency coefficients in NSCT domain represent the detailed components of the source images and have strong correlations among adjacent pixels, we use the sparse representation based fusion rule introduced in Sect. 2.1 to integrate the low frequency coefficients, the fusion rule based on improved PCNN developed in Sect. 2.2 to fuse each high frequency sub-bands of the source images. Having fused the frequency domain coefficients, the final fused image is produced by applying the

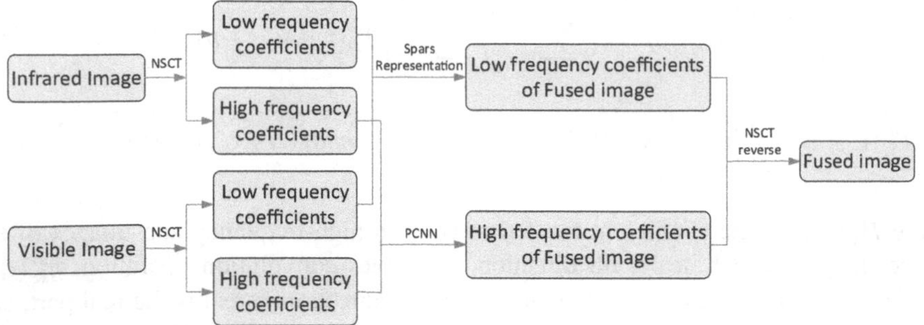

Fig. 2. Schematic diagram of the proposed image fusion method.

inverse NSCT on the merged sub-bands. The schematic diagram of the proposed image fusion method is shown in Fig. 2.

4 Fusion Results and Discussions

To evaluate the efficiency of the proposed algorithm, it is compared with four methods: NSCT (NSCT with average-maximum fusion rule), method in Ref. [6] (NSCT with region characteristics fusion rule), method in Ref. [7] (NSCT with fusion rule based on PCNN and sub-band coefficient filtering), method in Ref. [10] (NSCT with fusion rule based on PCNN and region features). For method of NSCT, decomposition level is 4, the number of directions are 2, 2, 3, 4 in the scales from coarser to finer. The fusion strategy used in the method of NSCT is average-selection rule for the low frequency and maximum-selection rule for the high frequency. For the methods in Ref. [6, 7, 10], the algorithm and parameter values are exactly the same as the original ones. Experimental environment: AMD Athlon(tm) 64 X2 Dual Core Processor 3800 + 2.00 GHZ, Memory 4G, 32 bit Win7 operating system, Matlab R2014a programming.

To evaluate the fused results of different methods as accurately as possible, objective and subjective evaluations are both employed. The objective indexes including standard deviation (SD), entropy of information (ENTR), average gradient (AG), spatial frequency (SF), and index $Q^{AB/F}$ are adopted. SD is used for measuring the image texture and dispersion degree of an image, the larger the dispersion degree is, the clearer the image would be. ENTR is used to judge the richness of the average information of an image, the higher the entropy is, the more information the image has. AG is used to describe the changing feature of image texture and the detailed information, larger values of the AG index correspond to higher spatial resolution. SF is used to describe the hierarchy and clarity of an image. And index $Q^{AB/F}$ is used to evaluate the level of the fusion algorithm in transferring input gradient information into the fused image. For all the five indexes, the larger the value, the better the fusion effect. The first four indexes are calculated by the following formulas:

$$SD = \sqrt{\frac{1}{M \times N}\sum_{m=1}^{M}\sum_{n=1}^{N}(F(m,n)-\bar{F})^2} \tag{11}$$

$$ENTR = -\sum_{i=0}^{L-1} P_i log_2 P_i \tag{12}$$

$$AG = \frac{1}{MN}\sum_{m=1}^{M}\sum_{n=1}^{N}\sqrt{\frac{\Delta F_x^2(m,n)+\Delta F_y^2(m,n)}{2}} \tag{13}$$

$$SF = \sqrt{RF^2 + CF^2} \tag{14}$$

where F(m,n) denotes gray value of the image, $\bar{F}$ is average gray value of the image, $M \times N$ is size of the image, ΔF_x and ΔF_y represent the difference of variable x and y, L is total grey level of the image, $P_0, P_1 \ldots, P_{L-1}$ is probability distribution of each level. In Eq. (14), RF and CF represents row and column frequency of the image, respectively, which are given as the follows:

$$RF = \sqrt{\frac{1}{M \times N}\sum_{m=1}^{M}\sum_{n=1}^{N}(F(m,n)-F(m,n-1))^2} \tag{15}$$

$$CF = \sqrt{\frac{1}{M \times N}\sum_{m=1}^{M}\sum_{n=1}^{N}(F(m,n)-F(m-1,n))^2} \tag{16}$$

The index $Q^{AB/F}$ is given by:

$$Q^{AB/F} = \frac{\sum_{n=1}^{N}\sum_{m=1}^{M}(Q^{AF}(n,m)W^A(n,m)+Q^{BF}(n,m)W^B(n,m))}{\sum_{n=1}^{N}\sum_{m=1}^{M}(W^A(n,m)+W^B(n,m))} \tag{17}$$

where $Q^{AF}(n,m) = Q_g^{AF}(n,m)Q_\alpha^{AF}(n,m)$, $Q_g^{AF}(n,m)$ and $Q_\alpha^{AF}(n,m)$ are the edge strength and orientation preservation values at location (n,m), respectively; $M \times N$ is the size of images. $Q^{BF}(n,m)$ is similar to $Q^{AF}(n,m)$. $W^A(n,m)$ and $W^B(n,m)$ reflect the importance of $Q^{AF}(n,m)$ and $Q^{BF}(n,m)$, respectively. The dynamic range of $Q^{AB/F}$ is [0 1], and it should be as close to 1 as possible.

The experimental tests are performed using a lot of infrared and visible images. Here, we use three examples to illustrate the fusion effect. Figure 3(a1)–(a3) are infrared image and Fig. 3(b1)–(b3) are visible image. Figures 4, 5 and 6 present fused

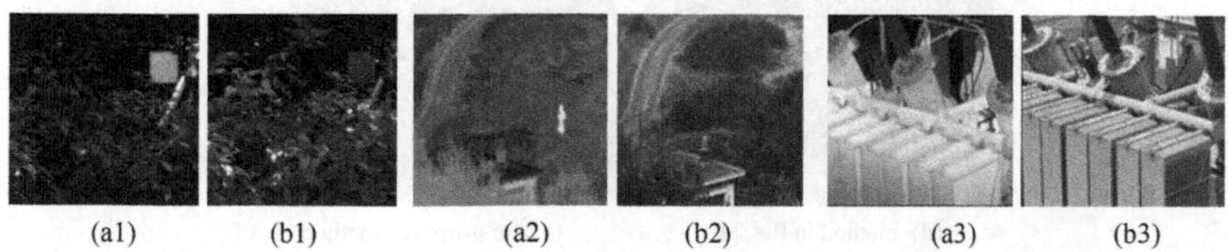

Fig. 3. Three groups of source images to be fused

images by the five multi-resolution methods. From Figs. 4, 5 and 6, it can be seen that the fused image of NSCT and method in Ref. [6] are low on contrast, fusion effect of method in Refs. [7, 10] are improved compared with the first two methods, but some

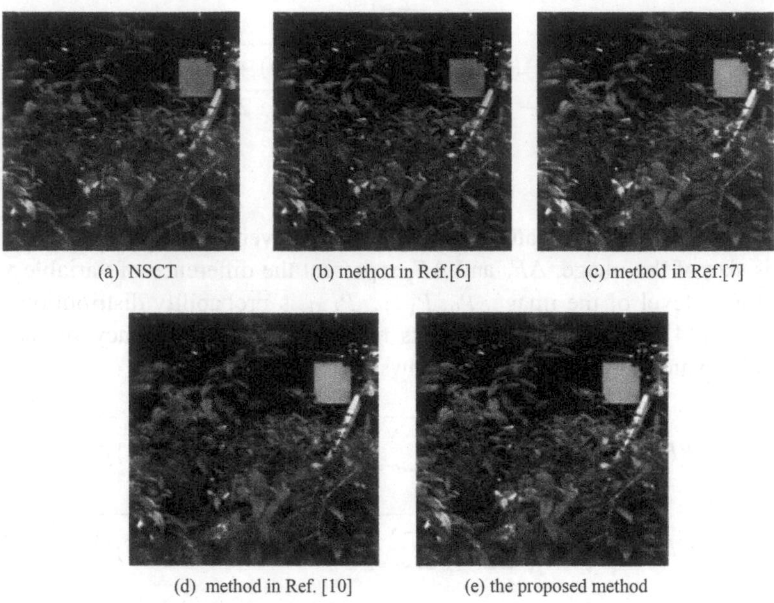

(a) NSCT (b) method in Ref.[6] (c) method in Ref.[7]

(d) method in Ref. [10] (e) the proposed method

Fig. 4. The first group fusion results of different methods

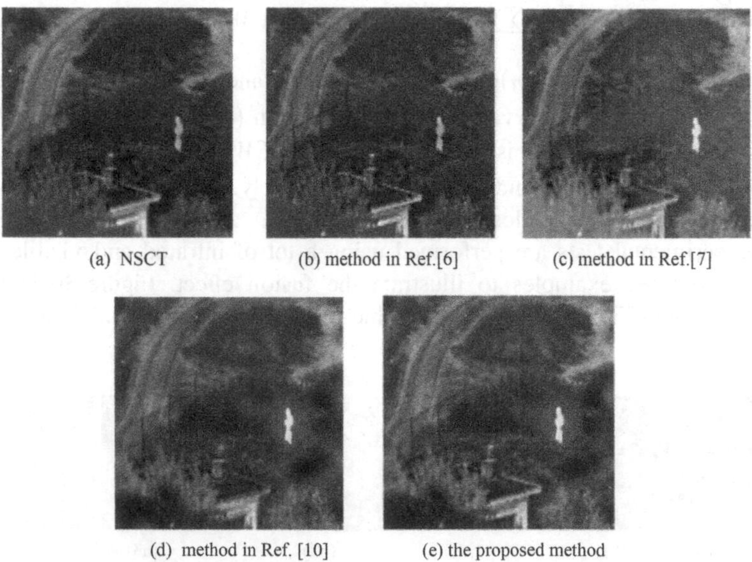

(a) NSCT (b) method in Ref.[6] (c) method in Ref.[7]

(d) method in Ref. [10] (e) the proposed method

Fig. 5. The second group fusion results of different methods

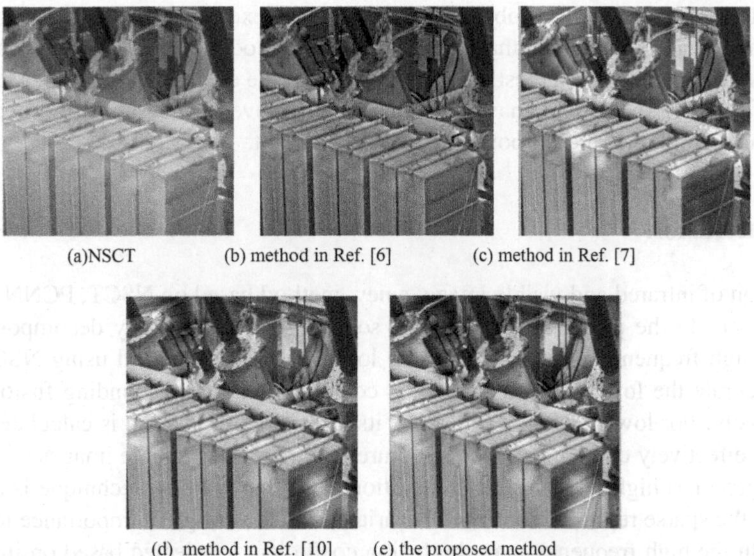

Fig. 6. The third group fusion results of different methods

details in the fused images are blurred, such as leaves and screws on the equipment. From Figs. 4(e), 5(e) and 6(e), we can see that fused image produced by the proposed method has higher clarity and contrast, clearer edge and richer texture than the other results. This demonstrates that the proposed method can transfer more useful information from the source images into the fused result, is an efficient fusion method.

Table 1. Results of objective evaluation indexes

Group	Method	SD	ENTR	AG	SF	$Q^{AB/F}$
1	NSCT	23.6696	6.3215	3.9102	7.7460	0.4127
	Method on Ref. [6]	24.6561	6.3873	5.1108	10.4508	0.4518
	Method on Ref. [7]	31.7406	6.9181	5.6551	10.8878	0.4566
	Method on Ref. [10]	31.0404	6.8962	5.3025	10.1793	0.4063
	Proposed method	**33.8520**	**7.0329**	**5.9779**	**11.5004**	**0.4574**
2	NSCT	33.1741	6.4844	7.7003	17.1117	0.6481
	Method on Ref. [6]	30.9550	6.0857	8.8707	20.1712	0.6638
	Method on Ref. [7]	39.4242	6.3763	8.8215	19.5976	0.6562
	Method on Ref. [10]	38.7752	6.6961	8.7040	18.9664	0.6217
	Proposed method	**41.4036**	**6.7828**	**9.4534**	**20.4756**	**0.6897**
3	NSCT	51.2651	7.5422	7.9844	16.9052	0.3775
	Method on Ref. [6]	46.1498	7.4614	13.5484	28.1221	0.5900
	Method on Ref. [7]	57.8401	7.6593	13.9834	29.1089	**0.6332**
	Method on Ref. [10]	56.7755	7.6505	13.5253	28.1996	0.5987
	Proposed method	**60.8869**	**7.6934**	**14.1860**	**29.4380**	0.6305

Table 1 lists the results of objective evaluation indexes of different fusion methods. From these values, we find that the proposed method outperforms the other four methods by acquiring the largest value for almost all the evaluation indexes. Moreover, indexes of SD, AG and SF have significantly improved. The subjective evaluation results also show that the proposed method is efficient.

5 Conclusion

For fusion of infrared and visible image, a new method based on NSCT, PCNN and SR is proposed. In the proposed method, the source images are firstly decomposed into several high frequency sub-bands and one low frequency sub-band using NSCT, and then integrate the low and high frequency coefficients by corresponding fusion rules, respectively. For low frequency sub-band, its sparse representation is calculated using OMP to effectively capture the deep structure features of the source images so that the fused image has higher resolution. In addition, a sliding window technique is adopted to make the sparse representation shift invariant, which is of great importance to image fusion. In the high frequency part, the fusion coefficient is selected based on improved PCNN by making full use of its characteristics synchronous excitation. The linking strength in convention PCNN has constant value, while in our improved PCNN model we give it the value of SML and Log-Gabor energy. The fused image generated by the proposed method not only preserves the scene information of the visible image, but also successfully integrates the target information of the infrared image, so it has both high contrast and rich details information. The fusion result of the proposed method outperforms the four other multi-resolution fusion method chosen in this paper in both visual quality and objective evaluation indexes.

Acknowledgments. This work is supported by the National Natural Science Foundation of China (No. 61571046, No. 61272026, 61571046, No. 61370193), Science and Technology Development Fund of Macao SAR (No. 097/2013/A3).

References

1. Xiang, T., Yan, L., Gao, R.: A fusion algorithm for infrared and visible images based on adaptive dual-channel unit-linking PCNN in NSCT domain. Infrared Phys. Technol. **69**, 53–61 (2015)
2. Do, M.N., Vetterli, M.: The contourlet transform: an efficient directional multiresolution image representation. IEEE Trans. Image Process. **14**(12), 2091 (2005)
3. Da, C.A., Zhou, J., Do, M.N.: The nonsubsampled contourlet transform: theory, design, and applications. IEEE Trans. Image Process. **15**(10), 3089–3101 (2006). A Publication of the IEEE Signal Processing Society
4. Li, H., Qiu, H., Yu, Z., Zhang, Y.: Infrared and visible image fusion scheme based on NSCT and low-level visual features. Infrared Phys. Technol. **76**, 174–184 (2016)

5. Das, S., Kundu, M.K.: NSCT-based multimodal medical image fusion using pulse-coupled neural network and modified spatial frequency. Med. Biol. Eng. Comput. **50**(10), 1105–1114 (2012)
6. Zhang, J.L., Zhao, E.Y.: Fusion method for infrared and visible light images based on NSCT. Laser Infrared **43**(3), 319–323 (2013)
7. Ikuta, C., Zhang, S., Uwate, Y., Yang, G.: A novel fusion algorithm for visible and infrared image using non-subsampled contourlet transform and pulse-coupled neural network. In: International Conference on Computer Vision Theory and Applications, pp. 160–164. IEEE (2014)
8. Zhang, G.M., Zhang, C.Z., Harvey, D.M.: Sparse signal representation and its applications in ultrasonic NDE. Ultrasonics **52**(3), 351–363 (2012)
9. Yu, N., Qiu, T., Bi, F., Wang, A.: Image features extraction and fusion based on joint sparse representation. IEEE J. Sel. Top. Sig. Process. **5**(5), 1074–1082 (2011)
10. Shen, C.: A new effective image fusion algorithm based on NSCT and PCNN. J. Inf. Comput. Sci. **12**(10), 4137–4144 (2015)
11. Liu, Y., Liu, S., Wang, Z.: A general framework for image fusion based on multi-scale transform and sparse representation. Inf. Fusion **24**, 147–164 (2015)
12. Wang, Z., Ma, Y.: Medical image fusion using m-PCNN. Inf. Fusion **9**(2), 176–185 (2008)
13. Yang, Y., Tong, S., Huang, S., Lin, P.: Log-gabor energy based multimodal medical image fusion in NSCT domain. Comput. Math. Methods Med. **2014**(2), 835481 (2014)

Methods for Eliminating the Complex Background of Pedestrian Images

Di Wu[1(✉)], Si-Jia Zheng[1], You-hua Zhang[2], and Zhi-peng Li[1]

[1] School of Electronics and Information Engineering, Tongji University, Shanghai 201804, China
wudi_qingyuan@163.com
[2] School of Information and Computer, Anhui Agricultural University, Hefei, Anhui, China

Abstract. In order to eliminate the complex background of pedestrian images in the re-identification task, this paper introduces grab-cut and co-segmentation algorithms to eliminate the complex background for the different pedestrian datasets. Specifically, we use grab-cut to handle small person dataset. For the bigger dataset, the designed co-segmentation algorithm which based on HSV color space is applied. Experimental results show that the proposed methods have a good effect.

Keywords: Complex background · Pedestrian images · Grab-cut · Co-segmentation

1 Introduction

As a basic task in multi-camera surveillance system, person re-identification aims to match pedestrians observed from non-overlapping cameras or across different time with a single camera [1–4], it has important applications in behavioral understanding, threat detection and video surveillance [1]. Recently, the task has gained much attention in computer vision field [5], despite the researchers make great efforts on this task, person re-identification remains a challenging problem. See Fig. 1 for some typical examples of pedestrians' images. Person's appearance can change significantly when large variations in view angle, illumination, and complex background are involved [1, 5], among those challenges, the complex background has the most influence on the re-identification effect, because of the complex background could directly affects the exploration of robust feature representations, which is an crucial component in person re-identification task. So far, an enormous amount of methods have been proposed to obtain robust feature representations, which are used to discriminatively represent the person image. However, those methods directly develop the features on the original person images, which contain some background clutter. Since complex background is not belonging to the any part of body, it only has negative effects on the feature representations. Despite lots of works about feature extraction notice the seriousness of this problem, hardly any relevant works have been done. Therefore, it is of great significance in eliminating the background clutter of person images in the re-identification task.

D.-S. Huang et al. (Eds.): ICIC 2017, Part I, LNCS 10361, pp. 432–441, 2017.
DOI: 10.1007/978-3-319-63309-1_40

Fig. 1. Some examples of person images

Since the color depth of the background in person images varies and the resolution of person images is very low, it is difficult to disregard the background. The existing method used in person re-identification to eliminate the background clutter is STEL (Stel component analysis) generative model [6]. Due to the complex scene and the low resolution in the image, this method cannot obtain a good effect. The performance of this method is shown as Fig. 2. From Fig. 2, we can see the pedestrians' heads are eliminated as the background noise. Since the head and shoulder could provide the most discriminative representation [1], it is important to retain the pedestrians' heads while eliminating the background clutter.

Based on the above description, aiming to the different characteristics of pedestrian datasets, in this paper, we introduce grab-cut [7] and designed co-segmentation algorithms to eliminate the background clutter for the different pedestrian datasets. Specifically, for small datasets with only two images for one person, like VIPeR [8], which is one of the most challenging dataset for the person re-identification task. The images in this dataset have various viewpoint and suffer from significant resolution changes across different camera views, which are improper to use co-segmentation algorithm. In this circumstance, we utilize grab-cut algorithm to deal with the dataset. For the bigger datasets, like CUHK01 [9], which is one of the largest published person re-identification datasets, consists of four different images for one person, we could use co-segmentation algorithm to segment the image sequence for one pedestrian.

The remainder of this paper is organized as follows: In Sect. 2, we review the grab-cut algorithm and introduce it to solve this conundrum for the small dataset. In Sect. 3, a suitable co-segmentation algorithm is designed for the large person

Fig. 2. Some segmentation results by stel model

re-identification dataset. In Sect. 4, we present our experimental results. Some suggestions regarding future work are given in Sect. 5.

2 Grab-Cut

Grab-Cut [7] is an efficient image segmentation algorithm to extract object from complex background. Different from the classic segmentation algorithm, the technique uses both texture and edge information to segment image. This method could achieve a high segmentation precision with a little interactive operation. Moreover, grab-cut can not only segment original grey image, but also segment colorful image. Based on the advantages discussed above, this method is suitable for segmenting the Pedestrian images which contain complex background.

The basic idea of the grab-cut algorithm is to use graph cut technology to implement segmentation, and to employ GMM (Gauss mixture model) to model the region item. The procedure of the algorithm is shown as follows:

Initialization

Obtained the trimap T through user selecting the target region (foreground pixels) TU, and outside the TU is background region (background pixels) TB; $\bar{T}_U = T_B$.

Initialize $l_n = 0$ when n $\in T_B$; $l_n = 1$ when n $\in T_U$.

According to the values of l_n, the foreground and background GMMs' parameters are initialized respectively.

Iterative minimization

(1) Allocate GMM component for each pixels:

$$k_n := \arg\min_{k_n} D_n(l_n, k_n, \theta, z_n)$$

(2) Learn GMM parameters from image z:

$$\theta := \arg\min_{\theta} \ U(L, K, \theta, Z)$$

(3) Use min cut to estimate the segmentation:

$$\min_{k} \ E(L, K, \theta, Z)$$

(4) Repeat above steps 1 to 3, until convergence.

As mentioned earlier, the size of VIPeR dataset for person re-identification task is fairly small, and the pedestrian images in this dataset contain complex background, which have a bad influence on the re-identification result. To overcome this vexed problem, we exploit grab-cut algorithm to eliminate the complex background of the VIPeR dataset. The experimental results are presented in Sect. 4.

3 Co-segmentation

Co-segmentation is an important weak supervised segmentation algorithm [10], unlike single image segmentation algorithms [11–15], the idea is to extract common object area from a group of images. This method can jointly segment several images based on the co-occurrence of objects in the images, it can obtain a semantic object area without any additional image label information. The algorithm has aroused more and more concern from scholars nowadays. The mathematical definitions of co-segmentation are as follows:

S represents a group of image which contains N images, S_k ($1 \leq \mathrm{k} \leq \mathrm{N}$) is the k_{th} image, $I_k(x, y)(1 \leq x \leq x_{max}, 1 \leq y \leq y_{max})$ is the pixel of the k_{th} image at (x, y), δ_k is the whole area of the k_{th} image, the segmented subarea sequence $\delta_k^i(k \geq 2, 1 \leq i \leq i_{max})$ must be meet the following conditions simultaneously:

(1) $\delta_k = \cup_{i=1}^{i_{max}} \delta_k^i$;

(2) $\delta_k^i \cap \delta_k^j = \emptyset \ (1 \leq i, j \leq i_{max})$;

(3) δ_k^i is a connected region;

(4) All pixels in the region δ_k^i have similar characteristics;

However, the diversity and weak information of the common object area lead co-segmentation task extremely challenging. As regards our Pedestrian images, the diversity and weak information does not exist. From Fig. 2, we can see the person (common object area) area is explicit. Furthermore, the bigger dataset consists of several different images for one person, so the method is applicable to the bigger dataset.

So far, there doesn't exist an effective universal one in image segmentation method, aiming at the illumination variance problem in person images, we use a color feature based foreground similarity measurement, which can segment common pedestrians from a group of images. Sections 3.1 and 3.2 describe the approach.

3.1 Color Feature and Foreground Similarity Measurement

Considering the person images in the dataset have dramatic variations caused by light changes, we chose the HSV color space as the color descriptor for the δ_k^i. The HSV color space has the property of illumination invariant, which would reduce the influence of light change on color stability. Then we apply the measurement method in [10]. Specifically, for each subarea δ_k^i, we use HSV color histogram (h) to represent it. Let $g(\delta_k^i)$ as the descriptor of δ_k^i, then $g(\delta_k^i) = h$. For the person image, the color value is a three (H, S, V) dimensional vector.

The region similarity f is measured by the similarity between a certain pixel (p) and the region δ_k^i, the formula is defined as $f(p, g(\delta_k^i)) = h(p)$. When the color value of the pixel p has a big proportion in the color histogram h, it shows that the region δ_k^i has a lot pixels which are same as the pixel p in color, in other word, the pixel p has a great similarity to the region. However, the pixel values of the similarity color exist tiny differences in practice, therefore, we set β as the upper limit of the pixel values. Based on β, the $f(p, g(\delta_k^i))$ can be redefined as:

$$f(p, g(\delta_k^i)) = \sum_{|p'-p| \leq \beta} h(p')$$

where the $|p' - p|$ is the distance of the two different color values, furthermore, $|p' - p|$ take the distance as the largest difference in H, S, and V color channels.

3.2 Energy Function

The definition of the energy function is written as:

$$E_k(C_k) = \mu Length(C_k) + vArea(\delta_k^i) - \gamma_k^i \int_{\delta_k^i} f[I_k(x, y), g(\delta_{1-k}^i)] dxdy - \gamma_k^0 \int_{\delta_k^0} f[I_k(x, y), g(\delta_k^0)] dxdy$$

Where C_k represents a curve in the image I_k, and $E_k(C_k)$ is the energy function of C_k. δ_k^0 and δ_k^i is the internal and external areas of the curve C_k, respectively. $Length(C_k)$ is the length of C_k, and $Area(\delta_k^i)$ is the area of region δ_k^i, μ, v, γ_k^i, γ_k^0 are the parameters and both of them are greater than zero.

For the foreground pixel $I_k(i,j)$ in the image I_k, there has $\mathrm{f}\left[I_k(i,j), \mathrm{g}\left(\delta^i_{1-k}\right)\right] > f\left[I_k(i,j), \mathrm{g}\left(\delta^0_k\right)\right]$, in contrast, for the background pixel $I_k(i,j)$, there has $\mathrm{f}[I_k(i,j), \mathrm{g}(\delta^i_{1-k})] < f\left[I_k(i,j), \mathrm{g}\left(\delta^0_k\right)\right]$, therefore, any interchange of foreground or background pixel segmentation results would lead to the energy value increase. In other word, when the curve can accurately segment the common objects, the energy function value achieves minimum. Accordingly, the co-segmentation problem can be expressed as:

$$C^*_k = argmin\ E_k(C_k)$$

From the energy function, we can see that the foreground and the background similarity is designed by the reward strategy, by manipulating the energy function, the level-set-based energy function is then given by:

$$\begin{aligned} E_k(\varphi_k) = \mu \int_{\sigma_k} \varepsilon(\varphi_k(x,y))|\nabla\varphi_k(x,y)|dxdy + v \int_{\sigma_k} H(\varphi_k(x,y))dxdy \\ - \gamma^i_k \int_{\sigma_k} \mathrm{f}\left[I_k(x,y), \mathrm{g}\left(\delta^i_{1-k}\right)\right] H(\varphi_k(x,y))dxdy \\ - \gamma^0_k \int_{\sigma_k} \mathrm{f}\left[I_k(x,y), \mathrm{g}\left(\delta^0_k\right)\right](1 - H(\varphi_k(x,y)))dxdy \end{aligned}$$

Where φ_k represents the zero-level set function, H(x) is the Heaviside function, $\varepsilon(x)$ is the 1-D function, $\varepsilon(\varphi_k(x,y))$ denotes C_k, $H(\varphi_k(x,y))$ is the inside region, $1 - H(\varphi_k(x,y))$ represents exterior region.

Finally, we utilize the energy function minimization method mentioned by [10] to minimize the energy function.

Based on the HSV color space and the reward strategy mentioned above, we perform the co-segmentation algorithm on the bigger person dataset, and the experimental results are shown in Sect. 4.

4 Experiments

In this section, we use two well-known datasets, i.e. VIPeR and CUHK01, both of them contain a set of pedestrian images. Among them, VIPeR is the most challenging dataset in the person re-identification arena owing to the huge variance and discrepancy between the images, moreover, the dataset is relatively small. To the above situation, we use grab-cut algorithm to eliminate the complex background for VIPeR dataset. For the CUHK01 dataset, the images are relatively simple and identical, we use the co-segmentation structure to tackle with it. Some experiment results are shown as follows:

As shown above, Fig. 3 is the results of grab-cut algorithm, we can see the segmentation results of pedestrian images are correct and accurate, compared with the STEL generative model, grab-cut can get more accurate results for segmentation. The grab-cut algorithm aims to separate the foreground and the background via a little interactive operation, and the algorithm uses both texture and edge information to segment image, so this method has strongly robustness and could achieve better segmentation effect under the complex background. Figure 4 shows the results of the co-segmentation algorithm. Here, we exhibit four sequences, each sequence contains eight images for one person, as shown above, the four images on the left side of the sequence are the original images from the dataset, and the rest four images are the segmentation results. From Fig. 4 we can see that the complex background is eliminated basically, meanwhile, we recognize the segmentation results exist some minor deficiency, such as the persons' legs and feet are eliminated as background, however, some exist work on person re-identification show that legs and feet provide the least reliable features, and contribute little to the person re-id task [1], therefore, our segmentation results could satisfy the person re-id task. Moreover, the experiments results already better than the previous method. In summary, the introduced algorithms could achieve the desired effect.

Fig. 3. Some experiment results in VIPeR dataset

Fig. 4. Some experiment results in CUHK01 dataset

5 Conclusions

This paper utilizes grab-cut and designed co-segmentation algorithms to segment the pedestrian images. According to the characteristics of different datasets, we use grab-cut to manage the small dataset, and use the designed co-segmentation algorithm handle the bigger dataset. For the designed co-segmentation, we use the HSV color feature to describe the foreground, which can segment the common person from a group of images. The segmentation results of experiments demonstrate the effective of our proposed methods. However, the grab-cut algorithm requires users' interaction and the designed co-segmentation algorithm exists the situation of inexact segmentation, the future work will focus on designing a more effective segmentation algorithm.

Acknowledgments. This work was supported by the grants of the National Science Foundation of China, Nos. 61472173, 61572447, 61373098, 61672382, 61472280, 61672203, 61402334, 61520106006, 31571364, U1611265, and 61532008, China Postdoctoral Science Foundation Grant, Nos. 2016M601646.

References

1. Cheng, D., Gong, Y., Zhou, S., et al.: Person re-identification by multi-channel parts-based CNN with improved triplet loss function. In: Proceedings of the IEEE Conference on Computer Vision and Pattern Recognition, pp. 1335–1344 (2016)
2. Xiao, T., Li, H., Ouyang, W., et al.: Learning deep feature representations with domain guided dropout for person re-identification. In: Proceedings of the IEEE Conference on Computer Vision and Pattern Recognition, pp. 1249–1258 (2016)
3. Li, W., Zhao, R., Xiao, T., et al.: DeepReID: deep filter pairing neural network for person re-identification. In: Proceedings of the IEEE Conference on Computer Vision and Pattern Recognition, pp. 152–159 (2014)
4. Wang, T., Gong, S., Zhu, X., et al.: Person re-identification by discriminative selection in video ranking. IEEE Trans. Pattern Anal. Mach. Intell. **38**(12), 2501–2514 (2016)
5. Matsukawa, T., Okabe, T., Suzuki, E., et al.: Hierarchical Gaussian descriptor for person re-identification. In: Proceedings of the IEEE Conference on Computer Vision and Pattern Recognition, pp. 1363–1372 (2016)
6. Jojic, N., Perina, A., Cristani, M., et al.: Stel component analysis: modeling spatial correlations in image class structure. In: Computer Vision and Pattern Recognition, pp. 2044–2051 (2009)
7. Rother, C., Kolmogorov, V., Blake, A.: Grabcut: Interactive foreground extraction using iterated graph cuts. ACM Trans. Graph. (TOG) **23**(3), 309–314 (2004). ACM
8. Gray, D., Tao, H.: Viewpoint invariant pedestrian recognition with an ensemble of localized features. In: Forsyth, D., Torr, P., Zisserman, A. (eds.) ECCV 2008. LNCS, vol. 5302, pp. 262–275. Springer, Heidelberg (2008). doi:10.1007/978-3-540-88682-2_21
9. Li, W., Wang, X.: Locally aligned feature transforms across views. In: Proceedings of the IEEE Conference on Computer Vision and Pattern Recognition, pp. 3594–3601 (2013)
10. Meng, F., Li, H., Liu, G., et al.: Image cosegmentation by in-corporating color reward strategy and active contour model. IEEE Trans. Cybern. **43**(2), 725–737 (2013)

11. Huang, D.S., Du, J.-X.: A constructive hybrid structure optimization methodology for radial basis probabilistic neural networks. IEEE Trans. Neural Netw. **19**(12), 2099–2115 (2008)
12. Huang, D.S.: Radial basis probabilistic neural networks: model and application. Int. J. Pattern Recogn. Artif. Intell. **13**(7), 1083–1101 (1999)
13. Wang, X.-F., Huang, D.S.: A novel density-based clustering framework by using level set method. IEEE Trans. Knowl. Data Eng. **21**(11), 1515–1531 (2009)
14. Wang, X.-F., Huang, D.S., Xu, H.: An efficient local Chan-Vese model for image segmentation. Pattern Recogn. **43**(3), 603–618 (2010)
15. Li, B., Huang, D.S.: Locally linear discriminant embedding: an efficient method for face recognition. Pattern Recogn. **41**(12), 3813–3821 (2008)

Information Security

Image Encryption Algorithms Based on Non-uniform Second-Order Reversible Cellular Automata with Balanced Rules

Kai Li, Mingyu Sun, Lvzhou Li(✉), and Juhua Chen

School of Data and Computer Science, Sun Yat-sen University,
Guangzhou 510006, People's Republic of China
lilvzh@mail.sysu.edu.cn

Abstract. In this paper, we propose an image encryption algorithm based on non-uniform second-order reversible cellular automata (CA) which consist of two-state cells and use the Moore neighborhood. Different from other existing algorithms based CA, our algorithm uses the balanced CA rules that have a higher randomness than unbalanced rules. More specifically, before encryption we first mix the pixels of the plain image and then group these pixels with 128 bits for each block. Then each block is encrypted based on non-uniform second-order reversible CA by an iterative method. Since different key are constructed and applied for each block, the properties of confusion and diffusion of the encryption algorithm are well satisfied, and our algorithm can avoid the problem in the Electronic Codebook Book (ECB) encryption method that two plaintext blocks that have identical content are encrypted into identical cipher-text blocks. Simulation results show that our algorithm performs well in several security measurements such as histogram, information entropy, correlation of adjacent pixels, and sensitivity.

Keywords: Image encryption · Cellular automata · Balanced rules · Security

1 Introduction

Today, the security is of great importance for the global networks which are characterized by an enormous growth of storage and transmission of digital information. Hence, public and private organizations have become increasingly dependent on cryptographic techniques to guarantee security and authenticity in many areas, especially in electronic commerce transactions.

We can refer to [1] for some image encryption methods used in various fields. However, the traditional encryption methods cannot achieve the ideal effect in encryption efficiency and performance, because of the strong correlation between pixels, large amounts of data, high redundancy and other characteristics of images. One promising technique for image encryption is by using cellular automata (CA). CA, as dynamic systems discrete in time and space, are a good choice for transmitting large numbers of data [2]. Due to its parallelism, ease to implement and simple hardware structures, it is used more and more widely in the encryption technology. More information can be found in [3, 4]. In symmetric cryptography, CA were first used by

D.-S. Huang et al. (Eds.): ICIC 2017, Part I, LNCS 10361, pp. 445–455, 2017.
DOI: 10.1007/978-3-319-63309-1_41

Wolfram to design the stream-based encryption system [5], followed by Habutsu et al. [6], Nandi et al. [7] and Gutowitz [8]. In the field of public-key cryptography, there are also some works based on CA, for example Guan [9].

Over the last decades, CA were used more often as a tool for image encryption. Abdo et al. [10] proposed an image encryption algorithm based on a special kind of periodic boundary cellular automata with unity attractors. In [11], Petre developed a high-performance encryption system that works according with the Programmable Cellular Automata (PCA) theory. In [12], Kamel proposed a block cipher algorithm based on second-order reversible CA constructed from one-dimensional CA with radius 3. In [13], a stream cipher system for color images was constructed based on two-dimensional CA.

In this paper we proposes an encryption algorithm for color images based on the balanced rules (to be defined) of two-dimensional Moore CA and the RGB channel mixing scheme. Simulation results are given to show the security of our encryption algorithm.

The remainder of the paper is organized as follows. Section 2 introduces the basis concept of CA rules. In Sect. 3 we present the encryption algorithm. In Sect. 4, experimental results and security analysis of the proposed encryption method are given.

2 Preliminaries

Cellular automata (CA) [3, 4], as dynamical systems discrete in space and time, are among the oldest models of natural computing, dating back over half a century. A CA consists of a number of cells organized in the form of a lattice. The next state of each cell depends on the current states of a set of cells, which includes the cell itself and some cells surrounding it, and is called its *neighborhood*.

In this paper we need two-dimensional CA (2D CA, for short), of which the cells are arranged in a two-dimensional grid. For 2D CA, two types of neighborhood modes are usually considered: the von Neumann neighborhood and the Moore neighborhood. Figure 1 shows the two types neighborhood modes. Consider a 2D CA consists of cells organized in a $N \times M$ gird, of which each cell has the state set S. Then the CA's configuration at time t can be represented by

$$C^t = \left(s^t_{1,1}, s^t_{1,2} \cdots, s^t_{1,M}, \cdots, s^t_{N,M}\right)$$

where $s^t_{a,b} \in S$ denotes the state of the cell at the position (a, b) at time t. Denote C the set of all configurations of a CA. Then its global transition function $F : C \to C$ is specified by the local update rule of each cell. Consider 2D CA with the Moore neighborhood. Each cell updates its state according to the rule $f : S^9 \to S$ defined by

$$f\left(s^t_{a-1,b-1}, s^t_{a-1,b}, s^t_{a-1,b+1}, s^t_{a,b-1}, s^t_{a,b}, s^t_{a,b+1}, s^t_{a+1,b-1}, s^t_{a+1,b}, s^t_{a+1,b+1}\right) = s^{t+1}_{a,b},$$

which means that the next state of the cell at (a, b) depends on the current states of its Moore neighborhood. In this paper, 2D CA are always assumed to be with the Moore neighborhood and with the cell state set $S = \{0, 1\}$. Then f is a Boolean function with nine variables and thus can be characterized by a 512-bit string. More specifically, f can be expressed by the followings equations:

$$f(000000000) = k_0, \ldots, f(bin(i)) = k_i, \ldots, f(111111111) = k_{511},$$

where $bin(i)$ indicates the binary notation of i (9 bits), and $k_i \in \{0, 1\}, i \in \{0, 1, \ldots, 511\}$. Then we can use the 512-bit string $k_{511} \cdots k_1 k_0$ to represent the rule f, and the decimal integer represented by the bit string is referred to as the "rule number". If we flip each bit in the bit-string representation of f, then we obtain a new rule, called the *complement* of f, denoted by $\bar{f}$. f is called a *balanced rule*, if its bit-string representation contains equal numbers of 0 s and 1 s. The encryption scheme using balanced rules can increase the confusion of the ciphertext.

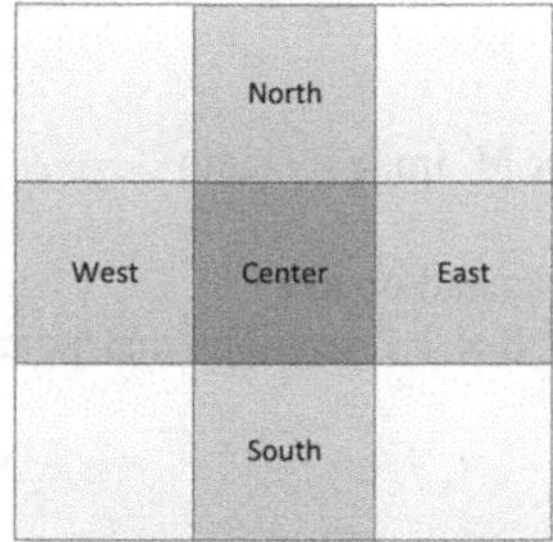

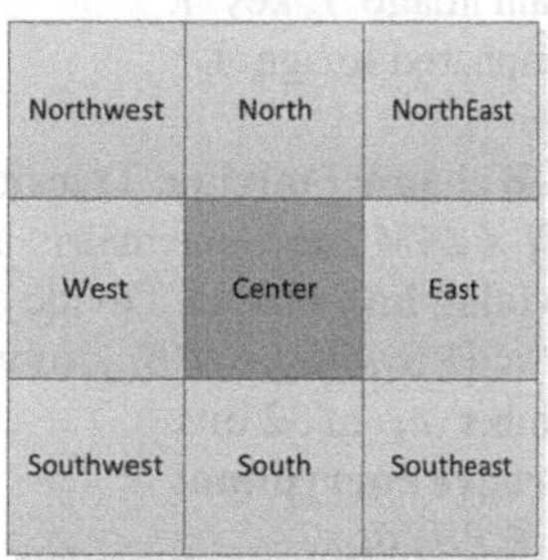

Fig. 1. The von Neumann neighborhood (left) and the Moore neighborhood (right) of 2D CA.

Reversible CA (RCA) means that every configuration has a unique successor and a unique predecessor. The second-order cellular automaton method invented by [16] allows us to construct a RCA from any CA which is not necessarily reversible. More specifically, we can build a second-order RCA based on the CA with the global transition function F using the following equation:

$$C^t = F\left(C^{t-1}\right) \oplus C^{t-2} \tag{1}$$

The defined RCA can then be reversed trivially using the following equation:

$$C^{t-2} = F\left(C^{t-1}\right) \oplus C^t \tag{2}$$

Then the corresponding local rule for each cell is as follows:

$$s_i^t = f\left(s_{\{i\}}^{t-1}\right) \oplus s_i^{t-2} \tag{3}$$

$$s_i^{t-2} = f\left(s_{\{i\}}^{t-1}\right) \oplus s_i^t \tag{4}$$

where $s_{\{i\}}^{t-1}$ denotes the states of cells in the neighborhood of cell i, at time t.

In the above, it is assumed that all cells of CA use the same rule to update they states. Such CA are called *uniform*. Researchers have also studied another kind of CA which are called *non-uniform* CA where different sets of cells can use different rules.

3 Encryption Algorithms Based on CA for Color Images

The decryption algorithm is to reverse each step in the encryption algorithm and thus we will omit the details. This section proposes an encryption algorithm for color images based on non-uniform second-order RCA. In the following, CA are assumed to be 2D CA with the Moore neighborhood and with the cell state set $S = \{0, 1\}$.

Inputs: plain image I, key K
Outputs: ciphered image I_c
Procedure:
step 1. RGB channel mixing: Transform the $N \times M$ image I into a $N \times 24M$ zero-one matrix I'.
step 2. Dividing into blocks: Divide I' into blocks and obtain $PB = \{PB_1, PB_2, \cdots PB_k\}$, of which P_i is 8×16 zero-one matrix and has a number n_i of 32 bits.
step 3. Iterative encryption:
while($PB_i \in PB$) **do**
$CB_i = Iencryption(PB_i, n_i, K)$.
step 4. Merging blocks:
return $I_c = (CB_1, CB_2, \cdots CB_k)$

Fig. 2. Image encryption algorithm based on RCA.

Note that the proposed cryptosystem is symmetric, meaning that the same key is required for encryption and decryption. The encryption algorithm is give in Fig. 2. More specifically the encryption algorithm consists of the following steps:

Step 1: **RGB channel mixing.** As we know, each pixel of a RGB image has three color channels: R (red), G (green) and B (blue). The RGB channel mixing procedure transforms a given $N \times M$ image into an $N \times M$ block matrix with each block being a 24-bit string as follows: For the pixel at (i, j), we obtain a 24-bit string which comprises the eight bits of each channel in the order of R, G and B. As a result, an $N \times M$ image I is transformed into a $N \times 24M$ matrix I' with each entry being 0 or 1 (called *zero-one* matrix in this paper).

Step 2: **Dividing into blocks.** The plain text matrix I' is divided into blocks with each block being a 8×16 zero-one matrix of 128 bits. Therefore, we obtain a set of plaintext blocks $PB = \{PB_1, PB_2, \cdots PB_k\}$, of which each block PB_i is assigned with a block number n_i of 32 bits.

Inputs: plain block PB_i, number n_i, key K
Outputs: ciphered block CB_i
Procedure:
divide PB_i into C_i^0 and C_i^1;
$K_i = K \oplus (n_i n_i n_i n_i)$;
j=2;
while ($j < L$) **do**
{$K_i << 8$;
SK_i = the leftmost 64 bits of K_i;
$C_i^j = F(C_i^{j-1}) \oplus C_i^{j-2}$ and $C_i^{j+1} = F(C_i^j) \oplus C_i^{j-1}$;
j=j+1;
}
return $CB_i = C_i^L C_i^{L+1}$.

Fig. 3. Procedure of the function Iencryption

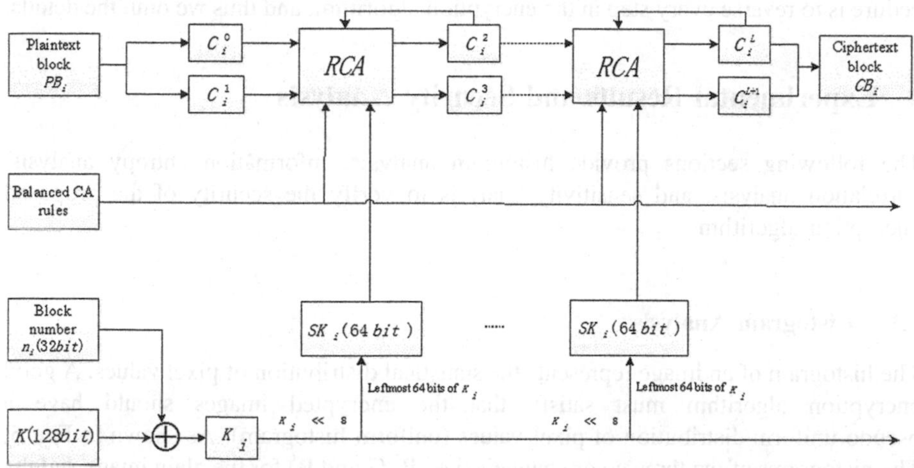

Fig. 4. Sketch of the function Iencryption

Step 3: Iterative encryption. Before encryption, there is a secret key shared by the communication participants, denoted by K, which is set to 128 bits. In addition, there is a set $S = \{R_1, R_2, \cdots, R_{64}\}$ of 64 balanced CA rules that is public. How to encrypt each block in PB is described in Fig. 3 and also the idea is sketched in Fig. 4. If the current block is PB_i, with the number n_i, then we perform the following operations:

1. PB_i is divided into two sub-blocks C_i^0 and C_i^1 with each being a 8×8 zero-one matrix, which are used as two initial configurations of a CA.
2. The key K is expanded into $K_i = K \oplus (n_i n_i n_i n_i)$.
3. Repeat the following iteration L (for example, let L = 80) times with j initially set to be 2:

(3.1) Update K_i by shifting K_i to left 8 bits, denoted by $K_i \ll 8$.[1]
(3.2) Let SK_i be the leftmost 64 bits of K_i.
(3.3) Update the two sub-blocks C_i^0 and C_i^1 according to Eq. (1) as follows: $C_i^j = F\left(C_i^{j-1}\right) \oplus C_i^{j-2}$ and $C_i^{j+1} = F\left(C_i^j\right) \oplus C_i^{j-1}$, where if the j th bit of SK_i is 0, then the j th cell of CA uses the rule $R_j \in S$; otherwise, the complement of R_j, denote by $\overline{R_j}$, is used. Note that all 64 cells in the CA use different rules. Thus, the CA is non-uniform.
(3.4) Update j to be $j = j + 1$.

4. As a result, $CB_i = C_i^{L-1} C_i^L$ is used as the ciphertext of PB_i.

Step 4: Merging blocks. After the above procedure, we obtain a set of ciphertext blocks $CB = \{CB_1, CB_2, \cdots CB_k\}$, and then we merge them in the order as the plaintext was divided into blocks, denoted by $I_c = (CB_1, CB_2, \cdots CB_k)$, which is as the result ciphertext.

Because each step in the encryption algorithm is reversible, the decryption procedure is to reverse every step in the encryption algorithm, and thus we omit the details.

4 Experimental Results and Security Analysis

The following sections provide histogram analysis, information entropy analysis, correlation analysis, and sensitivity analysis to verify the security of the proposed encryption algorithm.

4.1 Histogram Analysis

The histogram of an image represents the statistical distribution of pixel values. A good encryption algorithm must satisfy that the encrypted images should have a pseudo-uniform distribution of pixel values (uniform histogram). As shown in Fig. 5, The histograms of the three color channels (i.e., R, G and B) for the plain image and the encrypted image are given. One can see that the encrypted image has a uniform distribution of pixel values in contrast to that of the plain image, which indicates that the confusion of our algorithm is satisfied. Therefore, our encryption algorithm can resist statistical analysis attacks.

4.2 Information Entropy Analysis

According to Shannon's theory, entropy is one of the main measurements for randomness of information. For any message of n bit, the upper bound of the entropy is n. The entropy of a given raw data m can be defined as follows:

---

[1] For example, $01000 \ll 2 = 00001$.

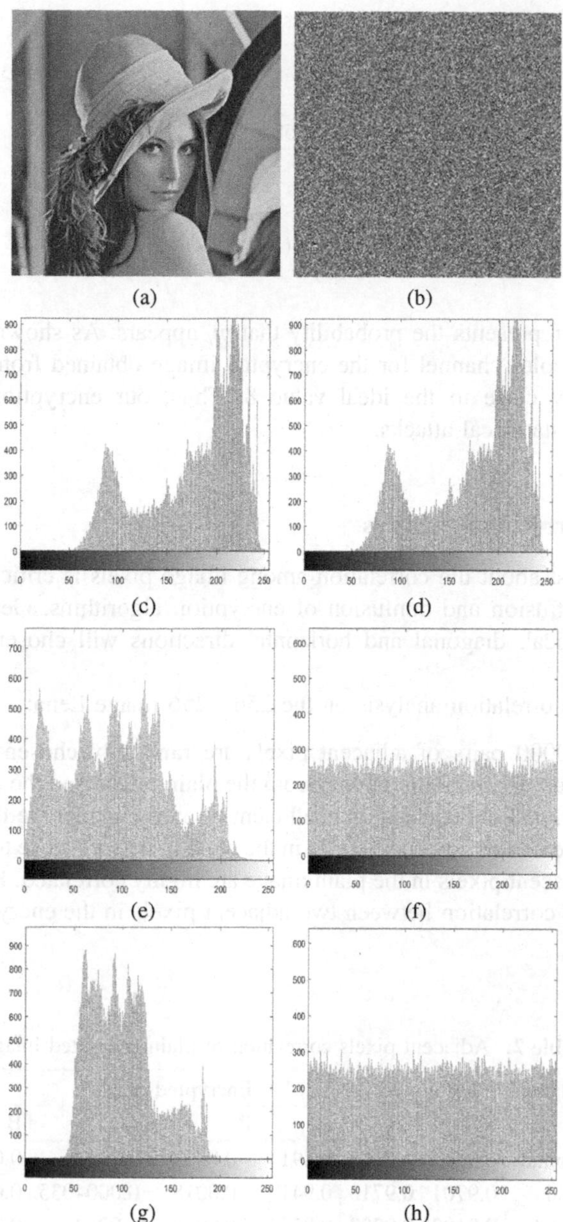

Fig. 5. Histograms for the plain image and the encrypted image: (a) plain image, (b) encrypted image, (c) histogram of red channel for the plain image, (d) histogram of red channel for the encrypted image, (e) histogram of green channel for the plain image, (f) histogram of green channel for the encrypted image, (g) histogram of blue channel for the plain image, (h) histogram of blue channel for the encrypted image. The X-axis represents pixel values of images, while the Y-axis denotes the number of pixels. For example, the point [Trial mode] in (h) stands for that in the blue channel of the encrypted image, there are 250 pixels having the value 50. (Color figure online)

Table 1. Entropy analysis of plain/cipher images

Image	Plain image			Cipher image		
	R	G	B	R	G	B
Lena	7.2477	7.5327	6.9584	7.9902	7.9902	7.9876

$$H(m) = \sum_{i=0}^{2^n-1} p(mi) \log_2 \frac{1}{p(mi)} \tag{5}$$

Where $p(m_i)$ represents the probability that m_i appears. As shown in Table 1, the entropy of each color channel for the encrypted image obtained from our encryption algorithm is very close to the ideal value 8. Thus, our encryption algorithm can effectively resist statistical attacks.

4.3 Image Correlation Analysis

The statistical test about the correlation among image pixels is critical for evaluating the quality of diffusion and confusion of encryption algorithms. Generally the pixel pairs in the vertical, diagonal and horizontal directions will chosen to analyze the correlation.

We perform correlation analysis on the 256 $*$ 256 image Lena:

(1) A set of 30000 pairs of adjacent pixels are randomly chosen (in the vertical, diagonal and horizontal directions) from the plain image and the encrypted image, respectively, and the correlation coefficient of each channel are then calculated in Table 2. The correlation coefficients in the plain image are close to 1, which means that two adjacent pixels in the plain image are highly correlated. Howcvcr, there is a negligible correlation between two adjacent pixels in the encrypted image.

Table 2. Adjacent pixels correlation of plain/encrypted images

Correlation	Plain image			Encrypted image		
	R	G	B	R	G	B
Horizontal	0.9415	0.9467	0.8914	−0.0062	−0.0024	0.0020
Vertical	0.9701	0.9713	0.9417	0.0018	−0.0004035	0.0062
Diagonal	0.9193	0.9203	0.8534	−0.0016	0.0064	0.0064

Table 3. Adjacent pixels correlation of plain and encrypted images at the same location

Image channel	Value of correlation
R	0.0011
G	0.0079
B	0.0020

(2) A set of 30000 pairs of pixels at the same location in the plain image and the encrypted image are randomly chosen, and the correlation coefficients for each channel are shown in Table 3. As can been seen that the correlation coefficients for three color channels are all close to zero, which means there is a trivial correlation between the plain image and the encrypted image.

4.4 Sensitivity Analysis

An encryption algorithm should satisfy that a small change in either the plain text or the key should result in a significant change in the ciphertext.

Sensitivity Analysis for Keys

(1) The sensitivity analysis is performed on a set of 1000 different plain blocks (128 bits) of the plain image *Lena*, and the averaged results are illustrated in Fig. 6. *key′* is a new key by flipping one bit from *key*. We use *key′* and *key* to encrypt the same plain block, respectively. The average change rate of the ciphertext owning to the one-bit difference between *key′* and *key* is calculated in the following:

$$d_i = (H(f(PB, key, n), (f(PB, key', n)))/128) \times 100\% \quad (6)$$

when H is the hamming distance between two 128-bit blocks and i represents the location of the different bit in key′ and key ($1 \leq i \leq 128$).
It can be observed from Fig. 6 that the average change rate is very high for each bit position modification, and so the key is very sensitive to modification. Furthermore, we perform the *NPCR* (Number of pixels change rate) analysis and *UACI* (Unified average changing intensity) analysis on the *Lena*. Then *NPCR* and

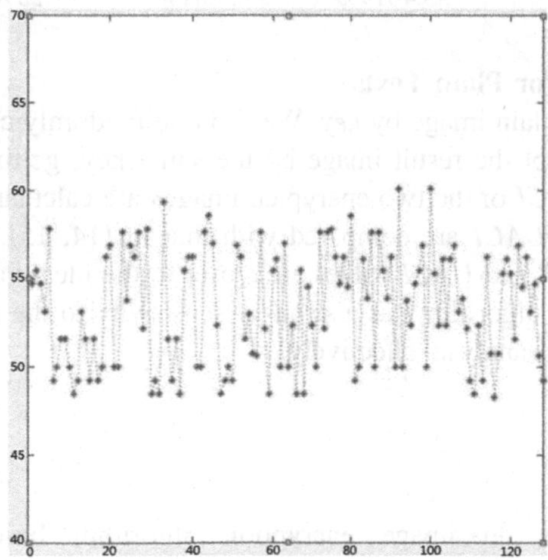

Fig. 6. Secret key sensitivity to elementary one-bit modification: X-axle denotes the position of the key's bit altered, and Y-axle denotes the average change rate (di)

Table 4. The average values of NPCR and UACI of three color channels

Plain image	NPCR	UACI
Lena	99.5043%	33.3348%

Table 5. Use the key and the key′ to decrypt the image encryption by the key

Plain image	Encryption by **key**	Decryption by **key**	Decryption by **key**′

(a)	(b)	(c)	(d)

UACI are shown in Table 4. It can be seen that the average values of *NPCR* and *UACI* are very close to the ideal values, respectively, which means that the two encrypted images are almost independent.

(2) After using *key* to encrypt the plain image, we use *key*′ to decrypt the encrypted image (*key*′ is a bit flip of the *key*). It shows in Table 5 that the image cannot be successfully decrypted by *key*′. It shows that decryption is also sensitive to small change of the key.

Table 6. NPCR/UACI values for plain text sensitivity

Lena (256 ∗ 256)	Our algorithm	Reference [14]	Reference [15]
NPCR	99.6037%	99.5863%	52.4832%
UACI	33.4619%	33.3861%	17.2570%

Sensitivity Analysis for Plain Texts

Firstly, we encrypt a plain image by *key*. We flip one randomly chosen bit in the plain image and then encrypt the result image by the same key, getting another encrypted image. *NPCR* and *UACI* of the two encrypted images are calculated, respectively. The results of *NPCR* and *UACI* are compared with that in [14, 15]. It can be seen from Table 6 that the *NPCR* and *UACI* values are closer to the ideal values than the ones in [14, 15]. It means our algorithm has a stronger sensitivity to the plain text, and thus it can against differential analysis effectively.

5 Conclusion

This paper proposes an image encryption algorithm based on non-uniform second-order reversible CA. Our algorithm uses the balanced CA rules that have a higher randomness than unbalanced rules. Since different keys are used to encrypt different plain blocks, our encryption algorithm can avoid the problem in the ECB encryption method that two different blocks with the same content result in the same ciphertext. Experimental results show that the proposed encryption scheme has a good performance in analysis of histogram, information entropy, correlation of adjacent pixels, and sensitivity.

Acknowledgement. This work is supported in part by the National Natural Science Foundation of China (Nos. 61472452, 61572532) and the National Natural Science Foundation of Guangdong Province of China (No. 2014A030313157), the Science and Technology Program of Guangzhou City of China (201707010194).

References

1. Schneier, B.: Applied Cryptography. Wiley, New York (1996)
2. Fuster-Sabater, A., Caballero-Gil, P.: On the use of cellular automata in symmetric cryptography. Acta. Appl. Math. **93**, 215–236 (2006)
3. Sarkar, P.: A brief history of cellular automata. ACM Comput. Surv. **32**, 80–107 (2000)
4. Kari, J.: Theory of cellular automata: a survey. Theor. Comput. Sci. **334**, 3–33 (2005)
5. Wolfram, S.: Cryptography with cellular automata. In: Williams, H.C. (ed.) CRYPTO 1985. LNCS, vol. 218, pp. 429–432. Springer, Heidelberg (1986). doi:10.1007/3-540-39799-X_32
6. Habutsu, T., Nishio, Y., Sasase, I., Mori, S.: A secret key cryptosystem by iterating a chaotic map. In: Davies, D.W. (ed.) EUROCRYPT 1991. LNCS, vol. 547, pp. 127–140. Springer, Heidelberg (1991). doi:10.1007/3-540-46416-6_11
7. Nandi, S., Kar, B.K., Chaudhuri, P.P.: Theory and applications of cellular automata in cryptography. IEEE Trans. Comput. **43**, 1346–1357 (1994)
8. Gutowitz, H.: Cryptography with dynamical systems. In: Goles, E., Boccara, N. (eds.) Cellular Automata and Cooperative Phenomena. Kluwer, Dordrecht (1993)
9. Guan, P.: Cellular automaton public-key cryptosystem. Complex Syst. **1**, 51–56 (1987)
10. Abdo, A.A., Lian, S.G., Ismail, I.A., Amin, M., Diab, H.: A cryptosystem based on elementary cellular automata. Commun. Nonlinear Sci. Numer. Simul. **18**, 136–147 (2013)
11. Anghelescu, P.: Encryption algorithm using programmable cellular automata. In: World Congress on Internet Security, pp. 233–239 (2011)
12. Faraoun, K.M.: Design of fast one-pass authenticated and randomized encryption schema using reversible cellular automata. Commun. Nonlinear Sci. Number Simul. **19**, 3136–3148 (2014)
13. Torres-Huitzil, C.: Hardware realization of a lightweight 2D cellular automata-based cipher for image encryption. In: IEEE Fourth Latin American Symposium on Circuits and Systems, pp. 1–4 (2013)
14. Wang, X., Jin, C.: Image encryption using game of life permutation and PWLCM chaotic system. Opt. Commun. **285**, 412–417 (2012)
15. Chen, R.J., Lai, J.L.: Image security system using recursive cellular automata substitution. Pattern Recogn. **40**, 1621–1631 (2007)
16. Wolfram, S.: Theory and applications of cellular automata. In: Advanced Series on Comples Systems, vol. 1, pp. 232–246 (1986)

Virtual Reality and Human-Computer Interaction

Multimodal Usability of Human-Computer Interaction Based on Task-Human-Computer-Centered Design

Yanbin Shi[1,2(✉)], Xiaoqi Li[1], Dantong Ouyang[2], and Hongtao Jiang[3]

[1] Militray Simulation Technological Research Institute, Aviation University of Air Force, Changchun 130022, China
shiyanbin_80@163.com

[2] Key Laboratory of Symbolic Computation and Knowledge Engineering of Ministry of Education, Jilin University, Changchun 130012, China

[3] Training Department, Aviation University of Air Force, Changchun 130022, China

Abstract. The cockpit of the aircraft provides the interface between the aircraft sensors and pilot, and enables the pilot to carry out the mission. Attention is concentrated on analyzing of the multimodal thinking about the interaction system's usability. Through comparing the human-centered and computer-centered design thinking in the human-computer interaction systematic design, "human-centered" emphasizes the unsubstitutability of human, but "computer-centered" prerequisite suppose computer can replace human. They all lost sight of the importance of the completing task. It is proposed that in order to improve the system usability, the designing of the multimodal interaction system should based on the task–human-computer centered design, that is consider synthetically human, computer and task in different develop phase.

Keywords: Multimodal · Human-computer interaction · Task-human-computer-centered design · Humanoid communication · Usability

1 Introduction

As the important component of the new generation information technology, the Internet of things [1], refers to Internet which links up everything, it includes two layers meaning. Firstly, the Internet of things is the extension and expansion on the basis of Internet. Secondly, the end-user and terminal of Internet is extended and expanded to any materials, exchanging information and communication. According to the definition of ITU (International Telecommunication Union), Internet of things mainly solves the interconnection between Thing to Thing, Human to Thing, and Human to Human. In the essence, the interconnecting among Thing to Thing and Human to Thing, the majority is for realizing interconnection between Human to Human. Where the interconnection between Thing to Thing is the main subject in the human engineering and Human-Computer Interface and Human-Computer Interaction.

While keep the pilot from the damage of the external environment condition, the cockpit of the aircraft also utilizes airborne bus to concentrate all kind terminals, and the cockpit display systems provide a visual presentation of the information and data

D.-S. Huang et al. (Eds.): ICIC 2017, Part I, LNCS 10361, pp. 459–468, 2017.
DOI: 10.1007/978-3-319-63309-1_42

from the aircraft sensors and systems to the pilot to enable the pilot to fly the aircraft safely and carry out the mission [2]. They are thus vital to the operation of any aircraft as they provide the pilot, whether civil or military, with primary flight information, navigation information, engine data, airframe data and warning information. The military pilot has also a wide array of additional information to view, such as: infrared imaging sensors, radar, tactical mission data, weapon aiming, threat warnings and so on. The detailed function diagram of the cockpit [3] is shown as Fig. 1.

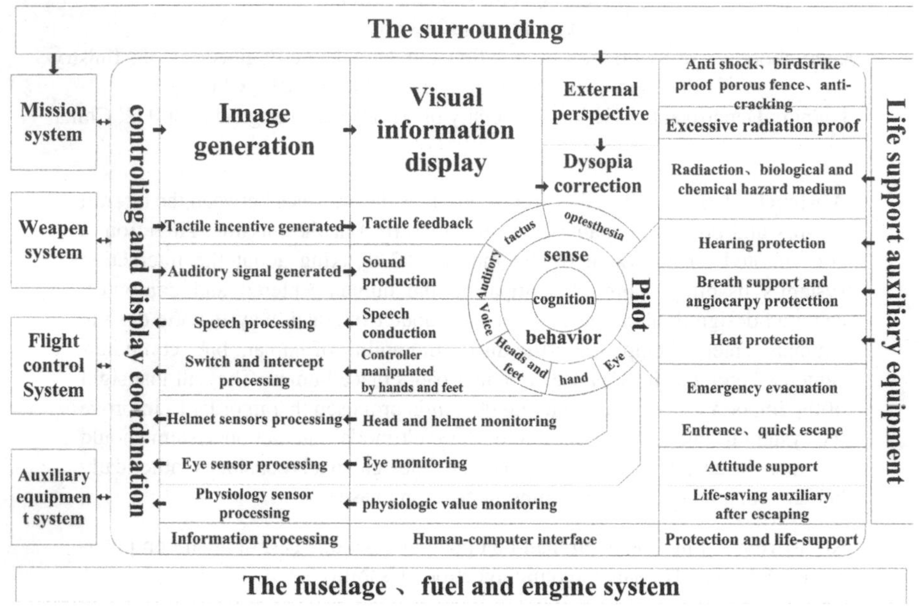

Fig. 1. Detailed function diagram of the cockpit

The pilot is able to rapidly absorb and process substantial amounts of visual information but it is clear that the information must be displayed in a way which can be readily assimilated, and unnecessary information must be eliminated to ease the pilot's task in high work load situations [4]. A number of developments have taken place to improve the pilot–display interaction and this is a continuing activity as new technology and components become available. The proven technique such as HUD (Head-up Display), HMD (Head Mounted Display), multi-function color displays, digitally generated color moving map displays.

Equally important and complementary to the cockpit display systems in the human-computer interaction are the means provided for the pilot to control the operation of the avionic systems and to enter data [5]. Again, this is a field where continual development is taking place. Multi-function keyboards and multi-function touch panel displays are now widely used. Speech recognition technology has now reached sufficient maturity for 'direct voice input' control to be installed in the new generation of military aircraft. Audio warning systems are now well established in both military and civil aircraft. The integration and management of all the display surfaces by audio/tactile inputs enables a very significant reduction in the pilot's workload to be achieved in the new generation of single seat fighter/strike aircraft.

With the diversification development of the human-computer interaction, it presents the humanoid communication, the research on the modal get more and more attention. Nowadays, the research on the modal not only include the traditional natural senses, such as visual, touch feeling and auditory sense, or through explicit operate by the keyboard and mouse, but also include the body language understanding for human computer interaction. So, the machine can understand the human's intention, attitude and mood, and realize the natural human-computer interaction.

The development of the science and technology, especially the sensor technology provides the important base for the multimodal human-computer interaction. But in the category of technology, in order to improve the usability, the systematic usability methods also are vital for collecting, or producing usability information. The mobile APP gives a powerful push to human-computer interaction's development, such as the proposition of the concept of the "micro-interaction" [6] further emphasized the human-computer interaction system "lets users use happily and easily" is fantastic system.

2 Task-Human-Computer-Centered Design

Along with the increasingly widespread way into people's work and life, human-computer interaction becomes more and more important. This trend becomes an indispensable factor for efficiency of work and the quality of life. The design of the cockpit HCI (human-computer interface) based on human-centered or computer-centered, which is controversial topic for a long time.

2.1 Computer-Centered Design

Traditional HCI is computer-centered, people have to learn and adapt to a series of rules designed for computer (machines). In order to get rid of the shackles of various rules, natural HCI emerges, which is human-centered, and computers are required to be able to understand and simulate human behaviors. Machine learning is the solid choice of achieving this goal. The process of HCI consists of three important stages, which are data input and output, manipulation of computers by human, and feedback of manipulation by computers, according to the order of occurrence.

2.2 Human-Centered Design

Such as the cockpit of the aircraft [7], which has close human-computer interface, the human-centered designer thinks human should be in the core position during designing the system, guaranteeing system can satisfy the demand under various operating position. In the area of design, the human-centered design already becomes major view, so the designers usually adopt this thinking undoubtedly. LAI [8] based on the expandence of human-centered design, introduced the behavioral picture and scenario based design rule, propose a kind of design procedure which is suitable for the complicated products system.

Human-centered design (HCD) is a design and management framework that develops solutions to problems by involving the human perspective in all steps of the problem-solving process. Human involvement typically takes place in observing the problem within context, brainstorming, conceptualizing, developing, and implementing the solution. Human-centered design is a creative approach to interactive systems development that aims to make systems usable and useful by focusing on the users, designing around their needs and requirements at all stages, and by applying human factors/ergonomics, usability knowledge, and techniques. This approach enhances effectiveness and efficiency, improves human well-being, user satisfaction, accessibility and sustainability; and counteracts possible adverse effects of use on human health, safety and performance.

Human-centered design already is utilized in multiple fields, including sociological sciences and technology. It has been noted for its ability to consider human dignity, access, and ability roles when developing solutions. Because of this, human-centered design may more fully incorporate culturally sound, human-informed, and appropriate solutions to problems in a variety of fields rather than solely product and technology-based fields. Typically, human-centered design is more focused on "methodologies and techniques for interacting with people in such a manner as to facilitate the detection of meanings, desires and needs, either by verbal or non-verbal means." In contrast, user-centered design is another approach and framework of processes which considers the human role in product use, but focuses largely on the production of interactive technology designed around the user's physical attributes rather than social problem solving.

In the last decades, mechatronic prostheses emerged and especially actuated ones increased the biomechanical functionality of their users.

2.3 Task-Centered Design

As an important aspect of the ergonomics, the human-computer interactions are being used in vast sum of products. However, human-computer interaction researches haven't been keeping pace with the development of software technology. The main performance is that design are usually handled by the programmers due to the fact that traditional engineers lack professional training in designing interaction which might result in apart from the function, the majority of products are lack of usability, aesthetics, fault tolerance. ZHOU [9] researched and analyzed the existing model of human-computer interaction, as well as discuss in depth the key problems. "Task-centered system design" comprehensive model focuses on functional task, and ignore the importance and ease to use software during the use of software, which dwindles "user centered design" has developed as a limited view of design. Instead of looking at a person's entire activity, it has primarily focused upon page-by-page analysis, screen-by-screen. And error messages don't contain explanations and offer alternative ways of proceeding from the message itself. Interactive methods can't improve effectively. Aiming the existing problems in the active design model, ZHOU puts forward a User Using Centered Design model by combing the user-centered and task-centered models, and introducing the latitude process mode and hierarchy classification method. The proposed model solves the problem of uncertain functional importance and no effective improvement over interactive way with respect to 3-D processes and important hierarchy classifications.

2.4 Task-Human-Computer-Centered Design

It can be seen from the development of the design thinking change from computer-centered design, human-centered design to task-centered design. The aim always is improve the usability of the system. When the technology can satisfy the human's demand, designers take more attention on the computer. On the contrary, when the technology cannot satisfy the human's demand, designers take more attention on the human. For the special system, for example, when consider the cockpit, in different time, some designers think it should be computer-centered design, some designers think it should be human-centered design, and some designers think it should be task-centered design. These designer proposed those different opinions based on different starting points and different work positions in different time, different technology condition. For the designing of the aircraft cockpit, attention must be concentrated on the human (pilot), computer and task, which should be human-computer-task-centered design (3D-centered design). That's means when design the cockpit human-computer interaction system, the designer must aiming at the task demand, utilizes the existing technologies, try best to satisfy the polit's need. These three dimensionalities are coordinate and unified, but it must be pointed out, in different time, the designer need pay more attention on one dimensionality (human, computer or task).

3 Multimodal Human-Computer Interaction

An important feature of human behavior is that some activities seem to require little or no mental effort, and so can be performed in parallel with other activities. Others seem to absorb all our mental capacity [10, 11]. In order to reduce the pilot's workload, the designing of the cockpit human-computer interaction system must use all kinds modal when the technological condition is feasible. Multimodal human-computer interaction refers to the "interaction with the virtual and physical environment through natural modes of communication", i.e. the modes involving the five human senses (touch, smell, taste, hearing and vision), that is humanoid communication. This implies that multi-modal interaction enables a more free and natural communication, interfacing users with automated systems in both input and output. Specifically, multimodal systems can offer a flexible, efficient and usable environment allowing users to interact through input modalities, such as speech, handwriting, hand gesture and gaze, and to receive information by the system through output modalities, such as speech synthesis, smart graphics and others modalities, opportunely combined. Then a multimodal system has to recognize the inputs from the different modalities combining them according to temporal and contextual constraints in order to allow their interpretation [12]. This process is known as multimodal fusion, and it is the object of several research works from nineties to now. The fused inputs are interpreted by the system. Naturalness and flexibility can produce more than one interpretation for each different modality (channel) and for their simultaneous use, and they consequently can produce multimodal ambiguity generally due to imprecision, noises or other similar factors. For solving ambiguities, several methods have been proposed. Finally the system returns to the user outputs through the various modal channels (disaggregated) arranged according to a consistent feedback (fission).

Two major groups of multimodal interfaces have merged, one concerned in alternate input methods and the other in combined input/output. The first group of interfaces combined various user input modes beyond the traditional keyboard and mouse input/output, such as speech, pen, touch, manual gestures, gaze and head and body movements. The most common such interface combines a visual modality (e.g. a display, keyboard, and mouse) with a voice modality (speech recognition for input, speech synthesis and recorded audio for output). However other modalities, such as pen-based input or haptic input/output may be used. Multimodal user interfaces are a research area in HCI.

The advantage of multiple input modalities is increased usability: the weaknesses of one modality are offset by the strengths of another. For example, on a mobile device with a small visual interface and keypad, a word may be quite difficult to type but very easy to say. In the cockpit of the military aircraft, because the pilot uses the HOTAS, the voice input can reduce workload prominently, facilitate finishing the task.

Multimodal input user interfaces have implications for accessibility. A well-designed multimodal application can be used by people with a wide variety of impairments. Visually impaired users rely on the voice modality with some keypad input. Hearing-impaired users rely on the visual modality with some speech input [13]. Other users will be "situationally impaired" (e.g. wearing gloves in a very noisy environment, driving, or needing to enter a credit card number in a public place) and will simply use the appropriate modalities as desired. On the other hand, a multimodal application that requires users to be able to operate all modalities is very poorly designed.

The second group of multimodal systems presents users with multimedia displays and multimodal output, primarily in the form of visual and auditory cues. Interface designers have also started to make use of other modalities, such as touch and olfaction. Proposed benefits of multimodal output system include synergy and redundancy. The information that is presented via several modalities is merged and refers to various aspects of the same process. The use of several modalities for processing exactly the same information provides an increased bandwidth of information transfer. Currently, multimodal output is used mainly for improving the mapping between communication medium and content and to support attention management in data-rich environment where operators face considerable visual attention demands.

An important step in multimodal interface design is the creation of natural mappings between modalities and the information and tasks. The auditory channel differs from vision in several aspects. It is omnidirection, transient and is always reserved. Speech output, one form of auditory information, received considerable attention.

The sense of touch was first utilized as a medium for communication in the late 1950s. It is not only a promising but also a unique communication channel. In contrast to vision and hearing, the two traditional senses employed in HCI, the sense of touch is proximal: it senses objects that are in contact with the body, and it is bidirectonal in that it supports both perception and acting on the environment.

4 The Task-Human-Computer-Centered Design Process Based on Usability

The usability of multimodal interaction system is the goal of making an interaction system fit the bodies and minds of its users in context. Its objectives are target values for things such as speed to perform representative tasks and number of errors allowable. These can be used to motivate designers and support resource allocation decisions. The target values can be selected to beat the competition or to meet the functional needs of well-defined tasks. The user of the aircraft cockpit is the pilot, it should request information in the order that the pilot is likely to receive it; it should make it easy to correct data that's often entered incorrectly; its hardware should fit in the space that pilots have available and look like it belongs there. Effective task and user (pilot) analysis requires close personal contact between members of the design team and pilot actually use interact system. Usually, both ends of the link between members of the design team and users are difficult to achieve, but for the aircraft cockpit, the ends of the links is clear and easy to get.

Understanding of the pilots' psychological and physiological condition, having an awareness of the human-computer interaction system's using environment is very important. As we all known, in the aircraft cockpit, the pilots must deal with large number information which not only include many kind sensor signals transmitted by the airborne bus, but also include the visual signal out of the cockpit. Especially for the military pilot, the interface contains weapon system information, flight information, and navigation information and so on. Until now, the visual sensor, hearing and touch sensor still are the most important interact modals in the cockpit. And at same time, the update speed of the interaction information is very quick, the human eye is a not a particularly good optical instrument, there are a number of visual problems which are specific to aviation, such as empty field and night myopia, perception time in high speed flight, dynamic visual acuity and depth perception. These are so arranged that related instruments are placed together and are viewed as a whole, rather than individually.

The electronic aids fitted to modern military aircraft greatly reduce pilot's mental and physical workload [2]. HUDs minimize the need to make large eye and head movements to take in the outside world and the cockpit interior, and facilitate shifts of attention between these sources of information. Although they reduce the overall level of aircrew workload, the monitoring component is increased since, in the event of a system malfunction or damage, the aircrew must be able to take over. Furthermore, as avionics technology advances, more and more systems tend to be crammed into the cockpit, so exacerbating the situation. Experimental evidence suggests that reversion from HUD to cockpit instruments is unlikely to be accomplished in less than 3 s and may take considerably longer.

The recognition of auditory signals is an integral part of tasks. Audition is more than the act of passive listening, and involves the interpretation by the brain of signals, often embedded in background noise. Generally, the auditory is considered as the strengthened or assisted means for the visual. Although there are some accessory equipments in the helmet help pilot to shielding the noise in cockpit, but in many aircraft, the noise still bring the serious interference to the pilot. The auditory signal should neither increase the communication pressure, nor interfere with communication, but need attract the pilot to notice, while offer the pilot a suggestion easy to understand the problems. Direct voice input control is a system which enables the pilot to enter data and control the operation of the aircraft's avionic systems by means of speech. The spoken commands and data are recognized by a speech recognition system which compares the spoken utterances with the stored speech templates of the system vocabulary. Feedback that the Direct voice input system has recognized the pilot's command correctly is provided visually on the HUD or HMD, and aurally by means of a speech synthesizer system. The pilot then confirms the correctly recognized command by saying 'enter' and the action is initiated.

After establishing a good understanding of the users and their tasks, a more traditional design process might abstract away from these facts and produce a general specification of the system and its user interface. The task-centered design process takes a more concrete approach. The designer should identify several representative tasks that the system will be used to accomplish. These should be tasks that users have actually described to the designers. The designer should analyze existing cockpit's human-computer interfaces that work for pilots and then according to the mission.

In fact the vast majority of current systems are multimodal, and unimodal systems represent a limiting case rather than the standard case. It can be seen from the Fig. 2, the interaction system in cockpit also is multimodal. The modals already include visual, auditory and tactile and so on. The designing of the cockpit in different historical stages should obey task-human-computer-centered. Figure 3 is task-human-computer-centered design spiral structure.

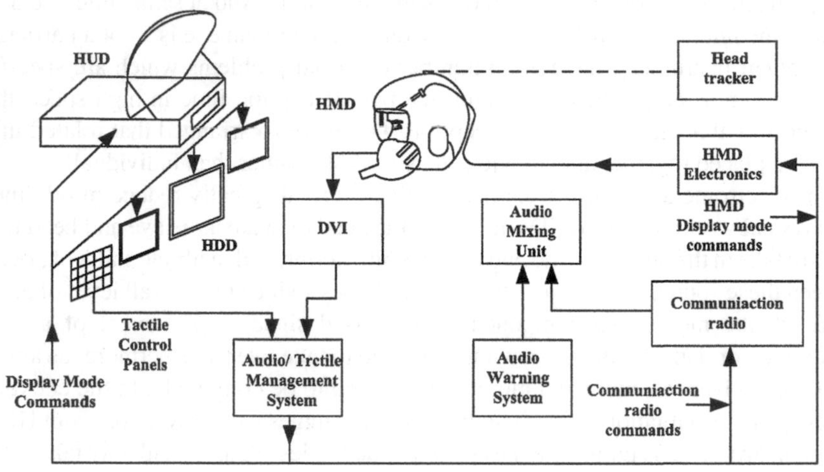

Fig. 2. Basic modal interaction system in cockpit

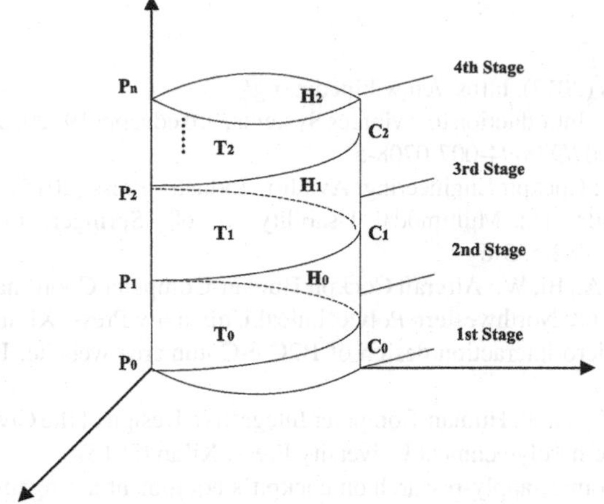

Fig. 3. 3D-centered design spiral structure

In the Fig. 3. P_i is the system developing stage, P_0 is the early stages of a simple design, which is 1st stage. The designing of the interaction system must figure out who's going to use the system to do what. So, it is task-centered. With the development of the research, choosing representative tasks for task-centered design, the designing will face the technology and human-computer coordination issues. So, the designer must solve these different problems, in different stage, the design based on human-centered or computer-centered. (T_0, C_0, H_0) is the first stage circle, T_i is task-centered, C_i is computer-centered, and H_i is human-centered in the ith stage. With i change from 0 to n, the human-computer interaction system will ripe gradually. Therefore, in order to improve the usability of the human-computer interaction system, the designing think should be based on task-human-computer-centered. The interface designers should not be a special group isolated from the rest of the system development effort.

5 Conclusion

Multimodal interaction is the future of the human-computer interaction. In cockpit, the pilots use many senses modals to accomplish the humanoid communication with airborne systems or sensors. In order to improve the system usability, it is suggested that the interface designing of the cockpit should base on task-human-computer-centered. It is proposed that using the task-human-computer-centered design spiral structure to complete cockpit designing.

Acknowledgment. This work is supported by the Education Department of Jilin Province "13th Five-Year" scientific research planning project (2016-514).

References

1. Internet of things (2017). https://en.wikipedia.org/
2. Collinson, R.P.G.: Introduction to Avionics Systems, 3rd edn, pp. 19–20. Springer, Heidelberg (2011). doi:10.1007/978-94-007-0708-5
3. Kong, Y., Qu, K.: Cockpit Engineering. Aviation Industry Press (2015)
4. Niels, O.B., Laila, D.: Multimodal Usability, p. 68. Springer, London (2010). doi: 10.1007/978-1-84882-553-6
5. Tang, Z., Zhang, A., Bi, W.: Aircraft Cockpit Human-Computer Coordination Design Theory and Methods., p. 68. Northwestern Polytechnical University Press, Xi'an (2015)
6. Qian, C.: The micro-interaction design of B2C e-Commerce website. Hubei Polytechnical University (2016)
7. Zhang, W., Ma, Z., Yu, J.: Human-Computer Integrative Design of the Civil Aircraft Cockpit., p. 68. Northwestern Polytechnical University Press, Xi'an (2015)
8. Ai, L.-Y.: Ergonomics-apply-research on cockpit's equipment arrangement. Aircr. Des. **32**, 78–80 (2012)
9. Lai, C., Xu, C.: Human-centered design of human-machine interaction complex product and system. Value Eng. 219–223 (2015)
10. Zhou, X.: Research of Human-Computer Interaction and Its Application in E-government Systems. Chongqing University, Chongqing (2006)
11. Luo, F., Sha, S.: An analysis of the connotation of the concept of computational thinking in order to calculate the angle of view: as the center. Comput. Knowl. Technol. **10**, 6307–6309 (2014)
12. Wang, H., Bian, T., Xue, C.: Layout design of display interface for a new generation fighter. Electro-Mech. Eng. **27**, 57–61 (2011)
13. Guo, Y., Gu, X., Shi, L.: Virtual Reality and Interaction Design. Wuhan University Press, Wuhan (2015)
14. Tan, H., Tan, Z., Jin, C.: Automotive Human Machine Interface Design. Publishing House of Electronics Industry, Beijing (2015)

Business Intelligence and Multimedia Technology

Double H.264 Compression Detection Scheme Based on Prediction Residual of Background Regions

Junjia Zheng[1], Tanfeng Sun[1,2(✉)], Xinghao Jiang[1,2], and Peisong He[1]

[1] School of Electronic Information and Electrical Engineering, Shanghai Jiao Tong University, Shanghai, China
{zhengjunjia,tfsun,xhjiang,gokeyhps}@sjtu.edu.cn

[2] National Engineering Lab on Information Content Analysis Techniques, Shanghai 036001, China

Abstract. Detection of double video compression plays an important role in video forensics. However, existing methods rarely focused on H.264 videos and are unreliable to provide detection results for static-background videos with fast moving foregrounds. In this paper, an effective double compression detection scheme based on Prediction Residual of Background Regions (PRBR) is proposed to overcome these limitations. Firstly, the mask of background regions in each frame is obtained by applying Visual Background Extractor (VIBE). VIBE is an efficient and robust background subtraction algorithm, which can distinguish the background and foreground of each frame at pixel level. Then, the PRBR feature is designed to characterize the statistical distribution of average prediction residual within the background mask. After that, the Jesen-Shannon Divergence is introduced to measure the difference between the PRBR features of the adjacent two frames. Finally, a periodic analysis method is applied to the final feature sequence for double H.264 compression detection and estimation of the first Group Of Pictures (GOP). Eighteen standard testing sequences captured by fixed cameras are used to establish the double compression dataset. Experiments demonstrate the proposed scheme can achieve better performance compared the-state-of-art methods.

Keywords: Double compression detection · Background subtraction · GOP estimation

1 Introduction

Due to the widely use of social media network and the development of video editing software, digital videos are easy to be accessed in daily life and can be tampered almost costlessly without professional knowledge. It has badly infringed the originality of videos, in which situation it is hard to judge the integrity and authenticity of videos [1]. The degrading general trust and the indispensable validation of video contents have called on the emergent research of video forensics, especially in the justice area. The research of video forensics can be classified into two basic categories, the active approaches and the passive approaches. Among the active approaches, most of the experts pay attention to digital watermarking. However, the active approaches have

D.-S. Huang et al. (Eds.): ICIC 2017, Part I, LNCS 10361, pp. 471–482, 2017.
DOI: 10.1007/978-3-319-63309-1_43

several limitations in practical applications, such as the extra requirements of hardware and software in recording process to embedding watermarking information [2]. In contrast, passive approaches can only depend on the traces left by tampering operations to verify the authenticity of the inquiry videos. Detection of double compressed videos is one of the most significant issues in video forensics. It is because the tampering process must be conducted on decompression domain and then tampered videos have to undergo the recompression process in most cases, for both intra-frame forgery and inter-frame forgery [1].

According to whether the GOP structure of the first and the second compression is matched or not, the detection of video double compression can be classified as two categories. For video double compression with the matched GOP structure, the properties of double compressed I-frames are applied to reveal the process of double compression, such as statistical patterns in the distribution of Discrete Cosine Transform (DCT) coefficients [3–6], Markov statistics [7, 8] and so on. For video double compression with non-aligned GOP structures (the GOP structure in the first compression is different from that in recompression), most of the existing methods [9–12] take the periodic artifacts left by I-P frames as the statistic feature, where I-P frames denote original I frames which are encoded as P frames in the second compression. [9, 10] achieved pretty good performance on detection of video double compression with non-aligned GOP structures, which take advantage of the variance of macroblock types in I-P frames and prediction residual distribution of I-P frames respectively. However, with the popularity of cameras and video surveillance, the valid detection of the originality of static-background videos becomes urgent. Both of the mentioned two algorithms lacks consideration of static-background videos with fast moving foregrounds. Although [12, 13] proposed an effective method focused on static-background videos or videos with no significant motion, they only analyzed the situation of MPEG-x videos specifically. To overcome such limitations, our work focuses on the static-background video double compression detection in H.264 standard.

In this paper, a background subtraction based approach is developed for double H. 264 compression detection. To the best of our knowledge, [9, 10] are the only two methods focused on H.264 videos for double compression detection with non-aligned GOP structures. And both of them are based on the analysis of the periodicity of I-P frames. However, the performance of the above two method is degraded due to the disturbance of the moving objects. To eliminate the effect of moving objects, background subtraction algorithm VIBE [14] is applied to extract the background mask of video sequence. The static or slow moving parts of the scene is called background while the moving objects of the scene is the foreground. VIBE is an efficient and effective background subtraction algorithm, which can distinguish the background and foreground of each frame at pixel level. Additionally, compared to the other background subtraction algorithms, it only takes the first frame as the reference frame and will not infringe the temporal clues for double compression detection. The prediction residual of background regions (PRBR) feature is calculated to represent the trace left by double compression. Database is constructed by all the known 18 YUV sequences with 352×288 resolution (CIF) which are captured by stationary cameras. Experiment results show better performance than the prior works.

The rest of the paper is arranged as follows. In Sect. 2, the disturbance of moving objects in prediction residual trace and the method to eliminate it are discussed. Section 3 presents the extraction of PRBR feature sequence in details. Then the framework of the proposed video double compression detection scheme is introduced in Sect. 4. Finally, the experimental results are given in Sect. 5 and the conclusion is drawn in Sect. 6.

2 Preliminary of the Proposed Model

In recent years, H.264 standard has become the most widely used video compression standard for online video transmission and data storage of mobile devices. In this work, we focus on the double compression detection of H.264 videos. There are three basic frames in H.264 videos, namely intra-coded frames (I frames), predictive-coded frames (P frames) and bi-directionally predictive-coded frames (B frames). Considering that the encoder used in this paper is based on the baseline profile for the H.264 standard, we do not consider the B frames in this work. And the number of reference frames is fixed to 1.

In the video double compression with non-aligned GOP structures, the first encoding GOP structure is different from the second one, which allows the possibility that I-frames in the first encoding will be compressed as P-frames in the secondary compression. As shown in Fig. 1, $F_1F_2F_3F_4F_5F_6F_7F_8F_9F_{10}$ is the initial frames of a raw video. The first and second GOP of video compression are set as 3 and 5. Then after twice compression the video sequence is structured as $I_1P_2P_3I_4P_5P_6I_7P_8P_9I_{10}$ and $I'_1P'_2P'_3P'_4P'_5I'_6P'_7P'_8P'_9P'_{10}$ respectively.

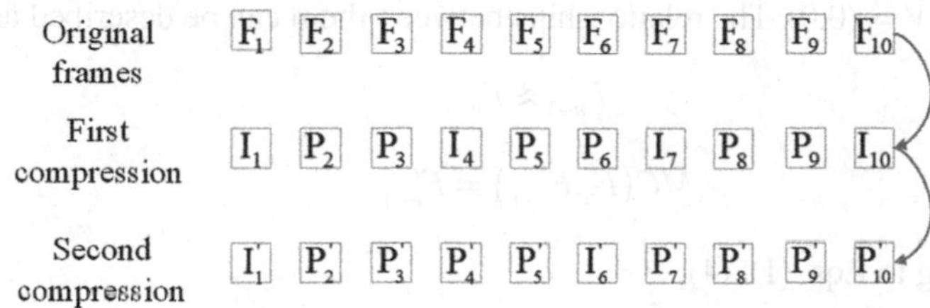

Fig. 1. An example of non-aligned double compression

As for P_3 is predicted from P_2 and I_4 is obtained by intra prediction, the difference between P_3 and I_4 is bigger than the difference between P_3 and P_2. In the similar way, the difference between P_3 and I_4 is bigger than the difference between I_4 and P_5. So the residual of P'_4 is relatively higher than P'_3 and P'_5, that is to say the residual of I-P frames peaks in the video sequence. This periodic peaks record the trace left by double compression.

The block scheme of H.264 compression using motion compression and transform based coding is shown in Fig. 2. P denotes the prediction residual of F_n and the difference between the current frame and the reference reconstructed frame is the quantization error introduced by quantization, which can be expressed as follows:

$$P_n = F_n - MC(F_n, F'_{n-1}) \quad (1)$$

$$F_{n-1} - F'_{n-1} = Q_E(F_{n-1}) \quad (2)$$

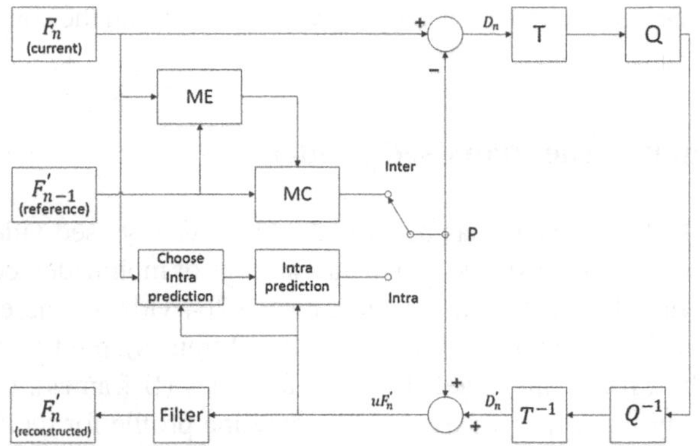

Fig. 2. Block scheme of H.264 compression

Where $MC(.)$ denotes the motion compensation operation; F_n denotes the nth frame; F'_{n-1} denotes the reconstructed frame of F_{n-1}; $Q_E(F_{n-1})$ denotes the quantization error of F_{n-1}.

- When the whole video sequence is still without moving objects, all of its frames are the same if noises are not taken into consideration. Then the best prediction block of the current frame is the block at the exact same location of the preference frame, namely $MV = (0,0)$. The relationships between them can be described as follows:

$$F_{n-1} \approx F_n \quad (3)$$

$$MC(F_n, F'_{n-1}) = F'_{n-1} \quad (4)$$

According to Eqs. (1)–(4),

$$P_n = F_n - F'_{n-1} = F_n - (F_{n-1} - Q_E(F_{n-1})) = F_n - F_{n-1} + Q_E(F_{n-1}) \approx Q_E(F_{n-1}) \quad (5)$$

Equations (5) shows that P_n is determined by quantization error of its referenced frame when the contents of video sequences are unchanged.

- When the video sequence is not still with moving objects, P_n is impacted by two main factors: the quantization error and the difference of video content between adjacent frames.

$$P_n = F_n - MC(F_n, F_{n-1} - Q_E(F_{n-1})) \quad (6)$$

In summary, the quantization error is the trace left by first compression, which allows us restoring the process of double compression. However, varying contents of video sequence influence the value of prediction residual, which disturbs the periodicity of I-P frames furthermore. In order to maintain the purity of periodicity, the disturbance of moving objects has to be eliminated. Here is a detailed example. Figure 3 shows the average prediction residual of a double compressed video sequence. The GOP structure of the first and second compression is 10 and 16 respectively. The bitrate for both compression is 300 kb/s. The video sequence named "bowing" is captured by stationary camera, which shows a man coming into scene and starting bowing from 50th frame to 210th frame. It is supposed to have a peak every ten frames for the periodicity of I-P frames mentioned above, namely 11th frame, 21th frame, 31th frame and so on. And the I-I frames (the common multiple of the first and second GOP) is found missing because only P frames are taken into consideration. However, what is shown in Fig. 3 is that the average prediction sequence begins to be noisy when moving objects come into being at 50th frame and the periodicity is obviously disturbed from 130th frame to 160th frame. In conclusion, the periodicity is more detectable in video sequences without moving objects.

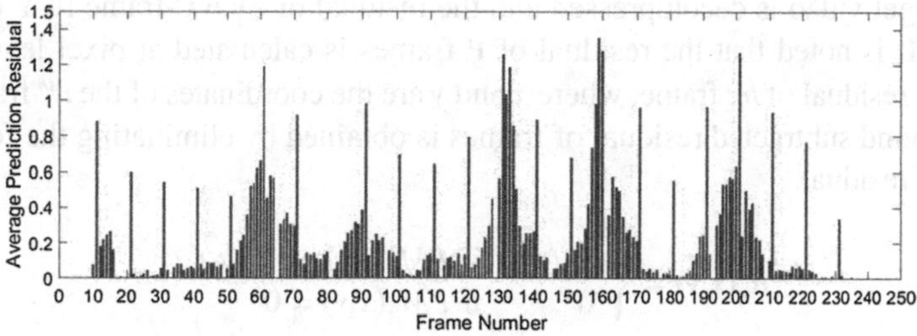

Fig. 3. The average prediction residual of bowing

In order to diminish or eliminate the influence brought by moving objects, a background subtraction algorithm is applied to extract the moving objects. The prediction residual in background region describe the periodicity of I-P frames. The common background subtraction algorithms often take several starting frames as reference frames. It results in the lacking consideration of the starting frames and effect the contribution of double compression to further forgery detection. For example, it infringes there liability of the inter frame tempering when the result of double compression is applied. Considering this consequence and the efficiency, VIBE [14] is used for background subtraction. Also the original code is appended in [14].

3 PRBR Feature Sequence Extraction

In this section, Prediction Residual of Background Region (PRBR) feature will be discussed in detail. As what is mentioned in Sect. 2, the prediction residual of I-P frames peaks periodically and the disturbance of moving objects weaken the periodicity. In order to strengthen the periodicity of the prediction residual of I-P frames and eliminate the influence brought by moving objects, only the prediction residual of background region will be considered. In common sense, the moving objects is regarded as foreground of the video.

In our implementation, VIBE [14] is employed to extract the foreground from the target video. VIBE acts excellent in both computation speed and detection rate. And it can extract the foreground of the video in pixel level with only one frame as the reference frame, which matches our experiments well. There are several parameters to be set. The first frame of the video is chosen as the reference frame because the first frame of the video is I frame and will not disturb the statistics of P frames. And we just employ the default setting of other parameters, such as the radius, the time subsampling factor and so on. Foreground of each frame is extracted at the pixel level.

$$FM_n(x,y) = \begin{cases} 0 \text{ if } (x,y) \in \textit{foreground area} \\ 1 \text{ if } (x,y) \in \textit{background area} \end{cases} \quad (7)$$

Where $FM_n(x,y)$ denotes the mask of background region and the index of row and column is x and y respectively in the nth frame.

The target video is decompressed and the residual of each P-frame in Y channel is extracted. It is noted that the residual of P frames is calculated at pixel level. $R_n(x,y)$ defines the residual of nth frame, where x and y are the coordinates of the nth frame. Then the foreground subtracted residual of frames is obtained by eliminating the foreground area of the residual.

$$R'_n(x,y) = \begin{cases} R_n(x,y) & \text{if } FM_n(x,y) = 1 \\ 0 & \text{if } FM_n(x,y) = 0 \end{cases} \quad (8)$$

Where $R'_n(x,y)$ defines the residual of nth frame, where x and y are the coordinates the nth frame. The experimental result shows that the number of $R'_n(x,y)$ bigger that 9 is less than 1%. So if value of $R'_n(x,y)$ is greater than 9, it will be truncated to 10, which is shown as follows.

$$R_{trn}(x,y) = \begin{cases} R'_n(x,y) & \text{if } R'_n(x,y) < 10 \\ 10 & \textit{otherwise} \end{cases} \quad (9)$$

Where $R_{trn}(x,y)$ denotes the truncating $R'_n(x,y)$. In order to reduce the disturbance of space distribution, the frequency distribution of residual to further describing the feature of each frame is applied, which is first proposed in [10].

$$P_m(k) = \frac{\sum R_{trn}(x,y) = k}{\sum FM_n(x,y)} \quad \mathrm{k} \in [0,10] \quad (10)$$

$P_m(k)$ defines the frequency of k in m^{th} frame and P_m defines the distribution of mth frame. Then the Jesen-Shannon divergence of each frame is calculated [10].

$$S_{JSD} = JSD(P_m \parallel P_{m-1}) + JSD(P_m \parallel P_{m+1}) \quad (11)$$

After the $S_{JSD}(m)$ is calculated for each frame of the target video, the median filter is used to eliminate noise in the sequence:

$$S_{MF}(m) = median(S_{JSD}(m-1), S_{JSD}(m), S_{JSD}(m+1)) \quad (12)$$

Then the noise can be eliminated according to the following equation:

$$S_{PRBR}(m) = max(S_{JSD}(m) - S_{MF}(m), 0) \quad (13)$$

Figure 4 shows the feature sequence of the same video with first GOP 10 and second GOP 16. (a) presents the feature sequence of background region and (b) presents the feature sequence of the whole frame. The peaks appear in a periodic pattern in the final sequence because of the fixed GOP size of the first and second compression. Basically, the sequence peaks every 10 frames for the first GOP size being 10 and is absent from peaks at the common multiple of first and second GOP size, namely 80th frame and 160th frame, which are the I frames in recompression process. There are periodical peaks every 10 frames in Fig. 4(a) while in Fig. 4(b) unexpected peaks(P-P frames, P frames of first compressed encoded as P frames in the second compression) appear such as 24th frame, 34th frame, etc and expected peaks(I-P frames) disappear such as 31th frame, 51th frame, etc. By comparison,

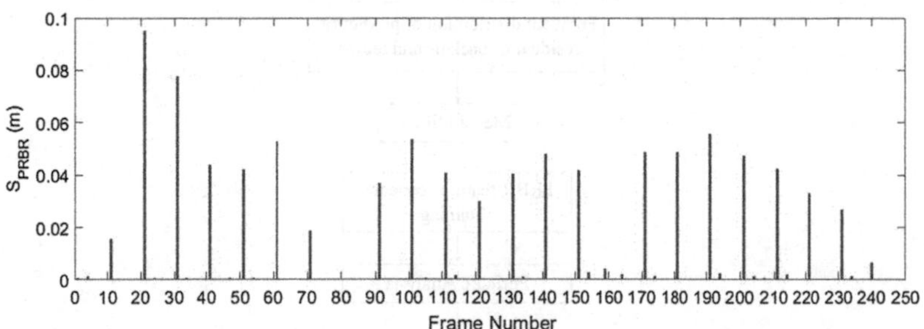

(a) Feature sequence of background region

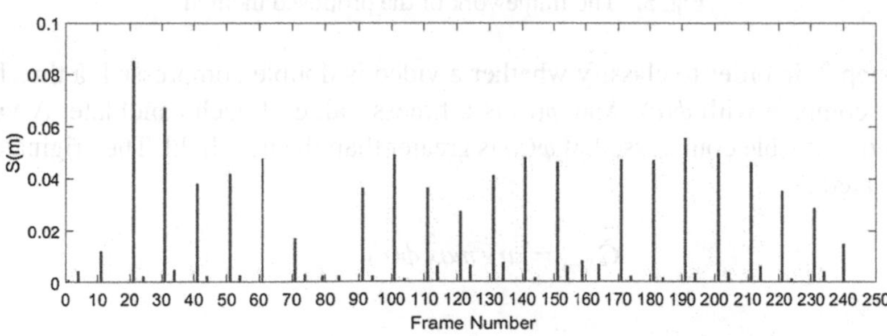

(b) Feature sequence of the whole region

Fig. 4. The effect of background subtraction on video feature sequence

sequence of (a) is more clear and less noisy than (b), which contributes to a better periodicity. It is demonstrated that PRBR feature achieves better periodicity.

4 Framework of the Proposed Scheme

The proposed algorithm for video double compression detection is based on time domain analysis of residual of P frames. Figure 5 shows the block scheme of the pro-posed algorithm for detecting tampered videos. The algorithm begins with the extraction of video frames. Then the residual of P frames are calculated and concurrently, the foreground of the each frame is extracted. To this intent, the prediction residual of background region is employed to calculate the feature sequence. The above two steps is described in details in the Sect. 3. Finally a periodic analysis method [10] is introduced to detect double H.264 compression and estimate the first GOP size.

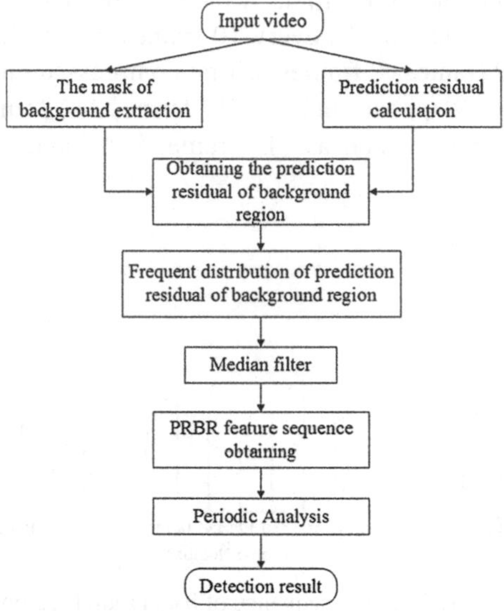

Fig. 5. The framework of the proposed method

In step 3, in order to classify whether a video is double compressed, a threshold is used to compare with $\phi(c)$. And $\phi(c)$ is a fitness value of each candidate. A video is regarded as double compressed, if $\phi(c)$ is greater than the threshold. The originate GOP is estimated as:

$$\hat{G}_{First} = arg\,max\,\phi(c) \tag{14}$$

5 Experiments

In this section, double H.264 compression detection and the first GOP estimation are both investigated. The experimental results of the proposed method are compared with those of VPF method in [9] and PRD method in [10] using the same dataset.

5.1 Dataset

All the known 18 YUV sequences with 352 × 288 resolution (CIF) which captured by stationary cameras are selected as source sequences. The names of these sequences and the address of the website where these sequences can be downloaded are provided in the Appendix A. Because these sequences have different frame numbers, the minimum frame number of all the sequences is considered, namely 240 frames. For those sequences which have more than 240 frames, only the first 240 frames are used. Then, all these sequences are encoded in constant bitrate (CBR) mode for realistic application. The GOP size of both first and second compression is fixed. The H.264 coding standard are used to generate both singly and double encoded videos, employing the FFmpeg3.2. Finally, with different combinations of GOP sizes and bitrates, 288 first encoding videos and 4608 s encoding videos are generated. The coding parameters are listed in Table 1.

Table 1. Parameters explanation for the double compression video dataset

Parameters	First encoding	Second encoding
Bitrate (kb/s)	{100, 300, 500, 700}	{100, 300, 500, 700}
GOP size	{10, 15, 30, 40}	{9, 16, 33, 50}
Sequence number	288	4608

The area under the receiver operating characteristic (ROC) curve, known as the AUC, is used to evaluate the performance of the detectors because it avoids the supposed subjectivity in the threshold selection process, when continuous probability derived scores are converted to a binary presence absence variable, by summarizing overall model performance over all possible thresholds.

5.2 Double H.264 Compression Detection

In this section, the performance of the proposed method in double H.264 compression detection is evaluated. AUC is used to evaluate the performance of the detectors. The results are compared with Vázquez-Padín et al. [9] and Chen Sheng et al. [10], because these two are the only prior work for the double H.264 compression detection with non-aligned GOP structures in the existing studying. All of the 288 singly encoded H.264 videos are chosen as negative samples while 288 doubly encoded videos are randomly selected from our dataset as positive samples. In order to cover all of the doubly encoded videos, the experiments are conducted 50 times and the average AUC is obtained. The results is shown in Table 2. The results in Table 2 shows that the proposed method has a better performance in the detection of double compression.

Table 2. Performance comparison results

Methods	Proposed	Chen [10]	Vázquez-Padín [9]
Results (AUC)	0.9516	0.9324	0.8898

5.3 The First GOP Estimation

The GOP size of the first encoding reveals the processing history of double compression which offers important information for further detection of video forgery. The performance of first GOP size estimation is analyzed from two aspects, different bitrate combinations and the overall accuracy. All the 4608 doubly compressed video sequences from our dataset are used in the experiment. Table 3 shows the accuracy rate and standard deviation comparison of the GOP estimation. It turns out that the accuracy rate of the proposed method is better than the other two methods and the standard deviation value of the proposed method is smaller than the other two methods. That is to say, the first GOP estimation of the proposed method tends to be more close to the correct GOP size than the other two methods, which shows the stability of the proposed method.

Table 3. Performance of GOP estimation comparison results

Methods	Proposed	Chen [10]	Vázquez-Padín [9]
Accuracy rate	0.9099	0.8997	0.8969

The results of different bitrates is shown in Table 4, compared with the previous two methods. b_1 is set as the first encoding bitrate and b_2 the second. From the Table 4, we can figure out that the detection accuracy shows better results when $b_1 < b_2$ or $b_1 = b_2$ and most of them almost get to 100%. Encoded with relatively low bitrates will leave more quantization error trace that make it easier to be detected. On the contrary, when $b_2 < b_1$ the traces of the first encoding will be easily concealed by the quantization in

Table 4. First GOP estimation accuracy comparison of different bitrates

b_1/b_2	Methods	100	300	500	700
100	**Proposed**	**98.96%**	100.00%	100.00%	100.00%
	Chen [10]	98.26%	100.00%	100.00%	100.00%
	Vázquez-Padín [9]	98.26%	100.00%	100.00%	100.00%
300	**Proposed**	**76.74%**	**100.00%**	99.65%	100.00%
	Chen [10]	74.31%	99.31%	**100.00%**	100.00%
	Vázquez-Padín [9]	76.39%	97.57%	99.65%	100.00%
500	**Proposed**	**59.03%**	92.71%	**99.31%**	100.00%
	Chen [10]	55.90%	**93.40%**	98.26%	100.00%
	Vázquez-Padín [9]	56.60%	90.63%	97.57%	99.65%
700	**Proposed**	**45.83%**	**87.50%**	**97.22%**	**99.31%**
	Chen [10]	40.28%	85.07%	96.88%	98.26%
	Vázquez-Padín [9]	44.79%	80.56%	95.14%	98.61%

the second compression, which make it harder to detect. The results turns out that the proposed method show better performance than the other two methods, especially when $b_2 < b_1$. For example, when $b_1 = 500k$, $b_2 = 100k(b_2 < b_1)$ the accuracy rate of the proposed method is 3.1% better than Chen and 2.4% better than Vázquez-Padín, and when $b_1 = 500k$, $b_2 = 500k(b_2 = b_1)$ the accuracy rate of the proposed method is 1.1% better than Chen and 1.8% better than Vázquez-Padín. And it proves the robust of our method to some extent.

6 Conclusions

In this paper, a prediction residual of background region based method is proposed to detect double H.264 compression. It's also a firm foundation for further forgery localization. It is found that the severe motion in video sequence will disturb stability of traces left by first compression. A background subtraction algorithm is employed to eliminate the disturbance of moving objects. Then the prediction residual of background region is calculated as the feature to detect double compression and estimate the first GOP size. Experimental results demonstrated that the proposed method is effective for double compression detection and first GOP estimation, especially for videos with severe motions, which is not well solved in prior works.

Acknowledgements. This work was supported by National Natural Science Foundation of China (No. 61572320, 61572321). Corresponding author is Prof. Tanfeng Sun, any comments should be addressed to tfsun@sjtu.edu.cn.

Appendix A

Akiyo, bowing, bridge-close, bridge-far, container, deadline, galleon, hall, ice, mother-daughter, news, news-announcer, pamphlet, paris, sign-irene, silent, students, washdc.

YUV address: http://trace.eas.asu.edu/yuv/index.html.

References

1. Milani, S., Fontani, M., Bestagini, P., et al.: An overview on video forensics. APSIPA Trans. Sig. Inf. Process. **1**, e2 (2012)
2. Tew, Y., Wong, K.S.: An overview of information hiding in H.264/AVC compressed video. IEEE Trans. Circuits Syst. Video Technol. **24**(2), 305–319 (2014)
3. Wang, W., Farid, H.: Exposing digital forgeries in video by detecting double MPEG compression. In: Proceedings of the 8th Workshop on Multimedia and Security, pp. 37–47. ACM (2006)
4. Wang, W., Farid, H.: Exposing digital forgeries in video by detecting double quantization. In: Proceedings of the 11th ACM Workshop on Multimedia and Security, pp. 39–48. ACM (2009)
5. Chen, W., Shi, Y.Q.: Detection of double MPEG compression based on first digit statistics. In: Kim, H.-J., Katzenbeisser, S., Ho, A.T.S. (eds.) IWDW 2008. LNCS, vol. 5450, pp. 16–30. Springer, Heidelberg (2009). doi:10.1007/978-3-642-04438-0_2

6. Sun, T., Wang, W., Jiang, X.: Exposing video forgeries by detecting MPEG double compression. In: 2012 IEEE International Conference on Acoustics, Speech and Signal Processing (ICASSP), pp. 1389–1392. IEEE (2012)
7. Jiang, X., Wang, W., Sun, T., Shi, Y.Q., Wang, S.: Detection of double compression in MPEG-4 videos based on Markov statistics. IEEE Signal Process. Lett. **20**(5), 447–450 (2013)
8. Ravi, H., Subramanyam, A., Gupta, G., Kumar, B.A.: Compression noise based video forgery detection. In: 2014 IEEE International Conference on Image Processing (ICIP), pp. 5352–5356. IEEE (2014)
9. Vazquez-Padin, D., Fontani, M., Bianchi, T., et al.: Detection of video double encoding with GOP size estimation. In: 2012 IEEE International Workshop on Information Forensics and Security (WIFS), pp. 151–156. IEEE (2012)
10. Chen, S., Sun, T.F., Jiang, X.H., He, P.S., Wang, S.L., Shi, Y.Q.: Detecting double H.264 compression based on analyzing prediction residual distribution. In: Shi, Y.Q., Kim, H.J., Perez-Gonzalez, F., Liu, F. (eds.) IWDW 2016. LNCS, vol. 10082, pp. 61–74. Springer, Cham (2017). doi:10.1007/978-3-319-53465-7_5
11. He, P., Jiang, X., Sun, T., et al.: Double compression detection based on local motion vector field analysis in static-background videos. J. Vis. Commun. Image Represent. **35**, 55–66 (2016)
12. He, P., Jiang, X., Sun, T., et al.: Detection of double compression in MPEG-4 videos based on block artifact measurement. Neurocomputing **228**, 84–96 (2017)
13. Aghamaleki, J.A., Behrad, A.: Inter-frame video forgery detection and localization using intrinsic effects of double compression on quantization errors of video coding. Sig. Process. Image Commun. **47**, 289–302 (2016)
14. Barnich, O., Van Droogenbroeck, M.: ViBe: a universal background subtraction algorithm for video sequences. IEEE Trans. Image Process. **20**(6), 1709–1724 (2011)

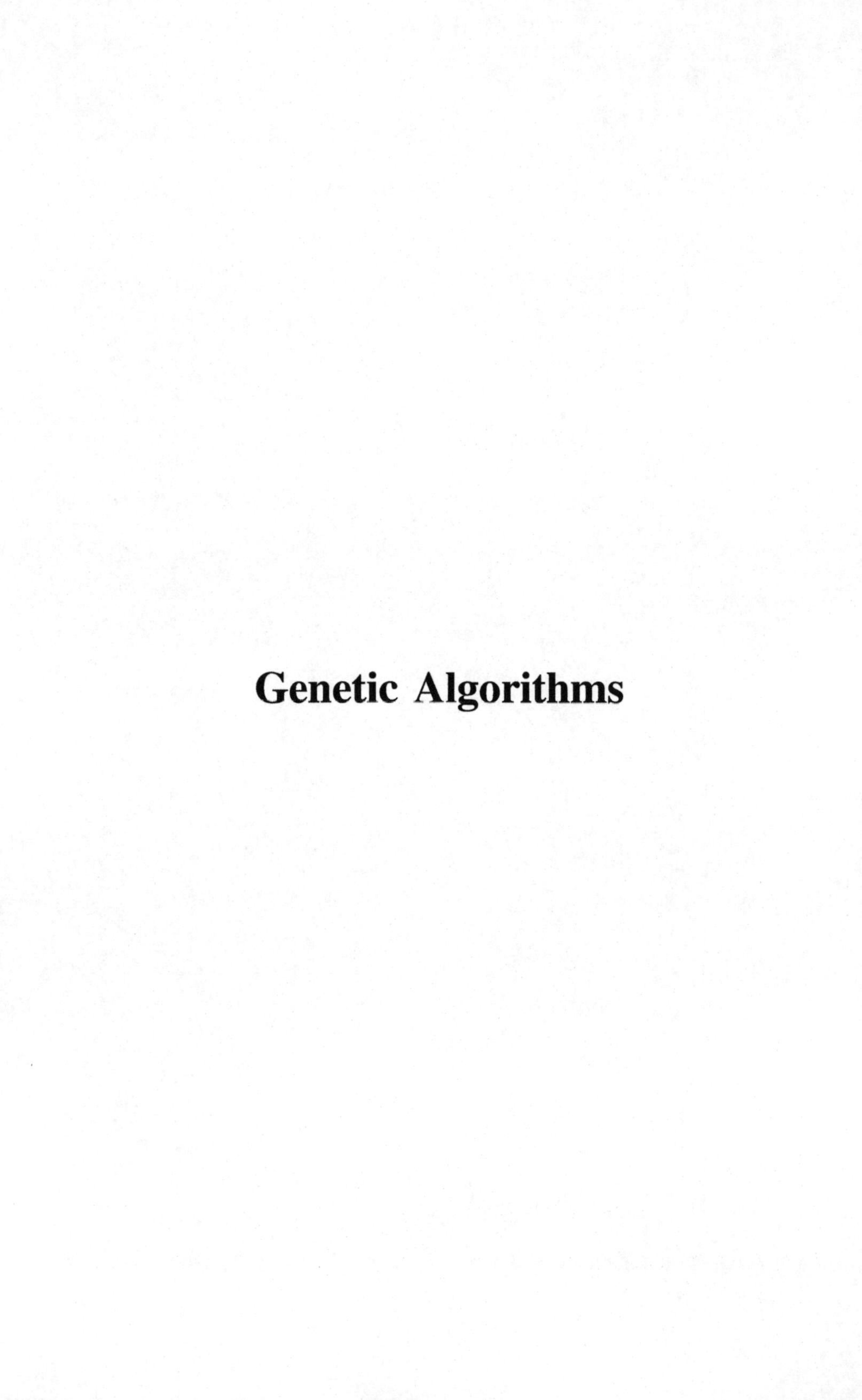

Genetic Algorithms

Research on Multi-UAV Loading Multi-type Sensors Cooperative Reconnaissance Task Planning Based on Genetic Algorithm

Ji-Ting Li(✉), Sheng Zhang, Zhan Zheng, Li-Ning Xing, and Ren-Jie He

College of Information System and Management, National University of Defense Technology, Changsha, HN, People's Republic of China
lijiting@nudt.edu.cn

Abstract. Unmanned Aerial Vehicle (UAV) has been playing an increasingly important role in modern military fields recently. The multi-UAV cooperative reconnaissance mission planning is one of the task allocation and resource scheduling problems in the field of multi-UAV co-operative control, which is full of challenges. In this paper, a multi-base, multi-target, multi-load and multi-UAV cooperative task model is established. Taking the actual battlefield situation into account, this paper built a confrontation scenario between the UAVs and radars. The objective function of the established model is the shortest route length of UAVs staying in the detection range of radars. This paper presented an improved genetic algorithm to address the problem scenario. The solving procedure consists of two steps. First of all, the route of UAVs that traverse targets within target group is considered as a Traveling Salesman Problem (TSP). Second, the route of UAVs that fly between different target groups is regarded as a Multiple Depot Vehicle Routing Problem (MDVRP). In addition, the working patterns of different sensors carried by UAVs are concerned. As a consequence, a more optimized route of UAVs is acquired. Finally, A simulated case is designed to verify the feasibility of our proposed algorithm.

Keywords: Multi-UAV · Mission planning · Genetic algorithm · TSP · MDVRP

1 Introduction

Unmanned Aerial Vehicle (UAV) is a new combat platform with autonomous capability and independent execution capability. It can not only perform non-offensive tasks such as military reconnaissance, surveillance, search and target orientation, but also can perform combat missions such as attack and target bombing and so on. With the rapid development of UAV technology, more and more UAVs will be used in the future battlefield. In a variety of applications of UAVs, arranging UAVs with different loads when they performing reconnaissance mission is an important application.

The implementation of reconnaissance missions by multi-UAV has been a research topic of concern to scholars all over the world. In the early days, Air Force Institute of

D.-S. Huang et al. (Eds.): ICIC 2017, Part I, LNCS 10361, pp. 485–500, 2017.
DOI: 10.1007/978-3-319-63309-1_44

Technology (AFIT) conducted a number of studies on multi-UAV collaborative reconnaissance mission planning with the background of 'Global Hawk' and 'Predator' unmanned reconnaissance aircraft. Ryan et al. [1] chose the 'Global Hawk' and 'Predator' of the US military as the objects of study, and transforms the task assignment problem of UAV into the Vehicle Routing Problem (VRP) model. The model regards the UAV as a vehicle, and reconnaissance target as a customer. Finally this paper used tabu search algorithm to solve the model. Vincent et al. [2] abstracted the multi-UAV collaborative reconnaissance problem into a multi-traveling salesman problem (MTSP), and used evolutionary algorithms to solve it. The paper [3] also modeled the same problem as a traveling salesman problem, and focused on the application of particle swarm optimization in solving TSP problems. Yang et al. [4] built a more practical model, namely the multi-base, multi-target, multi-UAV cooperative reconnaissance (M-M-MUCRM) model and designed a heuristic genetic algorithm to solve it.

While some studies have taken the situation in which multi-UAV carry different loads into account. Tian et al. [5] considered the type of sensor, indicating how to choose the sensor in different environments. Park et al. [6] mainly concerned about the function of sensors, and used them to build a multi-UAV information sharing structure. However, these papers remain to own the following deficiencies: (1) the previous literature rarely considered the working condition of the sensors for the UAV path planning. Most of them roughly dealt with the use of sensors, didn't consider the constraints related to sensors. So the path planning is not good enough. (2) And in the battlefield, the enemy's radar usually can detect and combat UAVs. How to make the UAVs effectively avoid them also need attention.

In this paper, we focus on the influence of multi-type load on the route planning of multi-UAV co-operation, and find a more optimized route. At the same time, this paper also considers the situation of radar defense in the real battlefield, and carries on the simulation solution of UAV reconnaissance task planning in this scene. In Sect. 2, the problem scenario we studied is introduced and the related mathematical model is established. After that in Sect. 3, a solving process based on genetic algorithm is put forward. Then in Sect. 4, simulation experiments are developed to validate the performance of our designed algorithms. Finally Sect. 5 displays our conclusion.

2 Scenario Description and Problem Modeling

2.1 Problem Scenario Description

A UAV combat troops are now equipped with a number of UAV bases and each base is equipped with a certain number of reconnaissance UAVs. The main function of reconnaissance UAV is to scout targets. Two kinds of sensors can be loaded on UAV. One is the imaging sensor, using the wide-area search mode to take images of targets, that is, once the targets fall within the sensor imaging bandwidth, can these targets be imaged. The imaging bandwidth of the sensor is generally 2 km, and the working principle of this kind of sensor is shown in Fig. 1.

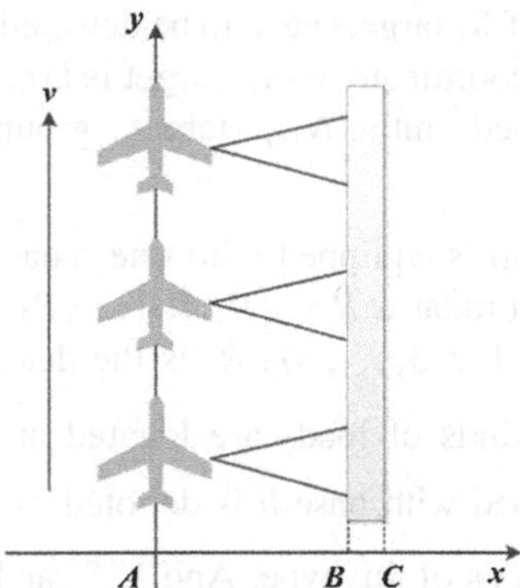

Fig. 1. The working pattern of imaging sensor

In Fig. 1, the length of BC is 2 km. The imaging sensor can only image on one side of UAV and is no longer adjusted during flight. In general, the target reconnaissance process needs some time to collect the required information. So the lateral distance AB must greater than a certain threshold AB_{min}, and there is also a maximum range limit AC_{max} of AC too. When the imaging sensor uses the wide area search mode to image the target, in order to ensure the imaging quality, the UAVs is required to make uniform linear motion.

Another kind of load is optical sensor, in order to achieve the required target recognition accuracy, the distance between optical sensor and target is under a required distance. And this kind of sensor can take photo instantly. Each UAV can only load one kind of sensor each time and in order to ensure the reconnaissance effect, two kinds of sensors are required to scout each target for at least once. Two different sensors have no more than 4 h of reconnaissance intervals for the same target. While the task is performed at the same time, multi-UAV can fly in formation, but a safe distance needed to maintain. After the completion of the task, the UAV will return to its original base.

According to the needs of combat missions, UAVs need to scout all targets in several target groups and each target group is equipped with a radar station. In the actual combat environment, once the reconnaissance UAVs fly into the radar detection range, the radar will turn on the air alert and search targets, then take appropriate measures, such as launch the missiles to destroy the UAV. So the longer the UAVs stay in the detection range of radar, the greater the possibility of its destruction. And the length of the UAV path directly determines the time it stays in radar detection range. Aiming at this complex confrontational task scene, it is necessary to develop a proper route and cooperative dispatching strategy for multi-UAV carrying multi-loads from multi-bases.

2.2 Mathematical Model

Notations

Base set: A total number of N_B bases are equipped with reconnaissance UAVs, and the base set is denoted as $B = \{B_1, B_2, \ldots, B_{N_B}\}$. The coordinate of one certain base is $(xb_i, yb_i), i = 1, 2, 3, \ldots N_B$.

Target set: Atotal number of N_T targets need to be detected, and the target set is denoted as $T = \{t_1, t_2, \ldots, t_{N_T}\}$. The coordinate of one target is $(xt_i, yt_i), i = 1, 2, 3, \ldots, N_T$. And all the targets are grouped into N_{TG} target groups, which is denoted as $TG = \{tg_1, tg_2, \ldots, tg_{N_T}\}$.

Radar set: Each target group is equipped with one radar station, so the total number N_R is equal to N_{TG}. The set of radar is $R = \{R_1, R_2, \ldots, R_{N_R}\}$, and the coordinate of one certain radar is $(xr_i, yr_i), i = 1, 2, 3, \ldots, N_R$. R_a is the detection radius of each radar.

UAV set: UAVs with Nv kinds of loads are located at N_B bases. The UAV set of loading k-type sensor equipped with base b is denoted as $V^{bk} = \left\{v_1^{bk}, \ldots, v_{N_b^k}^{bk}\right\}$. N_b^k is the maximum number of UAVs of this type. And V^{bk} can be null, which indicates that this base is not equipped with UAV carrying k-type load. In the problem discussed in this paper, $k \in \{1,2\}$, and 1-type load is imaging sensor while 2-type load is optical sensor. $NV = \Sigma_{b=1}^{N_B} \Sigma_{k=1}^{N_V} N_b^k$ is the number of all UAVs can be used.

The speed of UAV is υf. The takeoff time of one certain UAV is $T_0^{(b,k,i)}$ and the maximum flight time of UAV is T_{max}. The flying height of UAV is h. When a certain UAV from base b, carrying the k-type load is flying, the coordinates of it at the moment t is denoted as $\left(xv_{(t)}^{(b,k,i)}, yv_{(t)}^{(b,k,i)}\right), i = 1, 2, \ldots, N_b^k, T_0^{(b,k,i)} \leq t \leq T_{max}$.

rs_2 indicates the farthest distance from the target when optical sensor takes a picture. $TG_{(b,i)}^k$ indicates the target group set scouted by the i-th UAV from base b, carrying the k-type sensor.

$distance^k(t)$ denotes the sum of the distances of the UAVs loading the k-type sensor and scouting all targets in each target group.

$L_{(b,i)}^k$ denotes the edge set of the i-th UAV carrying the k-type sensor flying between the target groups.

$length(l)$ represents the length of each edge in the L set. To be more specific, the sketch map of the edge set $L_{(b,i)}^k$ is shown in Fig. 2.

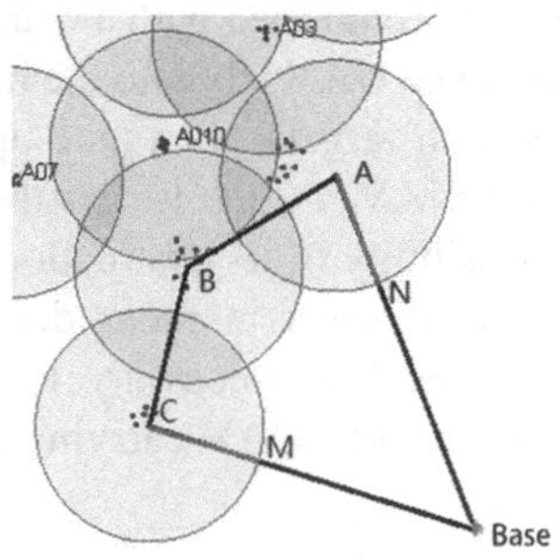

Fig. 2. The sketch map of the edge set $L_{(b,i)}^k$

As shown in Fig. 2, one certain UAV v_1^{bk} flies into target group C from M point, after scout all targets in group C, it flies to target group B and then flies to target group

A and returns to base. So $length(l) = L_{BC} + L_{BA}$ and the sum distance of UAV in the detection range of radar is $length(l) + 2r_a$.

Objective function

$$min\ Z = \sum_b \sum_i \sum_k \left(\sum_{l \in L(b,i)} length(l) + \sum_{t \in T(b,i)} distance^k(t) + 2r_a \right) \tag{1}$$

$$s.t.\ TG^k_{(b,i)} \subseteq \{1,2,3,\ldots,N_{TG}\}, k = 1,2 \tag{2}$$

$$\bigcup_{b,i} TG^k_{(b,i)} = \{1,2,3,\ldots,N_{TG}\}, k = 1,2 \tag{3}$$

Constraint conditions

$$\forall t, \forall v_i^{bk}, \forall v_j^{bk} \in V^{bk}, \left(\sqrt{(xv_{(t)}^{(b,k,i)} - xv_{(t)}^{(b,k,j)})^2 + (yv_{(t)}^{(b,k,i)} - yv_{(t)}^{(b,k,j)})^2} \leq 200\right) \tag{4}$$

$$\forall v_i^{bk} \in V^{bk}, \frac{\sum_{l \in L(p,q)} length(l) + \sum_{v \in V(p,q)} distance^k(v) + 2r_a}{vf} \leq T_{max} \tag{5}$$

$$\forall v_i^{bk}, v_j^{bk} \in V^{bk}, \left| T_0^{(b,k,i)} - T_0^{(b,k,j)} \geq T_{interval} \right| \tag{6}$$

Inequality (4) indicates that the flying distance between every two UAVs should bigger than 200 km. And since the average speed of the UAV is vf km/h, the maximum life time of T_{max} must meet inequality (5). Due to the limitation of the technical support of the base, the time interval between two UAVs' takeoffs from the same base must meet inequality (6).

2.3 Model Analyze

In order to solve the objective function, it is necessary to make further analyze whether the problem is applicable to some classical problem models. The objective function can be divided into two parts. When calculate

$$min \sum_b \sum_i \sum_k \left(\sum_{t \in T(b,i)} distance^k(t) \right) \tag{7}$$

the sub-problem can be modeled as a Traveling Salesman Problem (TSP). It can be described as finding a shortest path to traverse all targets in one certain target group.

And when calculate

$$min \sum_{b} \sum_{i} \left(\sum_{l \in L(b,i)} length(l) \right) \quad (8)$$

the sub-problem can be modeled as a Multiple Depot Vehicle Routing Problem (MDVRP). Mapped to this problem, the bases are regarded as depots, the targets are regarded as customers, and the UAVs are regarded as vehicles.

3 Algorithm Design

Based on the model analyze, a proper algorithm need to be designed to solve the problem scenario. As one of the classical evolutionary algorithms, genetic algorithm has a strong search ability and is widely used in solving TSP problems and performs well [7–9]. When solve the MDVRP problem, genetic algorithm also shows good performance [10–12]. So genetic algorithm was chosen to solve the problem proposed in this paper. And the overall pseudo-code of genetic algorithms designed in this paper is shown as follows.

```
1.FindOptimalSolution(B, TG, R,V^bk):
2.Initialize the optimal value to 0or infinity
3.solution = 0
4.Calulate the distance of UAV traversing targets in target group
5.For tg in TG:
6. For v_i^bk in V^bk:
7.   tspRoute = CalculateTSPbyGA(B,TG, V^bk )
8.   tspRoute = tspRoute  - L(CuttingEdge) –L(SensorWorkingPattern)
9. solution += tspRoute
10.Calculate the distance of UAVs flying between target groups
11.For v_i^bk in V^bk:
12. solution += CalculateVRPbyGA(F,n,TG)
13.Return solution
```

3.1 Genetic Algorithm for UAV Traversing Targets in a Target Group

This paper takes the working pattern of sensor into account, and the solving process of function (7) is shown as follows.

Genetic algorithm for solving TSP problem. This paper uses genetic algorithm to solve TSP problem, the brief solving process is shown in Fig. 3.

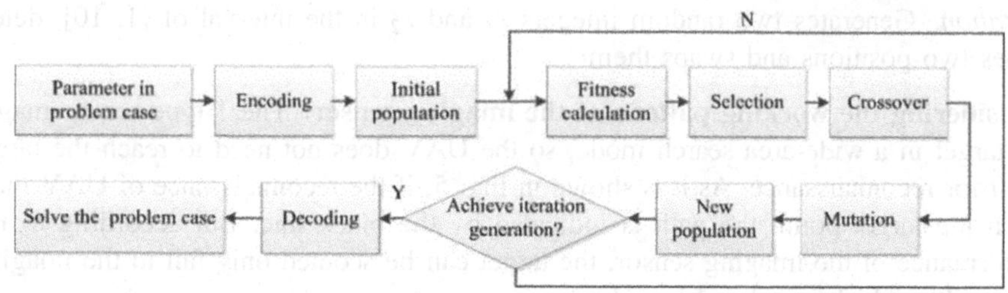

Fig. 3. The process of genetic algorithm

Encoding. Using integer encoding method, there are n targets in a target group, so the chromosome can be divided into n parts, and the order of genes indicate the order of targets under reconnaissance. For example, there are 10 targets in a target group, then 6|8|7|2|4|5|9|3|1|10 is a legal chromosome.

Population initialization. After completing the chromosome coding, an initial population must be generated as a starting solution. In this problem, the number of initialized populations depends on the number of targets.

Fitness function. Suppose that $|t_1|t_2|\cdots|t_i|\cdots|t_n|$ is a legal chromosome, and $D_{t_it_j}$ is the distance between target t_i and t_j, the fitness of the individual is:

$$fitness = \frac{1}{\sum_{i=1}^{n-1} D_{t_it_{i+1}} + D_{t_nt_1}} \tag{9}$$

Selection. Using roulette method to choose, the greater the individual fitness, the greater the probability of being selected.

Crossover. Suppose that the number of targets is 10, and the crossover method used in this paper is shown in Fig. 4.

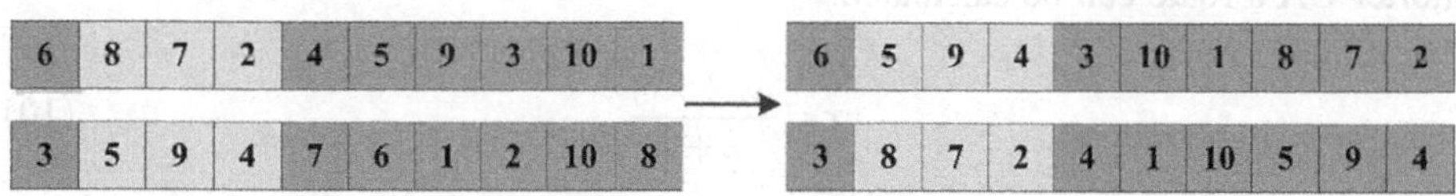

Fig. 4. The crossover operation in genetic algorithm

As it is shown in Fig. 4, two parent chromosome is a group, each of which repeats the following procedure: (1) Generates two random integers r_1 and r_2 in the [1, 10] interval, determining the two positions, crossing the intermediate data of these positions. (2) After crossing, there are duplicate target numbers in the same chromosome, non-repetitive numbers are retained. By using partial mapping, the conflicted numbers are eliminated, that is, mapping with the corresponding relationship of the middle segment.

Mutation. Generates two random integers r_1 and r_2 in the interval of [1, 10], determines two positions and swaps them.

Considering the working pattern of the imaging sensor. The 1-type sensor mages the target in a wide-area search mode, so the UAV does not need to reach the target point for reconnaissance. As it is shown in Fig. 5, if the reconnaissance of UAV must reach the target point, the path is indicated by the black line. But according to the performance of the imaging sensor, the target can be scouted only fall in the imaging bandwidth, which is colored in red.

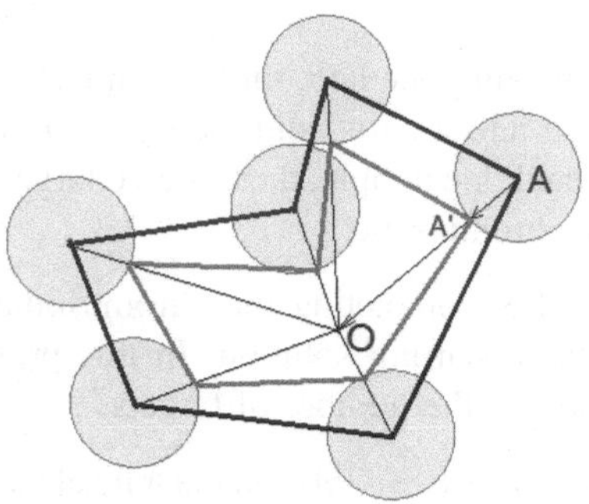

Fig. 5. The sketch map of UAV flight route when traversing targets in a target group (Color figure online)

As it is shown in Fig. 5, taking the target point as the center and inward extension of the distance α as the radius. Distance α is a certain value between the limits of AB and AC in Fig. 1. After that, we establish the center of gravity of the polygon, denoted as O in the figure, connect the point O and one certain target A, then the point A' is the new reconnaissance point and $\left|\overrightarrow{AA'}\right| = \alpha$. Set $\left|\overrightarrow{AA'}\right| = \lambda\left|\overrightarrow{A'O}\right|$ and the coordinates of A, A', O are $(x_1, y_1), (x, y), (x_0, y_0)$. The coordinates of A' are calculated as follows, then a shorter UAV route can be calculated.

$$x = \frac{x_1 + \lambda x_0}{\lambda + 1} \tag{10}$$

$$y = \frac{y_1 + \lambda y_0}{\lambda + 1} \tag{11}$$

$$\lambda = \frac{\left|\overrightarrow{AA'}\right|}{\left|\overrightarrow{A'O}\right|} = \frac{\alpha}{\left|\overrightarrow{AO}\right| - \alpha} \tag{12}$$

Cutting the edges. When a UAV traversing a target group, it usually chooses an entry point and an exit point, which are not coincide. Therefore, it is necessary to cut the edge of the UAV route in one target group. According to the law of military detection, when the UAV flies into a target group, it will detect the radar first, because only the first to understand the deployment of enemy's 'eyes', can you have a better grasp of the battlefield environment. Based on the working pattern of sensors introduced above, UAV can only detect the targets in one direction. Therefore, in each target group, when taking the radar as the entry point, it is necessary to calculate the total distance of UAV traversing all targets in clockwise direction and counterclockwise direction respectively. Then we can select the correct edge need to be cut.

Considering the working pattern of the optical sensor. The working pattern of the optical sensor is different from the imaging sensor. Optical sensor can take pictures of targets when the distance between UAV and targets does not exceed rs_2 km. While the UAV cruise flight height is h, so this paper will take the projection distance k as the effective range of optical sensor shooting range, as shown in Fig. 6.

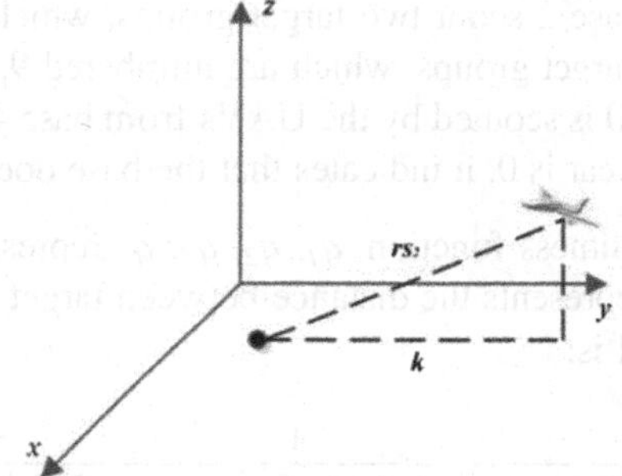

Fig. 6. The working pattern of optical sensor

Thus, when the target in the target group falls in the circle, of which center is the UAV and radius is k, the UAV can take photos of them all at once.

3.2 Genetic Algorithm for Calculating the UAV Flight Distance Between Target Groups

Encoding. According to the characteristics of this problem, this paper designed a two-part chromosome, that is, divided an individual chromosome into two parts. The length of the first part is n, which represents the feasible sequence of a group of UAV scouting target groups. The length of the rear part is m, which indicates the number of target groups should be scouted by each UAV base. The rear part and the front sequence corresponds in the same order from left to right. Thus, the length of the chromosome is $n + m$. To be more specific, an example of chromosome is shown in Fig. 7.

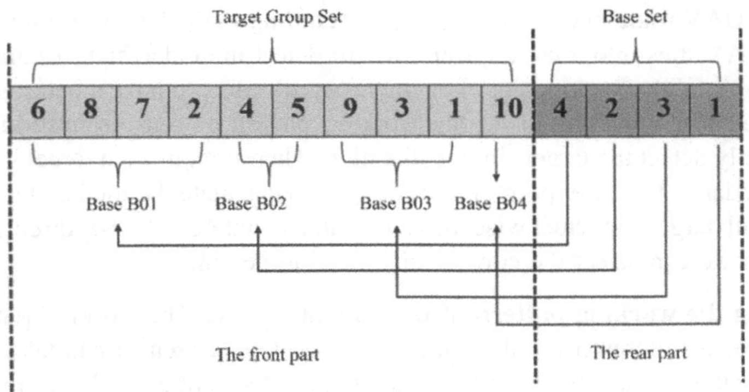

Fig. 7. The structure of two-part chromosome

In Fig. 7, the UAVs from base 1 scout four target groups, which are numbered 6, 8, 7, 2. And the UAVs from base 2 scout two target groups, which are numbered 4 and 5, while base 3 scout three target groups, which are numbered 9, 3 and 1. The remaining target groups numbered 10 is scouted by the UAVs from base 4. In particular, when the value of a position at the rear is 0, it indicates that the base does not dispatch the UAV.

Fitness function. In the fitness function, q_1, q_2, q_3, q_4 represent the value of the rear part coding. And $D_{t_it_{i+1}}$ represents the distance between target t_i and t_{i+1}. So the fitness function of the individual is:

$$fitness = \frac{1}{\sum_0^{q_1-1} D_{t_it_{i+1}} \sum_{q_1}^{q_1+q_2-1} D_{t_it_{i+1}} + \sum_{q_1+q_2}^{q_1+q_2+q_3-1} D_{t_it_{i+1}} + \sum_{q_1+q_2+q_3}^{n-1} D_{t_it_{i+1}} + \sum_{i \in B} 2r_a} \quad (13)$$

And it is worth mentioning that fitness function (9) and (13) is used to evaluate the performance of individuals in population of genetic algorithm. These two fitness functions are all parts of the objective function and are essential to it.

Crossover. The crossover operations of the front part and rear part are carried out separately. The method is as same as described in Sect. 3.1.

Mutation. The mutation operation of the front part is as same as the previous one. As shown in Fig. 8, in order to ensure the diversity of offspring chromosomes, the rear part of the chromosome carried out two kinds of mutation operations, which may occur in the same possibilities.

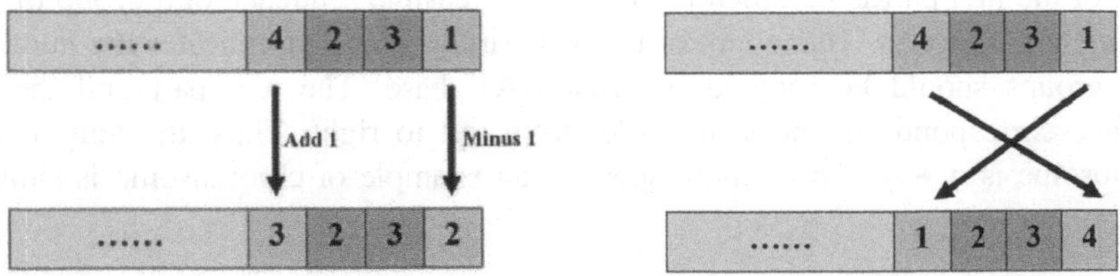

Fig. 8. The mutation operation in genetic algorithm based on two-part chromosome

4 Case Study

The case we studied in this paper is based on the problem description in Sect. 2. A simulated battlefield scenario is designed. The UAVs of attack side perform renaissance mission on the targets of defensive side, at the same time, in order to combat UAV, the defender also set the radar to detect the UAVs of attack side.

4.1 Case Generation

The coordinates of the bases and the number of UAVs equipped with each base is shown in Table 1.

Table 1. The situation of each UAV base

Name of base	Coordinates(Km)	The number of UAVs
Base B01	(368,319)	2
Base B02	(392,220)	2
Base B03	(392,275)	2
Base B04	(256,121)	2

Ten target groups of defensive side are generated and contain 68 targets. Each target group equipped with one radar station, and the detection radius of radar r_a = 70 km. The distribution of bases, targets and radars is shown in Fig. 9. The red circle is the detection range of radar station

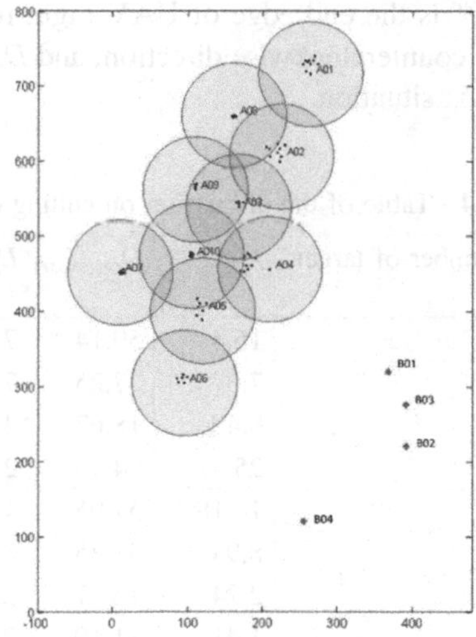

Fig. 9. The distribution of bases, targets and radars. The red circle is the detection range of radar station (Color figure online)

The flight parameters of UAV is shown in Table 2. And the parameter of genetic algorithm used in this paper is shown in Table 3. N_{pop} is the number of individuals in each generation, N_{gen} is the generation of both two genetic algorithms designed in this paper, p_c is the crossover probability, and p_m is the mutation probability

Table 2. The parameters of UAV

Parameter	Value (unit)	Parameter	Value (unit)
vf	200 km/h	AB_{min}	2 km
T_{max}	10 h	AC_{max}	8 km
h	1.5 km	rs_2	7.5 km

Table 3. The parameters of genetic algorithm

Parameter	N_{pop}	N_{gen}	p_c	p_m
Value	100	300	0.9	0.1

4.2 Results and Analyze

The proposed solving method is verified by using PYTHON code running on a PC with an Intel Core i7 processor operating at 3.30 GHZ. We repeated the experiment 5 times, and one of the results is shown as follows.

When performing cutting edge, the total distance of UAV traversing targets in each target group in clockwise direction and counterclockwise direction is shown in Table 4. $L_{end}^{clockwise}$ is the end edge of UAV flight routine when it traversing targets in a target group in clockwise direction, and $D_{clockwise}$ is the total distance after cutting edge under this situation. While $L_{end}^{counter}$ is the end edge of UAV flight routine when it traversing targets in a target group in counterclockwise direction, and $D_{counter}$ is the total distance after cutting edge under this situation.

Table 4. Table of the discussion on cutting edges

Name of target group	Number of targets	$L_{end}^{clockwise}$	$D_{clockwise}$	$L_{end}^{counter}$	$D_{counter}$
A01	10	16.4	59.14	7.21	68.33
A02	7	7.07	77.25	7.28	77.04
A03	5	8.49	18.67	4	23.16
A04	10	25	84.76	25.5	84.26
A05	7	11.18	57.95	12.04	57.09
A06	6	8.94	31.35	8.06	32.23
A07	6	2.24	16.77	2.83	16.18
A08	5	1.41	11.19	2.83	9.77
A09	5	7.28	9.40	2	14.68
A10	5	2.24	13.85	2	14.09

From Table 4, we can get the total length of route for UAV in clockwise direction is 380.33 km, while the total length of route for UAV in counterclockwise direction is 396.83 km. So UAV should fly in clockwise direction in each target group, and the imaging sensor should load on the left side of UAV. The final optimized distance of UAV traversing targets in each target group is shown in Table 5.

Table 5. Final optimized distance of UAV traversing targets in each target group

Name of target group	Final optimized distance (Km)	Name of target group	Final optimized distance (Km)
A01	47.40	A06	19.43
A02	56.65	A07	8.02
A03	10.21	A08	5.54
A04	70.06	A09	5.56
A05	43.63	A10	5.04

And Fig. 10. is the line chart compares the total distance of ignoring the working principle of imaging sensor and considering the working pattern of imaging sensor.

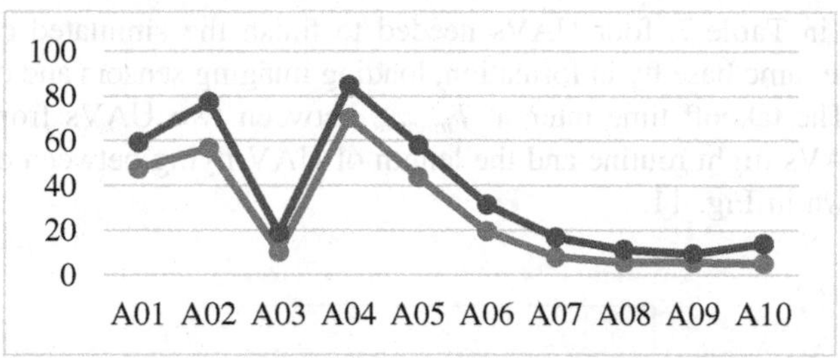

Fig. 10. The line chart compares the total distance of ignoring the working principle of imaging sensor and considering the working pattern of imaging sensor. The black line is the distance ignoring the working principle of imaging sensor, and the red line is the distance considering the working pattern of imaging sensor (Color figure online).

It is clearly that the length of route is shorter when taking the working pattern of imaging sensor into account. Considering the working pattern of optical sensor, the targets in target group A03,A07,A08,A09,A10 can be imaged once by optical sensor. And the optimized length of route for UAV carrying optical sensor traversing targets in each target group is shown in Table 6.

Table 6. The optimized length of route for UAV carrying optical sensor traversing targets in each target group

Name of target group	Final optimized distance (Km)	Name of target group	Final optimized distance (Km)
A01	47.40	A06	19.43
A02	56.65	A07	0
A03	0	A08	0
A04	70.06	A09	0
A05	43.63	A10	0

Then, the schedule table of UAVs is shown in Table 7.

Table 7. The schedule table of UAVs

UAV	Base	Flight route
v_1^{11}	B01	B01->A04->A03->A09->A08->A02->A01->B01
v_2^{12}	B01	B01->A04->A03->A09->A08->A02->A01->B01
v_1^{41}	B04	B04->A06->A05->A10->A07->B04
v_2^{42}	B04	B04->A06->A05->A10->A07->B04

As shown in Table 7, four UAVs needed to finish the simulated case. And two UAVs from the same base fly in formation, loading imaging sensors and optical sensors respectively. The takeoff time interval $T_{interval}$ between two UAVs from one base is 3 min. The UAVs flight routine and the length of UAV flying between different target groups is shown in Fig. 11.

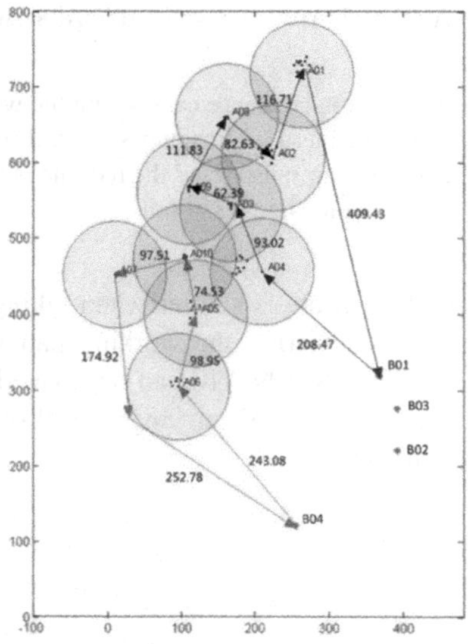

Fig. 11. The map of UAVs flight routine

And the minimum total distance of four UAVs staying in the detection range of radar is 2545.03 km. The schedule plan of UAVs of five experiments are same, and the total distances of five times experiments is shown in Table 8, and it verifies that the algorithms designed in this paper is relative stable.

Table 8. The results of five times experiments

Experiment	1	2	3	4	5
Total distance (km)	2545.03	2540.71	2545.03	2545.03	2540.71

5 Conclusion

Nowadays, multi-UAV cooperative control is one of the hot issues studied in the area of multi-robots system, which has great practical and academic influence in both military and civil aspect [13]. And this paper built a multi-base, multi-target, multi-load, multi-UAV cooperative task model and used genetic algorithm to solve the problem scenario. A simulated case is built and used the solving method proposed in this paper, we can get that the imaging sensor should be loaded on the left side of UAV, and when traversing targets in a target group, UAV should fly in clockwise direction. There are four UAVs needed to complete the simulated case and the average minimum total distances when UAVs stay in the detection range of radars is 2543.3 km. It is proved that the solving method proposed in this paper is feasible and stable.

References

1. Ryan, J.L., Bailey, T.G., Moore, J.T., Carlton, W.B.: Reactive tabu search in unmanned aerial reconnaissance simulations. In: Simulation Conference Proceedings, vol.1, pp. 873–879. DBLP (1999)
2. Vincent, P., Rubin, I.: A framework and analysis for cooperative search using UAV swarms. In: ACM Symposium on Applied Computing, pp. 79–86. DBLP (2004)
3. Seerest, B.R.: Traveling Salesman Problem for Surveillance Mission Using Particle Swarm Optimization. Master's thesis, Airforce Institution of Technology, Wright-Patterson Air Force Base, Ohio, USA (2001)
4. Yang, W.L., Lei, L., Deng, J.S.: Optimization and improvement for multi-UAV cooperative reconnaissance mission planning problem. In: International Computer Conference on Wavelet Active Media Technology and Information Processing, pp. 10–15. IEEE (2014)
5. Tian, J., Shen, L.: A multi-objective evolutionary algorithm for multi-UAV cooperative reconnaissance problem. In: King, I., Wang, J., Chan, L.-W., Wang, D. (eds.) ICONIP 2006. LNCS, vol. 4234, pp. 900–909. Springer, Heidelberg (2006). doi:10.1007/11893295_99
6. Park, C., Cho, N., Lee, K., Kim, Y.: Formation flight of multiple UAVs via onboard sensor information sharing. Sensors **15**(7), 17397 (2015)
7. Liu, F., Zeng, G.: Study of genetic algorithm with reinforcement learning to solve the TSP. Expert Syst. Appl. **36**(3), 6995–7001 (2009)
8. Wang, L.Y., Zhang, J., Li, H.: An improved genetic algorithm for TSP. In: International Conference on Machine Learning and Cybernetics, vol. 2, pp. 925–928. IEEE (2007)

9. Tao, Z.: TSP problem solution based on improved genetic algorithm. In: International Conference on Natural Computation, vol. 1, pp. 686–690. IEEE (2008)
10. Vaira, G., Kurasova, O.: Genetic algorithm for VRP with constraints based on feasible insertion. Informatica **25**(1), 155–184 (2014)
11. Zhang, Q., Gao, L.Q., Hu, X.P., Wu, W.: Research on multi-objective vehicle routing problem of optimization based on clustering analysis and improved genetic algorithm. Control Decis. **18**(4), 418–422 (2003)
12. Zou, T.: Genetic algorithm for multiple-depot vehicle routing problem. Comput. Eng. Appl. **40**(21), 82–85 (2004)
13. Tian, J., Shen, L.C.: Research on multi-base multi-UAV cooperative reconnaissance problem. Acta Aeronaut. Astronaut. Sinica **28**(4), 913–921 (2007)

Discriminative Motif Elicitation via Maximization of Statistical Overpresentation

Ning Li(✉)

Institute of Machine Learning and Systems Biology,
College of Electronics and Information Engineering, Tongji University,
No. 4800 Caoan Road, Shanghai 201804, China
jclining@126.com

Abstract. The Fisher Exact Test score (FETS) and its variants are based on the hypergeometric distribution. It's very natural to describe the enrichment level of TF binding site (TFBS) by it. And several widely used methods that discriminant motif discovery have choose them as the objective functions, for example, HOMER and DERME. Although the method is highly efficient and universal, FETS is a non-smooth and non-differentiable function. So it can not be optimized numerically. In order to solve the problem, the current methods that learn to optimize FETS either reduce the search set to discrete domain or introduce some external variables which will definitely hurt the precision, not to mention that to use the complete potential of input sequences for generate motifs. In this paper, we propose an approach that allows direct learning the motifs parameters in the continuous space use the FETS as the objective function. We find that when the loss function is optimized in a coordinate-wise mode, the cost function can be a piece-wise constant function in each resultant sub-problem. The process of finding optimal value is exactly and efficiently. Furthermore one key step in every iteration of optimize the FETS requires finding the most statistically significant scores among the tens of thousands of Fisher's exact test scores, which is solved efficiently by a 'lookahead' technique. Experiments on ENCODE ChIP-seq data testify the performance of the proposed method.

Keywords: Transcription factor (TF) · Motif · Fisher's exact test

1 Introduction

By binding to its genome target sequences and regulating gene expression patterns, transcription factor (TF) plays an important role in the transcriptional regulatory networks that control cell differentiation and development. In general, TF binding is a complex process, affected by a variety of factors, such as DNA shape, cofactor and chromatin status, etc. [1, 2]. In particular, TF is preferred to bind to similar short sequences (called TF binding sites, TFBS) across the DNA genome [3]. In order to unravel the complex mechanisms of genetic variation under different developmental and environmental conditions, the first step is to tap the potential nucleotide sequence pattern of TFBS, which is called the TFBS sequence motif [4].

D.-S. Huang et al. (Eds.): ICIC 2017, Part I, LNCS 10361, pp. 501–512, 2017.
DOI: 10.1007/978-3-319-63309-1_45

In the past ten years, high-throughput sequencing technology (such as ChIP-seq [5, 6]) is introduced, greatly increasing the data mining for sequence motifs available, which provides a huge opportunity for accurate classification of TFBS sequence motifs. However, it also poses new computational challenges. For example, a large number of currently available data may introduce a significant number of false positive sequences. Even for true positive sequences, the location of TFBS is not known. In addition, most of the computational complexity of the traditional sequential primitive mining algorithms is very large in the case of such large-scale data. So far, there have been many methods for heuristic primitives in high-throughput data mining sequence is proposed. Among all these methods, motif (discriminative motif discovery, DMD) algorithm provides a feasible solution for these problems [3, 7–10]. The traditional motif learning algorithm uses the statistical model to represent the non binding background sequence, which highlights the enrichment degree of foreground sequence data sets [8, 11]. The DMD method search can distinguish the foreground data set and the sequence pattern of the special collection of background sequence data sets. This method can help to distinguish the false positive, and mine the qualified motif signal. In addition, the DMD can also be used to study the context dependent regulatory activities by designing a negative case set in a specific way [2, 3, 8, 11].

In DMD, Fisher's exact test score (Fisher Exact Test score, FETS) is an important indicator of motif analysis [12]. The method is based on the hypergeometric distribution, and considers that the positive and negative cases are obtained from the same distribution. It is a statistical observation of the probability of such data (and even more extreme), thus providing a very natural method for quantifying the degree of enrichment of transcription factor binding site motif. As a result, FETS is often used as a sort of candidate motif in the study [12–14], as well as several widely used DMD methods as the objective function, such as DREME, Amadeus and HOMER [10].

Although it is very common, but as a discrete, non differentiable function, effective optimization method FETS cannot use the standard, such as gradient descent, is a serious challenge for the attempt by the optimization FETS motif learning algorithm. In order to scalability based on dealing with these difficulties without sacrificing, the algorithm selection will motif the search space in discrete space (such as DREME) or indirect scheme, namely the iterative threshold optimization to reconstruct motif (e.g. HOMER and Amadeus). In a word, none of these strategies can be used to obtain the motif parameters directly in a continuous space, which may be regarded as the accuracy of the motif mining and result in the deviation from the current experience.

In this paper, we have developed an algorithm called FSdO. The method can directly maximize the FETS of motifs. By analyzing the nature of the problem, we find that when the loss function of the results is optimized in the coordinate system, the cost function is piece-wise linear. The optimal value of the consumption function can be accurately identified. To further improve the practicability of the algorithm in the future, we propose an efficient "look-ahead" algorithm to solve the necessary steps in each iteration. In this step, you need to find the most statistically significant.

The reminder is arranged as follows: In Sect. 2, we discuss the background which is necessary; In Sect. 3, we present the FSdO method; In Sect. 4, we present experimental configurations and results.

2 Background

2.1 Problem Setting

In the general discriminant motif learning problem [3, 7–9], we have DNA sequences T as input, each string S ϵ T of length |S| defined over the DNA alphabet Σ = {A, C, G, T}. T is divided into tow sets, foreground set(Ƒ) and background set(Ɓ). Depending on the problem, the meaning of the positive case set and the negative case set may be different. For example, it can indicate whether a TF binding site exists in the collection [8].

Motif based on the different ways of expression, the traditional way can be divided into two kinds. Word-based [3, 4, 15] and position weight matrix [16] based on sequence expression.

Based on different representation of motif, the traditional ways can be divided into two kinds. That are word-based sequence expression and position weight matrix (PWM). Word-based approaches use the input data deterministic nucleotide strings as input that are overrepresented. In contrast, PWM-based methods have lower scalability. By representing motifs as PWMs, it have more modeling flexibility in which using product distribution of TFBS described the ambiguity in instance-level. Each product distribution is separately modeled as multinomial distribution on alphabet. Specifically, each entry in a PWM W ϵ $R^{|\sum| \times l}$ defined as [7, 17, 18]:

$$W_{[b,i]} = \log\left(\frac{p(b,i)}{p(b)}\right), \quad (1)$$

Where l is the length of motif, p(b,i) is the probability of letter b in the i-th position, and p(b) is the frequency of background in corresponding position. As in [7, 17], the match score between sequence l (s = s[1]s[2]…s[l]) with W is given blow:

$$f_W(s) = \sum_{i=1}^{l} \left(II\left(s_{[i]} = A\right) W_{[1,i]} + II\left(s_{[i]} = C\right) W_{[2,i]} + II\left(s_{[i]} = G\right) W_{[3,i]} + II\left(s_{[i]} = T\right) W_{[4,i]} \right) \quad (2)$$

2.2 Fisher Exact Test Score

Express foreground subset (background subset) as Ƒm (Bm) in which sequences that contain or not the motif W we interested. FETS can be represent as below [3]:

$$FETS_W = \sum_{k=|Fm|}^{\min(|F|,|Fm|+|Bm|)} \left(\frac{\binom{F}{k}\binom{|B|}{|Fm| + |Bm| - k}}{\binom{|F| + |B|}{|Fm| + |Bm|}} \right), \quad (3)$$

That is FETS is the probability that observing at least |Fm| motif occurrences in F, under the null assumption that such a occurrence is equal in foreground and background. In the representation, |*| is the cardinality of a set.

For the word-based approaches, when the occurrence of motif is well defined, the calculation is transparent. i.g. DREME. For PWM-based approaches, whether a sequence contains a motif is defined by a threshold b which is an additional site score. That is when the site score above the threshold it contains the motif [13]. We can rewrite |Fm|and |Bm| as below:

$$\begin{aligned} |Fm| &= \sum_{S \in F} II(f_W(S) - b > 0) \\ |Bm| &= \sum_{S \in B} II(f_W(S) - b > 0) \end{aligned} \tag{4}$$

Where $f_W(S)$ is the maximal matching score between W and the complete set of l-long sub-sequences of S. Let P denote that. And Ş can be obtained by using a sliding window which width is l and scan S sequence and its reverse complement $\overline{S}$ [19]:

$$\begin{aligned} & f_W(S) = \max_{s \in P(S)} (f_W(s)) \\ & P(S) = \left\{ S_{[i,i+1,\ldots,i+l-1]} | i = 1,2,\ldots,|S| - l + 1 \right\} \cup \left\{ \overline{S}_{[i,i+1,\ldots,i+l-1]} | i = 1,2,\ldots,|S| - l + 1 \right\} \end{aligned} \tag{5}$$

For both word-based and PWM-based methods, it is easy to see that FETS is the discontinuous function that is non-convex and can't be optimized numerically, because it can't be differentated. To solve this problem, a lot of heuristic method were used in previous approaches. For example, DREME, the classic word-based method, restricted the motif searching range in a discrete domain (IUPAC words). And optimize it by replacing one position of motif with more ambiguous alphabet. Therefore the FETS can be improved gradually. And for PWM-based methods, i.g. HOMER and Amaeus, although the details are different they all use the same strategy, namely iterative optimization [16, 20]. Tweak the score threshold thus the motif occurrences in foreground can be maximized. The method is also dependent on an external variable, indirect optimization parameters, even if the method has a more detailed representation than DREME. The results can't even guarantee the FETS is monotonically decreasing.

In view of the above observations, we propose a method that can efficiently optimize the (3) with a more general pattern.

3 Proposed Method

3.1 Encode Mode

In order to facilitate future discussion, firstly rewrite (2) as a simpler form. Let g(*) as a function that ACGT as e_i, $i \in [1,4]$, where $e_i \in R^4$ is the i-th natural basis. Then concatenate corresponding coding in each position of $s \in P(S)$ as a vector and appending an additional variable 1. So we can represent s into a (4l + 1)-dimensional vector as:

$$x = g(s) = \left[g\left(s_{[1]}\right)^T, g\left(s_{[2]}\right)^T, \ldots, g\left(s_{[l]}\right)^T, 1\right]^T \in R^{4l+1} \tag{6}$$

Based on (6), S can be converted to X:

$$\mathrm{X}(S) = \{x : x = g(s), s \in P(S)\} \tag{7}$$

Then, the W and b can be represent as:

$$w = \left[W_{[1,1]}, \ldots, W_{[4,1]}, \ldots, W_{[1,l]}, \ldots, W_{[4,l]}, -b\right]^T \tag{8}$$

Using (6) and (8), (2) and (5) can be express as:

$$\begin{aligned} f_W(s) - b &= w^T x = w^T g(s) \\ f_W(S) - b &= \max_{x \in X(S)} \left(w^T x\right) \end{aligned} \tag{9}$$

3.2 Problem Description

Based on (9), (3) can be express as FETSw, a function of w. Out method is similar to the coordinate descent in [21]. As in Algorithm 1. Explicitly, we start from an initial point, step by step to produce intermediate results until convergence. The process from wk to wk + 1 is the outer iteration. Specifically, there are 4 1 + 1 internal iteration for each outer iteration. We use it to calculate one variable update for each coordinate by solving following sub-problem:

$$\min_t L_t(t) \equiv FETS_w\left(w^k + t \cdot e_i\right) \tag{10}$$

Then choose the most objective decrease in the coordinate as the updating direction.

Algorithm

Input: Foreground set F, background set B, initial solution w0, iteration number k = 0.

Output: the optimized wk.

Repeat

Compute wk + 1, i for every i∈ [1,4l + 1] by solving (10)

$i_o = \underset{1 \le i \le 4l+1}{\arg\max} \left(FETS_w\left(w^k\right) - FETS_w\left(w^{k+1,i}\right)\right)$

$w^{k+1} = w^{k+1,i_o}, k = k+1$

Until convergence

3.3 Analysis

Using (4) (5) (9), $L_t(t)$ can be rewrite as:

$$L_t(t) = \sum_{k=|Fm(t)|}^{\min(|F|,|Fm(t)|+|Bm(t)|)} \left(\frac{\binom{|F|}{k}\binom{|B|}{|Fm(t)|+|Bm(t)|-k}}{\binom{|F|+|B|}{|Fm(t)|+|Bm(t)|}} \right) \tag{11}$$

Where

$$\begin{aligned} |Fm(t)| &= \sum_{S\in F} o_S(t) \\ |Bm(t)| &= \sum_{S\in B} o_S(t) \end{aligned} \tag{12}$$

$$o_S(t) = II\left(\max_{x\in X(S)}\left(\left(w^k + t\cdot e_i\right)^T x\right) \geq 0\right) \tag{13}$$

Occurrence function $o_s(t)$ returns 1 if the sequence contains the motif that is parametrized by wk + t*ei. In order to get best optimization of (10), we analyze it as follows.

We can know that vector (6) can be divided into two index set through analysis.

$$\begin{aligned} I_S &= \{x : x \in X(S),\ x_{[i]} = 1\} \\ O_S &= \{x : x \in X(S),\ x_{[i]} = 0\} \end{aligned} \tag{14}$$

The training set then can be divided into three parts:

$$\begin{aligned} A_1^f &= \{S|S \in T, I_S \neq \Theta, O_S \neq \Theta\} \\ A_1^f &= \{S|S \in T, I_S = \Theta, O_S \neq \Theta\} \\ A_1^f &= \{S|S \in T, I_S \neq \Theta, O_S = \Theta\} \end{aligned} \tag{15}$$

We assume that the training set is not empty, so that Is Os do not be considered. Based on (15), we find that fs(t) can also be divided into three corresponding types:

Type I

S ϵ A1f, so the fs(t) is a piece-wise linear function. And the ts is the break point:

$$f_S(t) = \begin{cases} \max\limits_{x\in O_S}\{x^T w^k\}, t \leq t_s \\ \max\limits_{x\in I_S}\{x^T w^k\} + t, t > t_s \end{cases} \tag{16}$$

Where

$$t_s = \max_{x \in O_S}\{x^T w^k\} - \max_{x \in I_S}\{x^T w^k\} \tag{17}$$

Type II
S ϵ A2f, so the fs(t) is a line:

$$f_S(t) = \max_{x \in O_S}\{x^T w^k\} \tag{18}$$

Type III S ϵ A3f, so the fs(t) is a line:

$$f_S(t) = \max_{x \in I_S}\{x^T w^k\} + t \tag{19}$$

By analyzing the $f_s(t)$, it is easy to know that gs(t) have three kinds of shapes also:

Scenario I

$$S \in A_1^o = \left\{ S | S \in A_1^f \; and \; \max_{x \in O_S}\{x^T w^k\} \leq 0, \; or \, S \in A_3^f \right\} \tag{20}$$

Then

$$o_s(t) = \begin{cases} 0, t \leq t_e \\ 1, t > t_e \end{cases} \tag{21}$$

Where

$$t_s = -\max_{x \in I_s}\{x^T w^k\} \tag{22}$$

Scenario II

$$S \in A_2^o = \left\{ S | S \in A_1^f \cup A_2^f \; and \; \max_{x \in O_S}\{x^T w^k\} > 0 \right\} \tag{23}$$

Then

$$o_s(t) = 1 \tag{24}$$

Scenario III

$$S \in A_3^o = \left\{ S | S \in A_2^f \; and \; \max_{x \in O_S}\{x^T w^k\} \leq 0 \right\} \tag{25}$$

Then

$$o_s(t) = 1 \tag{26}$$

3.4 Algorithm Overview

Through analysis we know the os(t) is constant in scenario II and scenario III, i.e.S ϵ A2o U A3o, and is piece-wise constant in scenario I. Combined with (11)–(13), the loss function L(t) is a function of |Fm(t)| and |Bm(t)| which means the simply sum of all os (t) with S ϵ F and S ϵ B. Therefore, L(t), |Fm(t)| and |Bm(t)| are all functions of t. And will change only at the break points of the occurrence functions.

So, we use the scheme write below to get the optimal solution of (11). Obviously, we need to record all the break points for every S. But in there, we only need to consider the scenario I, because other two don't have any break point so will not let any change of motif occurrences numbers. We collect the break points in foreground set and background set into sets LF and LB.

When all breakpoints are collected, sort them in ascending order. Then calculate every interval of break point by {(nF(ts), nB(ts))} then incrementally calculate the right value. The max number of interval is |F| + |B|, because there mostly has |F| + |B| break points. After all this process, we can get the minimal FETS from all these interval. Since the value in every interval is constant, we can sample a value from it as the solution.

3.5 Look-ahead Algorithm

There is a problem left in the above algorithm, that is it required to pick out the interval that have the minimized FETS value in ten thousands of interval. As in the earlier step, we have already calculate the value of |Fm| and |Bm| for each interval. The problem can be solved as below:

Let's see the standard approach for calculate the (3):

$$\begin{aligned} FETS &= \left(\frac{(|F|)!(B)!}{(|F|+|B|)!}\right) \sum_{k=|Fm|}^{\min(|F|,|Fm|+|Bm|)} \left(FETS_{sub}^{k}\right) \\ &\Rightarrow \sum_{k=|Fm|}^{\min(|F|,|Fm|+|Bm|)} \left(FETS_{sub}^{k}\right) \end{aligned} \tag{27}$$

Where $\left(FETS_{sub}^{k}\right)$ is defined as:

$$\frac{(|Fm|+|Bm|)!(|F|+|B|-|Fm|-|Bm|)!}{(k)!(|F|-k)!(|Fm|+|Bm|-k)!(|B|-|Fm|-|Bm|-k)!} \tag{28}$$

Cause that constant part of FETS don't depend on |Fm| or |Bm|, we can then discuss the subsequent. The (28) can be rewritten as:

$$\exp\left(\begin{array}{c} -\log\Gamma(k+1)-\log\Gamma(|F|-k+1)- \\ \log\Gamma(|Fm|+|Bm|-k+1)+\log\Gamma(|Fm|+|Bm|+1)- \\ \log\Gamma(|B|-|Fm|-|Bm|+k)+ \\ \log\Gamma(|F|+|B|-|Fm|-|Bm|+1) \end{array}\right) \tag{29}$$

Where $\Gamma(k+1) = (k)!$ Is the gamma function. As implemented in previous methods such as HOMER and DREME, using the log gamma function can make the factorials be efficiently calculated by continued fraction technique.

More over, we only interested in which point we get the minimal FETS, so we don't need to calculate all the FETS, some points can be excluded without calculate (27). As we go through the set P, we have the record of minimal FETS which was get by the elements we have checked. Each element that can be a minimum must satisfy a condition that must be lower than the threshold FETS*. Consequently, every point in P can be evaluated in an term by term manner. That is $\mathrm{FETS}_{sub}^{k+1}$ is calculated by the FETS_{sub}^{k} added the score. So the FETS_{sub}^{k} need to smaller than FETS*, i.e.:

$$\sum_{k=|Fm|}^{k}\left(FETS_{sub}^{k}\right) \leq FETS^{*} \tag{30}$$

4 Experiments

We use the 43 TFs in K562 cell line in the ENCODE [22] public ChIP-seq datasets to evaluate the performance. For each TF, positive set are collected by the highest intensity of 1000 peaks in the length of 100–500 nt [7]. And the negative set is generated use the "shuffle" function [23] make the positive set as the input, so can get the same first order of Markov properties.

Because of the complexity of the data, the exact location of the tie point is not known, so we use the same method as the previous measurement methods. And we use the "reference-free" cross-validation strategy [7, 16, 24, 25] to measure the accuracy. This is, for each TF, we divided the positive sequences and negative sequence into three sets of equal size. For every set(fold), we trained a PWM and evaluated it on the other fold. We considered the area under the receiver operating characteristic curve (AUC) as metric [26–28].

The cross-validation performance for each TF has put into the Tables 1, 2 and 3. In which we can see that almost all the cases FSdO have higher AUC than the other method.

Table 1. Statistical summary of improvements in AUC in cross validation

	DREME	HOMER
Higher AUC (% of TFs)	97.67	95.35
Higher AUC by 0.03(% of TFs)	81.40	76.75
Higher AUC by 0.07(% of folds)	32.56	34.88
Lower AUC (% of TFs)	2.33	4.65
Lower AUC by 0.03 (% of TFs)	0.00	0.00
Lower AUC by 0.07 (% of TFs)	0.00	0.00

Table 2. Statistical summary of improvements in negative logarithm in cross validation

	DREME	HOMER
Higher Negative Log FETS (% of TFs)	100.00	95.35
Higher Negative Log FETS by 3(% of TFs)	86.05	83.72
Higher Negative Log FETS by 7(% of folds)	65.12	69.77
Lower Negative Log FETS (% of TFs)	0.00	4.65
Lower Negative Log FETS by (% of TFs)	0.00	2.33
Lower Negative Log FETS by 7 (% of TFs)	0.00	0.00

5 Conclusion

In this paper, we develop a novel algorithm called FSdO to optimize the FETS of motifs. Compared with previous heuristic methods which rely on discrete search range or tuning of a external parameters. FSdO optimize the motif parameters directly in continuous space and making the process can be calculated numerically. Experimental evaluations on real ChIP-seq data showed the performance of FSdO. An interesting direction of future research would be to extend FSdO to optimize other metrics, such as AUC.

Acknowledgments. This work was supported by the grants of the National Science Foundation of China, Nos. 61532008, 61672203, 61402334, 61472282, 61520106006, 31571364, U1611265, 61472280, 61472173, 61572447, 61373098 and 61672382, China Postdoctoral Science Foundation Grant, Nos. 2016M601646.

References

1. Slattery, M., Zhou, T.Y., Yang, L., Machado, A.C.D., Gordan, R., Rohs, R.: Absence of a simple code: how transcription factors read the genome. Trends Biochem. Sci. **39**, 381–399 (2014)
2. Mason, M.J., Plath, K., Zhou, Q.: Identification of context dependent motifs by contrasting ChIP binding data. Bioinformatics **26**, 2826–2832 (2010)
3. Bailey, T.L.: DREME: motif discovery in transcription factor ChIPseq data. Bioinformatics **27**, 1653–1659 (2011)
4. Ichinose, N., Yada, T., Gotoh, O.: Large-scale motif discovery using DNA Gray code and equiprobable oligomers. Bioinformatics **28**, 25–31 (2012)

5. Furey, T.S.: ChIP-seq and beyond: new and improved methodologies to detect and characterize protein-DNA interactions. Nat. Rev. Genet. **13**, 840–852 (2012)
6. Zhu, L., Guo, W.L., Deng, S.P., Huang, D.S.: ChIP-PIT: enhancing the analysis of ChIP-seq data using convex-relaxed pair-wise interaction tensor decomposition. IEEE/ACM Trans. Comput. Biol. Bioinf. **13**, 55–63 (2016)
7. Patel, R.Y., Stormo, G.D.: Discriminative motif optimization based on perceptron training. Bioinformatics **30**, 941–948 (2014)
8. Yao, Z., MacQuarrie, K.L., Fong, A.P., Tapscott, S.J., Ruzzo, W.L., Gentleman, R.C.: Discriminative motif analysis of high-throughput dataset. Bioinformatics **30**, 775–783 (2013)
9. Agostini, F., Cirillo, D., Ponti, R., Tartaglia, G.: SeAMotE: a method for high-throughput motif discovery in nucleic acid sequences. BMC Genom. **15**, 925 (2014)
10. Heinz, S., Benner, C., Spann, N., Bertolino, E., Lin, Y.C., Laslo, P., et al.: Simple combinations of lineage-determining transcription factors prime cis-regulatory elements required for macrophage and B Cell Identities. Mol. Cell **38**, 576–589 (2010)
11. Maaskola, J., Rajewsky, N.: Binding site discovery from nucleic acid sequences by discriminative learning of hidden Markov models. Nucleic Acids Res. **42**, 12995–13011 (2014)
12. McLeay, R.C., Bailey, T.L.: Motif Enrichment Analysis: a unified framework and an evaluation on ChIP data. BMC Bioinform. **11**, 11 (2010)
13. Tanaka, E., Bailey, T.L., Keich, U.: Improving MEME via a twotiered significance analysis. Bioinformatics **30**, 1965–1973 (2014)
14. Liseron-Monfils, C., Lewis, T., Ashlock, D., McNicholas, P.D., Fauteux, F., Strömvik, M., et al.: Promzea: a pipeline for discovery of coregulatory motifs in maize and other plant species and its application to the anthocyanin and phlobaphene biosynthetic pathways and the Maize Development Atlas. BMC Plant Biol. **13**, 1–17 (2013)
15. Yu, Q., Huo, H.W., Vitter, J.S., Huan, J., Nekrich, Y.: An efficient exact algorithm for the motif stem search problem over large alphabets. IEEE-ACM Trans. Comput. Biol. Bioinform. **12**, 384–397 (2015)
16. Hartmann, H., Guthöhrlein, E.W., Siebert, M., Luehr, S., Söding, J.: P-value-based regulatory motif discovery using positional weight matrices. Genome Res. **23**, 181–194 (2013)
17. Pizzi, C., Rastas, P., Ukkonen, E.: Finding significant matches of position weight matrices in linear time. IEEE-ACM Trans. Comput. Biol. Bioinform. **8**, 69–79 (2011)
18. Valen, E., Sandelin, A., Winther, O., Krogh, A.: Discovery of regulatory elements is improved by a discriminatory approach. PLoS Comput. Biol. **5**, 8 (2009)
19. Colombo, N., Vlassis, N.: FastMotif: spectral sequence motif discovery. Bioinformatics **31**, 2623–2631 (2015)
20. Eden, E., Lipson, D., Yogev, S., Yakhini, Z.: Discovering motifs in ranked lists of DNA sequences. PLoS Comput. Biol. **3**, e39 (2007)
21. Hsieh, C.-J., Dhillon, I.S.: Fast coordinate descent methods with variable selection for non-negative matrix factorization. In: KDD, San Diego, CA, USA, pp. 1064–1072 (2011)
22. ENCODE-Project-Consortium: An integrated encyclopedia of DNA elements in the human genome. Nature **489**, 57–74 (2012)
23. Finn, R.D., Clements, J., Eddy, S.R.: HMMER web server: interactive sequence similarity searching. Nucleic Acids Res. **39**, W29–W37 (2011)
24. Simcha, D., Price, N.D., Geman, D.: The limits of De Novo DNA motif discovery. PLoS ONE **7**, 9 (2012)
25. Eggeling, R., Roos, T., Myllymäki, P., Grosse, I.: Inferring intramotif dependencies of DNA binding sites from ChIP-seq data. BMC Bioinformatics **16**, 1–15 (2015)

26. Huang, D.S., Zheng, C.H.: Independent component analysis-based penalized discriminant method for tumor classification using gene expression data. Bioinformatics **22**, 1855–1862 (2006)
27. Wang, B., Chen, P., Huang, D.S., Li, J.J., Lok, T.M., Lyu, M.R.: Predicting protein interaction sites from residue spatial sequence profile and evolution rate. FEBS Lett. **580**, 380–384 (2006)
28. Zhu, L., You, Z.H., Huang, D.S.: Increasing the reliability of protein-protein interaction networks via non-convex semantic embedding. Neurocomputing **121**, 99–107 (2013)

Biomedical Informatics Theory and Methods

DSD-SVMs: Human Promoter Recognition Based on Multiple Deep Divergence Features

Wenxuan Xu(✉), Wenzheng Bao, Lin Yuan, and ZhiChao Jiang

Institute of Machine Learning and Systems Biology,
College of Electronics and Information Engineering,
Tongji University, Shanghai, China
rifflexiansen@qq.com

Abstract. Accurate prediction and recognition of promoters remains a challenge in DNA sequence analysis. In this paper, the gene set firstly can be divided into two parts by CpG-island analysis. Then, in each part, a set of statistical divergence (SD) algorithms and sparse auto-encoders (SAEs) are integrated to optimize a series kinds of kmers and get multiple deep divergence features which compromises the merits of signal and context features. Extracted from the total possible combinations of kmers, the informative kmers can be selected by optimizing the differentiating extents of four sparse distributions based on promoter and non-promoters training samples. SAE in deep learning can convert the kmer feature based on SD into multiple deep divergence feature and reduce the dimension. Finally, multiple support vector machines and a bilevel decision model construct a human promoter recognition method called DSD-SVMs. Framework is flexible that it can integrate new features or new classification models freely. Experimental result shows the method has high sensitivity and specificity.

Keywords: Promoter recognition · Kmer · CpG-island · Sparse autoencoder · Statistical divergence · Support vector machine

1 Introduction

In genetics, a promoter is a region of DNA that initiates transcription of a particular gene and determines the direction, speed and accuracy of DNA transcription [1]. The promoter recognition contributed to the development of studying the regulation of human gene expression [2]. Bioinformatics and computer technology are combined to predict and recognize the promoter, which is low cost, less time-consuming and leads to more reliable results [26]. How to quickly and accurately recognize human promoter remains a big challenge in DNA sequence analysis at present.

How to extract the most discriminative feature to differentiate the categories of promoters from non-promoters is one of the key problems in promoter analysis. Signal, context and structure features are the three main types of features which can be used to analyze core-promoter regions essentially. CpG-island is widely used in many recognition and predict algorithms as a strong signal feature [3]. In addition, the context feature based on the unit DNA words called kmer is also used to recognize promoters

D.-S. Huang et al. (Eds.): ICIC 2017, Part I, LNCS 10361, pp. 515–526, 2017.
DOI: 10.1007/978-3-319-63309-1_46

which is statistically significant [4]. Information theory is more and more popular in many promoter recognition which can simplify the kmer feature extraction. The well-known Kullback-Leibler (KL) divergence in conditional entropy of SD [8] is widely applied [3, 9, 10]. And symmetrized divergence (also called the J divergence) and the Jensen–Shannon (JS) divergence are also the meaningful statistical measures between two probability distributions [11].

To get the most discriminative features, this paper introduces deep divergence, which combines sparse auto-encoder (SAE) and SD, to transform features in a deep way. SAE is a neural network model for unsupervised feature learning [14]. As an autoencoder, SAE is imposed with a sparseness constraint on hidden units and tries to learn an approximation to the identity function which minimizes an average distortion measure between inputs and outputs [15]. Typically, SAE is used for learning a representation of the raw data [16].

By integrating SAE, we propose a human promoter recognition method based on CpG-island analysis and deep divergence (DSD-SVMs). We firstly perform CpG-island analysis and divide the gene set into two parts and kmers statistical features are extracted in each part. Second, we apply a set of methods of SD to select the most informative and discriminative kmers. And then, SAE are combined with SD to get a meaningful representation, called deep divergence. Finally, three SVMs are independently adopted to classify these processed features. And a bilevel decision model (BD) is used to combine the outputs of three classifiers.

The contribution of this paper is applying CpG-island analysis to help extract features with higher separability and introducing deep divergence, which combines the advantage of statistical divergence and multiple SAEs to extract deep divergence feature, significantly reducing the kmer search space and feature dimension. In addition, a classification framework of multiple SVMs is presented and a decision model is used to integrate the prediction. The rest of this paper is organized as follows. Section 2 introduces deep divergence feature extraction. Section 3 introducccs CpG-island analysis. Section 4 proposes classifier ensemble recognition method based on the multiple SVMs and bilevel decision model, called DSD-MSVMs. We show experimental results in Sect. 5 and conclude this paper in Sect. 6.

2 DSD-SVMS: Deep Divergence Feature Extraction

In this paper, we focus on differentiating [–200, +50] bps around the transcription start point (TSS) which are defined by the DBTSS database [17] from other genomic regions, such as exons, introns, 3'UTRs and intragenic regions, and other alternative TSSs related to tissue specific gene expression are not considered.

Figure 1 shows the framework of DSD-SVMS. There are three stages in our framework. The first stage is CpG-island analysis. The second stage extracts deep divergence features for promoter recognition. The final stage is to learn these deep divergence features by the ensemble way. This section introduces the basic stage about feature extraction.

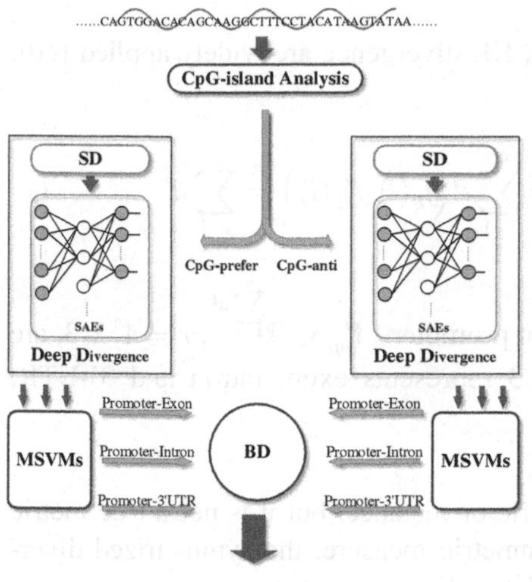

Fig. 1. Framework of DSD-MSVMs

Table 1. The feature dimensions in three stage of deep divergence feature extraction. The kmers feature dimension is 5460. The second column of each "Exon & Promoter", "Intron & Promoter" and "3'UTR & Promoter" shows the dimensions of features after SD. The third column shows the dimensions of features after deep divergence. "CpG-prefer" and "CpG-anti" represents two parts of dataset separated by CpG-island analysis

	Exon & Promoter			Intron & Promoter			3'UTR & Promoter		
	kmer	SD	SAEs	kmer	SD	SAEs	kmer	SD	SAEs
CpG-prefer									
KLD		276	220		115	110		118	100
JD	5460	554	443	5460	535	481	5460	517	413
JSD		4190	1000		4150	1000		4213	1000
CpG-anti									
KLD		431	344		531	477		535	428
JD	5460	569	512	5460	574	459	5460	557	445
JSD		4548	1000		4525	1000		4506	900

2.1 Kmer Feature

The context feature based on kmer can be used to recognize promoters which is statistically significant in their distribution [4]. Some experiments in the study [3, 5, 8] have shown that the statistic of kmer even may reveal fine details of unknown promoter characteristics and can help reduce the false positive rates while maintaining a relatively high sensitivity in promoter recognition [3, 5, 8, 9].

Let M be the set of total possible combinations of kmers, where $|M| = 4^k$. We get kmers probability distribution of M by counting the frequency of kmers at each site of the nucleotide sequence. Note that the tractable search space has a size $4^{k^{L-k+1}}$, where L is the length of sequence. Thus, the search space becomes more complex when n and L increase. To get a balance between the size of the tractable search space and the discriminant performance, we extract kmers $k \in \{1, 2, 3, 4, 5, 6\}$ for the following processing. Different from previous algorithms which use single kmer as feature, we combine all six kmers to form high dimensional input vectors.

2.2 Statistical Divergence

The concepts of bio-entropy and bioenergetics stem from the classical notions of thermodynamic entropy and weaved in the web of information theory (due to Shannon and Weaver) [12]. The relative entropy estimator methods based on statistical divergence are used to extract meaningful features to distinguish different regions of DNA sequences. Many promoter recognition algorithms make simplifications in the n-mer feature extraction based on the knowledge of information theory [3, 5, 8].

We apply the KL divergence, the J divergence, the JS divergence to select the most informative kmers as the features based on the probability distributions of promoter and non-promoters training samples.

Kullback-Leibler Divergence
Such as in conditional entropy aspects of SD, KL divergence are widely applied [10]. The KL divergence is defined as follows:

$$D_r\left(\mathbf{f}_p||\mathbf{f}_{np}^r\right) = \sum_{i=1}^{4^n} f_p(i)\ln\frac{f_p(i)}{f_{np}^r(i)} = \sum_{i=1}^{4^n} d\left(f_p(i), f_{np}^r(i)\right) = \sum_{i=1}^{4^n} d_i^r \tag{1}$$

where $\mathbf{f}_p \in \mathbb{R}^{\sum_{k=1}^{6} 4^k}$ is the probability density on promoters, $\mathbf{f}_{np}^r \in \mathbb{R}^{\sum_{k=1}^{6} 4^k}$, $r = 1, 2, 3$, are on three kinds of non-promoters, r = 1, 2, 3 represents exon, intron and 3'-UTR, respectively.

J Divergence
The KL divergence is often intuited as a metric or distance, but it is not a true metric because it is not symmetric. To obtain a symmetric measure, the symmetrized divergence (J divergence) can be used, which is defined as:

$$JD_r\left(\mathbf{f}_p||\mathbf{f}_{np}^r\right) = \frac{1}{2}D_r\left(\mathbf{f}_p||\mathbf{f}_{np}^r\right) + \frac{1}{2}D_r\left(\mathbf{f}_{np}^r||\mathbf{f}_p\right) = \sum_{i=1}^{4^n} d_i^r \tag{2}$$

where both $D_r\left(\mathbf{f}_p||\mathbf{f}_{np}^r\right)$ and $D_r\left(\mathbf{f}_{np}^r||\mathbf{f}_p\right)$ are the KL divergence.

Jensen-Shannon (JS) divergence
The Jensen-Shannon (JS) divergence is based on the KL divergence with some notable and useful differences, such as the JS divergence is symmetric and always has a finite value [10], which is defined as follows:

$$JSD_r\left(\mathbf{f}_p||\mathbf{f}_{np}^r\right) = \frac{1}{2}D_r\left(\mathbf{f}_p||\bar{\mathbf{f}}^r\right) + \frac{1}{2}D_r\left(\mathbf{f}_{np}^r||\bar{\mathbf{f}}^r\right) = \sum_{i=1}^{4^n} d_i^r \tag{3}$$

Now, we can select informative kmers based on (1), (2) or (3). First, we simply sort d_i^r in descending order and form a new vector $\mathbf{d}^r = \left[d_1^r, \cdots, d_{\sum_{k=1}^{6} 4^k}^r\right]^T \in \mathbb{R}^{\sum_{k=1}^{6} 4^k}$. The following optimization problem is defined to obtain the informative kmers:

$$\min_{m^r} \frac{\sum_{i=1}^{m^r} d_i^r}{\sum_{i=1}^{4^n} d_i^r} - \theta \tag{4}$$

where m^r is the number of the informative kmers, and $\theta > 0$ is a threshold, say 0.98. Let G^r be the set of the first m^r kmers. Generally given a gene $\mathbf{g} \in \mathbb{R}^L$, let the context features of $\mathbf{g}$ be $\mathbf{z}^r$, where $\mathbf{z}^r \in Z^r \subset \mathbb{R}^{m^r}$ and Z^r is the set of context features. And $\mathbf{z}^r$ is the probability density of G^r which is kmer features.

The kmer features are extracted from the total possible combinations of kmers with a large search space at each site of sequences, totally. And when k increases, the extraction of total possible combinations of kmers is overwhelmingly large. However, SD can reduce the feature dimension obviously and the computation can be simplified for maintaining the most information as show in Table 1. In addition, JSD can select the most kmers with the highest dimension and KL can select the least kmers with the lowest dimension.

2.3 Deep Divergence with SAEs

As mentioned before, SAE is a neural network model for unsupervised feature learning and used for learning a representation of the raw data. Especially, although SD can select the most informative kmers, the dimension is still very high. So, we apply multiple SAEs to deal with high dimensional divergence features while improve the separability in a deep way, which we called deep divergence.

In order to extract deep divergence features, we use three SAEs $h^{r}_{\mathrm{w,b}}(\bullet)$ to process the three kmer feature sets Z^r, respectively, $\mathbf{W}$ is the weight matrix and $\mathbf{b}$ is the bias vector. As we can see in Table 1, multiple SAEs can reduce the dimension evidently and transform the SD feature into more informative features. By separately training three SAEs on the obtained feature sets $\{\mathbf{z}_i^r\}_{i=1}^{N^r}$ and $N^r = |Z^r|$, we can have three new sets $\{\mathbf{x}_i^r\}_{i=1}^{N^r}$ where $\mathbf{x}_i^r \in \mathbb{R}^{l^r}$, $\mathbf{x}_i^r$ is the output of the last hidden layer for the rth SAE, l^r is the number of the hidden unit in the last hidden layer for the rth SAE.

3 DSD-SVMs: CpG-Island Analysis

The CpG-island is a region of DNA longer than 200 bp enriched with phosphodiesterase-linked C and G pairs [18], (Guanine: G, Cytosine: C), where the global frequency of GC content (C + G) is greater than 50% and the ratio of expected to observed CG dinucleotide (Obs/Exp) is greater than 60%. More specifically, CpG-islands can be found around promoters in about half of mammals [19]. CpG-island is an important feature for human promoter recognition. However, only about 70% of human gene promoters are associated with CpG-island [4]. So, different promoters with or without CpG-island has different characteristics and should be considered respectively.

Based on the analysis of two kinds of promoters, we call a sequence CpG-prefer, in which GC content is greater than 60%. Otherwise, a sequence is called CpG-anti. And all the genes can be divided into two parts, in which deep divergence feature extraction above is independently performed. Let the subset of CpG-prefer gene sequences be X_{pr}, $X_{pr} \in \mathbb{R}^{n_1}$, and the subset of CpG-anti gene sequences be X_{at}, $X_{at} \in \mathbb{R}^{n_2}$ and $n_1 + n_2 = n$.

4 DSD-MSVMs: Classifier Ensemble

As mentioned before, the final stage is to learn these deep divergence features by the ensemble way. In this section, we discuss the classifier ensemble.

4.1 Multiple Support Vector Machines

SVM is a universal learner based on statistical learning theory proposed by Vapnic et al. [23] and has been proved to be a good algorithm for promoter recognition as a supervised classification method [13]. Compared to other statistical or machine learning methods, SVM has a good performance in dealing with the large number of high-dimensional and complex data in most promoter recognition tasks [13, 26]. To implement SVM, we use the libsvm-3.17 toolbox written by Chih-Jen Lin (http://www.csie.ntu.edu.tw/~cjlin). We choose the radial basis function (RBF) kernel $k(\mathbf{x}_i, \mathbf{x}) = exp(-\gamma \parallel \mathbf{x}_i - \mathbf{x} \parallel^2)$ with a kernel parameter $\gamma > 0$. There are two parameters C and γ in SVMs. To obtain optimal parameters, we use grid search algorithm and 10-fold cross-validation. In addition, by training the parameters of an additional sigmoid function [22], label outputs can be mapped into probabilities.

We apply multiple SVMs (MSVMs) to process deep divergence features. In this paper, for each part of CpG-prefer and CpG-anti feature sets, we construct MSVMs based on $\left\{\mathbf{x}_i^r\right\}_{i=1}^{N^r}$. An unseen gene $\mathbf{g}$ can be firstly performed CpG-island analysis to determine whether it is CpG-prefer or CpG-anti. And then it will be described as three kmer features $\mathbf{z}^r$ according to G^r, and then be respectively represented as the deep divergence features $\mathbf{x}^r$ according to $h^r_{\mathbf{W},\mathbf{b}}(\mathbf{z}^r)$. We can get three label outputs $f^r(\mathbf{x}^r)$ and three possibility outputs $p^r(\mathbf{x}^r)$ from SVMs. Finally, we combine the outputs of three classifiers using the bilevel decision model algorithm.

4.2 Bilevel Decision Model

To determine whether $\mathbf{g}$ is the promoter or not, in this section, we introduce a bilevel decision model to combine these outputs. This decision model has two stage. Firstly, based on majority voting, we make the first prediction as follows:

$$\widehat{y}_1 = \begin{cases} \max\{p^r(\mathbf{x}^r)\}, & if \quad \sum_{r=1}^{3} \frac{(f^r(\mathbf{x}^r)+1)}{2} = 2 \\ \overline{p^r(\mathbf{x}^r)} & if \quad \sum_{r=1}^{3} \frac{(f^r(\mathbf{x}^r)+1)}{2} = 1 \\ \min\{p^r(\mathbf{x}^r)\}, & \quad Otherwise \end{cases} \tag{5}$$

In the second stage, we can get the final prediction according to majority probability rules:

$$\widehat{y} = \begin{cases} +1, & if\, \widehat{y}_1 \geq 0.5 \\ -1, & Otherwise \end{cases} \tag{6}$$

where $\widehat{y}$ is the estimated value for **g**. If $\widehat{y} = +1$, then **g** is a promoter. Otherwise **g** is a non-promoter.

5 Experiments and Results

All experiments are performed on the personal computer with a 2.5 GHz Intel(R) Core (TM) i5-2450 M CPU and 8 G bytes of memory. This computer runs on Windows10, with MATLAB R2016b compiler installed.

5.1 Datasets and Performance Evaluation

An experiment of a recognition algorithm using statistical pattern recognition methods requires a large number of promoters and non-promoters with accurate annotation. In this paper, we use 30,964 promoter sequences [−200, +50] bps around the TSSs from the DBTSS dataset to be the training and test sets because DBTSS provides the best combination of coverage and quality at present. We construct non-promoter sets by randomly extracting 10,000 exons and 10,000 introns with 251 bps in length from the EID database [22], and 10,000 3'UTR sequences with 251 bps in length from the UTRdb database [23]. We randomly select 4000 samples from the promoter and 2500 samples from exon, intron and 3'UTR sets, respectively.

Some evaluation measures proposed by Bajic [24] are used to evaluate our method, which are the sensitivity S_n, the specificity S_p and the averaged conditional probability ACP. These measures are defined as follows:

$$S_n = \frac{TP}{TP + FN}, S_p = \frac{TP}{TP + FP} \tag{7}$$

$$ACP = \frac{1}{4}\left(\frac{TP}{TP + FN} + \frac{TP}{TP + FP} + \frac{TN}{TN + FP} + \frac{TN}{TN + FN}\right) \tag{8}$$

where TP denotes the number of positive sample identified correctly, TN denotes the number of negative sample identified correctly, FP is the number of negative sample which is identified as positive samples and FN denotes the number of positive samples which are not to be identified correctly.

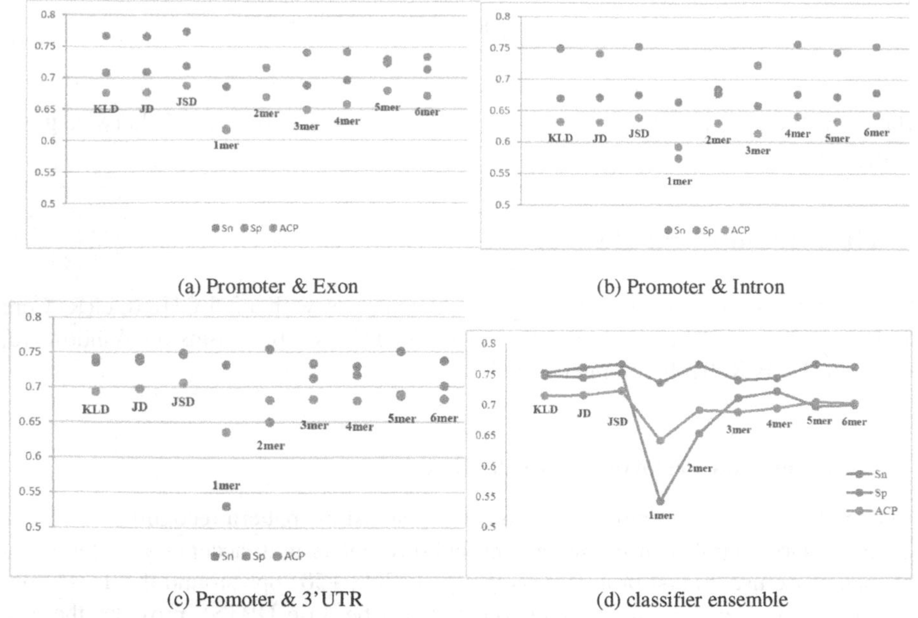

(a) Promoter & Exon (b) Promoter & Intron

(c) Promoter & 3'UTR (d) classifier ensemble

Fig. 2. Performances of promoter & exon, promoter & intron and promoter & 3'-UTR classifiers with or without deep divergence feature on the controlled datasets.

5.2 Effectiveness Evaluation of Deep Divergence Feature Extraction

In some methods [6–9], single kmer is used to perform feature extraction. However, DSD-MSVMs uses series kinds kmers to get more information and deep divergence can extract more meaningful features. So, experiments between DSD-MSVMs and single kmer shows the results in Fig. 2. Figure 2(a)–(c) shows the S_n, S_p and ACP of a sub-classifier for discriminating between promoters and exons, introns, and 3'UTRs, respectively. Figure 2(d) gives the S_n, S_p and ACP of combining three sub-classifiers. In Fig. 2(a)–(c), KLD, JD and JSD mean the performance of the single SVM only trained on the features extracted by deep divergence with the KL divergence, J divergence and JS divergence, respectively. 1mer, 2mer, 3mer, 4mer, 5mer and 6mer mean the performance of a single SVM trained on the single kmer feature. In Fig. 2(d), 1mer, 2mer, 3mer, 4mer, 5mer and 6mer mean the performance of combining three SVMs trained on the single kmer features. KLD, JD and JSD mean the performance of combining three SVM trained on the deep divergence features.

From Fig. 2(a)–(c), results show that S_n, S_p and ACP are all on the rise as k increases in all three sub-classifiers, but 4mer and 5mer have the better performance with $1/16$ and $1/4$ dimension of 6mer which indicates that high dimensional kmer features can improve recognition rate, however, the high sparsity can also weaken the classifier performance. So, DSD-MSVMs applies KLD, JD and JSD to select the most informative kmers and applies multiple SAEs to deal with high dimensional divergence features while improve the separability of promoter and non-promoters features by deep divergence.

We can see that KLD, JD and JSD has better sensitivity and specificity than single kmer according to Fig. 2(a)–(c), which indicates that the features extracted by deep divergence works well in all three sub-classifiers. Obviously, combining classifiers achieves much better performance on the sensitivity and the specificity, as shown in Fig. 2(d). First, the deep divergence feature is much better than the single kmer feature in distinguishing promoters from non-promoters. Second, ensemble learning outperforms the single classifier. Naturally, our method consisting of the deep divergence feature and ensemble learning achieves the best ACP in three SD methods. By the way, we can observe that JSD is the best among SD methods because the JSD can select the most informative kmers.

5.3 Effectiveness of CpG-Island Analysis

Figure 3 shows the S_n, S_p of DSD-MSVMs with or without CpG-island analysis. As we can see in Fig. 3 CpG-prefer(KLD), CpG-prefer(JD) and CpG-prefer(JSD) have the better S_n, S_p than CpG-anti(KLD), CpG-anti(JD) and CpG-anti(JSD) which indicates that CpG-island is a strong signal feature for promoter recognition and different promoters with or without CpG-island has different characteristics (Fig. 4).

In addition, DSD-MSVMs performs better in CpG-prefer part. The sensitivity and specificity of KLD + SAEs, JD + SAEs and JSD + SAEs are improved by CpG-island analysis compared with CpG(KLD), CpG(JD) and CpG(JSD). Table 2 shows the performance comparison in terms of the averaged conditional probability, which combines the sensitivity and the specificity. We can have the same conclusions as described above. So, CpG-island analysis can help extract features with higher separability and improve the performance of classifier.

Table 2. Comparison of AUC for DSD-MSVMs (KLD, JD and JSD) from gkm-SVM and SeqGL.

	DSD-MSVMs(KLD)	DSD-MSVMs(JD)	DSD-MSVMs(JSD)	SeqGL	gkm-SVM
AUC	0.7969	0.8022	0.8133	0.7631	0.7991

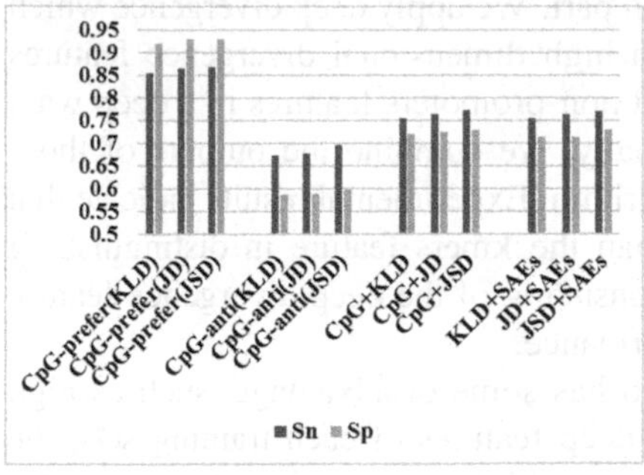

Fig. 3. Performances of DSD-MSVMs with or without CpG-island analysis.

Table 3. Comparison of DSD-MSVMs with or without CpG-island analysis.

	KLD	JD	JSD
ACP	CpG-prefer		
	0.7091	0.7402	0.7407
	CpG-anti		
	0.6604	0.6642	0.6782
	CpG		
	0.7028	0.7111	0.7213
	non-CpG		
	0.6998	0.7084	0.7179

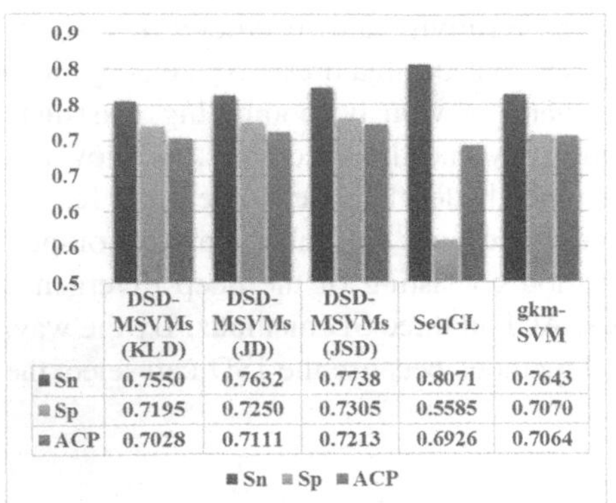

Fig. 4. Performance comparison of gkm-SVM, SeqGL and DSD-MSVMs (KLD, JD and JSD) on the controlled datasets.

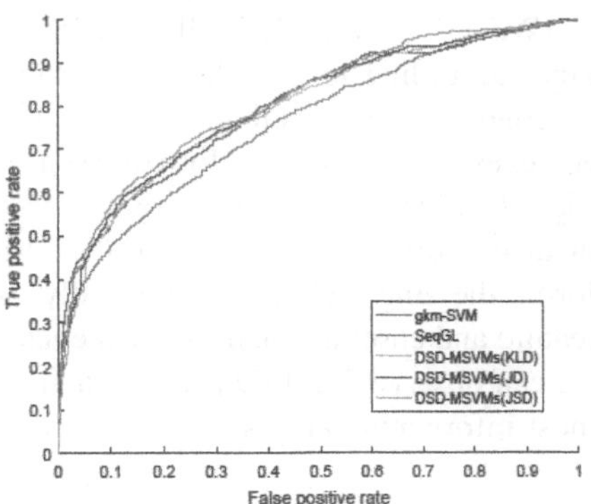

Fig. 5. Comparison of ROC curves for DSD-MSVMs (KLD, JD and JSD) from gkm-SVM and SeqGL.

5.4 Effectiveness of SD-MSAEs

In this section, we compare our method DSD-MSVMs with SeqGL and gkm-SVM proposed in [7, 8]. From Fig. 5, we can clearly see that DSD-MSVMs(JSD) can achieve the best ACP, followed by DSD-MSVMs(JD) and gkm-SVM. Although SeqGL has best sensitivity, its specificity is the lowest. DSD-MSVMs(KLD), DSD-MSVMs(JD) and DSD-MSVMs(JSD) are better than gkm-SVM and SeqGL on sensitivity and specificity. We computed the ROC curves and AUC for DSD-MSVMs (KLD, JD and JSD) from gkm-SVM and SeqGL based on average results (Fig. 5 and Table 3). Both ROC curves and AUC can also indicate that our method DSD-MSVMs for promoter recognition is highly effective.

6 Conclusion

This paper presents an DSD-MSVMs for promoter recognition. In DSD-MSVMs, we firstly perform CpG-island analysis and divide the gene set into two parts and kmers statistical feature extraction is performed in each part. We apply deep divergence which combines SD and multiple SAEs to deal with high dimensional divergence features while improve the separability of promoter and non-promoters features in a deep way. SVMs are used to classify these features. Finally, we combine the outputs of these classifiers using the bilevel decision model algorithm. Experimental results indicate that the deep divergence feature is much better than the kmers feature in distinguishing promoters from non-promoters. Our method consisting of the deep divergence feature and ensemble learning achieves the best performance.

Although DSD-MSVMs is effective, it also has some disadvantage, such as high time complexity. To get the most meaningful deep features of each training sets, the deep feature extraction algorithm needs complex multiple parameters optimization for each hidden layer and this process cost much time in most deep-learning algorithms.

The optimization of computational time will be the next focus of our research in the future work. Since the genetic data is very complex and high-dimensional, we only perform experiments on limited samples and limited kinds of features. Therefore, in the future research, more representative training data should be used to extracted features and more feature extraction methods should be considered.

Acknowledgment. This work was supported by the grants of the National Science Foundation of China, Nos. 61520106006, 31571364, U1611265, 61532008, 61672203, 61402334, 61472282, 61472280, 61472173, 61572447, 61373098 and 61672382, China Postdoctoral Science Foundation Grant, Nos. 2016M601646.

References

1. Bajic, V.B., Chong, A., Seah, S.H., et al.: An intelligent system for vertebrate promoter recognition. IEEE Intell. Syst. **17**(4), 64–70 (2002)
2. Fickett, J.W., Hatzigeorgiou, A.G.: Eukaryotic promoter recognition. Genome Res. **7**, 861–878 (1997)
3. Zeng, J., et al.: SCS: signal, context, and structure features for genome-wide human promoter recognition. IEEE/ACM Trans. Comput. Biol. Bioinf. **7**(3), 550–562 (2010)
4. Saxonov, S., Berg, P., Brutlag, D.L.: A genome-wide analysis of CpG dinucleotides in the human genome distinguishes two distinct classes of promoters. Proc. Natl. Acad. Sci. **103**(5), 1412–1417 (2006)
5. Werner, T.: The state of the art of mammalian promoter recognition. Brief Bioinform. **2014** (2014)
6. Setty, M., Leslie, C.S.: SeqGL Identifies Context-Dependent Binding Signals in Genome-Wide Regulatory Element Maps. PLoS Comput. Biol. **11**(5), e1004271 (2015)
7. Ghandi, M., Lee, D., Mohammad-Noori, M., et al.: Enhanced regulatory sequence prediction using gapped k-mer features. PLoS Comput. Biol. **10**(12), e1003711 (2014)
8. Vinga, S.: Information theory applications for biological sequence analysis. Brief. Bioinform. **15**(3), 376–389 (2014)
9. Zeng, J., Cao, X.Q., Yan, H.: Human promoter recognition using Kullback-Leibler divergence. In: IEEE International Conference on Machinc Lcarning and Cybernetics, pp. 3319–3325 (2007)
10. Zhao, X.Y., et al.: Promoter recognition based on the maximum entropy hidden markov model. Comput. Biol. Med. **51**(15), 73–81 (2014)
11. Neelakanta, P., et al.: Information-theoretic algorithms in bioinformatics and bio-/medical-imaging: a review. In: IEEE International Conference on Recent Trends in Information Technology, pp. 183–188 (2011)
12. Nielsen, F., Nock, R.: Sided and symmetrized Bregman centroids. IEEE Trans. Inf. Theory **55**(6), 2882–2904 (2009)
13. Anwar, F., et al.: Pol II promoter prediction using characteristic 4-mer motifs: a machine learning approach. BMC Bioinformatics **9**(1), 414 (2008)
14. Ng, A.: Sparse autoencoder. CS294A Lecture Notes for Stanford University (2011)
15. Baldi, P., Lu, Z.: Complex-valued autoencoders. Neural Netw. **33**(3), 136–147 (2014)
16. Ng, A., Ngiam, J., Foo, C.Y., Mai, Y., Suen, C.: UFLDL tutorial: building deep networks for classification. An online tutorial (2013)

17. Suzuki, Y., et al.: DBTSS, DataBase of Transcriptional Start Sites: progress report 2004. Nucleic Acids Res. **32**(Database issue D), 78–81 (2004)
18. Goddard, N.L., et al.: Sequence dependent rigidity of single stranded DNA. Phys. Rev. Lett. **85**(11), 2400–2403 (2000)
19. Liu, W., Kou, Q.B., Wei, L.H., et al.: Plant promoter recognition based on analysis of base bias and SVM. J. Liaoning Normal Univ. (2012)
20. Vapnik, V., Cortes, C.: Support vector networks. Mach. Learn. **20**(3), 273–297 (1995)
21. Platt, J.C.: Probabilistic outputs for support vector machines and comparisons to regularized likelihood methods. Adv. Large Margin Classif. **10**(4), 61–74 (1999)
22. Saxonov, S., Daizadeh, I., Fedorov, A., Gillbert, W.: EID: the exon-intron database—an exhaustive database of protein-coding intron-containing genes. Nucleic Acids Res. **28**(1), 185–190 (2000)
23. Pesole, G., Liuni, S., Grillo, G., et al.: UTRdb and UTRsite: specialized databases of sequences and functional elements of 5' and 3' untranslated regions of eukaryotic mRNAs. Update 2002. Nucleic Acids Res. **30**(1), 335 (2002)
24. Bajić, V.B.: Comparing the success of different prediction software in sequence analysis: a review. Brief. Bioinform. **1**(3), 214 (2000)
25. Zhu, L., Guo, W.L., Lu, C., Huang, D.S.: Collaborative completion of transcription factor binding profiles via local sensitive unified embedding. IEEE Trans. Nanobiosci. **99,** 1 (2016)
26. Liang, X., Zhu, L., Huang, DS.: Multi-task ranking SVM for image cosegmentaiton. Neurocomputing (2017)

MD-MSVMs: A Human Promoter Recognition Method Based on Single Nucleotide Statistics and Multilayer Decision

Wenxuan Xu(✉), Wenzheng Bao, Lin Yuan, and ZhiChao Jiang

Institute of Machine Learning and Systems Biology,
College of Electronics and Information Engineering,
Tongji University, Shanghai, China
rifflexiansen@qq.com

Abstract. The prediction and recognition of promoter in human genome play an important role in DNA sequence analysis. Nucleotide content is a multiple utility in bioinformatics details analysis. The single nucleotide statistics method based on nucleotide content can help extract features with higher separability and make decision. In this paper, a human promoter recognition method based on multiple gene features and multilayer decision, which is called MD-MSVMs, is proposed. In our method, we firstly perform single nucleotide analysis and divide the gene set into two parts. Secondly, the multiple gene features are extracted from each part, including CpG-island, n-mer and rigidity. And then, based on multiple features, multiple support vector machines and multilayer decision model are combined to construct a human promoter recognition framework, which is flexible and can integrate new feature extraction or new classification models freely. Experimental result shows that our method has better performance and helps understanding multiple features integrating.

Keywords: CpG-island · DNA rigidity · Human promoter recognition · Kullback-Leibler divergence · Nucleotide statistics · Support vector machines

1 Introduction

In genetics, a promoter is a region of DNA that initiates transcription of a particular gene [1]. It determines the direction, speed and accuracy of DNA transcription. The promoter recognition plays an important role in studying the regulation of human gene expression. Thus, it is an important task that how to quickly and accurately recognize human promoter at present. Since Fickett and Hatzigeorgiou published the first review paper on promoter recognition algorithms in 1997 [2], more and more researchers use the knowledge of bioinformatics to predict and recognize the promoter with the help of computer technology, which is low cost, less time-consuming and leads to more reliable results [27].

One of the key problems in promoter recognition is how to extract the most discriminative features to differentiate the categories of promoters from non-promoters. Signal, context and structure features are the three types of features which can be used to recognize core-promoter regions essentially. CpG-islands is widely used in many

D.-S. Huang et al. (Eds.): ICIC 2017, Part I, LNCS 10361, pp. 527–538, 2017.
DOI: 10.1007/978-3-319-63309-1_47

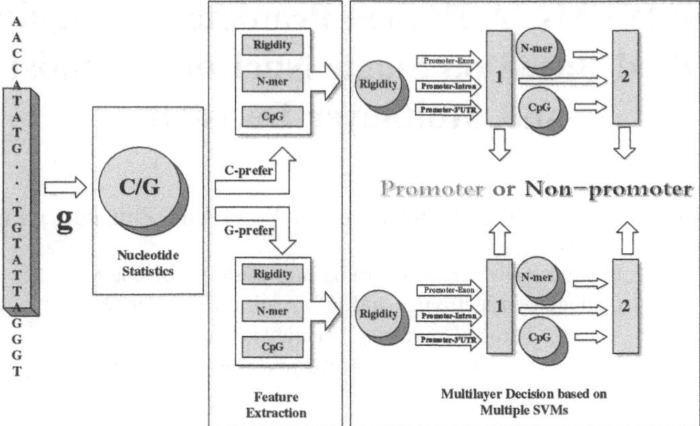

Fig. 1. Framework of MD-MSVMs

recognition algorithms as one of the signal features [3–5], but only about 70% of human promoters have a high CpG content [6]. N-mers feature based on unit nucleotide words are widely applied in promoter prediction and recognition, which belongs to the context feature [7]. Information theory can be used to simplify the extraction of n-mers, which has large search space, for example, Kullback-Leibler (KL) divergence [8]. As an important structure feature, which is derived from DNA three-dimensional structures [9], DNA Rigidity has been proposed to be an effective feature for promoter recognition [10]. Although can provide position information, it is still sequence dependent.

Selecting appropriate classification model and integrating multiple features can improve the prediction or recognition effect. A lot of models in machine learning can be applied as classifiers, such as support vector machine (SVM) [11], Markov model [12], relevance vector machine [13], linear and quadratic discriminant analysis [14] and neural network [15]. PromoterExplorer [16] applies cascade AdaBoost to integrate CpG-island, 5-mer and digitized DNA sequence feature, but it only combines all features to form high dimensional input vectors. SCS [4] combines CpG-island, n-mers and processed rigidity features and use decision trees to make the optimal classification decision, but it does not analyze sub classifier and improve the performance.

In this paper, a human promoter recognition method based on multiple gene features and multilayer decision, which is called MD-MSVMs, is proposed. We firstly perform single nucleotide analysis and divide the gene set into two parts. Secondly, the multiple gene features are extracted from each part, including CpG-island, n-mer and rigidity. And then, based on multiple features, multiple SVMs and multilayer decision model are combined to construct a human promoter recognition framework, which is flexible and can freely integrate new feature extraction or new classification models.

The contribution of this paper is applying single nucleotide statistics to help extract features with higher separability and make decision and combining multiple gene features and multilayer decision for promoter recognition. In addition, a classification system of multiple SVMs based on these features is presented. The rest of this paper is organized as follows. Section 2 introduces three kinds of features. Section 3 introduces single nucleotide statistics. Multilayer decision based on multiple SVMs is proposed in Sect. 4. We show experimental results in Sect. 5 and conclude this paper in Sect. 6.

2 MD-MSVMs: Feature Extraction

In this paper, we focus on differentiating [−200, +50] bps around the transcription start point (TSS) which are defined by the DBTSS database [17] from other genomic regions, and other alternative TSSs related to tissue specific gene expression are not considered. The present development trend of promoter recognition is considering the promoter, the coding exons and the introns of genomic regions at the same time because the properties of promoter regions are considerably different from those of other genomic regions, such as exons, introns, 3'UTRs and intragenic regions [8, 18].

Figure 1 shows the framework of our promoter recognition method, which combines single nucleotide statistics, multiple gene features and multilayer decision. There are three stages in our framework. The first stage is single nucleotide statistics. The second stage is feature extraction for promoter recognition. The last stage is to learn multiple features by the ensemble way and preform multilayer decision.

This section introduces the basic stage about feature extraction firstly, because the gene set will be divided into two parts after single nucleotide and the same feature extraction procedure will be performed in each part.

2.1 DNA Rigidity

DNA three-dimensional structure features are characterized by the local angular parameters (twist, roll, and tilt) as well as the translational parameters (shift, slide, and rise). Sequence-dependent DNA rigidity is an important physical property derived from DNA three-dimensional structure [19]. The general DNA rigidity patterns in human promoters have been examined and used for computational promoter prediction [20].

Statistical mechanics model can be used to obtain DNA rigidity profiles. Trinucleotide model [21] will be taken to calculate the rigidity features of human gene sequences. Trinucleotide parameter values of the trinucleotide model are provided in [21, Table 1]. We calculate the characteristic value of each site in sequence based on 6-mers. 6-mers rigidity values r are calculated by adding four overlapping trinucleotide parameter values:

$$r = \sum_{i=1}^{4} t_i \tag{1}$$

where i is the position index and t_i is the rigidity parameter of each trinucleotide at position i. We calculate the 6-mer rigidity values from the starting position of the sequence.

2.2 Context Feature

DNA sequences can be considered as the collections of documents consisting of the letter A (adenine), C (cytosine), G (guanine) and T (thymine). Context features based on the unit DNA words called n-mers are used to predict and recognize promoters which belong to the context features [7]. Specially, n-mers may reduce the false positive rates while maintaining a relatively high sensitivity in promoter recognition due to the biological significance in n-mers' distribution [4]. Thus, we extract n-mers feature from the datasets and use KL divergence to select the most informative and discriminative n-mers features for classification.

Let $\mathbf{f}_{pr}$ be the frequency of n-mers in promoters and $\mathbf{f}^a_{np}(a = 1, 2, 3)$ be respectively the frequency of n-mers in three kinds of non-promoters where a = 1 represents exon, a = 2 represents intron and a = 3 represents 3'-UTR. KL divergence is defined as follows:

$$D_{\mathrm{a}}\left(\mathbf{f}_p \| \mathbf{f}^a_{np}\right) = \sum_{i=1}^{4^n} f_p(i) \ln \frac{f_p(i)}{f^r_{np}(i)} = \sum_{i=1}^{4^n} d\left(f_p(i), f^a_{np}(i)\right) = \sum_{i=1}^{4^n} d^a_i \tag{2}$$

where $d^a_i = d\left(f_p(i), f^a_{np}(i)\right)$, $\mathbf{f}_p = [f_p(1), \cdots, f_p(4^n)]$ and $\mathbf{f}^a_{np} = \left[f^a_{np}(1), \cdots, f^a_{np}(4^n)\right]$.

To obtain the most m^a discriminative n-mers, we sort d^a_i, $i = 1, \cdots, 4^5$ in descending order and form a new vector $\mathbf{d}^a = \left[d^a_1, \cdots, d^a_{4^n}\right]^T \in \mathbb{R}^{4^n}$. We define the following optimization problem:

$$\min_{m^r} \frac{\sum_{i=1}^{m^r} d^r_i}{\sum_{i=1}^{4^n} d^r_i} - \theta \tag{3}$$

where $\theta > 0$ is a threshold, say 0.98. Let G^a be the set of the first m^a n-mers. And we can extract the feature $\mathbf{S}^a \in \mathbb{R}^{m^a}$ according to G^a.

2.3 CpG-Island

The CpG island is a region of DNA longer than 200 bp enriched with phosphodiesterase-linked C and G pairs [20], where the global frequency of GC content (C + G) is greater than 50% and the ratio of expected to observed CG dinucleotide (Obs/Exp) is greater than 60%. More specifically, it shows that CpG islands can be found around promoters in about half of mammals according to the statistics of DNA data available [22]. So, the CpG-island is an important feature for human promoter recognition. We use two important features including the global frequency of GC content (GC_con) and the ratio of expected to observed CG dinucleotide (o/e): $GC\_con = \frac{n_C + n_G}{L}$, $o/e = \frac{n_{CG} \times L}{n_C \times n_G}$, where L is the length of a DNA sequence and n_C, n_G and n_{CG} are numbers of C, G and CG in the DNA sequence, respectively.

3 MD-MSVMs: Nucleotide Statistics

It's particularly difficult to extract the common features, which have obvious biological significance, because the genetic data is very complex, high-dimensional and contains a huge amount of information. The composition of nucleotides in different DNA fragments is different. In addition, the composition of nucleotides at different positions in the same fragment is also different.

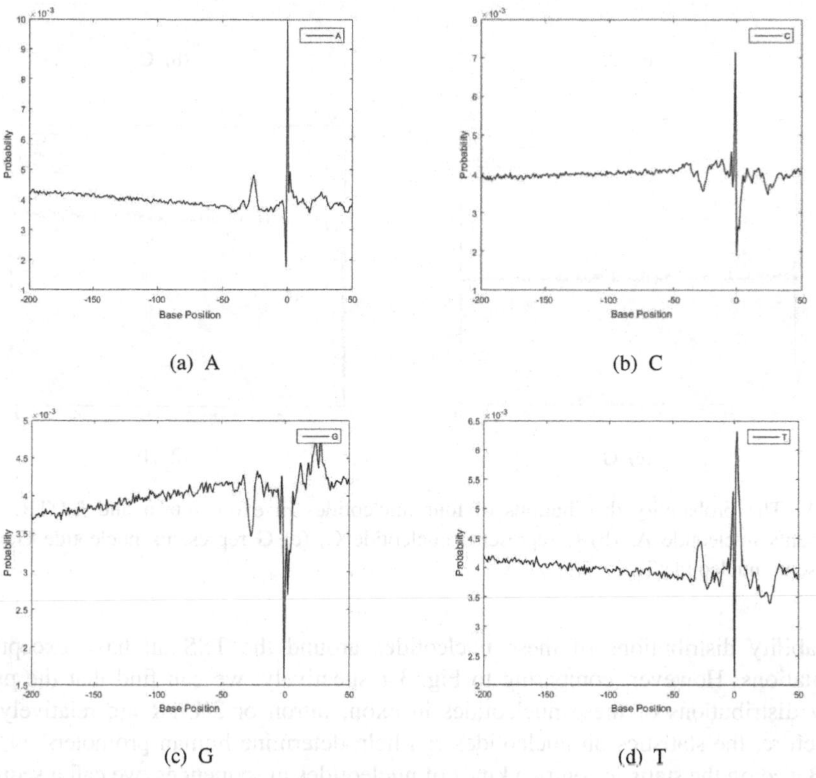

(a) A (b) C

(c) G (d) T

Fig. 2. The probability distributions of four nucleotides around the TSS. (a) A represents nucleotide A; (b) C represents nucleotide C; (c) G represents nucleotide G; (d) T represents nucleotide T.

We analyzed 30,964 promoter sequences, [−200, +50] bps, from the DBTSS. 10,000 exons and 10,000 introns with the same length from the EID database [23], and 10,000 3'-UTR sequences with 251 bps in length from the UTRdb database [24].

In promoters, we analyzed the probability distributions of four kinds of nucleotides in different base positions. In humans, the TSS is surrounded with a core-promoter region. We calculate the probability distributions of four nucleotides around TSS. At the same time, we analyzed the probability distributions of four kinds of nucleotides in different base positions with the same length, 251 bps. As the Fig. 2 shows, the

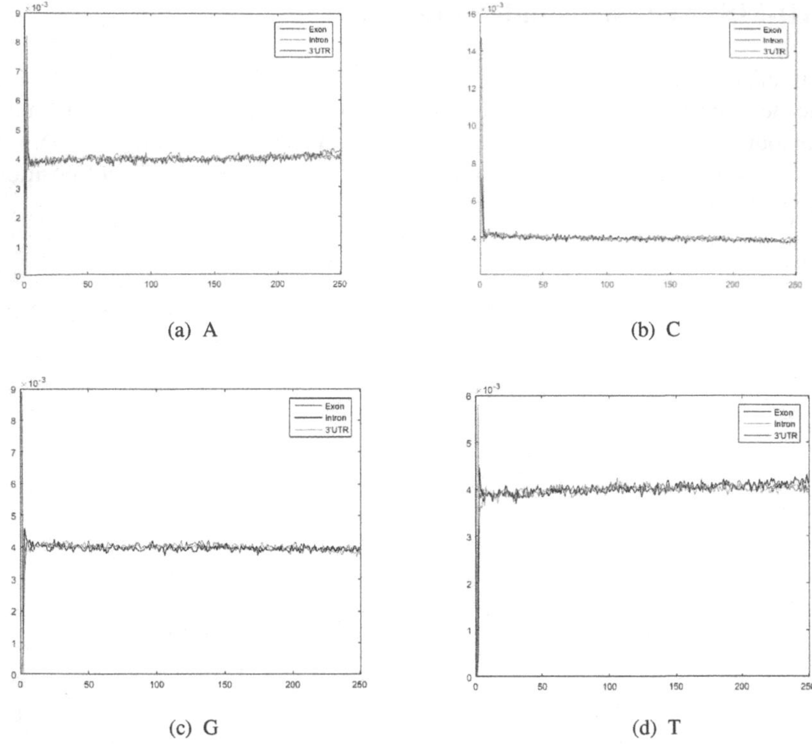

Fig. 3. The probability distributions of four nucleotides in exon, intron and 3'UTR. (a) A represents nucleotide A; (b) C represents nucleotide C; (c) G represents nucleotide G; (d) T represents nucleotide T.

probability distributions of these nucleotides around the TSS all have exceptional undulations. However, comparing to Fig. 3 respectively, we can find that the probability distributions of these nucleotides in exon, intron or 3'UTR are relatively flat. Therefore, the statistics on nucleotides can help determine human promoters.

Based on the statistics on two kinds of nucleotides in sequences, we call a sequence N_1-prefer, in which the content of N_1 is greater than N_2. Otherwise, a sequence is called N_2-prefer, where N_1 and N_2 are two kinds of nucleotides. According to this, all the genes can be divided into two parts, in which feature extraction above is independently performed.

To implement the classification of N_1-prefer and N_2-prefer genes, we use the ratio of contents of N_1 and N_2. In Sect. 5, the results show nucleotides C and G have better performance in MD-MSVMs. So, according to the contents of C and G in each sequence, the unknown gene will be divided into C-prefer and G-prefer parts.

4 MD-MSVMs: Multilayer Decision Based on Multiple SVMs

As mentioned before, our method has three stages. The last stage is to learn multiple features by the ensemble way and preform multilayer decision. In this section, we discuss the subsequent classifier ensemble. We apply multiple support vector machines (SVMs) as the classifiers to classify promoters from non-promoters based on processed features. Finally, we combine the outputs of multiple classifiers using the multilayer decision algorithm.

4.1 Multiple Support Vector Machines

SVM is a universal learner based on statistical learning theory proposed by Vapnic et al. [25] and has been proved to be a good algorithm for promoter recognition [11]. SVM can implement the structural risk minimization rule to achieve good generalization performance [28]. In SVM, kernel functions are used to map the original samples into a high-dimensional feature space, in which the original sample could be linearly separable.

To implement SVM, we use the libsvm-3.21 toolbox written by Chih-Jen Lin (http://www.csie.ntu.edu.tw/~cjlin). We choose the radial basis function (RBF) kernel $k(\mathbf{x}_i, \mathbf{x}) = \exp(-\gamma \parallel \mathbf{x}_i - \mathbf{x} \parallel^2)$ with a kernel parameter $\gamma > 0$. There are two parameters C and γ in SVMs. To optimize parameters for our method, we use the popular parameter optimization algorithms–grid search algorithm and 10-fold cross-validation to select the optimal parameters C and γ.

Multiple SVMs are constructed to process multiple features. For rigidity feature, we applied three SVMs: SVM$-$R^a, which used to differentiate between promoter and exon or intron or 3'UTR. We use the feature set of promoters and one kind of non-promoters to train the corresponding SVM. For context feature, we applied three SVMs: SVM$-$S^a corresponding to feature sets $\mathbf{S}^a$. For CpG feature, we train one SVM: SVM$-$PI.

4.2 Multilayer Decision

In this paper, we combine the outputs of multiple SVMs using the multilayer decision algorithm. And. In first layer, we have three outputs $f^a(\mathbf{R})$ from SVM$-$R^a, $a = 1, 2, 3$. To determine whether $\mathbf{g}$ is the promoter or not, we need to combine these outputs as follows:

$$\widehat{y}_1 = \sum_{a=1}^{3} \frac{(f^a(\mathbf{R}) + 1)}{2} \tag{4}$$

where $\widehat{y}_1$ is the estimated value for $\mathbf{g}$. If $\widehat{y}_1 = +3$, $\mathbf{g}$ is a promoter. If $\widehat{y}_1 = 0$, $\mathbf{g}$ is a non-promoter. Otherwise, it will be described as other two feature. We have three outputs $f^a(\mathbf{S}^a)$ from SVM$-$S^a and one output $f(\mathbf{PI})$ from SVM$-$PI. In next layer, we combine these outputs and $\widehat{y}_1$. Firstly, three outputs $f^a(\mathbf{S}^a)$ based on the majority voting rule. And it can be described as:

$$\widehat{y}_2 = \begin{cases} +1, & if \sum_{a=1}^{3} \frac{(f^a(\mathbf{S}^a)+1)}{2} \geq 2 \\ -1, & Otherwise \end{cases} \tag{5}$$

And then, $\widehat{y}_3 = \widehat{y}_1 + \widehat{y}_2 + f(\mathbf{PI})$, if $\widehat{y}_3 > 0$, $\mathbf{g}$ is a promoter. Otherwise, $\mathbf{g}$ is a non-promoter.

5 Experiments and Results

All experiments are performed on the personal computer with a 2.5 GHz Intel(R) Core (TM) i5-2450 M CPU and 4 G bytes of memory. This computer runs on Windows10, with MATLAB R2015b compiler installed.

5.1 Datasets and Performance Evaluation

We use 30,964 promoter sequences [−200, +50] bps around the TSSs from the DBTSS. Non-promoter sets are constructed by randomly extracting 10,000 exons and 10,000 introns with 251 bps in length from the EID database [23], and 10,000 3'-UTR sequences with 251 bps in length from the UTRdb database [24]. We randomly select 8000 samples from the promoter, exon, intron and 3'-UTR sets, respectively. Among 8000 samples, 4000 samples are considered as the training ones and the rest are test ones for each class. Thus, totally the positive and negative samples are unbalanced.

And some evaluation measures proposed by Bajic [26] are used to evaluate our method, which are the sensitivity S_n, the specificity S_p and the averaged conditional probability ACP. These measures are defined as follows:

$$S_n = \frac{TP}{TP + FN}, \quad S_p = \frac{TP}{TP + FP} \tag{6}$$

$$ACP = \frac{1}{4}\left(\frac{TP}{TP + FN} + \frac{TP}{TP + FP} + \frac{TN}{TN + FP} + \frac{TN}{TN + FN}\right) \tag{7}$$

where TP denotes the number of positive sample identified correctly, TN denotes the number of negative sample identified correctly, FP is the number of negative sample which is identified as positive samples and FN denotes the number of positive samples which are not to be identified correctly.

5.2 Efficiency Evaluation of Single Nucleotide Statistics

In the first step of MD-MSVMs, genes are divided into two categories according to single nucleotide statistics based on the ratio of contents of two nucleotides. In our

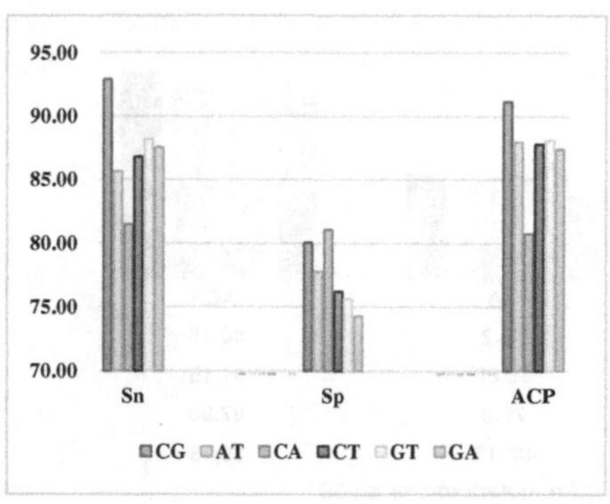

Fig. 4. Comparison of every combination preforming single nucleotide statistics in MD-MSVMs, including CG, AT, CA, CT, GT and GA.

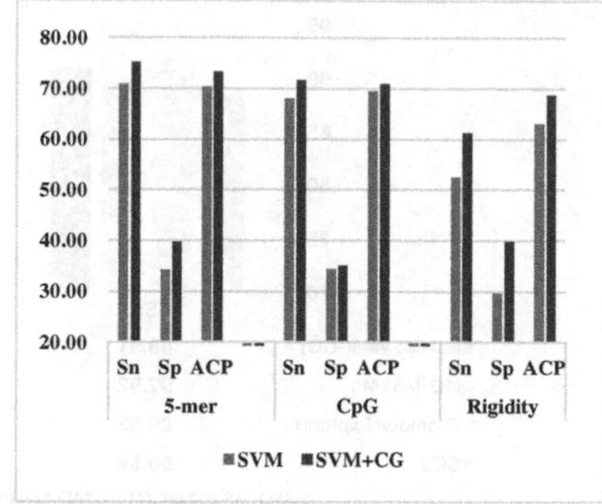

Fig. 5. Comparison of single feature, 5-mer, CpG and Rigidity models with and without single nucleotide statistics. For 5-mer, we use three SVMs and get the estimated value based on majority voting rule. For CpG and Rigidity, single SVM is constructed to get the result respectively.

method, we choose C and G. But there are 6 kinds of combinations of two nucleotides in total. We preform every combination in MD-MSVMs. As the Fig. 4 shows, although the Sp of CG is not the highest, the D-value between CG and the highest (CA) is only 1% and both Sn and ACP are highest among the six. In the figure, CG, AT, CA, CT, GT and GA represent the result from the MD-MSVMs, which using single nucleotide statistics based on two nucleotides of CG, AT, CA, CT, GT and GA.

And then, we observe the effect of single nucleotide statistics on single feature. Figure 5 shows the evaluation measures of 3 kinds of single features with (SVM + CG) and without (SVM) single nucleotide statistics, respectively. We can see that the performance with single nucleotide statistics is higher than that without single nucleotide statistics. For the whole method, as show in the Fig. 6, all the Sn, Sp and ACP are higher than the MD-MSVMs without performing single nucleotide statistics, which in figure is MD-MSVMs(-CG).

In a nutshell, sensitivity Sn, specificity Sp and ACP are all improved when applying single nucleotide statistics. Thus, it is necessary to perform single nucleotide statistics for promoter recognition.

5.3 Efficiency of MD-MSVMs

In Sect. 4. D, we show that MD-MSVMs based on single nucleotide statistics and multilayer decision are efficient by experiments. Here, we compare our method MD-MSVMs with methods proposed in [4, 16]. PromoterExplorer [16] applies cascade AdaBoost to integrate CpG-island, 5-mer and digitized DNA sequence feature, but it only combines all features to form high dimensional input vectors. SCS [4] combines CpG-island, n-mers

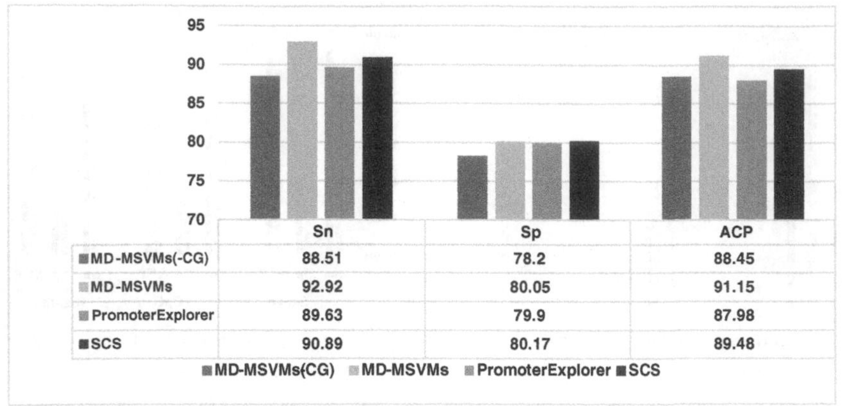

	Sn	Sp	ACP
■MD-MSVMs(-CG)	88.51	78.2	88.45
■MD-MSVMs	92.92	80.05	91.15
■PromoterExplorer	89.63	79.9	87.98
■SCS	90.89	80.17	89.48

Fig. 6. Comparison of MD-MSVMs(-CG), MD-MSVMs, PromoterExplorer and SCS.

and processed rigidity features and use decision trees to make the optimal classification decision, but it does not analyze sub classifier and improve the performance.

The comparison of MD-MSVMs(-CG), PromoterExplorer and SCS with our method is shown in Fig. 6. We can see clearly from these figures, SCS can achieve the best Sp, followed by MD-MSVMs and PromoterExplorer. SCS has the best Sp, 80.17% and MD-MSVMs is little lower than SCS but higher than PromoterExplorer. The ACP of PromoterExplorer is only about 87.98% and MD-MSVMs has the highest ACP, 91.15%. All the results can indicate that our method combining single nucleotide statistics and multilayer decision for promoter recognition with multiple SVMs is highly efficient.

6 Conclusion

This paper presents an MD-MSVMs for promoter recognition. In MD-MSVMs, we firstly perform single nucleotide analysis. And the multiple gene features are extracted from each part. Based on multiple features, multiple SVMs and multilayer decision model are combined to construct for human promoter recognition, which is flexible and can integrate new feature extraction or new classification models freely. Experimental results on the DBTSS dataset indicate that the single nucleotide statistics as statistical feature can help distinguish promoters from non-promoters. In addition, multilayer decision on multiple SVMs has been proved. Thus, our method consisting of single nucleotide statistics and multilayer decision achieves the best performance in these indexes.

Since the genetic data is very complex and high-dimensional, we only perform experiments on limited samples and limited kinds of features. Therefore, in the future research, more representative training data should be used to extracted features and more feature extraction methods should be considered.

Acknowledgment. This work was supported by the grants of the National Science Foundation of China, Nos. 31571364, U1611265, 61532008, 61672203, 61402334, 61472282, 61520106006, 61472280, 61472173, 61572447, 61373098 and 61672382, China Postdoctoral Science Foundation Grant, Nos. 2016M601646.

References

1. Bajic, V.B., Chong, A., Seah, S.H., et al.: An intelligent system for vertebrate promoter recognition. IEEE Intell. Syst. **17**(4), 64–70 (2002)
2. Fickett, J.W., Hatzigeorgiou, A.G.: Eukaryotic promoter recognition. Genome Res. **7**, 861–878 (1997)
3. Umesh, P., Dubey, J.K., Karthika, R.V., et al.: A novel sequence and context based method for promoter recognition. Bioinformation **10**(4), 175–179 (2014)
4. Zeng, J., Zhao, X.Y., Cao, X.Q., Yan, H.: SCS: signal, context, and structure features for genome-wide human promoter recognition. IEEE/ACM Trans. Comput. Biol. Bioinform. **7**(3), 550–562 (2010)
5. Deng, J., Liang, H., Zhang, R., et al.: Methylated CpG site count of dapper homolog 1 (DACT1) promoter prediction the poor survival of gastric cancer. Am. J. Cancer Res. **4**, 518–527 (2014)
6. Saxonov, S., Berg, P., Brutlag, D.L.: A genome-wide analysis of CpG dinucleotides in the human genome distinguishes two distinct classes of promoters. Proc. Natl. Acad. Sci. U.S.A. **103**, 1412–1417 (2015)
7. Huang, W.L., Tung, C.W., Liaw, C., Huang, H.L., Ho, S.Y.: Rule-based knowledge acquisition method for promoter prediction in human and Drosophila species. Sci. World J. **2014**, 1–14 (2014)
8. Vinga, S.: Information theory applications for biological sequence analysis. Brief. Bioinform. **15**(3), 376–389 (2014)
9. Fujii, S., Kono, H., Takenaka, S., Go, N., Sarai, A.: Sequence-dependent DNA deformability studied using molecular dynamics simulations. Nucleic Acids Res. **35**, 6063–6074 (2007)
10. Gan, Y., Guan, J., Zhou, S.: A comparison study on feature selection of DNA structural properties for promoter prediction. BMC Bioinform. **7**, 13–14 (2012)
11. Anwar, F., Baker, S.M., Jabid, T., Mehedi, H.M., Shoyaib, M., Khan, H., Walshe, R.: Pol II promoter prediction using characteristic 4-mer motifs: a machine learning approach. BMC Bioinform. **9**(1), 414–418 (2008)
12. Zhao, X.Y., Zhang, J., Chen, Y.Y., Li, Q., Yang, T., Pian, C., Zhang, L.Y.: Promoter recognition based on the maximum entropy hidden Markov model. Comput. Biol. Med. **51**, 73–81 (2014)
13. Li, Y., Lee, K.K., Walsh, S., Smith, C., Hadingham, S., Sorefan, K., Cawley, G., Bevan, M.W.: Establishing glucose- and ABA-regulated transcription networks in Arabidopsis by microarray analysis and promoter classification using a Relevance Vector Machine. Genome Res. **16**(3), 414–427 (2006)
14. Lu, J., Luo, L.: Prediction for human transcription start site using diversity measure with quadratic discriminant. Bioinformation **2**(7), 316–321 (2008)
15. Wang, J., Ungar, L.H., Tseng, H., Hannenhalli, S.: MetaProm: a neural network based meta-predictor for alternative human promoter prediction. BMC Genom. **8**, 374 (2007)
16. Xie, X., Wu, S., Lam, K., Yan, H.: PromoterExplorer: an effective promoter identification method based on the AdaBoost algorithm. Bioinformatics **22**, 2722–2728 (2006)

17. Suzuki, A., Wakaguri, H., Yamashita, R., Kawano, S., Tsuchihara, K., Sugano, S., Suzuki, Y., Nakai, K.: DBTSS as an integrative platform for transcriptome, epigenome and genome sequence variation data. Nucleic Acids Res. **43**(Database issue), D87–D91 (2014)
18. Zeng, J., Cao, X., Yan, H.: Human promoter recognition using Kullback-Leibler divergence. In: 2007 International Conference on Machine Learning and Cybernetics, vol. 6, pp. 3319–3325 (2007)
19. Goddard, N.L., Bonnet, G., Krichevsky, O., Libchaber, A.: Sequence dependent rigidity of single stranded DNA. Phys. Rev. Lett. **85**, 2400–2403 (2000)
20. Zeng, J., Zhu, S., Yan, H.: Towards accurate human promoter recognition: a review of currently used sequence features and classification methods. Brief. Bioinform. **10**, 498–508 (2009)
21. Brukner, I., Sanchez, R., Suck, D., Pongor, S.: Sequence-dependent bending propensity of DNA as revealed by DNase I. parameters for trinucleotides. EMBO J. **14**, 1812–1818 (1995)
22. Li, W., Kou, Q., Wei, L., Liu, J.: Plant promoter recognition based on analysis of base bias and SVM. J. Liaoning Normal Univ. (Natural Science Edition) **35**, 183–187 (2012)
23. Saxonov, S., Daizadeh, I., Fedorov, A., Gilbert, W.: EID: the exon-intron database — an exhaustive database of protein coding intron-containing genes. Nucleic Acids Res. **28**, 185–190 (2000)
24. Licciulli, Mignone, F., Gissi, C., Saccone, C.: F., Gissi, C., Saccone, C.: UTRdb and UTRsite: specialized databases of sequences and functional elements of 5' and 3' untranslated regions of eukaryotic mRNAs. Nucleic Acids Res. **30**, 335–340 (2002)
25. Vapnik, V., Cortes, C.: Support-vector networks. Mach. Learn. **20**(3), 273–297 (1995)
26. Bajic, V.B.: Comparing the success of different prediction programs in sequence analysis: a review. Brief. Bioinform. **1**(3), 214–228 (2000)
27. Zhu, L., Guo, W.L., Lu, C., Huang, D.S.: Collaborative completion of transcription factor binding profiles via local sensitive unified embedding. IEEE Trans. Nanobiosci. **PP**(99), 1 (2016)
28. Liang, X., Zhu, L., Huang, D.S.: Multi-task ranking SVM for image cosegmentaiton. Neurocomputing (2017)

A Novel Computational Method for MiRNA-Disease Association Prediction

Zhi-Chao Jiang(✉), Zhen Shen, and Wenzheng Bao

School of Electronics and Information Engineering,
Tongji University, Shanghai 201804, China
1531680@tongji.edu.cn

Abstract. Accumulating biological and clinical reports have indicated that disorders of microRNAs (miRNAs) are closely related with occurrence and development of various complex human diseases. Developing computational models to infer potential miRNA-disease associations has attracted increasing attention. In this paper, we developed the model of Improved Random Walk with Restart for MiRNA-Disease Association prediction (IRWRMDA) to identify potentially related miRNAs for investigated diseases. By taking advantages of known miRNA-disease association network and miRNA functional similarity network, IRWRMDA obtained reliable performance with AUC of 0.8208. What's more, Colon Neoplasms and Kidney Neoplasms were taken as case studies, where 45 and 43 out of the top 50 predicted miRNAs were successfully confirmed by recent clinical researches, respectively. It is anticipated that IRWRMDA would serve as an important biological resource for future experimental guidance.

Keywords: MicroRNA · Disease · Association prediction · Random walk with restart

1 Introduction

Recently, genome sequencing project has indicated that there are only about 20,000 protein-coding genes in the whole human genome and more than 98% of the human genome does not encode protein sequences [1–3]. Increasing evidences suggest that non-coding RNAs (ncRNAs) could serve as important modulators in a variety of biological processes [4, 5]. MicroRNAs (miRNAs) are a category of small single-stranded ncRNAs (containing ~22 nucleotides), which normally suppress gene expression and protein production of corresponding target messenger RNAs at post-transcription level [6, 7]. However, recent biological evidences have shown that miRNAs could also function as positive regulators in some specific cases [8, 9]. Based on experimental discovery and computational prediction, 28,645 miRNA entries were recorded in the latest version of miRBase, including more than 1,000 human miRNAs [10].

Some researchers developed machine learning-based models to prioritize candidate miRNAs for diseases of interest. For example, Xu *et al.* [11] linked expression profiles of miRNA and mRNA with target genes and then constructed the miRNA target-dysregulation network (MTDN). For each investigated miRNA, four kinds of network

D.-S. Huang et al. (Eds.): ICIC 2017, Part I, LNCS 10361, pp. 539–547, 2017.
DOI: 10.1007/978-3-319-63309-1_48

topological features (i.e. D_{out}, N_{miRNA}, $R_{pc\text{-}miRNA}$, $R_{tarpc\text{-}miRNA}$) were extracted from MTDN. After that, Support Vector Machine (SVM) classifier was trained based on these features to distinguish positive miRNA-disease associations from negative ones. As a well-known supervised learning algorithm, SVM demands not only positive samples but also negative samples for model training. However, it is really hard and even impossible to collect negative associations. In this study, they simply classed unlabeled miRNA-disease pairs (pairs without supporting association evidences) as negative training samples, which may lead to a confusing decision boundary. Furthermore, Chen *et al.* [12] developed a miRNA-disease association prediction model named Regularized Least Squares for MiRNA-Disease Association (RLSMDA) prediction. RLSMDA achieved excellent prediction performance by taking advantages of miRNA functional similarity network as well as disease semantic similarity network. More importantly, RLSMDA was designed in framework of semi-supervised learning, which did not demand information of negative miRNA-disease associations. RLSMDA could prioritize candidate miRNAs for diseases without any known related miRNAs, which provided meaningful guidance for experimental association discovery for new diseases. Most of existing models only considered binary miRNA-disease associations. As a matter of fact, associations between miRNAs and diseases could be grouped into multiple types based on association evidences from circulation, genetics, epigenetics and target [13]. Chen *et al.* [14] presented the model of Restricted Boltzmann Machine for Multiple MiRNA-Disease Association (RBMMMDA) prediction. RBMMMDA was the first proposed model to predict multiple miRNA-disease association types, which could provide more accurate knowledge of molecular basis of diseases in miRNA level. However, due to the divisions of known miRNA-disease associations, data support for each association type is far from meeting demand of model training. Data limitation seriously influenced final prediction performance of RBMMMDA.

In this paper, we developed a novel computational model of Improved Random Walk with Restart for MiRNA-Disease Association prediction (IRWRMDA) to infer potential associations between miRNAs and investigated diseases. We implemented leave-one-out cross validation (LOOCV) to evaluate the performance of IRWRMDA. As a result, IRWRMDA achieved AUC of 0.8202. In the case study of Conlon Neoplasms and Kidney Neoplasms, 45 and 43 out of top 50 predicted miRNAs were confirmed by experimental evidences.

2 Materials and Methods

2.1 MiRNA-Disease Associations

We downloaded 10368 high-quality miRNA-disease entries from Human microRNA Disease Database (HMDD, http://www.cuilab.cn/hmdd) [13]. After removing duplicated records, adjacency matrix ***A*** was then adopted to quantify the relationship between miRNAs and diseases. Meanwhile, to present the number of miRNAs and diseases investigated in the paper, variables *nm* and *nd* were respectively defined.

2.2 MiRNA Similarity Network

In previous study, Wang *et al.* [15] developed miRNA functional similarity calculation based on the assumption that miRNAs with similar functions are always associated with diseases with similar phenotypes. We derived miRNA functional similarity from http://www.cuilab.cn/files/images/cuilab/misim.zip and constructed miRNA functional similarity matrix *FM*, where functional similarity score between miRNA *m(i)* and *m(j)* is stored in entity *FM(i,j)*.

Based on the assumption that two miRNAs share more common associations with diseases, the bigger similarity score is, the miRNA similarity of known associations could be calculated using Jaccard measure as follows [16].

$$JM(\mathrm{i,j}) = \frac{D_{11}(\mathrm{i,j})}{D_{01}(\mathrm{i,j}) + D_{10}(\mathrm{i,j}) + \mathrm{D}_{11}(\mathrm{i,j}) + \varepsilon} \tag{1}$$

where D_{11} *(i,j)* represents the number of diseases with associations with both miRNA *i* and *j*. $D_{10}(i,j)$ is the number of diseases only having known associations with miRNA *i*, while $D_{01}(i,j)$ is the number of diseases only having known associations with miRNA *j*. To be clear, we here set ε to avoid denominator 0.

Finally,miRNA similarity matrix *SM* could be constructed as follows.

$$SM(\mathrm{i,j}) = FM(\mathrm{i,j}) * (1 + JM(\mathrm{i,j})) \tag{2}$$

2.3 Random Walk with Restart

Random walk with restart algorithm for investigated disease could be divided into three steps: (1) initialize probability of each miRNA, (2) implemented random walk on miRNA similarity matrix *SM*, (3) obtain final probability and rank candidate miRNAs [17].

For investigated disease *d*, we use vector *b* to denote the presence or absence of association between *d* and all miRNAs. $b(i) = 0$ represents there is no known association between *d* and miRNA *i*, while $b(i) = 1$ represents known association between *d* and miRNA *i*. The initial probability *pro(0)* could be defined as follows.

$$pro(0) = \frac{b}{\sum_{i=1}^{nm} b(\mathrm{i})} \tag{3}$$

We define vector *pro(k)* to demonstrate the probability between *d* and all miRNAs at step *k*, and the random walk could be iterate as Eq. (4).

$$pro(\mathrm{k}+1) = (1-\alpha) * SM * pro(\mathrm{k}) + \alpha * pro(0) \tag{4}$$

where α is the restart parameter which controls the contribution of source nodes. We stop the algorithm when the random walk is stable (i.e. the change between *pro(k + 1)* and *pro(k)* is less than 10^{-6} in *L1* norm).

When random walk is stopped, we could obtain final probability vector *pro(final)*. MiRNAs with high score in *pro(final)* is considered to be associated with investigated disease *d*.

3 Results

3.1 Performance Evaluation

We implemented leave-one-out cross validation on the experimentally verified miRNA-disease associations collected in HMDD database [13] to evaluate the prediction performance of IRWRMDA. We left out each known miRNA-disease association in turn for model testing while adopted other known miRNA-disease associations as training samples. Importantly, we only took miRNAs without known confirmed relevance with investigated disease as candidate samples. If the rank of test samples exceeds the given threshold, it would be considered as a successful prediction. By setting various thresholds, true positive rates (TPR, sensitivity) were calculated by counting percentages of the test samples with higher ranks than given thresholds. Meanwhile, false positive rates (FPR, 1-specificity), which denote the percentages of negative samples exceeding the given thresholds, were also obtained. To visualize the prediction ability of IRWRMDA, receiver-operating characteristics (ROC) curves were drawn. Area under ROC curve (AUC) was finally calculated as an essential performance evaluation criterion for RWRMDA. Generally, AUC = 1 demonstrates that the model achieves perfect performance. As a result, IRWRMDA obtained AUCs of 0.8208 in leave one out cross validation framework (Fig. 1).

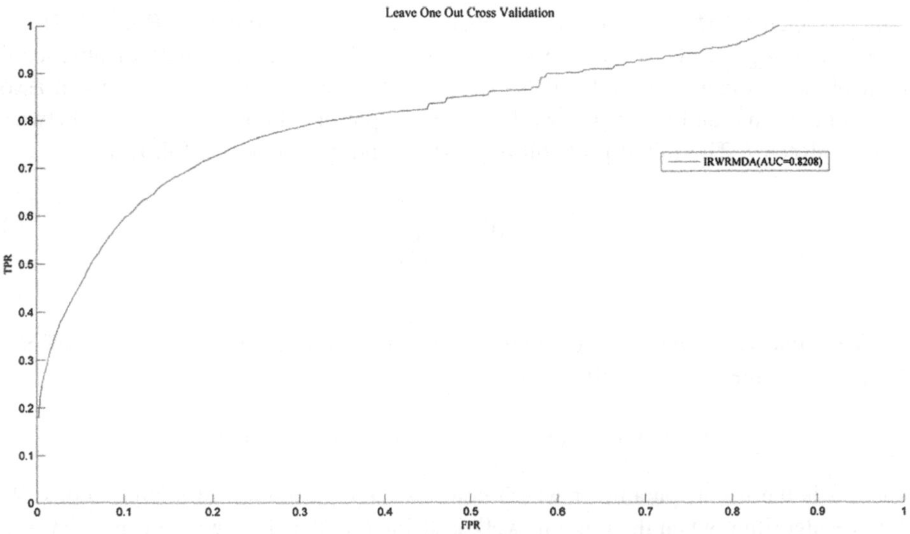

Fig. 1. IRWRMDA achieved AUCs of 0.8208 based on leave one out cross validation.

3.2 Case Study

Colon Neoplasms (CN) is the second most commonly diagnosed cancer in females and the third in males, with over 1,400,000 new cancer cases and 694,000 deaths annually [18]. Recently, death rates of CN have been decreasing in several developed countries, largely resulting from improved treatment techniques and increased awareness [19]. However, CN survival rates are far from satisfaction in developing countries because of limited resources and health infrastructure. With the development of high-throughput sequencing technologies, more and more CN-related miRNAs have been identified. For example, miR-145 was found serving as a robust inhibitor by down-regulating the IRS-1protein in colon cancer lines [20]. It is also found that mir-17 and miR-106a, which were always lost in CN tissues, took E2F1 as a target mRNA and suppressed neoplastic cells growth [21]. Taking CN as a case study to implemented SPYSMDA for potential miRNA prediction, 10 out of top 10, 18 out of top 20 and 45 out of top 50 predicted miRNAs were successfully confirmed based on evidences from dbDEMC

Table 1. Case study on colon neoplasms.

Top 1–25		Top 26–50	
miRNA	Evidence	miRNA	Evidence
hsa-mir-143	dbDEMC; miR2Disease	hsa-mir-221	dbDEMC; miR2Disease
hsa-mir-20a	dbDEMC; miR2Disease	hsa-let-7c	dbDEMC
hsa-mir-125b	dbDEMC	hsa-mir-29b	dbDEMC; miR2Disease
has-mir-223	dbDEMC; miR2Disease	hsa-mir-199a	unconfirmed
hsa-mir-155	dbDEMC; miR2Disease	hsa-mir-29a	dbDEMC; miR2Disease
hsa-mir-21	dbDEMC; miR2Disease	hsa-mir-192	dbDEMC; miR2Disease
has-mir-34a	dbDEMC; miR2Disease	hsa-mir-142	unconfirmed
hsa-mir-18a	dbDEMC; miR2Disease	hsa-mir-30c	dbDEMC; miR2Disease
hsa-mir-16	dbDEMC	has-mir-1	dbDEMC
hsa-mir-19a	dbDEMC; miR2Disease	hsa-let-7i	dbDEMC
hsa-mir-200b	dbDEMC	hsa-mir-203	dbDEMC; miR2Disease
hsa-mir-19b	dbDEMC; miR2Disease	hsa-mir-222	dbDEMC
hsa-mir-127	dbDEMC; miR2Disease	hsa-mir-32	dbDEMC; miR2Disease
hsa-mir-92a	unconfirmed	hsa-mir-34c	miR2Disease
hsa-mir-191	dbDEMC; miR2Disease	hsa-mir-137	dbDEMC; miR2Disease
hsa-mir-141	dbDEMC; miR2Disease	hsa-mir-183	dbDEMC; miR2Disease
hsa-mir-101	unconfirmed	hsa-let-7f	dbDEMC; miR2Disease
hsa-mir-146a	dbDEMC	hsa-mir-106b	dbDEMC; miR2Disease
hsa-mir-107	dbDEMC; miR2Disease	hsa-mir-224	dbDEMC; miR2Disease
hsa-mir-132	miR2Disease	hsa-mir-15b	miR2Disease
hsa-let-7e	dbDEMC	hsa-mir-150	unconfirmed
hsa-let-7b	dbDEMC; miR2Disease	hsa-mir-200c	dbDEMC; miR2Disease
hsa-mir-31	dbDEMC; miR2Disease	hsa-mir-24	miR2Disease
hsa-mir-125a	dbDEMC; miR2Disease	hsa-let-7 g	dbDEMC; miR2Disease
hsa-let-7a	dbDEMC; miR2Disease	hsa-mir-194	dbDEMC; miR2Disease

and miR2Disease (See Table 1). As for top 5 confirmed CN-related miRNAs, miR-20a, miR-125b and miR-155 were found up-regulated in CN tissues while miR-143 was found down-regulated [22–25]. Importantly, miR-125b (3rd in the prediction list) is directly involved in CN progression and could work as a reliable prognostic indicator for colon cancer patients [26].

According to latest cancer statistics report, more than 62,700 new cases and 14,240 deaths will be caused by Kidney Neoplasms (KN) in the United States in 2016 [27]. Patients with kidney cancer in advanced stage only had an 18% two-year survival rate. However, if kidney cancer could be diagnosed in early stage, five-year survival rate would increase to 95% [28]. Therefore, clinical therapy is in urgent need of sensitive markers for early KN detection. Recently, it is reported that miR-192, miR-194, miR-141, miR-200c and miR-215 could serve as reliable diagnostic biomarkers for its significant down-regulation in renal childhood neoplasms [29]. It is also found that clinical characteristic of KN was directly correlated to expression of miR-21 [30], up-regulation of which would lead to low survival rates of KN patients. Furthermore,

Table 2. Case study on kidney neoplasms.

Top 1–25		Top 26–50	
miRNA	Evidence	miRNA	Evidence
hsa-mir-429	dbDEMC	hsa-mir-203	dbDEMC
hsa-mir-200b	dbDEMC; miR2Disease	hsa-mir-1	dbDEMC
hsa-mir-205	unconfirmed	hsa-mir-92a	unconfirmed
has-mir-200a	dbDEMC	hsa-mir-7b	unconfirmed
hsa-mir-155	dbDEMC	hsa-mir-20a	dbDEMC; miR2Disease
hsa-mir-204	dbDEMC	hsa-mir-18a	dbDEMC
has-mir-210	dbDEMC; miR2Disease	hsa-mir-133a	unconfirmed
hsa-mir-16	dbDEMC	hsa-mir-30c	dbDEMC
hsa-mir-126	dbDEMC; miR2Disease	has-mir-34c	dbDEMC
hsa-mir-17	dbDEMC; miR2Disease	hsa-mir-106b	dbDEMC; miR2Disease
hsa-mir-29b	dbDEMC; miR2Disease	hsa-mir-196a	dbDEMC
hsa-mir-27a	dbDEMC; miR2Disease	hsa-mir-19b	dbDEMC; miR2Disease
hsa-let-7a	dbDEMC	hsa-mir-93	dbDEMC
hsa-mir-218	dbDEMC	hsa-mir-302c	unconfirmed
hsa-mir-146a	dbDEMC	hsa-let-7f	dbDEMC; miR2Disease
hsa-mir-10b	dbDEMC	hsa-mir-125b	unconfirmed
hsa-mir-145	dbDEMC	hsa-mir-182	dbDEMC; miR2Disease
hsa-mir-34b	dbDEMC	hsa-mir-29c	dbDEMC; miR2Disease
hsa-mir-29a	dbDEMC; miR2Disease	hsa-mir-452	dbDEMC;
hsa-mir-143	miR2Disease	hsa-mir-367	unconfirmed
hsa-mir-7	dbDEMC; miR2Disease	hsa-mir-302b	unconfirmed
hsa-mir-127	dbDEMC	hsa-let-7c	dbDEMC
hsa-let-7d	dbDEMC	hsa-mir-130b	dbDEMC
hsa-mir-15b	dbDEMC	hsa-mir-135a	unconfirmed
hsa-mir-19a	dbDEMC	hsa-mir-101	dbDEMC; miR2Disease

up-regulated miR-15a could be measured in both biopsy and urine samples from KN patients, and could be treated as an important maker for malignant renal cell carcinoma (RCC) identification [31]. We implemented SPYSMDA to identify potential KN-related miRNAs, and the results showed that 9 out of top 10 and 41 out of top 50 predicted miRNAs were confirmed by published experimental literatures (See Table 2). Evidences of top 5 confirmed miRNAs were concluded in these biological reports [32–34].

For further biological and clinical experiment validation, we prioritized and publicly released the prediction of all the unknown miRNA-disease pairs. It is anticipated that the candidate miRNA-disease pairs with higher ranks could offer valuable clues and would be confirmed by experimental observation in the near future.

4 Discussion

Based on the assumption that miRNAs with similar functions tend to be associated with diseases with similar phenotypes, we presented a novel computational model of IRWRMDA to prioritize candidate miRNA-disease pairs for further biological experiment validation. IRWRMDA achieved reliable prediction performance with AUCs of 0.8208 in leave one out cross validation framework. In addition, case studies of Colon Neoplasms and Kidney Neoplasms were implemented and 9, 9 out of top 10 as well as 45, 41 out of top 50 predicted miRNAs of the two human cancers were confirmed by recent experimental literatures.

According to the fact pointed out by recent literatures, it is not considerable to explain disease pathology only relying on binary miRNA-disease associations. In the future, we will integrate more data sources (such as miRNA expression data, disease-related miRNA-environmental factor interactions, and disease-related miRNA-target interactions) to enhance the robustness of SPYSMDA.

Acknowledgement. This work was supported by the grants of the National Science Foundation of China, Nos. 61472282, 61520106006, 31571364, U1611265, 61672203, 61402334, 61472280, 6 1532008, 61472173, 61572447, 61373098 and 61672382, China Postdoctoral Science Foundation Grant, Nos. 2016M601646.

References

1. Lander, E.S., Linton, L.M., Birren, B., Nusbaum, C., Zody, M.C., Baldwin, J., et al.: Initial sequencing and analysis of the human genome. Nature **409**, 860–921 (2001)
2. Claverie, J.M.: Fewer genes, more noncoding RNA. Science **309**, 1529–1530 (2005)
3. Birney, E., Stamatoyannopoulos, J.A., Dutta, A., Guigo, R., Gingeras, T.R., Margulies, E.H., et al.: Identification and analysis of functional elements in 1% of the human genome by the ENCODE pilot project. Nature **447**, 799–816 (2007)
4. Gutschner, T., Diederichs, S.: The hallmarks of cancer: a long non-coding RNA point of view. RNA Biol. **9**, 703–719 (2012)
5. Eddy, S.R.: Non-coding RNA genes and the modern RNA world. Nat. Rev. Genet. **2**, 919–929 (2001)

6. Bartel, D.P.: MicroRNAs: genomics, biogenesis, mechanism, and function. Cell **116**, 281–297 (2004)
7. Ambros, V.: The functions of animal microRNAs. Nature **431**, 350–355 (2004)
8. Jopling, C.L., Yi, M., Lancaster, A.M., Lemon, S.M., Sarnow, P.: Modulation of hepatitis C virus RNA abundance by a liver-specific MicroRNA. Science **309**, 1577–1581 (2005)
9. Vasudevan, S., Tong, Y., Steitz, J.A.: Switching from repression to activation: microRNAs can up-regulate translation. Science **318**, 1931–1934 (2007)
10. Griffiths-Jones, S., Saini, H.K., van Dongen, S., Enright, A.J.: miRBase: tools for microRNA genomics. Nucleic Acids Res. **36**, D154–D158 (2008)
11. Xu, J., Li, C.X., Lv, J.Y., Li, Y.S., Xiao, Y., Shao, T.T., et al.: Prioritizing candidate disease miRNAs by topological features in the miRNA target-dysregulated network: case study of prostate cancer. Mol. Cancer Ther. **10**, 1857–1866 (2011)
12. Chen, X., Yan, G.Y.: Semi-supervised learning for potential human microRNA-disease associations inference. Sci. Rep. **4**, 5501 (2014)
13. Li, Y., Qiu, C., Tu, J., Geng, B., Yang, J., Jiang, T., et al.: HMDD v2.0: a database for experimentally supported human microRNA and disease associations. Nucleic Acids Res. **42**, D1070–D1074 (2014)
14. Chen, X., Yan, C.C., Zhang, X., Li, Z., Deng, L., Zhang, Y., et al.: RBMMMDA: predicting multiple types of disease-microRNA associations. Sci. Rep. **5**, 13877 (2015)
15. Wang, D., Wang, J., Lu, M., Song, F., Cui, Q.: Inferring the human microRNA functional similarity and functional network based on microRNA-associated diseases. Bioinformatics **26**, 1644–1650 (2010)
16. Gu, C., Liao, B., Li, X., Li, K.: Network consistency projection for human miRNA-disease associations inference. Sci. Rep. **6**, 36054 (2016)
17. Chen, X., Liu, M.X., Yan, G.Y.: RWRMDA: predicting novel human microRNA-disease associations. Mol. BioSyst. **8**, 2792–2798 (2012)
18. Stewart, B., Wild, C.P.: World cancer report 2014. World (2016)
19. Jemal, A., Bray, F., Center, M.M., Ferlay, J., Ward, E., Forman, D.: Global cancer statistics. CA Cancer J. Clin. **61**, 69–90 (2011)
20. Shi, B., Sepp-Lorenzino, L., Prisco, M., Linsley, P., de Angelis, T., Baserga, R.: Micro RNA 145 targets the insulin receptor substrate-1 and inhibits the growth of colon cancer cells. J. Biol. Chem. **282**, 32582–32590 (2007)
21. Diaz, R., Silva, J., Garcia, J.M., Lorenzo, Y., Garcia, V., Pena, C., et al.: Deregulated expression of miR-106a predicts survival in human colon cancer patients. Genes Chromosomes Cancer **47**, 794–802 (2008)
22. Slaby, O., Svoboda, M., Fabian, P., Smerdova, T., Knoflickova, D., Bednarikova, M., et al.: Altered expression of miR-21, miR-31, miR-143 and miR-145 is related to clinicopathologic features of colorectal cancer. Oncology **72**, 397–402 (2007)
23. Chai, H., Liu, M., Tian, R., Li, X., Tang, H.: miR-20a targets BNIP2 and contributes chemotherapeutic resistance in colorectal adenocarcinoma SW480 and SW620 cell lines. Acta Biochim. Biophys. Sin. (Shanghai) **43**, 217–225 (2011)
24. Earle, J.S., Luthra, R., Romans, A., Abraham, R., Ensor, J., Yao, H., et al.: Association of microRNA expression with microsatellite instability status in colorectal adenocarcinoma. J. Mol. Diagn. **12**, 433–440 (2010)
25. Zhang, G.J., Xiao, H.X., Tian, H.P., Liu, Z.L., Xia, S.S., Zhou, T.: Upregulation of microRNA-155 promotes the migration and invasion of colorectal cancer cells through the regulation of claudin-1 expression. Int. J. Mol. Med. **31**, 1375–1380 (2013)
26. Nishida, N., Yokobori, T., Mimori, K., Sudo, T., Tanaka, F., Shibata, K., et al.: MicroRNA miR-125b is a prognostic marker in human colorectal cancer. Int. J. Oncol. **38**, 1437–1443 (2011)

27. Deng, S.-P., Zhu, L., Huang, D.S.: Mining the bladder cancer-associated genes by an integrated strategy for the construction and analysis of differential co-expression networks. BMC Genom. **16**(Suppl 3), S4 (2015)
28. Zheng, C.-H., Zhang, L., Ng, V.T.-Y., Shiu, S.C.-K., Huang, D.S.: Molecular pattern discovery based on penalized matrix decomposition. IEEE/ACM Trans. Comput. Biol. Bioinform. **8**(6), 1592–1603 (2011)
29. Senanayake, U., Das, S., Vesely, P., Alzoughbi, W., Frohlich, L.F., Chowdhury, P., et al.: miR-192, miR-194, miR-215, miR-200c and miR-141 are downregulated and their common target ACVR2B is strongly expressed in renal childhood neoplasms. Carcinogenesis **33**, 1014–1021 (2012)
30. Zaman, M.S., Shahryari, V., Deng, G., Thamminana, S., Saini, S., Majid, S., et al.: Up-regulation of microRNA-21 correlates with lower kidney cancer survival. PLoS ONE **7**, e31060 (2012)
31. von Brandenstein, M., Pandarakalam, J.J., Kroon, L., Loeser, H., Herden, J., Braun, G., et al.: MicroRNA 15a, inversely correlated to PKCalpha, is a potential marker to differentiate between benign and malignant renal tumors in biopsy and urine samples. Am. J. Pathol. **180**, 1787–1797 (2012)
32. Xiong, M., Jiang, L., Zhou, Y., Qiu, W., Fang, L., Tan, R., et al.: The miR-200 family regulates TGF-beta1-induced renal tubular epithelial to mesenchymal transition through Smad pathway by targeting ZEB1 and ZEB2 expression. Am. J. Physiol. Renal Physiol. **302**, F369–F379 (2012)
33. Huang, D.S., Zhang, L., Han, K., Deng, S., Yang, K., Zhang, H.: Prediction of protein-protein interactions based on protein-protein correlation using least squares regression. Curr. Protein Pept. Sci. **15**(6), 553–560 (2014)
34. Wang, G., Kwan, B.C., Lai, F.M., Chow, K.M., Li, P.K., Szeto, C.C.: Elevated levels of miR-146a and miR-155 in kidney biopsy and urine from patients with IgA nephropathy. Dis. Markers **30**, 171–179 (2011)
35. Huang, D.S., Yu, H.-J.: Normalized feature vectors: a novel alignment-free sequence comparison method based on the numbers of adjacent amino acids. IEEE/ACM Trans. Comput. Biol. Bioinform. **10**(2), 457–467 (2013)

Particle Swarm Optimization and Niche Technology

Petri Net Model and Its Optimization for the Problem of Robot Rescue Path Planning

Na Geng[1(✉)], Dun-wei Gong[2], and Yong Zhang[2]

[1] School of Electrical Engineering and Automation, Jiangsu Normal University, Xuzhou 221116, China
gengna@126.com

[2] School of Information and Control Engineering, China University of Mining and Technology, Xuzhou 221116, China

Abstract. Path planning is very important for the problem of robot rescue, establishing an appropriate model is helpful to obtain efficient solutions for the problem above. In this paper, Petri net model for the problem of robot rescue path planning is discussed, and the solutions of the above-mentioned problem are gained through optimizing the model. Firstly, taking the trapped man's life-strength into consideration, the Petri net model with time constraint is established; then, by employing Particle Swarm Optimization (PSO) algorithm, the best transition fired sequence is obtained by combining the problem's domain knowledge, designing the particle's fitness function, and giving the update strategy of particle's local and global best as well. Simulation results show that the established model is reasonable, and the proposed method is effective.

Keywords: Path planning · Robot rescue · Petri net · Particle Swarm Optimization

1 Introduction

Within a limited time, the robot completes as many tasks as possible, which is widespread in all aspects of life and industry, *e.g.*, after the disaster caused by earthquake, tsunamis and explosion, the robot needs to rescue as many trapped men as possible in a short time. The problem has close relationship with robot path planning, which can be attributed to the problem of robot path planning.

In view of the problem of robot rescue path planning, researchers have carried out lots of meaningful works [1–3]. The above works do well in rescuing, but rarely consider formulating the model of the problem to improve the rescue efficiency.

In order to solve the rescue planning problem, first, a model for the problem is needed. Considering that Petri net is a graphic model tool, it can not only characterize the system's structure, but also can accurately describe the events' relationships of sequence, concurrency and synchronization, as a result, it is suitable to describe the dynamic process of discrete events [4–7]. Recently, Petri net model is one of the most active technologies in dynamic system modeling for discrete events, and Petri net used in the field of robotic were extensively studied. The Petri net based on language was proposed by Ziparo *et al.* [8], which can be applied into different robots programming

D.-S. Huang et al. (Eds.): ICIC 2017, Part I, LNCS 10361, pp. 551–563, 2017.
DOI: 10.1007/978-3-319-63309-1_49

system behavior. Yu *et al.* [9] utilized the Petri net with time constraint to model and analyzes the implementation of the project. Since the problem of robot rescue path planning is a typical dynamic discrete event, therefore, Petri net is suitable to model the above problem.

In previous research, Dijkstra algorithm is often employed to optimal Petri net. However, when the number of the nodes is large, the algorithm becomes less efficiency and occupies a large space. Since PSO algorithm shows strong ability to deal with complex problems, thus, it can shorten the time of optimizing the Petri net effectively. Up till now, there are few outcomes in dynamic network model by employing PSO algorithm, so, it is essential to study the PSO algorithm to optimize Petri net model.

At present, there exist lots of researches applied PSO in the field of robot path planning [10–14]. However, PSO algorithm is rarely used in network optimization, even in dealing with the Petri net. So, it can surely expand the scope of application for PSO when it is employed to optimize the optimal transitions fired sequence of the Petri net.

2 Problem Description and Petri Net Model

2.1 Problem Description

In order to facilitate the description, assumptions are made as following:

(1) The robot goes along straight lines in a constant velocity in the rescue operation;
(2) There is no secondary disaster occurring in the rescue operation.

It is shown as Fig. 1, robot is denoted by a solid square, there have M trapped men (targets), and their locations are already known, they are denoted by solid circles, represented as $G\{G_1, G_2, G_3, \cdots, G_M\}$. Once the target's life-strength is smaller than the setting threshold value, there is no need to rescue them. The problem we study in this paper is described as follows: planning the robot's path to rescue the maximum targets in time constraint that each target has a limited lifespan.

In the process of rescuing, if target i is rescued, it must obey the following constraint:

$$\sigma_i > \Delta\sigma \tag{1}$$

where σ_i is the i-th target's life-strength, and it declines rapidly over time; $\Delta\sigma$ is the threshold set artificially according to experience. Referring to the calculation method in [15], the functional relationship between life-strength and rescue time of t is set as follows:

$$\sigma_i = \beta\sigma_{0i}e^{-0.037t} \tag{2}$$

where σ_{0i} is the initial life-strength of the i-th target; β is a constant; $t = \tau + L/v_r$, τ is the rescue time, we do not consider rescue process in detail, and each time it consumes time of τ; L is the distance between any two targets; v_r is the robot's velocity.

Without loss of generality, at time t, if the target's life-strength is smaller than $\Delta\sigma$, then, there is no necessity to rescue the target.

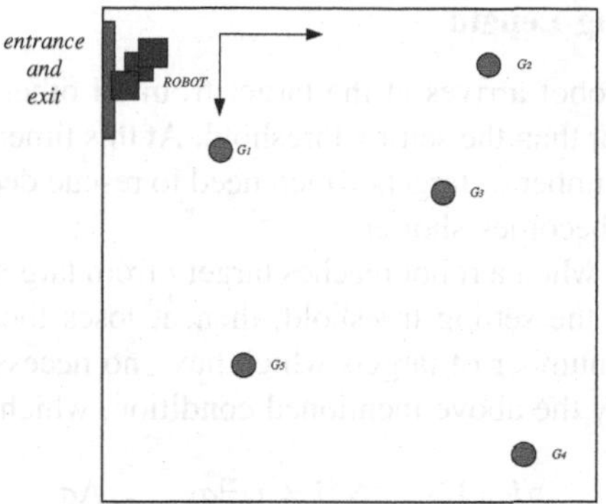

Fig. 1. Robot rescue environment

2.2 Petri Net Model

In this paper, multi robots will be used to complete the rescue operation, and the Petri net is appropriate to describe the problem.

According to the environment in Fig. 1, the Petri net is constructed as Fig. 2 shown, and each symbol has the following meaning: places represent the start points and the targets' locations, denoted as P_0 and $P = \{P_1, P_2, \cdots, P_5\}$; transitions, denoted as $T = \{t_1, t_2, \cdots, t_5\}$, represents the event that robot reaches to the target's location, the fire of transition means that target is rescued, and its fire condition is that the target's life-strength is larger than the setting threshold when the robot reaches to the target, *i.e.*, satisfy (1).

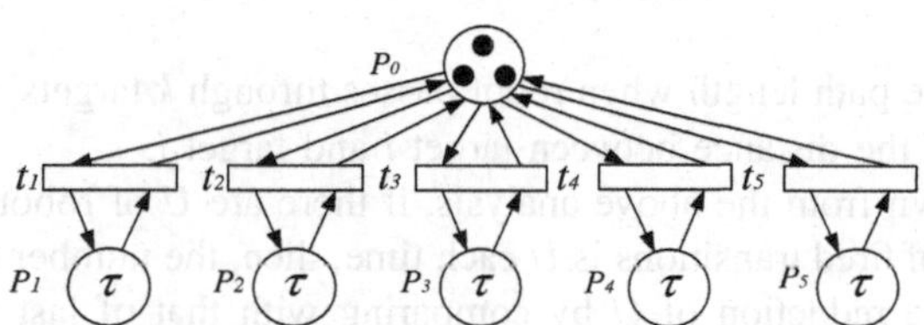

Fig. 2. Petri net for the rescue problem in Fig. 1

3 Particle Swarm Optimization

Since PSO algorithm has many advantages, we employ PSO algorithm to get the optimal transition fired sequence of Petri net, and gain the best rescue sequence. For this purpose, particle is encoded by integer, and particle after decoding is the transition fired sequence.

3.1 Particle Encoding Length

In real cases, when a robot arrives at the target from all other targets, all the target's life-strengths are smaller than the setting threshold. At this time, it needs to abandon the target, as a result, the number of targets which need to rescue declines, and the length of particle encoding also becomes shorter.

Easy to understand, when a robot reaches target i from target j, if the life-strength of target i is smaller than the setting threshold, then, it loses the necessity to rescue. In order to determine the number of targets which have no necessity to rescued, deleting the targets which satisfy the above mentioned condition, which is shown as following:

$$\forall j = 1, \cdots, N, j \neq i, \exists \sigma_{j \to i} < \Delta\sigma \tag{3}$$

Thereby the length of particle coding is reduced, in (3), $\sigma_{j \to i}$ is the life-strength of target i when robot reaches it from target j.

3.2 Particle Coding and Decoding Strategy

In order to obtain the optimal transition fired sequence of Petri net, the coding and decoding strategy of particle is needed. Referring to [16], integer encoding is used to determine the rescue sequence. Assume the encoding of a particle is $x = (x_1, x_2, \cdots, x_M)$, after decoding, the transitions fired sequence of Petri net is $X = (X_1, X_2, \cdots, X_M)$, then, the path length of its corresponding path is

$$L = \sum_{i=1, j=i+1}^{k} d(X_i, X_j), \; i < k, \; k < M \tag{4}$$

which indicates the path length when robot passes through k targets, where $d(X_i, X_j)$ is the path length of the distance between target i and target j.

It can be known from the above analysis, if there are U of robots to rescue M targets, the number of fired transitions is U each time, then, the number of transitions may fired would with a reduction of U by comparing with that of last time. A sequence contain M integers is generated randomly, *i.e.*, $\{x_1, \cdots, x_M\}$, then, the particle decoding method proposed in this section is shown as Fig. 3.

3.3 Fitness Function

The objective of robot rescue path planning problem is to rescue as many targets as possible, based on it, we take the number of rescued targets as the fitness function to estimate particles in this section.

Assume particle is $\{x_1, \cdots, x_i, \cdots, x_M\}$, and the transition sequence after decoding is $\{X_1, \cdots, X_i, \cdots, X_M\}$, and U robots take part in the rescue operation, according to the method in Sect. 3.2, the rescue sequence and its fitness function are calculated: firstly, the relationship between the life-strength and the setting threshold is compared, if the life-strength is smaller than or equal to the threshold, then, the number of rescued targets stays the same; otherwise, the number adds one. According to the above relationship, the number of rescued targets for each robot is calculated; then, after cumulative adding the numbers of rescued targets of all U robots, then, we can get the whole number of rescued targets, which is shown as following:

$$F=\sum_{j=1}^{U}\sum_{i=1}^{m_i}\max\{\text{sgn}(\sigma_i-\Delta\sigma),0\} \tag{5}$$

where m_i is the number of the targets that the i-th robot needs to rescue, $\text{sgn}(\cdot)$ is a sign function, if $\sigma_i-\Delta\sigma>0$, then, $sgn(\sigma_i-\Delta\sigma)=1$, which means the i-th target can be rescued; otherwise, $sgn(\sigma_i-\Delta\sigma)=-1$, which means the i-th target cannot be rescued.

Particle decoding method

Input: particle x
Output: rescue path S after decoding
begin
$x \leftarrow \{x_1, \ldots, x_i, \ldots x_M\}$;
$T \leftarrow \{t_1, t_2, \ldots, t_M\}$;
for i=1 to M **do**
$r_i \leftarrow Mod(x_i, length(x))$
$X=\{X \cap x_{r_i+1}\}$
$x=\{x / x_{r_i+1}\}$
i=i+1;
End For
Output X
End

Fig. 3. Particle decoding method

3.4 Global Best Solutions Update Strategy

For the problem in this paper, the optimal solutions may be more than one. Even if some optimal solutions have the same fitness value, their rescue sequences may be different, thus, the corresponding rescue paths are different. For the rescue environment in Fig. 1, if the robot can rescue two targets once, then, the rescue sequence may be target 2 and 3, or target 3 and 4, or some others paths. In this case, we keep the particle with the same fitness but different rescue in a set, and gbest is selected from the set randomly.

3.5 Particle Velocity and Location Updating Method

As the particle using integer encoding method, the update method of particle's velocity and location is different from CPSO, formula (6, 7) is used to update them:

$$v_{ij}(t+1) = \omega v_{ij}(t) + c_1 r_1 (p_{ij}(t) - x_i(t)) + c_2 r_2 (g_j(t) - x_{ij}(t)) \quad (6)$$

$$x_{ij}(t+1) = Int(x_{ij}(t) + v_{ij}(t+1)) \quad (7)$$

where $Int(\cdot)$ is the rounding function.

3.6 Performance Analysis

Firstly, the structure complexity of the Petri net is discussed. According to the analysis in [17], when Petri net is used to describe robot rescue operation, its structure complexity is small and acceptable.

Next, the complexity of PSO is discussed; it is mainly about the time complexity of this method. We can see that the time complexity of PSO is related to the number of the places of the Petri net, i.e., the number of targets, whereas has no relationship to the structure of the Petri net.

4 Simulations

In order to verify the effectiveness of the proposed method, we use MATLAB 7.0 to program about the problem of robot rescue path planning. The running environment is P_4 2.66 GHz, the RAM memory is 512 M. The parameters of PSO is setting as follows [18]: particle size is 20, the threshold of life-strength is 1.0, $\omega = \omega_{\max} - t/T_{\max} * (\omega_{\max} - \omega_{\min})$, $\omega_{\max} = 0.9$, $\omega_{\min} = 0.4$, $c_1 = 2$, $c_2 = 2$.

4.1 The Problem of Robot Rescue Shown in Fig. 1

Establishing the coordinate system shown in Fig. 1, and two robots are used to rescue the targets, the initial locations of the robots are (0, 0) and (0, 0). For the rescue environment in Fig. 1, the initial life-strength of each target is obtained by the sensors network, shown in Table 1. Setting the velocity of the robot as 3, according to formula (2), we can get the that, when all other targets and all robots reach target 5, the life-strength of target 5 is always lower than 1.0, as a result, before planning the rescue path, deleting target 5, then, the number of targets needs to rescue is four, shown in Fig. 4.

Table 1. Locations and initial life-strength of the targets

	1	2	3	4	5
(x,y)	(15,20)	(70,0)	(60,25)	(20,60)	(80,80)
σ_0	2.0	4.5	5.8	3.8	1.5

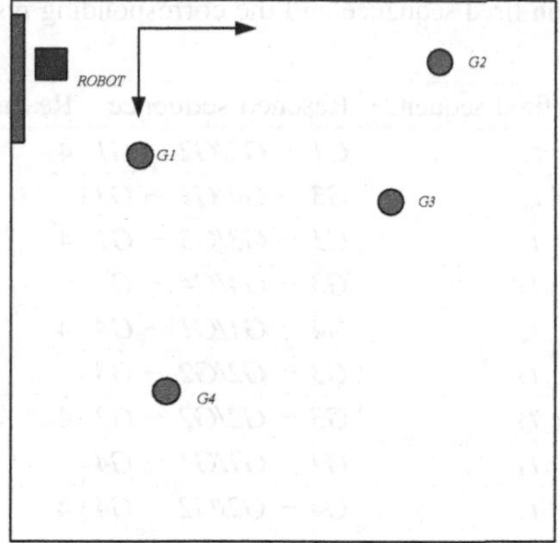

Fig. 4. Rescue environment after deleting target 5

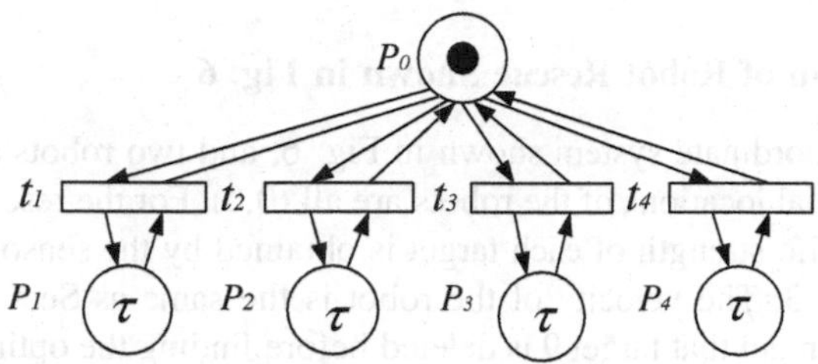

Fig. 5. Petri net for Fig. 4 after delete target 5

Abandoning rescuing target 5, reordering the remaining targets, and the rescue environment and the corresponding Petri net is shown in Fig. 5. Using the Petri net simulator Hierarchical Petri net Simulator [19] to verify the Petri net in Fig. 5, we can find that the Petri net in this paper, with the character of no deadlock, reachable, live, safe and bounded, is a correct Petri net. The transitions of the Petri net in Fig. 5 are $T = \{t_1, t_2, t_3, t_4\}$.

Using the PSO method proposed in this paper to optimal the Petri net in Fig. 5, and the results are shown in Table 2.

From Table 2, we can see that, the robots can rescue four targets at most for the rescue environment in Fig. 4, and there have 24 rescue paths. In the period of the targets' limited lifespan, go along any path in Table 2, the robots can complete rescuing all 4 targets except target 5.

Table 2. The best transition fired sequence and the corresponding rescue number of the targets for the Petri net in Fig. 5

	Transition fired sequence	Rescued sequence	Rescued number
1	$t_1 - t_2/t_2 - t_1$	$G1 - G2/G2 - G1$	4
	$t_4 - t_3/t_3 - t_4$	$G3 - G4/G4 - G3$	
2	$t_3 - t_1/t_1 - t_3$	$G1 - G3/G3 - G1$	4
	$t_3 - t_4/t_4 - t_3$	$G3 - G4/G4 - G3$	
3	$t_4 - t_1/t_1 - t_4$	$G4 - G1/G1 - G4$	4
	$t_3 - t_2/t_2 - t_3$	$G3 - G2/G2 - G3$	
4	$t_3 - t_2/t_2 - t_3$	$G3 - G2/G2 - G3$	4
	$t_4 - t_1/t_1 - t_4$	$G4 - G1/G1 - G4$	
5	$t_4 - t_2/t_2 - t_4$	$G4 - G2/G2 - G4$	4
	$t_3 - t_1/t_1 - t_3$	$G3 - G1/G1 - G3$	
6	$t_3 - t_4/t_4 - t_3$	$G3 - G4/G4 - G3$	4
	$t_2 - t_1/t_1 - t_2$	$G2 - G1/G1 - G2$	

4.2 The Problem of Robot Rescue Shown in Fig. 6

Establishing the coordinate system shown in Fig. 6, and two robots are used to rescue the targets, the initial locations of the robots are all (0,0). For the rescue environment in Fig. 6, the initial life-strength of each target is obtained by the sensors network, which is shown in Table 3. The velocity of the robot is the same as Sect. 4.1, according to formula (3), we can get that target 9 is deleted before finding the optimal path, then, the number of the targets need to rescue is ten.

Abandoning rescuing target 9, reordering the remaining targets, setting target 11 as target 9, Petri net for this scenario is similar to Fig. 5, the difference is that the number of tokens in P_0 adds to 2 and the number of transitions adds to ten. The transitions of the Petri net in Fig. 6 are $T = \{t_1, t_2, \cdots t_i, \cdots, t_9, t_{10}\}$.

In Fig. 7, different targets are represented by rectangles with different colors, each row represents a solution, each solution is divided into 3 parts, and each part indicates the rescue sequence for each robot. From Fig. 7, we can see that: (1) by the proposed method, we can get six solutions, and the three robots' rescued number are four, two and three, respectively, as a result, the whole rescued number is 9; (2) since target 10 is far away from each robot, in order to reach to the goal of rescue as many targets as possible, in the process of rescue, target 10 has to be abandoned.

4.3 Robot Rescue 100 Targets

Considering the situation that there are 100 targets, their x and y axis of each target's location are generated within [0, 100] randomly. Five robots are used to rescue these 100 targets, and the locations of the robots are all (0,0). According to the robots' locations, 12 targets have to abandon. Owning to space constraints in this paper, we omit their life-strength. Then, the five robots need to rescue the remaining 88 targets.

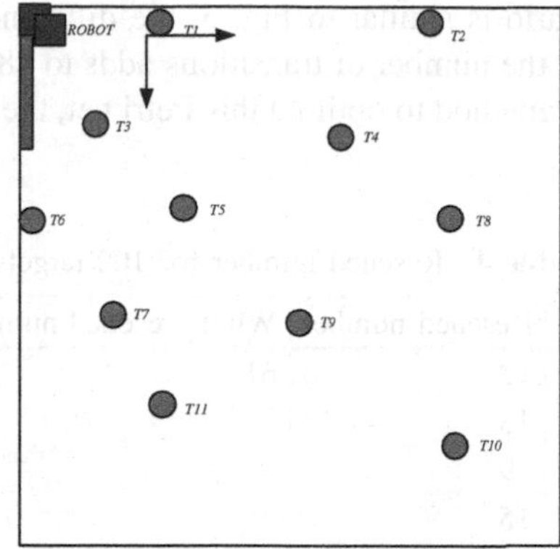

Fig. 6. Robot rescue environment

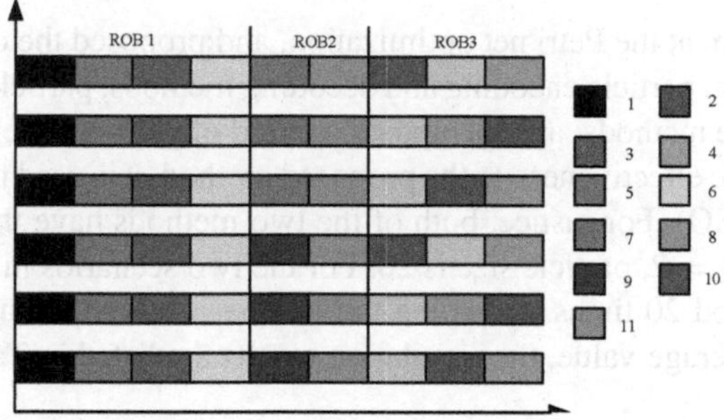

Fig. 7. Rescue sequence of the target rescued by three robots (Color figure online)

Table 3. Locations and initial life-strengths of the targets

	(x,y)	σ_0
1	(25,0)	2.0
2	(70,0)	4.5
3	(15,20)	5.8
4	(60,25)	3.8
5	(40,35)	8.0
6	(0,40)	2.5
7	(20,60)	4.3
8	(80,40)	6.7
9	(50,65)	1.2
10	(80,80)	7.4
11	(28,75)	10.9

Petri net for this scenario is similar to Fig. 5, the difference is that the number of tokens in P_0 adds to 5 and the number of transitions adds to 88, here, we do not give in detail. Using the proposed method to optimal this Petri net, the rescue number is shown as Table 4.

Table 4. Rescued number for 100 targets

	Rescued number	Whole rescued number
ROB1	12	61
ROB2	13	
ROB3	9	
ROB4	15	
ROB5	12	

4.4 Comparison with CPSO

In this paper, we aim at the Petri net optimization, and proposed the corresponding PSO algorithm, including: particle encoding and decoding methods, particle's local and global best solution update methods, and the update method of particle's velocity and location. In order to verify the effectiveness of the proposed method, it is used to compare with the common PSO (CPSO). For justice, both of the two methods have the same parameters setting, i.e., $c_1 = c_2 = 2$, particle size is 20. For the two scenarios in Sects. 4.1 and 4.2, running each method 20 times, recording the number of the optimal rescue paths, and calculating their average value, the simulation results are listed in Table 5.

Table 5. The number of rescue path using different methods

	Scenario 4.1	Scenario 4.2
Proposed method	7.8	16.5
CPSO	4.7	12.3

Table 6. Results of different coding and endecoding methods for scenario 4.1

	Rescued number	Running time	Feasible or not
Proposed method	4	0.869	Y
Common decoding method	4	0.556	N
Integer decoding with correct operator	4	1.150	Y

Table 7. Comparison results of different decoding method for scenario 4.2

	Proposed method	Integer coding method
Rescued number	9	9
Average runtime	15.68	17.89

From Table 5, we can see that the number of optimal paths is obviously larger than that of the CPSO. Since in the update phase, the proposed method takes the situation that the individuals which is the same as the global best solution into consideration, and keeps them in a set, while the CPSO does not use this strategy. For the rescue problem, if the secondary disaster occurrences or the current path is unfeasible, then a more path does a lot for the rescue operation.

Table 6 lists the comparison results for scenario 4.1, from the table, we can see that: (1) the proposed method can rescue all four targets, and the traditional decoding method also can achieve the goal, but it cannot guarantee that the gained solutions are feasible, some solutions exist the phenomenon of target occurrence repeatedly; (2) for the traditional method, the correct operator is performed on it to insure the solution is feasible, but it is time consumption, which is 1.150, is longer than that of the proposed method, which is 0.869, so the proposed method has advantage in running time and high efficiency.

Table 7 is the comparison results between the encoding methods in this paper and the common integer coding method for scenario 4.2. Through the simulation on rescuing ten targets, both of the two methods can rescue nine targets, while the running time of the proposed method is less than the integer encoding method, for the integer encoding method will generate some infeasible solutions, which needs to correct, while the proposed method doesn't need to, resulting in the less time consumption in algorithm running.

4.5 Comparison Between the Proposed Method and Dijkstra Algorithm

In addition, we did comparison between the proposed method and Dijkstra algorithm, aiming at the rescue environment with ten targets shown in Sect. 4.2, and the results are listed in Table 8.

Table 8. Comparison results between four methods when rescue 10 and 100 targets

	The proposed method	Dijkstra algorithm
10 target	9	9
100 targets	61	37

Table 8 reports that the proposed method performs better than the Dijkstra algorithm. When the number of recused targets is small, the results are same. If the number is large, the results are different, and the difference of these two methods is huge. So the proposed method is better than its counterpart.

5 Conclusions

For the problem of robot rescue path planning, the Petri net is used to establish the problem's model and integer PSO is employed to optimize the transitions sequence, and then we can get the optimal rescue path.

Note that, for the same problem of robot rescue path planning, different people constructs different Petri net model, while the Petri net constructed in this paper is one of them, and we optimal it to get the optimal transition fired sequence. If we use a more suitable method to construct a better Petri net for the problem, and optimal the rescue path based on the model, then, the effectiveness of solving the problem of rescue path planning would be further improved. Constructing a better Petri net for the problem of robot rescue path planning and solving the problem based on PSO is the further issue needs to study.

References

1. Geng, N., Sun, X.Y., Gong, D.W., Zhang, Y.: Solving robot path planning in an environment with terrains based on interval multi-objective PSO. Int. J. Rob. Autom. **31**(2), 100–110 (2016)
2. Tian, Z.J., Zhang, L.Y., Chen, W.: Improved algorithm for navigation of rescue robots in underground mines. Comput. Electr. Eng. **39**, 1088–1094 (2013)
3. Basilico, N., Amigoni, F.: Exploration strategies based on multi-criteria decision making for searching environments in rescue operations. Auton. Rob. **31**, 401–417 (2011)
4. Zhang, M.Q., Zhou, Y.: Review of application of Petri net theory in logistics management. Logist. Technol. **7**, 13–16 (2011). (In Chinese)
5. Marius, K., Cristian, M.: A Petri net based approach for multi-robot path planning. Discrete Event Dyn. Syst. **24**(4), 417–445 (2014)
6. Yasuda, G.: Discrete event behavior-based distributed architecture design for autonomous intelligent control of mobile robots with embedded petri nets. In: Taher Azar, A., Vaidyanathan, S. (eds.) Advances in Chaos Theory and Intelligent Control. SFSC, vol. 337, pp. 805–827. Springer, Cham (2016). doi:10.1007/978-3-319-30340-6_34
7. Farineli, A., Raeissi, M.M., Marchi, N., et al.: Interacting with team oriented plans in multi-robot systems. Auton. Agent. Multi-Agent Syst. **31**(2), 332–361 (2017)
8. Ziparo, V.A., Iocchi, L., Lima, P.U., et al.: Petri net plans: a framework for collaboration and coordination in multi-robot systems. Auton. Agent. Multi-Agent Syst. **23**, 344–383 (2011)
9. Yu, R.Q., Huang, Z.Q., Wang, L.: Modeling and analyzing project performance with timing constraint petri net. In: International Conference on Computer Engineering and Technology on Computer Engineering and Technology, Washington, D.C., USA, pp. 243–246. IEEE (2009)
10. Zhang, Y., Gong, D.W., Zhang, J.H.: Robot path planning in uncertain environment using multi-objective particle swarm optimization. Neurocomputing **103**, 172–185 (2013)
11. Deepak, B.B.V.L., Parhi, D.R., Raju, B.M.V.A.: Advance particle swarm optimization-based navigational controller for mobile robot. Arab. J. Sci. Eng. **39**, 6477–6487 (2014)

12. Katić, D., Ćosić, A., Šušić, M., Graovac, S.: An integrated approach for intelligent path planning and control of mobile robot in structured environment. In: Pisla, D., Bleuler, H., Rodic, A., Vaida, C., Pisla, A. (eds.) New Trends in Medical and Service Robots. MMS, vol. 16, pp. 161–176. Springer, Cham (2014). doi:10.1007/978-3-319-01592-7_12
13. Wang, X.W., Shi, Y.P., Yan, Y.X., Gu, X.S.: Intelligent welding robot path planning optimization based on discrete elite PSO. Soft Comput. **2016**, 1–13 (2016). doi:10.1007/s00500-016-2121-2
14. Bilbeisi, G., Al-Madi, N., Awad, F.: PSO-AG: a multi-robot path planning and obstacle avoidance algorithm. In: IEEE Jordan Conference on Applied Electrical Engineering and Computing Technologies, Amman, pp. 1–6 (2015)
15. Kuwata, Y., Takada S.: Rescue ability for earthquake casuality during the 1995 Kobe earthquake (2000). http://www.1ib.koheu.ac.jp/re.Pository/00231322.oaf
16. Gonzalo, M.J., Carlos, M., Cardona, J.L., Castro, A.L.: Petri nets and genetic algorithms for complex manufacturing systems scheduling. Int. J. Prod. Res. **50**(3), 791–803 (2012)
17. Wang, J.G.: Analysis of structure complication in software based on the Petri net. Math. Pract. Theor. **32**(5), 738–744 (2002). (In Chinese)
18. Shi, Y.H., Eberhart, R.C.: A modified particle swarm optimizer. In: Proceedings of the IEEE Congress on Evolutionary Computation (CEC 1998), Anchorage, AK, pp. 63–79 (1998)
19. HiPS: hierarchical petri net simulator (2012). http://sourceforge.net/projects/hips-tools/

Swarm Intelligence and Optimization

Aircraft Scheduling Considering Discrete Airborne Delay and Holding Pattern in the Near Terminal Area

K.K.H. Ng and C.K.M. Lee(✉)

Department of Industrial and Systems Engineering,
The Hong Kong Polytechnic University, Hung Hom, Hong Kong, China
Kkh.ng@connect.polyu.hk, ckm.lee@polyu.edu.hk

Abstract. In this paper, a constructive heuristic using the artificial bee colony algorithm is proposed to resolve the aircraft landing problem considering speed control for airborne delay and holding pattern in the near terminal area. Safety is a top priority in civil aviation management, and air traffic control has to consider handling air traffic promptly. The degree of conservatism in dealing with airborne and terminal traffic should be increased to maintain a high level of resilience for the runways system, enhance the robustness of landing schedule, and reduce the workload of air traffic controllers. The computational results show that the proposed algorithm can resolve the problem in a reasonable amount of time for practical usage.

Keywords: Aircraft landing problem · Discrete airborne delays · Holding pattern · Swarm intelligence · Artificial bee colony algorithm

1 Introduction

The Air Traffic Control (ATC) tower plays a major role in maintaining smooth air traffic and balancing the airborne and airport traffic. The growing demand for air transport increases the pressure on the efficiency of ATC, especially during peak hours. Most of the runways in the international airports are foreseen to reach the maximum runway capacity. Airport capacity expansion is urgently needed to avoid the consequences of exceeding capacity and enhance resilience on managing airport resources. Overcrowded air traffic is a serious issue in managing passengers' satisfaction and comfort and affects airport's reputation. Also, the authorities have an obligation to attempt to resolve safety and delay issues arising at the turnaround and terminal control. Such "alarms" become critical in the future, as most airports foresee a strong growth in the aviation sector. ATC require a robust delay and risk management system in handling daily air traffic and maintaining modern aviation safety standards. The airport capacity in controlling turnaround free-flow progress is limited by a scarce resource – the runway [1, 2]. The planning and construction of a new runway require a long lead-time. Besides the need for runway expansion, aviation authorities are seeking the computational intelligence to reduce the workload of the ATC tower and utilise the current airport resources.

D.-S. Huang et al. (Eds.): ICIC 2017, Part I, LNCS 10361, pp. 567–576, 2017.
DOI: 10.1007/978-3-319-63309-1_50

The Aircraft Sequencing and Scheduling Problem (ASSP) has been well studied in the current literature with different configurations and model objectives leading to variants of the ASSP model. The ASSP model includes the Aircraft Landing Problem [3, 4], Aircraft Take-off Problem [5, 6] and mixed-mode aircraft sequencing operation [7, 8]. Depending on the airport design configuration, the runway system can be heterogeneous or interdependent. As for heterogeneous runways systems, the flight can have a significant difference in the estimated time of arrival (ETA) when runways are located in different position [9]. An interdependent runway system refers to the aircraft scheduling operation using a pair of adjacent runways. The approaching flight generates vortices and may affect the nearby flights from other runways and trailing flights from the same runways [10]. The literature adopts a static approach without consideration of speed control and holding pattern.

In civil aviation, ATC is required to manage the air traffic and safety issues in airborne and airport traffic in the Terminal Manoeuvring Area (TMA). ATC need to determine the scheduled time of landing by priorities and the possible arrival time landing by priority and possible arrival time. Under free flow situation, the speed restriction can be cancelled by the ATC (e.g. flight CX710, no speed restrictions/resume normal speed). Flight speed needs to be maintained when there is a high volume of airborne traffic. ATC restricts the speed of the flight within the TMA to keep a safe distance between flights via speed command. Besides, ATC has the authority to command a particular flight speed during high traffic situations in near terminal area (e.g. Flight BER456, reduce speed to 210 knots.). Moreover, the engine type, flight weight and vertical altitude at standard pressure are also the factors affecting the possible upper bound and lower bound speeds. Therefore, the speed profile of each flight is a class-dependent and flight-level-dependent set.

In this research, we aim to reduce the ATC workload and maintain a high level of resilience for the runway system by the adopting the computational intelligence to obtain an aircraft scheduling solution considering discrete airborne delay and holding patterns. The complexity of the ASSP model is a Non-deterministic Polynomial hard (NP-hard) problem [11]. Therefore, swarm intelligence is proposed to reduce computational effort while ensuring the quality of the solution with close-to-optimal performance.

2 An Aircraft Scheduling Considering Discrete Airborne Delays and Holding Pattern

As mentioned in the previous section, aircraft scheduling considering discrete airborne delays and holding pattern is considered in the model. The runway is configured as an aircraft landing problem, in which the runways are solely for approaching flights only. The objective is to minimise the total tardiness (airborne delays and delay time caused by holding) of all flights. In order to avoid unnecessary workload and confusion in voice-communication-based command between ATC and pilots, the design of the system is intended to enhance the resilience of the aircraft schedule via speed control and the number of aeronautical holdings in the TMA, as shown in Fig. 1.

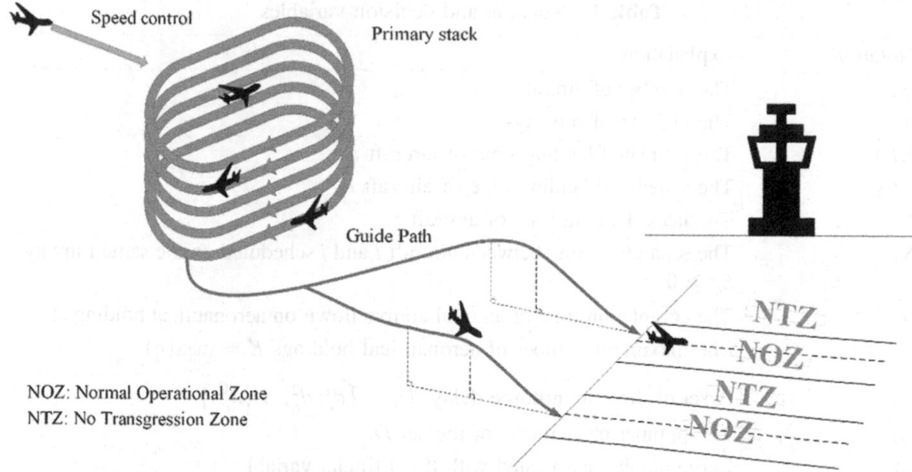

Fig. 1. The aircraft scheduling problem taking into account of speed control and the number of holding on fixed stack

2.1 Problem Formulation

Table 1 shows the notation and decision variables in the model. A flight is denoted as $i, i = (1, 2, \ldots, n)$ and the total number of flights is n. A multiple runway system is considered in the model. Each runway is denoted as $r, r = (1, 2, \ldots, m)$, where m is the maximum number of runways in the airport. The decision variable x_{ir} determines the runway assignment of each flight i, while y_{ijr} denotes the sequence of flight i and j (not necessarily immediately) on the same runway r.

The safety regulation enforced is that any pair of consecutive landings flights on the same runway must be separated with a time buffer – separation time S_{ij}, where flight i is the leading flight and flight j is the trailing flight. S_{ij} is a flight class-dependent value. The detailed separation requirement is shown as Table 2.

The estimated time of arrival ETA_i of flight i is a predetermined/roughly calculated arrival time found from the distance between the departure airport and destination airport. However, the formulation of ETA_i does not consider the air traffic pattern, queue length in a specific time interval and runway capacity of the destination airport. The final approaching time STA_i is usually assigned by ATC when the flight is ready to enter the TMA. The assigned airborne delay time d_i is from a set of discrete value $D_i = \left\{d_i^1, d_i^2, \ldots, d_i^{Q_i}\right\}$, where Q_i is the maximum element in the set. Besides controlling the speed of approaching flights, ATC arranges flights on the queue for landing by utilizing aeronautical holding in TMA when the airspace in TMA is congested. A completed time for an oval course flown in aeronautical holding is defined as h, and the cumulative number of completing holdings is denoted as t_i, where t_i is less than the maximum number of aeronautical holdings K. Therefore, the total holding time of flight i is calculated by $t_i \times h$.

Table 1. Notation and decision variables

Notations	Explanation
n	The number of aircraft
m	The number of runways
ETA_i	The estimated landing time of aircraft i
STA_i	The scheduled landing time of aircraft i
l_i	The latest landing time of aircraft i
S_{ij}	The separation time between aircraft i and j scheduled on the same runway, $S_{ij} \geq 0$
h	The completion time of an oval course flown on aeronautical holding stack
K	The maximum number of aeronautical holdings $K = \max(t_i)$
D_i	A set of discrete airborne delay, $D_i = \{d_i^1, d_i^2, \ldots, d_i^{Q_i}\}$
Q_i	The number of elements of the set D_i
M	Large number associated with the artificial variable
Decision variables	Explanation
t_i	The cumulative number of completing aeronautical holding(s) of aircraft i, $(i = 1, 2, \ldots, n)$
d_i	The airborne delay time of aircraft i, $(i = 1, 2, \ldots, n)$
x_{ir}	1, if aircraft i is assigned to runway r, $(r = 1, 2, \ldots, m)$ 0, otherwise
y_{ijr}	1, if aircraft i is scheduled to land before aircraft j on runway r 0, otherwise

Table 2. Separation time (in seconds) between two consecutive flights in aircraft landing problem

Separation time (sec)		Trailing flight (Arrivals)		
		SSF	MSF	LSF
Leading flight (Arrivals)	SSF	82	69	60
	MSF	131	69	70
	LSF	196	157	96

SSF = Small size flight; MSF = Medium size flight;
LSF = Large size flight

The model minimises the total delay from the estimated time of arrival in absolute value for the worst case directly. The completed mathematical formation for aircraft scheduling considering the airborne and holding pattern is shown as below:

$$\min f = \sum_{i=1}^{n} (|d_i| + ht_i) \quad (1)$$

s.t.

$$STA_i = ETA_i + d_i + ht_i, \forall i \tag{2}$$

$$STA_i \leq l_i, \forall i \tag{3}$$

$$STA_j - STA_i \geq S_{ij}y_{ijr} - My_{jir}, \forall i, j, i \neq j \tag{4}$$

$$y_{ijr} + y_{jir} \leq 1, \forall i, j, r, i \neq j \tag{5}$$

$$\sum\nolimits_{r=1}^{m} x_{ir} = 1, \forall i \tag{6}$$

$$x_{ir}, y_{ijr} \in \{0, 1\}, \forall i, r \tag{7}$$

$$d_i \in D_i = \left\{d_i^1, d_i^2, \ldots, d_i^{Q_i}\right\} \tag{8}$$

$$t_i \in \mathbb{Z}, 0 = \min(t_i) \leq t_i \leq \max(t_i) \tag{9}$$

The objective function (1) is used to minimise the airborne delay in absolute values and the total holding time of all flights. Constraint (2) computes the scheduled time of arrival STA_i of flight i by the sum of estimated time ETA_i of arrival of flight i, airborne time d_i via speed control and the total holding time ht_i. Constraint (3) limits the scheduled time of arrival STA_i before the latest time of arrival l_i. Constraint (4) guarantees that the scheduled time of arrival STA_j of flight j can only approach when the landing procedure of flight i is completed with separation time S_{ij}. Constraint (5) ensures that the landing sequence on the same runway r, either flight i before flight j or flight j before flight i. Each flight i is restricted to being assigned on only one runway r by constraint (6). Constraint (7) confirms that decision variables x_{ir}, y_{ijr} are binary numbers. Considering the travel time by speed up or slowdown is a discrete operation in aviation management, and the controlled airborne delay d_i is a discrete value from a set of speed profiles D_i of flight i by constraint (8). Constraint (9) denotes the minimum and maximum rounds of aeronautical holding in integer values.

3 Resolution Procedure for Aircraft Scheduling

3.1 Proposed MIP-Based Artificial Bee Colony Algorithm

The Artificial Bee Colony (ABC) algorithm is considered as a Swarm Intelligence based (SI-based) algorithm in optimisation problem [12]. The major advantage of using the ABC algorithm in optimisation problem is that the design of the algorithm focuses on balancing the exploration and exploitation during searching. Exploitation refers to the ability of searching from a known solution, while exploration refers to the ability of escaping from local optimal. Three major features foster the optimisation process effectively and efficiently. These include: decentralisation, self-organizing and collective behaviour [13]. The notation and the process flowchart of ABC algorithm are shown in Table 3 and Fig. 2 correspondingly. The ABC algorithm constructs the

Table 3. Notation of proposed artificial bee colony algorithm

Notations	Explanation
CS	The size of bee colony
SN	The number of candidate solutions
D	The dimension of an independent solution
$c_i, i = 1, 2, \ldots, SN$	The position of each solution in bee colony
$\text{fun}(c_i)$	The objective value of solution c_i
$\text{fit}(c_i)$	The fitness value of solution c_i
$Prob(c_i)$	The probability of an individual solution c_i among the entire colony in term of fitness value
$\overline{c_i}$	The neighbour solution of an individual solution c_i
$\text{trial}(c_i)$	The accumulated trial value of an individual solution c_i, which cannot be enhanced the quality of solution in terms of its objective value
limit	The maximum tolerance of $\text{trial}(c_i)$
p	Random number, $0 \leq p < 1$

aircraft scheduling and sequencing solution, while the mixed integer programming is involved in obtaining the discrete airborne delay and number of aeronautical holdings.

3.2 Constructive Heuristic in the Initialization Phase

Compared with randomised initialization, constructive heuristic provides a fairly good starting point for optimal searching from a promising solution region. Given a satisfactory initial solution with high quality, the algorithm is able to reduce the convergence time and computational burden. A simple constructive heuristic is proposed.

The initial solution is constructed from a sequential order of a set of ETA_i. The objective function is to minimize the airborne delay and holding delay. The construction from a First-come-first-serve sequence provides a good initial solution for further exploitation. In order to maintain diversity of the solution sets by constructive heuristic, a random runway assignment is considered from a sorted FCFS sequence in ascending order of ETA_i. The flight with the earlier time in ETA_i will be assigned an earlier position with a random runway assignment, and vice versa.

3.3 Employed Bee Phase

In each iteration, an employed bee performs neighbourhood search operators to generate a neighbourhood solution $\overline{c_i}$ from a known solution $c_i = 1, 2, \ldots, SN$. The greedy method is applied to obtain better solution quality by comparing the objective value of the known solution $fun(c_i)$ and neighborhood solution $\overline{fun(c_i)}$. Two operators are considered in this phase: The swap operator and the insert operator. The swap operator randomly selects two flights from different runways and performs swapping of the position of the two elements. The insert operator aims to reassign the randomly selected flight to another runway at a particular position. Any unsuccessful update in the candidate solution will be cumulated by the parameter $trial(c_i)$.

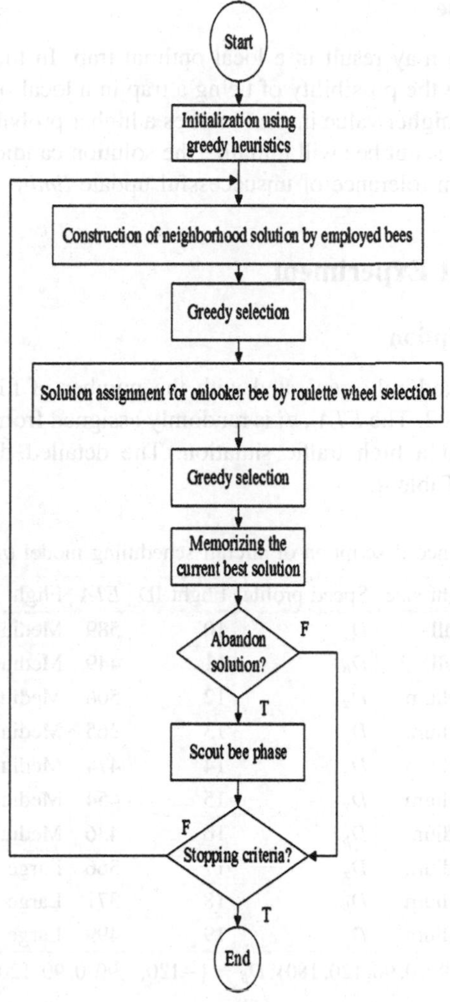

Fig. 2. The process flowchart of proposed artificial bee colony algorithm

3.4 Onlooker Bee Phase

The onlook bee further enhances the solution quality by the neighbourhood search operators. The selection criterion of candidate follows the fitness probability distribution using Eq. (10). The high value in fitness approximation $\text{fit}(c_i)$ denotes a high solution quality in terms of the objective value across the population. The selective probability of each solution is calculated by Eq. (11).

$$\text{fit}(c_i) = 1/(1 + \mathit{fun}(c_i)), \forall \text{i} \tag{10}$$

$$\text{Prob}(c_i) = \mathit{fit}(c_i) / \sum\nolimits_{i=1}^{SN} \mathit{fit}(c_i), \forall \text{i} \tag{11}$$

3.5 Scout Bee Phase

Excessive exploitation may result in a local optimal trap. In the ABC algorithm, the scout bee will evaluate the possibility of being a trap in a local optimal by considering the parameter *trial*. A higher value in *trial* implies a higher probability of being trapped in a local optimal. The scout bee will initialize the solution candidate when the $trial(c_i)$ is excess the maximum tolerance of unsuccessful update *limit*.

4 Computational Experiment

4.1 Instance Description

The test instance is randomly generated with the number of flights n is 20, and the number of runways r is 2. The $ETA_i, \forall i$ is randomly assigned from a uniform interval of $[240, 600]$ to represent a high traffic situation. The detailed description of the test instance is shown as Table 4.

Table 4. Instance description of aircraft scheduling model ($n = 20, m = 2$)

Flight ID	*ETA*	Flight size	Speed profile	Flight ID	*ETA*	Flight size	Speed profile
0	491	Small	D_α	10	589	Medium	D_α
1	375	Small	D_β	11	449	Medium	D_β
2	371	Medium	D_α	12	506	Medium	D_β
3	388	Medium	D_γ	13	265	Medium	D_β
4	534	Medium	D_α	14	474	Medium	D_β
5	371	Medium	D_α	15	454	Medium	D_β
6	327	Medium	D_β	16	436	Medium	D_β
7	291	Medium	D_α	17	366	Large	D_α
8	424	Medium	D_β	18	371	Large	D_β
9	491	Medium	D_γ	19	499	Large	D_α

$D_\alpha = \{-180, -120, -90, 0, 90, 120, 180\}; D_\beta = \{-120, -90, 0, 90, 120\}; D_\gamma = \{-90, 0, 90\};$

4.2 Effectiveness of the Proposed Algorithm

To measure the effectiveness of the proposed ABC algorithm, an exact method by *IBM ILOG CPLEX Optimization Studio 12.6.3* and the original ABC algorithm are also applied as a baseline for comparison. The optimal solution by the exact method is shown in Fig. 3. The configuration of the computational environment is Intel Core i7 3.60 GHz CPU and 16 GB RAM under Window 7 Enterprise 64-bit operating system. The algorithms are written in C# language with visual studio 2015. In our preliminary study, the parameters of the proposed ABC algorithm are set as follows: $\text{CS} = 40, \text{SN} = CS/2$, $limit = SN \times m \times n$. The computational time of exact method by *IBM ILOG CPLEX* is limited to an hour, while the ABC algorithm was given a maximum computational time of 300 s for resolution. Each algorithm is repeated 10 times to obtain the average performance. Table 5 indicates that proposed ABC algorithm obtain a fairly good approximation solution compared with the original ABC algorithm.

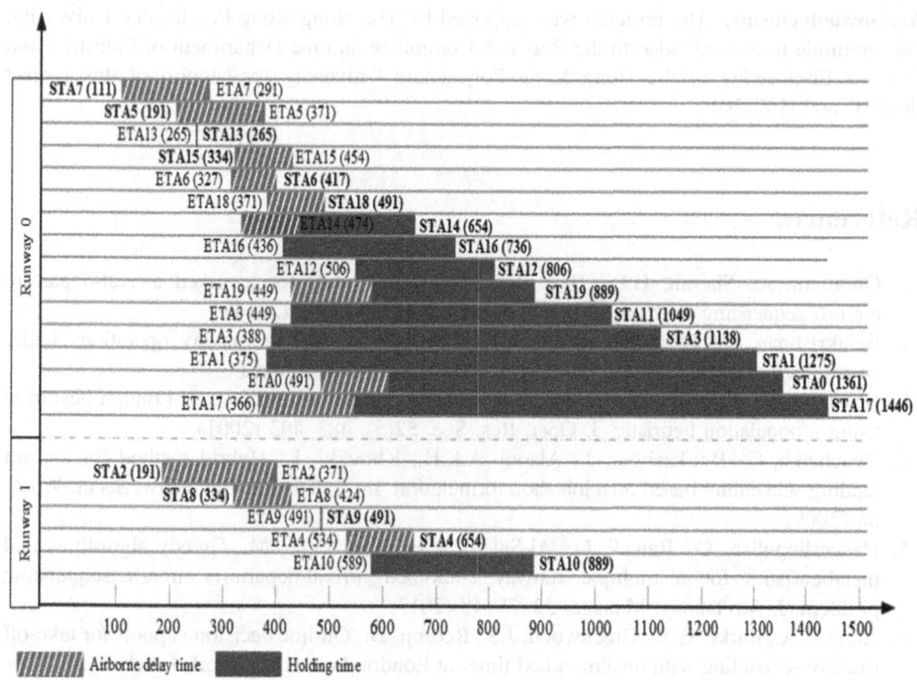

Fig. 3. Gantt charts for the solution obtained by exact method with one-hour computational limit

Table 5. Computational performance by exact method and ABC algorithm

CPU = 3600 s	CPU = 300 s					
MIP w/CPLEX	ABC algorithm			Proposed ABC algorithm		
Optimal	Avg obj	Best obj	Gap (Opt.)	Avg obj	Best obj	Gap (Opt.)
6990	9546	8430	36.57%	7785	7500	11.37%

Avg obj = Average objective value; Best obj = Best objective value; Gap (Opt.) = Deviation between average objective value and optimal by *ILOG CPLEX*

5 Conclusion

Due to the increase of air traffic, the workload is increasingly affecting the resilience of ATC. During heavy air traffic situations and in a dynamic environment, the landing schedule may be adjusted from time-by-time. Pilots may be confused when ATC provides an excessive voice-communication-control on flight speed and landing time by rescheduling. To reduce the workload of ATC, a swarm intelligence algorithm is proposed to solve the aircraft scheduling problem considering discrete airborne delays and holding patterns for daily operation.

Acknowledgements. The research was supported by The Hong Kong Polytechnic University. Our gratitude is also extended to the Research Committee and the Department of Industrial and Systems Engineering of the Hong Kong Polytechnic University for support of this project (RU8H) and (I-ZE3B).

References

1. Ghoniem, A., Sherali, H.D., Baik, H.: Enhanced models for a mixed arrival-departure aircraft sequencing problem. Inf. J. Comput. **26**(3), 514–530 (2014)
2. Balakrishnan, H., Chandran, B.G.: Algorithms for scheduling runway operations under constrained position shifting. Oper. Res. **58**(6), 1650–1665 (2010)
3. Beasley, J.E., Sonander, J., Havelock, P.: Scheduling aircraft landings at London heathrow using a population heuristic. J. Oper. Res. Soc. **52**(5), 483–493 (2001)
4. Bencheikh, G., Boukachour, J., Alaoui, A.E.H., Khoukhi, F.: Hybrid method for aircraft landing scheduling based on a job shop formulation. Int. J. Comput. Sci. Netw. Secur. **9**, 78–88 (2009)
5. Hancerliogullari, G., Rabadi, G., Al-Salem, A.H., Kharbeche, M.: Greedy algorithms and metaheuristics for a multiple runway combined arrival-departure aircraft sequencing problem. J. Air Transp. Manage. **32**, 39–48 (2013)
6. Atkin, J.A., Burke, E.K., Greenwood, J.S., Reeson, D.: On-line decision support for take-off runway scheduling with uncertain taxi times at London heathrow airport. J. Sched. **11**, 323–346 (2008)
7. Bennell, J.A., Mesgarpour, M., Potts, C.N.: Airport runway scheduling. 4OR Q. J. Oper. Res. **9**, 115 (2011). https://link.springer.com/article/10.1007/s10288-011-0172-x
8. Lieder, A., Stolletz, R.: Scheduling aircraft take-offs and landings on interdependent and heterogeneous runways. Transp. Res. Part E Logist. Transp. Rev. **88**, 167–188 (2016)
9. Liu, Y.H.: A genetic local search algorithm with a threshold accepting mechanism for solving the runway dependent aircraft landing problem. Optim. Lett. **5**, 229–245 (2011)
10. Beasley, J.E., Krishnamoorthy, M., Sharaiha, Y.M., Abramson, D.: Scheduling aircraft landings—the static case. Transp. Sci. **34**, 180–197 (2000)
11. Bianco, L., Dell'Olmo, P., Giordani, S.: Scheduling models and algorithms for TMA traffic management. In: Bianco, L., Dell'Olmo, P., Odoni, A.R. (eds.) Modelling and Simulation in Air Traffic Management. Transportation Analysis, pp. 139–167. Springer, Heidelberg (1997)
12. Karaboga, D.: An idea based on honey bee swarm for numerical optimization. Technical report-tr06, Erciyes University, Engineering Faculty, Computer Engineering Department (2005)
13. Zhang, S., Lee, C., Choy, K., Ho, W., Ip, W.: Design and development of a hybrid artificial bee colony algorithm for the environmental vehicle routing problem. Transp. Res. Part D Transp. Environ. **31**, 85–99 (2014)

Minimization of Makespan Through Jointly Scheduling Strategy in Production System with Mould Maintenance Consideration

Xiaoyue Fu[1], Felix T.S. Chan[1], Ben Niu[2(✉)], S.H. Chung[1], and Ying Bi[3]

[1] Department of Industrial and Systems Engineering, The Hong Kong Polytechnic University, Kowloon, Hong Kong
[2] College of Management, Shenzhen University, Shenzhen, China
drniuben@gmail.com
[3] School of Engineering and Computing Science, Victoria University of Wellington, Wellington, New Zealand

Abstract. Job shop scheduling problem with machine maintenance has attracted the attention of many scholars over the past decades. However, only a limited number of studies investigate the availability of injection mould which is important to guarantee the regular production of plastic industry. Furthermore, most researchers only consider the situation that the maintenance duration and interval are fixed. But in reality, maintenance duration and interval may vary based on the resource age. This paper solves the job shop scheduling with mould maintenance problem (JSS-MMP) aiming at minimizing the overall makespan through a jointly schedule strategy. Particle Swarm Optimization Algorithm (PSO) and Genetic Algorithm (GA) are used to solve this optimization problem. The simulation results show that under the condition that the convergence time of two algorithms are similar, PSO is more efficient than GA in terms of convergence rate and solution quality.

Keywords: Jointly scheduling · PSO · GA · Machine maintenance · Mould maintenance

1 Introduction

To maintain high productivity and profitability during production, it is important for manufacturers to have reliable production plans and schedules. However, resource breakdown or failure often occurs which interrupts the current production schedule. Maintenance planning is essential to minimize the occurrence of system failures and to improve the reliability of production schedules [1]. Because of the internal conflict and relationship between production scheduling and resource maintenance, it is necessary to make a compromise decision which can both maximize the productivity and maintain resource reliability. Recently, more and more scholars and researchers pay attention to the production scheduling with maintenance problem which means integrating the production schedule with maintenance plans to harmonize both activities [2–4]. However, in most of these studies, only the availability of machine is considered. In fact,

D.-S. Huang et al. (Eds.): ICIC 2017, Part I, LNCS 10361, pp. 577–586, 2017.
DOI: 10.1007/978-3-319-63309-1_51

the availability of the injection mould is also one of the important factors to guarantee the regular production for some industries, such as the plastic industry [5].

Production scheduling problem with mould maintenance consideration under time-dependent deteriorating maintenance schemes was firstly investigated by Wong et al. [6] and a genetic algorithm was proposed to solve this problem. Furthermore, they [7] proposed a jointly scheduling method to solve the production-maintenance scheduling model with multiple resources and maintenance tasks. Then, Ben Niu, et al. [8] studied the production scheduling problem with mold and machine maintenance consideration through structure-redesign-based bacterial foraging optimization algorithm. So far, research on production scheduling problem with mold maintenance is still lacking and more algorithms need to be explored to solve the problem efficiently.

PSO is proposed by Kennedy and Eberhart [9], which is an intelligent optimization algorithm inspired by the social behavior of animals. Due to the attractive features of easy implementation, fast convergence and a few parameters, PSO has been applied in a wide variety of optimization problems including job shop scheduling problem [10–12]. GA is firstly proposed by Holland in the field of machine learning [13]. As one of the most popular algorithms, GA is known as a suitable and effective method to deal with scheduling problem [14, 15]. It is a stochastic search algorithm which imitates the mechanism of natural genetic evolution in biological systems. In most cases, a specific chromosome is required to be designed to represent a solution of the individual problem and the performance of GA will be affected by the genetic operator [16].

This research deals with the job shop scheduling with mould maintenance problem (JSS-MMP) to minimize the overall makespan, PSO is designed to solve the JSS-MMP and GA is used as the comparison to illustrate the effectiveness of the proposed PSO.

2 Problem Description

In the JSS-MMP firstly proposed by Wong et al. [6], jobs ($j = 1, 2, \ldots J$) are allocated to injection machines ($m = 1, 2, \ldots M$) with corresponding injection moulds ($n = 1, 2, \ldots N$). Each job can be distributed on different machines (but not all machines) with a specific mould and a specific operating time. Since in a real production system, less maintenance time is needed if earlier maintenance task is performed [17], we assume that the relationship between maintenance time and resource age can be fitted by a piecewise linear function. The hypothetical maintenance scheme can be seen from Table 1 and Fig. 1. Specifically, age is defined as the cumulated operating time of a resource. If a machine or mould age reaches the maximum age, maintenance has to be conducted for the particular machine after the completion of the current job. After each maintenance task, machine or mould age will be reset to 0, which means that the condition of the machine or mould is assumed to be as good as new after maintenance (perfect maintenance). Our target is to decide job scheduling and maintenance plan on machine and mould aiming at minimizing the makespan.

Table 1. Maintenance time based on machine/mould age

Machine age	Maintenance time	Mould age	Maintenance time
$0 < A_1 <= 190$	160	$0 < A_2 <= 130$	160
$190 < A_1 <= 430$	$160 + A_1/3 + 40 - 70$	$130 < A_2 <= 250$	$160 + A_2/2 + 4 - 70$
$430 < A_1 <= 600$	$310 + A_1/3 - 150$	$250 < A_2 <= 400$	$310 + A_2/2 - 150$
$600 < A_1$	720	$400 < A_2$	620

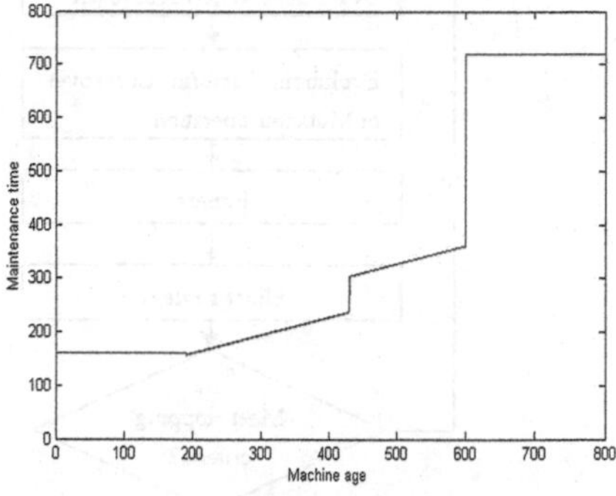

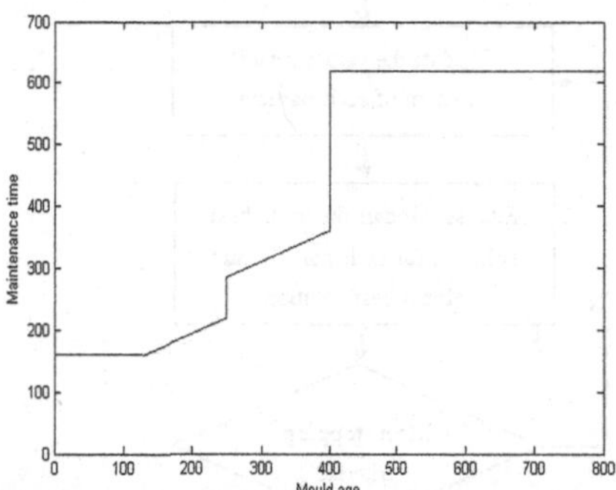

Fig. 1. Maintenance time based on machine/mould age

The assumptions and constraints considered in this paper are summarized as follows:

1. All jobs, machines, and moulds are available for processing at time zero.
2. Set-up times are included in processing times.
3. A job will be operated according to its order quantity (a batch) without splitting.
4. No interruption is allowed when a job is processing.
5. After each maintenance task, machine or mould age will be reset to 0.
6. Each job can only be performed on one machine with one mould.
7. Each machine can carry out only one job at a time unit.
8. Each mould can carry out only one job at a time unit.

3 Optimization Methodology

3.1 PSO

The Process of PSO. The flowchart of PSO and its application in JSS-MMP can be seen from Fig. 2. The details can be described as follows:

Step 1: Initialization. Initialize a population of particles with random positions and velocities on $4 * J$ (J is the number of jobs) dimensions in the searching space.

Step 2: Fitness. Measure the fitness of each particle in the population, and find the local best solution for all particles and global best solution.

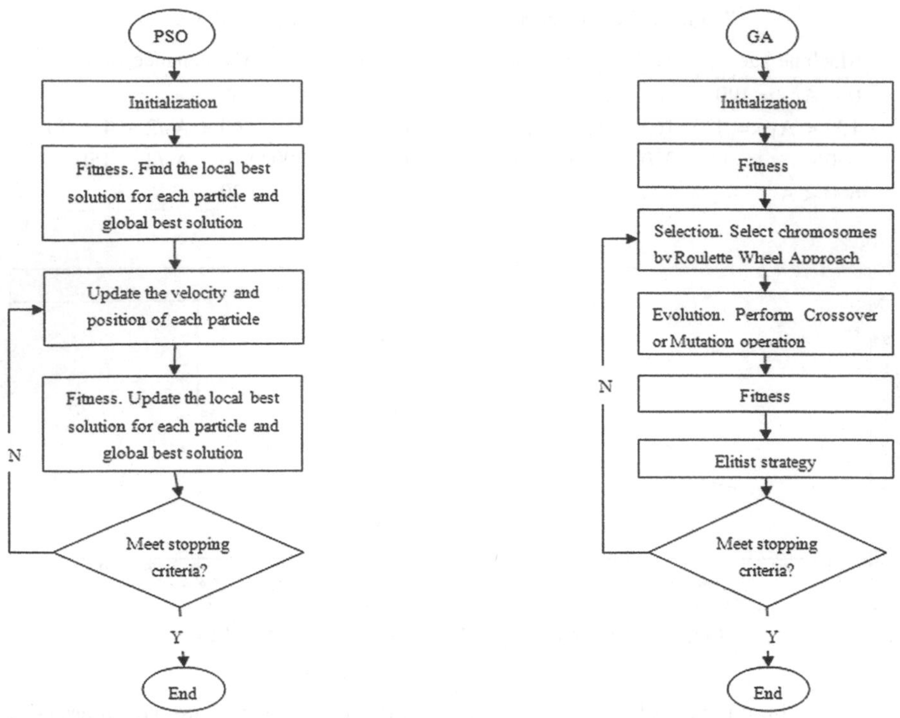

Fig. 2. Flowchart of PSO and GA

Step 3: Update the velocity and position of each particle. The velocity and position are updated according to Eqs. 1–3.

Step 4: Fitness. Update the optimal value of each particle. P_{ibest} is the best value for particle i during the iteration process, and P_i^r is the current fitness of particle i. If $p_i^r < P_{ibest}$, then set $P_{ibest} = p_i^r$; otherwise, P_{ibest} retains. Update the optimal value of the population. Define g_{best} as the best value of the particle population, and $g_{best} = \min(P_{ibest}), (i = 1, 2, \ldots P_{e_{num}})$. If $g_{best}^r < g_{best}$, then set $g_{best} = g_{best}^r$; otherwise, g_{best} retains.

Step 5: Termination. Set $r = r + 1$, and check the condition that if $r <= r_{max}$, then go to Step 3, start a new iteration; otherwise, terminate the algorithm and calculate minimum makespan.

$$V_{ij}(r+1) = W \times V_{ij}(r) + C_1 \times R_1 \times \left(P_{ij}(r) - X_{ij}(r)\right) + C_2 \times R_2 \times \left(P_{gj}(r) - X_{ij}(r)\right) \tag{1}$$

$$X_{ij}(r+1) = X_{ij}(r) + V_{ij}(r+1) \tag{2}$$

$$W = W_{max} - \frac{W_{max} - W_{min}}{r_{max}} \times r \tag{3}$$

Encoding and Decoding of Particle. Since each particle represents a candidate position (i.e. solution), every particle includes the information of job allocation and maintenance plan. For the problem containing (J) jobs, the dimension of the particle is 4 * J which is divided into four groups, respectively representing the job sequence (J), the corresponding machine sequence (M), the corresponding machine maintenance (AM), the corresponding mould maintenance (OM). To transfer the continuous position vector of the particle into a suitable scheduling solution for this problem, random key representation [18] and the smallest position value (SPV) rule [19] are used. Figure 3 shows an example of a particle after decoding.

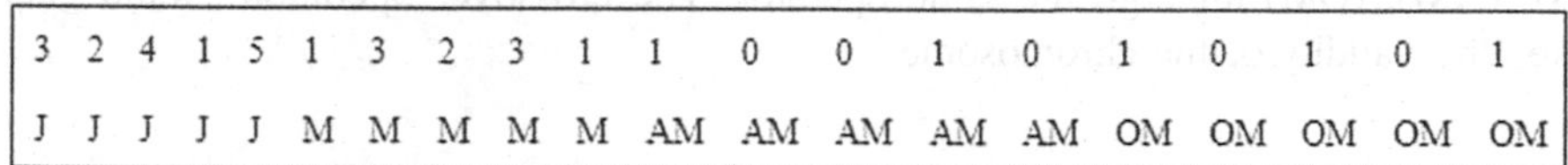

3	2	4	1	5	1	3	2	3	1	1	0	0	1	0	1	0	1	0	1
J	J	J	J	J	M	M	M	M	M	AM	AM	AM	AM	AM	OM	OM	OM	OM	OM

Fig. 3. An example particle after decoding

There are 5 jobs and 3 machines in this example, so this example particle contains 20 dimensions which can be divided into 4 groups. The first and the second group represent the sequence of the jobs (J) and the corresponding machine sequence for the jobs (M). The third and the fourth group represent the machine maintenance (AM) and the mould maintenance (OM). If the AM parameter is denoted as 1, the machine will be maintained after the appointed job is processed by the machine and if the OM parameter is denoted as 1, the mould will be maintained. Otherwise, the machine and the mould will not be maintained. From Fig. 3, we can know that job 3 is distributed on machine 1 and machine 1 will be maintained after processing job 3 and the injection mould on machine 1 will also be maintained. Job 2 is allocated to machine 3, while the machine 3 will not be maintained since the corresponding M parameter is 0, and the injection mould on machine 3 will not be maintained.

3.2 GA

The Process of GA. The flowchart of GA and its application in JSS-MMP can be seen from Fig. 2 and the details can be described as follows:

Step 1: Initialization. Initialize a pool of chromosomes with 4 * J (J is the number of jobs) genes randomly in the chromosomes pool, each chromosome represents a candidate solution. The decoding of the chromosome is the same as the particle in PSO.

Step 2: Fitness. Calculate the makespan and fitness of each chromosome in the pool.

Step 3: Selection. Select chromosomes by Roulette Wheel Approach.

Step 4: Evolution. Perform Crossover or Mutation operation.

Step 5: Fitness. Calculate the makespan and fitness of each chromosome in the pool.

Step 6: Elitist Strategy. Record the best chromosome.

Step 7: Termination. Set $r = r + 1$, and check the condition that if $r < r_{max}$, then go to Step 3; otherwise, terminate the algorithm and calculate minimum makespan.

Crossover Operator. To create new offspring chromosomes, some segments of two parent chromosomes are exchanged hoping to get better chromosomes through replacing the weaker segments with the stronger segments from others.

In our algorithm, two chromosomes are randomly selected as one pair of parent chromosomes that undergo crossover operator. The crossover rate is 0.4. Figure 4 shows an example of the crossover operator. If there are 5 jobs in the problem which means 4 * 5 genes in a chromosome and 2 genes in each group will undergo crossover operator. The first gene in every chromosome is the initial position to begin the crossover operator. The gene in the job group is the conclusive gene. Once any gene in the job group (J) undergo the crossover operator, the corresponding gene in the group (M), (AM), (OM) undergo the same operator. The crossover operation should guarantee the validity of the chromosome.

Parent 1: **3 2** 5 4 1 1 3 2 3 1 1 0 0 1 0 1 0 1 0 1

Parent 2: **1 4** 5 3 2 2 3 1 2 3 0 1 1 0 0 0 1 0 1 1

Step 1:

Offspring 1: **1 4** 0 0 0 **2 3** 0 0 0 **0 1** 0 0 0 **0 1** 0 0 0

Offspring 2: **3 2** 0 0 0 **1 3** 0 0 0 **1 0** 0 0 0 **1 0** 0 0 0

Step 2:

Offspring 1: 1 4 3 2 5 2 3 1 3 2 0 1 1 0 0 0 1 1 0 1

Offspring 2: 3 2 1 4 5 1 3 2 3 1 1 0 1 1 0 1 0 1 0 1

Fig. 4. An examplc of crossover

Mutation Operator. The purpose of mutation is to increase the diversity of chromosomes. To avoid repetition with crossover operator, only machines used by jobs are changed. The gene in the M group is chosen stochastically. And the number of genes undergo the mutation operator is decided by mutation rate. In this paper, we set the mutation rate as 0.5, which means that 2 genes in the M gene group are changed. An example of mutation can be seen from Fig. 5.

Parent: 3 2 5 4 1 1 **3** 2 3 **1** 1 0 0 1 0 1 0 1 0 1

After mutation:

Offspring: 3 2 5 4 1 1 **2** 2 3 **3** 1 0 0 1 0 1 0 1 0 1

Fig. 5. An example of mutation

4 Numerical Example

In the proposed PSO, we set $C_1 = C_2 = 2, V_{max} = 1, W_{max} = 0.9, W_{min} = 0.4$.

To show the effectiveness of PSO, four datasets are tested. GA is used as a comparison. The Parameters of the instances are shown in Table 2. The operation time of moulds is randomly generated between 30 and 55 units of time. The batch size of jobs is randomly generated between 2 and 6 units. The initial size of the population is 50 and the number of iteration is 1000. Each algorithm will be run 10 times in order to measure the deviation of the solution obtained and the convergence time. The results obtained are shown in Table 3. The convergence curves for each instance can be seen from Figs. 6 and 7

Table 2. Parameters of instances

Data set number	Job number	Machine number	Mould number
1	20	3	4
2	35	5	10
3	50	8	14
4	65	9	15

Table 3. Results comparison of PSO and GA

Data set	Algorithm	Average	Min	Max	St. dev	CPU time
1	PSO	1622	1502	1714	72	120.75 s
1	GA	1749	1645	1884	73	115.70 s
2	PSO	2953	2736	3137	115	160.20 s
2	GA	3165	2865	3266	129	150.81 s
3	PSO	2862	2644	3091	138	210.54 s
3	GA	3201	3000	3354	111	196.62 s
4	PSO	3861	3758	4097	104	289.94 s
4	GA	4238	3962	4411	120	268.14 s

From the results, we can see that for those four different cases, the convergence time of PSO and GA are similar, but the Average makespan, Min makespan and Max makespan obtained by PSO are all smaller than the corresponding results obtained by GA. Furthermore, PSO always converges faster than GA. From all the results, we can conclude that PSO is more suitable to solve these four cases in terms of the convergence rate and solution quality.

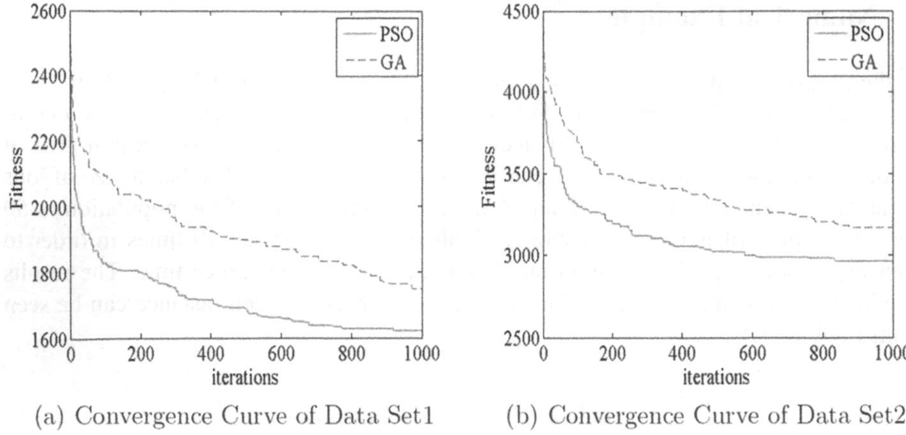

(a) Convergence Curve of Data Set1 (b) Convergence Curve of Data Set2

Fig. 6. Convergence curve of Data Set 1 and 2

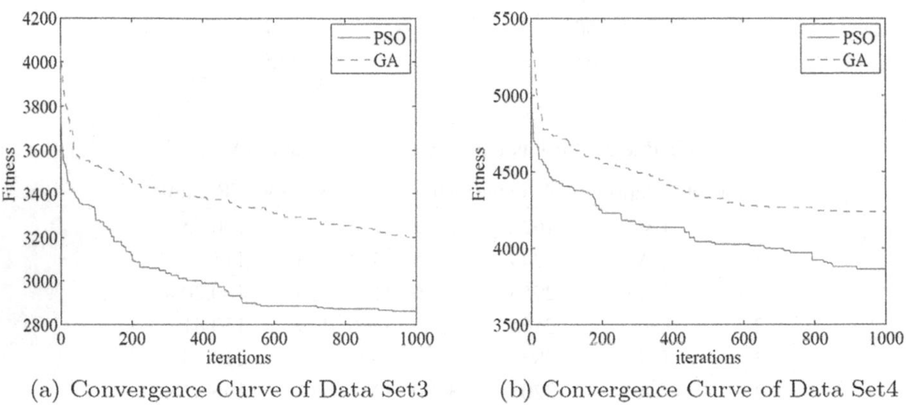

(a) Convergence Curve of Data Set3 (b) Convergence Curve of Data Set4

Fig. 7. Convergence curve of Data Set 3 and 4

5 Conclusion

In this paper, job shop scheduling with mould maintenance problem (JSS-MMP) is studied, which integrates mould maintenance into the traditional job shop scheduling problem. Maintenance duration and interval are subject to the age of resource. To minimize the overall makespan, a jointly scheduling strategy is used. Particle Swarm Optimization Algorithm (PSO) is designed to solve this optimization problem and Genetic Algorithm (GA) is the comparison algorithm. From simulation results of four instances with different sizes, we can know that PSO is more efficient to deal with the integrated problem than GA in terms of convergence rate and solution quality. Imperfect maintenance will be considered in further research related to job shop scheduling problem and more efficient algorithms based on PSO will be explored to solve the problem.

Acknowledgments. The work described in this paper was supported by grants from the Research Grants Council of the Hong Kong Special Administrative Region, China (Project No. PolyU 15201414). The Natural Science Foundation of China (Grant No. 71471158, 71571120, 71271140); and under student account code RUKH.

References

1. Rajkumar, M., Asokan, P., Vamsikrishna, V.: A GRASP algorithm for flexible job-shop scheduling with maintenance constraints. Int. J. Prod. Res. **48**(22), 6821–6836 (2010)
2. Berrichi, A., Amodeo, L., Yalaoui, F., Chatelet, E., Mezghiche, M.: Bi-objective optimization algorithms for joint production and maintenance scheduling: application to the parallel machine problem. J. Intell. Manuf. **20**(4), 389–400 (2008)
3. Chan, T.S., Chan, L.Y., Chung, S.H., Finke, G., Timari, M.K.: Solving distributed FMS scheduling problems subject to maintenance: genetic algorithms approach. Robot. Comput. Integr. Manuf. **22**(56), 493–504 (2006)
4. Chung, S.H., Lau, H.C.W., Ho, G.T.S., Ip, W.H.: Optimization of system reliability in multi-factory production networks by maintenance approach. Expert Syst. Appl. **36**(6), 10188–10196 (2009)
5. Menges, G., Michaeli, W., Mohren, P.: How to make Injection Moulds, 3rd edn. Hanser Publishers, Munich (2001)
6. Wong, C.S., Chan, F.T.S., Chung, S.H.: A genetic algorithm approach for production scheduling with mould maintenance consideration. Int. J. Prod. Res. **50**(20), 5683–5697 (2011)
7. Wong, C.S., Chan, F.T.S., Chung, S.H.: A joint production scheduling approach considering multiple resources and preventive maintenance tasks. Int. J. Prod. Res. **51**(3), 883–896 (2013)
8. Niu, B., Bi, Y., Chan, F.T.S., Wang, Z.X.: SRBFO algorithm for production scheduling with mold and machine maintenance consideration. In: Huang, D.-S., Jo, K.-H., Hussain, A. (eds.) ICIC 2015. LNCS, vol. 9226, pp. 733–741. Springer, Cham (2015). doi:10.1007/978-3-319-22186-1_73
9. Kennedy, J., Eberhart, R.C.: Particle swarm optimization. In: Proceedings of IEEE International Conference on Neural Networks, Piscataway, NJ, USA, vol. 4, pp. 1942–1948 (1995)
10. Abdelhakim, A., Brahim, B., Mourad, B.: Branch-and-bound and PSO algorithms for no-wait job shop scheduling. J. Intell. Manuf. **27**(3), 679–688 (2016)
11. Liu, L.L., Zhao, G.P., Ou'Yang, S.S., Yang, Y.J.: Integrating theory of constraints and particle swarm optimization in order planning and scheduling for machine tool production. Int. J. Adv. Manuf. Tech. **57**(14), 285–296 (2011)
12. Shao, X.Y., Liu, W.Q., Liu, Q., Zhang, C.Y.: Hybrid discrete particle swarm optimization for multi-objective flexible job-shop scheduling problem. Int. J. Adv. Manuf. Tech. **67**(9), 2885–2901 (2013)
13. Holland, J.H.: Adaptation in Natural and Artificial Systems. University of Michigan Press, Ann Arbor (1975)
14. Jia, Z.Y., Lu, X.H., Yang, J.Y., Jia, D.F.: Research on job-shop scheduling problem based on genetic algorithm. Int. J. Prod. Res. **49**(12), 3585–3604 (2011)
15. Li, X.Y., Gao, L.: An effective hybrid genetic algorithm and tabu search for flexible job shop scheduling problem. Int. J. Prod. Eco. **174**, 93–110 (2016)

16. Chan, T.S.F., Choy, K.L., Bibhushan, K.L.: A genetic algorithm-based scheduler for multiproduct parallel machine sheet metal job shop. Expert Syst. Appl. **38**(7), 8703–8715 (2011)
17. Mosheiov, G., Sidney, J.B.: Scheduling a deteriorating maintenance activity on a single machine. J. Oper. Res. Soc. **61**(5), 882–887 (2010)
18. Bean, J.C.: Genetic algorithms and random keys for sequencing and optimization. ORSA J. Comput. **6**(2), 154–160 (1994)
19. Tasgetiren, M.F., Liang, Y.C., Sevkli, M., Gencyilmaz, G.: A particle swarm optimization algorithm for makespan and total flowtime minimization in the permutation flowshop sequencing problem. Eur. J. Oper. Res. **177**(3), 1930–1947 (2007)

Chaotic Optimization of Tethered Kites for Wind Energy Generator

Junfang Li[1,2(✉)], Mingwei Sun[2], Zenghui Wang[3], and Zengqiang Chen[1]

[1] College of Computer and Control Engineering, Nankai University, Tianjin 300350, China
wendyljf@sina.com

[2] College of Electrical and Electronic Engineering, Tianjin University of Technology, Tianjin 300384, China

[3] Department of Electrical and Mining Engineering, University of South Africa, Florida 1710 South Africa

Abstract. Transduction of the stable wind energy into electrical energy at high altitudes is an innovative eco-friendly electricity generation strategy. Effective trajectory optimization can maximize the power generation for the traction phase and the recovery phase of the high-altitude wind power generator. The offline receding horizon control or the fast nonlinear model predictive control was previously employed to realize effective trajectory optimization, however it is time-consuming and lacks of adaptability and flexibility to varying system configurations. A receding horizon optimization method for the tethered kite generator based on an online searching scheme is proposed to improve the flexibility of the system. The nonlinear optimization problem can be approximately reformulated to a univariate receding horizon sub-optimal issue in a short interval in four phases with different objectives. By using uniform sampling and chaotic searching approaches, the sub-optimal solution, subject to the physical constraints, can be sought online. The simulation results demonstrate the effectiveness of the proposed method.

Keywords: Wind power generator · Trajectory optimization · Online searching scheme

1 Introduction

Nowadays, about 80% of world's electricity is produced from thermal plants using fossil sources such as oil, gas and coal [1]. The economical, geopolitical and environmental problems related to such sources are becoming more critical. Today, the worldwide concern about these problems has led to increasing interest in technologies for the generation of renewable electrical energy. Although the wind can generate the eco-friendly energy, the wind turbines are hard to be generalized extensively because

This work was supported by the Natural Science Foundation of China Under Grants of 61573197, 61273138, 61573199, and the Tianjin Natural Science Foundation (Grant No. 14JCYBJC18700, 13JCYBJC17400).

D.-S. Huang et al. (Eds.): ICIC 2017, Part I, LNCS 10361, pp. 587–599, 2017.
DOI: 10.1007/978-3-319-63309-1_52

that they require heavy towers, foundations and huge blades, which can impose severe impacts on environment and investment. These factors contribute to the uncompetitive wind electric price all over the world. More importantly, wind farms generally occupy an area as large as 200 times a thermal plant of the same power, exhibiting another serious environmental problem [1]. From the technical point of view, the surface wind is quite unstable with remarkable fluctuations both in the speed and in the direction, which leads to fluctuant power output as a danger for the stability of the entire electric network. However, winds normally increase with altitude above the ground [1, 4], and the wind at higher altitudes is more constant with strong intensities. For example, the wind speed at an altitude 800 m doubles the wind speed at the altitude 100 m, and the generated power is 8 times greater. Therefore, several high-altitude wind energy generator concepts have been established [2, 3] since the pioneering philosophy proposed in [4], such as the Laddermill [5], the pumping mall, and the KiteGen [6]. The prototype of the KiteGen has already come out [6–10]. Furthermore in Germany, the high-altitude kite generator, SkySails [11–14], has put into commercial operation for the oceanic carrier ships to serve as the supplementary energy such that up to 2 MW of the main engine's propulsion power can be replaced instead. In addition, Argatov et al. [15, 16] carried out quantitative investigations on the energy conversion efficiency, which provides an insightful guideline for future research and development. Until now, high-altitude wind power generator demonstrates an appealing prospect.

The mechanism of high-altitude wind power generator is simple. Here the KiteGen [6–10] is looked as an example to show the mechanism of the high-altitude wind power generator. The kite is attached to two lines and the MEMS as well as the GPS are installed on the kite to measure its attitude, velocity and position in real time. The generator/motor is placed on the ground. The control algorithm calculates the variation rate of the line to manipulate the kite according to its states and the effective wind speed. The generation mode can be achieved when the lines are protracted such that the generator is operational with the pull force from the kite. Otherwise, when the lines are wrapped back due to the length limit, the motor must work and the electric power is consumed. The net energy gain between these two phases should be maximized.

As a power generation system, it is clear that the economical criterion, i.e. the power output per hour, is very crucial for its development even existence. The KiteGen system has a unique characteristic that the maximal net power generation is obtained by implementing an optimal trajectory planning subject to certain process constraints. Unlike the familiar trajectory planning problem for conventional flight aircrafts or cruise missiles, this one is on the basis of optimal power generation and remains open [17–28]. At present, there are two popular methods to deal with the trajectory optimization problem of the tethered kites subject to the constraints: pseudo-spectral method (PSM) [21] and model predictive control (MPC) approach [6–10, 12, 14, 17, 19, 23–26]. PSM is emerging as an efficient technique to solve the optimal trajectory problems, but as a kind of nonlinear programming scheme, the selection of reasonable initial values is very difficult, especially for the tethered kite system. In many cases, the selection of initial values, highly depending on existing empirical knowledge, are more difficult than solving the optimal trajectory algorithm itself. However, MPC is a sub-optimal alternative to eliminate the implementation difficulty for the optimal control in practice based on the receding horizon strategy. Since 1980th, extensive MPC implementations have been reported in process industry and have achieved considerable economic benefits [29–31]. One of the superiority of MPC is that it can handle various forms of process constraints straightforwardly, which is a difficult task for many other control algorithms. The linear model based MPC theory is rather mature with many successful applications [31]. By contrast, the nonlinear MPC is quite complicated both in theory and in practice and is always an open topic [29]. Milanese et al. presented an explicit MPC, which has been applied to the nonlinear systems subject to constraints by using set membership approximation technique [32]. In this method, the sufficiently dense operating points in terms of the states distributed within the possible range are selected and the solution to each point for a specified MPC problem can be obtained by using appropriate optimization tools offline. The interpolation strategy is employed among these sampling points with respect to the practical states when implementing a fast MPC algorithm online. Such approach has been used for the trajectory optimization of the KiteGen system [7, 9]. Although this explicit method can be readily applicable in practice, the solutions to tens of thousands of sampling points must be obtained in advance and stored prior to the application. Therefore, the intolerable computational complexity is unavoidable with extensive storage requirement otherwise the sparse sampling may lead to severe performance degradation. For example, over 10000 sampling points are needed for the KiteGen optimization which includes 6 states. Moreover, such result relies on the concrete values of lift and drag coefficients of the kite, and the optimal solution by using the nominal setup may deviate from anticipation, causing considerable performance deterioration. Therefore, when the aerodynamic coefficients change remarkably, the original solution may be invalid and the offline calculation should be repeated. In such a way, the adaptability, flexibility and extensibility are all severely limited, particularly in the initial R&D phase to investigate the reasonable system configurations. Hence, the explicit fast MPC has inherent pitfalls in reality and an online strategy is urgently needed.

This paper presents an online receding horizon trajectory optimization method for the KiteGen system to enhance its adaptability and flexibility. The entire generation cycle is divided into 4 stages with different cost functions to perform MPC optimization. By fixing the control horizon at 1, the performance cost can be described by a univariate function with respect to the control variable. Different sampling rates are specified for the command updating and the MPC solving to raise calculation efficiency. The optimization problem subject to process constrains can be reformulated as a sequencing problem with two priority levels. By combining the uniform sampling with a chaotic search around the local minimal points on specific intervals, the control variable can be optimized online. The effectiveness of the proposed method is validated by numerical simulations.

2 Problem Formulation

2.1 Dynamics of KiteGen System

The dynamic model of the KiteGen system can refer to [7] or [17]. There are two lines attaching the two edges of the kite, respectively. Assume that there is no kite

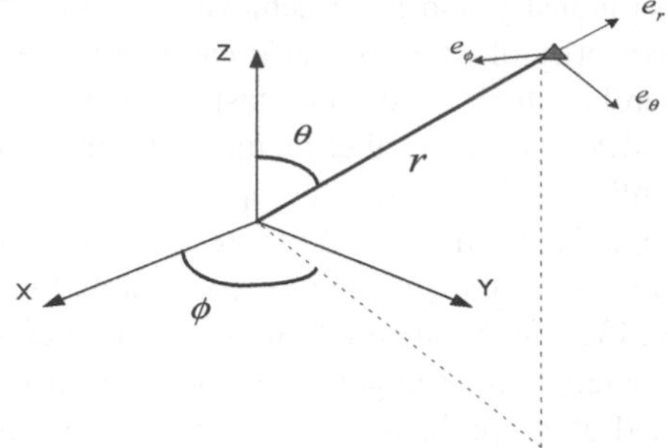

Fig. 1. Spherical and local coordinate system.

deformation. The distance between the two lines fixing points at the kite is d and the length difference of the two lines is Δl which is also the control variable manipulating the trajectory of the kite. A fixed Cartesian system (X, Y, Z) is established with X axis aligned with the nominal wind speed vector direction which is represented as

$$\boldsymbol{v}_0 = [\, v_x(Z) \quad 0 \quad 0]^{\mathrm{T}} \tag{1}$$

where $v_x(Z)$ is a prior known function describing the wind nominal speed at a certain altitude Z. The practical wind speed vector is

$$\boldsymbol{v}_l = \boldsymbol{v}_0 + \boldsymbol{v}_t \tag{2}$$

where $\boldsymbol{v}_t$ is not supposed to be known and accounts for wind measured turbulence. The dynamic model of the KiteGen system is

$$\begin{cases} r\ddot{\theta} - r\sin\theta\cos\theta\dot{\phi}^2 + 2\dot{\theta}\dot{r} = \frac{F_\theta}{m} \\ r\sin\theta\ddot{\phi} + 2r\cos\theta\dot{\phi}\dot{\theta} + 2\sin\theta\dot{\phi}\dot{r} = \frac{F_\phi}{m} \\ \ddot{r} - r\dot{\theta}^2 - r\sin^2\theta\dot{\phi}^2 = \frac{F_r}{m} \end{cases} \tag{3}$$

In this system, the kite position is given by its distance r from the origin and also by its elevation angle of θ and azimuth angle of ϕ as shown in Fig. 1, and m is the mass of the kite. The external forces F_θ, F_ϕ and F_r include the contributions of gravitational force mg, aerodynamic force $\vec{F}^{\mathrm{aer}}$ and force F^{c} exerted by the kite on the lines. They can be expressed in the local coordinates as

$$F_\theta = mg\sin\theta + F_\theta^{\mathrm{aer}} \tag{4}$$

$$F_\phi = F_\phi^{\mathrm{aer}} \tag{5}$$

$$F_r = -mg\cos\theta + F_r^{\mathrm{aer}} - F^{\mathrm{c}} \tag{6}$$

The aerodynamic forces in the local coordinate system are

$$F_\theta^{\text{aer}} = F_l((\boldsymbol{e}_l \times \boldsymbol{e}_t)\cdot \boldsymbol{e}_\theta) + F_d(\boldsymbol{e}_l \cdot \boldsymbol{e}_\theta) \tag{7}$$

$$F_\phi^{\text{aer}} = F_l((\boldsymbol{e}_l \times \boldsymbol{e}_t)\cdot \boldsymbol{e}_\phi) + F_d(\boldsymbol{e}_l \cdot \boldsymbol{e}_\phi) \tag{8}$$

$$F_r^{\text{aer}} = F_l((\boldsymbol{e}_l \times \boldsymbol{e}_t)\cdot \boldsymbol{e}_r) + F_d(\boldsymbol{e}_l \cdot \boldsymbol{e}_r) \tag{9}$$

The lift force is

$$F_l = \frac{1}{2}\rho C_l A ||\boldsymbol{v}_e||^2 \tag{10}$$

and the drag force is

$$F_d = \frac{1}{2}\rho C_d A ||\boldsymbol{v}_e||^2 \tag{11}$$

where ρ is the air density, A is the characteristic area of the kite, and C_l and C_d are the lift and drag coefficients, respectively. The effective wind speed is

$$\boldsymbol{v}_e = \boldsymbol{v}_l - \dot{\boldsymbol{p}} \tag{12}$$

The position of the kite is

$$\boldsymbol{p} = \begin{bmatrix} X \\ Y \\ Z \end{bmatrix} = \begin{bmatrix} r\sin\theta\cos\phi \\ r\sin\theta\sin\phi \\ r\cos\theta \end{bmatrix} \tag{13}$$

The basis vectors in (7)–(9) are defined as

$$\boldsymbol{e}_l = \frac{\boldsymbol{v}_e}{||\boldsymbol{v}_e||} \tag{14}$$

$$\boldsymbol{e}_\theta = [\cos\theta\cos\phi \quad \cos\theta\sin\phi \quad -\sin\theta]^{\mathrm{T}} \tag{15}$$

$$\boldsymbol{e}_\phi = [-\sin\phi \quad \cos\phi \quad 0]^{\mathrm{T}} \tag{16}$$

$$\boldsymbol{e}_r = [\sin\theta\cos\phi \quad \sin\theta\sin\phi \quad \cos\theta]^{\mathrm{T}} \tag{17}$$

$$\boldsymbol{e}_t = \boldsymbol{e}_w(-\cos\psi\sin\eta) + \boldsymbol{e}_o(\cos\psi\cos\eta) + \boldsymbol{e}_r(\sin\psi) \tag{18}$$

where

$$\boldsymbol{e}_o = \boldsymbol{e}_r \times \boldsymbol{e}_w \tag{19}$$

$$e_w = \frac{v_e^p}{||v_e^p||} \tag{20}$$

$$v_e^p = e_\theta(e_\theta \cdot v_e) + e_\phi(e_\phi \cdot v_e) = v_e - e_r(e_r \cdot v_e) \tag{21}$$

and

$$\eta = \arcsin(v_e \cdot e_r \frac{\tan\psi}{||v_e^p||}) \tag{22}$$

Here, ψ is just the control variable defined as

$$\psi = \arcsin(\frac{\Delta l}{d}) \tag{23}$$

and it is necessary to satisfy

$$\left| v_e \cdot e_r \frac{\tan\psi}{||v_e^p||} \right| \leq 1 \tag{24}$$

When the line is protracted, the power output of the generator is

$$P = \dot{r}F^c \tag{25}$$

When the line is wrapped back, the energy consumption by the motor is

$$P = -\dot{r}F^c \tag{26}$$

2.2 Constraints

Several constraints should be considered to guarantee the safety of the system as well as the feasibility of the control system. To ensure that the kite is unable to impact the ground, it is desired that

$$\theta(t) < \bar{\theta} < \pi/2 \tag{27}$$

The constraints on the control variable are

$$|\psi(t)| \leq \bar{\psi} \tag{28}$$

$$|\dot{\psi}(t)| \leq \bar{\dot{\psi}} \tag{29}$$

2.3 Cost Function

The general objective of the KiteGen system is to generate maximal power. Since the net power output only exists in the generation phase, this phase should be as long as possible and the cost function is maximizing the power output. When r reaches a value $\bar{r}$, it is necessary to wrap the lines back in order to start a new cycle again. During this phase, the least amount of energy should be spent such that the net energy gain of the entire cycle can be maximized. In the transition phase between these two stages, the pitch and azimuth angles should meet the specific requirements to ensure sufficient use of wind energy subsequently.

The entire cycle can be divided into 4 Sections.

(1) Generation phase
This phase is initialized provided that the following conditions are satisfied

$$0 < \theta_{\mathrm{I,min}} \le \theta(t) \le \theta_{\mathrm{I,max}} < \pi/2 \tag{30}$$

$$|\phi(t)| \le \bar{\phi}_{\mathrm{I}} < \pi/2 \tag{31}$$

$$r_{\mathrm{I,min}} \le r(t) \le r_{\mathrm{I,max}} \tag{32}$$

implying a symmetrical region down the wind. The cost function is

$$J(t_k) = \max \int_{t_k}^{t_k+T_p} \dot{r}(\tau) F^c(\tau) \mathrm{d}\tau \tag{33}$$

where T_p is the prediction horizon. For simplicity and without loss of generality, $\dot{r}(t)$ can be chosen as a fixed value. When the line length reaches its maximal value or

$$r(t) = \bar{r} \tag{34}$$

this phase terminates and the recovery phase gradually starts.

(2) Recovery phase I
The objective of this stage is to change $\dot{r}(t)$ from a positive fixed value smoothly to 0. Meanwhile, θ is to decrease and $|\phi|$ is to increase so that the energy consumption can be saved effectively. Thus, the cost function can be defined as

$$J(t_k) = \min \int_{t_k}^{t_k+T_p} (\theta^2(\tau) + (|\phi(\tau)| - \pi/2)^2) \mathrm{d}\tau \tag{35}$$

When the following requirements are met, the second recovery phase begins

$$|\phi(t)| \ge \phi_{\mathrm{II}} \tag{36}$$

$$\theta(t) \le \bar{\theta}_{\mathrm{II}} \tag{37}$$

(3) Recovery phase II
This is the primary part of the recovery phase. In this stage, $\dot{r}(t)$ changes from 0 linearly to a negative constant and stays at it and the cost function is

$$J(t_k) = \min \int_{t_k}^{t_k+T_p} |\dot{r}(\tau)| F^c(\tau) \mathrm{d}\tau \tag{38}$$

When (32) is satisfied, the third recovery phase initializes.

4) Recovery phase III
In this stage, $\dot{r}(t)$ changes from a negative constant linearly to 0 in a fast way and stays at it. The objective of this phase is to wait for the satisfaction of (30) and (31) in order to initialize the generation phase once again and its cost function is

$$J(t_k) = \min \int_{t_k}^{t_k+T_p} (|\theta(\tau) - \theta_1| + |\phi(\tau)|) \mathrm{d}\tau \tag{39}$$

where

$$\theta_1 = (\theta_{\mathrm{I,min}} + \theta_{\mathrm{I,max}})/2 \tag{40}$$

These four phases will alternate in order repeatedly.

3 Receding Horizon Control Implementation

3.1 Basic Philosophy

The receding horizon optimization used in MPC is a sub-optimal strategy. Consider the following dynamics

$$\dot{\boldsymbol{x}} = f(\boldsymbol{x}, \boldsymbol{u}) \tag{41}$$

where $\boldsymbol{x} \in \mathrm{R}^n$ is the state, and $u \in \mathrm{R}^m$ is the control variable. The objective function at time t_k is defined as

$$\begin{cases} J(t_k) = \min \sum_{i=1}^{N_y} p(\boldsymbol{x}(k+i|k), \boldsymbol{u}(k+i-1|k)) \\ \text{s.t. } \boldsymbol{x}(k+i|k) \in \boldsymbol{X}, \boldsymbol{u}(k+i-1|k) \in U \end{cases} \tag{42}$$

where $p(\cdot,\cdot)$ is a given penalty functional, $x(k+i|k)$ and $u(k+i-1|k)$ are the i-step ahead state and control predictions respectively, N_y is the prediction horizon or the effective range of (42), and $\boldsymbol{X}$ and $\boldsymbol{U}$ are the feasible domains of state and control variable respectively. Obviously, it is a N_y- dimensional nonlinear programming problem. At each sampling interval, solve the optimal control vector as

$$U^*_{\mathrm{opt}} = [\boldsymbol{u}^*(k|k), \cdots, \boldsymbol{u}^*(k+N_y-1|k)]^T \tag{43}$$

Only the first component of this vector, $u^*(k|k)$, is implemented. The procedure is repeated in the next sampling interval.

3.2 Numerical Optimization Method

Except rare linear systems, there is no analytical solution to the general MPC problem. Numerical solving approaches are normally utilized to handle nonlinear MPC problems. A control horizon, N_u, can be introduced to reduce the computational complexity, that is when $i > N_u$, the control variable remains constant or

$$\Delta \boldsymbol{u}(k+i-1|k) = 0, \quad i > N_u \tag{44}$$

In this paper, letting $N_u = 1$ can considerably reduce the optimization procedure to a one-dimensional problem.

As the KiteGen optimization problem contains process constraints, there are naturally two-level priorities, i.e. the cost function should be optimized subject to the satisfaction of constraints. The major constraint on the system is to avoid impacting ground. Therefore, the number of constraints violation in every receding horizon, N_c, is considered as an important index: the smaller N_c is, the higher priority the corresponding control variable has. With the same priority, the values of the cost function then are compared with each other.

A uniform sampling method in combination of a locally fine Tent chaotic searching approach are employed here to determine the optimal control variable on a specific interval.

At each sampling instant k, denote the feasible interval for the control variable as $[\psi_{\min}(t_k), \psi_{\max}(t_k)]$ based on the constraints of (28) and (29). Select N_s sampling points uniformly distributed on this interval and denote this sequence as $\psi_s(k,i)(i = 1, 2, \cdots, N_s)$. For each $i(1 \le i \le N_s)$ and the given interval constant T_c, calculate the related N_c and $J(\psi_s(k,i))$ from $t_0 = kT_c$ to $t_f = (k+N_y-1)T_c$ by using numerical integration and then conduct sequencing for these results by using the two-level priority rule mentioned above. Next, the local optimal points can be determined accordingly and their corresponding intervals can also be confined with a number of N_l. Define the sequence number of the sole sampling point in such each interval as $i_l(l = 1, 2, \cdots, N_l)$.

For each interval $[\psi_{i_l-1}, \psi_{i_l+1}]$ containing a local optimal point, the chaotic searching scheme is used to achieve a fine search further. Chaos is a unique characteristic of specific nonlinear dynamic systems, which exhibits inherent randomness in a determinate system. Several features may accompany chaos, such as ergodicity y, fractals and strange attractors. Among these ones, ergodicity, that is to traverse a complete set within a given range without any repetition from a possibility point of view, is an appealing mechanism to be used in global optimization process in order to escape local minimal trap according to this intrinsic property. Up to now, chaotic optimization has received extensive attention in academia as an innovative optimization technique and a lot of successful applications have been reported. Chaotic optimization has perfect global searching capability with simple implementation structure, which has a promising prospect. The Tent mapping [33] is employed to achieve the fine search objective here, which is defined as

$$x_{n+1} = 1 - 2|x_n - 0.5|, \quad 0 \leq x_0 \leq 1 \tag{45}$$

It can be proved that the Tent mapping can traverse the interval of [0,1] with uniform probability density distribution from an irrational initial number. The procedure of chaotic optimization is as follows:

(1) Select an initial irrational number of $x_0 (0 < x_0 < 1)$ and let $n = 0$ as well as specify the expected iterative number of N_i.
(2) Calculate x_{n+1} according to (45) and project it on the corresponding physical interval of the control variable as

$$\psi_{n+1} = \psi_{i_l - 1} + (\psi_{i_l + 1} - \psi_{i_l - 1}) x_{n+1} \tag{46}$$

by using a linear carrier function. Next, calculate $N_c(\psi_{n+1})$ and $J(\psi_{n+1})$. Then $n = n + 1$.
(3) When $n < N_i$, go to (2), otherwise go to (4).
(4) Arrange the order of the dual-element sequence $(N_c(\psi_i), J(\psi_i))$ by using the two-level priority criterion and determine ψ_i related to the minimal one.

Then the fine searching can be realized on a specific interval. This operation is performed for each interval containing the local minimum. Select the minimal one among them and the boundary values to determine the global minimum on the entire possible interval of $[\psi_{\min}(t_k), \psi_{\max}(t_k)]$ at time k.

4 Numerical Examples

Take the example of [7] to carry out the simulations. The wind speed is 6 m/s. The mass of the kite is 2.5 kg, the characteristic area is 5 m^2. The lift coefficient $C_l = 1.2$, and the drag coefficient $C_d = 0.15$. The constraints on the control variable are $\bar{\psi} = 4°$ and $\bar{\dot{\psi}} = 20°/\mathrm{s}$. $\dot{r} = 0.5$ m/s during the generation phase, and $\dot{r} = -2.3$ m/s during the recovery phase. The trajectory optimization interval is $T_c = 0.1$s and the prediction horizon is $T_p = 25$. The parameters of process constraints are set as follows: $\theta_{I,\min} = 35°$, $\theta_{I,\max} = 45°$, $\bar{\phi}_I = 5°$, $r_{I,\min} = 95$ m, $r_{I,\max} = 105$ m, $\bar{r} = 280$ m, $\phi_{II} = 30°$, $\bar{\theta}_{\mathrm{II}} = 60°$ and $\bar{\theta} = 80°$ [7]. The number of uniform sampling is $N_s = 100$ and the iterative number of chaotic optimizations at each sub-interval containing a local minimal point is $N_i = 20$. The initial value of Tent mapping is an irrational number of $x_0 = 1/\sqrt{2}$.

The simulation results are shown in Figs. 2, 3, 4 and 5. According to Fig. 2, the kite is flying down the wind with a repetitive circular mode in the generation phase where the distance between two neighboring circles is quite small such that a long generation

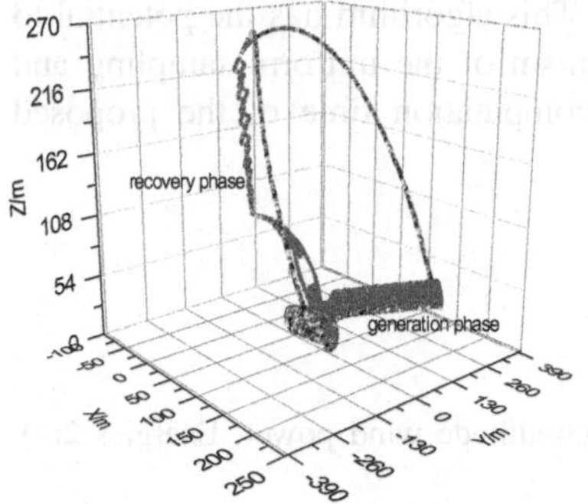

Fig. 2. Three-dimension trajectory.

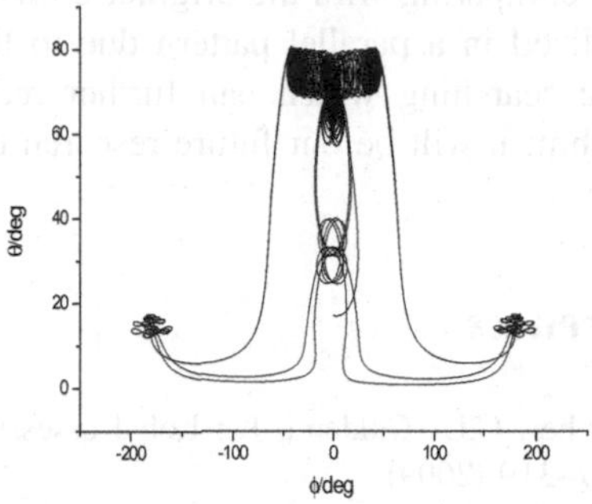

Fig. 3. Phase plot between pitch and azimuth angle.

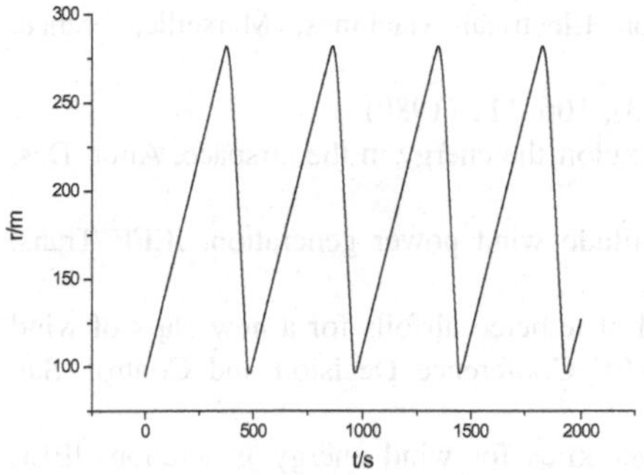

Fig. 4. Variation of line length.

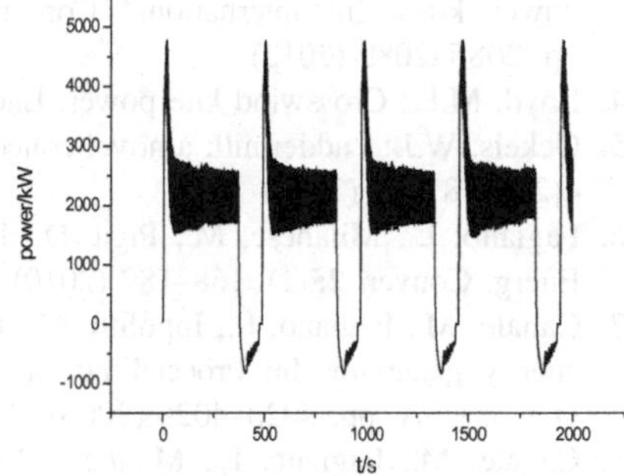

Fig. 5. Generated and dissipated power.

period can be maintained. On the other hand, the recovery phase is rather short to reduce energy consumption. The phase plane of pitch and yaw angles are shown in Fig. 3. The variation of the line length is shown in Fig. 4. From Fig. 5, it can be seen that the power output is far more than the power consumption. The net power is 754 kJ, 3% higher than the result of [7]. In the generation phase, the average power output is about 2 kW. In the recovery phase, the power consumption is around 250 W.

5 Conclusion

This paper presented an online trajectory optimization for a tethered kite generation system based on the receding horizon strategy, which can reduce the computational complexity and enhance the flexibility of the original offline algorithm. Moreover, the proposed method can provide a fast and convenient tool at the initial R&D stage to optimize the system configurations. In addition, this algorithm would have strong adaptability in conjunction with identification of aerodynamic coefficients using the power output. The effectiveness of the proposed approach was demonstrated by the numerical simulations, which showed the proposed method can achieve more energy output comparing with the original offline method. This algorithm has the potential to be realized in a parallel pattern due to the mechanism of the uniform sampling and chaotic searching which can further reduce the computation time of the proposed algorithm; it will be our future research topic.

References

1. Archer, C.L., Caldeira, K.: Lobal assessment of high-altitude wind power. Energies **2**(2), 307–319 (2009)
2. Lansdorp, B., Ockels, W. J.: Comparison of concepts for high-altitude wind energy generation with ground based generator. In: The 2nd RE Asia Conference, Beijing, China (2005). 28(2), pp. 253–263
3. Ahmed, M., Hably, A., Bacha, S.: High altitude wind power systems: a survey on flexible power kites. In: International Conference on Electrical Machines, Marseille, France, pp. 2085–2089 (2012)
4. Loyd, M.L.: Crosswind kite power. Energy **4**(3), 106–111 (1980)
5. Ockels, W.J.: Laddermill: a novel concept to exploit the energy in the airspace. Aircr. Des. **4**(2–3), 81–97 (2001)
6. Fagiano, L., Milanese, M., Piga, D.: High-altitude wind power generation. IEEE Trans. Energ. Conver. **25**(1), 168–180 (2010)
7. Canale, M., Fagiano, L., Ippolito, M.: Control of tethered airfoils for a new class of wind energy generator. In: Proceedings of 45th IEEE Conference Decision and Control, San Diego, USA, pp. 4020–4026 (2006)
8. Canale, M., Fagiano, L., Milanese, M.: Power kites for wind energy generation. IEEE Control Syst. Mag. **27**(6), 25–38 (2007)
9. Canale, M., Fagiano, L., Milanese, M.: High altitude wind energy generation using controlled power kites. IEEE Trans. Control Syst. Technol. **18**(2), 279–293 (2010)
10. Fagiano, L., Milanese, M., Piga, D.: Optimization of airborne wind energy generators. Int. J. Robust Nonlinear Control **22**(18), 2055–2083 (2012)
11. Erhand, M., Strauch, H.: Control of towing kites for seagoing vessels. IEEE Trans. Control Syst. Technol. **21**(5), 1629–1640 (2013)
12. Canale, M., Fagiano, L., Milanese, M., Razza, V.: Control of tethered aircrafts for sustainable marine transportation. In: Proceedings of 2010 IEEE International Conference on Control Application Part of 2010 IEEE Multiconference System Control, Yokohama, Japan, pp. 1904–1909 (2010)
13. Dadd, G.M., Hudson, D.A., Shenoi, R.A.: Determination of kite forces using three-dimensional flight trajectories for ship propulsion. Renew. Energy **36**(10), 2667–2678 (2011)
14. Fagiano, L., Milanese, M., Razza, V., Bonansone, M.: High-altitude wind energy for sustainable marine transportation. IEEE Trans. Intell. Transp. Syst. **13**(2), 781–791 (2012)
15. Argatov, I., Rautakorpi, P., Silvennoinen, R.: Estimation of the mechanical energy output of the kite wind generator. Renew. Energy **34**(6), 1525–1532 (2009)
16. Argatov, I., Silvennoinen, R.: Energy conversion efficiency of the pumping kite wind generator. Renew. Energy **35**(5), 1052–1060 (2010)
17. Diehl, M.: Real-time optimization for large scale nonlinear processes. Ph.D. dissertation, Dept. Electr. Eng., Heidelberg Univ., Heidelberg, Germany (2001)

18. Ahmed, M., Hably, A., Bacha, S.: Power maximization of a closed-orbit kite generator system. In: Proceedings of 50th IEEE Conference on Decision Control and European Control Conference, Orlando, FL, USA, pp. 7717–7722 (2011)
19. Houska, B., Diehl, M.: Optimal control for power generating kites. In: Proceedings of 9th European Control Conference, Kos, Greece, pp. 3560–3567 (2007)
20. Furey, A., Harvey, I.: Evolution of neural networks for active control of tethered airfoils. In: Proceedings of 9th European Conference on Artificial Life, pp. 746–756 (2007)
21. Williams, P., Lansdorp, B., Ockels, W.: Optimal crosswind towing and power generation with tethered kites. J. Guidance Control Dyn. **31**(1), 81–93 (2008)
22. Williams, P., Lansdorp, B., Ockels, W.: Nonlinear control and estimation of a tethered kite in changing wind conditions. J. Guidance Control Dyn. **31**(3), 793–799 (2008)
23. Ilzhoefer, A., Houska, B., Diehl, M.: Nonlinear MPC of kites under varying wind conditions for a new class of large scale wind power generators. Int. J. Robust Nonlinear Control **17**(17), 1590–1599 (2007)
24. Houska, B., Diehl, M.: Robustness and stability optimization of power generating kite systems in a periodic pumping mode. In: Proceedings of IEEE Multiconference System Control, Yokohama, pp. 2172–2177 (2010)
25. Novara, C., Fagiano, L., Milanese, M.: Direct data-driven inverse control of a power kite for high altitude wind energy conversion. In: Proceedings of IEEE International Conference Control Application, Denver, USA, pp. 240–245 (2011)
26. Ferreau, H.J., Houska, B., Geebelen, K., Diehl, M.: Real-time control of a kite-model using an auto-generated nonlinear MPC algorithm. In: Proceedings of 18th IFAC World Conference, Milan, Italy, pp. 2488–2493 (2011)
27. Gillis, J., Goos, J., Geebelen, K., Swevers, J., Diehl, M.: Optimal periodic control of power harvesting tethered airplanes: how to fly fast without wind and without propeller. In: Proceedings of American Control Conference, Fairmont Queen Elizabeth, Montréal, Canada, pp. 2527–2532 (2012)
28. Baayen, J.H., Ockels, W.J.: Tracking control with adaption of kites. IET Control Theory Appl. **6**(2), 182–191 (2012)
29. Henson, M.A.: Nonlinear model predictive control: current status and future directions. Comput. Chem. Eng. **23**(2), 187–202 (1998)
30. Morari, M., Lee, J.H.: Model predictive control: past, present and future. Comput. Chem. Eng. **23**(4–5), 667–682 (1999)
31. Qin, S.J., Badgwell, T.A.: A survey of industrial model predictive control technology. Control Eng. Pract. **11**(7), 733–764 (2003)
32. Canale, M., Fagiano, L., Milanese, M.: Set membership approximation theory for fast implementation of model predictive control laws. Automatica **45**(1), 45–54 (2009)
33. Song, Y., Chen, Z.Q., Yuan, Z.Z.: New chaotic PSO-based neural network predictive control for nonlinear process. IEEE Trans. Neural Netw. **18**(2), 595–600 (2007)

A Machine Vision Method for Automatic Circular Parts Detection Based on Optimization Algorithm

Wenyan Wang[1], Kun Lu[1], Rui Hong[1], Peng Chen[3], Jun Zhang[4], and Bing Wang[1,2(✉)]

[1] School of Electrical and Information Engineering, Anhui University of Technology, Ma'anshan 243002, China
wangbing@ustc.edu
[2] Key Laboratory of Metallurgical Emission Reduction and Resources Recycling, Ministry of Education, Ma'anshan 243002, China
[3] College of Electrical Engineering and Automation, Anhui University, Hefei, Anhui, China
[4] Institute of Health Sciences, Anhui University, Hefei, Anhui, China

Abstract. Circle Hough transform (CHT) is the most commonly used method to inspect circular shapes for its advantage in strong robustness. However, it requires large amounts of storage and computing power, which cannot meet the requirements of real-time processing. To overcome this deficiency, this paper presents a novel circle detection method based on adaptive artificial fish swarm algorithm (AAFSA) by determining the circle center and the radius of circular parts. A new fitness function had been developed to evaluate the similarity of a candidate circle with a real circle. Based on the fitness values, a batch of encoded candidate circles is modified through the AAFSA in order that they can match with the actual circles on the edge map. Experiments results show our proposed method can accurately detect circular parts by search the optimum values in the parameter space. Compared to other popular approaches, i.e., the CHT, the least square method (LS) and the random sample consensus (RANSAC), the proposed method achieved a remarkable improvement in both accuracy and speed of circular parts detection.

Keywords: AAFSA · Circular parts detection · Computer vision · Hough transform

1 Introduction

With the development of industrial techniques, mechanical processing show a higher demand in machining precision, speed and automation level, which caused that traditional manual detection methods cannot meet the requirements of product quality and quantity. Computer vision-based measurement methods featured with easy operation, strong capacity of resisting disturbance, and high precision has been widely used in the measurement of different workpieces. Detecting and localizing circular parts from a digital image is a significant and basic work in the field of mechanical manufacturing, especially for some industrial applications with high automatic level [1, 2].

D.-S. Huang et al. (Eds.): ICIC 2017, Part I, LNCS 10361, pp. 600–611, 2017.
DOI: 10.1007/978-3-319-63309-1_53

Lots of industrial parts are designed in circular, such as bearing, flange, friction plate and oil seal. Usually the center and the radius of circular objects can be extracted to determine the quality of the circular parts. Currently, many methods are utilized to inspect circles based on least-squares or multiple chords for the advantages of high real-time and easy implementation [3]. However, these approaches are suffering from many shortcomings, such as poor stability and robustness, the noise sensibility, low precision and so on. Therefore, some optimization-based methods had been adopted to address this problem, which is one of classical strategies in application area [4–12].

The Hough transform (HT) technique has been widely used in the field of machine vision for detecting straight lines and circles in a digital image, which was first put forward by Duda and Hart [13, 14]. In HT-based methods, a set of feature points in image space had been transformed into a parameter space where a set of accumulated votes can be obtained by an edge detection strategy. For each image feature point, votes over all parameter combinations are accumulated in an accumulator array. By searching for the accumulator array with large numbers of votes, the candidate of the target shape, i.e., circle, can be extracted. It can be found that HT is robust to noise, but huge memory and computation cost is a big problem in real-time application [15, 16].

Some studies had been presented to improve the performance of circle detection by developing new computational strategies for HT algorithm. Xu et al. proposed a randomized HT (RHT) where randomly selects three edge points are selected randomly and mapped into one point, which can reduce computational time and storage requirement significantly [17]. Chen and Chung improved Xu's method by randomly selecting four pixels. But the detection accuracy of RHT method is heavily dependent on the number of edge pixels. Therefore, it is hard to detect circles well in an image. Yip et al. came up with parallel edge points to determine the parameters of an ellipse and a circle [18], which reduces the parameter dimension, but it will be invalid for circles with partial occlusion due to the failure of simultaneous appearance of both parallel edge points. Kimme et al. identified circles by means of two-dimensional accumulator [19] and Atherton et al. accelerated the processing using Hough transform filter [20]. However, the performance of these approaches are subject to the limitations of prior knowledge, such as radius range. If there is no enough information available, the detection process will need long time and large amounts of space resources. Moreover, some circles will not be successfully located in complicated images that have several circles with different radius, since circles with large radius will conceal the small ones.

To address these problems, this work proposed a new adaptive artificial fish swarm algorithm optimized Hough transform (AAFSA-HT) detect the circular parts automatically. Firstly, AAFSA-HT finds the single pixel parts edge in the image by an edge detection method. Then, each combination of center and radius has been seen as an artificial fish of a candidate circle over the edge image. Further, a novel fitness function is developed to evaluate the difference of a candidate circle with a real circle. With guidance of the values of fitness function, the encoded candidate circles are modified by the optimization of the AAFSA algorithm to make them matching with the actual circles on the edge map. AAFSA-HT method can also detect multiple circles with different radius and partial occlusion. The experimental results we achieved show the efficiency and the effectiveness of our proposed approach.

The framework of the rest of this paper is organized as follows. Section 2 gives a brief introduction of AFSA while Sect. 3 provides detailed steps for how to implement the proposed method. The applications of this approach to several images of industrial parts are demonstrated in Sect. 4. Section 5 makes a summary about this research work.

2 Artificial Fish Swarm Algorithm

The artificial fish swarm algorithm (AFSA) is a new global optimization algorithm based on the simulation of fish behaviors, such as moving, prey and swarming [21]. Generally, in a body of water, a fish is capable of seeking out the area with more food through individual searching or tagging along with other fish. Therefore, the higher the food concentration of a water area is, the more the fish will be. Therefore, the basic ideal of AFSA is to construct a set of artificial fish in a body of water, and mimic their foraging behaviors to find the place with the most abundant food. Previous studies demonstrated that this kind methodology is an effective and simple optimization strategy with the advantages of the parallel computing, quick convergence and robustness, and has many applications, such as NP problem, numerical analysis, image processing, transportation system, and etc. In this work, an adaptive artificial fish swarm algorithm (AAFSA) is presented to improve the searching accuracy of AFSA.

2.1 Definitions

In AFSA, the behaviors of a fish to find the water area with more food had been transformed to be a optimization problem, where the water are can be seen as a solution space, food concentration is the objective function, and each fish is one solution for this optimization problem. For a set of artificial fish $\{X_i\}$ ($i = 1, 2, \ldots, n$ where n is the number of fish), the *i*th fish can be denoted as a vector $X_i = (x_i^1, x_i^2, \ldots, x_i^p)$, where p is the length of dimension., and the food concentration in its location can be expressed as $Y_i = f(X_i)$, Y indicates the fitness function value. The distance between two artificial fishes can be expressed by $d_{ij} = ||X_i - X_j||$. In AFSA, the status of artificial fishes can be updated by four behaviors, i.e., random moving, preying, swarming and following.

If the current status of artificial fish is X, The random moving behavior (*AF-move*) of artificial fish can be described as follows:

$$X_i^{'} = X_i + \alpha_r V_i \tag{1}$$

where $X_i^{'}$ is the new status of artificial fish set X_i, V_i denotes the visual range of the ith fish, and α_r is random ratio fallen in an interval [0, 1].

The preying behavior (*AF-prey*) will occur if a fish change its location from X_i to X_j if the food concentration Y_j is bigger than Y_i:

$$X_i^{'} = X_i + \alpha_p h \frac{X_j - X_i}{||X_j - X_i||}, \text{ if } Y_i < Y_j \tag{2}$$

where h is the step a fish move, and α_r is random ratio fallen in an interval [0, 1].

The following behavior (*AF-follow*) will occur if the best status in the visual range of a fish X_i^{bst} is better than the current status X_i, which can be described as follows:

$$X_i^{'} = X_i + \alpha_f h \frac{X_i^{bst} - X_i}{||X_i^{bst} - X_i||}, \; Y_i^{bst} > \delta \cdot n_f \cdot Y_i \tag{3}$$

where X_i^{bst} is the best status the *i*th fish can find in its visual range, Y_i^{bst} is the fitness value of, n_f is the number of artificial fishes in the *i*th fish's visual range, and δ is a coefficient of artificial fish congestion degree.

The swarming behavior (*AF-swarm*) will happen if the status in the center of the ith fish's visual range is better than its current status:

$$X_i^{'} = X_i + \alpha_s h \frac{X_i^c - X_i}{||X_i^c - X_i||}, \; Y_i^c > \delta \cdot n_f \cdot Y_i \tag{4}$$

where Y_i^c is the fitness value of the status of the center in the *i*th fish visual range X_i^c, and α_s is a random ratio fallen in an interval [0, 1].

2.2 Adaptive Strategy for Artificial Fish Swarm Algorithm

In the AFSA described above, the visual range V_i, and moving step h are constant during the whole optimization process, which will cause the algorithm easily to fall into local optimum, and results in slow convergence speed and fitness value shock in the late period [22, 23]. To improve the convergence performance of AFSA, this work adopts a self-adaption strategy adjust the visual range V_i, and moving step h. In the early stage, both of them are set a bigger value to get a good global search capability. With the iterations of the algorithm, their values will decrease gradually to raise the local search ability. Therefore, the change of the visual range and move step from *j*th iteration to the next can be shown as the following:

$$V_i^j = V_i^{j+1} \cdot (1 - \frac{N_p}{N_m}) \tag{5}$$

$$h_i^j = \alpha_h h_i^{j+1} \tag{6}$$

where N_p and N_m denote iterated numbers and maximum iteration numbers respectively; α_h denotes the attenuation coefficient whose value falls between 0 and 1.

3 The Hough Transform Based on AAFSA

In this research, an edge detection method is firstly applied to generate a one-pixel-width contour for the circular parts in the original images. Here, a classical Sobel operator is adopted to implement edge detection. All the edge points are then stored in a vector array $I=\{(x_1, y_1), (x_2, y_2), \ldots, (x_n, y_n)\}$, where $(x_i, y_i)(i = 1, 2, \ldots, n)$

is the coordinate of *i*th edge point in the edge map and *n* is the total number of edge points. Normally, a circle in R^2 can be described by the following equation:

$$(x-a)^2 + (y-b)^2 = r^2 \tag{7}$$

where parameters *a* and *b* represent the center of the circle, and *r* stands for the circle radius. Accordingly, we use the combination of these three parameters as the artificial fish of candidate circles over the edge image. To calculate the fitness value of each individual *C* = (*a*, *b*, *r*) in the population, a new fitness function *Y*(*C*) is derived as shown in Eq. (8). The primary function of *Y*(*C*) is to accumulate the number of desired edge pixels that are located on the circle candidate *C*.

$$Y(C) = \sum_{i=1}^{n} E(C, x_i, y_i) \tag{8}$$

where $E(C, x_i, y_i)$ is to determine whether an edge pixel lies on the circle *C*, and can be defined as:

$$E(C, x_i, y_i) = \begin{cases} 1 & \text{if } \left| \sqrt{(x_i - a)^2 + (y_i - b)^2} - r \right| \leq \varepsilon \\ 0 & \text{otherwise} \end{cases} \tag{9}$$

where ε is the compensation of image digitization and quantification, which can is a threshold for detection error. Through lots of trials and analyses, we noticed that taking $\varepsilon = 3$ is suitable enough for this practical application.

3.1 Algorithm Implementation

Our purpose is to maximize Y(C) because a larger value of Y(C) inidicates a better response of the circularity operator. Each step of the whole detection algorithm is presented as follows:

1) Parameter settings: Set up the population size N, the congestion factor M_{th}, the perceiving range V, the maximum iteration times N_m.
2) Fish swarm initialization: A total number of N artificial fishes are generated to constitute the initial fish group and they are all randomly distributed in the search space.
3) Fitness calculating: The fitness value of each artificial fish is calculated through Eqs. (8) and (9), and the optimal one and its status are recorded in the bulletin.
4) Behavior assessment: Implementing artificial fish behaviors, i.e., AF-move, AF-prey, AF-swarm, and AF-follow, and then updating the fitness value and fish position according to Eqs. (8) and (9).
5) Updating the bulletin board: For each iteration, bulletin board will be updated if the fitness value of behaviors is larger than the records in bulletin board. Otherwise keep the bulletin board is unchanged.

6) Updating Visual and Step: The visual range and the move step are changed according to the Eqs. (5) and (6).
7) Output: The optimal solution is acquired when the cut-off condition is reached, such as the maximum iteration number or other termination change of two consecutive steps is smaller than a pre-defined threshold. Otherwise go to step (4).

3.2 Multiple Circle Detection

There are many cases that multiple circles should be detected in a single situation, and some circle detection method has been employed to find multiple circles. N. Dong et al. proposed a Chaotic Hybrid Algorithm (CHA) to detect multiple circles using a combination optimization methods based on particle swarm optimization and genetic algorithm [24, 25]. They identify multiple species within a population and determine the neighborhood best for each species. However, the exact number of circles and the distance between their centers are provided in advance. Our algorithm is also applied to multiple circle detection with slight modification. Firstly, the AAFSA algorithm is implemented over the edge-only image until convergence is achieved for the first circle detection. Then this shape is masked on the original edge-map. The AAFSA circle detector will be implemented again on the masked edge image. The procedure is repeated until the fitness value reaches a minimum predefined threshold M_{th}. Finally, a validation over all detected circles is performed by analyzing continuity of the detected circumference segments.

4 Simulation and Discussion

Several experiment tests have been developed to evaluate the performance of circle detection in different situation, i.e., single circle part detection, noised circle image detection, and multiple circle parts detection. All of the test images are collected from practical applications.

Table 1 shows the experimental parameters which have been experimentally determined and remained unchanged through all experiments.

Table 1. Parameter setup for AAFSA

Iterations	V	Try_number	N	δ	α
50	60	10	40	0.2	0.95

4.1 Single Circle Detection

As shown in Fig. 1(a), 552 × 552 image of cutting disk with incompletely close edge is acquired to verify the performance of the proposed algorithm. For the purpose of comparison, we apply the Circle Hough transform (CHT), the least square method (LS), the random sample consensus (RANSAC) and our proposed adaptive artificial fish swarm algorithm-Hough transform (AAFSA-HT) to this circular parts with incompletely close edge individually. Figure 1(a)–(d) displays the detection results after applying the four methods to such image. The experiment was repeated for 50 times and the average detection results are shown in Table 2.

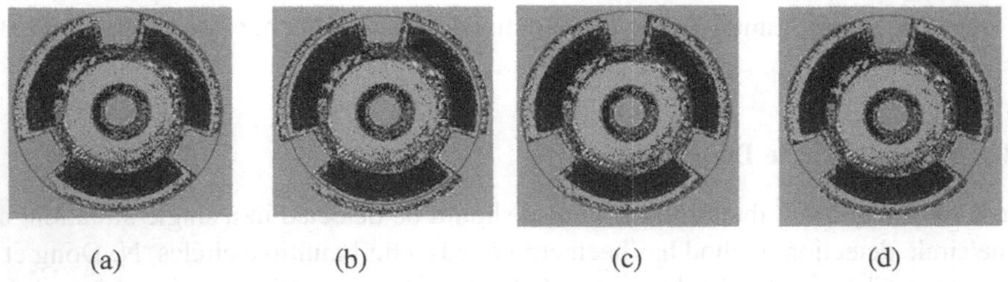

Fig. 1. (a) The detection result of proposed method. (b) The detection result of LS. (c) The detection result of CHT. (d) The detection result of RANSAC.

Table 2 shows the average running time and value of detected center and radius by using CHT, LS, RANSAC and the proposed method 50 times separately. From the table, we can see that the running speed of the algorithm in this paper is significantly improved by comparison with CHT and the detected parameters are relatively close to the ideal values, which can meet the requirements of precision and real-time detection.

Table 2. The estimated circle parameters of cutting disk

Parameters	Ideal value	The proposed method	LS	CHT	RANSAC
a	274.3301	274.5171	273.1492	274.0000	270.6586
b	277.5047	276.3899	278.2079	277.0000	272.9031
r	198.7120	198.0348	183.3936	199.0000	193.8937
Running time (s)	–	1.8752	0.5122	38.5633	6.2838

Figure 2 is the iterative curve of fitness function of adaptive artificial fish swarm algorithm. As is shown in the graph, we can observe that the proposed method has been convergence at about 8th times, which proves the speediness and validity of the method.

4.2 Circle Detection with Noise

Another test is carried out to assess the ability of resisting noise of our algorithm. In order to evaluate the accuracy of the test results, two related parameters (*ES* and *SR*) are defined. At first, the circle in the original edge map is detected manually and the parameters is indicated as (x_{GT}, y_{GT}, r_{GT}). The center and radius of the detected circle found by algorithm can be expressed as (x_D, y_D, r_D). Therefore, we can define an error score (*ES*) in the following way,

$$ES = \eta(|x_{GT} - x_D| + |y_{GT} - y_D|) + \mu|r_{GT} - r_D| \tag{10}$$

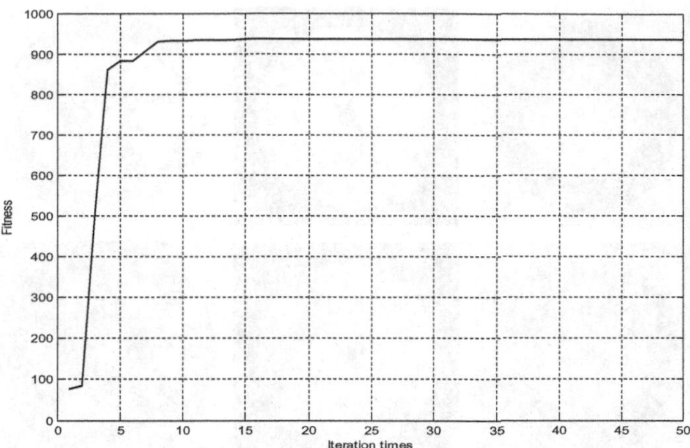

Fig. 2. Fitness function iterative curve of adaptive artificial fish swarm algorithm

The first term signifies the shift of center of the detected circle by comparing with the benchmark circle and the second term accounts for the difference between their radii. η and μ are two weights associated with each term in Eq. (10) and they are chosen according to the required accuracy as $\eta = 0.05$ and $\mu = 0.1$. Such particular choice of parameters ensures that radii difference gets priority to the difference of the centers of the manually detected circle and the machine-detected circle. If the *ES* is less than 1, the algorithm gets a success, otherwise we say that it has failed to detect the edge-circle. Note that for $\eta = 0.05$ and $\mu = 0.1$, $ES < 1$ indicates the maximum difference of radius tolerated is 10 pixels while the maximum mismatch for the center location can be up to 20 pixels. From this consideration, success rate (*SR*) can be defined as percentage of reaching success in certain number of trials.

In this test, the percentage of density of the salt & pepper noise was increased from 0 to 40 percent and the success rate of the proposed method was counted. Figure 3 shows the result of applying the proposed algorithm on three noisy images. The circular objects was detected correctly when 30% salt & pepper noise was added into the image. And we draw the graph about the success rate versus noise density for oil seal seen in Fig. 4. We can see from the graph that the noise with less than 35% density has no effect on the algorithm's accuracy and the algorithm resists against noise until this degree. In contrast the CHT loses robustness against noise from 25%. After this threshold, the success rate decreases exponentially for both algorithm but sharper for CHT, and it failed totally when the noise density exceeds 55% for the AAFSA and 30% for the CHT. Hence, the noise resistance performance of the AFSA-based method was superior to CHT. Table 3 shows the averaged error score for the AFSA and CHT, for the three circular parts shown by Fig. 3.

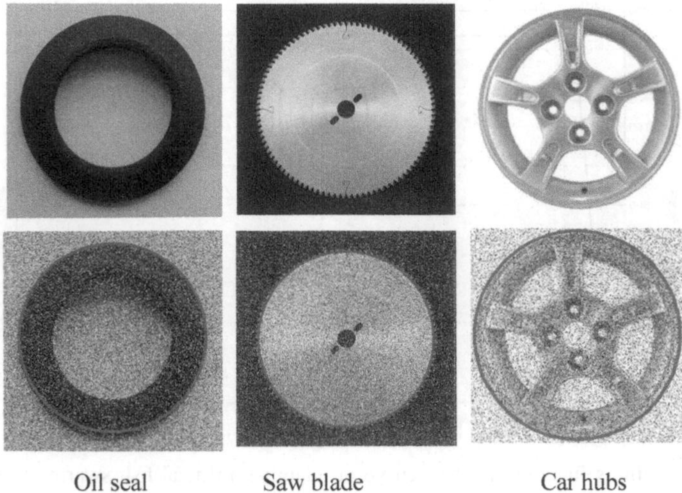

Fig. 3. Detection result of the proposed method for 30% salt & pepper noise

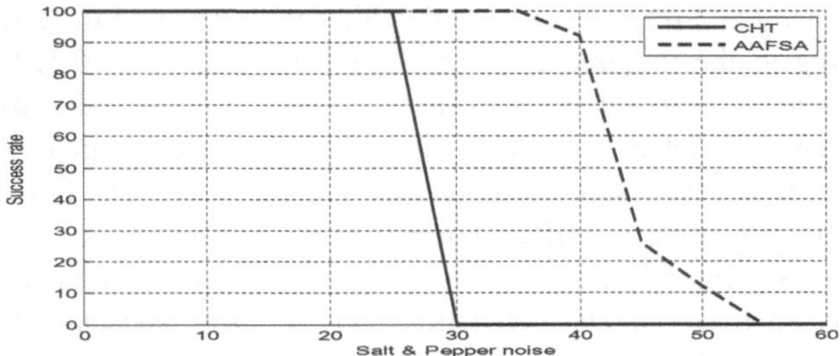

Fig. 4. Resistance of the proposed method and the CHT against salt & pepper noise

Table 3. The ES for the different noise density

Noise density (%)		Image		
		Oil seal	Saw blade	Car hubs
AFSA	0	0.2767	0.2319	0.3603
	10	0.3237	0.1603	0.2175
	20	0.3874	0.0688	0.4186
	30	0.4118	0.0638	0.3026
	40	0.5897	0.3532	0.5389
CHT	0	0.15	14.1	0.15
	10	0.15	14.15	0.1
	20	0.15	14.1	16.3
	30	18.2	14.2	21.05
	40	19.2	14.15	10.5

4.3 Multiple Circle Detection

The applications of the proposed method to the multiple circle detection in industrial parts are demonstrated in Fig. 5. Figure 5(a) shows the two flange gaskets with partial occlusion and Fig. 5(c) displays the two intersecting rubber gaskets. The detected results are shown in Fig. 5(b) and (d) respectively, in which the multiple circles are located accurately.

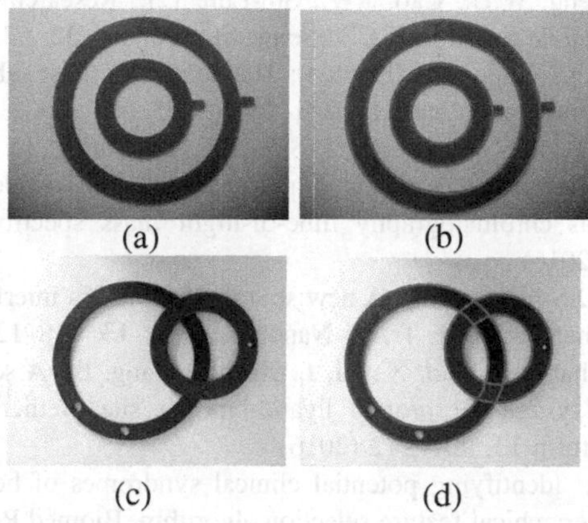

Fig. 5. Some mechanical parts with circular shapes. (a) Flange gasket with partial occlusion. (b) The detected circles of (a). (c) Two intersecting rubber gaskets. (d) The detected circles of (c).

5 Conclusions

In order to improve the speed and precision of circular parts detection, an improved Hough transform method is proposed in this paper, which is based on the adaptive artificial fish swarm algorithm. Firstly, we present the brief introduction of the AFSA and then apply an optimal AFSA to Hough transform successfully. The measured value of circular parts can be accurately obtained by using AAFSA to search the optimum parameters in the parameter space. The algorithm consumes less computing resources for it does not need to traverse the whole search space. Lastly, we make the simulations according to our setting parameters and compare with CHT, LS and RANSAC. Experimental results indicate that the proposed method has a remarkable improvement in both accuracy and speed of circular parts detection, which has important significance in the field of mechanical manufacturing and production.

Acknowledgement. This work is supported National Natural Science Foundation of China under Grant Nos. 61472282 and 61672035, Natural Science Foundation of Anhui Province under Grant No.1508085MF129,and the Key Laboratory of Metallurgical Emission Reduction & Resources Recycling (Anhui Uinversity of Technology, Minisitry of Education) under No. KF17-02.

References

1. Davies, E.R.: Design of cost-effective systems for the inspection of certain food products during manufacture. In: Proceeding of. 4th Conference on Robot Vision and Sensory Controls, pp. 437–446 (1984)
2. Davies, E.R.: Radial histograms as an aid in the inspection of circular objects. In: IEEE Proceedings of the Control Theory and Applications [see also IEEE Proceedings-Control Theory and Applications], vol. 132, pp. 158–163 (1985)
3. Jin, T.U., Sheng-Deng, W.U., Cao, J.W., Shi-Lun, L.I.: Research on measure method of diameter of partial circle based on CCD. Semicon. Technol. **32**, 573–574 (2007)
4. Mi, J.X., Huang, D.S., Wang, B., Zhu, X.J.: The nearest-farthest subspace classification for face recognition. Neurocomputing **113**, 241–250 (2013)
5. Wang, B., Shen, H., Fang, A., Huang, D.S., Jiang, C., Zhang, J., Chen, P.: A regression model for calculating the second dimension retention index in comprehensive two-dimensional gas chromatography time-of-flight mass spectrometry. J. Chromatogr. A **1451**, 127–134 (2016)
6. Wang, B., Huang, D.S., Jiang, C.J.: A new strategy for protein interface identification using manifold learning method. IEEE Trans. Nanobioscience **13**, 118–123 (2014)
7. Chen, P., Hu, S., Zhang, J., Gao, X., Li, J., Xia, J., Wang, B.: A sequence-based dynamic ensemble learning system for protein ligand-binding site prediction. IEEE/ACM Trans Comput Biol Bioinform **13**, 901–912 (2016)
8. Ji, Z.W., Wang, B.: Identifying potential clinical syndromes of hepatocellular carcinoma using PSO-based hierarchical feature selection algorithm. Biomed Res. Int. **2014**, 12 (2014)
9. Zhao, Y., Zhang, J., Wang, B., Kim, S.H., Fang, A., Bogdanov, B., Zhou, Z., McClain, C., Zhang, X.: A method of calculating the second dimension retention index in comprehensive two-dimensional gas chromatography time-of-flight mass spectrometry. J. Chromatogr. A **1218**, 2577–2583 (2011)
10. Chen, P., Liu, C., Burge, L., Li, J., Mohammad, M., Southerland, W., Gloster, C., Wang, B.: DomSVR: domain boundary prediction with support vector regression from sequence information alone. Amino Acid **39**, 713–726 (2010)
11. Wang, B., Chen, P., Wang, P., Zhao, G., Zhang, X.: Radial basis function neural network ensemble for predicting protein-protein interaction sites in hetero complexes. Protein Pept Lett **17**, 1111–1116 (2010)
12. Xia, S., Chen, P., Zhang, J., Li, X.P., Wang, B.: Utilization of rotation-invariant uniform LBP histogram distribution and statistics of connected regions in automatic image annotation based on multi-label learning. Neurocomputing **228**, 11–18 (2017)
13. Hough, P.V.C.: Method and means for recognizing complex patterns (1962)
14. Duda, R.O., Hart, P.E.: Use of the Hough transformation to detect lines and curves in pictures. Commun. ACM **15**, 11–15 (1972)
15. Chien, C.F., Cheng, Y.C., Lin, T.T.: Robust ellipse detection based on hierarchical image pyramid and Hough transform. J. Opt. Soc. Am. A: **28**, 581–589 (2011)
16. Teng, A.Z., Kim, J.H., Kang, D.J.: Ellipse detection: a simple and precise method based on randomized Hough transform. Optical Eng. **51**, 84 (2012)
17. Xu, L., Oja, E., Kultanen, P.: A new curve detection method: Randomized Hough transform (RHT). Pattern Recogn. Lett. **11**, 331–338 (1990)
18. Yip, R.K.K., Tam, P.K.S., Leung, D.N.K.: Modification of hough transform for circles and ellipses detection using a 2-dimensional array. Pattern Recogn. **25**, 1007–1022 (1992)
19. Kimme, C.: Finding circles by an array of accumulators. Commun. ACM **18**, 120–122 (1975)

20. Atherton, T.J., Kerbyson, D.J.: Size invariant circle detection. Image Vis. Comput. **17**, 795–803 (1999)
21. Li, X.L., Shao, Z.J., Qian, J.X.: An optimizing method based on autonomous animate: Fish swarm algorithm. Syst. Eng. Theor. Pract. **22**, 32–38 (2002)
22. Gao, Y., Guan, L., Wang, T.: Optimal artificial fish swarm algorithm for the field calibration on marine navigation. Measurement **50**, 297–304 (2014)
23. Yong, P., Tang, G.L., Xue, Z.C.: Optimal operation of cascade reservoirs based on improved artificial fish swarm algorithm. Syst. Eng. Theor. Pract. **31**, 1118–1125 (2011)
24. Dong, N., Wu, C.H., Ip, W.H., Chen, Z.Q., Chan, C.Y., Yung, K.L.: An opposition-based chaotic GA/PSO hybrid algorithm and its application in circle detection. Comput. Math. Appl. **64**, 1886–1902 (2012)
25. Kelly, M., Levine, M.: Finding and Describing Objects in Complex Images: Advances in Image Understanding, pp. 209–225. IEEE Computer Society Press, Baco Raton (1997)

Univariate Gaussian Model for Multimodal Inseparable Problems

Geng Zhang[1], Yangmin Li[2], Bingxiao Ding[3], and Yun Li[1,4(✉)]

[1] School of Computer Science and Network Security, Dongguan University of Technology, Dongguan 523808, China
zggyd@163.com

[2] Department of Industrial and Systems Engineering, The Hong Kong Polytechnic University, Hung Hom, Hong Kong
yangmin.li@polyu.edu.hk

[3] Faculty of Science and Technology, Department of Electromechanical Engineering, University of Macau, Macao, China
bingxding@hotmail.com

[4] School of Engineering, University of Glasgow, Glasgow G12 8LT, UK
Yun.Li@ieee.org

Abstract. It has been widely perceived that a univariate Gaussian model for evolutionary search can be used to solve separable problems only. This paper explores whether and how the univariate Gaussian model may also be used to solve inseparable problems. The analysis is followed up with experimental tests. The results show that the univariate Gaussian model stipulates no inclination towards separable problems. Further, it is revealed that the model is not only an efficient but also an effective method for solving multimodal inseparable problems. To verify its relative convergence speed, a restart strategy is applied to a univariate Gaussian model (the univariate marginal distribution algorithm) on inseparable problems. The results confirm that the univariate Gaussian model outperforms the five peer algorithms studied in this paper.

Keywords: Univariate model · Inseparable problem · Random sampling · Evolutionary computation

1 Introduction

Optimization has been an active area of research for decades. One challenging problem in this area is the unconstrained optimization problem such as a D-dimensional minimization problem:

$$Min \quad f(X), \quad X = [x_1, x_2, \ldots x_d \ldots, x_D] \tag{1}$$

where $X = [x_1, x_2, \ldots, x_d \ldots, x_D]$ is the vector of variables to be optimized, D is the number of variables.

One effective approach to this problem is the estimation of distribution al- gorithms (EDAs) [1–3], which is a population-based optimization algorithm and has attracted an

D.-S. Huang et al. (Eds.): ICIC 2017, Part I, LNCS 10361, pp. 612–623, 2017.
DOI: 10.1007/978-3-319-63309-1_54

increasing attention during the last decade. Its principle is to explore the space of potential solutions by building and sampling promising candidates ongoing. The most important step of an EDA is the construction of an explicit probabilistic model that attempts to capture the probability distribution of the promising solutions by using tree-structured or Bayesian networks [2]. However, it is challenging to retrieve information needed from the population when interactions among the variables increase [3]. Reference [4] extends a simple univariate model EDA from boolean to continuous search spaces. Since the univariate model EDA builds the probabilistic model with all the variables sampled separately, it is regarded that the univariate model has a preference to separable problems and is inadequate to inseparable continuous problems [2, 6]. However, experimental tests [7] have shown that a univariate model such as that used in the univariate marginal distribution algorithm continuous domains (UMDAc) [8] is not restricted to separable problems and it can be effective in dealing with certain inseparable problems.

This paper studies the above phenomenon. In Sect. 2, we study correlation among multiple variables in UMDAc in solving multimodal inseparable problems. Section 3 analyzes effectiveness of this model in dealing with multimodal inseparable problems. Based on the analyses, Sect. 4 applies a simple restart strategy to the UMDAc in order to verify its performance in comparison with other peer algorithms. Section 5 conducts benchmark tests quantitatively to compare the performance of the RUMDAc with that of peer algorithms in the literature. Conclusions are drawn in Sect. 6.

2 Univariate Gaussian Model

EDAs are iterative optimization algorithms that explore the space of candidate solutions by sampling an explicit probabilistic model constructed from promising solutions found that far at iteration. EDAs typically work with a population of candidate solutions to the problem, starting with a population generated from a uniform distribution throughout the admissible solution space [2]. Then, individuals in the population are ranked according to their performance. Based on the ranking, a number of most promising solutions are selected to construct a probabilistic model to estimate the probability distribution of the selected solutions. New solutions are then obtained by sampling the distribution encoded by this probabilistic model. These new solutions are then incorporated back into the old population. This process is iterated until a termination criterion is met.

A critical step in EDAs is the construction of a probability model that captures the features of the promising solutions. When the univariate model of UMDAc is extended to solve problems of a continuous search space, the probability distribution that is used to model each vector element is expressed as single Gaussian distribution defined as follows:

$$N(u_d, \sigma_d) = \frac{1}{\sqrt{2\Pi}\sigma_d} e^{-\frac{(x_d-u_d)^2}{2\sigma_d^2}} \tag{2}$$

where $N(u_d, \sigma_d)$ stands for a Gaussian distribution of vector element x_d with mean u_d and standard deviation σ_d. The mean value u_d and standard deviation σ_d are defined as

$u_d = \frac{1}{N}\sum_{n=1}^{N} x_{dn}$ and $\sigma_d = \sqrt{\frac{1}{N}\sum_{n=1}^{N}(x_{dn} - u_d)^2}$. Then the new solutions are sampled under this Gaussian distribution. Figure 1 shows the pseudocode of UMDAc that adopts the univariate Gaussian model.

```
Define the optimization problem and initialize M candidate solutions
Repeat:
    Evaluate and rank the M solutions
    Select best N solutions sorted as X_1 to X_N
        for each dimension d ∈ [1, D] :
            The dth data set is summarized as mean u_d and standard deviation σ_d
        endfor
    Obtain M new solutions based on the mean u and standard deviation σ
Until Stopping criterion is met
```

Fig. 1. Pseudocode of UMDAc with a univariate Gaussian model

Since the univariate model EDA assumes that there is no correlation among variables and builds the probabilistic model with all the variables are sampled separately, it is treated as the simplest model to be implemented among EDAs. In addition, the simple univariate model keeps a lower level of computational complexity than other complex EDAs. In UMDAc, The complexity of model estimation and solution sampling is $O(DN)$ and $O(DM)$, respectively [7].

However, it is widely perceived that the univariate model has a preference to separable problems and is inadequate to inseparable problems [2, 6]. The probabilistic model cannot be built accurately when dealing with inseparable problems, whereas the fundamental task of EDAs is to search the global optimum of the given optimization problem, rather than to model the structure of the optimization accurately [9]. Meanwhile, experimental test shows that the univariate EDAs such as UMDAc are actually very effective to deal with some multimodal inseparable problems.

In order to evaluate the influence of the correlation to univariate Gaussian model, Table 1 lists the comparison results of UMDAc and two optimization algorithms on 30 dimensional Rastrigin's function and Rotated Rastrigin's function, which have been tested 30 times. The expressions of Rastrigin's function and Rotated Rastrigin's function [10] are listed in Table 2 of Sect. 5. The two variants of particle swarm optimization (PSO) algorithms [11] are cooperative particle swarm optimization-hybrid (CPSO-H_k) [12] and comprehensive learning particle swarm optimization (CLPSO) [10], which have an inclination to separable problems. Here we set the parameter k = D in CPSO-H_k to facilitate the algorithm for solving separable problems. The other parameters of CPSO-H_k and CLPSO are set according to the corresponding reference [10, 12]. As for UMDAc, we set the parameters M = 200 and N = 40 and the termination criterion is a maximum number of 200,000 fitness evaluations (FEs).

From Table 1, we can see that the UMDAc has no advantage on Rastrigin's function when compared with some algorithms that are inclined to separable problems. However, it performs the best on Rotated Rastrigin's function. The results indicate that the UMDAc is less affected by the correlation among the multiple variables.

Table 1. Comparison results of UMDAc, CPSO-H_k and CLPSO on 30 dimensional Rastrigin's function and rotated Rastrigin's function

Functions	CPSO-H_k	CLPSO	UMDAc
Rastrigin's function	2.24e − 012 ± 1.67e − 012	**5.78e − 013 ± 4.45e − 013**	4.21e + 000 ± 2.44e + 000
Rotated Rastrigin's function	6.35e + 001 ± 4.55e + 001	3.94e + 001 ± 5.23e + 000	**7.35e + 000 ± 3.21e + 000**

3 Working Principle of the Univariate Gaussian Model

In order to explain this phenomenon, this section analyzes the working principle of the univariate Gaussian model in the continuous search space. Figures 2, 3, 4 and 5 illustrate the entire process in detail. Figure 2 shows the contour map of an inseparable 2-D Rotated Rastrigin's function. The different colors stand for different fitness values of $f(X)$. It is clear that there are many local optima among which global optimum is located. Figure 3 shows an initialization of M sampling points (candidate solutions) inside the search area and N relatively better ones are chosen. The initialization is performed by uniformly sampling on each dimension x_d. Since better candidate solutions represents better search areas, the chosen N better candidate solutions need to be generalized to form a new search area as shown in Fig. 4. The generalization of the chosen better candidate solutions is expressed as mean value $u = [u_1, \ldots, u_d, \ldots u_D]$ and standard deviation $\sigma = [\sigma_1, \ldots, \sigma_d, \ldots, \sigma_D]$. As for each dimension x_d, the generalization is summarized as the dth mean value and the dth standard deviation σ_d. The mean value u_d and standard deviation σ_d are defined as $u_d = \frac{1}{N}\sum_{n=1}^{N} x_{dn}$ and $\sigma_d = \sqrt{\frac{1}{N}\sum_{n=1}^{N}(x_{dn} - u_d)^2}$.

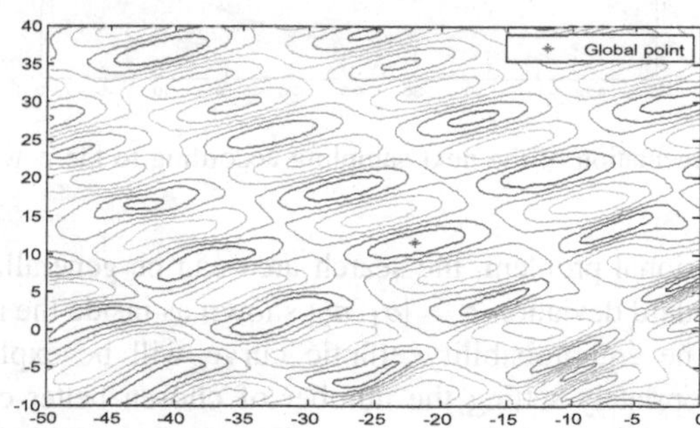

Fig. 2. Contour map of the 2-D rotated Rastrigin's function.

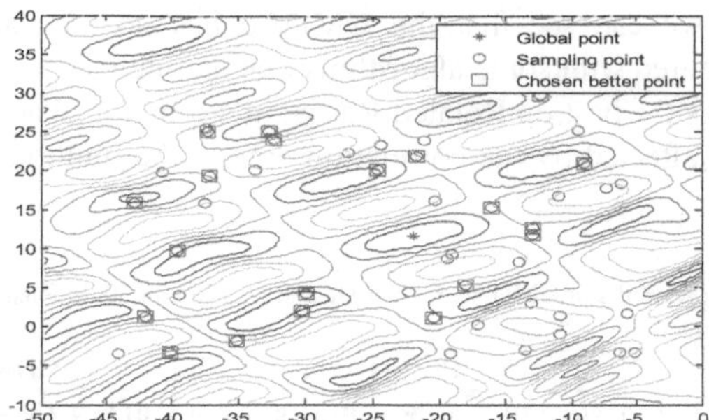

Fig. 3. Sampling a certain number of M candidate solutions in the search area and selecting N better ones.

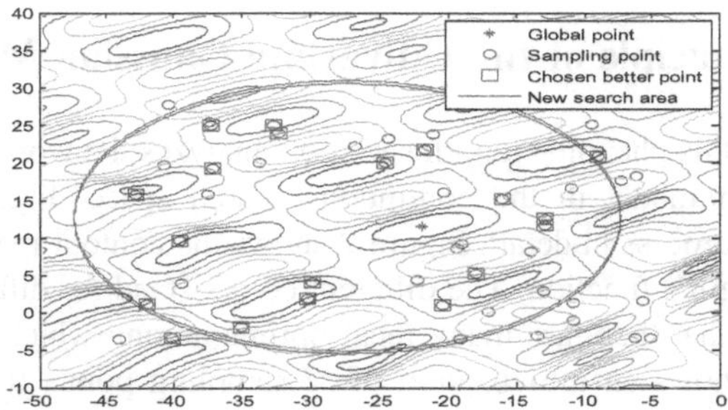

Fig. 4. Generalization of the chosen better points as the new search area which is inside the red ellipse. (Color figure online)

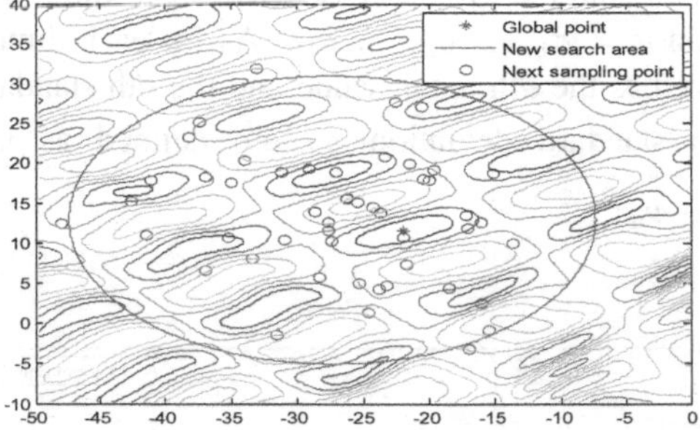

Fig. 5. Implementation of the next sampling according to the new search area.

As a two dimensional problem, the search area can be generalized as mean value $u = [u_1, u_2]$ and standard deviation $\sigma = [\sigma_1, \sigma_2]$ shown as inside the red equiprobability ellipse in Fig. 4. The equiprobability elliptic curve will be explained in the next paragraph. By appropriately setting the number of chosen better candidate solutions N which is smaller than the total number of candidate solutions M, the new search area is smaller than the original one. Based on the new search area, random sampling M candidate solutions is implemented as shown in Fig. 5. The M new candidate solutions are obtained by sampling on each dimension x_d with normal distribution according to the dth mean value u_d and the dth standard deviation σ_d. That is, the normal distribution on each dimension x_d of the new candidate solutions follows:

$$x_d \sim u_d + \sigma_d \cdot N(0,1) \quad for \quad d = 1, \ldots, D \tag{3}$$

where $\sim$ denotes the same distribution on the left and right side, $N(0,1)$ is the standard normal distribution.

If the same process of sampling and selecting is implemented on smaller search area, the search area will get smaller and smaller and finally converge to a point. Based on the above analysis, the whole process is as follows: (1) a certain number of M candidate solutions are created by sampling and N better ones are selected; (2) the N better ones are generalized to be a new promising search area because better ones represent that the areas around them are better; (3) based on the new search area, the process of sampling new candidate solutions and selecting is repeated until some criteria are met.

The procedure of univariate Gaussian model is simple because it only contains the process of sampling and selecting and there are no other learning strategies attached. A distinctive feature is that better candidate solutions are generalized to be a promising search area, which prevents the premature convergence of candidate solutions and makes random sampling possible. Since the D-dimensional better solutions are only used to confine search space, they can be treated as D 1-dimensional data sets and the complex multivariate normal distribution is ignored.

As for each 1-dimensional data set, it is suitable to summarize the data as normal distribution expressed in terms of mean and standard deviation because of the random sampling. For example, assuming in Fig. 4 there are N better candidate solutions $X_1 = [x_{11}, x_{21}], \ldots, X_n = [x_{1n}, x_{2n}], \ldots, X_N = [x_{1N}, x_{2N}]$ with mean value $u = [u_1, u_2]$, these 2-dimensional candidate solutions can be treated as two 1-dimensional data sets $[x_{11}, \ldots x_{1n}, \ldots x_{1N}]$ and $[x_{21}, \ldots, x_{2n}, \ldots x_{2N}]$. Since each data set is summarized in normal distribution, the joint distribution of two dimensional case forms equiprobability elliptic curve as described in Fig. 4. Based on mean value and standard deviation of these two data sets $[u_1, \sigma_1]$ and $[u_2, \sigma_2]$, two new 1-dimensional data sets $[x_{new11}, \ldots, x_{new1m}, \ldots x_{new1M}]$ and $[x_{new21}, \ldots, x_{new2m}, \ldots x_{new2M}]$ can be created according to Eq. (3).

The next sampling can be implemented by recombining these two data sets as M new candidate solutions $X_{new1} = [x_{new11}, x_{new21}], \ldots, X_{newm} = [x_{new1m}, x_{new2m}], \ldots, X_{newM} = [x_{new1M}, x_{new2M}]$ as shown in Fig. 5. The red ellipse in Figs. 4 and 5 shows the new search area where the next candidate solutions are mainly focused.

The UMDAc in Fig. 3 starts the random initialization according to uniform distribution on each vector variable x_d. Then the univariate Gaussian distribution is applied to randomly sample each vector variable x_d. Based on the above analysis, we present two aspects to explain why the univariate Gaussian model can be used for solving inseparable problems.

1. Using univariate Gaussian distribution for sampling is similar to the initialization step, the only difference is the initialization step adopts uniform distribution on each vector element. If the initialization is treated as random sampling to inseparable problems, so does the univariate Gaussian distribution. The random sampling can also be observed from Figs. 3 and 5. To sum up, the working principle of univariate Gaussian model is to do random sampling on smaller promising search area, which is formed by better selected candidate solutions. Since the number of selected solutions is smaller than the total number of solutions, the promising search area is reduced and therefore better solutions can always be found. Therefore, the univariate Gaussian model can optimize both separable and inseparable problems.
2. Since inseparable problems contain strong interactions among vector variables, it is inadequate to the method of optimizing each vector with only one variable has been changed [14]. However, the univariate Gaussian model that adopts independent sampling on each vector variable is totally different. The major difference relies on the evaluation of the candidate solutions. References [12–14] use a divide-and-conquer strategy which divides the search space into several smaller ones, similar to a dimensionality reduction method to fix certain variables while optimizing the other. Therefore, it is suitable for separable problems if the division of the search space is accurate. In spite that the variables are sampled separately in univariate Gaussian model, the candidate solutions are evaluated with all the vector variables having been changed according to the mean u and the standard deviation σ and there are no variables fixed in the whole process just as described above. Hence, the univariate Gaussian model gives no preference to separable problems and can be applied to solve inseparable problems.

4 Restart Strategy for the Univariate Gaussian Model

In order to evaluate the univariate model's search performance on inseparable problems, a restart strategy is adopted for the UMDAc considering its high convergence speed [9]. Figure 6 shows the pseudocode of the restart UMDAc (RUMDAc). The restart is determined by the maximum value of the standard deviation σ. If the maximum value of σ is smaller than a predefined threshold (0.001 in Fig. 6), it means the algorithm has converged and the search stops. Therefore, a restart UMDAc should be implemented to find better results. A restart is launched with the model of which the mean value is the same as the previous one and the standard deviation σ is set to be the absolute value of the half search range multiplied by the random vector $rand(1, D)$.

5 Experimental Results and Comparisons

5.1 Test Functions

In order to test the univariate Gaussian model's performance in different multimodal inseparable problems, we choose two separable functions and six inseparable functions of 30 dimensions. Details of these functions are given in Table 2. The minimum value

```
Define the optimization problem and initialize M candidate solutions
Repeat:
    Evaluate and rank the M solutions
    Select best N solutions sorted as X_1 to X_N
    for each dimension d ∈ [1, D] :
        The dth data set is summarized as mean u_d and standard deviation σ_d
    endfor
    Obtain M new solutions based on the mean u and standard deviation σ
    if the maximum value of σ is smaller than 0.001
        Reinitialize the  as σ =0.5*abs(search range)*rand(1,D)
    endif
Until Stopping criterion is met
```

Fig. 6. Pseudocode of the restart UMDAc (RUMDAc)

for all the test functions is set to 0. $f_1(X)$ and $f_2(X)$ are two separable functions, which can be used to compare the results with that of corresponding rotated functions $f_5(X)$ and $f_8(X)$ to show whether the algorithm has a preference to separable functions. The correlation elements among $f_3(X)$ are in sequence and each element has two correlation elements that are close to it. The correlation elements of the other five rotated functions are created randomly by an orthogonal matrix *M*. The new rotated vector $Y = M X$, which is obtained through the original vector *X* left multiplied by orthogonal matrix *M*, performs like the high correlation vector, because all elements in vector *Y* will be affected once one element in vector *X* changes.

5.2 Algorithms Compared

In the following part, we briefly describe five algorithms in the literature to compare with RUMDAc. The first, the fully informed particle swarm (FIPS), one of the variants of PSO [11], uses all neighbor particles to influence the flying velocity [15]. The second, the CPSO-H_k, uses *k* subcomponent swarms to search each subcomponent separately and is combined with the original PSO to improve the performance on multimodal problems [12]. The third, the CLPSO, is proposed for solving multimodal problems and its learning strategy is to use all the other particles' best historical information to update one particle's velocity [10]. The fourth algorithm is the harmony search (HS), a popular algorithm inspired from music improvisation [16]. The fifth is the covariance matrix adaptation- evolution strategy (CMA-ES), which uses the adaptation of covariance matrix to learn appropriate mutation distribution and increase the searching ability [17, 18].

The parameters of the above algorithms are adopted according to their corresponding references. As for RUMDAc, increasing *M* can improve the searching performance to most of the test functions. But it consumes more FEs to converge and the

Table 2. Test functions

Name	Test function	Search range
Rastrigin	$f_1(X) = \sum_{i=1}^{D} x_i^2 - 10\cos(2\Pi x_i) + 10$	$[-5.12, 5.12]^D$
Schwefel	$f_2(X) = 418.9829 \times D - \sum_{i=1}^{D} x_i \sin(\lvert x_i \rvert^{\frac{1}{2}})$	$[-500, 500]^D$
Rosenbrock	$f_3(X) = \sum_{i=1}^{D-1} 100(x_{i+1} - x_i^2)^2 + (x_i - 1)^2$	$[-2, 2]^D$
Rot. Ackley	$f_4(X) = \sum_{i=1}^{D-1} 100(x_{i+1} - x_i^2)^2 + (x_i - 1)^2$ where: $Y = M \cdot X$, M is the orthogonal matrix	$[-32, 32]^D$
Rot. Rastrigin	$f_5(X) = \sum_{i=1}^{D} x_i^2 - 10\cos(2\Pi x_i) + 10$ where: $Y = M \cdot X$, M is the orthogonal matrix	$[-5.12, 5.12]^D$
Rot. Noncon. Rastrigin	$f_6(X) = \sum_{i=1}^{D} x_i^2 - 10\cos(2\Pi x_i) + 10$ where: z_i is expressed as: $z_i = \begin{cases} y_i & \lvert y_i \rvert < \frac{1}{2} \\ \frac{round(2y_i)}{2} & \lvert y_i \rvert > = \frac{1}{2} \end{cases} \quad for \quad i = 1, 2, \ldots, D$ $Y = M \cdot X$, M is the orthogonal matrix	$[-5.12, 5.12]^D$
Rot. Weierstrass	$f_7(X) = \sum_{i=1}^{D} (\sum_{k=0}^{kmax} [a^k \cos(2\Pi b^k (y_i + 0.5))]) - D \sum_{k=0}^{kmax} [a^k \cos(\Pi b^k)]$ where: $Y = M \cdot X, \quad a = 0.5, \quad b = 3, \quad kmax = 20,$ M is the orthogonal matrix	$[-0.5, 0.5]^D$
Rot. Schwefel	$f_8(X) = 418.9829 \times D - \sum_{i=1}^{D} x_i \sin(\lvert Z_i \rvert^{\frac{1}{2}})$ where z_i is expressed as: $z_i = \begin{cases} y_i \sin(\lvert y_i \rvert^{\frac{1}{2}}) & if \quad \lvert y_i \rvert < 500 \\ 0 & otherwise \end{cases} \quad for \quad i = 1, 2, \ldots, D$ $Y = Y' + 420.96.,\quad Y' = M * (X - 420.96)$. M is an orthogonal matrix	$[-500, 500]^D$

computational complexity is increased correspondingly as the complexity of solution sampling is $O(DM)$. In general, smaller N can obtain better results to most of the test functions. However, it will cause the algorithm premature convergence if the value N is too small. Therefore, based on experimental test, the parameters of RUMDAc are roughly set as $M = 200$ and $N = 40$. The stopping criterion is measured by the maximal number of FEs, which is set 200,000 for 30 dimensions.

5.3 Results and Discussions

In order to compare RUMDAc fairly with the five peer algorithms, all algorithms are implemented and run 30 times on the eight test functions. Table 3 shows the means and variances of the six algorithms on the eight test functions with dimensions D = 30. The best result for each test function is shown in bold. From Table 3, we can see that the FIPS performs well on some multimodal inseparable functions like f_4 and f_7. In addition, the algorithm has not been affected too much by rotation, which shows that the algorithm has no preference to separable problems. However, the poorly performance of FIPS on f_5 and f_6 demonstrates the algorithm is highly dependent on different test functions.

By comparing the results of separable functions f_1 and f_2 with that of corresponding inseparable functions f_5 and f_8, the three algorithms, CPSO-H_k, CLPSO and HS, have strong preference to separable problems. The CMA-ES obtains the best results on functions f_3 and f_7. Meanwhile, it works well on all the inseparable functions and the

Table 3. Results for the 30 dimensional problems.

Functions	FIPS	CPSO-H_k	CLPSO
f_1	7.34e + 001 ± 1.92e + 001	**6.45e−012 ± *3.89e-012***	5.99e−010 ± *4.50e−010*
f_2	2.07e + 003 ± 1.03e + 003	3.12e + 002 ± *2.10e* + 002	**1.22e−012 ± *7.51e−013***
f_3	2.39e + 001 ± *8.03e* + 000	2.18e + 001 ± *7.97e* + 000	1.74e + 001 ± 1.68e + 001
f_4	2.87e−004 ± *2.74e−004*	1.63e + 000 ± *1.80e* + 000	**2.35e−005 ± *2.78e−005***
f_5	1.09e + 002 ± *3.11e* + 001	8.53e + 001 ± *6.88e* + 001	3.45e + 001 ± *5.82e* + 000
f_6	1.27e + 002 ± *4.23e* + 001	7.42e + 001 ± *3.71e* + 001	3.73e + 001 ± 1.12e + 001
f_7	7.72e−002 ± *3.58e−002*	1.31e + 001 ± *6.02e* + 000	4.26e + 000 ± 1.35e + 000
f_8	2.69e + 003 ± 1.40e + 003	3.74e + 003 ± 1.69e + 003	2.18e + 003 ± *8.03e* + 002
Functions	**HS**	**CMA-ES**	**RUMDAc**
f_1	1.36e + 000 ± *5.43e−001*	4.59e + 000 ± 2.49e + 000	1.23e + 000 ± *3.60e−001*
f_2	7.65e−006 ± *3.61e−006*	3.81e + 003 ± *1.29e* + 003	1.28e + 003 ± *3.85e* + 002
f_3	3.03e + 001 ± *5.11e* + 000	**1.33e−026 ± *6.39e−027***	2.64e + 001 ± *2.44e* + 000
f_4	1.59e + 000 ± *7.53e−001*	1.93e + 001 ± *7.45e−001*	1.86e−003 ± *2.31e−004*
f_5	1.95e + 001 ± *1.26e* + 001	6.70e + 000 ± 2.88e + 000	**1.59e + 000 ± *6.87e-001***
f_6	2.43e + 001 ± *7.48e* + 000	1.15e + 001 ± *7.53e* + 000	**9.02e + 000 ± *6.31e* + 000**
f_7	9.31e + 000 ± *3.35e* + 000	**0.00e + 000 ± *0.00e* + 000**	8.72e−001 ± *2.35e−002*
f_8	1.85e + 003 ± *5.38e* + 002	3.41e + 003 ± *9.87e* + 002	**1.14e + 003 ± *4.73e* + 002**

algorithm has not been affected by rotation. The RUMDAc, simply adopting the restart strategy to UMDAc, has not shown superiority on separable problems and deceptive function f_3 when compared with other algorithms such as CPSO-H_k and CMA-ES. However, the algorithm has got great advantage on multimodal inseparable functions and obtains three best results on f_5, f_6 and f_8. It implies that, when dealing with complex multimodal inseparable functions with many local optima, the covariance matrix can be easily misled by local optima. Whereas, the simple structure RUMDAc, by reducing the search space gradually, is easier to jump out of local optima and therefore obtains better results on these test functions.

By analyzing the results of 30 dimensional problems, we select four algorithms, CPSO-H_k, CLPSO, CMA-ES and RUMDAc, that obtain best solutions ever among the eight test functions. Table 4 summaries their preference for different functions. The CPSO-H_k and CLPSO are suitable solutions for separable functions. The univariate model, similar to the CMA-ES, is less affected by correlation among variables, but they have different advantages: the CMA-ES relies on the covariance matrix to explore correlation among variable and therefore is more effective for deceptive functions such as f_3; the univariate model like RUMDAc focuses on random sampling and has the advantage in solving multimodal inseparable functions with many local optima which is very intractable for the CMA-ES to explore the correlation. However, the simple sampling method does not have the ability to learn the correlation among variables. Therefore, its searching ability is inferior to CMA-ES when the correlation among variables is simple and can be perfectly explored such as the deceptive function f_3.

Table 4. CPSO-H_k, CLPSO, CMA-ES and RUMDAc's preference for different functions.

CPSO-H_k	CLPSO	CMA-ES	RUMDAc
Separable functions	Separable functions	Deceptive multimodal functions	Multimodal inseparable functions with many local optima

6 Conclusion

This paper has analyzed the working principle of the univariate Gaussian model in a multimodal evolutionary search space for multivariate inseparable problems. The results show that the univariate Gaussian model has no inclination towards separable problems in a continuous search space, despite the perception otherwise. The univariate Gaussian model relies on random sampling on smaller search spaces to find better solutions. Experimental tests verify that it is an efficient and effective method for solving multimodal inseparable problems. By simply adopting a restart strategy, the RUMDAc is capable of outperforming the five peer algorithms in the literature in solving inseparable problems. Therefore, the univariate Gaussian model is very competitive for solving inseparable problems. Based on the working principle of the univariate Gaussian model discovered herein, future work includes using univariant sampling to improve the searching performance of evolutionary algorithms in solving inseparable problems.

References

1. Mühlenbein, H., Paaß, G.: From recombination of genes to the estimation of distributions I. binary parameters. In: Voigt, H.M., Ebeling, W., Rechenberg, I., Schwefel, H.P. (eds.) PPSN 1996. LNCS, vol. 1141, pp. 178–187. Springer, Heidelberg (1996). doi:10.1007/3-540-61723-X_982
2. Hauschild, M., Pelikan, M.: An introduction and survey of estimation of distribution algorithms. Swarms Evol. Comput. **1**(3), 111–128 (2011)
3. Echegoyen, C., Zhang, Q., Mendiburu, A., Santana, R., Lozano, J.A.: On the limits of effectiveness in estimation of distribution algorithms. In: Proceeding of IEEE Congress Evolutionary Computation, pp. 1573–1580 (2011)
4. Sebag, M., Ducoulombier, A.: Extending population-based incremental learning to continuous search spaces. In: Eiben, A., Bäck, T., Schoenauer, M., Schwefel, H.P. (eds.) PPSN 1998. LNCS, vol. 1498, pp. 418–427. Springer, Heidelberg (1998). doi:10.1007/BFb0056884
5. Baluja, S.: Population-based incremental learning: A method for integrating genetic search based function optimization and competitive learning in Technical report CMU-CS-94-163, Carnegie Mellon University, Pittsburgh, PA (1994)
6. Kabán, A., Bootkrajang, J., Durrant, R.J.: Toward large-scale continuous EDA: a random matrix theory perspective. Evol. Comput. **24**(2), 255–291 (2016)
7. Dong, W., Chen, T., Tino, P., Yao, X.: Scaling up estimation of distribution algorithms for continuous optimization. IEEE Trans. Evol. Comput. **17**(6), 797–822 (2011)

8. Larrañaga, P., Etxeberria, R., Lozano, J.A., Peña, J.M.: Optimization in continuous domains by learning and simulation of Gaussian networks. In: Proceeding of 2000 Genetic and Evolutionary Computation Conference Workshop Program, pp. 201–204 (2000)
9. Wang, Y., Lin, B.: A restart univariate estimation of distribution algorithm: sampling under mixed Gaussian and l′evy probability distribution. In: Proceeding of IEEE Congress Evolutionary Computation, pp. 3917–3924 (2008)
10. Liang, J.J., Qin, A.K., Suganthan, P.N., Baskar, S.: Comprehensive learning particle swarm optimizer for global optimization of multimodal functions. IEEE Trans. Evol. Comput. **10**(3), 281–295 (2006)
11. Kennedy, J., Eberhart, R.C.: Particle swarm optimization. In: Proceeding of IEEE International Conference on Neural Networks, pp. 1942–1948 (1995)
12. van den Bergh, F., Engelbrecht, A.P.: A cooperative approach to particle swarm optimization. IEEE Trans. Evol. Comput. **8**(3), 225–239 (2004)
13. Zhang, G., Li, Y.: Cooperative particle swarm optimizer with elimination mechanism for global optimization of multimodal problems. In: Proceeding of IEEE Congress Evolutionary Compution, pp. 210–217 (2014)
14. Zhang, G., Li, Y.: A memetic algorithm for global optimization of multimodal nonseparable problems. IEEE Trans. Cybern. **46**(6),1375–1387 (2016)
15. Mendes, R., Kennedy, J., Neves, J.: The fully informed particle swarm: Simpler, maybe better. IEEE Trans. Evol. Comput. **8**(3), 204–210 (2004)
16. Lee, K., Green, Z.W.: A new meta-heuristic algorithm for continuous engineering optimization: harmony search theory and practice. Comp. Meth. Appl. Mech. Engng. **194**, 3902–3933 (2005)
17. Hansen, N.: The CMA evolution strategy: a tutorial. http://www.lri.fr/hansen/cmatutorial.pdf
18. Hansen, N., Müller, S., Koumoutsakos, P.: Reducing the time complexity of the derandomized evolution strategy with covariance matrix adaptation (CMA-ES). Evolut. Comput. **11**(1), 1–18 (2003)

Whale Swarm Algorithm for Function Optimization

Bing Zeng, Liang Gao(✉), and Xinyu Li

Huazhong University of Science and Technology, Wuhan, China
gaoliang@hust.edu.cn

Abstract. Increasing nature-inspired metaheuristic algorithms are applied to solving the real-world optimization problems, as they have some advantages over the classical methods of numerical optimization. This paper proposes a new nature-inspired metaheuristic called Whale Swarm Algorithm for function optimization, which is inspired from the whales' behavior of communicating with each other via ultrasound for hunting. The proposed Whale Swarm Algorithm is compared with several popular metaheuristic algorithms on comprehensive performance metrics. According to the experimental results, Whale Swarm Algorithm has a quite competitive performance when compared with other algorithms.

Keywords: Whale Swarm Algorithm · Ultrasound · Nature-inspired · Metaheuristic · Function optimization

1 Introduction

Nature-inspired algorithms are becoming powerful in solving numerical optimization problems, especially the NP-hard problems such as the travelling salesman problem [1], vehicle routing [2], classification problems [3], routing problem of wireless sensor networks (WSN) [4] and multiprocessor scheduling problem [5], etc. These real-world optimization problems often probably come with multiple global or local optima of a given mathematical model (i.e., objective function). And if a point-by-point classical method of numerical optimization is used for this task, the classical method has to try many times for locating different optimal solutions in each time [6], which will take a lot of time and work. Therefore, using nature-inspired metaheuristic algorithms to solve these problems has become a hot research topic, as they are easy to implement and can converge to the global optima with high probability. In this paper, we propose a new nature-inspired metaheuristic called Whale Swarm Algorithm (WSA) for function optimization, based on the whales' behavior of communicating with each other via ultrasound for hunting. Here, a brief overview of nature-inspired metaheuristic algorithms is presented.

Genetic Algorithm (GA) was initially proposed by Holland to solve the numerical optimization problem [7], which simulates Darwin's genetic choice and natural elimination biology evolution process and has opened the prelude of nature-inspired metaheuristic algorithms. It mainly utilizes selection, crossover and mutation operations on the individuals (chromosomes) to find the global optimum as far as possible. In

D.-S. Huang et al. (Eds.): ICIC 2017, Part I, LNCS 10361, pp. 624–639, 2017.
DOI: 10.1007/978-3-319-63309-1_55

GAs, the crossover operator that is utilized to create new individuals by combining parts of two individuals significantly affects the performance of a genetic system [8]. Until now, lots of researchers have proposed diverse crossover operators for different optimization problems. For instance, Syswerda proposed order based crossover operator (OBX) for permutation encoding when dealing with schedule optimization problem [9]. A detailed review of crossover operators for permutation encoding can be seen from reference [10]. Mutation is another important operator in GAs, which provides a random diversity in the population [11], so as to prevent premature convergence of algorithm. Michalewicz proposed random (uniform) mutation and non-uniform mutation [12] for numerical optimization problems. And polynomial mutation operator proposed by Deb is one of the most widely used mutation operators [13]. A comprehensive introduction to mutation operator can be seen from [14]. In a word, it is very important to choose or design appropriate select, crossover and mutation operators of GAs, when dealing with different optimization problems.

Storn and Price proposed Differential Evolution (DE) algorithm for minimizing possibly nonlinear and non-differentiable continuous space functions [15]. It also contains three key operations, namely mutation, crossover and selection, which are different from those of GAs. First of all, a donor vector, corresponding to each member vector of population called target vector, is generated in the mutation phase of DE algorithm. Then, the crossover operation takes place between the target vector and the donor vector, wherein a trial vector is created by selecting components from the donor vector or the target vector with the crossover probability. The selection process determines whether the target or the trial vector survives in the next generation. If the trial vector is better, it replaces the target vector; otherwise remaining the target vector in the population. Since put forward, DE algorithm has gained increasing popularity from researchers and engineers in solving lots of real-world optimization problems [16, 17] and various schemes have been proposed for it [18]. The general convention used to name the different DE schemes is "DE/*x*/*y*/*z*", where DE represents "Differential Evolution", *x* stands for a string indicating the base vector need to be perturbed, for example, it can be set as "*best*" and "*rand*", *y* denotes the number of difference vectors used to perturb *x*, and *z* represents the type of crossover operation which can be binomial (*bin*) or exponential (*exp*) [18]. Some popular existing DE schemes are DE/*best*/1/*bin*, DE/*best*/1/*exp*, DE/*rand*/1/*bin*, DE/*best*/2/*exp*, E/*rand*/2/*exp*, etc.

Particle Swarm Optimization (PSO) is a swarm intelligence based algorithm proposed by Kennedy and Eberhart, which is inspired by social behavior of bird flocking [19]. PSO algorithm has been applied to solve lots of complex and difficult real-world optimization problems [20, 21], since it was put forward. In the traditional PSO algorithm, each particle moves to a new position based on the update of its velocity and position, where the velocity is concerned with its cognitive best position and social best position. Until now, there are lots of PSO variants are proposed for different optimization problems. For instance, Shi and Eberhart introduced a linear decreasing inertia weight into PSO (PSO-LDIW) [22], which can balance the global search and local search, for function optimization. Zhan et al. proposed Adaptive PSO (APSO) [23] for function optimization, which enables the automatic control of parameters to improve the search efficiency and convergence rate, and employs an elitist learning strategy to jump out of the likely local optima. Qu et al. proposed Distance-based Locally

Informed PSO (LIPS) that eliminates the need to specify any niching parameter and enhance the fine search ability of PSO for multimodal function optimization [6], etc.

In addition to the above, there are large amounts of other nature inspired algorithms such as Ant Colony Optimization (ACO) [24], Bees Swarm Optimization (BSO) [25] and Big Bang-Big Crunch (BB-BC) [26], etc. A comprehensive review of nature inspired algorithms is beyond the scope of this paper. A detailed and complete reference on the motif can be seen from [27, 28].

The rest of this paper is organized as follows. Section 2 describes the proposed WSA in sufficient detail. The experiment setup is presented in Sect. 3. Section 4 presents the experimental results performed to evaluate the proposed algorithm. The last section is the conclusions and topics for further works.

2 Whale Swarm Algorithm

First of all, this section introduces the behavior of whales probably, especially the behavior of whales hunting. Then, the details of Whale Swarm Algorithm are presented.

2.1 Behavior of Whales

Whales with great intellectual and physical capacities are completely aquatic mammals, and there are about eighty whale species in the vast ocean. They are social animal and live in groups. Such as pregnant females gather together with other female whales and calves to enhance defense capabilities. And sperm whales are often spotted in groups of some 15 to 20 individuals, as shown in Fig. 1. The whale sounds are beautiful songs in the ocean and their sound range is very wide. Until now, scientists have found 34 species of whale sounds, such as whistling, squeaking, groaning, longing, roaring, warbling, clicking, buzzing, churring, conversing, trumpeting, clopping, etc. These sounds made by whales can often be linked to important functions such as their migration, feeding and mating patterns. What's more, a large part of sounds made by whales are ultrasound beyond the scope of human hearing. And whales determine foods azimuth and keep in touch with each other from a great distance by the ultrasound.

Fig. 1. The swarm of sperm whales.

When a whale has found food source, it will make sounds to notify other whales nearby of the quality and quantity of food. So each whale will receive lots of notifications from the neighbors, and then move to the proper place to find food based on these notifications. The behavior of whales communicating with each other by sound for hunting inspire us to develop a new metaheuristic algorithm for function optimization problems. In the rest of this section, we will discuss the implementation of Whale Swarm Algorithm in detail.

2.2 Whale Swarm Algorithm

To develop whale swarm inspired algorithm for solving function optimization problem, we have idealized some hunting rules of whale. For simplicity in describing our new Whale Swarm Algorithm, the following four idealized rules are employed: (1) all the whales communicate with each other by ultrasound in the search area; (2) each whale has a certain degree of computing ability to calculate the distance to other whales; (3) the quality and quantity of food found by each whale are associated to its fitness; (4) the movement of a whale is guided by the nearest one among the whales that are better (judged by fitness) than it, such nearest whale is called the "better and nearest whale" in this paper.

Iterative Equation. As we know, both radio wave and light wave are electromagnetic waves, which can propagate without any medium. If propagating in water, they will attenuate quickly due to the large electrical conductivity of water. Whereas, sound wave is one kind of mechanical wave that needs a medium through which to travel, whether it is water, air, wood or metal. And ultrasound belongs to sound wave, whose transmission speed and distance largely depends on the medium. For instance, ultrasound travels about 1450 m/s in water, which is faster than that (about 340 m/s) in air. What's more, some ultrasound with pre-specified intensity can travel about 100 m underwater, but can only transmit 2 m in air. That is because the intensity of mechanical wave is continuously attenuated by the molecules of medium, and the intensity of ultrasound traveling in air is attenuated far more quickly than that in water. The intensity ρ of ultrasound at any distance d from the source can be formulated as follows [29].

$$\rho = \rho_0 \cdot e^{-\eta \cdot d} \tag{1}$$

where, ρ_0 is the intensity of ultrasound at the origin of source, e denotes the natural constant. η is the attenuation coefficient, which depends on both the physico-chemical properties of medium and the characteristics of ultrasound itself (such as the ultrasonic frequency) [29].

As we can see from Eq. 1, ρ decreases exponentially with the increment of d when η is constant, which means that the distortion of message conveyed by the ultrasound transmitted by a whale will occur with a great probability, when the travel distance of ultrasound gets quite far. So a whale will not sure whether its understanding about the message send out by another whale is correct, when that whale is quite far away from it.

Thus, a whale would move negatively and randomly towards its better and nearest whale which is quite far away from it.

Based on the above, it can be seen that a whale would move positively and randomly towards its better and nearest whale which is close to it, and move negatively and randomly towards that whale which is quite far away from it, when hunting food. Thus, some whale swarms will form after a period of time. Each whale moves randomly towards its better and nearest whale, because random movement is an important feature of whales' behavior, like the behavior of many other animals such as ant, birds, etc., which is employed to find better food. These rules inspire us to find a new position iterative equation, wishing the proposed algorithm to avoid falling into the local optima quickly and enhance the population diversity and the global exploration ability, as well as contribute to locating multiple global optima. Then, the random movement of a whale **X** guided by its better and nearest whale **Y** can be formulated as follows.

$$x_i^{t+1} = x_i^t + \text{rand}\left(0,\ \rho_0 \cdot e^{-\eta \cdot d_{\mathbf{X},\mathbf{Y}}}\right) * \left(y_i^t - x_i^t\right) \tag{2}$$

where, x_i^t and x_i^{t+1} are the i-th elements of **X**'s position at t and $t + 1$ iterations respectively, similarly, y_i^t denotes the i-th element of **Y**'s position at t iteration. $d_{\mathbf{X},\mathbf{Y}}$ represents the Euclidean distance between **X** and **Y**. And $\text{rand}\left(0,\ \rho_0 \cdot e^{-\eta \cdot d_{\mathbf{X},\mathbf{Y}}}\right)$ means a random number between 0 and $\rho_0 \cdot e^{-\eta \cdot d_{\mathbf{X},\mathbf{Y}}}$. Based on a large number of experiments, ρ_0 can be set to 2 for almost all the cases.

As mentioned previous, the attenuation coefficient η depends on both the physico-chemical properties of medium and the characteristics of ultrasound itself. Here, for function optimization problem, those factors that affect η can be associated to the characteristics of fitness function, including the function dimension, range of variables and distribution of peaks. Therefore, it is important to set appropriate η value for different objective functions. For engineer's convenience in application of WSA, the initial approximate value of η can be set as follows, based on a large number of experimental results. First of all, we should make $\rho_0 \cdot e^{-\eta \cdot (d_{\max}/20)} = 0.5$, i.e., $2 \cdot e^{-\eta \cdot (d_{\max}/20)} = 0.5$, since ρ_0 is always set to 2, wherein $d_{\max}$ denotes the maximum distance between any two whales in the search space that can be formulated as $d_{\max} = \sqrt{\sum_{i=1}^{n} \left(v_i^U - v_i^L\right)^2}$, n is the dimension of fitness function, v_i^L and v_i^U represent the lower limit and upper limit of the i-th variable respectively. This equation means that if the distance between whale **X** and its better and nearest whale **Y** is $d_{\max}/20$, the part $\rho_0 \cdot e^{-\eta \cdot d_{\mathbf{X},\mathbf{Y}}}$ of Eq. 2 that affects the moving range of whale **X** should be set to 0.5. Next, we can get that $\eta = -20 \cdot \ln(0.25)/d_{\max}$. Then, it is easy to adjust η to the optimal or near-optimal value based on this initial approximate value.

Equation 2 shows that a whale will move towards its better and nearest whale positively and randomly, if the distance between them is small. Otherwise, it will move towards its better and nearest whale negatively and randomly, which can be illustrated with Fig. 2 when the dimension of fitness function is equal to 2. In Fig. 2, the red stars denote the global optima, the circles represent the whales and the rectangular regions signed with imaginary lines are the reachable regions of whales in current iteration.

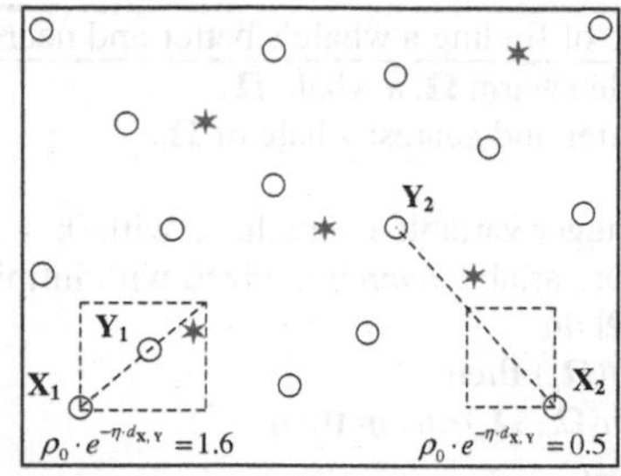

Fig. 2. Sketch map of a whale's movement guided by its better and nearest whale.

General Framework of WSA. Based on the above rules, the general framework of the proposed WSA can be summarized as shown in Fig. 3, where $|\mathbf{\Omega}|$ in line 6 denotes the number of members in $\mathbf{\Omega}$, namely the swarm size, and $\mathbf{\Omega}_i$ in line 7 is the *i*-th whale in $\mathbf{\Omega}$. It can be seen from Fig. 3 that those steps before iterative computation are some initialization steps, including initializing configuration parameters, initializing individuals' positions and evaluating each individual, which are similar with most other metaheuristic algorithms. Here, all the whales are randomly assigned to the search area. Next come the core step of WSA: whales move (lines 5–13). Each whale needs to move for the better food via group cooperation. First of all, a whale should find its better and nearest whale (lines 7), as shown in Fig. 4, where $f(\mathbf{\Omega}_i)$ in line 5 is the fitness value of whale $\mathbf{\Omega}_i$ and $dist(\mathbf{\Omega}_i, \mathbf{\Omega}_u)$ in line 6 denotes the distance between $\mathbf{\Omega}_i$ and $\mathbf{\Omega}_u$. If its better and nearest whale exists, then it will move under the guidance of the better and nearest whale (lines 9 in Fig. 3). As described above, the framework of WSA is fairly simple, which is convenient for applying WSA in solving the real-world optimization problems.

```
The general framework of Whale Swarm Algorithm
Input: An objective function, the whale swarm Ω.
Output: The global optima.
 1: begin
 2: Initialize parameters;
 3: Initialize whales' positions;
 4: Evaluate all the whales (calculate their fitness values);
 5: while termination criterion is not satisfied do
 6:     for i=1 to |Ω| do
 7:         Find the better and nearest whale Y of Ω_i;
 8:         if Y exists then
 9:             Ω_i moves under the guidance of Y according to Eq. 2;
10:             Evaluate Ω_i;
11:         end if
12:     end for
13: end while
14: return the global optima;
15: end
```

Fig. 3. The general framework of WSA.

```
The pseudo code of finding a whale's better and nearest whale
Input: The whale swarm Ω, a whale Ω_u.
Output: The better and nearest whale of Ω_u.
begin
Define an integer variable v initialized with 0;
Define a float variable temp initialized with infinity;
for i=1 to |Ω| do
    if f(Ω_i)<f(Ω_u) then
        if dist(Ω_i, Ω_u)<temp then
            v=i;
            temp=dist(Ω_i, Ω_u);
        end if
    end if
end for
return Ω_v;
end
```

Fig. 4. The pseudo code of finding a whale's better and nearest whale.

3 Experimental setup

The proposed WSA and other algorithms compared are all implemented with C++ programming language by Microsoft visual studio 2015 and executed on the PC with 2.3 GHz Intel core i7 3610QM processor, 8 GB RAM and Microsoft Windows 10 operating system. The source code of the proposed WSA can be download from the website "https://drive.google.com/open?id=0B8qyXW8tulBqeGFUUmtKOV9sYWc". In addition to the GA with elitism, non-uniform arithmetic crossover and basic bit mutation strategies, DE/best/1/bin [30] and PSO with inertia weight [22], the following 4 popular multimodal optimization algorithms are also compared with WSA: The locally informed PSO (LIPS) [6], Speciation-based DE (SDE) [31], The original crowding DE (CDE) [32] and Speciation-based PSO (SPSO) [33].

In this paper, we utilize the number of function evaluations as the stopping criterion of these algorithms to test their performance.

3.1 Test Functions

To verify the performance of the proposed WSA, the comparative experiments were conducted on twelve benchmark test functions, which are taken from the studies of Deb [34], Michalewicz [12], Li [35] and Thomsen [32]. Test functions F1–F8 are multimodal (F1–F6 and F7–F10 are low and high dimensional multimodal functions respectively) that have multiple global or local optima. F11 and F12 are high dimensional unimodal functions with 100 dimension. Basic information of these test functions are summarized in Table 1. For functions F2–F6, the objective is to locate all the global optima, while for the rest the target is to escape the local optima (if they have) to hunt for the global optimum. And all test functions are minimization problems. It can be seen from Table 1, F7, F9, F11 and F12 all get the global optima at (0, 0, …, 0), and

Table 1. Test functions.

Fn.	Function name/Dimensions	Expression	Ranges	No. of global optima	Minimum value
1	Uneven Increasing Minima/1D	$f(\mathbf{X}) = -\exp\left(-2\log(2)\cdot\left(\frac{x_1-0.08}{0.854}\right)^2\right)\cdot\sin^6\left(5\pi\left(x_1^{3/4}-0.05\right)\right)$	[0,1]	1	–1
2	Uneven Minima/1D	$f(\mathbf{X}) = -\sin^6\left(5\pi\left(x_1^{3/4}-0.05\right)\right)$	[0,1]	5	–1
3	Himmelblau's function/2D	$f(\mathbf{X}) = (x_1^2+x_2-11)^2+(x_1+x_2^2-7)^2-200$	$[-6,6]^2$	4	–200
4	Six-hump camel back/2D	$f(\mathbf{X}) = 4\left(\left(4-2.1x_1^2+\frac{x_1^4}{3}\right)x_1^2+x_1x_2+(-4+4x_2^2)x_2^2\right)$	[–1.9,1.9] [–1.1,1.1]	2	–4.126514
5	Inverted Shubert function/2D	$f(\mathbf{X}) = \prod_{i=1}^{2}\sum_{j=1}^{5} j\cos((j+1)x_i+j)$	$[-10,10]^2$	18	–186.7309
6	Branin RCOS/2D	$f(\mathbf{X}) = \left(x_2-\frac{5.1}{4\pi^2}x_1^2+\frac{5}{\pi}x_1-6\right)^2+10\left(1-\frac{1}{8\pi}\right)\cos(x_1)+10$	[–5,10] [0,15]	3	0.397887
7	Rastrigin/100D	$f(\mathbf{X}) = \sum_{i=1}^{D}(x_i^2-10\cos(2\pi x_i)+10)$	$[-100,100]^{100}$	1	0
8	Schwefel/100D	$f(\mathbf{X}) = D\cdot 418.9829-\sum_{i=1}^{D}x_i\sin\left(\sqrt{\lvert x_i\rvert}\right)$	$[-500,500]^{100}$	1	0
9	Griewank/100D	$f(\mathbf{X}) = -\prod_{i=1}^{D}\cos\left(\frac{x_i}{\sqrt{i}}\right)+\sum_{i=1}^{D}\frac{x_i^2}{4000}+1$	$[-100,100]^{100}$	1	0
10	Rosenbrock/100D	$f(\mathbf{X}) = \sum_{i=1}^{D-1}\left(100(x_{i+1}-x_i^2)^2+(x_i-1)^2\right)$	$[-15,15]^{100}$	1	0
11	Sphere/100D	$f(\mathbf{X}) = \sum_{i=1}^{D}x_i^2$	$[-100,100]^{100}$	1	0
12	Zakharov/100D	$f(\mathbf{X}) = \sum_{i=1}^{D}x_i^2+\left(\frac{1}{2}\sum_{i=1}^{D}ix_i\right)^2+\left(\frac{1}{2}\sum_{i=1}^{D}ix_i\right)^4$	$[-5,10]^{100}$	1	0

F10 gets the global optimum at (1, 1, …, 1), which are located near the middle of the feasible region. As we know, some algorithms are efficient in optimizing the functions whose optima are near the middle of feasible region, especially near zero, but perform badly when the optima are not near the middle of feasible region. For the sake of fairness, we shifte F7–F12, and the shift data of them are randomly generated within specified range.

3.2 Parameters Setting

Although the global optima of these test functions can be obtained by the method of derivation, they should still be treated as black-box problems, i.e., the known global optima of these test functions cannot be used by the algorithms during the iterations, so as to compare the performance of these algorithms. The fitness value and the number of global optima of each function are listed in Table 1. Here, we use fitness error $\mathcal{E}_f$, i.e., level of accuracy, to judge whether a solution is a real global optimum, i.e., if the difference between the fitness of a solution and the known global optimum is lower than $\mathcal{E}_f$, this solution can be considered as a global optimum. In our experiments, the fitness error $\mathcal{E}_f$, population size and maximal number of function evaluations for WSA and the 7 algorithms compared are listed in Table 2. It is worth to note that a function which has more optima or higher dimension requires a larger population size and more number of function evaluations.

Table 2. Test functions setting.

Fn.	ε_f	Population size	No. of function evaluations
F1	0.01	100	10000
F2	0.000001	100	10000
F3	0.05	100	10000
F4	0.001	100	10000
F5	0.05	300	100000
F6	0.002	200	20000
F7	0.001	100	500000
F8	0.001	100	500000
F9	0.001	100	500000
F10	0.001	100	500000
F11	0.001	100	500000
F12	0.001	100	500000

The user-specified control parameters of WSA, i.e., attenuation coefficient η, for these test functions are set as shown in Table 3.

Table 3. Parameter setting of WSA for test functions.

Parameter	F1	F2	F3	F4	F5	F6	F7	F8	F9	F10	F11	F12
η	40	40	1.55	5.5	0.6	1.5	5E–3	2.2E–3	2E–3	6.5E–2	5E–3	6.5E–2

The parameters value of the 7 algorithms compared are set as the same as those in their reference source respectively. Table 4 shows the setting of main parameters of these algorithms. The parameter species radius r_s of SDE and SPSO for these test functions are listed in Table 5.

Table 4. Setting of parameters of the algorithms.

Algorithms	Parameters
GA	P_c=0.95, P_m=0.05
DE	P_c=0.7, F=0.5
PSO	ω=0.729844, c_1=2, c_2=2
LIPS	ω=0.729844, *nsize*=2~5
SDE	P_c=0.9, F=0.5, m=10
CDE	P_c=0.9, F=0.5, CF=population size
SPSO	χ=0.729844, φ_1=2.05, φ_2=2.05

1. P_c: crossover probability; P_m: mutation probability; 2. F: scaling factor; 3. ω: inertia weight; c_1, c_2: acceleration factor; 4. *nsize*: neighborhood size; 5. m: species size; 6. CF: crowding factor; 7. χ: constriction factor; φ_1, φ_2: coefficient.

Table 5. Species radius setting for test functions.

	F1	F2	F3	F4	F5	F6	F7	F8	F9	F10	F11	F12
SDE	0.05	0.05	1	0.5	1	1	4	10	4	1	4	1
SPSO	0.01	0.01	1	0.2	1.2	2	800	4000	800	120	800	60

3.3 Performance Metrics

To compare the performance of WSA and other 7 algorithms, we have conducted 25 independent runs for each algorithm on each test function. And the following five metrics are used to measure the performance of all the algorithms.

(1) Success Rate (SR) [33]: the percentage of runs in which all the global optima are successfully located using the given level of accuracy.
(2) Average Number of Optima Found (ANOF) [36]: the average number of global optima found over 25 runs.
(3) Quality of optima found: the mean of fitness values of optima found over 25 runs.
(4) Convergence rate: the rate of an algorithm converging to the global optimum over function evaluations.

4 Experimental Results and Analysis

This section presents and analyzes the results of comparative experiments. All the algorithms are run under the experiment setup shown in the previous section.

4.1 Success Rate

The success rates of all the algorithms on each test function are presented in Table 6, in which the numbers within parentheses denote the ranks of each algorithm. If the success rates of any two algorithms on a test function are equal, they have the same ranks over this test function. The last row of this table shows the total ranks of algorithms, which are the summation of individual ranks on each test function. As we can see from Table 6, for multimodal functions F1–F10, the success rate of WSA on F3 is only a little bit lower than that of LIPS, but is far greater than those of other algorithms. Only LIPS can achieve nonzero success rates on F5, and no algorithm can achieve nonzero success rates on the four high dimensional multimodal functions F7–F10. What's more, WSA can achieve 100% success rates on test functions F1, F2, F4 and F6, which are much higher than those gained by most of other algorithms. Therefore, it can be seen that WSA has a very competitive performance on dealing with multimodal functions with respect to other algorithms. And for high-dimensional unimodal functions F11–F12, all the algorithms cannot achieve nonzero success rates on F12. However, WSA can achieve 100% success rate on F11, while the success rates of other algorithms on F11 are 0. Therefore, it can be concluded that WSA also has better performance than other algorithms on success rate when solving unimodal functions. It also can be seen that the better performance of WSA on success rate can be

Table 6. SR and ranks (in parentheses) of algorithms on F1–F12.

Fn.	GA	DE	PSO	CDE	SDE	SPSO	LIPS	WSA
F1	0.96 (6)	**1 (1)**	**1 (1)**	**1 (1)**	**1 (1)**	0.32 (8)	0.64 (7)	**1 (1)**
F2	0 (4)	0 (4)	0 (4)	0.04 (3)	0.32 (2)	0 (4)	0 (4)	**1 (1)**
F3	0 (5)	0 (5)	0 (5)	0.08 (4)	0.72 (3)	0 (5)	**1 (1)**	0.8 (2)
F4	0 (7)	0 (7)	0.08 (5)	0.08 (5)	0.60 (3)	0.12 (4)	**1 (1)**	**1 (1)**
F5	0 (2)	0 (2)	0 (2)	0 (2)	0 (2)	0 (2)	**0.76 (1)**	0 (2)
F6	0 (4)	0 (4)	0 (4)	0.04 (3)	0.64 (2)	0 (4)	0 (4)	**1 (1)**
F7	**0 (1)**	**0 (1)**	**0 (1)**	**0 (1)**	**0 (1)**	**0 (1)**	**0 (1)**	**0 (1)**
F8	**0 (1)**	**0 (1)**	**0 (1)**	**0 (1)**	**0 (1)**	**0 (1)**	**0 (1)**	**0 (1)**
F9	**0 (1)**	**0 (1)**	**0 (1)**	**0 (1)**	**0 (1)**	**0 (1)**	**0 (1)**	**0 (1)**
F10	**0 (1)**	**0 (1)**	**0 (1)**	**0 (1)**	**0 (1)**	**0 (1)**	**0 (1)**	**0 (1)**
F11	0 (2)	0 (2)	0 (2)	0 (2)	0 (2)	0 (2)	0 (2)	**1 (1)**
F12	**0 (1)**	**0 (1)**	**0 (1)**	**0 (1)**	**0 (1)**	**0 (1)**	**0 (1)**	**0 (1)**
Total rank	35	30	28	25	20	34	25	14

supported by the total rank of WSA which is much smaller than those gained by other algorithms. The better performance of WSA is due to its novel iteration rules based on the behavior of whales hunting, including that the random movement of a whale is guided by its better and nearest whale, and its range of movement depends on the intensity of the ultrasound received as shown as Eq. 2, which have a great contribution to the maintenance of population diversity and the enhancement of global exploration ability, so as to locate the global optimum (optima).

As some algorithms cannot obtain nonzero success rates on some multimodal functions, the metric ANOF is used to test the performance of those algorithms on locating multiple global optima. Table 7 presents the ANOF of all the algorithms over functions F2–F6 which have multiple global optima. As can be seen from this table, for these multimodal functions, the ANOF of WSA on test functions F2, F4 and F6 are much higher than those obtained by most of other algorithms, which echoes the 100% success rates of WSA on these functions as shown in Table 6. And the ANOF of WSA on F3 is only a little bit lower than that of LIPS, but is much higher than those of other algorithms, which is similar to the case of success rates of algorithms on this function. The ANOF of WSA on F5 is 6.76, which is much higher than those of other algorithms but the multimodal optimization algorithms LIPS and SDE. Therefore, the results of Table 7 further demonstrate the outstanding performance of WSA on finding multiple global optima with respect to other algorithms when solving multimodal functions.

4.2 Quality of Optima Found

In this section, WSA compares the performance with other algorithms in terms of the accuracy of optima found. The mean and standard deviation of fitness values of optima found by all the algorithms on each test function over 25 runs are listed in Table 8. Here, the ranks of algorithms are based on the mean of fitness values of optima found over the test functions. As we can see from Table 8, for low-dimensional multimodal functions F1–F6, WSA, DE and PSO achieve the best accuracy on F1. And DE algorithm ranks the best on F2–F6. But DE algorithm cannot get nonzero success rates on F2–F6 as shown in Table 6, and gains worse ANOF than most of other algorithms as shown in Table 7, which mean that DE algorithm has a poor performance on locating multiple global optima though it can achieve a few of multiple global optima with high accuracy, when solving low-dimensional multimodal functions. Whereas, WSA achieves very good accuracy over F2–F6, on the premise of keeping excellent SR and ANOF as shown in Tables 6 and 7. And for high-dimensional multimodal functions F7 and F8, WSA only performs a little bit worse than LIPS and SDE, but outperforms other algorithms. What's more, for high-dimensional multimodal functions F9–F10 and high-dimensional unimodal functions F11–F12, WSA achieves the best accuracy when compared with all the other algorithms. Particularly, the mean of fitness values of optima found by WSA on F11 over 25 runs is **2.60E–09** and the standard deviation is **6.68E–09**, which are far better than those obtained by all the other algorithms. Therefore, it can be concluded that WSA also performs better than most of other algorithms in terms of the quality of optima found. The outstanding performance of WSA on the quality of optima found is also due to its novel iteration rules, which contribute significantly to enhancing the local exploitation ability.

Table 7. ANOF and ranks (in parentheses) of algorithms on F2–F6.

Fn.	GA	DE	PSO	CDE	SDE	SPSO	LIPS	WSA
F2	0.96 (6)	1.08 (5)	1.64 (4)	2.60 (3)	3.28 (2)	0.08 (7)	0.04 (8)	**5** **(1)**
F3	0.08 (8)	1 (6)	1.16 (5)	2.40 (4)	3.56 (3)	0.96 (7)	**4** **(1)**	3.8 (2)
F4	0.36 (8)	1 (6)	1.08 (4)	0.64 (7)	1.60 (3)	1.04 (5)	**2** **(1)**	**2** **(1)**
F5	0.24 (8)	2.16 (4)	1.44 (5)	0.80 (7)	7.32 (2)	1.04 (6)	**17.72** **(1)**	6.76 (3)
F6	0 (7)	1 (5)	1.16 (3)	1.08 (4)	2.56 (2)	0 (7)	0.96 (6)	**3** **(1)**
Total rank	37	26	21	25	12	32	17	8

4.3 Convergence Rate

Based on the previous, it can be seen that WSA has a quite competitive performance when compared with other algorithms, in terms of the location of multiple global optima and the quality of optima found. To further demonstrate the superiority of WSA, it is compared with other algorithms on F3 in terms of convergence rate in this section. The convergence curves of all the algorithms on F3 are depicted in Fig. 5, in which the abscissa values denote function evaluations and the ordinate values represent the average fitness values of population over 25 runs. As can be seen from Fig. 5, WSA converges slower than DE, LIPS and SDE in the early iterations. However, in the mid and later iterations, WSA converges faster than LIPS and SDE, and it can achieve a better value than LIPS and SDE do. Although DE algorithm can converge to the global optimum (–200), it can only locate one of the four global optima in a single run, as shown in Table 7. Therefore, it can be concluded that WSA has better performance in terms of convergence rate than other algorithms on the premise of keeping good SR and ANOF.

Table 8. Quality of optima found and ranks (in parentheses) of algorithms on F1–F12.

Fn.	Measure	GA	DE	PSO	CDE	SDE	SPSO	LIPS	WSA
F1	Mean	3.03E–03	**8.91E–04**	**8.91E–04**	3.47E–03	1.31E–03	3.19E–02	1.17E–02	**8.91E–04**
	Std.	8.66E–03	**0.00E+00**	**7.06E–13**	2.68E–03	1.79E–03	2.38E–02	1.34E–02	**1.54E–16**
		(5)	**(1)**	**(1)**	(6)	(4)	(8)	(7)	**(1)**
F2	Mean	3.80E–07	**1.10E–11**	8.63E–08	4.09E–07	7.08E–08	1.62E–04	1.08E–04	3.69E–09
	Std.	9.68E–07	**5.37E–11**	1.27E–07	5.11E–07	1.27E–07	2.49E–04	1.53E–04	1.32E–08
		(5)	**(1)**	(4)	(6)	(3)	(8)	(7)	(2)
F3	Mean	6.36E–01	**7.03E–13**	2.26E–03	2.33E–02	5.83E–03	5.02E–03	2.09E–03	5.11E–04
	Std.	5.52E–01	**3.22E–12**	5.77E–03	1.55E–02	6.67E–03	1.24E–02	2.90E–03	1.96E–03
		(8)	**(1)**	(4)	(7)	(6)	(5)	(3)	(2)
F4	Mean	1.18E–02	**1.86E–07**	2.08E–05	1.22E–03	9.46E–05	5.00E–04	1.58E–05	4.18E–05
	Std.	2.31E–02	**0.00E+00**	8.38E–05	9.65E–04	1.97E–04	3.43E–04	3.64E–05	8.50E–05
		(8)	**(1)**	(3)	(7)	(5)	(6)	(2)	(4)
F5	Mean	2.01E + 00	**9.22E–08**	4.68E–03	7.63E–02	1.40E–03	4.79E–02	3.05E–03	1.15E–03
	Std.	2.21E + 00	**3.13E–07**	7.48E–03	9.61E–02	2.50E–03	4.33E–02	1.88E–03	1.76E–03
		(8)	**(1)**	(5)	(7)	(3)	(6)	(4)	(2)
F6	Mean	1.27E–01	**3.58E–07**	9.58E–05	1.55E–03	3.18E–04	1.59E–01	1.30E–03	3.35E–05
	Std.	1.51E–01	**0.00E+00**	2.35E–04	1.11E–03	4.32E–04	1.07E–01	1.39E–03	1.33E–04
		(7)	**(1)**	(3)	(6)	(4)	(8)	(5)	(2)
F7	Mean	1.66E + 05	2.41E+04	1.42E+05	6.07E+04	**6.13E+02**	6.14E+03	1.62E+03	5.39E+03
	Std.	6.56E + 03	5.70E+03	1.67E+04	2.91E+03	**3.94E+01**	4.85E+02	3.08E+02	1.26E+03
		(8)	(5)	(7)	(6)	**(1)**	(4)	(2)	(3)
F8	Mean	3.65E + 04	3.33E+04	2.39E+04	3.82E+04	2.30E+04	2.79E+04	**1.98E+04**	2.36E+04
	Std.	1.28E + 03	7.51E+02	2.35E+03	6.44E+02	8.40E+02	1.27E+03	**9.64E+02**	1.12E+03
		(7)	(6)	(4)	(8)	(2)	(5)	**(1)**	(3)
F9	Mean	5.33E + 01	8.50E+00	4.05E+01	1.30E+01	1.13E+00	2.35E+00	1.16E+00	**1.00E+00**
	Std.	2.19E + 00	2.49E+00	6.28E+00	5.52E–01	5.14E–03	1.52E–01	6.51E–02	**8.54E–14**
		(8)	(5)	(7)	(6)	(2)	(4)	(3)	**(1)**
F10	Mean	8.53E+07	5.19E+06	2.82E+07	1.36E+06	5.40E+03	6.84E+04	1.29E+04	**9.81E+01**
	Std.	7.06E+06	1.78E+06	1.89E+07	2.24E+05	3.59E+02	1.51E+04	7.20E+03	**7.92E–01**
		(8)	(6)	(7)	(5)	(2)	(4)	(3)	**(1)**
F11	Mean	1.78E+05	2.71E+04	1.41E+05	5.61E+04	5.28E+02	2.14E+04	1.70E+05	**2.60E–09**
	Std.	8.82E+03	7.66E+03	1.70E+04	2.92E+03	2.24E+01	2.61E+03	1.43E+04	**6.68E–09**
		(8)	(4)	(6)	(5)	(2)	(3)	(7)	**(1)**
F12	Mean	6.49E+02	1.26E+03	2.28E+03	1.77E+03	3.33E+02	1.63E+03	5.65E+02	**2.76E+02**
	Std.	3.89E+02	1.25E+02	2.50E+02	1.03E+02	3.37E+01	8.51E+02	5.57E+01	**4.74E+01**
		(4)	(5)	(8)	(7)	(2)	(6)	(3)	**(1)**
Total rank		84	37	59	76	36	67	47	23

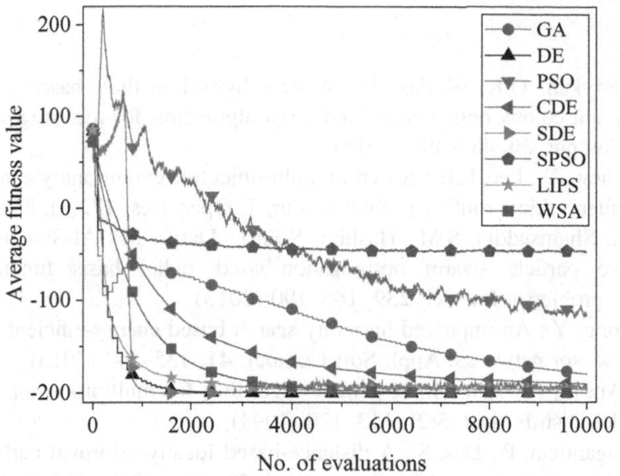

Fig. 5. The convergence graph of different algorithms on F3.

5 Conclusions

A new swarm intelligence based metaheuristic called Whale Swarm Algorithm, inspired by the whales' behavior of communicating with each other via ultrasound for hunting, is proposed for function optimization in this paper. The innovations of the iterative equation of WSA consist of two parts: the random movement of a whale is guided by its better and nearest whale; and its range of movement depends on the intensity of ultrasound received, which contribute significantly to the maintenance of population diversity, the avoidance of falling into the local optima quickly and the enhancement of global exploration ability, so as to locate the global optimum/optima. And the novel iteration rules also have a great contribution to the enhancement of local exploitation ability, especially when some whales have gathered around a same peak, so as to improve the quality of optima found. WSA is compared with several popular metaheuristic algorithms on four performance metrics. The experimental results show that WSA has a quite competitive performance when compared with other algorithms, in terms of the location of multiple global optima, the quality of optima found and convergence rate.

In the future, we will focus on the following aspects:

(1) Utilizing WSA to solve multi-objective optimization problems.
(2) Modifying WSA to deal with real-world optimization problems, especially the discrete optimization problems and the NP-hard problems.

Acknowledgment. This research was supported by the National Natural Science Foundation of China (NSFC) (51421062) and the National Key Technology Support Program (2015BAF01B04).

References

1. Mahi, M., Baykan, Ö.K., Kodaz, H.: A new hybrid method based on particle swarm optimization, ant colony optimization and 3-opt algorithms for traveling salesman problem. Appl. Soft Comput. **30**, 484–490 (2015)
2. Tan, K.C., Chew, Y., Lee, L.H.: A hybrid multi-objective evolutionary algorithm for solving truck and trailer vehicle routing problems. Eur. J. Oper. Res. **172**(3), 855–885 (2006)
3. Qasem, S.N., Shamsuddin, S.M., Hashim, S.Z.M., Darus, M., Al-Shammari, E.: Memetic multiobjective particle swarm optimization-based radial basis function network for classification problems. Inf. Sci. **239**, 165–190 (2013)
4. Zeng, B., Dong, Y.: An improved harmony search based energy-efficient routing algorithm for wireless sensor networks. Appl. Soft Comput. **41**, 135–147 (2016)
5. Hou, E.S., Ansari, N., Ren, H.: A genetic algorithm for multiprocessor scheduling. IEEE Trans. Parallel Distrib. Syst. **5**(2), 113–120 (1994)
6. Qu, B.Y., Suganthan, P., Das, S.: A distance-based locally informed particle swarm model for multimodal optimization. IEEE Trans. Evol. Comput. **17**(3), 387–402 (2013)
7. Holland, J.: Adaptation in Artificial and Natural Systems. The University of Michigan Press, Ann Arbor (1975)
8. Gen, M., Cheng, R.: Genetic Algorithms and Engineering Optimization. Wiley, New York (2000)
9. Syswerda, G.: Schedule optimization using genetic algorithms. In: Davis, L. (ed.) Handbook of Genetic Algorithms. Van Nostrand Reinhold, New York (1991)
10. Kellegöz, T., Toklu, B., Wilson, J.: Comparing efficiencies of genetic crossover operators for one machine total weighted tardiness problem. Appl. Math. Comput. **199**(2), 590–598 (2008)
11. Whitley, L.D.: Foundations of Genetic Algorithms 2. Morgan Kaufmann, San Mateo (1993)
12. Michalewicz, Z.: Genetic Algorithms + Data Structures = Evolution Programs. Springer Science & Business Media, Heidelberg (2013)
13. Deb, K.: Multi-objective optimization using evolutionary algorithms. John Wiley & Sons, Chichester (2001)
14. Deep, K., Thakur, M.: A new mutation operator for real coded genetic algorithms. Appl. Math. Comput. **193**(1), 211–230 (2007)
15. Storn, R., Price, K.: Differential Evolution - A Simple and Efficient Adaptive Scheme for Global Optimization Over Continuous Spaces. ICSI, Berkeley (1995)
16. Qing, A.: Dynamic differential evolution strategy and applications in electromagnetic inverse scattering problems. IEEE Trans. Geosci. Remote Sens. **44**(1), 116–125 (2006)
17. Gao, Z., Pan, Z., Gao, J.: A new highly efficient differential evolution scheme and its application to waveform inversion. IEEE Geosci. Remote Sens. Lett. **11**(10), 1702–1706 (2014)
18. Das, S., Suganthan, P.N.: Differential evolution: a survey of the state-of-the-art. IEEE Trans. Evol. Comput. **15**(1), 4–31 (2011)
19. Kennedy, J., Kennedy, J.F., Eberhart, R.C., Shi, Y.: Swarm Intelligence. Morgan Kaufmann, San Francisco (2001)
20. Liu, M., Xu, S., Sun, S.: An agent-assisted QoS-based routing algorithm for wireless sensor networks. J. Netw. Comput. Appl. **35**(1), 29–36 (2012)
21. Cheng, R., Jin, Y.: A social learning particle swarm optimization algorithm for scalable optimization. Inf. Sci. **291**, 43–60 (2015)

22. Shi, Y., Eberhart, R.: A modified particle swarm optimizer. In: The 1998 IEEE International Conference on Evolutionary Computation Proceedings, IEEE World Congress on Computational Intelligence, pp. 69–73. IEEE, Anchorage (1998)
23. Zhan, Z.H., Zhang, J., Li, Y., Chung, H.S.H.: Adaptive particle swarm optimization. IEEE Trans. Syst. Man Cybern. Part B Cybern. **39**(6), 1362–1381 (2009)
24. Dorigo, M.: Optimization, learning and natural algorithms. Ph. D. thesis, Politecnico di Milano, Italy (1992)
25. Drias, H., Sadeg, S., Yahi, S.: Cooperative bees swarm for solving the maximum weighted satisfiability problem. In: Cabestany, J., Prieto, A., Sandoval, F. (eds.) IWANN 2005. LNCS, vol. 3512, pp. 318–325. Springer, Heidelberg (2005). doi:10.1007/11494669_39
26. Erol, O.K., Eksin, I.: A new optimization method: big bang-big crunch. Adv. Eng. Softw. **37** (2), 106–111 (2006)
27. Zang, H., Zhang, S., Hapeshi, K.: A review of nature-inspired algorithms. J. Bionic Eng. **7**, 232–237 (2010)
28. BoussaïD, I., Lepagnot, J., Siarry, P.: A survey on optimization metaheuristics. Inf. Sci. **237**, 82–117 (2013)
29. Majumdar, S., Kumar, P.S., Pandit, A.: Effect of liquid-phase properties on ultrasound intensity and cavitational activity. Ultrason. Sonochem. **5**(3), 113–118 (1998)
30. Price, K.V.: An introduction to differential evolution. In: New ideas in optimization. McGraw-Hill Ltd, Maidenhead (1999)
31. Li, X.: Efficient differential evolution using speciation for multimodal function optimization. In: Proceedings of the 7th Annual Conference on Genetic and Evolutionary Computation, pp. 873–880. ACM, Washington (2005)
32. Thomsen, R.: Multimodal optimization using crowding-based differential evolution. In: Congress on Evolutionary Computation, CEC 2004, pp. 1382–1389. IEEE, Portland (2004)
33. Li, X.: Adaptively choosing neighbourhood bests using species in a particle swarm optimizer for multimodal function optimization. In: Deb, K. (ed.) GECCO 2004. LNCS, vol. 3102, pp. 105–116. Springer, Heidelberg (2004). doi:10.1007/978-3-540-24854-5_10
34. Deb, K.: Genetic algorithms in multimodal function optimization, Clearinghouse for Genetic Algorithms, Department of Engineering Mechanics, University of Alabama (1989)
35. Li, J.P., Balazs, M.E., Parks, G.T., Clarkson, P.J.: A species conserving genetic algorithm for multimodal function optimization. Evol. Comput. **10**(3), 207–234 (2002)
36. http://www.ntu.edu.sg/home/EPNSugan/index_files/CEC2015/CEC2015.htm

The Container Truck Route Optimization Problem by the Hybrid PSO-ACO Algorithm

Yi Liu[1(✉)], Mao Feng[1], and Sabina Shahbazzade[2]

[1] Hangzhou Dianzi University, Hangzhou 310018, China
liuyi@hdu.edu.cn
[2] University of California, Berkeley, USA

Abstract. This paper mainly research on the container tuck route optimization problem with the integrated loading and unloading operation. Considered the disperse-stacking of containers in yards and the loading/unloading operations of each berth, the objective function of scheduling problem is the optimal rout of the container truck. In order to solve this problem, the hybrid swarm intelligence algorithm (PSO-ACO) is proposed, which combined the particle swarm optimization algorithm with the ant colony optimization algorithm. The hybrid swarm intelligence algorithm takes advantage of strong local search ability of ant colony optimization algorithm and the ACO's pheromone taxis, which can avoid the particle swarm optimization algorithm fall in the local optimum during the convergence. The results show that the mathematical model and hybrid algorithm have effective, reliability and stability in solving the container truck scheduling problem.

Keywords: Container truck route · Scheduling · Particle swarm optimization algorithm · Ant colony optimization algorithm

1 Introduction

1.1 Brief Introduction

The seaport container terminals is the key role in the maritime transportation. The container truck perform the horizontal transportation between the container terminal and the container yard, which the non-automated container terminal yard has the flexible and large quantity transport equipment. The container truck scheduling problem can improve the operation efficiency of the container terminal.

1.2 Literature Review

Now, some scholars have researched the truck scheduling problems of the container terminal yard: Kim studied the integer programming method to solve the truck path optimization problem between the transfer crane and container yard [1]. Ebru, Thin-Yin, et al. established the mathematical model of the truck scheduling problem in the non-automated port, and used the heuristic algorithm to obtain the shortest operation time in the container port [2]. Bish, Chen, et al. used the heuristic algorithm to

D.-S. Huang et al. (Eds.): ICIC 2017, Part I, LNCS 10361, pp. 640–648, 2017.
DOI: 10.1007/978-3-319-63309-1_56

solve the truck scheduling problem in order to short the waiting time of the ship [3]. According to the characteristics model of truck dynamic path, Nishimura used the genetic algorithm to solve the dynamic scheduling truck problem [4]. Lin Shih Wei and Yu Vincent F used the simulated annealing (SA) heuristic to solving the truck and trailer routing problem with time windows (TTRPTW). The results of computational experiments indicate that the proposed SA heuristic is capable of consistently producing quality solutions [5]. Derigs U, Pullmann M, Vogel U proposed computational experience with a simple and flexible hybrid approach which is based on local search and large neighborhood search as well as standard metaheuristic control strategies to obtain the optimal path of the truck [6]. Ji Ming jun and Le Zhi hong considered the truck route optimization model for the shortest time of the crane operation and truck transport [7]. Wei Hong Gai, Zhu Jing established the container truck path optimization model based on the coordination of the loading and unloading operation of the ship and used the branch and bound method to solve truck optimal path and the optimal allocation scheme of the container [8]. Zheng Qing Cheng established the container terminal handling process scheduling model and designed the Q-learning scheduling algorithm in order to short the quay waiting time [9].

In order to improve loading/unloading operation efficiency of the truck resource, this paper establish the minimum loading and unloading operation time of container truck route scheduling model, which the operating time of quay cranes and yard cranes are considered. This paper propose the hybrid swarm intelligence algorithm combined the particle swarm optimization algorithm (PSO) with ant colony optimization (ACO) algorithm. At last, the small-scale and large-scale examples are adopted to examine the performance of the proposed hybrid swarm intelligence algorithm.

2 Model Formulation

The shortest operation time of truck scheduling problem is the truck route optimization problem to finish the transportation from the import to export containers, under the condition of container storage operation sequence have been identified. The following notations are also used in the model formulation:

Let m denote the number of containers in the import container area, n denote the number of containers in the export container area, k is the number of import berths, L is the number of export berths, Ci be the quantity of containers assigned to stockpile in the I import container area, Dj be the quantity of containers assigned to stockpile in the J export container area, Ba be the quantity of import containers through the a import berth, Hb be the quantity of export containers through the b export berth. Xaijb, Yai, Zjb are decision variables. Xaijb be the transportation number of trucks on the path of the “Working Group”, which is the transportation number on the path: the a import berth → the I import container area → the J export container area → the b export berth, and Yai be the transportation number of trucks on the path: the a import berth → the I import container area → the a import berth, and Zjb be the transportation number of trucks on the path: the b export berth → the j export container area the b export berth.

The truck scheduling problem for container truck at port container wharfs can be formulated as follows:

$$J = \min \sum_{i=1}^{m}\sum_{j=1}^{n}\sum_{a=1}^{k}\sum_{b=1}^{l}[(t_{ai}+t_i+t_{ij}+t_j+t_{jb}+t_b+t_{ba}+t_a)\times X_{aijb} + (t_{ai}+t_i+t_{ia}+t_a)\times Y_{ai}+(t_{bj}+t_j+t_{jb}+t_b)\times Z_{jb}] \tag{1}$$

$$s.t.\quad X_{aijb}, Y_{ai}, Z_{jb} \in Z \tag{2}$$

$$\begin{Bmatrix} \sum_{a=1}^{k} B_a > \sum_{b=1}^{l} H_b, \sum_{i=1}^{m}\sum_{a=1}^{k} Y_{ai} = \sum_{a=1}^{k} B_a - \sum_{b=1}^{l} H_b, \sum_{j=1}^{n}\sum_{b=1}^{l} Z_{jb} = 0 \\ \sum_{a=1}^{k} B_a < \sum_{b=1}^{l} H_b, \sum_{i=1}^{m}\sum_{a=1}^{k} Y_{ai} = 0, \sum_{j=1}^{n}\sum_{b=1}^{l} Z_{jb} = \sum_{b=1}^{l} H_b - \sum_{a=1}^{k} B_a \\ \sum_{b=1}^{l} H_b = \sum_{a=1}^{k} B_a, \sum_{j=1}^{n}\sum_{b=1}^{l} Z_{jb} = \sum_{i=1}^{m}\sum_{a=1}^{k} Y_{ai} = 0 \end{Bmatrix} \tag{3}$$

$$\sum_{i=1}^{m}\sum_{j=1}^{n}\sum_{a=1}^{k}\sum_{b=1}^{l} X_{aijb} = \min\left\{\sum_{a=1}^{k} B_a, \sum_{b=1}^{l} H_b\right\} \tag{4}$$

$$\sum_{j=1}^{n}\sum_{a=1}^{k}\sum_{b=1}^{l} X_{aijb} + \sum_{b=1}^{l} Z_{jb} = D_j, j = 1,2,\ldots,n \tag{5}$$

$$\sum_{j=1}^{n}\sum_{a=1}^{k}\sum_{b=1}^{l} X_{aijb} + \sum_{a=1}^{k} Y_{ai} = C_i, i = 1,2,\ldots,m \tag{6}$$

$$\sum_{i=1}^{m}\sum_{j=1}^{n}\sum_{b=1}^{l} X_{aijb} + \sum_{i=1}^{m} Y_{ai} = B_a, a = 1,2,\ldots,k \tag{7}$$

$$\sum_{i=1}^{m}\sum_{j=1}^{n}\sum_{a=1}^{k} X_{aijb} + \sum_{j=1}^{n} Z_{jb} = H_b, b = 1,2,\ldots,l \tag{8}$$

Equation (1) represents the objective function of the truck transportation shortest time problem that is to complete all the import and export containers. Constraints (2) represent the Xaijb, Yai, Zjb are integers. Constraints (3) represent the computation of truck travel time of "Working Group". Constraints (4) represent the single truck travel time to transport import or export container, $\sum_{j=1}^{n}\sum_{b=1}^{l} Z_{jb}, \sum_{i=1}^{m}\sum_{a=1}^{k} Y_{ai}$ at least one equals 0 and constraints (5) ensure that the truck must transport container from berth to the import container area i and stockpiling, constraints (6) ensure that the truck must transport container from the export container area j to berth. Constraints (7) represent that the imported container from the A imported berth must be transported by trucks from the A import berth to the imported container area. Constraints (8) represent that the exported container from the B exported berth must be transported by trucks from the exported container area to the B export berth.

3 The Hybrid Swarm Intelligence Algorithm

3.1 PSO Presentation

American social psychologist James Kennedy and electronic engineer Russell Eberhart put forward the Particle Swarm Optimization (PSO) theory in 1995 [10]. The PSO algorithm was derived from bird flock behaviour which can be regarded as the movement of particle swarm with no quality and volume. The bird flock can change its flying speed and adjust flying direction in order to make whole colony move into a better environment. Compared with other algorithms, PSO algorithm has many merits. One is that the simple algorithm flow can be easily achieved, the other is that few parameters need to be adjusted. Its algorithm [11] can be described as follows:

Total number of particles is n, each flies in a D-dimensional space. The velocity and location for the ith particle is represented as Vi = (Vi1, Vi2,…, Vin) and Xi = (Xi1, Xi2,…, Xin), respectively. Its best previous position is recorded and represented as Pi = (Pi1, Pi2,…, Pin), which is also called Pbest. The index of the best Pbest is represented by the symbol g, i.e. Pg, which is also called Gbest. At each step, the particles are manipulated according to the following equations:

$$V_i(t+1) = wV_i(t) + c_1 rand(P_{best}(t) - X_i(t)) + c_2 rand(G_{best}(t) - X_i(t)) \quad (9)$$

$$X_i(t+1) = X_i(t) + V_i(t+1) \quad (10)$$

where w is inertia weight, c1and c2 are acceleration constants, rand() is a random value from 0 to 1. In order to reduce the feasibility of particle departed from searching space, velocity was restricted in Vmax. If the particle accelerate its speed and exceed Vmax, in the same dimension, the speed will be given Vmax = k·Vmax, $(0.1 \leq k \leq 1.0)$.

In the (9) equation, the first part denotes particle initial speed, the second part means the "cognition" part that particles can think by themselves, the third part means the "social" part that particles can share the information and cooperate one another. The particles can imitate one another. "cognition" reinforced random activity will possibly appear in the future, which mean the particle will be encouraged to mini-deviation.

The PSO algorithm can keep and utilize the information of position and speed simultaneously in the evolutional course. Yet other kinds of algorithm, such as genetic algorithm merely use the position information. Therefore, the PSO algorithm has more advantage and feasibility than others in this optimization problem.

3.2 The Hybrid Swarm Intelligence Algorithm

The scheduling problem of container truck according the operation time can be split into two steps: The first step is to make distribution of all containers to each trucks, The second step is to optimization the routes of each truck to accomplish the task.

Step 1. Initialization the parameters of particle swarm algorithm: The number of virtual tasks is n, The total number of truck is M, The two dimensional particle swarm vector will be constructed, and the each dimension length is n that means the number of the task. The components X_i in the first dimension means the mapping task allocation, which value are the real numbers within [1, M + 1) according to the distributed container truck. The components Y_i in the second dimension means the sequence of vehicle route within the [Y_min, Y_max]. The position of particle will be updated by two-dimensional velocity vector, and the decoding rules of particle as following:

Firstly, The value X_{ij} of first dimension X_i should been Integer processed within [1, M] in order to make the j task corresponded to the l truck, which means the particle position vector corresponded to the task allocation plan.

Secondly, The task J are corresponded to the each truck l (l = 1, 2,..., M), And make the corresponded sequence of Y_{ij} by the Y_i, The truck path L will be determined ty the sequence results of the corresponded task (Tables 1 and 2).

For example, there is the scheduling problem, which has the n = 10 missions and M = 4 vehicles. The particle position vector of I are as following:

Table 1. Vector of particle position

Mission	1	2	3	4	5	6	7	8	9	10
X_i	1.8	4.1	1.5	3.8	2.6	3.3	2.9	1.8	3.7	4.9
Y_i	4.13	1.67	3.53	5.56	0.19	2.67	6.6	3.25	1.78	3.36

After the integral operation on the first dimension vector, the task allocations are as follows:

Table 2. The truck and task allocation

Mission	1	2	3	4	5	6	7	8	9	10
X_i	1	4	1	3	2	3	2	1	3	4

In the ascending order of the second dimension vector, the truck routing path are as follows:

Table 3. The rout path of truck

Truck	Path
$\frac{W_{max}-W_{min}}{iner_{max}}$	
1	0-1-3-8-0
2	0-5-7-0
3	0-9-6-4-0
4	0-2-10-0

The two-dimensional structure of particle swarm optimization algorithm has efficiency in solving the truck scheduling process, which take into account vehicle allocation and the corresponding path optimization.

Step 2. Construction and definition the Initial value of the two-dimensional particle velocity vector according to the two-dimensional position vector.

Step 3. The objective function of the model is as the fitness function, and calculate the fitness value and Initial the individual extreme according to the initial position of each particle vector, and then find the smallest as the initial global extreme value point of whole particle swarm.

Step 4. Adaptive inertia weight: Inertia weight is an very important parameter of the particle swarm optimum algorithm. The larger value of inertia weight can enlarger the search area of PSO, while the small one enhance the search capacity of PSO in the local area. Therefore, the improved PSO has the ability of local search and self-adaptive changing area by setting the value of inertia weight from larger one to small one. The function of adaptive inertia weight is as following:

$$W = W_{\max} - *iner \tag{11}$$

where $W_{\min}$ is the original value of inertia weight, $W_{\max}$ is the finally value of inertia weight, $iner_{\max}$ is the maximum iterative time and $iter$ is the current iterative time.

Step 5. Use the adaptive inertia weight method to update the velocity and position vector. To the first dimension vector, if the updated values are out of the bounds, the value will be set by the boundary. There is no limit to the second dimension vector.

Step 6. Evaluation of each particle by fitness function, update the Pbest, Gbest by Eqs. (9) and (10), make the particle search at new position respectively.

Step 7. If the maximum number of cycles is satisfied, then break to step8 and transmit the results of truck assignment to the ant colony optimization algorithm [12], else jump to the step 4 continue to make loop.

Step 8. Initialize the parameters of ant colony algorithm, and established the taboo search matrix and candidate list. antk(k = 1,2,..., m) determines the direction of the path based on the pheromone concentration. The transformation probability of $\mathrm{p}_{\mathrm{ij}}^{\mathrm{k}}(\mathrm{t})$ present the probability that ant k transfer from position i to j at t time.

$$\mathrm{p}_{\mathrm{ij}}^{\mathrm{k}}(\mathrm{t}) = \begin{cases} (\tau_{ij}^{\alpha}(t)\eta_{ij}^{\beta}) / \sum (\tau_{ij}^{\alpha}(t)\eta_{ij}^{\beta})\ if\ j \in S_i^k \\ 0 \quad otherwise \end{cases} \tag{12}$$

$\tau ij(t)$ is the information pheromone of path ij at t time, ηij is inspire information $\eta ij = 1/dij$, dij present the distance from ith city to jth city, α and β are constant, which present the importance of pheromone concentration and the inspired information separately, s_{ik} present the neighborhood of k ant near the point i, that is set of visited nodes.

Step 9. The ants variable $\mathrm{M_k}$ randomly distribute on the $\mathrm{N_k}$ nodes, (k = 1, 2,..., M). firstly, each ant select the transfer nodes in candidate list according to the state transfer formula. If there are no access nodes in the candidate list, then the ant colony optimization algorithm random choose the node as the next access node, else use the local pheromone update strategy to update the pheromone on the selected path.

Step 10. After every ant find the loop rout, use the 2-OPT local algorithm to optimize the path, The final results will be the path that ants has found. After calculating the length of all paths, The algorithm will compare the current shortest path with the all paths found. If there has the path better than the current shortest path, then the algorithm will update the pheromone on the current shortest path according to the global updating strategy.

Step 11. If the fitness value has been achieved and meet the end conditions, then the final results will be output, else jump to the Step 9 continue to make loop (Fig. 1).

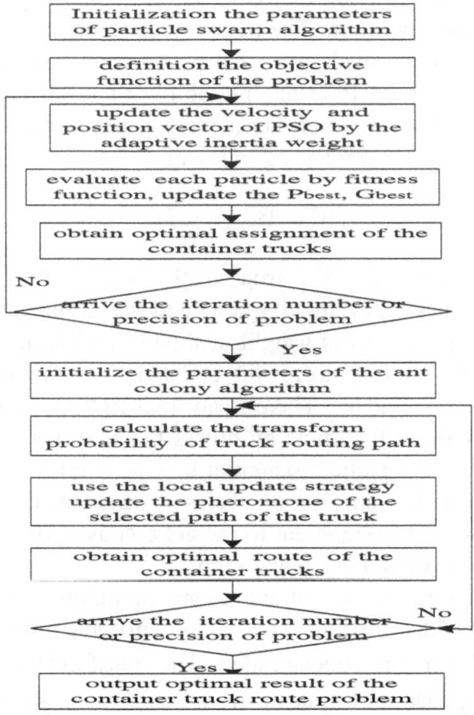

Fig. 1. The flow chart of PSO-ACO algorithm

4 Computational Experiments

Supposed there were 3 import container yards and 4 export container yards, which are used to stack dispersedly the loaded and unloaded containers. Sometime, The ships will anchor the A, B berth. In order to loaded 430TEU on the ship, the A berth will open 2 quays. In order to uploaded 655TEU from the ship, the B berth will open 3 quays. The distances between each quays to the container yards are as following, the plan container number in each yards (TEU) are show (Table 4). The hybrid swarm intelligence algorithm optimize the routing and driving distance of the container trucks (Tables 5 and 6).

Table 4. The distance from quay to the export and import container yards (km)

The distance	U_1	U_2	U_3	L_1	L_2	L_3	L_4
A berth NO. 1 quay to L				0.65	1.50	0.75	1.55
A berth NO. 2 quay to L				0	1.35	0.65	1.50
U to B berth NO. 1 quay	2.55	2.75	2.90				
U to B berth NO. 2 quay	2.60	2.75	2.36				
U to B berth NO. 3 quay	2.80	2.70	2.15				

Table 5. The container number of each yards (TEU)

The import container yards	U_1	U_2	U_3	SUM	The export container yards	L_1	L_2	L_3	L_4	SUM
TEU	215	210	230	655	TEU	100	120	130	80	430

Table 6. The optimizing route of the container truck

Transportation path	Loaded container	Unloaded container
B berth NO. 1 quay → NO. 1 imported container yard→ NO.1 exported container yard → A berth NO.1 quay	89	89
B berth NO. 1 quay → NO. 1 imported container yard→ NO.2 exported container yard → A berth NO.1 quay	58	58
B berth NO. 1 quay → NO. 1 imported container yard→ NO.2 exported container yard → A berth NO.2 quay	63	63
B berth NO. 1 quay → NO. 3 imported container yard→ NO.4 exported container yard → A berth NO.1 quay	9	9
B berth NO. 2 quay → NO. 3 imported container yard→ NO.3 exported container yard → A berth NO.2 quay	155	155
B berth NO. 2 quay → NO. 3 imported container yard→ NO.4 exported container yard → A berth NO.1 quay	66	66
B berth NO. 3 quay → The NO. 2 import container yards	0	190
B berth NO. 3 quay → The NO. 3 import container yards	0	17

The parameters of the PSO are as follows: the number of particles is 30, learning gene $C_1 = C_2 = 1.49$, $W = 0.729$, iterative number is 5 times. The parameters set of ACO algorithm are as following: $\alpha = 1$, $\beta = 2$, $\tau ij(t) = 0.1$, $\eta ij = 0.3$, $s_{ik} = 0.7$. After these algorithms run 500 times respectively, the solutions, run-time, and the rate of the best solution are showed on Table 3.

The Table 3 show that the transportation distance of truck on the model is moor less than that the independent of loading and unloading from the truck. The total distance saves 20.28%, greatly reduce the truck empty transportation distance. The hybrid swarm intelligence algorithm improve the efficiency of the container truck, so that save the energy consumption of the wharf.

5 Conclusion

This paper construct the integer programming model of the truck routing optimization problem. The objective of the problem is the minimum process time of the loaded and unloaded containers. The hybrid swarm intelligence algorithm(PSO-ACO) is proposed to handle this problem, which take advantage of strong local search ability of ACO and the global optimum search ability of PSO. Results of the simulation show that the hybrid swarm intelligence algorithm has high converging speed, stability and suitable for solving container truck scheduling problem.

Acknowledgment. The project supported by the zhejiang provincial natural science foundation of China (Foundation No. LY14G010006).

References

1. Kim, K.H.: An optimal routing algorithm for a transfer crane in port container terminals. Transp. Sci. **33**(1), 17–33 (1999)
2. Bish, E.K., Leong, T., Li, C., et al.: Analysis of a new vehicle scheduling and location problem. Naval Res. Logist. **48**, 363–385 (2001)
3. Bish, E.K., Chen, F.Y., Leong, Y.T., et al.: Dispatching vehicles in a mega container terminal. OR Spectr. **27**(4), 491–506 (2005)
4. Nishimura, E., Akio, I., Stratos, P.: Yard trailer routing at a maritime container terminal. Transp. Res. E **41**(1), 53–76 (2005)
5. Lin, S., Wei, Y., Vincent, F., Lu, C.: A simulated annealing heuristic for the truck and trailer routing problem with time windows. Expert Syst. Appl. **38**(12), 15244–15252 (2011)
6. Derigs, U., Pullmann, M., Vogel, U.: Truck and trailer routing problems, heuristics and computational experience. Comput. Oper. Res. **40**(2), 536–546 (2013)
7. Ji, M., Jin, Z.: A united optimization of crane scheduling and yard trailer routing in a container terminal. J. Fudan Univ. Nat. Sci. **46**(4), 476–480 (2007)
8. Hong, G., Zhu, J.: Operation priority strategy of container port truck path optimization based on. Chin. water Transp. **12**, 70–72 (2012)
9. Qing, C., Zhong, Z.: A scheduling model and Q-learning algorithm for yard trailers at container terminals. J. Harbin Eng. Univ. **29**(1), 1–4 (2008)
10. Cao, Q.-K., Zhao, F.: Port trucks route optimization based on GA-ACO. Syst. Eng. Theor. Pract. **33**(7), 1820–1828 (2013)
11. Chen, T.Y., Chi, T.M.: On the improvements of the particle swarm optimization algorithm. Adv. Eng. Softw. **41**(2), 229–239 (2010)
12. Kaur, N., Sharma, J.P.: Mobile Sink and ant colony optimization based energy efficient routing algorithm. Int. J. Comput. Appl. **121**(1), 23–31 (2015)

A New Adaptive Firefly Algorithm for Solving Optimization Problems

Wenjun Wang[1], Hui Wang[2,3(✉)], Jia Zhao[2,3], and Li Lv[2,3]

[1] School of Business Administration, Nanchang Institute of Technology, Nanchang 330099, China

[2] Jiangxi Province Key Laboratory of Water Information Cooperative Sensing and Intelligent Processing, Nanchang Institute of Technology, Nanchang 330099, China

huiwang@whu.edu.cn

[3] School of Information Engineering, Nanchang Institute of Technology, Nanchang 330099, China

Abstract. Firefly algorithm (FA) is an efficient optimization technique, which has been widely used to solve various engineering problems. However, FA is sensitive to its control parameters. Recently, a memetic FA (called MFA) is proposed to improve the sensitivity of FA. To further enhance the performance of MFA, this paper proposes a new method to adaptively adjust the step factor. Experiments on several benchmark problems show that our approach is superior to the standard FA, MFA, and some other improved FAs.

Keywords: Firefly algorithm · Dynamic · Adaptive parameter · Optimization

1 Introduction

Optimization problems exist in many engineering fields. With the growth of problem complexity, stronger optimization algorithms are required. In the past decades, several new optimization techniques have been proposed by the inspiration of swarm intelligence, such as artificial bee colony (ABC) [1–4], bat algorithm (BA) [5–7], firefly algorithm (FA) [8–10], cuckoo search (CS) [11, 12], fruit fly optimization (FFO) [13], and artificial plant optimization algorithm [14, 15]. FA is a new optimization technique originally invented by Prof. Yang [8]. In FA, every firefly moves toward new positions and find potential solutions by the attraction of other brighter fireflies. Some latest researches proved that FA is an efficient optimization tool.

However, the standard FA still has some drawbacks. For instance, FA is sensitive to the control parameters, and the convergence speed is slow. To tackle these issues, several improved strategies have been designed. Recently, a memetic FA (MFA) was proposed, in which two improved strategies are employed [16]. First, the step factor α is dynamically decreased based on an empirical model. Second, the attractiveness β is constrained into a predefined range [0.2, 1.0]. Moreover, the factor α is multiplied by the length of search interval. Gandomi et al. [17] presented a chaotic FA, in which different chaotic maps were used to adjust the step factor α and the light absorption coefficient γ. Besides Gandomi's work, some researchers also combined FA with chaos

D.-S. Huang et al. (Eds.): ICIC 2017, Part I, LNCS 10361, pp. 649–657, 2017.
DOI: 10.1007/978-3-319-63309-1_57

to obtain a good performance. In [18], an adaptive selection method was used to choose the parameter α from a candidate set. In [19], Wang et al. investigated the relationships between convergence and the parameter α. Results show that α should tend to zero when FA converges to a solution. Based on this principle, Wang et al. designed two dynamic methods adjust α, in which α is gradually decreased based on different models as the generation increases [19, 20].

In this paper, we present a new adaptive firefly algorithm (AFA), which is an enhanced version of MFA. In AFA, we combine MFA with a new adaptive parameter strategy to dynamically adjust the step factor α. Thirteen famous test functions arc used for performance verification. Results show that AFA is superior to the standard FA, MFA [16], chaotic FA (CFA) [17], and FA with random attraction (RaFA) [9].

2 Firefly Algorithm

For two fireflies X_i and X_j, their attractiveness β is defined as follows [21].

$$\beta(r_{ij}) = \beta_0 e^{-\gamma r_{ij}^2} \tag{1}$$

where β_0 is the attractiveness at $r = 0$, γ is the light absorption coefficient, and r_{ij} is the distance between X_i and X_j. The distance r_{ij} is computed as follows [21].

$$r_{ij} = \left\| X_i - X_j \right\| = \sqrt{\sum_{d=1}^{D} (x_{id} - x_{jd})^2} \tag{2}$$

where D is the problem size.

When X_j is brighter (better) than X_i, X_i will move towards X_j because of the attraction. In the standard FA, this movement is defined as follows [21].

$$x_{id}(t+1) = x_{id}(t) + \beta(r_{ij}) \cdot \left(x_{jd}(t) - x_{id}(t)\right) + \alpha\left(rand - \frac{1}{2}\right) \tag{3}$$

where x_{id} and x_{jd} are the dth dimensions of X_i and X_j, respectively, $\alpha \in [0, 1]$ is called step factor, and *rand* is a random value within [0, 1].

3 Proposed Approach

Recently, a memetic firefly algorithm (MFA) was designed to enhance the performance of FA [16]. The MFA made three improvements. Firstly, the step factor α is dynamically updated as follows.

$$\alpha(t+1) = \left(\frac{1}{9000}\right)^{\frac{1}{t}} \alpha(t) \tag{4}$$

where t indicates the generation index. We can find that the value of α decreases with the growth of t.

Secondly, the definitions of the attractiveness β is modified. The new β is calculated as follows.

$$\beta(r_{ij}) = \beta_{\min} + (\beta_0 - \beta_{\min})e^{-\gamma r_{ij}^2} \quad (5)$$

where $\beta_{\min}$ is the minimum value of the attractiveness β. The β is constrained into the range $[\beta_{\min}, \beta_0]$. In [16], $\beta_{\min}$ and β_0 are equal to 0.2, and 1.0, respectively.

Thirdly, the step factor α is multiplied by the length of the search range by the suggestions of [21]. Then, the new movement equation is modified as follows.

$$x_{id}(t+1) = x_{id}(t) + \beta(r_{ij}) \cdot (x_{jd}(t) - x_{id}(t)) + \alpha(t) \cdot s_d \cdot \left(rand - \frac{1}{2}\right) \quad (6)$$

where s_d is the length of the search interval of the dth dimension.

Based on MFA, we propose an enhanced version by employing a new adaptive parameter strategy to adjust the step factor α. In our approach AFA, Eq. (4) is modified as follows.

$$\alpha(t+1) = \left(1 - \frac{FEs}{MaxFEs}\right)^m \alpha(t) \quad (7)$$

where $m > 0$, *FEs* represents the number of fitness evaluations, *MaxFEs* indicates the maximum value for the *FEs*, and t is the generation number. When $0 < m < 1$, a small

Algorithm 1: The Proposed AFA

```
Begin
    Randomly initialize all fireflies in the population;
    FEs=N;
    while FEs <= MaxFEs do
        Update the parameter α according to Eq. (7);
        for i=1 to N do
            for j=1 to i do
                if f(X_j)<f(X_i) then
                    Move X_i towards X_j according to Eq. (6);
                    Evaluate the new solution X_i;
                    FEs++:
                end if
            end for
        end for
    end while
End
```

Fig. 1. The framework of our approach AFA.

value is added to the weighed term of $\alpha(t)$. It avoids that $1 - FEs/MaxFEs = 0$. In our experiments, m is set to 0.5. In fact, Eq. (7) is a general version of our previous work [19].

The framework of AFA is given in Fig. 1. Compared to MFA, AFA only modifies the updating strategy of the step factor α. Therefore, AFA has the same complexity with MFA.

4 Experimental Study

4.1 Experimental Setup

In the experiment, thirteen benchmark functions are utilized for performance verification [22–26]. All test functions are minimization problems. Table 1 gives a brief description of these functions. More detailed descriptions of these functions can be found in [27–30].

Table 1. A brief descriptions of test functions.

Functions	Search interval	Global minimum
Sphere (f_1)	$[-100, 100]$	0
Schwefel 2.22 (f_2)	$[-10, 10]$	0
Schwefel 1.2 (f_3)	$[-100, 100]$	0
Schwefel 2.21 (f_4)	$[-100, 100]$	0
Rosenbrock (f_5)	$[-30, 30]$	0
Step (f_6)	$[-100, 100]$	0
Quartic with noise (f_7)	$[-1.28, 1.28]$	0
Schwefel 2.26 (f_8)	$[-500, 500]$	0
Rastrigin (f_9)	$[-5.12, 5.12]$	0
Ackley (f_{10})	$[-32, 32]$	0
Griewank (f_{11})	$[-600, 600]$	0
Penalized 1 (f_{12})	$[-50, 50]$	0
Penalized 2 (f_{13})	$[-50, 50]$	0

In the comparison, AFA is compared with four FAs. The related FAs are presented as follows.

- FA
- MFA [16]
- CFA [17]
- RaFA [9]
- Proposed AFA

The parameters N and $MaxFEs$ are equal to 20 and $5.0E+05$, respectively. In the standard FA, the parameters α, β_0, and γ are set to 0.2, 1.0, and $\gamma = 1/\Gamma^2$, respectively. For MFA, RaFA, and AFA, the initial α, β_{min}, β_0, and γ are set to 0.5, 0.2, 1.0, and 1.0, respectively. The parameter m used in AFA is set to 0.5. Besides AFA, RaFA is also an improved version of MFA.

4.2 Results

Table 2 presents the results of AFA, FA, MFA, CFA, and RaFA on thirteen test functions, where "Mean" is the mean best fitness value over on 30 runs. From the results, AFA outperforms the standard FA on 11 functions, while AFA achieves worse solutions on two functions f_3 and f_7. For function f_3, AFA is trapped into local optima and can hardly obtain reasonable solutions. For function f_7, the standard FA is a little better than AFA. Compared to MFA, the proposed adaptive parameter strategy helps AFA to achieve significant improvements, especially for f_1, f_2, and f_{10}–f_{13}. CFA outperforms AFA on three functions f_3, f_5 and f_8. For function f_6, AFA, CFA, RaFA, and MFA find the same solution. RaFA is better than AFA on 6 functions, while AFA achieves more accurate solutions than RaFA on 6 functions.

Table 2. Results for different FAs.

Functions	FA mean	MFA mean	CFA mean	RaFA mean	AFA mean
f_1	5.14E−02	1.56E−05	3.27E−06	5.36E−184	5.36E−184
f_2	1.07E+00	1.85E−03	8.06E−04	8.76E−05	1.73E−07
f_3	1.26E−01	5.89E−05	1.24E−05	4.91E+02	8.57E+01
f_4	9.98E−02	1.73E−03	8.98E−04	2.43E+00	1.97E−04
f_5	3.41E+01	2.29E+01	2.06E+01	2.92E+01	2.69E+01
f_6	5.24E+03	0.00E+00	0.00E+00	0.00E+00	0.00E+00
f_7	7.55E−02	1.30E−01	9.03E−02	5.47E−02	7.87E−02
f_8	9.16E+03	4.94E+03	4.36E+03	5.03E+02	4.62E+03
f_9	4.95E+01	6.47E+01	5.27E+01	2.69E+01	4.08E+01
f_{10}	1.21E+01	4.23E−04	4.02E−04	3.61E−14	2.19E−14
f_{11}	2.13E−02	9.86E−03	7.91E−06	0.00E+00	7.78E−16
f_{12}	6.24E+00	5.04E−08	8.28E−09	4.50E−05	1.22E−25
f_{13}	5.11E+01	6.06E−07	1.69E−07	8.25E−32	8.36E−22

Figures 2 and 3 show the convergence graphs on some unimodal functions and multimodal functions, respectively. As seen, both RaFA and AFA converges faster than FA, MFA, and CFA.

To compare the optimization performance of the five FA variants on the whole test set, we calculate the mean rank values by the Friedman test. Table 3 gives the mean rank values of the five algorithms. The highest rank is marked in boldface. It is obvious that AFA obtains the highest rank. It demonstrates that AFA is the best one among AFA, FA, CFA, MFA, and RaFA.

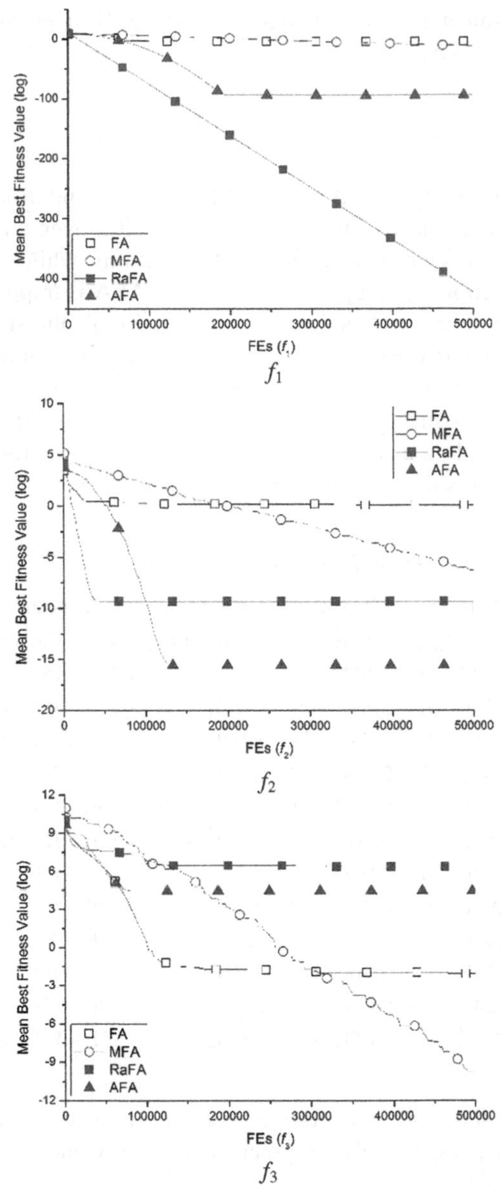

Fig. 2. The convergences processes of different FAs on some unimodal functions.

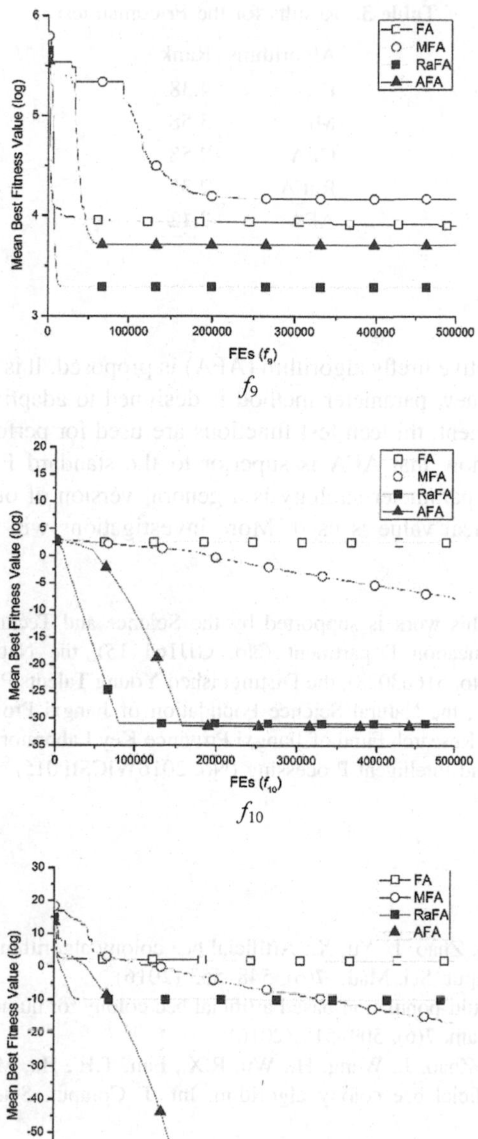

Fig. 3. The convergences processes of different FAs on some multimodal functions.

Table 3. Results for the Friedman test.

Algorithms	Rank
FA	4.38
MFA	3.58
CFA	2.58
RaFA	2.35
AFA	**2.12**

5 Conclusions

In this paper, an adaptive firefly algorithm (AFA) is proposed. It is an enhanced version of MFA. In AFA, a new parameter method is designed to adaptively change the step factor. In the experiment, thirteen test functions are used for performance verification. Simulation results show that AFA is superior to the standard FA, MFA, CFA, and RaFA. The adaptive parameter strategy is a general version of our previous work. In this paper, an empirical value is used. More investigations will be conducted in the future work.

Acknowledgement. This work is supported by the Science and Technology Plan Project of Jiangxi Provincial Education Department (No. GJJ161115), the National Natural Science Foundation of China (No. 61663028), the Distinguished Young Talents Plan of Jaingxi Province (No. 20171BCB23075), the Natural Science Foundation of Jiangxi Province (No. 20171BAB 202035), and the Open Research Fund of Jiangxi Province Key Laboratory of Water Information Cooperative Sensing and Intelligent Processing (No. 2016WICSIP015).

References

1. Sun, H., Wang, K., Zhao, J., Yu, X.: Artificial bee colony algorithm with improved special centre. Int. J. Comput. Sci. Math. **7**(6), 548–553 (2016)
2. Yun, G.: A new multi-population-based artificial bee colony for numerical optimization. Int. J. Comput. Sci. Math. **7**(6), 509–515 (2016)
3. Lv, L., Wu, L.Y., Zhao, J., Wang, H., Wu, R.X., Fan, T.H., Hu, M., Xie, Z.F.: Improved multi-strategy artificial bee colony algorithm. Int. J. Comput. Sci. Math. **7**(5), 467–475 (2016)
4. Lu, Y., Li, R.X., Li, S.M.: Artificial bee colony with bidirectional search. Int. J. Comput. Sci. Math. **7**(6), 586–593 (2016)
5. Cai, X., Gao, X.Z., Xue, Y.: Improved bat algorithm with optimal forage strategy and random disturbance strategy. Int. J. Bio-Inspired Comput. **8**(4), 205–214 (2016)
6. Xue, F., Cai, Y., Cao, Y., Cui, Z., Li, F.: Optimal parameter settings for bat algorithm. Int. J. Bio-Inspired Comput. **7**(2), 125–128 (2015)
7. Cai, X., Wang, L., Kang, Q., Wu, Q.: Bat algorithm with Gaussian walk. Int. J. Bio-Inspired Comput. **6**(3), 166–174 (2014)
8. Yang, X.S.: Nature-Inspired Metaheuristic Algorithms. Luniver Press, Beckington (2008)
9. Wang, H., Wang, W.J., Sun, H., Rahnamayan, S.: Firefly algorithm with random attraction. Int. J. Bio-Inspired Comput. **8**(1), 33–41 (2016)

10. Wang, H., Wang, W.J., Zhou, X.Y., Sun, H., Zhao, J., Yu, X., Cui, Z.: Firefly algorithm with neighborhood attraction. Inf. Sci. **382–383**, 374–387 (2017)
11. Cui, Z., Sun, B., Wang, G., Xue, Y.: A novel oriented cuckoo search algorithm to improve DV-Hop performance for cyber-physical systems. J. Parallel Distrib. Comput. **103**, 42–52 (2017)
12. Wang, G.G., Gandomi, A.H., Yang, X.S., Alavi, A.H.: A new hybrid method based on krill herd and cuckoo search for global optimization tasks. Int. J. Bio-Inspired Comput. **8**(5), 286–299 (2016)
13. Zhang, Y.W., Wu, J.T., Guo, X., Li, G.N.: Optimising web service composition based on differential fruit fly optimisation algorithm. Int. J. Comput. Sci. Math. **7**(1), 87–101 (2016)
14. Cui, Z., Fan, S., Zeng, J., Shi, Z.Z.: APOA with parabola model for directing orbits of chaotic systems. Int. J. Bio-Inspired Comput. **5**(1), 67–72 (2013)
15. Cui, Z., Fan, S., Zeng, J., Shi, Z.Z.: Artificial plant optimisation algorithm with three-period photosynthesis. Int. J. Bio-Inspired Comput. **5**(2), 133–139 (2013)
16. Fister Jr., I., Yang, X.S., Fister, I., Brest, J.: Memetic firefly algorithm for combinatorial optimization. arXiv preprint arXiv:1204.5165 (2012)
17. Gandomi, A.H., Yang, X.S., Talatahari, S., Alavi, A.H.: Firefly algorithm with chaos. Commun. Nonlinear Sci. Numer. Simul. **18**(1), 89–98 (2013)
18. Roy, A.G., Rakshit, P., Konar, A., Bhattacharya, S., Kim, E., Nagar, A.K.: Adaptive firefly algorithm for nonholonomic motion planning of car-like system. In: IEEE Congress on Evolutionary Computation (CEC 2013), pp. 2162–2169. IEEE, Cancun (2013)
19. Wang, H., Zhou, X.Y., Sun, H., Yu, X., Zhao, J., Zhang, H., Cui, L.Z.: Firefly algorithm with adaptive control parameters. Soft. Comput. 1–12 (2016). doi:10.1007/s00500-016-2104-3
20. Wang, H., Cui, Z.H., Sun, H., Rahnamayan, S., Yang, X.S.: Randomly attracted firefly algorithm with neighborhood search and dynamic parameter adjustment mechanism. Soft. Comput. 1–15 (2016). doi:10.1007/s00500-016-2116-z
21. Yang, X.S.: Engineering Optimization: An Introduction with Metaheuristic Applications. Wiley, Hoboken (2010)
22. Wang, H., Wu, Z.J., Rahnamayan, S., Liu, Y., Ventresca, M.: Enhancing particle swarm optimization using generalized opposition-based learning. Inf. Sci. **181**(20), 4699–4714 (2011)
23. Wang, H., Rahnamayan, S., Sun, H., Omran, M.G.H.: Gaussian bare-bones differential evolution. IEEE Trans. Cybern. **43**(2), 634–647 (2013)
24. Guo, Z.L., Wang, S.W., Yue, X.Z., Yin, B.: Enhanced social emotional optimisation algorithm with elite multi-parent crossover. Int. J. Comput. Sci. Math. **7**(6), 568–574 (2016)
25. Wang, H., Liu, Y., Li, C.H., Zeng, S.Y.: A hybrid particle swarm algorithm with Cauchy mutation. In: IEEE Swarm Intelligence Symposium (SIS 2007), pp. 356–360. IEEE, Honolulu (2007)
26. Yu, G.: An improved firefly algorithm based on probabilistic attraction. Int. J. Comput. Sci. Math. **7**(6), 530–536 (2016)
27. Wang, H., Sun, H., Li, C.H., Rahnamayan, S., Pan, J.S.: Diversity enhanced particle swarm optimization with neighborhood search. Inf. Sci. **223**, 119–135 (2013)
28. Wang, H., Wu, Z.J., Rahnamayan, S., Sun, H., Liu, Y., Pan, J.S.: Multi-strategy ensemble artificial bee colony algorithm. Inf. Sci. **279**, 587–603 (2014)
29. Zhou, X.Y., Wu, Z.J., Wang, H., Rahnamayan, S.: Gaussian bare-bones artificial bee colony algorithm. Soft. Comput. **20**(3), 907–924 (2016)
30. Zhou, X.Y., Wang, H., Wang, M.W., Wan, J.Y.: Enhancing the modified artificial bee colony algorithm with neighborhood search. Soft. Comput. **21**(10), 2733–2743 (2017)

A Hybrid Algorithm of Adaptive Particle Swarm Optimization Based on Adaptive Moment Estimation Method

Yan Jiang(✉) and Fei Han

School of Computer Science and Communication Engineering, Jiangsu University, Zhenjiang, Jiangsu, China
crystal_jyan@163.com, hanfei@ujs.edu.cn

Abstract. Particle swarm optimization (PSO) algorithm is a promising swarm intelligence optimization technology. It has been applied to a variety of complex optimization problems due to its outstanding global search ability. However, it suffers from premature convergence and slow convergence rate. Motivated by adaptive moment estimation (Adam) method, which is computationally efficient, little memory-required and also appropriate for non-stationary objectives, a hybrid algorithm combining adaptive PSO with a modified Adam method (AdamPSO) is proposed in this paper. Adaptive particle swarm optimization (APSO) is first used to perform stochastic and rough search. In the solution space obtained by APSO, Adam method is then used to perform further search, which may establish a new solution space. Depending on the fitness value of particles, the position of each particle switches alternately between APSO and Adam. The experimental results on six well-known benchmark functions show that our proposed algorithm gets better convergence performance compared to other five classical PSOs.

Keywords: Particle swarm optimization · Adaptive particle swarm optimization · Adaptive moment estimation

1 Introduction

The particle swarm optimization (PSO) algorithm is a global search strategy that can efficiently handle arbitrary optimization problems. PSO was first proposed by Kennedy and Eberhart in 1995, referred to a relatively heuristic search method whose mechanics were inspired by biological populations, such as bird flocks or fish schools [1]. Like other random search algorithms, there is also certain degree of premature phenomenon and slow convergence rate in PSO algorithm.

In order to improve the performance of traditional PSO algorithm, many improved PSOs have been proposed in recent years. It has been discussed in [2] that inertia weight has predictive effects on search ability of PSO algorithm. Clerc and Kennedy proposed the constriction factor approach PSO (CPSO), which ensured the convergence of the search procedures and could achieve higher-quality solutions [3, 4]. Riget proposed Attractive and Repulsive Particle Swarm Optimizer (ARPSO) to avoid premature convergence of particle swarm optimization [5]. Passive congregation PSO

D.-S. Huang et al. (Eds.): ICIC 2017, Part I, LNCS 10361, pp. 658–667, 2017.
DOI: 10.1007/978-3-319-63309-1_58

(PSOPC) introduced passive congregation for preserving swarm integrity to transfer information among individuals of the swarm [6]. A new hybrid optimization algorithm that combined the PSO algorithm with semi-deterministic search was proposed, the algorithm adaptively searched local minima with random search realized by APSO and performed global search with semi-deterministic search realized by DGPSOGS [7, 8]. Random search and semi-deterministic search were integrated into the ARPSO [9]. These improved algorithms above have improved the ability of local search, but they are still trapped into local minima and some of them require more time to find the best solution.

Adam is an algorithm for first-order gradient-based optimization of stochastic objective functions. Adam is computationally efficient, has little memory requirements and is well suited for problems that are large in terms of data and/or parameters [10]. In this paper, the Adam method is introduced in the APSO, which can improve the performance effectively. Firstly, each particle updates its position by the APSO. In the solution space obtained by APSO, then Adam method is used to calculate position of each particle. Finally, select the position which has better fitness value as the current position. The experimental results on six well-known benchmark functions show that our proposed algorithm gets better convergence performance compared to other five classical PSOs.

The rest of the article is organized as follows: Sect. 2 introduces particle swarm optimization algorithm and adaptive moment estimation method. Section 3 describes the AdamPSO in detail. Section 4 presents the test functions and the discussion of the experimental results. Section 5 the conclusions are given.

2 Preliminaries

2.1 Basic Particle Swarm Optimization

PSO algorithm is an evolutionary computation technique in searching for the best solution by simulating the movement of birds flocks [1]. Supposing in a D dimension objective search space, the total number of particles is N, and the swarm is $S = (X_1, X_2, \ldots, X_N)$, each particle represents a position in the D dimension, the position X_i of ith particle can be expressed as vector $X_i = (X_{i1}, X_{i2}, \ldots, X_{iD})$. Accordingly, flying velocity can be expressed as vector $V_i = (V_{i1}, V_{i2}, \ldots, V_{iD})$. $P_i = (P_{i1}, P_{i2}, \ldots, P_{iD})$ represents the best position of ith particle searching until now and $P_g = (P_{g1}, P_{g2}, \ldots, P_{gD})$ represents the best position of the total particle swarm searching until now. According to the literature [1], the formulas were described as:

$$V_i(t+1) = V_i(t) + c1 * rand() * (P_i(t) - X_i(t)) + c2 * rand() * (P_g(t) - X_i(t)) \quad (1)$$

$$\mathrm{X}_i(t+1) = \mathrm{X}_i(t) + V_i(t+1) \quad (2)$$

In Eq. (1), *c1*, *c2* are learning factor with positive values; *rand()* is a random number in the range [0,1].

In order to improve the performance of PSO, adaptive particle swarm optimization (APSO) was proposed by Shi and Eberhart in 1998 [2]. The velocity formula of APSO can be described as follows:

$$V_i(t+1) = W(t) * V_i(t) + c1 * rand() * (P_i(t) - X_i(t)) + c2 * rand() * (P_g(t) - X_i(t)) \quad (3)$$

W in Eq. (3) is the inertial weight that makes a tradeoff between the global and local search abilities. *W(t)* is defined as:

$$W(t) = W_{\max} - t * (W_{\max} - W_{\min})/N_{PSO} \quad (4)$$

$W_{\min}, W_{\max}$ in Eq. (4) are the initial inertial weight and the final inertial weight respectively. Accordingly, N_{PSO} is the maximum iterations.

2.2 Adaptive Moment Estimation Method

Adam is an algorithm for first-order gradient-based optimization of stochastic objective functions. The method is based on adaptive estimates of lower-order moments of the gradients. "Ada" originates from "adaptive", this method changes the learning rate over time according to gradients before. The meaning of moment in probability theory is described as: If a random variable *X* obeys a certain distribution, the first moment of *X* is *E(x)* and the second moment of *X* is $E(x^2)$. Adam algorithm computes individual adaptive learning rates for different parameters from estimates of first and second moments of the gradients. The magnitudes of parameter updates in Adam method are invariant to rescaling of the gradient, the step size of each learning iteration parameter has a certain range, the Adam method does not require a stationary objective, it works with sparse gradients and naturally performs a form of step size annealing [10]. Adam method takes the advantages of AdaGrad, which works well with sparse gradients [11], and RMSProp, which works well in on-line and non-stationary settings [12]. In summary, firstly, gradient g of parameters is computed. Secondly, Adam method is designed to compute two moment estimates: the first moment of *g (E(g))*, which helps avoid disordered moving and prevents settling into local optima, and the second moment of $g(E(g^2))$, which guarantees a upper bound of step size. Then, unbiased estimate of *E(g)* and $E(g^2)$ should be calculated. Finally, we can obtain the resulting parameter θ_t.

The flowchart of the Adam algorithm was shown in Fig. 1:

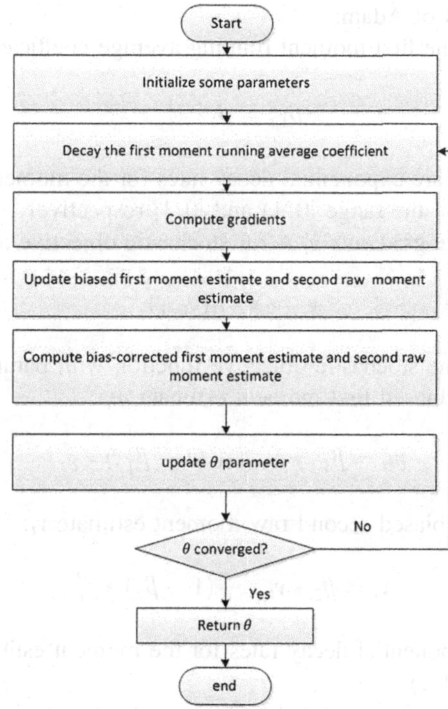

Fig. 1. The flowchart of Adam method

3 The Proposed Hybrid Algorithm

Although APSO is a population-based stochastic optimization technique and is easy to implement without complex evolutionary operations, its particles always track the current global best position or their own best position which is easy to get in local minima [13]. Therefore, in this paper, an APSO based on Adam method hybrid algorithm is proposed to improve the search ability of the population. APSO is used to perform stochastic and rough search. In the solution space obtained by APSO, Adam method is used to perform further search, which may establish a new solution space.

The detailed steps of the proposed method AdamPSO are described as follows:

Step 1: Initialize all particles randomly. Set swarm number N, the maximum iteration N_{PSO}, inertia weight and learning factors;

Step 2: Evaluate pbest of each particle and gbest of all particles;

Step 3: Each particle updates its position by the following formulas:

$$X' = X(t) + V_{apso} \tag{5}$$

In Eq. (5), V_{apso} represents the velocity update of the ith particle obtained by Eq. (3).

Step 4: In the solution space obtained by APSO ($\theta_0 = X'$), each particle updates its position by the method of Adam:

Substep 1: Decay the first moment running average coefficient $\beta_{1,t}$:

$$\beta_{1,t} = \beta_1 \lambda^{t-1} \tag{6}$$

β_1 and λ in Eq. (6) are exponential decay rates for the moment estimates. β_1 and λ are a random number in the range $[0, 1)$ and $[0, 1]$ respectively.

Substep 2: Compute gradients g_t w.r.t. stochastic objective at time step t:

$$g_t = \nabla_\theta f_t(\theta_{t-1}) \tag{7}$$

$f(\theta)$ in Eq. (7) is the stochastic objective function with parameters θ:

Substep 3: Update biased first moment estimate m_t:

$$m_t = \beta_{1,t} * m_{t-1} + (1 - \beta_{1,t}) * g_t \tag{8}$$

Substep 4: Update biased second raw moment estimate v_t:

$$v_t = \beta_2 * v_{t-1} + (1 - \beta_2) * g_t^2 \tag{9}$$

β_2 in Eq. (9) is exponential decay rates for the moment estimates. β_2 is a random number in the range $[0, 1)$.

Substep 5: Compute bias-corrected first moment estimate $\hat{m}_t$:

$$\hat{m}_t = m_t / (1 - \beta_{1,t}) \tag{10}$$

Substep 6: Compute bias-corrected second raw moment estimate $\hat{v}_t$:

$$\hat{v}_t = v_t / (1 - \beta_{2,t}) \tag{11}$$

Substep 7: Update parameter θ_t:

$$\theta_t = \theta_{t-1} - \alpha * \hat{m}_t / (\sqrt{\hat{v}_t} + \varepsilon) \tag{12}$$

α in Eq. (12) is the step size.

Substep 8: If θ_t does not converge, execute Substep 1 to Substep 7, otherwise return θ_t (resulting parameters).

Each particle updates its position by the following formulas:

$$X''(t) = \theta_t \tag{13}$$

Step 5: Calculate the fitness value of X' and X'', compare the fitness value of X' with the fitness value of X'', select the better fitness value as the current position.

$$X_i(t+1) = \begin{cases} X' & f(X') < f(X'') \\ X'' & f(X') >= f(X'') \end{cases} \quad (14)$$

Step 6: Evaluate end condition of iteration, if sustainable then output the result, if unsustainable then turn to Step 2.

4 Experimental Results and Discussion

In order to study algorithm performance, tests are conducted on six well-known benchmark functions in the following experiments: Sphere, Ellipsoid, Ackley, Griewangk, Rastrigin and Rosenbrock. Sphere and Ellipsoid are unimodal functions which have a single local minimum, while Ackley, Griewank, Rastrigin and Rosenbrock are multimodal functions which have a lot of local minima. Griewank and Rastrigin are used to test the ability of the proposed algorithm to escape from local minima, while Rosenbrock is used to test the ability of the proposed algorithm to navigate flat regions with small gradients. The description of the benchmark functions and their global optimums are listed in Table 1.

Table 1. Test functions used in the experiment

Test function	Equation	Search space	Global minima
Sphere(F1)	$\sum_{i=1}^{n} x_i^2$	$(-100, 100)^n$	0
Ellipsoid(F2)	$\sum_{i=1}^{n} i \cdot x_i^2$	$(-100, 100)^n$	0
Ackley(F3)	$-a \cdot e^{-b\sqrt{\frac{\sum_{i=1}^{n} x_i^2}{n}}} - e^{\frac{\sum_{i=1}^{n} \cos(c \cdot x_i)}{n}} + a + e^1$	$(-32, 32)^n$	0
Griewank(F4)	$1 + \frac{\sum_{i=1}^{n} x_i^2}{4000} - \prod_{i=1}^{n} \cos\left(\frac{x_i}{\sqrt{i}}\right)$	$(-600, 600)^n$	0
Rastrigin(F5)	$10n + \sum_{i=1}^{n} (x_i^2 - 10\cos(2\pi x_i))$	$(-5.12, 5.12)^n$	0
Rosenbrock(F6)	$\sum_{i=1}^{n-1} (100 \cdot (x_i - x_{i-1}^2)^2 + (1 - x_i)^2)$	$(-2, 2)^n$	0

In this section, experiments are conducted to compare AdamPSO with 5 peer algorithms: APSO, CPSO, SPSO, ARPSO and PSOPC. Each algorithm is executed 20 independently runs over the six benchmark functions in 10, 20 and 30 dimensions respectively. The initial swarm is the same for all the seven algorithms. The swarm size is 40. The decaying inertia weight is used which starts at 0.9 and ends at 0.4. Parameters c1 and c2 for AdamPSO, APSO are equal to 1.962. Parameters c1 and c2 for ARPSO are set as 2.0 and 1.5, the diversity threshold d_{low} and d_{high} for ARPSO are set as $5e^{-6}$ and 0.25, respectively. Parameters c1 and c2 for CPSO are both set as 2.05. The number of maximum iterations is 5000. Good default setting for the Adam algorithm are $\alpha = 0.001$, $\beta_1 = 0.9$, $\beta_2 = 0.999$, $\varepsilon = 10^{-8}$ and $\lambda = 1 - 10^{-8}$

Figure 2 presents the convergence characteristics in the term of the best fitness value of the median run of each algorithm for each test function. Without loss of the generality, F1 to F6 are all in ten dimensions. Figure 2(a) and (b) show the performance of each algorithm on unimodal and convex function, whereas Fig. 2(c)–(f) show the performance of each algorithm on multimodal function. From Fig. 2(a) and (b), all algorithms perform very well except PSOPC and CPSO. PSOPC and CPSO have not

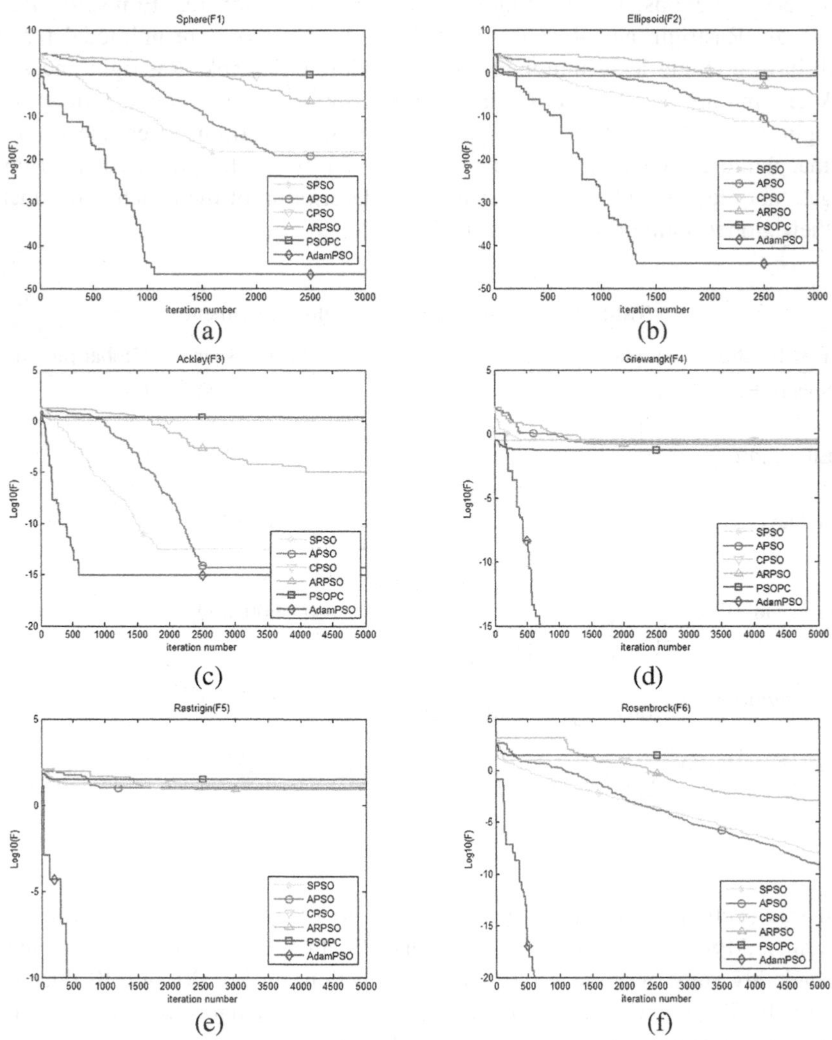

Fig. 2. Best solution versus iteration number for the six test functions by using six PSOs (a) Sphere (b) Ellipsoid (c) Ackley (d) Griewank (e) Rastrigin (f) Rosenbrock

Table 2. Mean best solution for the six test functions by using six PSOs

Test functions	D	SPSO	APSO	CPSO	ARPSO	PSOPC	AdamPSO
Sphere(F1)	10	3.08e−19	1.30e−19	5.86e−01	1.99e−08	4.58e−01	1.00e−48
	20	3.41e−11	2.93e−17	1.17e+03	3.70e−07	4.05e+00	5.59e−48
	30	2.65e−09	1.39e−15	5.24e+03	1.24e+03	7.21e+00	7.60e−54
Ellipsoid(F2)	10	3.01e−12	8.95e−16	1.22e+01	1.16e+03	9.44e−01	2.53e−46
	20	1.46e−07	3.43e−12	1.85e+03	2.69e+03	5.80e+00	7.64e−47
	30	1.25e−02	6.11e−05	6.11e+03	4.51e−02	1.58e+01	6.09e−44
Ackley(F3)	10	4.99e−13	2.11e−14	2.74e+00	2.35e−05	2.06e+00	7.64e−47
	20	1.16e−08	1.69e−13	9.67e+00	1.53e−04	3.12e+00	1.00e+00
	30	3.65e−07	9.86e−01	1.18e+01	1.03e+00	3.60e+00	2.00e+00
Griewank(F4)	10	1.77e−01	1.87e−01	3.79e−01	1.44e−01	5.48e−02	0
	20	6.95e−02	8.02e−02	1.40e+01	6.44e−02	2.10e−01	0
	30	1.89e−02	1.97e−02	3.69e+01	3.38e+01	3.01e−01	0
Rastrigin(F5)	10	1.51e+01	8.26e+00	1.73e+01	5.07e+00	2.88e+01	0
	20	4.41e+01	3.66e+01	7.97e+01	2.23e+01	9.55e+01	0
	30	7.16e+01	6.39e+01	1.45e+02	4.19e+01	1.85e+02	0
Rosenbrock(F6)	10	1.99e−01	5.98e−01	6.66e+00	7.63e−04	5.83e+01	0
	20	6.85e−01	3.36e−01	6.30e+01	2.02e+02	4.07e+02	0
	30	1.66e+01	1.10e+01	2.40e+02	3.81e+02	1.22e+03	0

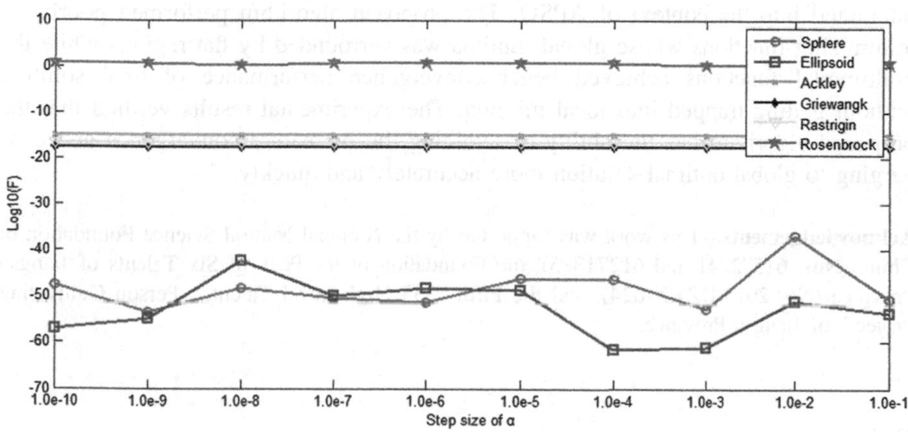

Fig. 3. Mean best solution for the six test functions by using different step sizes

been changed nearly after 560 and 1190 iterations. From Fig. 2(c), all algorithms converge to local minima obviously, the proposed algorithm can obtain high-quality solutions than others. AdamPSO, CPSO, PSOPC, SPSO, APSO and ARPSO converge to local minima around 600, 650, 750, 1830, 2550 and 4090 iterations. Figure 2(d) and (e), PSOPC, SPSO, APSO, ARPSO and CPSO converge to local minima obviously. From Fig. 2(f), PSOPC and CPSO converge to local minima around 790 and 1300 iterations, whereas ARPSO, SPSO and APSO do not converge to local minima obviously. It is delightful that the proposed algorithm has an obvious downward trend with the iteration number in the six figures, that is to say, the proposed algorithm has the ability of avoiding the premature convergence.

Table 2 shows mean best solution for the six test functions on different dimensions by using six PSOs. It is delightful that the proposed algorithm has better convergence accuracy than other PSOs in all cases. Although SPSO, APSO and ARPSO obtain better convergence accuracy than AdamPSO on the function F3 in twenty dimensions as well as SPSO, APSO, CPSO and ARPSO obtain better convergence accuracy than AdamPSO on the function F3 in thirty dimensions, it requires more iterations and the proposed algorithm sometimes can find the global minima.

Figure 3 shows the relationship between step size α in the AdamPSO and the best solutions for six test functions. α is selected from $[10^{-10}, 10^{-1}]$. It can be found that two suitable values of the step size for the function F1 are 10^{-3} and 10^{-9}. 10^{-3} and 10^{-4} are the two suitable values of the step size for the function F2. Different step sizes have no effect on the function F3 and F4, All the step sizes are suitable values for the function F5 except 10^{-2}, 10^{-4} and 10^{-7}. 10^{-3} is the only suitable value of the step size for the function F6. So we select 10^{-3} as the value of the step size eventually.

5 Conclusions

In order to enhance the search ability of APSO, the modified Adam method was integrated into the context of APSO. The proposed algorithm performed poorly on minimizing functions whose global minima was surrounded by flat regions while the multimodal functions achieved better convergence performance of final solution without getting trapped into local minima. The experimental results verified that the proposed algorithm has the ability of avoiding the premature convergence and converging to global optimal solution more accurately and quickly.

Acknowledgements. This work was supported by the National Natural Science Foundation of China (Nos. 61572241 and 61271385), the Foundation of the Peak of Six Talents of Jiangsu Province (No. 2015-DZXX-024), and the Fifth “333 High Level Talented Person Cultivating Project” of Jiangsu Province.

References

1. Kennedy, J., Eberhart, R.C.: Particle swarm optimization. In: IEEE International Conference on Neural Networks, vol. 4, pp. 1942–1948 (1995)
2. Shi, Y., Eberhart, R.C.: A modified particle swarm optimizer. Comput. Intell. **6**(1), 69–73 (1998)
3. Clerc, M.: The swarm and the queen: towards a deterministic and adaptive particle swarm optimization. In: Evolutionary Computation, pp. 1951–1957. IEEE, Washington (1999)
4. Corne, D., Dorigo, M., Glover, F.: New Ideas in Optimization, pp. 379–387. McGraw Hill, London (1999)

5. Riget, J., Vesterstrom J.S.: A diversity-guided particle swarm optimizer - the ARPSO. Evalife Technical report (2002)
6. He, S., et al.: A particle swarm optimizer with passive congregation. Bio Syst. **78**(1–3), 135–147 (2004)
7. Han, F., Liu, Q.: A diversity-guided hybrid particle swarm optimization. In: Huang, D.-S., Gupta, P., Zhang, X., Premaratne, P. (eds.) ICIC 2012. CCIS, vol. 304, pp. 461–466. Springer, Heidelberg (2012). doi:10.1007/978-3-642-31837-5_67
8. Liu, Q., Han, F.: A hybrid attractive and repulsive particle swarm optimization based on gradient search. In: Huang, D.-S., Jo, K.-H., Zhou, Y.-Q., Han, K. (eds.) ICIC 2013. LNCS, vol. 7996, pp. 155–162. Springer, Heidelberg (2013). doi:10.1007/978-3-642-39482-9_18
9. Han, F., Liu, Q.: A diversity-guided hybrid particle swarm optimization. Neurocomputing **137**(4), 234–240 (2014)
10. Kingma, D., Ba, J.: Adam: a method for stochastic optimization. Comput. Sci. (2014)
11. Duchi, J., Hazan, E., Singer, Y.: Adaptive subgradient methods for online learning and stochastic optimization. J. Mach. Learn. Res. **12**(7), 2121–2159 (2011)
12. Tieleman, T., Hinton, G.: Lecture 6.5 - RMSProp, COURSERA: neural networks for machine learning. Technical report (2012)
13. Wang, L.: Intelligent Optimization Algorithms with Applications. Tsinghua University Press, Beijing (2001)

An Improved Multi-swarm Particle Swarm Optimization Based on Knowledge Billboard and Periodic Search Mechanism

Pan-pan Du(✉) and Fei Han

School of Computer Science and Communication Engineering, Jiangsu University, Zhenjiang, Jiangsu, China
jstdupanpna@163.com, hanfei@ujs.edu.cn

Abstract. Multi-swarm particle swarm optimization has faster convergence rate, wider range of search, and higher convergence accuracy. However, the information among sub-swarms is not updated in time, which may decrease the search ability of the multiple swarms. An improved multi-swarm particle swarm optimization based on the periodic search mechanisms and the knowledge billboard (KBMPSO) is proposed. The swarm is divided into several sub-swarms using the improved K-means method. In a search cycle, one sub-swarm searches collaboratively and the remaining sub-swarms search independently. When the particles evolve independently to a certain generation, the global best value is periodically updated. The information stored in the knowledge billboard can help the sub-swarm jump out the local optimum. The KBMPSO algorithm will exchange the information between the adjacent sub-swarms every fixed number of generations. Once the sub-swarm is trapped into the local optimum during the search process, it will affect the convergence effect of its adjacent sub-swarm. Introducing the knowledge billboard to the sub-swarm during its searching avoids the sub-swarm trapping into the local optimum. To effectively keep the balance between the global exploration and the exploitation, the particle takes advantage of the shared information which stored on the knowledge billboard. In the simulation studies, several benchmark functions are conducted to verify the superiority of the KBMPSO algorithm.

Keywords: Particle swarm optimization · Multi-swarm · Periodic shared · Improved K-means · Knowledge billboard

1 Introduction

Particle swarm optimization (PSO) is a swarm-intelligence optimization algorithm. It is motivated from the simulation of the behavior of a flock of birds, firstly introduced by Kennedy and Eberhart [1, 2]. PSO has attracted a lot of attentions from researchers around the world owning to the advantages of the PSO, such as few parameters to adjust and easy to implement. However, particles swarm optimization will be easy to congregate to local optimum due to the loss of the diversity of swarms. The convergence accuracy of the algorithm is limited. PSO has difficulties in controlling the balance between exploration and exploitation. In order to improve the ability of

D.-S. Huang et al. (Eds.): ICIC 2017, Part I, LNCS 10361, pp. 668–678, 2017.
DOI: 10.1007/978-3-319-63309-1_59

exploration and exploitation, a number of variations to the standard PSO had been proposed. Many strategies and methods were applied to particle swarm optimization, which could improve the convergence precision of the algorithm. Introduced the inertia weight into the original PSO can balance the global and local search abilities [3]. Comparing inertia weights and constriction factors in particle swarm optimization (CPSO) was proposed [4]. Constriction factor provided a method to choose the inertia weight and acceleration constants. It was concluded that the best approach was using the constriction factor while limiting the maximum velocity to the dynamic range of the maximum position. In particle swarm optimization algorithm for random inertia weight (RPSO) [5], the velocity of the particles was only a function of the current position of the particles in the SPSO. We could try to set the inertia weight to a random number. Multi-swarm cooperative particle swarm optimization (MCPSO) [6] was based on a master–slave model, in which a population consists of one master swarm and several slave swarms. The slave swarms executed a single PSO or its variants independently to maintain the diversity of particles, while the master swarm evolved based on its own knowledge and the knowledge of the slave swarms. Competitive and cooperative co-evolutionary particle swarm optimization (CCPSO) [7] was proposed, which was based on the co-evolution of competing and cooperating species. This method was very useful to solve complex problems. Competitive and cooperative particle swarm optimization with information sharing mechanism for global optimization problems was proposed (CCPSO-ISM) [8]. Dynamic particle swarm optimization (DPSO) [9, 10] was a modified version of the PSO, which utilized a different update mechanism for the velocity of the particles. Multi-swarm particle swarm optimization (MPSO) [11] divided the swarm into several small sub-swarms which could increase the diversity of the swarm. However, the particle swarm updated its position and velocity information too frequently. The computational time of this method would be longer. Dynamic multi-swarm particle swarm optimizer based on K-means clustering and its application (IKMPSO) was proposed [12]. The swarm was divided into several sub-swarms using the K-means clustering. In order to increase the information exchange of sub-swarms, the sub-swarm is dynamically constructed, and the velocity of each particle is adjusted by clustering center that it belongs to and all particles in its neighborhood including itself. Dynamic multi-swarm particle swarm optimizer with cooperative learning strategy was proposed [13]. Introducing the cooperative learning strategy to DMS-PSO could achieve a good balance between the exploration and exploitation abilities.

When sub-swarms search independently to a certain generation, the global best value is periodically updated. During the search process, all of sub-swarms search the global optimum based on its own knowledge and the knowledge of all the swarms. Based on the above idea, this paper proposes the multi-swarm particle swarm optimization based on the periodic search mechanisms and the knowledge billboard (KBMPSO). The swarm is decomposed into several sub-swarms, each sub-swarm independently evolves, and the global best position is updated when the cycle is reached. During the search process, information searched by all the sub-swarms will be sent to the knowledge billboard. The knowledge billboard will be used to store the information which has been searched by all sub-swarms, such as the diversity of the swarms, best position and fitness value. The knowledge billboard judges the search status of each sub-swarm based on the collected information. The periodic shared strategy helps to share information among sub-swarms in the search process. This paper uses the improved K-means to divide the swarms into several sub-swarms. This method can evenly distribute the particles to the sub-swarms and increase the diversity of the swarm. In KBMPSO, each stage, one sub-swarm searches collaboratively and other sub-swarms search independently. This strategy can increase the diversity of the swarm, but also can improve the convergence of the algorithm. When the sub-swarm searches collaboratively in the space area, the previous sub-swarm adjacent to it will give it a good position to guide its searching. The direction of the particles in each sub-swarm can be changed quickly and fly toward the best position. We can judge the search status of the sub-swarm based on the information stored in the knowledge billboard. Once the previous sub-swarm adjacent to it was trapped in the local optimum, the information in the knowledge billboard will be used to help it jump out.

The rest of the article is organized as follows: Sect. 2 presents a general introduction of PSO and MPSO. Section 3 describes the KBMPSO in detail. Section 4 presents the test functions and the discussion of the experimental results. Section 5 the conclusions and future works are given.

2 Preliminaries

2.1 Standard Particle Swarm Optimization

Particle swarm optimization algorithm was a new optimization search algorithm which is similar to the genetic algorithm. Particle swarm optimization algorithm is different from the genetic algorithm which has the mutation mechanism. Once the particles were trapped into local optimum, the standard particle swarm algorithm is difficult to jump out the local optimum by increasing the number of iterations. In PSO, a potential solution for a problem is considered as a bird, which is named particle here, flies through a D-dimensional search space and adjusts its position according to the previous best position and the global best position. The velocity and location update strategy of the particle are presented as the Eqs. (1) and (2):

$$v_{id}(t+1) = v_{id}(t) + c_1 * r_1 * (p_{id}(\mathrm{t}) - x_{id}(t)) + c_2 * r_2 * (p_{gd}(\mathrm{t}) - x_{id}(t)) \quad (1)$$

$$x_{id}(t+1) = x_{id}(t) + v_{id}(t+1) \quad (2)$$

Assume that the dimension of the search space is D, and the number of the swarm is N, the swarm describes as $S = (x_1, x_2, x_3 \cdots x_N)$; each particle represents a position in the D dimension; $x_i = (x_{i1}, x_{i2}, \cdots, x_{iD})$ represents the location of the ith particle in the D-dimensional space. $v_i = (v_{i1}, v_{i2}, \cdots, v_{iD})$ represents the velocity of the ith particle in the D-dimension space. ω represents the inertia weight. r_1 and r_2 are random values between 0 and 1. c_1 and c_2 are acceleration constants. $p_i = (p_{i1}, p_{i2}, \cdots p_{id}, \cdots p_{iD})$ is the best previous position of the ith particle $p_g = (p_{g1}, p_{g2}, \cdots, p_{gd}, \cdots, p_{gD})$ is the best position among all the particles in the swarm. Shi and Eberhart [3] brought the inertia weight to the PSO, which is called the standard PSO (SPSO). Inertia weight ω is an important parameter. The larger inertia weight enhances the global search capability of the algorithm, while the smaller inertia weight enhances the local search capability of the algorithm. $\omega = 0.9 - \frac{t}{MaxIter} \times 0.5$, MaxIter is the maximum number of iterations, t is the contemporary iterations. With the increase of the number of iterations t, the ω linearity decreases, and the experimental results show that there is a good convergence performance. The velocity and position update strategy of the particle are presented as the Eqs. (3) and (2):

$$v_{id}(t+1) = \omega * v_{id}(t) + c_1 * r_1 * (p_{id}(\mathrm{t}) - x_{id}(t)) + c_2 * r_2 * (p_{gd}(\mathrm{t}) - x_{id}(t)) \quad (3)$$

2.2 Multi-swarm Particle Swarm Optimization

As the process of searching, the diversity of the swarm will be lost. Losing diversity may lead the particle of the swarm to be trapped into the local optimum which called "premature convergence". Introduced the sub-swarms to particle swarm optimization could increase the diversity of the swarm and improve the convergence performance of the algorithm. The swarm was divided into several sub-swarms. Each sub-swarm represented a solution of the target. All sub-swarms not only searched their best solutions individually but also shared the information. It has been proved that multi-swarms PSO algorithm was superior to single-swarm PSO algorithm in terms of convergence efficiency and convergence accuracy [14]. This strategy of sub-swarms called collaboration strategy. Island model and Neighborhood model are two models of the cooperative evolution strategy. Different cooperative models have different strategies for the multi-swarm particle optimization. In the process of the evolution of the standard PSO, the particle will update the individual of historical best position and group of historical best position for each generation. Once the particle trapped into local optimum, the algorithm will stop searching and affect the global convergence. If particle periodically update the global best value which can not only make full use of the historical best position but also use the global best position of all sub-swarms. The process of the MPSO is given as Fig. 1.

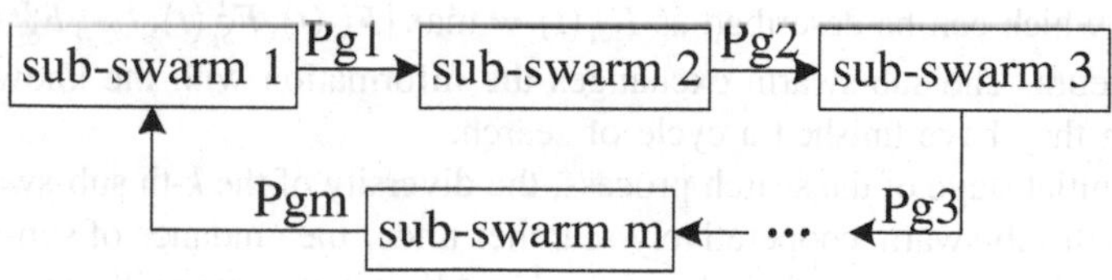

Fig. 1. The process of the MPSO

In the MPSO, each sub-swarm updates its position and velocity using the Eqs. (2) and (3) in the first period. In the second period, sub-swarm 1 will guide sub-swarm 2 to search the space. The second sub-swarm updates its position and velocity position using the Eqs. (2) and (4), PG represents the global best solution found by the previous sub-swarm adjacent. Other sub-swarms search the space independently and update the position and velocity using the Eqs. (2) and (3) during this period. All sub-swarms search the space like this. Each sub-swarm will be guided to search by its adjacent before it.

$$v_{id}(t+1) = \omega v_{id}(t) + c_1 * r_1 * (p_{id}(t) - x_{id}(t)) + c_2 * r_2 * (p_{gd}(t) - x_{id}(t)) + c_3 * r_3 * (PG - x_{id}(t)) \quad (4)$$

3 The Proposed Algorithm

In MPSO algorithm, sub-swarms exchange information they searched too frequently. Sub-swarm does not take full advantage of the best solution which searched by other sub-swarms. Once the sub-swarm is trapped in the local optimum, it will affect other sub-swarm adjacent after it. In order to improve the convergence efficiency and convergence accuracy of the algorithm, we introduce the periodic sharing mechanism and the knowledge billboard to the MPSO. The paper proposes an improved multi-swarm particle swarm optimization, which is based on the knowledge billboard and the periodic shared strategy. The swarm is divided into some sub-swarms using the improved initial cluster center K-means method. Using the optimal binary tree idea improved the initial clustering center K-means algorithm, which can increase the diversity of the swarm. In the first period, all the sub-swarms will search the space and find their own best solutions which will be used to affect their neighborhoods. Information sharing strategy is introduced to the sub-swarms during the period. It can help the sub-swarm jump out the local optimum. During the process of searching, search ability, best position and its fitness value will be stored on the knowledge billboard [15], which will be used to guide the searching for all sub-swarms.$C_{S_k}(t)$ is the search ability which can be described as Eq. (5).

$$C_{S_k}(t) = |F_{S_k}(t) - F_{S_k}(t - \Delta T)| \quad (5)$$

$p_{S_k}(t)$ is the best position of the kth sub-swarm which can be described as $p_{S_k}(t) = \arg\max\{F_{S_k}^1(t), F_{S_k}^2(t), \cdots, F_{S_k}^{N_k}(t)\}$ and $F_{S_k}(t)$ is best fitness value of the kth sub-swarm which can be described as $F_{S_k}(t) = \max\{F_{S_k}^1(t), F_{S_k}^2(t), \cdots, F_{S_k}^{N_k}(t)\}$. ΔT is interval algebra. The sub-swarm exchanges the information with the knowledge billboard when they have finished a cycle of search.

At the initial stage of the search process, the diversity of the *k*-th sub-swarm is very high. The *k*-th sub-swarm cooperatively searches under the guidance of sub-swarm *k*-1. And other sub-swarms search independently. All sub-swarms will save the search information to the knowledge billboard. As the search process progresses, the diversity of the swarm will be reduced and the capacity of the exploitation will also be reduced. If the *k*-th sub-swarm falls into the local optimum, the information searched by the sub-swarm k-1 cannot help it to jump out of local optimum. The information stored on the knowledge billboard can help sub-swarms to jump out of the local optimum.

The detail steps of the multi-swarm particle swarm optimization based on the knowledge billboard and the periodic shared strategy (KBMPSO) is as follow:

Step1: Initialize velocity and position of all particles randomly and all parameters;
Step2: Calculate pbest of each particle and gbest of all particles;
Step3: Divide the swarm into some sub-swarms using the improved K-means method;
Step4: Deal the iteration steps according to the period R;
Step5: Each sub-swarm search the space using the Eqs. (2) and (3) to update their position and velocity in the first period;
Step6: The sub-swarm will update its position and velocity by the Eqs. (2) and (4), which will be affected by its previous sub-swarm in the next period. Other swarms will update their position and velocity by the Eqs. (2) and (3) independently;
Step7: Introduce the information sharing strategy to step6 avoids the sub-swarm trapping in the local optimum. When the sub-swarm is trapped in the local optimum, PG will be updated as follows

$$PG = \begin{cases} Pg & \text{Sub - swarm has strong search ability} \\ \frac{\sum_{i=1}^{K} Pg_i}{K} & \text{Sub - swarm falls into local optimum} \end{cases}$$

Step8: Satisfy the requirements of the accuracy, repeat the steps from step4 to step7.

4 Experimental Results and Discussion

This paper chooses five benchmark functions which are depicted in Table 1 to test the performance of the proposed algorithm. Five algorithms of the PSOs (SPSO, RPSO, CPSO, MPSO and IKMPSO) are implemented for comparing their performance with the KBMPSO algorithm. All parameters configuration are referred to the corresponding reference. The parameters used for SPSO, RPSO CPSO were recommended in [3–5]. The parameter configuration of the MPSO and IKMPSO are based on the suggestion in the references [11, 12]. All of the programs of the PSOS are executed in MATLAB 2012a environment on an Intel Core(TM) i3-2350 2.30 GHZ CPU. The number of particles is set to 60 and the population dimension is set to 10 for all PSOS. Meanwhile each benchmark function is run 30 times and the maximal iteration is set at 1000. The inertia weight w decreases linearly from 0.9 to 0.4 for all PSOs according to the literature [10]. This method can achieve a good balance between the exploration ability and the exploitation ability. The acceleration constants c1 and c2 are set to 2.0 in all PSOs. For MPSO, IKMPSO and KBMPSO, c1 and c2 are set to 2.0, while c3 is set to 2.05. The swarms are divided into five sub-swarms for MPSO, IKMPSO and KBMPSO. Each benchmark function which is conducted runs 30 times.

The experiment results (the fitness values in terms of the best, worst and mean for each algorithm for 10 dimensions) are listed in Table 2. The bold numbers in Table 2 indicate the relatively better results after running 30 runs on each functions F1 $\sim$ F5 (Fig. 2).

The first function (F1) is an unimodal function, which has only one local optimal point. The local optimal value is the global optimal value. All algorithms can search for

Table 1. Five benchmark functions

Functions	Mathematical representation	Search range	Global minima
Sphere (F1)	$\sum_{i=1}^{n} x_i^2$	$(-100,100)^n$	0
Ackley (F2)	$-20\exp(-0.2\sqrt{\frac{1}{n}\sum_{i=1}^{n} x_i^2}) - \exp(\frac{1}{n}\sum_{i=1}^{n}\cos(2\pi x_i) + 20 + e)$	$(-32,32)^n$	0
Griewank (F3)	$1 + \sum_{i=1}^{n} x_i^2/4000 - \prod_{i=1}^{n}\cos(\frac{x_i}{\sqrt{i}})$	$(-100,100)^n$	0
Rosenbrock (F4)	$\sum_{i=1}^{n-1} (100(x_{i+1} - x_i^2)^2 + (x_i - 1)^2)$	$(-100,100)^n$	0
Rastrigrin (F5)	$10n + \sum_{i=1}^{n} (x_i^2 - 10\cos(2\pi x_i))$	$(-5.12,5.12)^n$	0

Table 2. Results on five benchmarks for 10-D

Functions	Solutions	SPSO	CPSO	RPSO	MPSO	IKMPSO	KBMPSO
Sphere(f1)	best	1.8082e–017	1.6297e–024	3.2041e–021	9.2581e–032	4.9912e–039	6.7862e–078
	mean	2.6407e–013	2.5154e–017	6.8572e–012	7.9554e–028	2.8777e–034	5.6743e–045
	worst	8.1527e–008	6.0619e–019	2.3795e–011	7.1568e–021	8.5796e–030	7.5674e–037
Ackley(f2)	best	2.7384e–001	1.6046e–001	1.8046e–001	6.7400e–002	5.9400e–002	8.8818e–012
	mean	1.4539e–000	1.1565e–000	1.7565e–000	1.3200e–000	1.4998e–000	2.8873e–001
	worst	3.5187e–000	9.8828e–000	9.6528e–000	5.4425e–000	1.1466e+001	1.1972e+001
Griewank(f3)	best	2.9600e–002	2.7000e–002	2.4600e–002	0	0	0
	mean	8.3600e–002	2.0680e–002	1.5520e–002	8.789e–002	6.636e–002	2.2047e–016
	worst	2.6810e–001	9.0800e–001	8.7900e–001	2.1088e+000	2.6208e–001	1.7454e–012
Rosenbrock(f4)	best	1.0560e–001	6.1000e–002	4.3780e–001	4.4526e–004	3.0638e–004	0
	mean	5.8515e+001	1.5974e+001	2.5381e+002	8.4158e–003	1.3221e–002	2.8748e+000
	worst	8.5307e+002	2.0250e+002	3.2367e+001	2.7832e–001	3.200e–002	7.5135e+001
Rastrigrin(f5)	best	7.8337e–010	2.9849e+000	1.9899e+000	2.8777e–034	0	0
	mean	3.7228e+000	4.0793e+001	1.9900e+001	7.8433e–017	2.8612e–15	3.7529e–023
	worst	6.9647e+000	7.4532e+001	8.5567e+000	3.9870e–015	3.7870e–011	3.6955e–016

global optimizations as long as there is enough iteration. We can conclude that all algorithms converge to the optimum quickly and have a better convergence performance from Fig. 3(a). The curves in Fig. 3(a) shows that the convergence speed of MPSO, IKMPSO and KBMPSO is faster than other algorithms. The KBMPSO algorithm searched the best solution after running 100 iterations. Dividing the swarm into multiple sub-swarms can increase the diversity of the swarm. The convergence performance of the KBMPSO is better than the other algorithms. Introducing the cycle sharing mechanism and knowledge billboard to the KBMPSO helps the sub-swarm exchange the information in time and finds a better solution. Ackley function (F2) is a continuous function of the exponential function superimposed on a moderately enlarged cosine. The search of this function is very complicated, because in the process of searching, the swarms will inevitably fall into the local optimum. Unfortunately, KBMPSO does not perform well on the Ackley function. The results in Table 2 show

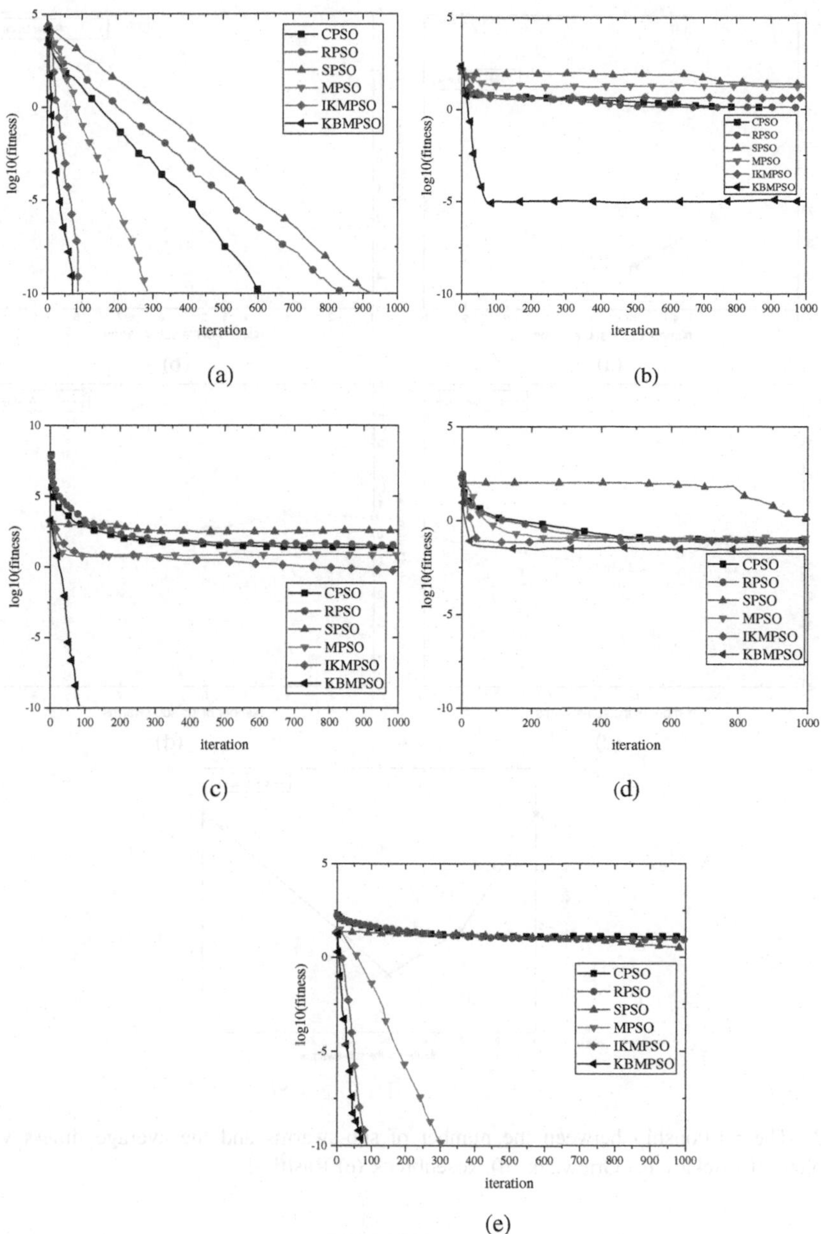

Fig. 2. Best solution versus iteration number for the five test functions by using six PSOs (a) Sphere (b) Ackley (c) Griewank (d) Rosenbrock (e) Rastrigin

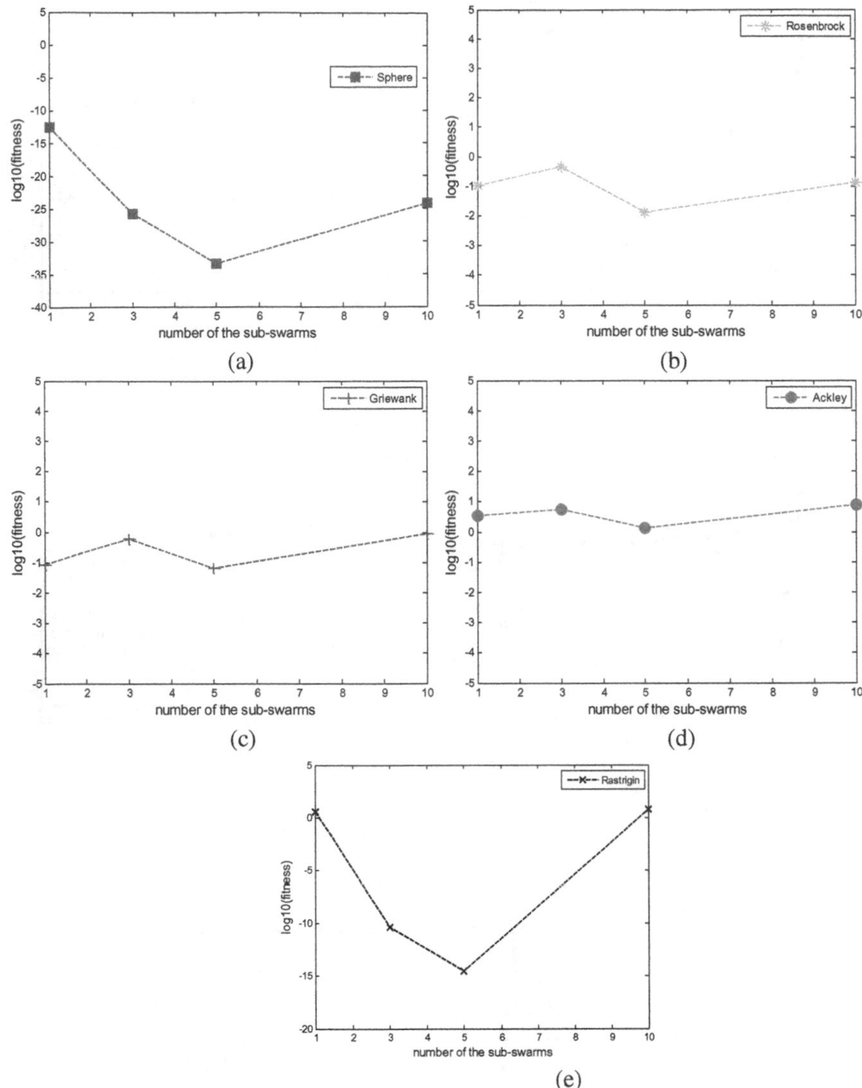

Fig. 3. The relationship between the number of sub-swarms and the average fitness values (a) Sphere (b) Ackley (c) Griewank (d) Rosenbrock (e) Rastrigrin

that there is no much difference between KBMPSO and other optimization algorithms on the average best fitness value for Ackley functions. The curve in Fig. 3(b) shows the convergence performance is not good. Griewank function (F3) exhibits a pattern similar to the function (F5). Both RPSO and CPSO have a fast convergence speed and can achieve a higher convergence accuracy. The results in Table 2 show that both MPSO and KBMPSO can find the best optimum. The curve in Fig. 3(c) shows the KBMPSO convergence performance is better than MPSO. Rosenbrock function (F4) which also called banana function, is a classic optimization problem. The global optimum is inside a long, narrow, parabolic shaped valley. To search the valley is trivial and convergence to the global optimum is difficult. All algorithms convergence performances are not good from the Fig. 3(d). However, the KBMPSO algorithm is better than other algorithms. The information stored in the knowledge billboard helps the sub-swarms to jump out the local optimum. From the curve in Fig. 3(d), we can conclude that the KBMPSO convergence performance is better than other algorithms after running 300 iterations. Rastrigin function (F5) is a multi-modal function. The curves in Fig. 3(e) shows that the SPSO convergence performance is a little bad but eventually better compared to RPSO and CPSO. MPSO method clearly performed best and gave consistently a near-optimum result. The KBMPSO convergence performance and the convergence trend are better than the MPSO. During the search, KBMPSO can find the best optimum result. The KBMPSO algorithm can be greatly improved the performance of the PSOs. However, the convergence accuracy of the KBMPSO is worse than DMS-PSO-CLS that proposed in the literature [13]. The number of iterations of the KBMPSO is less than the DMS-PSO-CLS.

The curves in Fig. 3 records the average of the fitness values searched by the KBMPSO algorithm on the five benchmark test functions 30 times. The results show that the number of sub-swarms is too small or too big, the convergence accuracy of the algorithm is not good. From curves in Fig. 3(a $\sim$ e), we can conclude that the KBMPSO algorithm has the highest convergence precision when the number of sub-swarms is set to five.

5 Conclusions

In the paper, we presented a multi-swarm particle swarm optimization algorithm based on the periodic search mechanisms and the knowledge billboard. The periodic search mechanism makes information among sub-swarms update in time. The information stored on the knowledge billboard can help sub-swarm to jump out the local optimum. The new strategy makes the particles have more exemplars to learn from and a larger potential space to fly. The periodic search mechanisms and the knowledge billboard can improve the convergence performance of PSO.

Five benchmark functions are used to test the performance of the KBMPSO algorithm. The KBMPSO can achieve superior features both in the convergence accuracy and convergence rate. Further work may focus on improving the performance of the KBMPSO algorithm.

Acknowledgements. This work was supported by the National Natural Science Foundation of China (Nos. 61572241 and 61271385), the Foundation of the Peak of Six Talents of Jiangsu Province (No. 2015-DZXX-024), and the Fifth “333 High Level Talented Person Cultivating Project” of Jiangsu Province.

References

1. Eberchart, R.C., Kennedy, J.: A new optimizer using particle swarm theory. In: Proceedings of the 6th International Symposium on Micromachine and Human Science, Nagoya, Japan, pp. 39–43 (1995)
2. Eberchart, R.C., Kennedy, J.: Particle swarm optimization. In: Proceedings of the IEEE International Conference on Neural Networks, Perth, Australia, pp. 1942–1948 (1995)
3. Zhang, Q.L., Li, X., Tran, Q.A.: A modified particle swarm optimization algorithm. In: International Conference on Machine Learning and Cybernetics, vol. 5, pp. 2993–2995 (2005)
4. Eberhart, R.C., Shi, Y.: Comparing inertia weights and constriction factors in particle swarm optimization. In: Proceedings of IEEE International Congress on Evolutionary Computation, San Diego, CA, vol. 1, pp. 84–88 (2000)
5. Ebehart, R.C., Shi, Y.: Tracking and optimizing dynamic systems with particle swarms. In: Proceedings of IEEE Congress on Evolutionary Computation, Seoul, Korea, vol. 1, pp. 94–97 (2001)
6. Niu, B., Zhu, Y.L.: MCPSO: a multi-swarm cooperative particle swarm optimizer. Appl. Math. Comput. **185**, 1050–1062 (2007)
7. Goh, C.K., Tan, K.C., Liu, D.S.: A competitive and cooperative co-evolutionary approach to multi-objective particle swarm optimization algorithm design. Eur. J. Oper. Res. **202**, 42–44 (2010)
8. Li, Y., Zhan, Z.H., Lin, S., Zhang, J., Luo, X.: Competitive and cooperative particle swarm optimization with information sharing mechanism for global optimization problems. Inf. Sci. **293**(3), 370–382 (2015)
9. Chatterjee, A., Siarry, P.: Nonlinear inertia weight variation for dynamic adaptation in particle swarm optimization. Comput. Oper. Res. **31**, 859–871 (2006)
10. Zhao, S.Z., Suganthan, P.N., Pan, Q.K.: Dynamic multi-swarm particle swarm optimizer with harmony search. Expert Syst. Appl. **38**, 3735–3742 (2011)
11. Niu, B., Zhu, Y., He, X.: Multi-population cooperative particle swarm optimization. In: Capcarrère, M.S., Freitas, A.A., Bentley, P.J., Johnson, C.G., Timmis, J. (eds.) ECAL 2005. LNCS, vol. 3630, pp. 874–883. Springer, Heidelberg (2005). doi:10.1007/11553090_88
12. Liu, Y.M., Sui, C.L., Zhao, Q.Z.: Dynamic multi-swarm particle swarm optimizer based on k-means clustering and its application. Control Decis. **26**(7), 1019–1025 (2011)
13. Xu, X., Tang, Y., Li, J., Hua, C., Guan, X.: Dynamic multi-swarm particle swarm optimizer with cooperative learning strategy. Appl. Soft Comput. **29**, 169–183 (2015)
14. Zhao, S.Z., Liang, J.J., Suganthan, P.N.: Dynamic multi-swarm particle swarm optimizer with local search for large global optimization. In: Proceedings of IEEE Swarm Intelligence Symposium, Hong Kong, pp. 3845–3852 (2008)
15. Cui, Z.H., Zeng, J.C.: Particle Swarm Optimization, 1st edn. Science Press, Beijing (2011)

Independent Component Analysis

Quantile Kurtosis in ICA and Integrated Feature Extraction for Classification

Md Shamim Reza and Jinwen Ma(✉)

Department of Information Science, School of Mathematical Sciences and LMAM, Peking University, Beijing 100871, China
shamim@pku.edu.cn, jwma@math.pku.edu.cn

Abstract. As an effective statistic in independent component analysis (ICA), kurtosis can provide valuable information for testing normality, determining features shape and ordering independent components of feature extraction in classification analysis. However, it may lead to the poor performance in certain situations so that the quantile kurtosis has been developed. In this paper, we propose a robust quantile measure of kurtosis in ICA for feature extraction. Moreover, we also present a feature extraction method which integrates the extracted features of principal component analysis (PCA), linear discriminant analysis (LDA), ICA and random forest algorithm (RFA) together. For the ICA based feature extraction, independent components are sorted according to the proposed quantile kurtosis. The experimental results show that our integrated feature extraction method, especially with the help of the proposed quantile kurtosis, outperforms the others.

Keywords: Quantile · Kurtosis · Normality · ICA · PCA · LDA · RFA

1 Introduction

Over the last few decades, Independent Component Analysis (ICA) has been found to be a very convenient and effective tool which helps us to extract representative features in classification analysis. Actually, ICA was originally proposed by Jutton and Herault [1] for solving the blind source separation (BSS) problem. But now, many investigations have found that it can serve as an effective feature extraction method of improving the classification performance in both supervised classifications [2–4], and unsupervised classifications [5–7].

In pattern classification, ICA is useful as a dimension preserving transformation because it produces statistically independent components. In fact, it could be directly used for feature extraction [8–10]. In earlier studies, Kwak et al. [11] even showed that ICA could outperform the PCA and LDA feature extraction methods on face recognition. More recently, Reza and Ma [12] successfully applied the ICA and PCA integrated feature extraction method with support vector machine (SVM), naive Bayes, decision tree (C5.0) and multi-layer perception (MLP), respectively, to the supervised classification of some UCI (The University of California at Irvine) machine learning databases.

D.-S. Huang et al. (Eds.): ICIC 2017, Part I, LNCS 10361, pp. 681–692, 2017.
DOI: 10.1007/978-3-319-63309-1_60

The fundamental restriction of ICA is that independent components (IC's) must be non-Gaussian, and we cannot determine the order of the obtained independent components [13]. This two ambiguities of ICA are the main obstacle for extracting representative feature in classification analysis. Clearly, the Principal Components (PC's) can be sorted according to the related eigenvalues, but there is no reasonable measure to order the independent components (IC's) [12, 13]. However, the past studies have shown that non-Gaussian IC's are sometimes significant to classification and kurtosis statistic can be considered as a measure of non-gaussianity as well as for sorting the IC's [14, 15].

Given this emerging concentration of kurtosis in ICA, all of the previous work concerning kurtosis in ICA has used the classical measures of kurtosis [14, 15]. Usually, classical measures of kurtosis are based on the sample average and very sensitive to outliers. In order to overcome this problem, Moors [20] proposed a quantile kurtosis alternatively, but this quantile kurtosis is not so robust to ordering independent components. In this paper, we propose an improved quantile measure of kurtosis to sort the independent components and compare their performances with four other kurtosis measures that are found in the recent statistics literature.

In the classification problem, there may happen irrelevant features that affect the learning process and thus lead to an unsatisfactory result [16]. Additionally, as the dimension of feature space becomes very large, the classification method requires many attributes to find out the association of the features, which triggers slow training and testing in both of supervised and unsupervised learning algorithm. Some of the feature extraction techniques such as ICA, PCA, LDA, random forest algorithm (RFA) and wrapper method can be directly used for feature extraction, but they cannot guarantee to generate the useful information individually [11, 12, 17]. In our earlier work, we proposed two integrated feature extraction methods only based on ICA and PCA, which outperformed the others adoptive methods [12]. In this paper, we further integrate with LDA, PCA, ICA, and RFA feature based on some statistical criterions to generate a more representative feature for the purpose of improving the classification performance.

The remaining of this paper is organized as follows. Section 2 gives a review of classical and quantile measures of kurtosis. Section 3 presents our proposed robust method of quantile kurtosis and the experimental results to demonstrate its performance. In Sect. 4, we propose our integrated feature extraction method for classification. Section 5 summarizes the experimental results and comparisons. Finally, we conclude briefly in Sect. 6.

2 Review of Kurtosis Measures

Pearson [18] originally introduced kurtosis as a measure of how flat the top of a symmetric distribution is in comparison with a normal distribution of the same variance. This conventional measure can be formally defined as the standardized fourth population moment about the mean.

$$K_1 = \frac{E(x-\mu)^4}{\left(E(x-\mu)^2\right)^2} - 3 = \frac{\mu_4}{\sigma^4} - 3 \quad (1)$$

Since the conventional measures of kurtosis are essentially based on sample averages, they are sensitive to outliers. Moreover, the impact of outliers is greatly amplified in the conventional measures of kurtosis due to the fact that they are raised to the third and fourth powers [19].

To overcome of conventional measure of kurtosis, Moors [20] proposed a quantile kurtosis alternative to K_1. The quantity of Moors kurtosis is

$$K_2 = \frac{(E_7 - E_2) + (E_3 - E_1)}{(E_6 - E_2)} \tag{2}$$

where E_i is i-th octile; that is $E_i = F^{-1}(i/8)$. For Gaussian independent components, Moor's quantile kurtosis is equal to 1.23. One advantage of the quantile measures of kurtosis is that it doesn't depend on the first moment and second moment. So the measure is not affected by outliers.

While investigating how to test light-tailed distributions against heavy-tailed distributions, Hogg [21] found that the following measure of kurtosis performs better than the traditional measure in detecting heavy-tailed distributions:

$$K_3 = \frac{U_\alpha - L_\alpha}{U_\beta - L_\beta} \tag{3}$$

Where U_α (L_α) is the average of the upper (lower) α quantile defined as:

$$U_\alpha = \frac{1}{\alpha} \int_{1-\alpha}^{1} F^{-1}(y)dy, \text{ and } L_\alpha = \frac{1}{\alpha} \int_{0}^{\alpha} F^{-1}(y)dy$$

for $\alpha \epsilon (0,1)$. According to Hoggs simulation experiments, $\alpha = 0.05$ and $\beta = 0.5$ gave the most satisfactory results. For the normal distribution, Hogg kurtosis value is 2.54. Hence, the centered Hogg coefficient is given by:

$$K_3 = \frac{U_{0.05} - L_{0.05}}{U_{0.5} - L_{0.5}} - 2.54 \tag{4}$$

Another interesting measure based on quantiles has been used in Crow and Siddiqui [22], which is given by

$$K_4 = \frac{F^{-1}(1-\alpha) + F^{-1}(\alpha)}{F^{-1}(1-\beta) + F^{-1}(\beta)} \tag{5}$$

where $\alpha, \beta \epsilon (0, 1)$. Their choices for α and β are 0.025 and 0.25 respectively. For these values, we obtain $F^{-1}(0.975) = -F^{-1}(0.025) = 1.96$ and $F^{-1}(0.75) = -F^{-1}(0.25) = -0.68$ for N (0,1) and the coefficient is 2.91.

3 Robust Quantile Kurtosis

In this study, we take an attempt to propose a robust measure of kurtosis which based on quantile and modification of moor's kurtosis estimator. In statistics, quantiles are points taken at regular intervals from the cumulative distribution function (CDF) of a random variable. It generalization of the idea of the median, where median is the value which splits data into two equal parts. Similarly, a quantile partitions the data into other proportions. Dividing ordered data into essentially equal-sized data subsets is the motivation for q-quantiles. We have used specialized 16-quantiles are called hexadeciles where ordered data are divided into 16 equal sizes. The proposed measure is given by

$$K_5 = \frac{(E_{15} - E_9) + (E_7 - E_1)}{(E_{15} - E_1)} \tag{6}$$

Where E_i is i-th hexadeciles; that is $E_i = F^{-1}(i/16)$. We easily to calculate that $E_1 = -E_{15} = -1.53$, $E_7 = -E_9 = -0.16$ for N (0,1) and therefore the coefficient of kurtosis is 0.8975. Hence, the centered coefficient is given by:

$$K_5 = \frac{(E_{15} - E_9) + (E_7 - E_1)}{(E_{15} - E_1)} - 0.8975 \tag{7}$$

Since our measure based on hexadeciles, it covers a wide range of data and doesn't depend on sample mean and variance. So it's less affected by outlier and more robust than the classical measure of kurtosis. In the next section, we will discuss the robustness of an estimator.

3.1 Qualitative Robust Index

Qualitative robustness, influence function, and breakdown point are three main concepts to judge an estimator from the viewpoint of robust estimation. Nasser et al. [23] have proposed a definition of finite-version qualitative robustness, their estimator with finite breakdown point equal to zero should have empirically lower QRI whereas estimators with high breakdown point should have higher QRI. They proposed two versions of SQRI (SQRI-1 and SQRI-2):

$$SQRI - I = \frac{1}{1 + max_j\left|\hat{\theta} - \hat{\theta}(j)\right|} \tag{8}$$

$$SQRI - II = \frac{1}{1 + max_{i \neq j}\left|\hat{\theta}(i) - \hat{\theta}(j)\right|} \tag{9}$$

It is easy to prove (i) It's maximum value is 1. (ii) It's minimum value is zero or above zero. The more SQRI the estimator is more qualitative robust.

3.2 Datasets and Results

To find out qualitative robust measure of kurtosis by using SQRI-1 and SQRI-2 method, we have used USGS (United States Geological Survey) data, counting earthquake by yearly contains 112 observations (1900-2011) and considered magnitude range of earthquake 7.0 to 9.9. In our experiment, we have taken four datasets, where three from UCGS, earthquake data and one set of melanoma skin cancer data. Sample data structure are given below:

- Data-1: 60 sample have drowned from 112 observations of UCGS data.
- Data-2: 60 sample have drowned from 112 observations and 5 samples have drawn from the Cauchy distribution with parameter 2.
- Data-3: 60 sample have drowned from 112 observations and 5 samples have drowned from the student-t distribution with degrees of freedom 2.
- Data-4: 37 sample have taken from melanoma skin cancer incidence data during the year (1936 – 1972). Source: R data package (lattice).

In descriptive statistics, the box plot is a convenient way of graphically displaying variation in samples of a statistical population without making any assumptions about the underlying statistical distribution. Figure 1, we have taken 60 samples in 1000 times from earthquake data and calculate each of kurtosis estimators. The graph shows that propose kurtosis estimator (K_5) distribution is more consistent than others. To check

Table 1. Results comparison for SQRI-1 of six kurtosis estimators.

Kurtosis estimators	Data-1	Data-2	Data-3	Data-4
K_1	0.725	0.825	0.874	0.952
K_2	0.891	0.852	0.837	0.944
K_3	0.849	0.935	0.918	0.946
K_4	0.698	0.782	0.807	0.955
K_5 (proposed)	**0.925**	**0.942**	**0.931**	**0.968**

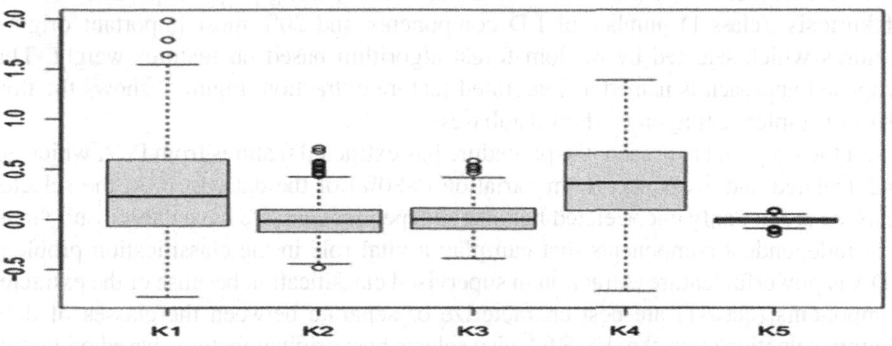

Fig. 1. Boxplot of different kurtosis estimators, Number of replication, N = 1000.

numerically of a robust estimator, we have used two SQRI for all kurtosis estimators which result are displayed in Tables 1 and 2.

Table 2. Results comparison for SQRI-2 for six kurtosis estimators.

Kurtosis estimators	Data-1	Data-2	Data-3	Data-4
K_1	0.575	0.726	0.810	0.905
K_2	0.853	0.830	0.767	0.902
K_3	0.812	0.879	0.882	0.904
K_4	0.602	0.662	0.735	0.917
K_5 (proposed)	**0.905**	**0.922**	**0.894**	**0.926**

Notes and Comments: In our experiments, we consider bootstrap techniques to find θ, where the number of replication N = 1000. The results show that the proposed method successfully chooses the best robust estimator (K_5) because it's all value closest to 1. According to SQRI index, the more SQRI value conveys high break down point and indicating as the more robust estimator. So this robust kurtosis estimator can provide useful information in ICA as well as to sorts of independent components and extract representative features in a classification problem, in details refer to Sect. 4.

4 Integrated Feature Extraction Paradigm

Although, the performances of PCA, ICA, and LDA are powerful in the field of data visualization and blind source separation. For classification problem, feature extraction technique of LDA performances is good if certain assumptions are hold in data [24], but PCA and ICA are not as good as expected [11, 25]. To overcome the problem, we propose a feature extraction method, which integrates with LDA, PCA, ICA, and a feature selection technique random forest algorithm (RFA) to represent significant feature sets for classification problem.

The idea of the proposed feature extraction is very simple. In the proposed approach, LDA, ICA, PCA, and feature selection algorithm have applied to the original data individually, we then retain those PC's that can explain at least 80% of the total variation, most sub-Gaussian IC's (kurtosis < 0) are ordered by using propose quantile measure of kurtosis, (class-1) number of LD components and 20% most important original features which selected by random forest algorithm based on features weight. This proposed approach is named as integrated feature extraction. Figure 2 shows the flow chart of implementing on the four databases.

In the proposed approach, the procedure has extracted features from PCA which are uncorrelated and gives maximum variation (>80%) of the data. In ICA, the selected features are not only uncorrelated but also independent and we have chosen sub-Gaussian independent components that can play a vital role in the classification problem. LDA is powerful feature extraction in supervised classification because of the extracted components (class-1) are best characterize or separate between the classes of data. Feature selection algorithm FS-RFA also selects best original features based on feature

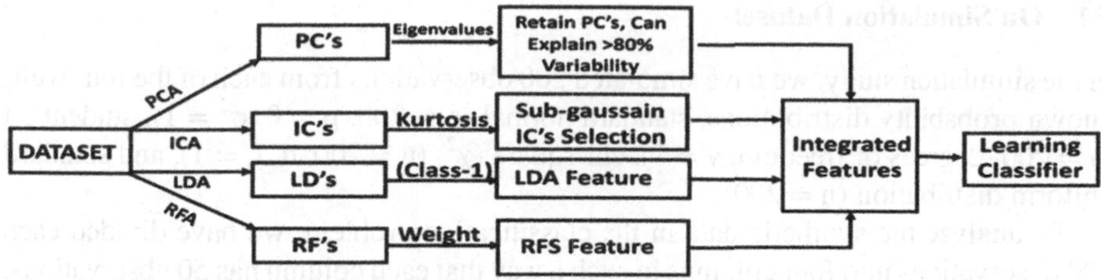

Fig. 2. Flow chart for implementing integrated feature extraction.

weights of class information. Finally, we have been integrated a significant features sets to learning classifiers.

5 Experimental Framework and Results

In this section, our proposed feature extraction approach is tested on a simulated dataset, and three real datasets from UCI database [26], namely Satellite, Ionosphere, and Sonar datasets, respectively.

In order to test the efficiency of the proposed feature extraction methods, we select the most significant number of original attributes by using random forest algorithm (FS-RFA), which is available in R package, FSelector [27]. In random forest algorithm FS-RFA, first, employs a weight function to generate weights for each feature. To select significance of weight, the algorithm use mean decrease accuracy, besides it selects an optimum number of subset feature through the statistical function chi-square and information gain. Finally, the procedure sorts a top most dominant original subset of features. To apply ICA for feature extraction, training data was transformed to zero mean and unit variance by using fastICA algorithm [13], PCA and LDA were applied directly over data.

In classifier system, we have used multi-layer perception (MLP), support vector machine (SVM), decision tree (C5.0), and naive Bayes classifier. In our experiment, we have driven 10-fold cross validation in getting the performance as follows: The observations have been divided randomly into 10 disjoint fold or sets. For each experiment, 9 of these fold is used as training data, while 10th set observation is reserved for testing. The experiment is repeated 10 times in such a way that every fold appears once as a part of a test set.

To show the effectiveness of our method, we have compared the performances of the proposed methods with PCA, LDA, ICA, IPCA, IC-PC and FS-RFA. Feature extraction techniques, IPCA and IC-PC we have already been proposed our earlier work [12]. All the experiments conduct here are implemented in R-studio software with different R-CRAN (The Comprehensive R Archive Network) packages of machine learning [27], and run on a 4 core GPU system.

5.1 On Simulation Dataset

In the simulation study, we have simulated 200 observations from each of the four well-known probability distributions, standard normal (n = 200, $\mu = 0$, $\sigma^2 = 1$), student's t (n = 200, degrees of freedom, v = 1), chi-square (χ^2) (n = 200, d. f. = 1), and standard uniform distribution (n = 200).

To analyze the synthetic data in the classification problem, we have divided each 200 observations into four columns in such a way that each column has 50 observations. An additional column has also been inserted to input class label. As for example of the Gaussian distribution (mean = 0, variance = 1), the generated 200 observations have been divided into four columns, where each column contains 50 observations, then each of the first 50 observations has been labeled by 1 in the additional column. Similarly, for chi-square (χ^2) distribution, 200 generated observations were divided into four columns and inserted the class label 2 and so on. Finally, we have combined the observations to obtain a data frame that contains 5 features including one class label attribute each with 200 observations.

In simulated data, first 3 PC's can explain 84.23% of the total variation, then we have applied ICA algorithm on 3 PC's to construct IPCA feature. In IC-PC, we have united first 3 PC's and one sub-Gaussian (kurtosis < 0) IC's. To make propose integrated feature, we have combined first 3 PC's with sub-Gaussian IC's (kurtosis < 0), one LD component and one most important original feature which obtained by using random forest algorithm. Table 3 shows the classification performances of different classifiers. The classification accuracy is obtained by using 10-fold cross-validation and we found that our integrated feature extraction methods improvement in all the classification performances in a certain degree.

Table 3. Classification accuracy (%) for simulated data (parentheses are the number of PC's & IC's respectively)

Features	SVM	Naïve Bayes	C5.0	MLP
Original	62.5	66.5	69.0	51.5
FS-RFA	64.0	69.0	65.0	56.0
PCA	65.0 (3)	65.5 (3)	68.5 (3)	56.5 (3)
LDA	61.5	66.5	66.0	51.5
ICA	62.0	67.5	65.0	52.0
IPCA	66.0 (3)	69.0 (3)	65.5 (3)	56.0 (3)
IC-PC	66.0 (3,1)	72.0 (2,2)	**72.0 (3,4)**	57.0 (3,4)
Integrated	**67.5**	**72.5**	71.8	**58.5**

LDA performances are not so good on simulated data because, under the assumption of LDA, the distribution of samples in each class are normal and homoscedastic. But our simulated data includes normal and three other classes (student's t, chi-square, and uniform) apart from normal, which may influence LDA performances.

5.2 On Satellite Dataset

The original Landsat data for this database was generated from data purchased from NASA by the Australian Centre for Remote Sensing and used for research at the Centre for Remote Sensing, University of New South Wales, Australia. These data have been taken from the UCI Repository of Machine Learning Databases [26].

Data frame with 36 inputs, one target on 6435 observations. The database consists of the multi-spectral values of pixels in 3×3 neighborhoods in a satellite image and the classification associated with the central pixel in each neighborhood. The aim is to predict this classification, given the multi-spectral values.

In satellite data, only first 4 PC's (out of 36) can explain 92.02% of the total variation. All of the classifiers performs well when we integrate with first 7 PC's, one most sub-Gaussian IC's that ordered by our proposed quantile kurtosis, (6-1) class i.e. 5 LD components and most two important original variables that obtained from random forest algorithm (RFA). Table 4 shows that the performances of integrated and IC-PC feature extraction method have achieved 100% classification accuracy of the decision tree (C50) classifier. The IC-PC (7,2) method integrated only 7 PC's and 2 most sub-Gaussian IC's features out of 36 features and achieved 100% accuracy.

Table 4. Classification accuracy (%) for satellite data (parentheses are the number of PC's & IC's respectively)

Features	SVM	Naïve Bayes	C5.0	MLP
Original	80.34	79.12	84.61	83.31
FS-RFA	85.12	72.69	83.12	81.18
PCA	87.88 (7)	82.18 (7)	85.85 (7)	83.02 (7)
LDA	87.13	84.55	84.97	84.04
ICA	86.20	80.26	79.37	83.68
IPCA	87.44 (5)	82.79 (6)	85.68 (7)	83.80 (7)
IC-PC	87.56 (5,1)	82.64 (7,1)	**100 (7,2)**	83.73 (7,1)
Integrated	**89.29**	**85.75**	**100**	**84.16**

5.3 On Ionosphere Dataset

These data have been taken from the UCI Repository of Machine Learning Databases [26]. This radar data was collected by a system in Goose Bay, Labrador. The targets were free electrons in the ionosphere. "good" radar returns are those showing evidence of some type of structure in the ionosphere. "bad" returns are those that do not; their signals pass through the ionosphere. Data frame with 351 observations on 35 independent variables, some numerical and 2 nominal, and one last defining the class. This dataset is often used to test and compare the performances of various classification algorithms.

In Ionosphere data, first 11 PC's can explain 80% of the total variation, while original feature number is 35. The classification accuracy of the four classifiers are displayed in Table 5. It can be seen that propose integrated features (11 PC's, 1 IC's, 1 LD component, and 1 original attribute) perform better than others. The cross validation classification accuracy of SVM classifier exceeds than others in the past work on this dataset. In Ionosphere data, the naive Bayes classifier also performs better than others because of our proposed feature extraction method produced uncorrelated and independent features which coincide the assumptions of naive Bayes classification techniques.

Table 5. Classification accuracy (%) for ionosphere data (parentheses are the number of PC's & IC's respectively)

Features	SVM	Naïve Bayes	C5.0	MLP
Original	94.87	78.89	88.91	89.77
FS-RFA	94.87	91.73	89.75	90.31
PCA	96.01 (11)	90.32 (11)	89.73 (11)	87.46 (11)
LDA	89.45	88.89	88.89	89.45
ICA	93.73	89.74	84.33	87.22
IPCA	95.72 (11)	92.89 (11)	87.76 (11)	87.17 (11)
IC-PC	95.43 (11,3)	90.60 (11,2)	90.59 (11,1)	87.18 (11,2)
Integrated	**96.29**	**95.73**	**91.46**	**91.73**

5.4 On Sonar Dataset

This is the data set used by Gorman and Sejnowski in their study of the classification of sonar signals using a neural network. The task is to train a network to discriminate between sonar signals bounced off a metal cylinder and those bounced off a roughly cylindrical rock. These data have been taken from the UCI Repository of Machine Learning Databases [26], and data frame with 208 observations on 61 variables, all numerical and one (the class) nominal.

In sonar data, only first 14 PC's (out of 60 PC's) can explain 81.19% of the total variation. We have then compared proposed feature extraction approach with PCA, LDA, ICA, IC-PC and IPCA. In Table 6 show that most of the cases, LDA and our extracted feature (15 PC's, 2 IC's, one LD components, and 3 original variables) outperforms the others.

Table 6. Classification Accuracy (%) for Sonar data (Parentheses are the number of PC's & IC's respectively)

Features	SVM	Naïve Bayes	C5.0	MLP
Original	84.59	66.38	73.09	83.17
FS-RFA	83.62	73.07	81.73	80.78
PCA	85.02 (16)	75.97 (9)	76.50 (11)	82.19 (15)
LDA	88.46	85.22	**89.90**	**90.86**
ICA	79.83	60.64	59.66	75.02
IPCA	85.02 (16)	69.26 (16)	75.50 (15)	80.71 (15)
IC-PC	86.93 (16,3)	75.48 (16,2)	76.50 (16,2)	84.09 (15,1)
Integrated	**90.83**	**89.45**	86.50	87.45

6 Conclusion

We have proposed a robust measure of quantile kurtosis for ICA, and a novel integrated feature extraction approach based on ICA, PCA, LDA and RFA for supervised classification. We have tested our proposed integrated feature extraction method on synthetic and real datasets in comparison with the other existing feature extraction methods. It is demonstrated by the experimental results that the extracting features, PCA, ICA, even LDA, doesn't perform well individually. But if we fuse these features by using some statistical criterions, it generates a more representative and dominant feature set for the classifier. In most of the cases, integrated feature extraction method can improve the classification performances substantially. The findings of this work clearly show that dimensionality reduction and integrated feature extraction by fusing some different kinds of features together are necessary and effective in the field of data mining and pattern recognition.

Acknowledgements. This work was supported by the Natural Science Foundation of China for Grant 61171138.

References

1. Jutten, C., Herault, J.: Blind separation of sources, part 1: an adaptive algorithm based on neuro mimetic architecture. Sig. Process. **24**, 1–10 (1991)
2. Kwak, N., Choi, C.-H., Choi, J.Y.: Feature extraction using ICA. In: Dorffner, G., Bischof, H., Hornik, K. (eds.) ICANN 2001. LNCS, vol. 2130, pp. 568–573. Springer, Heidelberg (2001). doi:10.1007/3-540-44668-0_80
3. Yu, S.N., Chou, K.T.: Integration of independent component analysis and neural networks for ECG beat classification. Expert Syst. Appl. **34**, 2841–2846 (2008)
4. Fan, L., Poh, K.L., Zhou, P.: A sequential feature extraction approach for naive Bayes classification of microarray data. Export Syst. Appl. **36**, 9919–9923 (2009)
5. Kapoor, A., Bowles, T., Chambers, J.: A novel combined ICA and clustering technique for the classification of gene expression data. In: Proceedings of IEEE International Conference on Acoustics, Speech, and Signal Processing, vol. 5, pp. 621–624 (2005)
6. Kwak, N.: Feature extraction for classification problems and its application to face recognition. Pattern Recogn. **41**, 1701–1717 (2008)
7. Hyvarinen, A., Oja, E., Hoyer, P., Hurri, J.: Image feature extraction by sparse coding and independent component analysis. In: Proceedings of 14th International Conference on Pattern Recognition, August 1998
8. Back, A.D., Trappenberg, T.P.: Input variable selection using independent component analysis. In: Proceedings of International Joint Conference on Neural Networks, July 1999
9. Yang, H.H., Moody, J.: Data visualization and feature selection: new algorithms for non-gaussian data. In: Advances in Neural Information Processing Systems, vol. 12 (2000)
10. Yang, T.-Y., Chen, C.C.: Data Visualization by PCA, LDA, and ICA, ACEAT-493 (2015)
11. Kwak, N., Choi, C.: Feature extraction based on ICA for binary classification problems. IEEE Trans. Knowl. Data Eng. **15**, 1374–1388 (2003)
12. Reza, M.S., Ma, J.: ICA and PCA integrated feature extraction for classification. In: The 13 IEEE International Conference on Signal Processing, pp. 1083–1088 (2016). doi:10.1109/ICSP.2016.7877996

13. Hyvarinen, A., Oja, E.: Independent component analysis: algorithms and applications. Neural Netw. **4–5**(13), 411–430 (2000)
14. Reza, M.S., Nasser, M., Shahjaman, M.: An improved version of kurtosis measure and their application in ICA. Int. J. Wirel. Commun. Inf. Syst. **l**(1), 6–11 (2011)
15. Scholz, M., Gibon, Y., Stitt, M., Selbig, J.: Independent component analysis of starch-deficient pgm mutants. In: Proceedings of the German Conference on Bioinformatics, pp. 95–104 (2004)
16. Cios, K.J., Pedrycz, W., Swiniarski, R.W.: Data Mining Methods for Knowledge Discovery. Kluwer Academic Publishers, Boston (1998). Chap. 9
17. Fan, L., Poh, K.L., Zhou, P.: Partition-conditional ICA for Bayesian classification of microarray data. Export Syst. Appl. **37**, 8188–8192 (2010)
18. Pearson, K.: Skew variation, a rejoinder. Biometrika **4**, 169–212 (1905)
19. Kim, T.H., White, H.: On more robust estimation of skewness and kurtosis: simulation and application to the S\&P500 index, Department of Economics, UCSD (2003)
20. Moors, J.J.A.: A quantile alternative for kurtosis. The Stat. **37**, 25–32 (1988)
21. Hogg, R.V.: More light on the kurtosis and related statistics. J. Am. Stat. Assoc. **67**, 422–424 (1972)
22. Crow, E.L., Siddiqui, M.M.: Robust estimation of location. J. Am. Stat. Assoc. **62**, 353–389 (1967)
23. Nasser, M., Hamzah, N.A., Alam, A.: Qualitative robustness in estimation. PJSOR **8**(3), 619–634 (2012). Statistics in the Twenty-First Century, Special Volume
24. Oh, J., Kwak, N., Lee, M., Choi, C.H.: Generalized mean for feature extraction in one-class classification problems. Pattern Classif. **46,** 3328–3340 (2013)
25. Martinez, A.M., Kak, A.C.: PCA versus LDA. IEEE Trans. Pattern Anal. Mach. Intell. **23**(2), 228–233 (2001)
26. The University of California, Irvine (UCI) Machine Learning Repository. http://www.ics.uci.edu/~mlearn. Accessed 11 May 2017
27. The Comprehensive R Archive Network. https://cran.r-project.org. Accessed 11 May 2017

Compressed Sensing and Sparse Coding

Similarity Matrix Construction Methods in Sparse Subspace Clustering Algorithm for Hyperspectral Imagery Clustering

Qing Yan[1], Yun Ding[1], Jing-Jing Zhang[1], Li-Na Xun[1], and Chun-Hou Zheng[2(✉)]

[1] College of Electrical Engineering and Automation, Anhui University, Hefei, Anhui 230601, China

[2] College of Computer Science and Technology, Anhui University, Hefei, Anhui 230601, China

zhengch99@126.com

Abstract. The clustering of hyperspectral images is a challenging task because of the high dimensionality of the data. The sparse subspace clustering (SSC) algorithm is one of the popular used clustering algorithm for high dimensionality data. But, SSC has not considered the spectral and spatial information fully, so it is not satisfied for Hyperspectral Imagrery (HSI) clustering. In this paper, a novel similarity matrix construction methods are proposed which combined the high spectral correlation and rich spatial connection. Firstly, we utilize the cosine similarity of sparse representation vector to construct a novel similarity matrix. Then, the similarity matrix based on Euclidean distance of the sparse representation vector can connect spectral correlation with spatial information. Several experiments on HSIs demonstrated that the proposed algorithms are effective for hyperspectral images (HSIs) clustering.

Keywords: High dimensionality · Sparse subspace clustering · Similarity matrix · Hyperspectral imagery

1 Introduction

Hyperspectral imagery (HSI) can record hundreds of spectral bands for each pixel, and each pixel contains values that correspond to the detailed spectrum of reflected light. In essence, the existing HSI recognition approaches can be divided into two categories i.e., supervised classification and unsupervised clustering [1]. Supervised classification techniques require the availability of a training set for training the classifier, and identify unlabeled pixels by the classifier. Unsupervised clustering is just exploiting label information conveyed by the unlabeled data, without requiring any labeled training sample set. So it is more suitable for hyperspectral data analysis [2]. Clustering methods of hyperspectral data mainly incorporate K-means [3], FCM, Spectral clustering, etc. K-means algorithm is an iterative method. It has expensive computational

Q. Yan et al.—These authors are contributed equally to the paper as first authors.

D.-S. Huang et al. (Eds.): ICIC 2017, Part I, LNCS 10361, pp. 695–700, 2017.
DOI: 10.1007/978-3-319-63309-1_61

cost, and the effect of clustering is closely related to initial randomly centroids. Another classic clustering approach is FCM [4], which considers each cluster as a fuzzy set. Since HSI contains high-dimensional data and possess complex structure. FCM may not make full use of enough information of HSI.

Spectral clustering is a popular clustering method for HSI clustering. For the past few years, the sparse subspace clustering (SSC) [5] algorithm has been proposed for image clustering. Directly applying SSC to HSIs usually failed to make full use of the high spectral connection and abundant spatial information which will affect the performance of HSI clustering. To solve this problem, Zhang et al. [6] raises spatial information SSC (SSC-S) algorithm which design a new sparse model to obtain sparse coefficient matrix by adding spatial information, which consider the high spectral correlation and rich spatial information of the HSIs in the SSC model to obtain a more accurate coefficient matrix. We consider using the whole sparse representation vector to structure similarity matrix to add spatial information optimization in the SSC model.

2 Related Work of the SSC Algorithm in HSI Field

Sparse subspace clustering (SSC) is a novel framework for data clustering based on spectral clustering. High-dimensional data usually lies in a union of several low-dimensional subspaces, which allows sparse representation of high-dimensional data with an appropriate dictionary. Firstly, let we review the content of SSC algorithm. Let $\{S_l\}_{l=1}^{n}$ be an array of n linear subspaces of IR^D of dimensions $\{d_l\}_{l=1}^{n}$. Given a collection of N data points $\{y_i\}_{i=1}^{N} \in Y$ that lie in the union of the n subspaces. Thereby, each data point $y_i \in S_l$ can be written as

$$\min \| C \|_1 \ \text{s.t.}\ \mathrm{Y} = YC,\ \mathrm{diag(C)} = 0 \tag{1}$$

Where $C \triangleq [c_1\, c_2 \cdots c_N] \in R^{N \times N}$ is a matrix whose i th column corresponds to the sparse representation of y_i and $\mathrm{diag(C)} \in R^N$ is the vector of the diagonal elements of C. Now we can build the similarity matrix W as Eq. (2). $W \in R^{N \times N}$ is a symmetric nonnegative similarity matrix.

$$W_{ij} = |C_{ij}| + |C_{ji}| \tag{2}$$

Finally, applying spectral clustering to the similarity matrix W and getting the clustering result of the data: $Y_1, Y_2, \cdots, Y_n$.

3 The Cosine-Euclidean Similarity Matrix Construction

Firstly, we recognize the significance of extracting spectral information from complex HSI structure. In nature, Wu et al. [2] raise the similarity matrix construction which is linked closely with spectral information based on the cosine similarity between the sparse representation vectors. And they also proposed the cosine similarity algorithm, which has achieved excellent performance in the general data. We logically extend this idea to SSC. In particular, Wu et al. only concentrate on cosine similarity and neglect spatial connection in building similarity matrix. So, we improve the construction of similarity matrix in SSC by adding the spatial information.

Let c_i and c_j are the sparse representation vectors of data objects y_i and y_j. Theoretically, if the y_i and y_j are similar, then we consider that their sparse representation vectors, c_i and c_j are also similar. Since cosine measure is often used as similarity measure of two vectors. The spectral similarity matrix is defined as follows.

$$\cos_{ij} = \max\left\{0, \frac{c_i \cdot c_j}{\| c_i \|_2 \times \| c_j \|_2}\right\} i, j = 1, 2, \cdots MN \tag{3}$$

where c_i and c_j are both from Eq. (4) by solving l_1 program and MN represent the number of pixel points. The value of cos range from 0 to 1. The Eq. (3) carries plenty of spectral information of HSI pixel with high correlation. But, in fact, the spatial information is also very important for clustering. So we add more spatial information in the construction of similarity matrix. So the Euclidean distance of pixel spectral value can reflect the local spatial information to some extent. Euclidean distance formula is defined as follows:

$$d_{ij} = \| y_i - y_j \| \; i, j = 1, 2, \cdots, MN \tag{4}$$

The value of spectral similarity matrix number ranges from 0 to 1. Consequently, Euclidean distance needs to be normalized as (9)

$$d_{ij}^* = \frac{d_{ij} - d_{\min}}{d_{\max} - d_{\min}} \tag{5}$$

where $d_{\min}$ and $d_{\max}$ represents the minimum value and maximum value of d_{ij} respectively. If two spectral vectors have the smaller Euclidean distance and the larger spectral cosine angle at the same time, we consider they are similar. So, we define the Cosine-Euclidean Similarity Matrix (abbreviated as CE) by combining Euclidean distance with spectral cosine angle as:

$$W_{CE}^* = \cos_{ij} - d_{ij}^* \tag{6}$$

4 Experimental Work and Analysis

In this section, we conduct a set of experiments to further evaluate the effectiveness of proposed algorithms for HSI SSC, SSC-S, K-means, FCM, Nystrom are used as benchmarks.

4.1 Experimental Datasets

We conduct experiments on two hyperspectral imagery datasets, which include Pavia Centre scene. A brief description of the data sets is listed as follows.

Pavia Centre scene: Pavia Centre is a 1096*1096 pixels image, and the number of spectral bands is 102. The geometric resolution is 1.3 m. Pavia Centre image ground truths differentiate 8 classes. The Pavia Centre scene data set is shown as abroad black strips (Fig. 1(a), (b)).

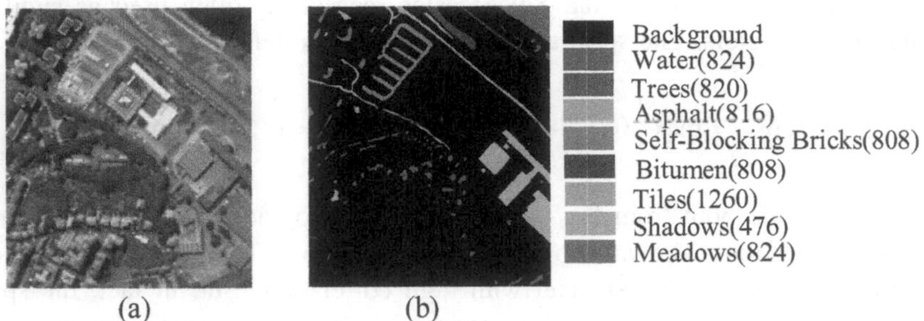

Fig. 1. Part of Pavia centre scene. (a) The HSI in false color (RGB 3,65,101) (b) Ground truth

4.2 Evaluation Metric

The clustering results usually is estimated kappa coefficient (KC), overall accuracy (OA), producer's accuracy (PA). These evaluation indicator values range from 0 to 1, and a bigger value indicates a better clustering result.

4.3 Experimental Results and Analysis

We use our proposed method CE, compared with SSC, SSC-S, K-means, FCM, and Nystrom, to analyze the Pavia Centre scene.

The visual clustering performance of these methods on the Pavia Centre image is shown in Fig. 2(a)–(e), and the corresponding quantitative analysis of the clustering results is listed in Tables 1 and 2. From Fig. 2, it can be seen that the visual clustering effect of the Self-Blocking Bricks, Tiles in (e) are better than the others and more closer to the true ground. From the quantitative evaluation of the clustering results in Table 1, the producer's accuracy (PA) precision of the Self-Blocking Bricks and a Tiles of the Pavia Centre scene are 75.71% and 63.71% by using our proposed algorithms. The clustering results of other four methods included K-means, FCM, Nystrom SSC ans SSC-S algorithms, are also list in Table 1. It can be clearly seen that our proposed algorithms obtain a better visual cluster map than these five methods by the stander of effectively discriminating the two classes of the Self-Blocking Bricks and Tiles. It is also consistent with the quantitative analysis. The main reason is that we have added spectral correlation of sparse representation vector of the objective instance and Euclidean distance of HSIs pixels with spatial correlation in sparse similarity matrix construction. Which can ensure that the signals with higher correlation are preferentially selected in the sparse representation process. Thereby, it can build a more accurate similarity matrix, which is critical for yielding higher cluster accuracy. Moreover, the proposed method with CE algorithm gets the highest precision, with the best OA of 68.44% and KC of 0.6000 listed in the Table 2. The CE method can obtain an OA improvement of 12.44% compared with SSC-S algorithm. In summary, the clustering effect of our proposed algorithm for HSIs data clustering achieves better performance, both in visually and quantitatively.

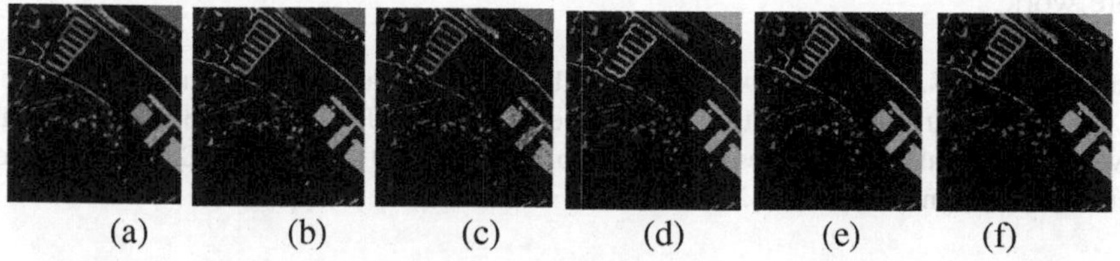

(a) (b) (c) (d) (e) (f)

Fig. 2. Cluster maps of the different methods on the Pavia centre scene. (a) K-means (b) FCM (c) Nystrom (d) SSC (e) SSC-S (f) CE

Table 1. The quantitative analysis on Pavia Centre scene

Evaluation	PA					
Method Class	kmeans	Fcm	Nystrom	SSC	SSC-S	CE
Water	69.05	69.15	**76.69**	0	69.15	6.49
Trees	73.48	**91.22**	20.79	88.83	77.07	81.82
Asphalt	6.19	26.57	2.52	**31.27**	1.22	0
Self-Blocking Bricks	58.55	60.92	31.21	30.13	64.32	**75.71**
Bitumen	66.06	**91.68**	9.28	71.08	75.16	76.77
Tiles	4.79	41.23	38.21	42.39	9.12	63.71
Shadows	67.98	**74.97**	53.38	71.53	66.32	**63.61**
Meadows	0	0	31.00	**99.28**	0	99.06

Table 2. The overall quantitative evaluation on Pavia Centre scene

Method Evalution	kmeans	Fcm	Nystrom	SSC	SSC-S	CE
KC	0.5148	0.5308	0.3054	0.5290	0.4589	**0.6000**
OA (%)	61.08	60.59	40.43	62.21	56.00	**68.44**

5 Conclusion

In this paper, we proposed the new similarity matrix construction method for making full use of the spectral correlation and spatial correlation of HSIs information. In view of the complex spatial and spectral structure of hyper spectral image, our method acquires similarity matrix based on whole sparse representation vector, which is covered spectral information sparse representation vectors and spatial connection of sparse representation objects. The extensive experimental results, compared with the traditional methods, consistently demonstrate that our proposed approach exhibit better performance than the state-of-the-art methods.

However, the proposed algorithm still has the room of improvement. For example, the more spatial and shape information of HSI in SSC model need to be taken into consideration on similarity matrix construction. We will address the question in our future work.

Acknowledgments. This work was supported by the National Science Foundation for China (No.61602002), AnHui University Youth Skeleton Teacher Project (E12333010289), Anhui University Doctoral Scientific Research Start-up Funding (J10113190084), China Postdoctoral Science Foundation (2015M582826).

References

1. Chen, Y., Nasrabadi, N.M., Tran, T.D.: Hyperspectral image classification via kernel sparse representation. IEEE Trans. Geosci. Remote Sens. **51**(1), 217–231 (2013)
2. Wu, S., Feng, X., Zhou, W.: Spectral clustering of high-dimensional data exploiting sparse representation vectors. Neurocomputing **135**(8), 229–239 (2014)
3. Filho, A.G.S., et al.: Hyperspectral images clustering on reconfigurable hardware using the k-means algorithm. In: Proceedings of the Symposium on Integrated Circuits and Systems Design, SBCCI 2003 (2003)
4. Niazmardi, S., Homayouni, S., Safari, A.: An improved FCM algorithm based on the SVDD for unsupervised hyperspectral data classification. IEEE J. Sel. Top. Appl. Earth Observ. Remote Sens. **6**(2), 831–839 (2013)
5. Elhamifar, E., Vidal, R.: Sparse subspace clustering: algorithm, theory, and applications. IEEE Trans. Pattern Anal. Mach. Intell. **35**(11), 2765–2781 (2013)
6. Zhang, H., et al.: Spectral-spatial sparse subspace clustering for hyperspectral remote sensing images. IEEE Trans. Geosci. Remote Sens. **54**(6), 3672–3684 (2016)

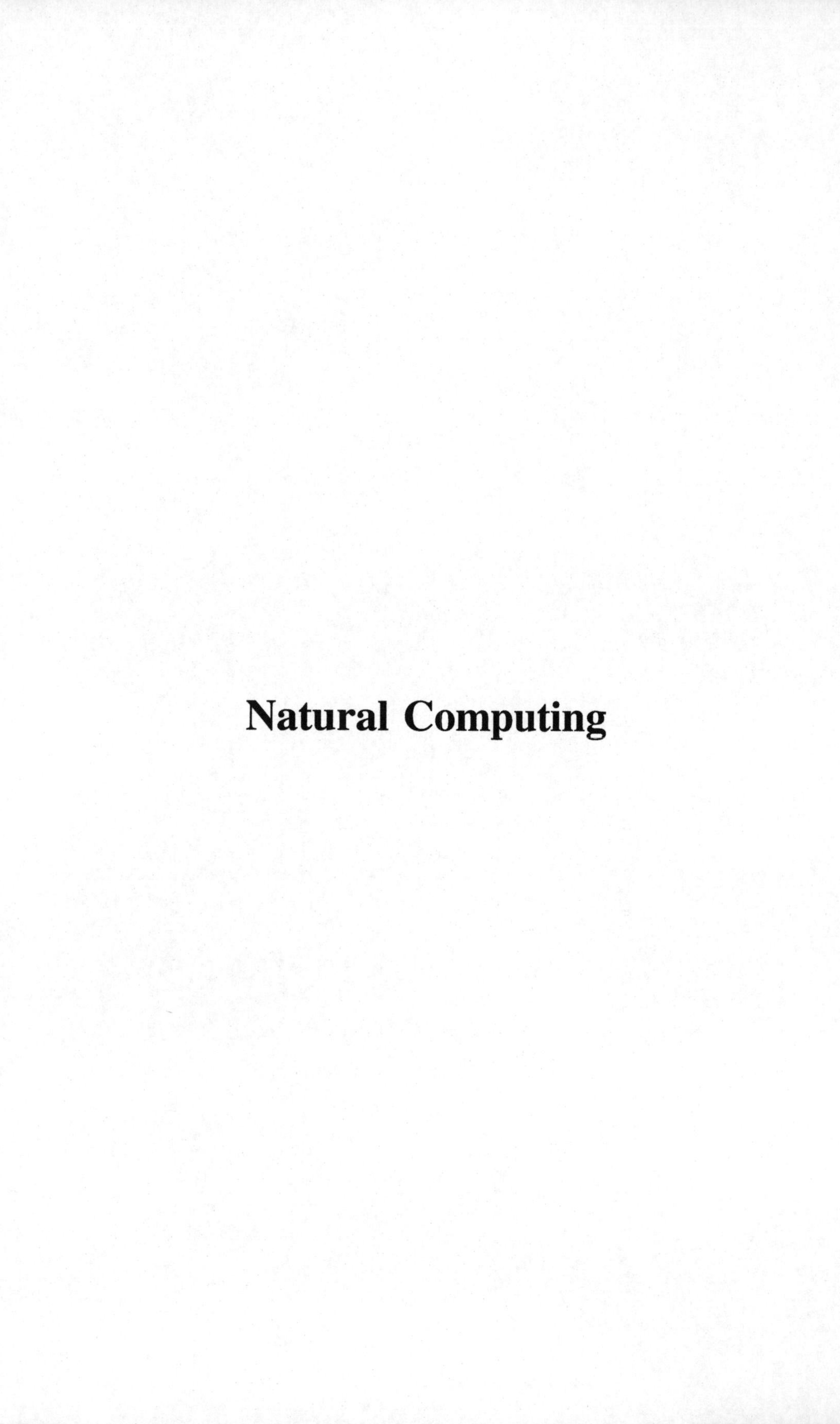

Natural Computing

A Fast Optimization Algorithm for *K*-Coverage Problem

Jingwen Pei[1], Maomao[1], and Jiayin Wang[2,3,4(✉)]

[1] Department of Computer Science and Engineering, University of Connecticut, Storrs, CT 06269, USA

[2] School of Management, Xi'an Jiaotong University, Xi'an, China
wangjiayin@mail.xjtu.edu.cn

[3] Institute of Data Science and Information Quality, Xi'an Jiaotong University, Xi'an, China

[4] Shaanxi Engineering Research Center of Medical and Health Big Data, Xi'an Jiaotong University, Xi'an, Shaanxi 710049, China

Abstract. *K*-coverage optimization is widely used in healthcare environments, which minimizing the number of directional receivers that guarantee a given region is covered *k* times. As *K*-coverage optimization is NP-hard, a commonly used approach for finding the optimal solution is integer linear programming. However, it can be slow for many practical instances. In this article, we propose an exact dynamic programming algorithm based on tree-decomposition. A probability-distribution density function is introduced to describe the relative importance of various areas and compute the minimal optimized sub-structures. We also show that this algorithm can be easily extended to provide exact solutions for different coverage ratio requirements. When compared with the ILP approach and greedy algorithm, our algorithm provides accurate solutions without scarifying too much on efficiency.

Keywords: K-coverage problem · ILP · Tree decomposition algorithm

1 Introduction

Benefiting from the wireless networking, many types of wireless equipments are widely used in clinical practices. The information collected by medical devices transmits via wireless communications. To reduce the interference, it is suggested that, for a particular region, an optimal design prefers to use multiple directional antenna receivers corresponding to the same number of the types of equipments due to less channel disturbance. And thus, the design of receivers that cover a given region becomes a fundamental problem. Different from omni-directional sensors, directional sensors are characterized by an orientation and a working angle such that the field of view is fan-shaped or can be described as a sector. The work of Ai et al. [1] first studied this network coverage problem using directional sensors, which has since attracted much subsequent interests. Since some types of sensors (e.g. video and ultrasonic sensors) can be directional, research on directional sensors has been actively studied [2–4]. In this article, we address the directional sensor deployment problem in the pervasive healthcare environment.

D.-S. Huang et al. (Eds.): ICIC 2017, Part I, LNCS 10361, pp. 703–714, 2017.
DOI: 10.1007/978-3-319-63309-1_62

Patients carrying medical recording devices are able to move freely, and thus, the coverage problem focuses on regional coverage (i.e. covering a set of pre-defined regions) rather than target coverage. A traditional sensor network, e.g. those used for surveillance, cannot be re-charged during its lifetime, and this energy savings is usually a concern in deployment. Because the sensors in a healthcare environment must operate continually, there is less need to consider energy conservation.

We develop a tree decomposition algorithm that minimizes the number of required sensors for the desired level of region coverage. To the best of our knowledge, no previous research work has used tree decomposition-based optimization to solve the coverage problem in a sensor network, whose results from a decomposed tree are able to have the exact results as ILP, which is known to provide the optimal result. Moreover, the use of tree decomposition largely outperforms the ILP approach in terms of efficiency, and the optimal solution also satisfies the higher requirements specific for healthcare environment, such as area ratio.

In previous work, the existing coverage solutions primarily employ two approaches: integer linear programming (ILP) and the greedy algorithm (GA). ILP can obtain the exact solution. Ai et al. [1] first presented an ILP formulation for the coverage problem with directional sensors. In their work, the objective function maximizes the number of targets covered and also imposes a penalty intended to minimize the number of sensors by multiplying the number of activated sensors by a positive penalty coefficient ε ($\varepsilon \leq 1$). Considering unequal sensing ranges and angles of view, Osais et al. [5] extended the ILP model by aiming to minimize the "cost" of sensor placement, where the cost refers to the base station setup and the type of sensor instead of maximal coverage.

However, the coverage problem is known to be *NP*-hard [1] and ILP can be slow to solve in large-scale applications. They also designed two greedy-based algorithms: a centralized greedy algorithm (CGA) that searches for a (sensor, orientation) pair where the number of covered targets is maximized, and a distributed greedy algorithm (DGA) in which sensors gather data from a small area localized around their positions and make decisions based on the priority level of their neighbors within $2 \times R$ (R is the radius of sensing range). Moreover, in other approaches, the CGA assigns a weight for each orientation and target (Chen et al. [6]) and chooses the orientations with the larger weights. Cai et al. [7] designed a distributed algorithm known as the DCS-Dist. Unlike DGA, in DCS-Dist, each target (instead of a sensor) is labeled with a priority, which indicates the number of directions for the sensors covered. In this approach, a given sensor node looks for uncovered targets with the highest priorities while assessing messages received from its neighbors. Two new direction-optimizing algorithms, the greedy direction-adjusting (GDA) algorithm and the equitable direction-optimizing (EDO) algorithm, were proposed by Wen et al. [8]. The GDA optimizes the directions according to the amount of targets covered, whereas the EDO algorithm adjusts the directions of the nodes to cover the critical targets and allocates sensing resources fairly among nodes to minimize the coverage differences between nodes. Cheng et al. [9] have proposed a distributed scheduling algorithm known as Dgreedy. For each directional sensor, the Dgreedy algorithm chooses the least overlapped direction as its working direction. Each sensor is assigned a unique priority for placement of the sensing neighbors in an established order.

There are also other proposed methods to solve this coverage problem. Wang et al. [10] proposed a genetic algorithm approach to find the minimum subset of sensors. Li et al. [11] proposed a greedy approximation algorithm for a solution of the problem based on the boundary Voronoi diagram.

As mentioned previously, the problem in this paper is based on a pervasive healthcare environment, which represents a unique application area. The medical sensors carried by patients collect the medical information online. All information is transported to a data center via a wireless network, and patients are guaranteed protection at all times. To build this network, a set of receivers must be deployed to transport the information. The problem that subsequently arises involves the optimization of the deployment of those receivers. The receivers should cover all patients and satisfy certain requirements. Each receiver has a fixed working area for each orientation and can cover the patients in nearby areas. To make this clear, we use targets to denote the patients. Patients are moveable. So the receivers must cover regions instead of the targets themselves. Thus, in this article, we design a new framework for the region-coverage problem using the fewest receivers.

2 Problem Statement

In this section, we formulate the maximum region coverage in high-density area with the minimum sensors (MCHMS in short) problem.

We focus on the use of directional antenna receivers (sensors), since they are more robust and may be able to decrease the number of signal conflicts. Different choices of orientation can provide different sensing sectors. A directional sensor can be described by four parameters (position, orientation, angle, and radius). In this work, position refers to the location of the sensor, denoted as vector (x, y); orientation represents the direction chosen by the sensor in the form of a unit direction vector; and the angle and radius describe the size of the sensing region. For simplicity, instead of using a continuous orientation, we assume that every sensor has only a finite number of orientations. Each of the eight orientations has the same angle and mutually disjoint sensing sectors which is fan-shaped, and a combination of all orientations forms a full circular view.

In a pervasive healthcare environment such as a hospital, the entire area is usually divided into different functional regions. The functional regions contain patient rooms, nurses' stations, doctors' offices, and other regions. As a result, sensors are not useful in locations where patients have already received care, such as the nurses' station and the device room. Furthermore, patients are highly likely to appear in narrowly spaced areas, such as aisles or entrances. Traditional methods tend to cover larger areas quite well while ignoring small but important areas. To ensure that the problem description is practical, we introduce a probability density function $P(x, y)$ for an area to indicate the chance of target occurrence at position (x, y).

Definition: $P(x, y)$ is the probability density function in the target region and describes the probability density of a patient occurrence at one point (x, y) whose calculation on the entire target region is 1. The higher the probability a patient will appear at a point, the larger the value of $P(x, y)$.

Moreover, due to the complex structure of the healthcare environment, we may be constrained by a limited region in which candidate locations, such as in the aisles, are set up. In fact, to facilitate the installment of devices, random choice of candidate locations is not allowed in the target region. Additionally, we also need to consider other factors, such as the location of the routers that direct the sensors. As a result, we chose a series of candidate locations in the target region using an orderly process instead of scattering them randomly.

According to the above consideration, we conclude the following assumptions: (1) We assume the availability of the necessary information for the healthcare environment in our study. For example, if we are to build a sensor network in a hospital, we assume that the necessary information on the building structure and its layout is given. (2) We assume that sensors can only be located along the boundaries of the served areas. (3) We assume that the sensors are homogenous and contain identical detection ability. In a hospital, deployment of different types of devices is quite rare. (4) In the beginning, we assume that every location contains a sensor, and the relevant personnel must decide which sensors will be activated. When the final decision has been made, we determine where to set up the sensors and which orientations to select.

We define the problem of deploying (sensor, orientation) pairs to minimize the number of sensors for coverage of the required area ratio. More specifically, the areas with higher priority are covered first. In particular,

Given: The target area A_c, and a set of candidate locations of directional sensors, denoted as $S = \{s_1, s_2, \cdots, s_n\}$. Each sensor has N orientations, the angle of each coverage sector is α where $N\alpha = 2\pi$. A collection of parts of A_c, $F = \{A_{ij} | A_{ij} \subseteq A_C\}$, the element of which denotes the area covered by sensor i with its orientation j. Also, we have a coverage ratio κ. A probability density function $P(x, y)$.

Objective: Select a number of candidate locations at which to set up sensors and determine their orientations. Minimize the number of sensors used while the required coverage ratio κ is satisfied.

Preliminary Model: To reach coverage ratio κ, there exist many combinations of A_{ij}. Our task is to find a subset $Z \subseteq F$ to minimize the number of elements in Z and maximize the weight $\iint_s P(x, y) dxdy$, where $S = \bigcup_{A_{ij} \subseteq Z} A_{ij}$ with the constraint that with the same i at most one A_{ij} can be chosen.

We consider the problem on a 2D level in which the areas requiring supervision are spread or scattered in a bounded area.

3 Methods

A main difficulty of the coverage problem is the overlapping areas and irregular shapes of A_C. For simplicity, we propose to use of spatial discretization to solve the problem. We discretize the target region A_C into a 2D grid with equal intervals. Each point at an intersection represents the small square to its bottom left. We assign the weight as $\iint_s P(s) ds$ to the red point. The weight is the probability of an occurrence in the shaded S, denoted $\omega(x, y)$, and (x, y) is the position of the red point.

3.1 Integer Linear Programming Formulation

We first give an ILP formulation to solve the problem, which uses the following parameters:

- $\omega(x,y)$ denotes the probability that a target exists in the area on the bottom-left of (x,y).
- $T(x,y)$ is a binary variable, where $T(x,y) = 1$, if point (x,y) is covered by at least one receiver; otherwise, $T(x,y) = 0$.
- $s(i,j)$ is a binary variable equal to 1 if sensor i uses the j th direction; otherwise, $s(i,j) = 0$.
- $A_{i,j}^{(x,y)}$ is a binary variable equal to 1 if target (x,y) can be covered by the j th direction of sensor i; otherwise, $A_{i,j}^{(x,y)} = 0$.
- κ is the required coverage ratio.
- t is the number of targets to be covered.
- ρ is the penalty coefficient of the objective function.

The integer linear programming model is stated as:

$$\max\left(\sum_{(x,y)}\omega(x,y)T(x,y) - \rho\sum_{(i,j)}s(i,j)\right)$$

s.t.

$$\frac{\sum_{(i,j)}A_{i,j}^{(x,y)}s(i,j)}{S_{total}} \le T(i,j) \le \sum_{(i,j)}A_{i,j}^{(x,y)}s(i,j)$$

$$\sum_{j}s(i,j) \le 1, \sum_{(x,y)}T(x,y) \ge \kappa t$$

where $\sum_{(x,y)}\omega(x,y)T(x,y)$ maximizes the weight of the deployment, while $\rho\sum_{(i,j)}s(i,j)$ minimizes the number of sensors being used. In this formulation, a penalty coefficient ρ is imposed, whose value can be adjusted slightly based on the importance of the second objective; it is used to balance these two components, which is our main goal.

3.2 Tree Decomposition

Tree decomposition performs well for solving *NP*-hard problems, such as the maximum independent set problem [12, 13]. This method is based on a decomposed tree derived from the original, and this allows a more efficient algorithm. The sparser the graph is, the smaller the tree width and the more efficient the algorithm will be.

Note that the target areas are usually scattered across the entire available area in a pervasive healthcare environment, which makes the graph less dense such that sensors and targets in one area have little to no interaction with those in other areas located farther away. Additionally, to improve the algorithm, we consider decision-making among a small number of sensors rather than all sensors. For these reasons, the tree decomposition is a sensible approach.

In this process, TD combines sensors characterized by close relationships into a node or a tree. Each pair of neighboring nodes is linked by the common targets that they share. Using dynamic programming, we can first locally optimize the deployment of a node and subsequently enlarge and enhance the deployment step-by-step in a larger area.

The following is a rough outline of the algorithm procedure:

1. Establish a graph with the sensors.
2. Map the graph into a tree using the tree decomposition principle.
3. Perform dynamic programming on the tree and determine the deployment.
4. Choose the most desirable deployment strategy according to the most applicable root, and trace back to the bottom of the tree for detail.

Based on the sensor candidate locations, we construct a graph denoted as $G(S,E)$ in which S is the set of vertices of the graph and E is the set of edges. The set of vertices is equal to the sensor set S, and each vertex in the graph represents a candidate sensor location. Each edge reflects the relationship between two sensors. We sum up all of the target weights $\sum_{(x,y)\in T(i,j)\wedge(x,y)\in T(i',j')} \omega(x,y)$ in the overlapping area of each pair of sensors i and i' for use in the decision whether to create an edge. In this representation, $I(i,i')$ is the collection of targets covered by sensors i and i'. If $\sum_{(x,y)\in I(i,i')} \omega(x,y) > 0$, then we create an edge between sensors i and i', denote as $e(i,i')$. We subsequently assign a weight to each edge with this value.

After constructing the graph, we now consider the task of mapping. Let $(X = \{X_i | i \in S\}, Tr(S,E))$ denotes a tree decomposition of $G(S,E)$, where $X = \{X_1, X_2, \cdots, X_n\}$ is a collection of subsets of S and node set of tree Tr, $I = \{1, 2, \cdots, n\}$. If the tree satisfies the following properties, then this tree can be considered as the tree decomposition of the graph.

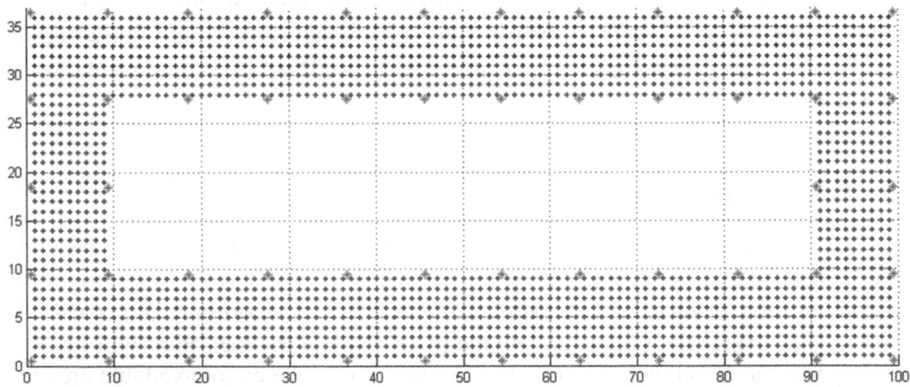

Fig. 1. Simulation: red points represent 52 candidate locations and blue points represent 2106 targets scattered in the target region. (Color figure online)

Property 1: $\bigcup_i X_i = S$
Property 2: $\forall e(i, i') \in E, \exists X_k,\ s.t.(i, i')$ belongs to X_k
Property 3: If $i \in X_k \wedge i \in X_l$, then all nodes in the path between X_k and X_l contain i.

An important property of a graph, which helps estimate of efficiency of algorithm operates on tree decomposition is tree-width. The width of a tree decomposition is $\max_{i \in S}|X_i| - 1$. The tree-width $Tw(G)$ of a graph is the minimum width among all its possible tree decompositions. A single graph might be decomposed into more than one tree. Different methods contribute to tree decompositions with different shape and the corresponding tree-width. Take a simple case for example, one node with of all of the vertices contained in it forms the simplest tree of this graph. Regardless of the method chosen, the tree will have the same properties. The goal is a tree decomposition that has a small tree width.

The lemma below guarantees our following dynamic programming operating accurately and properly.

Lemma: The intersection of each two descendant nodes with the same parent node is empty; otherwise, the intersection must be the subset of their parent node.

Proof: Suppose two descendant node of a parent node X_k are X_{k1} and X_{k2}.

When $X_{k1} \cap X_{k2} = \phi$, it is proved.

Otherwise, according to property 3, all the nodes in their path contains $X_{k1} \cap X_{k2}$. Obviously, the parent node links its two descendant nodes. Thus, we have $X_{k1} \cap X_{k2} \subseteq X_k$.

We first randomly choose a node among X as the root of the tree. For a node $X_k = \{s_{k1}, s_{k2}, \cdots, s_{km}\}$ of a tree, let X_j denote one of its descendants, and D_i denote the union of X_j, that is $D_i = \bigcup_j X_j$, and $I = X_i \bigcap D_i$. Let Ψ represents a deployment set that contains the elements of the (sensor, orientation) pair where $\Psi_{X_i} = \{(s_{i1}, o_{i1}), (s_{i2}, o_{i2}), \cdots, (s_{im}, o_{im})\}$. As each directional sensor has different choice of orientations or even being inactivated, Ψ_{X_i} has $(\mathrm{N}+1)^{\mathrm{m}}$ different choices. But each two different Ψ_{X_i} must have same number of elements which means Ψ_{X_i} must contains all the deployment information about the elements in node $\mathrm{X_i}$. Let T_Ψ denotes a subset of target sets whose elements are covered under deployment of Ψ.

Let Ψ_i denotes the deployment set that contains the information of elements in I. Given Ψ_{X_i}, we have fixed set Ψ_i. For each choice of a deployment set Ψ_{X_i} let $A(\Psi_i, k, i)$ denote the largest summation of weight of targets that are covered by the sub-tree from node X_i when only activating k directional sensors under deployment Ψ_i.

With all the variants set up, here comes the process of dynamic programming. When we operate the algorithm throughout the tree, it is the bottom-up traversal which starts at the bottom layer of tree and ends up at root node. Subjectively, tree decomposition divides the problem into several connected sub-problems (tree nodes). Every time when we reach the next layer of a tree decomposition, algorithm obtains the solution of the sub-tree from current layer. During the process, we obtain two different conditions of $A(\Psi_i, k, i)$ which can be stated as follows:

When we encounter a leaf node of the tree, we enumerate all possible deployment schemes Ψ_{X_i} and assign each corresponding $A(\Psi_i, k, i)$ with the value of $\sum_{(x,y)\in T_{\Psi_{X_i}}} w(x, y)$. Here the value of k has only one condition that is determined by Ψ_{X_i}.

When we encounter an internal node, we also enumerate all possible deployment schemes, but assign $A(\Psi_i, k, i)$ with the following formulation.

$$A(\Psi_i, k) = \sum_{(x,y)\in T_{\Psi_{X_i}}} w(x, y) + \sum_j (B(\Psi_i, \Psi_{X_j}, k_j) - \sum_{(x,y)\in T_{\Psi_i \cap \Psi_{X_j}}} w(x, y))$$

$$\sum_j B(\Psi_i, \Psi_{X_j}, k_j,) = \max(\sum_j A(\Psi_j, k_j, j))$$

such that

① Ψ_j is partially determined by Ψ_i

② $k_i + \sum_j (k_j - k_{ij}) = k$

Here Ψ_i can determine a subset of $\Psi_{X_i}$and Ψ_j. k_i is the number of active directional sensor under Ψ_i. Similarly k_j is for Ψ_j. k_{ij} is for $\Psi_i \cap \Psi_j$. This means that we divide all choices of the descendants of X_i into small groups by considering the k value, and each group is a combination of choices of descendants with the same k. As a result, we choose the option with the largest weight summation in each group and use it to decide the value of $A(\Psi_i, k, i)$.

Using the calculation of all $A(\Psi_i, k, i)$ via bottom-up traversal of the tree, we finally obtain all of the optimal results under each possible k. Moreover, from the root node, we could not only get the optimal result under required coverage ratio, but also the optimal results satisfying other coverage ratios. It actually contains the results for coverage ratio from 0% to the best.

In this problem, we are able to take advantage of the coverage ratio. By applying such a constraint to the algorithm, memory space can be conserved. After each calculation of $A(\Psi_i, k, i)$, we simply estimate whether each deployment would contribute to the final decision. If we determine that one option will not contribute to the optimal choice, then that choice is deleted. The estimation can be stated simply as

$$N_{deployment} - N_1 \cdot k \geq (1 - \kappa) \cdot N$$

where N is the total number of targets, N_1 is the largest number of targets that one sensor can cover, $N_{deployment}$ is the number of all targets that can be covered by only those sensors in the particular deployment, and k is the number of sensors activated under such deployment.

4 Experiments and Results

In this section, we compare the performance of the tree decomposition approach with those of the integer linear programming and the greedy algorithm. For simplicity, we abbreviate integer linear program as ILP, greedy algorithm as GA, dynamic programming based on tree decomposition as TD. ILP is intended to provide the optimal result of the deployment but suffers from low efficiency. The greedy algorithm is an approximate algorithm that only provides a feasible result. We first fix the candidate locations and targets in a uniform area and subsequently implement these three methods to identify the influence of each parameter. Finally, we apply the methods to an actual healthcare dataset and evaluate the performance of the tree decomposition approach.

In our problem, the target region and the candidate locations are previously fixed and the outcome is the number of sensors. Because the covered region is determined by the features of the directional sensors and the coverage ratio requirement, we perform a number of experiments with respect to three parameters: the number of orientations, the sensing range and the coverage ratio requirement. Other than the positions of the sensors that are going to be decided, these three parameters describe the properties of the sensors. We consider an example hospital complex, where the building structure is in the rectangular-ambulatory plane frame. The target region consists of a 37 by 100 rectangular area with a width of 10. Along the boundary of target region, we set up 52 candidate locations at a uniform interval of 9. In the target region, 2106 targets are distributed equally in the grid. The probability that one target will appear in any one place is the same. Next, we identify the influence of various parameters on the different methods.

To identify the effect of the number of orientations, we fix the sensing range and coverage ratio and change the number of orientations from 2 to 8. This formulation represents the most commonly encountered situation because additional orientations will make the covered region too narrow and result in additional sensor demand.

In our first experiment, we fix the sensing range to R = 10 and the coverage ratio to 100%. As the number of orientations increases, the number of sensors used increases correspondingly. When N = 2 and N = 4, the coverage ratio is able to exceed 99%. However, when N = 8, no solution exists that will satisfy the coverage ratio requirement, and the largest area we can cover is approximately 94% overall by ILP and TD while GA could only provides 80% coverage. Here, the coverage ratio of GA is listed in the parenthesis behind its outcome to denote the case that it cannot reach the coverage ratio of ILP and TD (same with the other tables in this paper).

From the results of the experiment, we observe that the outcomes of the integer linear programming and tree decomposition approaches are the same, but the results from the greedy algorithm use more than 50% of the sensors or cannot attain a desirable coverage ratio (Table 1).

Table 1. Comparison of ILP, GA and TD for influence of the number of orientations

Number of orientation	ILP	GA	TD	Coverage ratio
N = 2	32	49	32	99%
N = 4	52	52	50	99%
N = 8	52	52(80%)	52	94%

To evaluate the influence of the sensing range, we use a sensor with 2 orientations and require a coverage ratio of 100%. If the sensing range is not enough to cover the entire area, then the system will use all of the candidate locations, but the coverage ratio will increase with the increases in the sensing range. When the 100% coverage ratio is satisfied, the number of sensors selected decreases gradually.

Because the sensing range directly affects the relationship between sensors, the graph will be slightly altered due to the increase in the number of edges. When the sensing range increases to be greater than 15, the efficiency of the tree decomposition approach seems to decrease significantly because the tree width resulting from tree decomposition is closely related to the complexity of the graph. Despite the lower efficiency, the result from the tree decomposition approach remains the same as that from integer linear programming (Table 2).

Table 2. Comparisons of ILP, GA and TD for the influence of sensing range with fixed N = 2

Radius length	ILP	GA	TD	Coverage ratio
5	52	52	52	79%
6	52	52(89%)	52	90%
7	52	52(93%)	52	96%
8	48	52(96%)	48	99%
9	32	52(98%)	34	99%
10	32	49	32	99%
11	16	48	16	100%
12	16	48	16	100%
13	16	48	16	100%
14	16	50	16	100%
15	10	50	10	100%

As the coverage ratio requirement decreases, the number of sensors activated will be affected. In this case, we fix the number of orientations at N = 4 and the sensing range to R = 10 and decrease the ratio requirement from 100% to 80% to observe the difference.

With the decrease in the ratio, the number of selected sensors also decreases. Note that not only is the tree decomposition approach much more efficient than integer linear programming, but it only requires one operation to provide all optimal solutions satisfying the different coverage ratio requirements. Therefore, with the number of orientations and the required sensing ranges, tree decomposition is able to provide the optimal solution in any feasible zone. Meanwhile, ILP requires additional time-consuming operations to obtain a solution for a different coverage ratio (Table 3).

Table 3. Comparison of ILP, GA and TD for the influence of coverage ratio

Coverage ratio	ILP	GA	TD
100%	50	52	50
95%	30	40	34
90%	27	39	25
85%	24	37	24
80%	22	36	24

5 Conclusion

In this paper, we study the optimization of deployment of directional sensors for a network coverage problem. We first propose the MCHMS problem to characterize the healthcare background and state our objectives. Next, we present the optimal solution found by integer linear programming and approximate solutions found by the greedy algorithm. Additionally, we propose the use of dynamic programming with tree decomposition to improve the efficiency as well as accuracy. Finally, we compare three methods within a uniform background to systematically evaluate the properties of the tree decomposition approach. Furthermore, we apply the method using a real healthcare dataset.

The tree decomposition approach is much more efficient than ILP when the tree width is relatively small. When the graph is quite dense, which leads to a larger tree width, tree decomposition also suffers from low efficiency. However, in a healthcare environment, the target areas are usually scattered. Alternatively, we could also adjust the candidate locations and identify a smaller tree decomposition to render our method more tractable. Furthermore, tree decomposition provides an optimal result for a feasible zone. If we control the properties of the directional sensors and measure the target region, tree decomposition may offer different exact solutions according to different coverage requirements.

Acknowledgement. This work is supported by the National Science Foundation of China (Grant No: 81400632), Shaanxi Science Plan Project (Grant No: 2014JM8350) and the Fundamental Research Funds for the Central Universities (XJTU).

References

1. Ai, J., Abouzeid, A.A.: Coverage by directional sensors in randomly deployed wireless sensor networks. J. Comb. Optim. (JCO), **11**(1), 21–41 (2006)
2. Tao, D., Ma, H., Liu, L.: Coverage-enhancing algorithm for directional sensor networks. In: Cao, J., Stojmenovic, I., Jia, X., Das, S.K. (eds.) MSN 2006. LNCS, vol. 4325, pp. 256–267. Springer, Heidelberg (2006). doi:10.1007/11943952_22
3. Zhao, J., Zeng, J.: An electrostatic field-based coverage-enhancing algorithm for wireless multimedia sensor networks. In: Proceedings of International Conference on Wireless Communications, Networking and Mobile Computing (WiCom), Beijing, China, pp. 1–5 (2009)
4. Fusco, G., Gupta, H.: Selection and orientation of directional sensors for coverage maximization. In: Proceedings of IEEE International Conference on Sensor, Mesh and Ad Hoc Communications and Networks (SECON), Rome, Italy, pp. 1–9 (2009)
5. Osais, Y., St-Hilaire, M., Yu, F.: Directional sensor placement with optimal sensing range, field of view and orientation. In: Proceedings of the IEEE International Conference on Wireless and Mobile Computing (WIMOB), Avignon, France, pp. 19–24 (2008)
6. Chen, U., Chiou, B., Chen, J., Lin, W.: An adjustable target coverage method in directional sensor networks. In: Asia-Pacific Conference on Services Computing (APSCC), Los Alamitos, CA, USA, pp. 174–180 (2008)

7. Cai, Y., Lou, W., Li, M.: Cover set problem in directional sensor networks. In: Proceedings of IEEE International Conference on Future Generation Communication and Networking (FGCN), Washington, DC, USA, pp. 274–278 (2007)
8. Wen, J., Fang, L., Jiang, J., Dou, W.: Coverage optimizing and node scheduling in directional wireless sensor networks. In: Proceedings of IEEE International Conference on Wireless Communications, Networking and Mobile Computing (WiCom), Dalian, China, pp. 1–4 (2008)
9. Cheng, W., Li, S., Liao, X., Changxiang, S., Chen, H.: Maximal coverage scheduling in randomly deployed directional sensor networks. In: Proceedings of International Conference on Parallel Processing Workshops (ICPPW), Xi-An, China, p. 68 (2007)
10. Wang, J., Niu, C., Shen, R.: Priority-based target coverage in directional sensor networks using a genetic algorithm. Comput. Math Appl. **57**(11–12), 1915–1922 (2009)
11. Li, J., Wang, R., Huang, H., Sun, L.: Voronoi based area coverage optimization for directional sensor networks. In: Proceedings of International Symposium on Electronic Commerce and Security (ISECS), Nanchang, China, vol. 1, pp. 488–493
12. Rooij, J.M.M., Bodlaender, H.L., Rossmanith, P.: Dynamic programming on tree decompositions using generalised fast subset convolution. In: Fiat, A., Sanders, P. (eds.) ESA 2009. LNCS, vol. 5757, pp. 566–577. Springer, Heidelberg (2009). doi:10.1007/978-3-642-04128-0_51
13. Zhao, J., Malmberg, R., Cai, L.: Rapid ab initio RNA folding including Pseudoknots via graph tree decomposition. J. Math. Biol. **56**, 145–159 (2008)

Intelligent Computing in Computer Vision

Human Carrying Baggage Classification Using Transfer Learning on CNN with Direction Attribute

Wahyono(✉) and Kang-Hyun Jo(✉)

Graduate School of Electrical Engineering, University of Ulsan,
Daehak-road 93, Ulsan 680-749, Korea
wahyono@islab.ulsan.ac.kr, acejo@ulsan.ac.kr

Abstract. Human carrying baggage classification is one of the important stages in identifying the owner of unattended baggage for a vision-based intelligent surveillance system. In this paper, an approach to classifying human carrying baggage region on surveillance video is proposed. The proposed approach utilized transfer learning strategy under convolution neural network with human pose direction attribute. For this purpose, we first constructed convolution neural network with the target including the presence of baggage and viewing direction of the human region. The network kernels are then fine-tuned to learning a new task in verifying whether the human carrying baggage or not. Rather than using the entire human region as input to the network, we divided the region into several sub-regions and assign them as a channel of the input layer. In the experiment, the standard public dataset is re-annotated with direction information of human pose to evaluate the effectiveness of the proposed approach.

Keywords: Human carrying baggage classification · Convolution neural network · Transfer learning · Direction attribute · Region division

1 Introduction

Detecting human carrying baggage is considered as important tasks for preventing theft detection, criminal behavior identification, and bombing prevention in an intelligent surveillance system. Specifically, it could be also used as a prior stage for identifying the owner of unattended baggage detection [1]. Conceptually, detecting human carrying baggage can be done by analyzing the human appearance. The baggage carried by the human can cause the changes in appearance, so it can be distinguished from the human without one.

In general, a method for detecting human carrying baggage could be divided into two main tasks: human detection and carried object identification. The standard approach for human detection is applying human detector which combines visual feature extraction, such as HOG [2], LBP [3] and machine learning for classification, such as support vector machine [4] and random forest [5]. The candidates of human regions are extracted by either sliding window or background subtraction. After human regions are extracted, the process is continued by applying carried object detector for classifying the region into either human with or without baggage. Damen et al. [6] proposed a

D.-S. Huang et al. (Eds.): ICIC 2017, Part I, LNCS 10361, pp. 717–724, 2017.
DOI: 10.1007/978-3-319-63309-1_63

method for detecting carried object in a human region by constructing spatial-temporal template from a sequence of walking human region. The template was matched with view-specific exemplars generate offline to find the best match. The temporal protrusion between them was detected as carried object. Tzanidou [7] extended the system by combining color information to extract more accurate carried object location. However, the method assumes that parts of the carried objects are protruding from the body silhouettes. Due to its dependency on a protrusion, the method cannot detect non-protruding carried object. To solve protruding problem, in our previous work [8], we developed human-baggage detector by modeling the human region into several body parts, such as head, torso, leg, and baggage parts. The carried object detection was done by combining feature extraction and training on each part and combined them using mixture model. Without assuming the carried object as protrusion part, an author from [9] proposed an ensemble of contour exemplars of humans with different standing and walking poses. The carried object is detected by analyzing the contour alignment between hypothesis mask and contour exemplars. However, aforementioned methods are mostly depended on human detection results which make the methods are not practical as we should implement both human and carried object detector. Recently, the utilization of convolution neural network (CNN) has been proven to be effective for classification task application [12, 13]. Thus, in this work, we present a framework for classifying human carrying object which can be directly applied on all possible candidate regions. It is based on convolution neural network with transfer learning strategy using human viewing direction attribute.

2 Transfer Learning on CNN with Viewing Direction Attribute

As shown in Fig. 1, our CNN architecture is composed of two convolution layers which are followed by sub-sampling layer using max-pooling strategy. For faster training, the rectified linear units (ReLu) is utilized as the hidden unit between convolution and pooling layers. In training, the bias is initially set to 0.1 and the weights are learned by using stochastic gradient descent with momentum [14] and a mini-batch strategy and a learning rate of 0.000005. The target output of our network is activated using Softmax function. The value of $P(c_r|x)$ is considered as response sample and final classification is determined by thresholding as follows:

$$C(x) = \begin{cases} 1 & P(c_p|x) \geq T_c \\ 0 & otherwise \end{cases} \tag{1}$$

where c_p is positive class and T_c is classification threshold which $C(x) = 1$ classifies the sample x as human carrying baggage class, and $C(x) = 0$ classifies the sample x as the human without baggage region.

In the fact that the baggage is usually only located on certain part of the body, we modified the input layer of CNN by dividing the region into several sub-regions, such as top, middle, and bottom sub-images. Each sub-image is assigned as a channel to form a new 3-channels image. As shown in Fig. 2, it can be observed that the human carrying

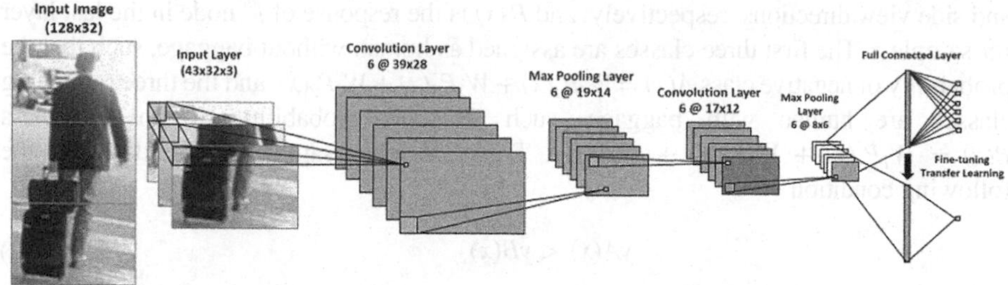

Fig. 1. Network architecture of the proposed CNN.

baggage samples (2nd, 4th, and 6th columns) produce reddish input layer since the baggage is usually darker and located in either middle or bottom sub-regions. These sub-regions cause the small values of intensity in the blue and green channels, so the input layer becomes more reddish.

Fig. 2. Samples of modified input layer obtained by region division. (Top) original input, (bottom) modified input. (Best viewed in color).

The candidate regions may contain human in different viewing direction which is either front, back, side views, since the camera mounted in fixed angle and location. Thus, the viewing direction of human is considered as part of the target. The utilization of human attributes including viewing direction has been proven effective to improve the accuracy of person re-identification [10, 11]. We increase the number of the target becomes six different classes, such as front, back, and side views without baggage, and front, back, side views with baggage.

It is inevitable that different view direction may have a different separable level. For instance, in front view direction, it is visually more difficult to identify the presence of the baggage comparing from back and side view directions when the person carrying a back-pack. Thus, we proposed a new loss function in training with a target including viewing direction. Formally, we define W_1, W_2, and W_3 be the weighting factors for the front, back, and side view directions, respectively, and $P_i(x)$ is the response of ith node in the last layer for sample x. The first three classes are assigned as human without baggage, such that the probability of negative class $A(x) = W_1P_1(x) + W_2P_2(x) + W_3P_3(x)$, and the three remaining classes are human with baggage, such that the probability of positive class $B(x) = W_1P_4(x) + W_2P_5(x) + W_3P_6(x)$. Therefore, each sample x should satisfy the following condition

$$yA(x) < yB(x) \tag{2}$$

where $y\epsilon\{1, -1\}$ is the target sample. Thus, the total loss function is defined as follows

$$L = \sum_{i=1}^{N} \left[y_i\left(A(x_i) - B(x_i)\right) + CE\right] \tag{3}$$

where N is the number of samples, and CE is the cross entropy loss.

As shown in Fig. 1, after performing training on convolution neural network with six different target units considering the viewing direction, we fine-tuned the learned weights using a new target which is only consisted of two different classes, such as the human with and without baggage. The obtained weights in the first training are used for as initial weight for the second training.

3 Experiments

The proposed system was implemented using MATLAB on PC equipped with Core i7-4770 CPU running at 3.40 GHz and 8 GB RAM. Precision, recall, and F-measures were used as evaluation protocols. Public databases are collected to form new data which is carefully chosen from sub-part of PETA data [15] to construct the new dataset for the task of human carrying baggage classification. The dataset consists of 2673 images (see Table 1) and is randomly distributed into training and testing sets as much as 75% and 25%, respectively. Each image is newly labeled with the presence of baggage and viewing direction (e.g. front, back, and side views). Figure 3 shows selected samples in the experiment.

Table 1. Dataset distribution

Viewing direction	Without baggage	Carrying baggage
Front view	330	562
Back view	318	605
Side view	325	533
Total	972	1700

Fig. 3. Several samples of images in our dataset: Human without baggage (the first sixth images) and human carrying baggage (the remaining images).

3.1 Variation of Our Proposed and Baseline Methods

The proposed method has been implemented with various configuration of steps in order to validate the effectiveness of each part, which are defined as follows:

HOG + SVM. Histogram of oriented gradient (HOG) [2] was utilized in feature extraction over an entire human region, and support vector machine (SVM) trains and cross-validates feature for classification.

HOG + RF. Based on HOG feature extraction method, the random forest (RF) classifier is used to train the training sub-set with 100 random trees.

Standard CNN. We straightforward implemented convolution neural network (CNN) as shown in Fig. 1 without region division and viewing attribute.

CNN + DA + SVM. We modified the standard CNN by setting the target class into six classes considering viewing direction (CNN + DA). The full connected layer outputs are assigned as feature maps which are trained using SVM.

CNN + DA + RF. Full connected layer extracted in the network with direction attribute is trained using random forest classifier with 100 random trees.

CNN + DA + TL. In the first training, we trained the CNN by setting the output layer with six different classes. The learned weights are used to initialize the network weight in the second training which only has two classes output layer.

CNNR. The standard CNN was modified using different input layer by dividing the region into three sub-regions, referred as CNN with region division (CNNR).

CNNR + DA + TL. This is the full proposed framework which covers region division as an input layer, considering viewing direction attribute of human region, and transfer learning strategy.

3.2 Results

First, we conduct an experiment to find optimal threshold Tc, as shown in Eq. (1). In the experiment, it is found that if the threshold is set to 0.5, the most configuration obtained the highest f-measure. Thus, this threshold value is used for comparison with state-of-the-art methods. Table 2 provided summaries of the comparison results between the proposed method and its variation with the baseline methods under 5 experiment trials. It can be observed that all variation of proposed methods outperforms to the CNN of 0.70, HOG + SVM [2] of 0.61, and HOG + RF [5] of 0.75. Among five variations of the proposed method, CNNR + DA + TL achieves the highest f-measure value of 0.91 which significantly improves the performance of CNN as much as 21%.

Furthermore, it can be verified that the utilization of viewing direction attribute (DA) achieves average improvement over CNN and HOG feature base classifiers around 4%. The modified input layer improves the performance of CNN as much as 6%. The combination of DA and transfer learning (TL) successfully increase the CNN accuracy with more than 10% improvement. Next, we analyze the response performances of the proposed and other methods using receiving operation characteristic (ROC). As shown Fig. 4, the proposed method and its variation achieves better results comparing to the standard CNN in term of AUC value. Among variation of proposed methods, CNN + DA + RF achieves the best result, while our full framework improves the CNN around 6%. Overall, it can be confirmed that our method achieved better performances in general compared to the baseline methods and gains significant improvement for standard CNN. However, our proposed method still obtained several wrong classifications. The false positives are mainly caused by occlusion with other objects and human appearance which is quite similar with human carrying baggage. On the other hand, the false negatives are mostly obtained from human carrying backpack captured in the front

Table 2. Comparison results (Tc = 0.5)

Method	Av. Precision	Av. Recall	Av. F-measure
HOG + SVM	0.96	0.45	0.61
HOG + KF	0.95	0.62	0.75
CNN	0.93	0.55	0.70
CNN + BA + SVM	0.86	0.50	0.63
CNN + DA + RF	0.90	0.69	0.78
CNN + BA + TL	0.95	0.68	0.79
CNNR	0.94	0.64	0.76
CNNR + DA + TL	0.93	0.90	0.91

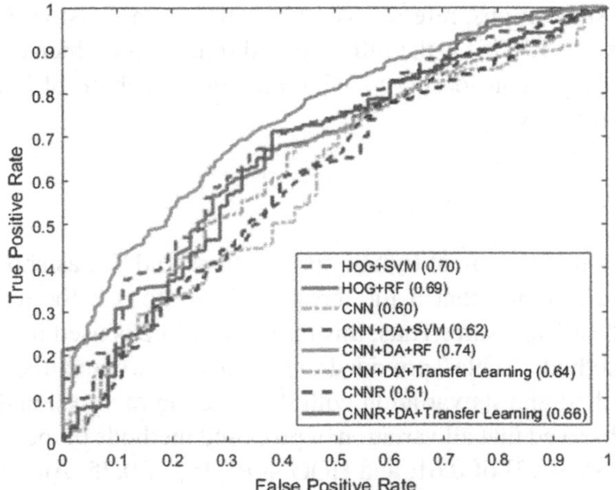

Fig. 4. Receiving operation characteristic. Bracket values represent the area under curve.

direction which is sometimes difficult to be identified. Thus, improvement is required to solve this problem.

4 Conclusion

A framework for classifying human carrying baggage has been successfully implemented with significant improvement of baseline methods. It is based on CNN with transfer learning strategy using viewing direction attribute. Instead of using the entire region as input feed of the network, the candidate region is divided equally into several subregions to form new input region. Based on the experimental result, the utilization of human viewing direction attribute (DA) and transfer learning (TL), as well as modification of input layer, achieve significant improvement over standard convolution neural network around 10%, 4%, and 6%, respectively.

Acknowledgment. This research was supported by the MSIP (Ministry of Science, ICT and Future Planning), Korea, under the Grand Information Technology Research Center support program (IITP-2017-2016-0-00318) supervised by the IITP (Institute for Information & communications Technology Promotion).

References

1. Wahyono, Filonenko, A., Jo, K.H.: Unattended object identification for intelligent surveillance systems using sequence of dual background difference. IEEE Trans. Ind. Inf. **12**(6), 2247–2255 (2016)
2. Dalal, N., Triggs, B.: Histograms of oriented gradients for human detection. In: IEEE CVPR, pp. 886–893 (2005)
3. Satpathy, A., Jiang, X., Eng, H.L.: Human detection using discriminative and robust local binary pattern. In: 2013 IEEE International Conference on Acoustics, Speech and Signal Processing, pp. 2376–2380, May 2013
4. Cortes, C., Vapnik, V.: Support-vector networks. Mach. Learn. **20**(3), 273–297 (1995)
5. Ho, T.K.: Random decision forest. In: Proceedings of the 3rd International Conference on Document Analysis and Recognition, pp. 14–16 (1995)
6. Damen, D., Hogg, D.: Detecting carried objects from sequences of walking pedestrians. IEEE Trans. Pattern Anal. Mach. Intell. **34**(6), 1056–1067 (2012)
7. Tzanidou, G., Zafar, I., Edirisinghe, E.A.: Carried object detection in videos using color information. IEEE Trans. Inf. Forensics Secur. **8**(10), 1620–1631 (2013)
8. Wahyono, Hariyono, J., Jo, K.H.: Body part boosting model for carried baggage detection and classification. Neurocomputing **228,** 106–118 (2017)
9. Ghadiri, F., Bergevin, R., Bilodeau, G.-A.: Carried object detection based on an ensemble of contour exemplars. In: Leibe, B., Matas, J., Sebe, N., Welling, M. (eds.) ECCV 2016. LNCS, vol. 9911, pp. 852–866. Springer, Cham (2016). doi:10.1007/978-3-319-46478-7_52
10. Layne, R., Hospedales, T.M., Gong, S.: Towards person identification and re-identification with attributes. In: Fusiello, A., Murino, V., Cucchiara, R. (eds.) ECCV 2012. LNCS, vol. 7583, pp. 402–412. Springer, Heidelberg (2012). doi:10.1007/978-3-642-33863-2_40
11. Lin, Y., Zheng, L., Zheng, Z., Wu, Y., Yang, Y.: Improving person re-identification by attribute and identity learning. arXiv:1703.07220, March 2017

12. He, X., Wang, G., Zhang, X.-P., Shang, L., Huang, Z.-K.: Leaf classification utilizing a convolutional neural network with a structure of single connected layer. In: Huang, D.-S., Jo, K.-H. (eds.) ICIC 2016. LNCS, vol. 9772, pp. 332–340. Springer, Cham (2016). doi: 10.1007/978-3-319-42294-7_29
13. Wang, Z., Jiang, P., Zhang, X., Wang, F.: Natural scene digit classification using convolutional neural networks. In: Huang, D.-S., Jo, K.-H. (eds.) ICIC 2016. LNCS, vol. 9772, pp. 311–321. Springer, Cham (2016). doi:10.1007/978-3-319-42294-7_27
14. Rumelhart, D.E., Hinton, G.E., Williams, R.J.: Learning representations by backpropagating errors. Nature **323**(6088) (1986)
15. Deng, Y., Luo, P., Loy, C.C., Tang, X.: Pedestrian attribute recognition at far distance. In: Proceedings of ACM Multimedia (ACM MM), pp. 1–4 (2014)

Fully Combined Convolutional Network with Soft Cost Function for Traffic Scene Parsing

Yan Wu, Tao Yang, Junqiao Zhao(✉), Linting Guan, and Jiqian Li

College of Electronics and Information Engineering, Tongji University, Shanghai 201804, China
{yanwu,zhaojunqiao}@tongji.edu.cn

Abstract. Autonomous car has achieved unprecedented improvement in object detection because of the high performance of deep convolutional neural networks, and now researches are devoted to more complex traffic scene parsing. In this paper, we present a novel traffic scene parsing algorithm by learning a fully combined convolutional network (FCCN). Our network improves the upsampling layer of a fully convolutional network, we add five unpooling layers after the final convolution layer, and each unpooling layer is corresponded to a former pooling layer. We then combine each pair of pooling and unpooling layers, add convolution layers after the combined layer. Since we find it is still hard to learn fine details or edge features of target objects, we propose a soft cost function for further improvement. Our cost function adds soft weights on different target objects. The weight of background is set as constantly one, and the weights for target objects are calculated dynamically, which should be larger than two. We evaluate our work on CamVid datasets. The results show that our FCCN achieves a considerable improvement in segmentation performance.

Keywords: Semantic segmentation · Combined convolutional network · Soft cost function · Traffic scene parsing

1 Introduction

In recent years, autonomous car has drawn great attention of both the academic and the business sectors. However, it is still a tough task for cars to fully understand what they are seeing, in other words, scene parsing. Rather than image classification and object detection, scene parsing predicts and labels every single pixel in an image, which is the focus of high-level interpretation task in image understanding [1].

Thanks to millions of labeled images, state-of-the-art image understanding methods are mostly based on the deep convolutional neural network (CNN). In the past ten years, CNN has shown its great success in image classification, object detection. CNN has also been introduced to scene parsing recently since the emerging of massive pixel-wise labeled datasets. In CVPR 2015, Jonathan Long et al. proposed the fully convolutional networks (FCNs) [2], which was a completely novel idea for image segmentation. FCNs brought up an end-to-end learning method by transforming fully connected layers into convolution layers. These convolution layers had a kernel size of 1×1, and a one-step

D.-S. Huang et al. (Eds.): ICIC 2017, Part I, LNCS 10361, pp. 725–731, 2017.
DOI: 10.1007/978-3-319-63309-1_64

upsampling layer was added after the final convolution layer. FCNs also skipped the final prediction layer with lower layers to reduce noise in output maps. Though a great improvement in image segmentation was achieved by FCNs, edge segmentation for target objects were still coarse.

In this paper, we propose a fully combined convolutional network (FCCN). We evaluate our model on CamVid dataset [3]. A considerable improvement is achieved in mean-IoU compared to our baseline. As a conclusion, our main contributions are:

- We adopt a layer-by-layer unpooling method, which achieves a better performance than one-step upsampling methods like FCNs.
- We combine every pooling layer with its corresponded unpooling layer. As a result, detailed edge information of target objects is better learned.
- We propose a new soft cost function, each object class is assigned a dynamic weight based on its area. And the weights on background are suppressed by low weight to help our net learn more target features.

2 Related Work

After the proposed FCNs, recently works on segmentation task can mainly be divided into three directions: improving the upsampling method [4–6], fine-tuning a more effective deep classification network [7], and proposing other new segmentation network [8–10]. Lin et al. [7] applied deep residual net [11] to image segmentation and achieved state-of-the-art performance. Paszke presented a real time convolution net by deploying operations of low latency [9], Mostajabi et al. [12] proposed a feed-forward segmentation with zoom-out features. A novel dilated convolution module proposed by Yu and Koltun [8] was widely applied recently. Dilated convolution could systematically aggregate multi-scale contextual information.

Recent works, which are similar to us, are SegNet [4], deconvolution network [5] and U-Net [6]. All of these three methods upsampled their feature maps by sequent layers. However, different from the decoder path in [4], the deconvolution operation in [5] and the 2×2 convolution ("up-convolution") in [6], we use bilinear interpolation to realize upsampling. More significantly, we do a combination between every pooling layer and corresponding unpooling layer, add one convolution layer after each combined layer. We implement the combination layers with a sum operation, rather than concatenation layers in U-Net, In addition, we add only one convolution layer after each upsampling operation in our network, while the others added more than two. In this way, we make our network more efficient and achieve a higher prediction.

3 Method

3.1 Fully Combined Convolutional Network

In FCNs, the upsampling operation is a one-step operation, which means the heat map is directly upsampled to output with the same size the input image. The main shortcoming of this method is that the final segmentation predictions will loss essential details after upsampling. To improve this, Long et al. [2], skipped fine feature maps in low layers, yielding several new models, such as fcn16s, fcn8s. This method enabled the model making local predictions while respecting the global structure.

To learn more feature information, we take another way to make up the details of edges between different target objects. Rather than the one-step upsampling in FCNs, we divide the upsampling layer in FCNs into several double sized unpooling layers. And each unpooling layer is designed to match to a previous pooling layer, both keep the same size of feature map. We do a sum operation on these matched pooling and unpooling feature maps. After each unpooling operation, we also add a convolution layer to help extract further features. By matching each unpooling layer to the former pooling layer, our network depicts a combined structure, as shown in Fig. 1. There are more than one convolution layer before a pooling operation, followed by dropout and non-linearity RELU layers, just like our baseline FCNs. As is shown in Fig. 1, in the left downsampling part, there are at least two convolution layers before a max pooling operation. While we add only one convolution layer after an unpooling layer in the right part. Thus, the cubic blocks in the right half is relatively "thinner" than that in the left half. We have tried to keep the number of convolution layers equivalent in both halves, but neither removing layers in the left nor adding layers in the right could improve segmentation performance. The number above the convolution cubic Fig. 1 is the channel of feature map in that layer.

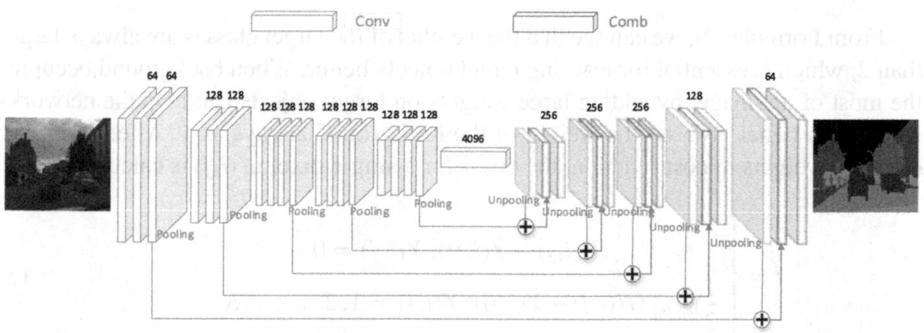

Fig. 1. Different colors of cubic blocks represent different operations, in which orange cubes is convolution layers, green cubes is pooling layers, yellow cubes is unpooling layers, and blue cubes is combination layers. A combination layer has two inputs, the last unpooling layer and corresponding pooling layer. (Color figure online)

We achieve significant improvements by modifying the upsampling structure of FCNs. However, we find that the output predictions, especially the edge features of objects are still coarse. To resolve it, we build a connection between each unpooling layer and its former corresponding pooling layer. In fact, we do a sum operation on each pair of corresponding feature maps. We have tried two ways of combination: combining followed by unpooling and unpooling followed by combining. We choose the second way since it brings out a better performance.

3.2 Soft Cost Function

The existing semantic segmentation networks adopt equal weights on target objects and the background in cost function. However, in most case, it is more important to extract more target information rather than the background. In this paper, we proposed a dynamic weighting strategy for the targets part in cost function.

We define L as the final cost, which is a (one-half) squared-error cost function between prediction output H and lable Y, we then define the overall cost function to be:

$$L = \sum_{i=1}^{m} \sum_{j=1}^{n} \frac{1}{2} \| H(i,j) - Y(i,j) \|^2 \quad (1)$$

In Formula (1), m and n represents the width and height of input image. Before we calculate the cost function, we firstly count the number of background pixels cb and the number of pixels of every single target class ct_i in the label. We then dynamically calculate the proportions between target classes and background in each image, and as a result we assign weight w_i to target class ct_i. We denote N to the number of classes, then we get:

$$w_i = max\left(\frac{cb + \sum_{i=1}^{N} ct_i}{ct_i}, 2 \right) \quad (2)$$

From Formula (2), we can see that the weights of the target classes are always larger than 2, which is essential for learning target objects better. When background occupies the most of an image, by adding large weights on target objects can help the networks learn them better. We set the weight of the background always as 1. After assigning different weights in cost function, the cost l for a single pixel in (i,j) is calculated as:

$$l = \begin{cases} \frac{1}{2}(H(i,j) - Y(i,j))^2 \; Y(i,j) = 0 \\ \frac{1}{2} w_{Y(i,j)} (H(i,j) - Y(i,j))^2 \; Y(i,j) = 1, 2, \ldots\ldots, N \end{cases} \quad (3)$$

Finally, the overall cost function is the sum of the costs l for all pixels.

$$L = \sum_{i=1}^{m} \sum_{j=1}^{n} l \quad (4)$$

4 Experiments

All experiments in this paper are implemented based on MatCovNet, a convolution neural network integration matlab toolbox [13]. We trained our network on two GeForce GTX TITAN X graphics cards, one Intel(R) Core(TM) i7-4790 K CPU processor with 8 cores and 16 G physical memory. Each image takes about 0.36 processing seconds when evaluation. In our experiments, we adopt a batch size of 10 images for training, and finish training after 50 epochs, and the learning rate is 0.0001. We initialize the weight with random numbers chosen from a normal distribution.

We evaluate our methods on CamVid dataset, which consists of 367 image training sets, 101 image validation sets, and 233 test sets of pictures. There are 11 categories of buildings, trees, sky, cars, and so on, plus 12 categories in the background. This is a section of continuous real traffic scene video. The resolution of each image is of 960*720. The original image is too large to train, so we scale the resolution into half, 480*360, in our experiment. As we start our work based on FCNs, we choose its performance on CamVid as our baseline, relying on the standard mean intersection-over-union metric.

There are three main steps in our work: basic combined network, then with combination of pooling layer and unpooling layer, and finally with more weights on target objects in cost function. Table 1 shows the segmentation performance of our work in each step, with a comparison to baseline. The first row is the result of FCNs, the next three rows show the result of our basic combined network, combining pooling layer and unpooling layer, adding soft weights in cost function based on combined network respectively. From Table 1, it can be seen that after adding weights on target object, we increase our result by more than 8 percentage points compared to FCNs.

Table 1. Results of our works on CamVid Dataset, compared to baseline FCNs.

Model	Mean IoU (%)	Mean pix.acc (%)	Pixel acc (%)
FCNs [2]	57.0	88.0	–
FCCN basic	58.23	84.45	65.83
FCCN + comb	60.76	85.91	70.35
FCCN + comb + SW	65.79	88.74	88.26

Table 2 shows Quantitative results on CamVid Dataset. Compared with other works recently on CamVid, we achieve the highest score on class Car, Road, Fence and Sidewalk, especially on Sidewalk and Fence, we achieve an improvement of more than 12 and 31 percentage points. For the mIoU performance, our performance is also competitive to the state-of-the-art. Figure 2 shows visual samples of segmentation results on CamVid, we display original images, ground truths, predictions of FCNs and ours in four columns. In FCNs, there exist many fragmenting artefacts inside a single object, while we make a significant improvements in this aspect. We also achieve a more obvious edge segmentation.

Table 2. Quantitative results on CamVid Dataset (%), consisting of 11 traffic scene categories.

Method	Building	Tree	Sky	Car	Sign	Road	Pedestrian	Fence	Pole	Sidewalk	Bicyclist	mIoU
Superparsing [14]	70.4	54.8	83.5	43.3	25.4	83.4	11.6	18.3	5.2	57.4	8.9	42
Liu and He [15]	66.8	66.6	90.1	62.9	21.4	85.8	28	17.8	8.3	63.5	8.5	47.2
SegNet [4]	81.3	72	**93**	81.3	14.8	93.3	62.4	31.5	**36.3**	73.7	42.6	47.7
Tripathi [16]	74.2	67.9	91	66.5	23.6	90.7	26.2	28.5	16.3	71.9	28.2	53.2
DilatedNet [8]	82.6	**76.2**	89.9	84	46.9	92.2	56.3	35.8	23.4	75.3	55.5	65.3
Kundu [17]	**84**	77.2	91.3	85.6	**49.9**	92.5	**59.1**	37.6	16.9	76	**57.2**	**66.1**
Ours	79.7	77.2	85.7	**86.1**	45.3	**94.9**	45.8	**69.0**	25.2	**86.2**	52.9	65.79

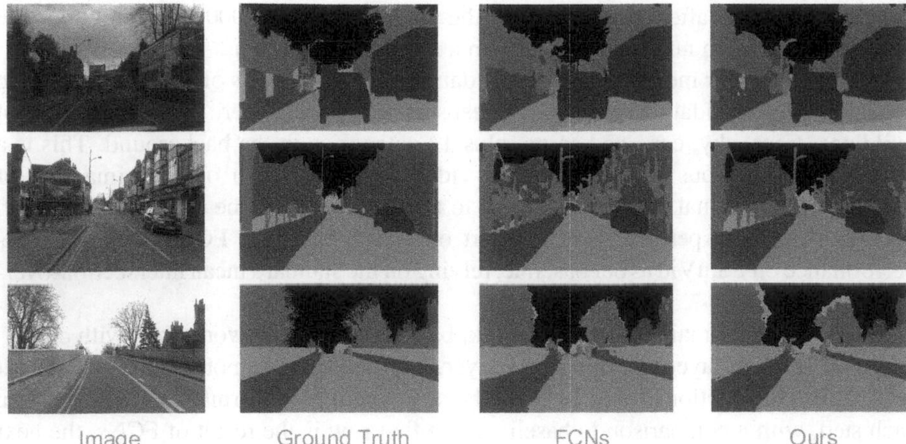

Fig. 2. Visual results on CamVid dataset. First column is original images, second is ground truths, and last two columns show segmentations of FCNs and ours. Predictions in FCNs are quite rough, while we have clearer segmentation inside a single object and more obvious edges

5 Conclusion

We proposed a fully combined convolutional network for traffic scene parsing in this paper. There are two paths in our network, downsampling for feature sampling and upsampling for feature restoring. We set combinations in corresponding pooling and unpooling layers, thus we can learn more detailed feature information. We also proposed a novel soft cost function for learning, we add adaptive weights on each single target object, which help our network to pay more attention on those small objects. Experiments on scene parsing datasets CamVid show that our FCCN outperforms FCNs and state of the art methods. As a conclusion, we hope our work, especially our method on cost function, can be helpful to those who are devoted to image semantic segmentation and scene parsing.

Acknowledgements. This work was supported by the Fundamental Research Funds for the Central Universities (20143436).

References

1. Gao, J., Xie, Z., Zhang, J., Wu, K.W.: Image semantic analysis and understanding: a review. Pattern Recog. Artif. Intell. **23**(2), 191–202 (2010)
2. Long, J., Shelhamer, E., Darrell, T.: Fully convolutional networks for semantic segmentation. IEEE Trans. Pattern Anal. Mach. Intell. **39**(4), 640–651 (2017)
3. Brostow, G.J., Fauqueur, J., Cipolla, R.: Semantic object classes in video: a high-definition ground truth database. Pattern Recogn. Lett. **30**(2), 88–97 (2009)
4. Badrinarayanan, V., Alex, K., Roberto, C.: Segnet: A deep convolutional encoder-decoder architecture for image segmentation. arXiv preprint arXiv:1511.00561v3 (2016)

5. Noh, H., Hong, S., Han, B.: Learning deconvolution network for semantic segmentation. In: Proceedings of the IEEE International Conference on Computer Vision, pp. 1520–1528 (2015)
6. Ronneberger, O., Fischer, P., Brox, T.: U-Net: convolutional networks for biomedical image segmentation. In: Navab, N., Hornegger, J., Wells, W.M., Frangi, A.F. (eds.) MICCAI 2015. LNCS, vol. 9351, pp. 234–241. Springer, Cham (2015). doi:10.1007/978-3-319-24574-4_28
7. Lin, G., Milan, A., Shen, C., Reid, I.: RefineNet: Multi-Path Refinement Networks with Identity Mappings for High-Resolution Semantic Segmentation. arXiv preprint arXiv: 1611.06612v3 (2016)
8. Yu, F., Koltun, V.: Multi-scale context aggregation by dilated convolutions. arXiv preprint arXiv:1511.07122v3 (2016)
9. He, K., Zhang, X., Ren, S., Sun, J.: Deep residual learning for image recognition. In: Proceedings of the IEEE Conference on Computer Vision and Pattern Recognition, pp. 770–778 (2016)
10. Paszke, A., Chaurasia, A., Kim, S., Culurciello, E.: ENet: A Deep Neural Network Architecture for Real-Time Semantic Segmentation. arXiv preprint arXiv:1606.02147v1 (2016)
11. Chen, L.C., Yang, Y., Wang, J., Xu, W., Yuille, A.L.: Attention to scale: Scale-aware semantic image segmentation. arXiv preprint arXiv:1511.03339v2 (2016)
12. Mostajabi, M., Yadollahpour, P., Shakhnarovich, G.: Feedforward semantic segmentation with zoom-out features. In: Proceedings of the IEEE Conference on Computer Vision and Pattern Recognition, pp. 3376–3385 (2015)
13. Vedaldi, A., Lenc, K.: Matconvnet: convolutional neural networks for matlab. In: Proceedings of the 23rd ACM International Conference on Multimedia, pp. 689–692. ACM (2016)
14. Tighe, J., Lazebnik, S.: SuperParsing: scalable nonparametric image parsing with superpixels. In: Daniilidis, K., Maragos, P., Paragios, N. (eds.) ECCV 2010. LNCS, vol. 6315, pp. 352–365. Springer, Heidelberg (2010). doi:10.1007/978-3-642-15555-0_26
15. Liu, B., Xuming, H.: Multiclass semantic video segmentation with object-level active inference. In: Proceedings of the IEEE Conference on Computer Vision and Pattern Recognition, pp. 4286–4294 (2015)
16. Tripathi, S., Belongie, S., Hwang, Y., Nguyen, T.: Semantic video segmentation: exploring inference efficiency. In: SoC Design Conference (ISOCC), pp. 157–158. IEEE (2015)
17. Kundu, A., Vibhav V., Vladlen, K.: Feature space optimization for semantic video segmentation. In: Proceedings of the IEEE Conference on Computer Vision and Pattern Recognition, pp. 3168–3175 (2016)

Computational Intelligence and Security for Image Applications in Social Network

Pedestrian Detection Based on Fast R-CNN and Batch Normalization

Zhong-Qiu Zhao[1(✉)], Haiman Bian[1], Donghui Hu[1], Wenjuan Cheng[1], and Hervé Glotin[2]

[1] School of Computer and Information, Hefei University of Technology, Hafei, China
z.zhao@hfut.edu.cn

[2] Systems and Information Sciences Lab, LSIS CNRS and University of Sud-Toulon Var - La Garde, Toulon, France

Abstract. Most of the pedestrian detection methods are based on hand-crafted features which produce low accuracy on complex scenes. With the development of deep learning method, pedestrian detection has achieved great success. In this paper, we take advantage of a convolutional neural network which is based on Fast R-CNN framework to extract robust pedestrian features for efficient and effective pedestrian detection in complicated environments. We use the Edge-Boxes algorithm to generate effective region proposals from an image, as the quality of extracted region proposals can greatly affect the detection performance. In order to reduce the training time and to improve the generalization performance, we add a batch normalization layer between the convolutional layer and the activation function layer. Experiments show that the proposed method achieves satisfactory performance on the INRIA and ETH datasets.

Keywords: Pedestrian detection · Batch normalization · Deep learning

1 Introduction

In recent years, pedestrian detection has become an important branch of object detection and has attracted wide attention in computer vision [4–6, 30, 32–34]. In real life, it is a key problem in automotive safety, video surveillance, smart vehicles and intelligent robotics. Because of the diversity of pedestrian body pose, object occlusions, clothing, lighting change and complicated backgrounds in the video sequence or image, the pedestrian detection is still a challenging task in computer vision.

In pedestrian detection, feature extraction is an important factor to influence the performance. Many features have been proposed by researchers for pedestrian detection, such as Haar-like features [29], Integral Channel Features (ICF) [6], Histogram of Oriented Gradients (HOG) [4], Local Binary Pattern (LBP) [19], Dense SIFT [28] etc. The proposed Deformable Part Based Model (DPM) [8] which is based on HOG, has maken a breakthrough in pedestrian detection. However, these features are hand-craft and are considered to be low-level. They use the low-level information while the high-level information is usually very important for pedestrian detection [25].

D.-S. Huang et al. (Eds.): ICIC 2017, Part I, LNCS 10361, pp. 735–746, 2017.
DOI: 10.1007/978-3-319-63309-1_65

Recently, with the development of deep learning techniques, deep neural networks have been successfully applied in object recognition tasks [9–11, 35]. Deep neural networks can achieve more excellent results due to their capability to learn discriminative features from raw pixels. Many researchers have applied deep learning techniques to pedestrian detection. Sermanet *et al.* [25] proposed a two layers convolutional model which adopts convolutional sparse coding to pre-train convolutional neural network for pedestrian detection. Chen *et al.* [2] proposed a pre-trained Deep Convolutional Neural Network (DCNN) to learn features from ACF [5] detector. These features are then fed to a SVM classifier. Ouyang *et al.* [21] proposed a joint deep model that jointly learns four key components in pedestrian detection: feature extraction, deformation handling, occlusion handling and classifier. Li *et al.* [17] proposed a SAF R-CNN which takes scales of pedestrians into account in pedestrian detection. Fast R-CNN [9] is one of the best detection approaches based on deep leaning. It determines regions of interest(RoI) in an image using superpixel method, and then extracts features using convolutional neural network model. The extracted features are then passed to softmax layer and bounding box regressor layer which are trained for each object, and finally the pedestrian and its position are detected.

It is well known that training a deep neural network is complicated due to the fact that *internal covariate shift* [14] can degrade the efficiency of training. Internal covariate shift is a change of distribution of the inputs. Which happens during training the feed-forward neural networks, by changing the parameters of a layer. Batch normalization (BN) [14] is a technique to control the distributions of feed-forward neural network activations, thereby reducing internal covariate shift. So deep neural networks trained with batch normalization can converge faster and generalize better. In this paper, we use Fast R-CNN architecture based on batch normalization(BN) to extract robust pedestrian features.

In pedestrian detection task, another important stage is locating the potential windows which may contain pedestrians. As an exhaustive and traditional method, sliding window method has two main shortcomings. First, it needs searching every possible position in an image. Second, it may produce many redundancy windows which affect the quality of detection. To improve the detection efficiency, the approaches of region proposals have been proposed to produce high quality regions, in which the most popular method is Selective Search [27]. It can get high quality of regions by using image segmentation. However, its speed is very slow and the detection performance is sensitive to the image quality. And the Selective Search method of extracting candidate windows is infeasible [2] but cannot provide precise localization of pedestrians. To produce more precise localized candidate windows, we employ EdgeBoxes [36] method, which is a fast and effective one [13]. It is believed that the edge information can precisely describe the object, so the EdgeBoxes method can extract higher quality of the candidate windows than the Selective Search method. Moreover, the EdgeBoxes method actually performs better in terms of the average intersection over union(IoU) across all images in the set [12].

In this paper, our main contributions are summarized as follows:

(1) In order to make the pedestrian detection more efficient, we use the EdgeBoxes method which can obtain low-redundancy and high quality of candidate windows.

(2) We use the Fast R-CNN architecture to extract robust features for pedestrian detection. In order to reduce the training time and prevent the gradients that explode or vanish when training the Fast R-CNN, we add the batch normalization(BN) layer into the Fast R-CNN architecture.

The rest of this paper can be organized as follows. In Sect. 2, we introduce the related work of pedestrian detection and briefly introduce the batch normalization layer. In Sect. 3, we introduce our pedestrian detection approach. In Sect. 4, we present our experiment results on two benchmark datasets, and some analyses as well. In Sect. 5, we conclude our work and discuss the possible advances to our model in the future.

2 Related Work

In this section, we review the related work of two important stages of pedestrian detection, viz. extracting region proposals and extracting features, respectively. And, we also briefly introduce the batch normalization layer.

Region Proposals: In an image, the position of a pedestrian can be anywhere and its size can be arbitrary, so it is necessary to search the whole image to localize the pedestrians. Traditional methods employ sliding window method to find the possible locations on the whole image. The sliding window method can get almost all potential locations through exhaustive search, but it is computational and may produce many redundancy windows. Some region proposals methods were proposed to overcome this problem [12], in which the most popular method is Selective Search [27]. It is a combination of exhaustive search and segmentation method. Wang *et al.* [31] proposed a region proposal fusion algorithm to fusion the Selective Search and BING [3] method. Nam *et al.* [20] proposed LDCF method to extract regions for pedestrian detection. The EdgeBoxes method has been shown as a state-of-the-art region proposal system [13, 36].

Extracting features: Many models based on hand-crafted features have been utilized for pedestrian detection [4, 6, 19, 29, 34]. Wang *et al.* [34] utilized the combination of HOG and LBP to handle the partial occlusion of pedestrian. Felzenswalb *et al.* [19] further improved the detection performance by combining the HOG with a deformable part model. Dollár *et al.* proposed the Integral Channel Features(ICF) [6] and Aggregated Channel Features(ACF) [5] which efficiently extract histograms and haar features. Chen *et al.* [1] used a multi-order context representation to take advantage of co-occurrence contexts of different objects. Deep learning models can be trained end-to-end to extract robust features from the given images. Sermanet *et al.* [25] proposed a convolutional neural network model which has two convolutional layers pre-trained by sparse coding. Ouyang *et al.* [21] proposed a joint deep model which contains a deformation layer to model mixture poses information. Tian *et al.* [26] jointly utilized semantic tasks to optimize pedestrian detection. More recently, some excellent works have been successfully applied in object detection [9–11], and the performance of pedestrian detection is largely improved [2, 17, 31]. Fast R-CNN [9] is one of the best detection approaches among them. It consists of two steps for object detection, in which the first step is to use non-deep learning methods to extract thousands of region proposals, and the second step is to use convolutional neural networks classifiers to classify categories at those locations.

Batch Normalization: Training a deep neural network is complicated due to the fact that changing the parameters of a layer will affect the distribution of the inputs to the succedent layers. As a result, the succedent layers are continually adapted to fit the shifting input distribution and unable to learn effectively. This phenomenon is called *Internal Covariate Shift* (ICS) [14]. The ICS slows down the course of training and needs careful parameter initialization. Batch normalization(BN) [14] is a technique to accelerate training and to improve generalization capability. Its basic idea is to standardize the internal representations inside the network to make the network converging faster and generalizing better, inspired by the way to whiten the network input to improve performance. Batch normalization layer can be applied to convolutional neural networks directly between the convolutional layer and the activation function layer, for example ReLU etc. It uses the local mean and variance computed over the minibatch x, and then corrects with a learned variance and bias term, namely γ and β:

$$BN(x;\gamma,\beta) = \beta + \gamma\frac{(x - E[x])}{(Var[x] + \epsilon)^{1/2}} \tag{1}$$

where ϵ is a small positive constant to improve numerical stability.

3 Our Approach

In this section, we will introduce the details of our approach for pedestrian detection. Firstly, we will introduce the proposal generation algorithm. Then, we will describe the pedestrian detection architecture. Finally, we will briefly introduce the whole pedestrian detection process.

3.1 Pedestrian Proposals

We utilize the EdgeBoxes as our proposal generation algorithm. The basic idea of EdgeBoxes is that it generates and scores the proposal based on the edge map of the given image.

Specifically, firstly, a structured edge detector is utilized to generate an edge map where each pixel contains a magnitude and orientation information of the edge. Secondly, the edges are grouped together by a greedy algorithm. Then, the affinity between two edge groups is computed. This is the input to the next step for the score computation. Thirdly, for a given bounding box, the score of the bounding box is computed based on the edge groups entirely inside the box. Finally, as the EdgeBoxes detector may generate many overlapping candidate pedestrian windows, the non-maximal suppression is used to filter out these overlapping bounding boxes by the following criterion:

$$OverlapArea = \frac{(IntersectionArea)}{(UnionArea)} \tag{2}$$

If two bounding boxes overlap by more than 50%, then the bounding box with the highest score is selected.

Compared with the Selective Search which is the most popular proposal method, EdgeBoxes is much faster. For EdgeBoxes, the average runtime is about 0.3 s per image while Selective Search is about 10 s per image. In the case of accuracy on the PASCAL VOC 2007 dataset, the mAP of Selective Search is 31.7% which is slightly smaller than the mAP of EdgeBoxes 31.8% [13], and on the INRIA dataset, the recall rates of Selective Search and EdgeBoxes are 23% and 93% respectively by our experiment. In brief, the EdgeBoxes can not only reduce the runtime complexity but also increase the accuracy. Therefore, we choose the EdgeBoxes as our proposal generation algorithm.

3.2 The Architecture for Pedestrian Detection

The architecture of our proposed Fast R-CNN based on batch normalization method is illustrated in Fig. 1. It has 7 layers. As batch normalization layer can normalize each scalar feature independently layer by layer, it avoids the gradient vanish and gradient explode problem. Batch normalization can lead to a faster convergence during the training and can improve the generalization performance as a regularization technique. In our proposed architecture, the batch normalization layer is placed between each convolutional layer and the succedent ReLU layer, as illustrated in Fig. 2.

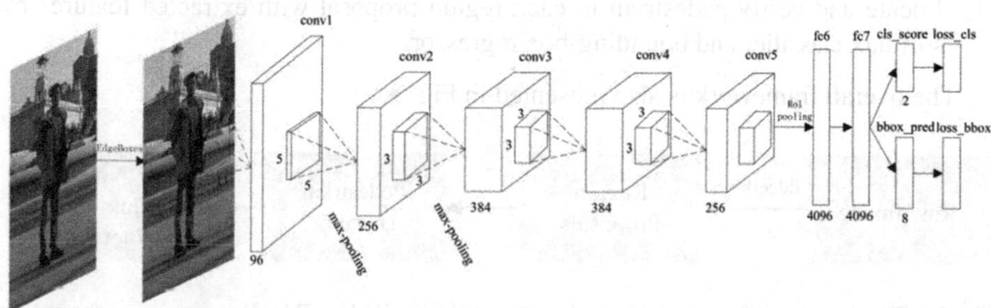

Fig. 1. The architecture of our proposed pedestrian detection model, where the batch normalization layer is placed between any one convolutional layer and the succedent activation function layer (such as ReLU layer, and details can be seen in Fig. 2).

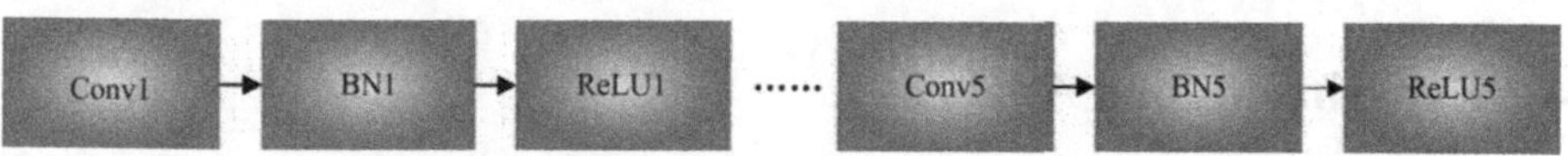

Fig. 2. The first five convolutional layers of our proposed model. The batch normalization (BN) layer is placed between each convolutional layer (Conv) and the succedent ReLU layer. In the first two convolutional layers, the max-pooling layer is placed to follow the ReLU layer.

In the first convolutional layer, there are 96 kernels of size 11 × 11 with a stride of 4 pixels and in the following max-pooling layer, the kernel size is 3 × 3. The second convolutional layer takes as input the output of the first convolutional layer, and there are 256 kernels of size 5 × 5 with a stride of 2 pixels. In the following max-pooling layer, the kernel size is the same as the first layer. For the next two (the third and the fourth) convolutional layers, the corresponding pooling layers are omitted and both contain 384 kernels. In the fifth convolutional layer, there are 256 kernels and the following pooling layer is a region of interest (RoI) pooling layer. The RoI pooling layer is utilized to pool the feature maps of each input object proposal into a fixed-length feature vector which is then fed into the fully connected layers. The next two layers are fully connected layers, both of which contain 4096 nodes. At the end of the architecture there are two output layers which produce two output vectors for per object proposal. Specifically, one is a softmax layer, which outputs classification scores over K object classes plus a "background" class. The other is a bounding-box regressor layer, which outputs four real-valued numbers for each of the K object classes. These 4 × K values encode refined bounding boxes for each class.

3.3 Pedestrian Detection Process

The entire pedestrian detection process can be summarized as follows:

(1) Run the EdgeBoxes algorithm to generate region proposals.
(2) Input the whole image and a number of region proposals to our proposed model.
(3) Extract the features of region proposals with convolutional neural networks.
(4) Locate and verify pedestrian in each region proposal with extracted features by softmax classifier and bounding-box regressor.

The overall framework is also presented in Fig. 3.

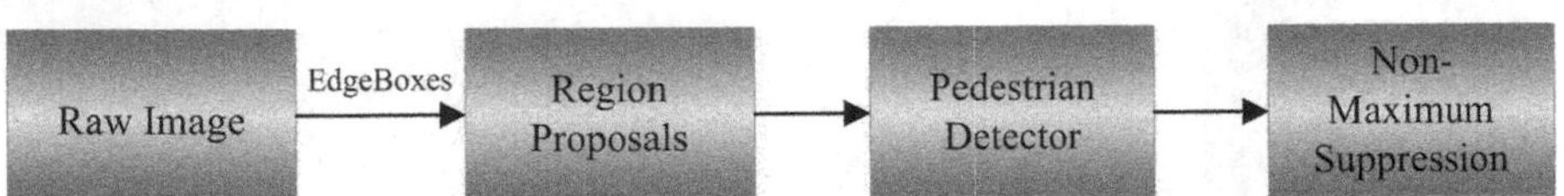

Fig. 3. The overview of our pedestrian detection system. (1) Use EdgeBoxes to generate region proposals. (2) Pass the region proposals to the pedestrian detector which is our proposed model. The detector will produce softmax scores for each proposal bounding box. (3) For each pedestrian class, use the non-maximum suppression (NMS) independently to greedily merge the overlapped proposals.

4 Experiments and Analysis

In this section, we evaluate our proposed pedestrian detection method on two popular benchmark datasets, INRIA and ETH. We follow the evaluation protocols proposed in [7]. The log average miss rate is used to summarize the detector performance. It is computed by averaging the miss rate at 9 FPPI (false positives per image) rates (0.05,

0.10, 0.20, 0.30, 0.40, 0.50, 0.64, 0.80, 1) which are evenly spaced in the log-space ranging from 10^{-2} to 10^{0}. In one test image, if a detected bounding box BB_{dt} can significantly cover most area of the body window BB_{gt} of the ground-truth pedestrian, the detection window is considered to properly detect the pedestrian.

$$\alpha_0 = \frac{area(BB_{dt} \cap BB_{gt})}{area(BB_{dt} \cup BB_{gt})} \geq \theta \tag{3}$$

where α_0 is the area of overlap between BB_{dt} and BB_{gt}, and θ is the threshold which is set to be 0.5. And the experiments are run on an Ubuntu 14.04 64 bit system with a single 4 GB memory NVIDIA GeForce GTX 980 GPU.

4.1 INRIA Dataset

The INRIA dataset is a very popular dataset for pedestrian detection. The annotations of pedestrians are of high quality in diverse conditions and poses. The details of INRIA dataset are listed in Table 1.

Table 1. The details of INRIA dataset

Statistic	Training	Testing
Positive images	614	288
Negative images	1218	453
Annotated windows	1237	589

Our proposed model is implemented based on the publicly Caffe platform [15] and the CaffeNet model is taken as our initial model which has been pre-trained on the ImageNet dataset. In order to use the Fast R-CNN method, we replace the last max-pooling layer of the CaffeNet with the RoI pooling layer to pool the feature maps of each region proposal into a fixed resolution, *i.e.* 6×6. We also replace the final fully connected layer and softmax layer with two sibling fully connected layers, *i.e.* softmax layer and bounding-box regressor layer.

In order to expand positive example set, we map each region proposal to the ground-truth which has a high intersection over union (IoU). If the IoU is lager than 0.5, then we will label the selected region proposal as a positive sample for pedestrian class. The rest region proposals, whose IoUs are smaller than 0.1, are regarded as negative instances. The proposed model is trained with the Stochastic Gradient Descent (SGD) method with the momentum of 0.9 and the weight decay factor of 0.0005. Since batch normalization layer is added, a large learning rate is used in our experiment. We set the initial global learning rate as 0.01 and the dropout ratio as 0.1 in the fc6 and fc7 layers. The whole process of training is 40000 iterations, the training time of our model (FRCNN + BN) is 0.094 s per iteration and that of Fast R-CNN is 0.153 s per iteration. The average time to process an image using our model is about 0.5 s, while the average time of the Fast R-CNN is about 0.6 s.

The overall experimental results are reported with the DET curves (miss rate versus FPPI) in Fig. 4. We compare our proposed model with existing methods, including two classical methods VJ [29], HOG [4] and other state-of-the-art methods such as ConvNet [25], ACF [5], VeryFast [23], FRCNN [9], Roerei [24], Sketch Tokens [16] and Spatial-Pooling [22]. We also give the digitized results of the miss rate at 0.1 FPPI.

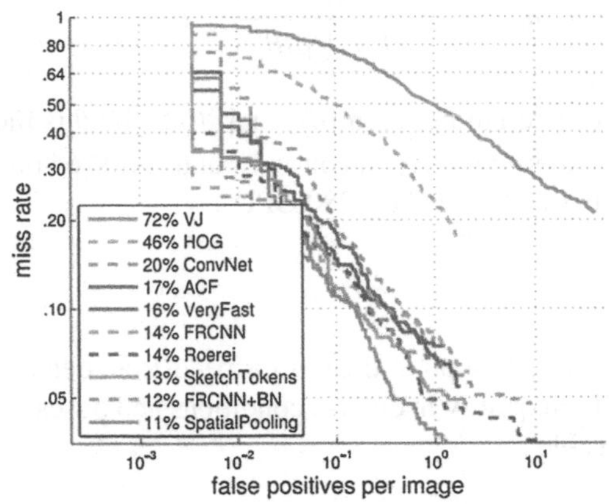

Fig. 4. Comparison of miss rate versus false positives per image between different methods on the INRIA dataset. Our model is denoted as FRCNN + BN. We obtain the miss rate of 12% at 0.1 FPPI.

From Fig. 4, we can see that our method outperforms most of the state-of-the-art methods. The miss rate of our method is 12%, which has an improvement of 8% over ConvNet and an improvement of 2% over FRCNN, but a slight decrease over the SpatialPooling method.

4.2 ETH Dataset

The ETH dataset derives from the binocular vision of pedestrian dataset. It is obtained by a pair of on-board camera and it gives the annotations of pedestrians. The set01, set02 and set03 are currently the most used three subsets. The details of these three subsets can be seen in Table 2.

Table 2. The details of ETH dataset

Statistic	set01	set02	set03
Positive images	999	446	354
Negative images	0	0	0
Annotated windows	8466	3472	2225

In order to evaluate the generalization performance of our proposed method, we train our proposed method on the INRIA dataset, and then apply the trained model to the ETH dataset. The performance evaluation is the same as that on the INRIA dataset. The average time to process an image using our model (FRCNN + BN) is about 0.4 s, while the average time of the Fast R-CNN is about 0.5 s. We compare our model with other methods, such as ConvNet [25], JointDeep [21], FRCNN [9], SDN [18], TA-CNN [26], VJ [29], HOG [4], ACF [5], VeryFast [23]. The comparison results are shown in Fig. 5, from which we can see that our method outperforms VJ, HOG, VeryFast, ACF,

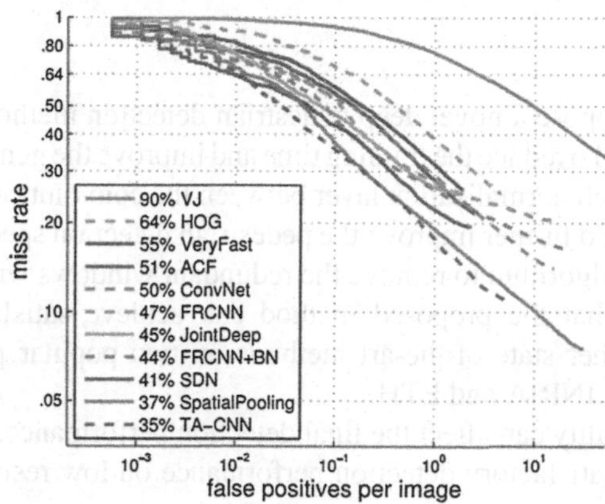

Fig. 5. Comparison of miss rate versus false positives per image between different methods on the ETH dataset. Our method is denoted as FRCNN + BN. We obtain the miss rate of 44% at 0.1 FPPI.

Fig. 6. Detection examples from INRIA dataset (a, b, c) and ETH dataset (d, e, f).

ConvNet, FRCNN and JointDeep but underperforms SDN, SpatialPooling and TA-CNN methods.

From Fig. 5, we can also see that the performance on the ETH dataset is worse than that on the INRIA dataset. One reason may lie in that the image quality of ETH dataset is much worse than that of the INRIA dataset, and the image resolution of the ETH dataset is lower. Another reason may be that we use an external dataset to train the model for the ETH dataset.

Finally, we also give some of our detection examples in Fig. 6.

5 Conclusions

In this paper, we propose a novel deep pedestrian detection method based on the Fast R-CNN framework. To reduce the training time and improve the generalization performance, we add the batch normalization layer between the convolutional layer and activation function layer. To further improve the pedestrian detection speed and accuracy, we use the EdgeBoxes algorithm to remove the redundant windows with poor quality. The experiments show that the proposed method can achieve satisfactory performance comparable with other state-of-the-art methods on two popular pedestrian detection benchmark datasets, INRIA and ETH.

As the image quality can affect the final detection performance, using our proposed method to achieve satisfactory detection performance on low resolution image is still an open issue.

References

1. Chen, G., Ding, Y., Xiao, J., Han, T.: Detection evolution with multi-order contextual co-occurrence. In: IEEE Conference on Computer Vision and Pattern Recognition, pp. 1798–1805 (2013)
2. Chen, X., Wei, P., Ke, W., Ye, Q., Jiao, J.: Pedestrian detection with deep convolutional neural network. In: Jawahar, C.V., Shan, S. (eds.) ACCV 2014. LNCS, vol. 9008, pp. 354–365. Springer, Cham (2015). doi:10.1007/978-3-319-16628-5_26
3. Cheng, M.M., Zhang, Z., Lin, W.Y., Torr, P.: Bing: binarized normed gradients for objectness estimation at 300fps. In: Proceedings of the IEEE Conference on Computer Vision and Pattern Recognition, pp. 3286–3293 (2014)
4. Dalal, N., Triggs, B.: Histograms of oriented gradients for human detection. In: IEEE Computer Society Conference on Computer Vision and Pattern Recognition, vol. 1, pp. 886–893 (2005)
5. Dollár, P., Appel, R., Belongie, S., Perona, P.: Fast feature pyramids for object detection. IEEE Trans. Pattern Anal. Mach. Intell. **36**(8), 1532–1545 (2014)
6. Dollár, P., Tu, Z., Perona, P., Belongie, S.: Integral channel features. In: British Machine Vision Conference 2, p. 5 (2009)
7. Dollár, P., Wojek, C., Schiele, B., Perona, P.: Pedestrian detection: an evaluation of the state of the art. IEEE Trans. Pattern Anal. Mach. Intell. **34**(4), 743–761 (2012)
8. Felzenszwalb, P.F., Girshick, R.B., McAllester, D., Ramanan, D.: Object detection with discriminatively trained part-based models. IEEE Trans. Pattern Anal. Mach. Intell. **32**(9), 1627–1645 (2010)

9. Girshick, R.: Fast r-cnn. In: Proceedings of the IEEE International Conference on Computer Vision, pp. 1440–1448 (2015)
10. Girshick, R., Donahue, J., Darrell, T., Malik, J.: Rich feature hierarchies for accurate object detection and semantic segmentation. In: Proceedings of the IEEE conference on computer vision and pattern recognition, pp. 580–587 (2014)
11. He, K., Zhang, X., Ren, S., Sun, J.: Spatial pyramid pooling in deep convolutional networks for visual recognition. IEEE Trans. Pattern Anal. Mach. Intell. **37**(9), 1904–1916 (2015)
12. Hosang, J., Benenson, R., Dollár, P., Schiele, B.: What makes for effective detection proposals? IEEE Trans. Pattern Anal. Mach. Intell. **38**(4), 814–830 (2015)
13. Hosang, J., Benenson, R., Schiele, B.: How good are detection proposals, really? arXiv preprint arXiv:1406.6962 (2014)
14. Ioffe, S., Szegedy, C.: Batch normalization: Accelerating deep network training by reducing internal covariate shift. arXiv preprint arXiv:1502.03167 (2015)
15. Jia, Y., Shelhamer, E., Donahue, J., Karayev, S., Long, J., Girshick, R., Guadarrama, S., Darrell, T.: Caffe: Convolutional architecture for fast feature embedding. In: ACM International Conference on Multimedia, pp. 675–678 (2014)
16. Joseph, J.L., C. Lawrence, Z., Dollár, P.: Sketch tokens: a learned mid-level representation for contour and object detection. In: IEEE Conference on Computer Vision and Pattern Recognition, pp. 3158–3165 (2013)
17. Li, J., Liang, X., Shen, S., Xu, T., Yan, S.: Scale-aware fast r-cnn for pedestrian detection. arXiv preprint arXiv:1510.08160 (2015)
18. Luo, P., Tian, Y., Wang, X., Tang, X.: Switchable deep network for pedestrian detection. In: Proceedings of the IEEE Conference on Computer Vision and Pattern Recognition, pp. 899–906 (2014)
19. Mu, Y., Yan, S., Liu, Y., Huang, T., Zhou, B.: Discriminative local binary patterns for human detection in personal album. In: IEEE Conference on Computer Vision and Pattern Recognition, pp. 1–8 (2008)
20. Nam, W., Dollár, P., Han, J.H.: Local decorrelation for improved pedestrian detection. In: Advances in Neural Information Processing Systems, pp. 424–432 (2014)
21. Ouyang, W., Wang, X.: Joint deep learning for pedestrian detection. In: Proceedings of the IEEE International Conference on Computer Vision, pp. 2056–2063 (2013)
22. Paisitkriangkrai, S., Shen, C., Hengel, A.: Strengthening the effectiveness of pedestrian detection with spatially pooled features. In: Fleet, D., Pajdla, T., Schiele, B., Tuytelaars, T. (eds.) ECCV 2014. LNCS, vol. 8692, pp. 546–561. Springer, Cham (2014). doi: 10.1007/978-3-319-10593-2_36
23. Rodrigo, B., Markus, M., Radu, T., Luc, V.G.: Pedestrian detection at 100 frames per second. In: IEEE Conference on Computer Vision and Pattern Recognition, pp. 2903–2910 (2012)
24. Rodrigo, B., Markus, M., Tinne, T., Luc, V.G.: Seeking the strongest rigid detector. In: IEEE Conference on Computer Vision and Pattern Recognition, pp. 3666–3673 (2013)
25. Sermanet, P., Kavukcuoglu, K., Chintala, S., LeCun, Y.: Pedestrian detection with unsupervised multi-stage feature learning. In: Proceedings of the IEEE Conference on Computer Vision and Pattern Recognition, pp. 3626–3633 (2013)
26. Tian, Y., Luo, P., Wang, X., Tang, X.: Pedestrian detection aided by deep learning semantic tasks. In: IEEE Conference on Computer Vision and Pattern Recognition, pp. 5079–5087 (2015)
27. Uijlings, J.R., van de Sande, K.E., Gevers, T., Smeulders, A.W.: Selective search for object recognition. Int. J. Comput. Vision **104**(2), 154–171 (2013)
28. Vedaldi, A., Gulshan, V., Varma, M., Zisserman, A.: Multiple kernels for object detection. In: IEEE International Conference on Computer Vision, pp. 606–613 (2009)

29. Viola, P., Jones, M.J.: Robust real-time face detection. Int. J. Comput. Vision **57**(2), 137–154 (2004)
30. Viola, P., Jones, M.J., Snow, D.: Detecting pedestrians using patterns of motion and appearance. Int. J. Comput. Vision **63**(2), 153–161 (2005)
31. Wang, B., Tang, S., Zhao, R., Liu, W., Cen, Y.: Pedestrian detection based on region proposal fusion. In: Multimedia Signal Processing (MMSP), pp. 1–6. IEEE (2015)
32. Wang, M., Gao, Y., Lu, K., Rui, Y.: View-based discriminative probabilistic modeling for 3d object retrieval and recognition. IEEE Trans. Image Process. **22**, 1395–1407 (2013)
33. Wang, M., Liu, X., Wu, X.: Visual classification by l1-hypergraph modeling. IEEE Trans. Knowl. Data Eng. **27**, 2564–2574 (2015)
34. Wang, X., Han, T.X., Yan, S.: An hog-lbp human detector with partial occlusion handling. In: IEEE International Conference on Computer Vision, pp. 32–39 (2009)
35. Zhao, Z.-Q., Xie, B.-J., Cheung, Y.-M., Wu, X.: Plant leaf identification via a growing convolution neural network with progressive sample learning. In: Cremers, D., Reid, I., Saito, H., Yang, M.-H. (eds.) ACCV 2014. LNCS, vol. 9004, pp. 348–361. Springer, Cham (2015). doi:10.1007/978-3-319-16808-1_24
36. Zitnick, C.L., Dollár, P.: Edge boxes: locating object proposals from edges. In: Fleet, D., Pajdla, T., Schiele, B., Tuytelaars, T. (eds.) ECCV 2014. LNCS, vol. 8693, pp. 391–405. Springer, Cham (2014). doi:10.1007/978-3-319-10602-1_26

Neural Networks: Theory and Application

Dynamic Analysis and Simulation for Two Different Chaos-Like Stochastic Neural Firing Patterns Observed in Real Biological System

Huijie Shang[1], Rongbin Xu[1], and Dong Wang[1,2](✉)

[1] School of Information Science and Engineering, University of Jinan, Jinan 250022, China
ise_wangd@ujn.edu.cn

[2] Shandong Provincial Key Laboratory of Network Based Intelligent Computing, University of Jinan, Jinan 250022, China

Abstract. In the present investigation, two neural firing patterns generated in the experimental neural pacemaker and simulated in the stochastic Chay model but not in the deterministic model, lying between period 2 bursting and period 3 bursting, were identified to be stochastic by the inter-event intervals (IEIs) analysis, although they exhibit chaos-like characteristics with the results of the inter-spike intervals (ISIs) analysis. The case I pattern was a transition between a string of period 2 burst and a string of period 3 burst, while between a single period 2 burst and a single period 3 burst in case II pattern. Two patterns were generated near a bifurcation point from period 2 bursting to period 3 bursting under the influence of noise by coherence resonance with the result of the bifurcation analysis. The Case I has exhibited coexistence of 2 bursting and period 3 bursting near the bifurcation while case II with no coexistence.

Keywords: Stochastic neural firing pattern · Period adding bifurcation · Coherence resonance · Chaos

1 Introduction

With the rapid development of nonlinear dynamics and neuroscience, the opinion that information in a neural system is transmitted not only by firing rate but also by interspike intervals (ISIs) including the dynamic structures has been widely recognized [1–4]. The ISI series revealed lots of different firing patterns, such as the periodic, chaotic, and stochastic firing patterns were observed both in the theoretical model and biological experiment [5–8]. In these previous researches, noise was found important to generate new stochastic firing patterns [9–12]. Periodic adding bifurcation scenarios were paid much attention as the special transition rules that there are not only chaotic patterns but also stochastic patterns in these bifurcation scenarios near the bifurcation point. However, there may be no lack of misjudgments previous. The present paper discusses two neural firing patterns in period adding bifurcation scenarios without chaos observed in neural pacemaker experiment and theoretical stimulation by Chay model. Next, we analyzed and compared these two patterns in two cases.

D.-S. Huang et al. (Eds.): ICIC 2017, Part I, LNCS 10361, pp. 749–757, 2017.
DOI: 10.1007/978-3-319-63309-1_66

2 Experimental Model and Results

Biological experiment was finished on experimental neural pacemakers formed at the injured site of adult rat [2], and we record the time intervals between the successive spikes seriatim as ISIs series. When experiment decreased the *extra-cellular* calcium concentration ($[Ca^{2+}]o$) gradually from 1.2 mmol/L to zero, we could see that period adding bifurcation scenarios or reverse period adding bifurcation scenarios without chaos from period 1 bursting to period 3 bursting or reverse cases, as shown in Fig. 1. Here the characteristics of the firing pattern lying between period 2 bursting and period 3 bursting were studied particularly.

The multimode characteristics of IEIs series are similar to those of the on-off pattern [6] caused by coherence resonance. The peaks in IEI histogram (IEIH) exhibited no significant characteristics but with a higher first peak, as shown in Fig. 2(a). First return map of the IEI series is composed of two groups of points, as shown in Fig. 2(b). NPE of this IEI series was nearly 1.0 when predicated from step 1 to step 10, as shown in Fig. 2(c), implying that there exists uncertain structures in the IEI series and such a firing pattern may not result from a dynamics of deterministic chaos.

In some other experimental units, another new neural firing pattern occurred from period 2 bursting to period 3 bursting which behavior is a stochastic transition. We can wee the same evaluation method in Fig. 3. This result indicates case II firing pattern may also result from a stochastic dynamics mechanism.

If period 3 burst was chosen as an event, the analysis on IEIs series above could get similar results.

3 Theoretical Model

Chay model, a mathematical neuronal model whose dynamics was verified to be very close to the experimental neural pacemaker in a series of previous studies [2, 3, 6, 7], was used to carry out the numerical simulation. Its deterministic form is given as follows:

$$\frac{dV}{dt} = g_I m_\infty^3 h_\infty (v_I - V) + g_{kv}(v_k - V)n^4 + g_{kc}\frac{C}{1+C}(v_k - V) + g_l(v_l - V) \tag{1}$$

$$\frac{dn}{dt} = \frac{n_\infty - n}{\tau_n} \tag{2}$$

$$\frac{dC}{dt} = \frac{m_\infty^3 h_\infty (v_c - V) - k_c C}{\tau_c} \tag{3}$$

where V is the membrane potential, n is the probability of potassium channel activation and C is the dimensionless intracellular concentration of calcium ion ([Ca2+]i). $\tau_n = 1/(\lambda_n(\alpha_n + \beta_n))$ and the explicit interpretation of Chay model is given as [15]. vc is the reversal potential of calcium ion, chosen as the bifurcation parameter, corresponding to the adjustment of [Ca2+]o in the experiment. λn is chosen as the conditional parameter.

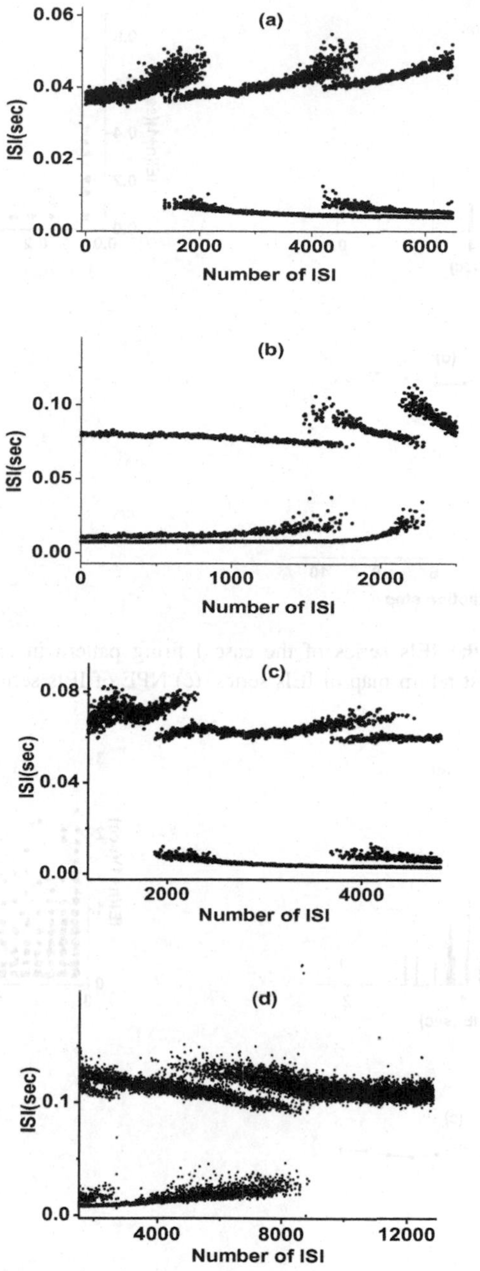

Fig. 1. Period adding bifurcation scenario from period 1 bursting to period 3 bursting or the reverse cases in the experiment when $[Ca^{2+}]_o$ was changed gradually. (a) Period adding bifurcation scenario with case I firing pattern. (b) Reverse period adding bifurcation scenario with case I firing pattern. (c) Period adding bifurcation scenario with case II firing pattern. (d) Reverse period adding bifurcation scenario with case II firing pattern.

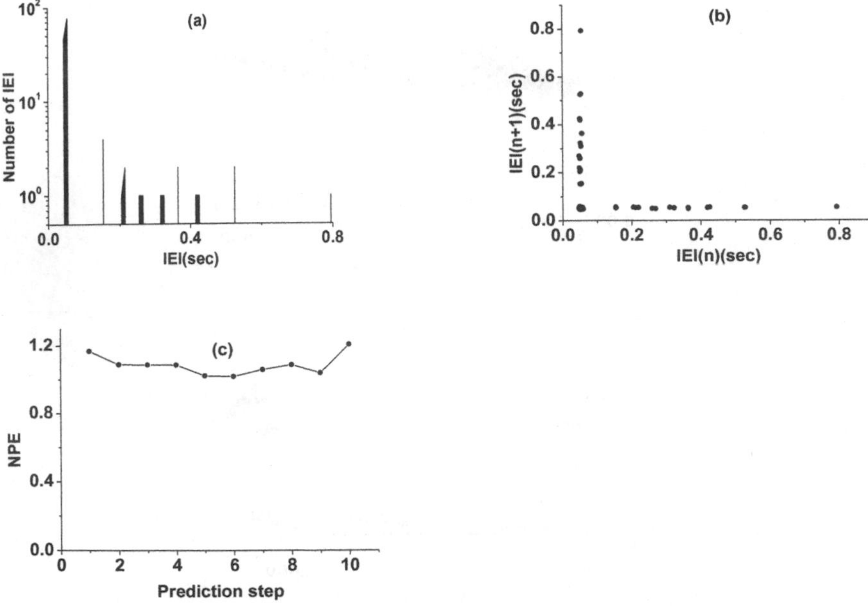

Fig. 2. Analysis on the IEIs series of the case I firing pattern in experiment. (a) The IEIs histogram. (b) The first return map of IEIs series. (c) NPE of IEIs series.

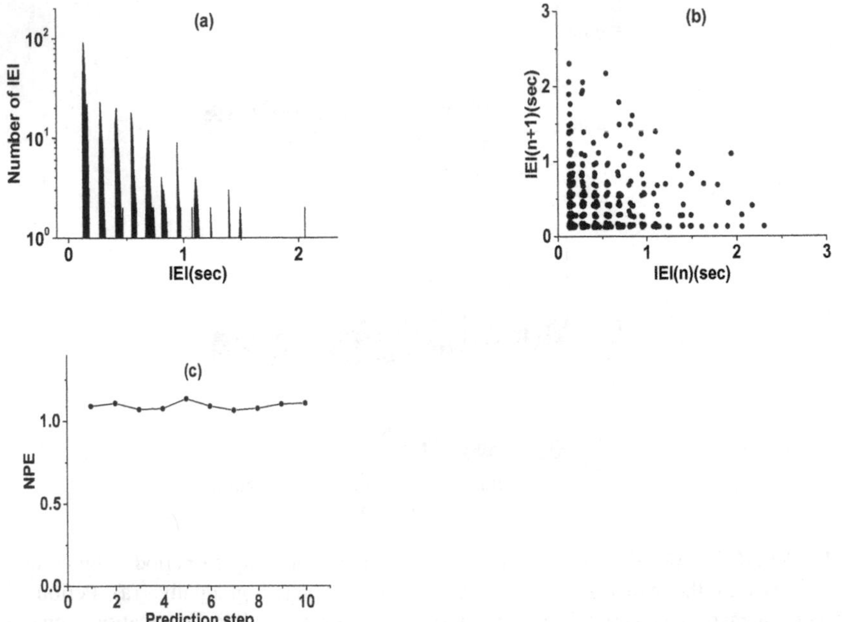

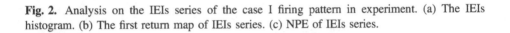

Fig. 3. The case II 2-3 alternation firing pattern in experiment. (a) The IEIs histogram. (b) The return map of IEIs series. (c) NPE of IEI series.

In this paper, the parameters are as follows: gI = 1800 mS/cm2, gkv = 1700 mS/cm2, gkc = 10 mS/cm2, gl = 7 mS/cm2, vI = 100 mV, vk = –75 mV, vl = 40 mV, τc = 100/27, kc = 3.3/18.

A stochastic factor, a Gaussian white noise, $\xi(t)$, reflecting fluctuation of environments, was directly added to the right side of Eq. (1) with other two equations unchanged to form the stochastic Chay model. The stochastic factor possessed the statistical properties as $<\xi(t)>=0$; $<\xi(t), \xi(t')> \ = 2D\delta(t-t')$; where D is the noise density and δ is the Dirac δ-function.

Both of the deterministic and stochastic Chay models are solved by Mannella numerical integrate method [14] with integration time step being 10-3 s. Upstrokes of the voltage reached the amplitude of –25.0 mV are counted as spikes.

4 Then Numerical Simulation Results

In the deterministic Chay model, when λn = 230, vc from 480 to 175 mV, and λn = 240, vc from 486 to 231 mV, we care the dynamics near the bifurcation between period 2 bursting and period 3 bursting. When λn = 230, the period 2 bursting was changed into period 3 bursting and the bifurcation point is vc ≈ 222.31 mV, vice versa. When

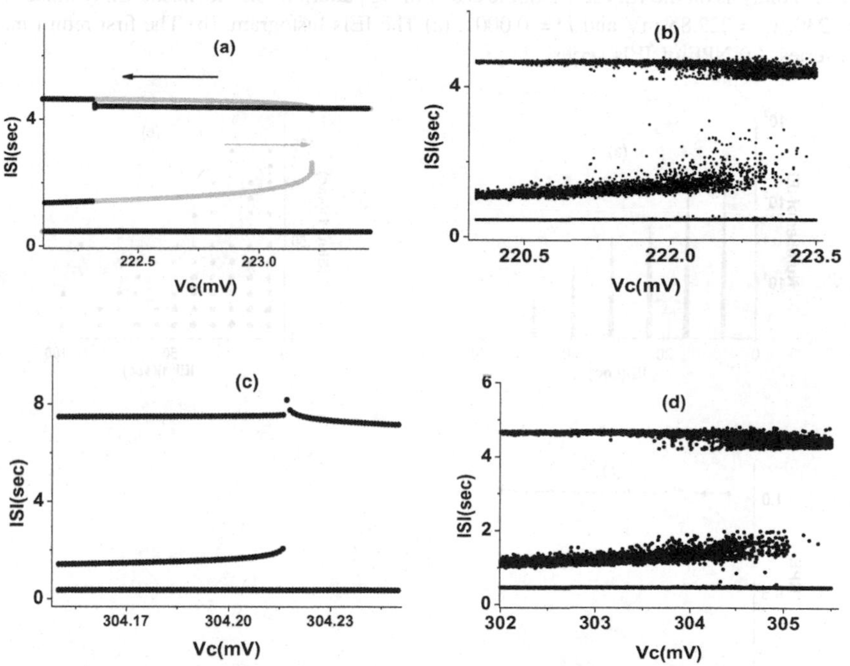

Fig. 4. Period adding bifurcation scenario from period 1 bursting to period 3 bursting in Chay model. (a) Deterministic model ($\lambda_n = 230$), the gray points represented the reverse period adding bifurcation scenario with v_c increased. (b) Stochastic model ($\lambda_n = 230$, $D = 0.0001$). (c) Deterministic model ($\lambda_n = 240$). (d) Stochastic model ($\lambda_n = 240$, $D = 0.0001$).

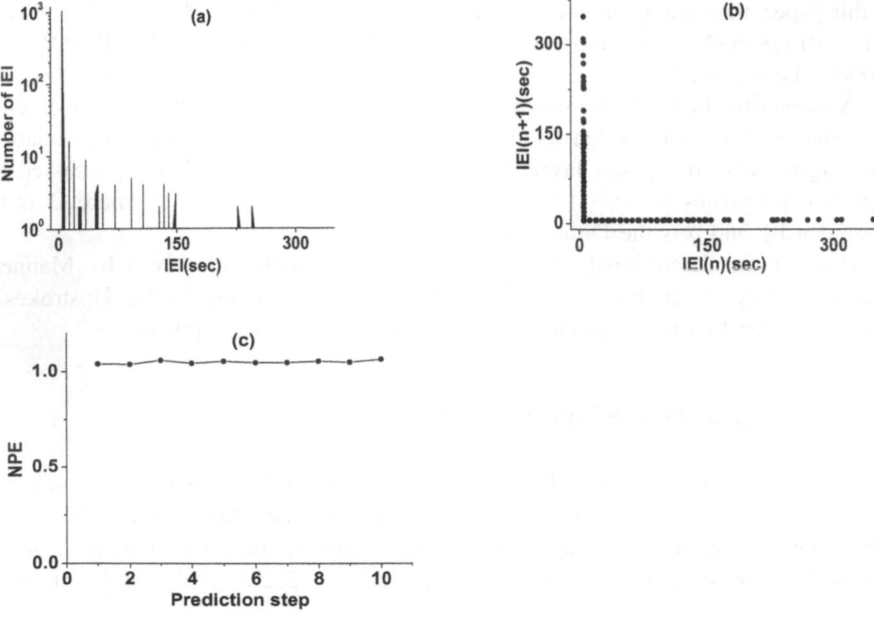

Fig. 5. Analysis on the IEIs series of the case I firing pattern in the stochastic Chay model when $\lambda_n = 230$, $v_c = 222.85$ mV and $D = 0.0001$. (a) The IEIs histogram. (b) The first return map of IEIs series. (c) NPE of IEIs series.

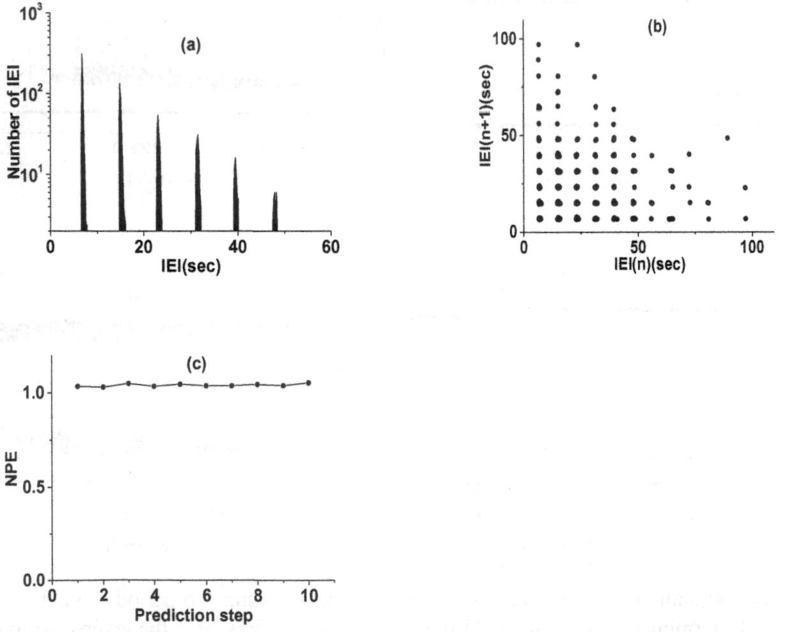

Fig. 6. The case II 2-3 alternation firing pattern in the stochastic Chay model when $\lambda_n = 240$, $v_c = 304.21$ mV and $D = 0.0001$. (a) The firing trains. (b) The IEIs histogram. (c) The first return map of IEIs series. (d) NPE of IEIs series.

vc was between 222.31 mV and 223.23 mV, the period 2 bursting and period 3 bursting coexist, forming an overlap composed of two bursts, as shown in Fig. 4(a). No coexist of period 2 bursting and period 3 bursting appeared, as shown in Fig. 4(c). In the deterministic Chay model, there are no other firing patterns.

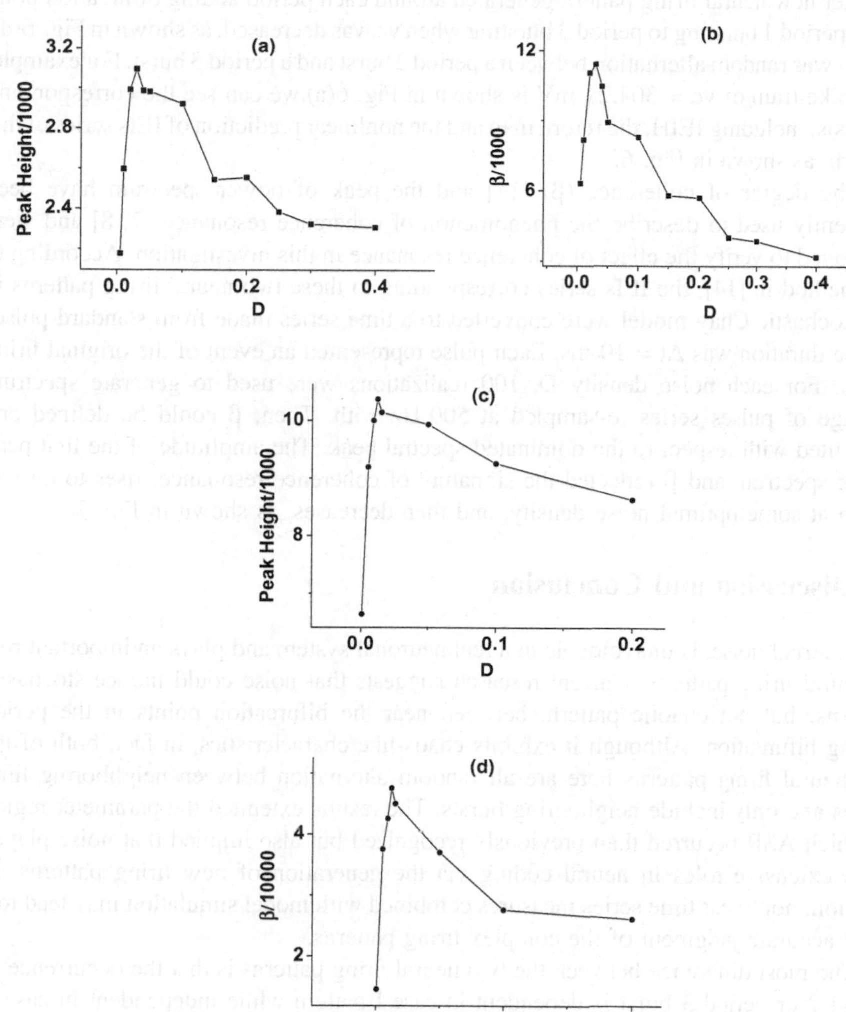

Fig. 7. Measurement on coherence resonance of case I ($\lambda_n = 230$, $v_c = 222.85$, $D = 0.0001$) and case II 2–3 ($\lambda_n = 240$, $v_c = 304.21$ mV, $D = 0.0001$) alternation firing pattern in the stochastic Chay model. (a) The relationship between amplitudes of dominant peaks in power spectrum and noise density in case I pattern. (b) The relationship between coherence degree (β) and noise density. (c) The relationship between amplitudes of dominant peaks in power spectrum and noise density in case II pattern. (d) The relationship between coherence degree (β) and noise density.

The chaos-like characteristics of the results are very similar to those of case I neural firing pattern discovered in experiment mentioned above. The IEIs series also exhibited the on-off pattern characteristics. The histogram, the return map and the nonlinear prediction of IEIs series of this 2–3 alternation firing pattern were similar to those of the experiments as well, as shown in Fig. 5.

Still in the stochastic Chay model, when λn = 240 and D = 0.0001, there was another new neural firing pattern generated around each period adding bifurcation point from period 1 bursting to period 3 bursting when vc was decreased, as shown in Fig. 6(d), which was random alternation between a period 2 burst and a period 3 burst. For example, the spike train of vc = 304.21 mV is shown in Fig. 6(a).we can see the corresponding analysis, including IEIH, the return map and the nonlinear prediction of IEIs series of this pattern, as shown in Fig. 6.

The degree of coherence (β) [11] and the peak of power spectrum have been frequently used to describe the phenomenon of coherence resonance [7, 8] and were employed to verify the effect of coherence resonance in this investigation. According to the method in [14], the IEIs series corresponding to these two neural firing patterns in the stochastic Chay model were converted to a time series made from standard pulses whose duration was Δt = 10 ms. Each pulse represented an event of the original firing series. For each noise density D, 100 realizations were used to generate spectrum average of pulses series re-sampled at 500 Hz with. Then, β could be defined and computed with respect to the dominated spectral peak. The amplitude of the first peak of the spectrum and β reflected the signature of coherence resonance: rises to a maximum at some optimal noise density, and then decreases, as shown in Fig. 7.

5 Discussion and Conclusion

The internal noise, is unavoidable in a real neuronal system and plays an important role in neural firing patterns. Current research suggests that noise could induce stochastic patterns, but not chaotic pattern, between near the bifurcation points in the period adding bifurcation. Although it exhibits chaos-like characteristics, in fact, both of the two neural firing patterns here are all random alternation between neighboring limit cycles and only include neighboring bursts. The results extended the parameter region in which ASR occurred than previously recognized but also implied that noise played more extensive roles in neural coding via the generation of new firing patterns. In addition, nonlinear time series measures combined with model simulation may lead to a more accurate judgment of the complex firing patterns.

The most difference between the two neural firing patterns is that the occurrence of period 2 or period 3 burst is dependent in case I pattern while independent in case II pattern. By the bifurcation analysis, we considered that it is caused mainly by the different bifurcation. Case I was generated near the bifurcation point with the coexistence of period 2 bursting and period 3 bursting while case II with no coexistence. In fact, this is a manifestation of the nonlinear characteristic of a system. The relationship between the two neural firing pattern should be further studied using bifurcation theory.

Acknowledgments. This research was supported by the National Key Research And Development Program of China (No. 2016YFC0106000), the Natural Science Foundation of China (Grant No. 61302128), and the Youth Science and Technology Star Program of Jinan City (201406003).

References

1. Braun, H.A., Wissing, H., Schafer, K., Christian, Hirsch M.: Oscillation and noise determine signal transduction in shark multimodal sensory cells. Nature **367**, 270–273 (1994)
2. Ren, W., Hu, S.J., Zhang, B.J., Wang, F.Z., Gong, Y.F., Xu, J.X.: Period-adding bifurcation with chaos in the inter-spike intervals generated by an experimental neural pacemaker. Int. J. Bifurcat. Chaos **7**, 1867–1872 (1997)
3. Gu, H.G., Yang, M.H., Li, L., Liu, Z.Q., Ren, W.:Chaotic and ASR induced firing pattern in experimental neural pacemaker. Dyn. Contin. Discrete Impulsive Syst. (Ser. B: Appl. Algorithms) **11a**, 19–24 (2004)
4. Yang, M.H., An, S.C., Gu, H.G., Liu, Z.Q., Ren, W.: Understanding of physiological neural firing patterns through dynamical bifurcation machineries. NeuroReport **17**(10), 995–999 (2006)
5. Jia, B., Gu, H.G., Xue, L.: A basic bifurcation structure from bursting to spiking of injured nerve fibers in a two-dimensional parameter space. Cogn. Neurodyn. **11**(2), 189–200 (2017)
6. Li, Y.Y., Gu, H.G.: The distinct stochastic and deterministic dynamics between period-adding and period-doubling bifurcations of neural bursting patterns. Nonlinear Dyn. **87**(4), 2541–2562 (2016)
7. Zhao, Z.G., Jia, B., Gu, H.G.: Bifurcations and enhancement of neuronal firing induced by negative feedback. **86**(3), 1549–1560 (2016)
8. Yang, M.H., Liu, Z.Q., Li, L., Xu, Y.L., Liu, H.J., Gu, H.G., et al.: Identifying distinct stochastic dynamics from chaos: a study on multimodal neural firing patterns. Int. J. Bifurcat. Chaos **19**, 453–485 (2009)
9. Jia, B., Gu, H.G.: Identifying type I excitability using dynamics of stochastic neural firing patterns. Cogn. Neurodyn. **6,** 485–497 (2012). doi:10.1007/s11571-012-9209-x. PMID: 24294334
10. Gu, H.G., Zhang, H.M., Wei, C.L., Yang, M.H., Liu, Z.Q., Ren, W.: Coherence resonance induced stochastic neural firing at a saddle-node bifurcation. Int. J. Mod. Phys. B **25**, 3977–3986 (2011)
11. Gu, H.G., Zhao, Z.G., Jia, B., Chen, S.G.: Dynamics of On-off neural firing patterns and stochastic effects near a sub-critical hopf bifurcation. PLoS ONE **10**(4), e0121028 (2015)
12. Xing, J.L., Hu, S.J., Xu, H., Han, S., Wan, Y.H.: Subthreshold membrane oscillations underlying integer multiples firing from injured sensory neurons. NeuroReport **12**, 1311–1313 (2011)
13. Chay, T.R.: Chaos in a three-variable model of an excitable cell. Physica D **16**, 233–242 (1985)
14. Mannella, R., Palleschi, V.: Fast and precise algorithm for computer simulation of stochastic differential equations. Phys. Rev. A **40**, 3381–3386 (1989)
15. Wiesenfeld, K., Pierson, D., Pantazelo, E.: Stochastic resonance on a circle. Phys. Rev. Lett. **72**(14), 2125–2129 (1994)

Smooth Multi-instance Learning for Object Detection

Dayuan Li[1(✉)], Zhipeng Li[1], and Youhua Zhang[2]

[1] School of Electronics and Information Engineering, Tongji University, Caoan Road 4800, Shanghai 201804, China
1531811@tongji.edu.cn

[2] School of Information and Computer, Anhui Agricultural University, Changjiang West Road 130, Hefei, Anhui, China
zhangyh@ahau.edu.cn

Abstract. The problem of object localization is one of the key problems in computer vision applications. Recently, multiple-instance learning (MIL) is a kind of machine learning framework which receiving a set of instances that are individually labeled. This framework has been verified that will get good effect in object localization in images. In this paper, we propose a novel method to handle the classical MIL problem. We preprocess images with superpixel techniques to speed up the whole procedure of training our model and regard the positiveness of instance as a continuous variable. The softmax model is used to bring a bridge between instances and bags and jointly optimize the bag label and instance label in a unified framework. At last, the model is trained by iterative weakly supervised training method. The extensive experiments demonstrate that out method achieves superior performance on various MIL benchmarks. The state-of-the-art results of object discovery on Pascal VOC datasets further confirm the advantages of the proposed method.

Keywords: Multiple-instance learning · Object localization · Superpixel · Smooth

1 Introduction

Over the last decades, object localization in image has been researched not only in computer vision but also in pattern recognition [3, 4, 26], and significant progress has bend made due to the development of technology in object localization and the availability of ever larger and more comprehensive datasets. Object localization is to identify all interesting objects in an image and it is much more challenging to localize the object when compared with image classification that is just to figure out whether this image contains the object we interested. The development of neural network theory [28, 29, 31] also improve the performance in object localization compared with traditional methods. Since the multiple-instance learning (MIL) problem was introduced by Dietterich et al. in 1997 [6] for the task of drug activity prediction, a large number of researches try to solve object localization problem with MIL techniques emerged in recent years [7–10]. In multiple-instance learning, the label of bag is given but the label

D.-S. Huang et al. (Eds.): ICIC 2017, Part I, LNCS 10361, pp. 758–767, 2017.
DOI: 10.1007/978-3-319-63309-1_67

of instances are unknown. One bag contains many instances. A bag is labeled as positive when at least one instance in this bag is positive and a bag will labeled as negative if none of the instances is positive in the binary classification task.

Experiments with the same detector but different proposal instances will result in very different result. A discriminative proposal instances plays a significant role in MIL problem. In last decades, many researchers have made significant progress. Latent SVM which is also called multiple instance SVM was used to image categorization [11]. M. Sapienza proposed a method to learn discriminative space–time action parts from weakly labelled videos [12]. R. Hong annotated image by multiple-instance learning with discriminative feature mapping and selection [13]. All the methods mentioned above regard instance selection and model learning as two separated procedures but we proposed a unified framework to calculate and optimize the label of instance and bag by smooth the discrete instance label and gradient descent.

In this paper, we adopt SLIC (Simple Linear Iterative Clustering) [14] to simplify the image as superpixels and utilize Edgebox [15] to propose plenty of windows as instance proposals. The probability of the instance label to be positive is represented by a Probit model. The relation between bag label and instance labels, which is also called MIL constraints, is the softmax function. The procedure of training our classifier follows the iterative weakly supervised training algorithm with a tradeoff between efficiency and accuracy. Our method is simple but effective. In the experiments, our approach leads to better localization on the training images using the PASCAL VOC 2007 dataset.

2 Related Work

Many multi-instance learning algorithms have been developed during the past decade. For example, DD, EM-DD, MIBoosting, miSVM and miGraph, etc. DD [16] finds regions in the instance space with instances from many different positive bags and few instances from negative bags to solve MIL. EM-DD [17] developed DD with expectation maximization (EM). MIBoosting [18] is a boosting approach for MIL assumed that all instances contribute equally and independently to a bag's label. miSVM [19] is an improved SVM designed for MIL. It searches max-margin hyperplanes to separate positive and negative instances in bags. miGraph [20] employs graph kernels to solve MIL problems which regards instances as a non-i.i.d.

Object localization in visual recognition is a common application scenarios for weakly supervised object discovery, such as object class discovery [5, 30], face recognition [21, 27], object detection [22], image segmentation [25]. Chunhui Gu's team [23] formed visual clusters from the data that are tight in appearance and configuration spaces and trained individual classifiers for each component. Furthermore, they trained a second classifier that operates at the category level by aggregating responses from multiple components. Our method is different from the existing object discovery methods, which divides an image with super-pixels and extracts instance features with the color, shape and texture of that super-pixel. We also smooth the typical MIL problem and trains our model with iterative weakly supervised training algorithm.

3 Bounding Boxes Selecting

The superpixels can capture the local structural cues of an image and speed up the processing. We adopt SLIC (Simple Linear Iterative Clustering) [14] to simplify the image as superpixels. SLIC is an approach generates superpixels according to the color similarity of pixels in an image. In this approach, one five dimensional vector is used to represent one pixel, color vector of this pixel in CIELAB color space and its position in one image. For an image with N pixels, the number of superpixels in this image is K and therefore the approximate interval of cach superpixel is $S = \sqrt{N/K}$. The distance in this 5D space is represented by D_s:

$$\begin{aligned} d_{lab} &= \sqrt{(l_k - l_i)^2 + (a_k - a_i)^2 + (b_k - b_i)^2} \\ d_{xy} &= \sqrt{(x_k - x_i)^2 + (y_k - y_i)^2} \\ D_s &= d_{lab} + \frac{m}{S} d_{xy} \end{aligned} \tag{1}$$

where d_{lab} is the lab distance and d_{xy} is the plane distance. D_s is the sum of the lab distance and the *xy* plane distance normalized by the grid interval *S*. The variable *m* introduced in D_s is defined to control the compactness of a superpixel and it ranges from 1 to 20. The greater the value of m is, the more compact the cluster will be. We often set its value to 10 by experience.

We utilize Edgebox [15] to capture plenty of windows as object proposals and extract features by color, texture and SIFT. Edges provide a sparse yet informative representation of an image and millions of candidate boxes can be evaluated in a fraction of a second.

4 Smooth MIL

Firstly, we give some notions about MIL to simplify our demonstration and make it easier to understand. Given a set of bags $\mathbf{B}_1, \cdots, \mathbf{B}_m$ with labels $Y_I \in \{0, 1\}$. Each bag $\mathbf{B}_I$ consists of many instance $\mathbf{x}_i = \{\mathbf{x}_{i1}, \cdots, \mathbf{x}_{ij}\}$, where j denotes the number of instances in the bag $\mathbf{B}_I$. The label Y_I of $\mathbf{B}_I$ is defined by this: $y_{ij} \in \{0, 1\}$ is the label of instance $\mathbf{x}_{ij}$, if $Y_I = 0$, then $y_{ij} = 0$ for all $i, j \in \mathbf{I}$. On the other hand, if $Y_I = 1$, then at least one instance $\mathbf{x}_{ij} \in \mathbf{B}_I$ is a positive instance.

In classical MIL problems, the labels of instances and bags are discrete and can only be 0 or 1. We smooth the instance label y_i to be a continues variable in the range of [0,1], which represents the probability of $\mathbf{x}_{ij}$ being positive, denoted as $p_{ij}.\mathbf{x}_{ij} \in \mathbf{R}^{d \times 1}$ is the j-th instance in the i-th bag. p_{ij} is given by a Probit model.

$$p_{ij} = \Pr(y_{ij} = 1 | \mathbf{x}_{ij}; \mathbf{w}) = \Phi(\mathbf{w}^T \mathbf{x}_{ij}), \tag{2}$$

where Pr denotes probability, and Φ is the cumulative distribution function of the standard normal distribution. $\mathbf{w}$ is the weight vector of the linear model that needs to be optimized through our formulation. On the other hand, we adopt the softmax model to bring a bridge between these two kinds of labels:

$$P_i = \frac{p_{ij}}{\sum_{j=1}^{m_i} p_{ij}} \tag{3}$$

where m_i is the number of instances in $\mathbf{B}_i$. Assuming that one instance in the bag is predicted as positive, then we can find $P_i = 1$ according to Eq. (2). P_i will equal to 0 if and only if all instances in the bag are predicted as zero.

We smooth the labels of bags and instances according to Eqs. (2) and (3) so that the problem becomes more tractable because we can differentiate all parts in Eq. (3). Our MIL objective function is given as follows with considering the instance-level loss, bag-level loss, and model regularization.

$$\begin{aligned} \min_{\mathbf{w}} \frac{\beta}{n} \sum_{i=1}^{n} & \underbrace{-\{Y_i \log P_i + (1 - Y_i)\log(1 - P_i)\}}_{\text{cost item for i-th bag}} \\ & + \frac{1}{n}\sum_{i=1}^{n}\frac{1}{m_i}\sum_{j=1}^{mi}\underbrace{\max(0, [\mathrm{m_O} - \mathrm{sgn}(p_{ij} - p_O)\mathbf{w}^T\mathbf{x}_{ij}])}_{\text{cost item for ij-th instance}} + \frac{\lambda}{2}\|\mathbf{w}\|^2 \end{aligned} \tag{4}$$

where sgn is the sign function; m_O is the margin parameter in SVM model used to separate the positive instances and negative instances distant from the hyper line in the feature space; $\mathrm{p_O}$ is a threshold defined before the training to determine positive or instance.

We can optimize Eq. (3) with gradient descent and the partial derivative of the cost item for bag with respect to $\mathbf{w}$ is derived as

$$\frac{\partial \mathcal{L}_{bagi}}{\partial \mathbf{w}} = \frac{\partial \mathcal{L}_{bagi}}{\partial P_i} \cdot \sum_{j=1}^{m_i} \frac{\partial P_i}{\partial p_{ij}} \frac{\partial p_{ij}}{\partial \mathbf{w}} \tag{5}$$

Where $\frac{\partial \mathcal{L}_{bagi}}{\partial P_i}$ and $\frac{\partial P_i}{\partial p_{ij}}$ is given by

$$\frac{\partial \mathcal{L}_{bagi}}{\partial P_i} = -\{\frac{Y_i}{P_i} - \frac{(1 - Y_i)}{1 - P_i}\} = -\frac{Y_i - P_i}{P_i(1 - P_i)} \tag{6}$$

$$\frac{\partial P_i}{\partial p_{ij}} = \frac{1}{p_{ij} + \sum\limits_{t \in [1, m_i] \setminus j} p_{it}} - \frac{1}{(p_{ij} + \sum\limits_{t \in [1, m_i] \setminus j} p_{it})^2} \tag{7}$$

As for the final expression of partial derivative of $\mathcal{L}_{bagi}$ with respect to $\mathbf{w}$ is

$$\frac{\partial \mathcal{L}_{bagi}}{\partial \mathbf{w}} = -\frac{Y_i - P_i}{P_i(1 - P_i)} \sum_{j=1}^{m_i} \frac{1}{p_{ij} + \sum\limits_{t \in [1, m_i] \setminus j} p_{it}} - \frac{1}{(p_{ij} + \sum\limits_{t \in [1, m_i] \setminus j} p_{it})^2} \frac{\partial p_{ij}}{\partial \mathbf{w}} \tag{8}$$

And then we can find the partial derivative of $\mathcal{L}_{insij}$ with respect to $\mathbf{w}$ is

$$\frac{\partial \mathcal{L}_{insij}}{\partial \mathbf{w}} = -[\text{sgn}(p_{ij} - p)\mathbf{w}^T\mathbf{x}_{ij} < m].\text{sgn}(p_{ij} - p)\mathbf{x}_{ij} \quad (9)$$

where $[\text{sgn}(p_{ij} - p)\mathbf{w}^T\mathbf{x}_{ij} < m]$ can only be one or zero which mean its argument is true or false. We update the weight vector using a varied learning rate $s_t = \frac{1}{t\lambda}$, and then $\mathbf{w}_{t+1} = \mathbf{w}_t - s_t \cdot \frac{\beta}{n} \cdot \Delta d$, where Δd is the partial derivative of the objective function Eq. (3) with respect to $\mathbf{w}$.

5 Iterative Weakly Supervised Training

We initialize positive and negative windows with Edgebox. In each iteration from 1 to T, we divide all instances in one bag into 10 folders and select top 50% instances in all folders according to scores provided by Edgebox. This procedure can reduce the time of training and do not decrease accuracy a lot. After the training, re-localizing positive windows with this detector. We perform hard-negative mining on some instances that are hard to classify. The variable T is set by experience. A hard negative is that when we get a bunch of false positives, we correct them to negatives and add them to training set. When we retrain our classifier, it should perform better with this extra knowledge, and not make as many false positives.

The whole procedure is shown below:

Algorithm 1 —Iterative weakly supervised training

1. Initialization: capture plenty of windows as object proposals.
2. For iteration t = 1 to T
   a) Divide all instances in one bag into 10 folders and select top 50% instances in all folders according to scores provided by Edgebox
   b) Train the model with all datasets.
   c) Re-localize positive windows using this detector
   d) Perform hard-negative mining using this detector
3. Return final detector and object windows in train data

6 Results

In this section, we perform some experiments on MIL benchmarks with our method and compare our results with other state-of-the-art methods. Experiments are carried out on a desktop machine with Intel core i7-3770 CPU and 16 GB RAM. Given a set of images, we adopt SLIC to simplify the image as superpixels and utilize Edgebox to propose plenty of windows as instance proposals. This strategy turns the object discovery problem into a well-defined MIL problem in which an image is a bag, an object proposal is an instance, and the label of image is used as the label of a bag. We extract shape, color and texture features of the instance as input of our model we proposed before. All features are extracted from original images not images processed after superpixel. Furthermore, we divide all instances in one bag into 10 folders and select top 50% instances in all folders according to scores provided by Edgebox. The model is trained by the iterative weakly supervised training mentioned in Sect. 5. At last, after the model is learnt, we report the object proposals with maximal value predicted by our model as the detected object. The results evaluated via CorLoc measure, which is the percentage of the correct location of objects under the Pascal criteria (intersection over union (IoU) > 0.5 between detected bounding boxes and the ground truth) (Fig. 1).

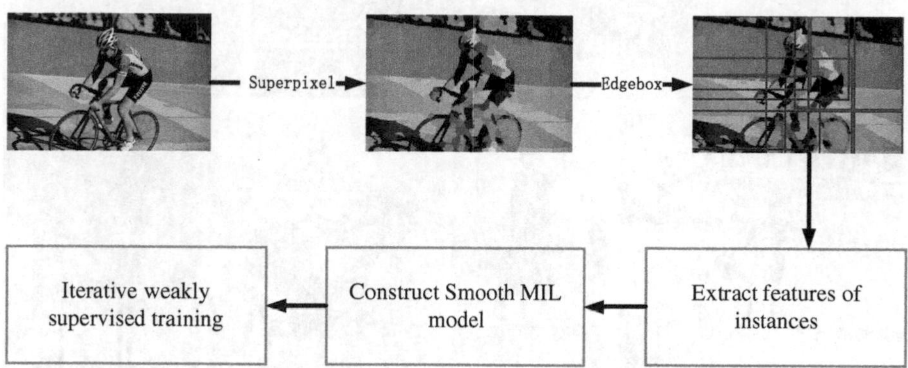

Fig. 1. The pipeline of our training procedure. The first picture is the original picture; the second one is the result of SLIC super-pixel division and the third one is the result of instance proposals provided by Edgebox.

6.1 Pascal

Pascal provides standard image data sets for object class recognition and the Pascal 2006 and 2007 datasets are extremely popular and challenging in computer vison. Two subsets are taken from Pascal 2006 and 2007 train + val dataset, which are divided into various of class and view combinations following the protocol of [24]. There are 2047 images divided into 45 class combinations in Pascal07 while total 2184 images from 33 class in Pascal06. We select all images that contain at least one object instance not marked as truncated or difficult in the ground truth.

The parameters are given via $T = 2000$, $\lambda = 0.0017$, $\beta = 5$, $p = 0.5$ and $m = 1.5$ on Pascal07 dataset, while $T = 2000$, $\lambda = 0.0017$, $\beta = 7$, $p = 0.5$ and $m = 1.48$ on Pascal06 dataset. The detectors and CorLoc are shown in Fig. 2 and Table 1, respectively.

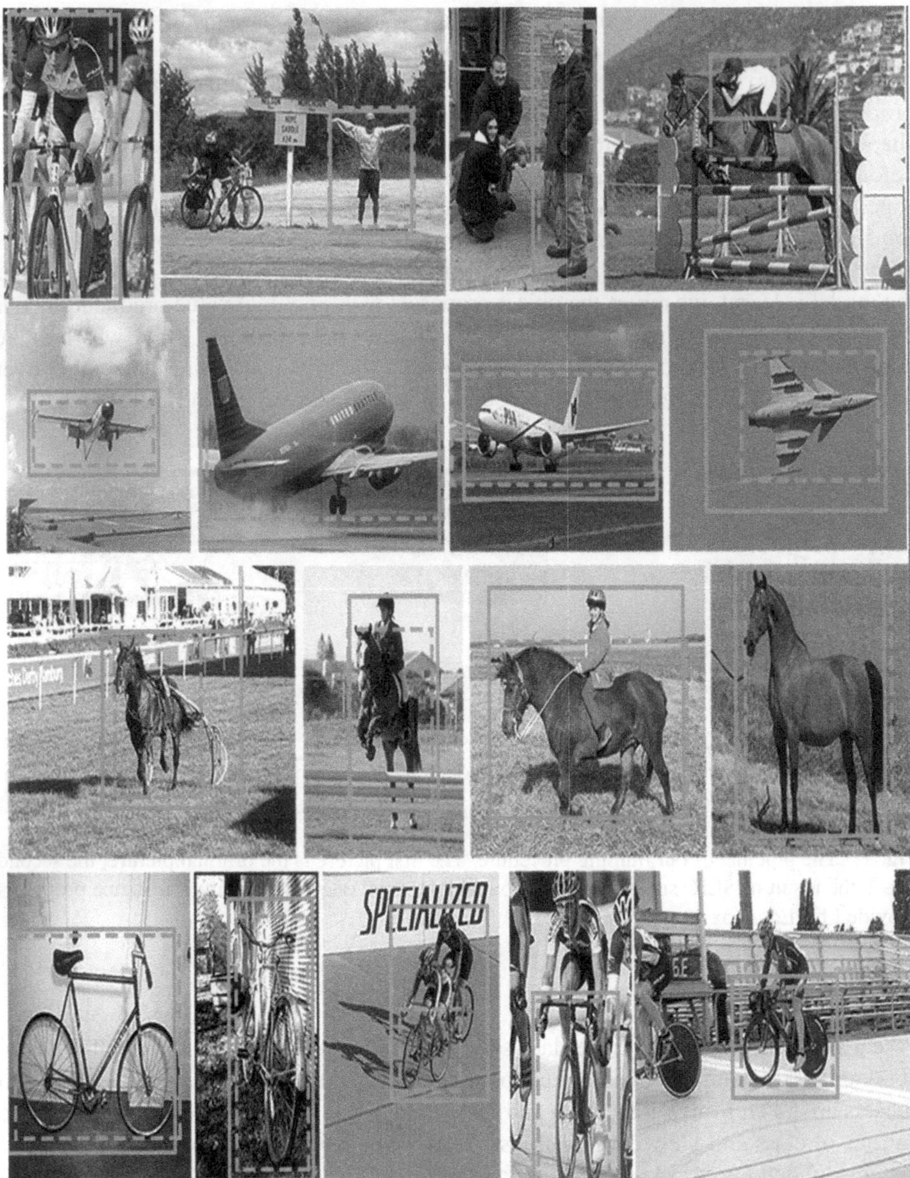

Fig. 2. Results of object discovery on several class on Pascal07 set. Each row denotes one class. These classes are, from top to people, aeroplanes, horses and bicycles. The solid green rectangle denotes the matched ground truth; the dashed green rectangle denotes the matched detection. (Color figure online)

Table 1. Object localization results evaluated via CorLoc on Pascal06 and Pascal07.

Dataset	Ours	Multi-fold	ADMM [1]	MIForests [2]	bMCL [5]
Pascal06	**52**	/	43	36	45
Pascal07	**37**	37	27	25	31

6.2 Drug Activation Prediction

Some molecules with particular shapes can bind well to a target protein and these molecules are very meaningful in pharmacy. Our task is to predict whether a new molecule can bind well to the target protein. Normally, a molecule always exhibits multiple shapes and a good molecule will bind well if at least one of its shapes is right, while a poor molecule will not bind well if none of its shapes can bind. Therefore, our task can be regarded as a MIL problem.

Musk dataset is a popular benchmark for drug prediction. The goal is to learn to predict whether new molecules will be musks or non-musks. Both MUSK1 and MUSK2 are composed of molecules (bags) in multiple conformations (instances). MUSK1 describes a set of 92 molecules of which 47 are judged by human experts to be musks and the remaining 45 molecules are judged to be non-musks and MUSK2 describes a set of 102 molecules of which 39 are judged by human experts to be musks and the remaining 63 molecules are judged to be non-musks.

We set $T = 2000$, $\lambda = 0.04$, $\beta = 1.8$, $p = 0.5$, $m = 0.5$ in our model and our results are compared with miSVM and MISVM proposed in [19] in Table 2. All three methods adopt linear kernel function. We can find from the result that the accuracy of our method improved about 10% when comparing with miSVM in MUSK2 dataset. Furthermore, on MUSK1 dataset, miSVM, MISVM and our method achieves a similar accuracy.

Table 2. Average prediction accuracy (%) on MUSK datasets.

Dataset	Ours	miSVM	MISVM
MUSK1	**80.4**	78.0	80.4
MUSK2	**82.7**	70.2	77.5

7 Conclusions

In this paper, we propose a novel method for solving MIL problem and apply it for object localization. In typical MIL problem, the labels of bag and instance are discrete but we smooth this problem into a convex problem and solve it efficiently. Our model is trained in a multi-fold way and the features of instances in image are extracted with color, shape and texture. Beside of object discovery, other recognition tasks, such as image classification, text categorization and automatic image annotation can also be solved by this method.

Acknowledgments. This work was supported by the grants of the National Science Foundation of China, Nos. 61520106006, 31571364, U1611265, 61532008, 61672203, 61402334, 61472282, 61472280, 61472173, 61572447, 61373098 and 61672382, China Postdoctoral Science Foundation Grant, Nos. 2016M601646.

References

1. Wang, X., Zhang, Z., Ma, Y., Bai, X., Liu, W., Tu, Z.: Robust subspace discovery via relaxed rank minimization. Neural Comput. **26**(3), 611–635 (2014)
2. Leistner, C., Saffari, A., Bischof, H.: MIForests: multiple-instance learning with randomized trees. In: Daniilidis, K., Maragos, P., Paragios, N. (eds.) ECCV 2010. LNCS, vol. 6316, pp. 29–42. Springer, Heidelberg (2010). doi:10.1007/978-3-642-15567-3_3
3. Chen, X., Shrivastava, A., Gupta, A.: Enriching visual knowledge bases via object discovery and segmentation. In: Proceedings of the IEEE Conference on Computer Vision and Pattern Recognition, pp. 2027–2034 (2014)
4. Tang, K., Joulin, A., Li, L.-J., Fei-Fei, L.: Co-localization in real-world images. In: 2014 IEEE Conference on Computer Vision and Pattern Recognition, pp. 1464–1471. IEEE (2014)
5. Zhu, J.-Y., Wu, J., Xu, Y., Chang, E., Tu, Z.: Unsupervised object class discovery via saliency-guided multiple class learning. IEEE Trans. Pattern Anal. Mach. Intell. **37**(4), 862–875 (2015)
6. Dietterich, T.G., Lathrop, R.H., Lozano-Pérez, T.: Solving the multiple instance problem with axis-parallel rectangles. Artif. Intell. **89**(1), 31–71 (1997)
7. Wu, J., Yu, Y., Huang, C., Yu, K.: Deep multiple instance learning for image classification and auto-annotation. In: Proceedings of the IEEE Conference on Computer Vision and Pattern Recognition, pp. 3460–3469 (2015)
8. Wei, X.-S., Wu, J., Zhou, Z.-H.: Scalable multi-instance learning. In: 2014 IEEE International Conference on Data Mining, pp. 1037–1042. IEEE (2014)
9. Oquab, M., Bottou, L., Laptev, I., Sivic, J.: Is object localization for free?-weakly-supervised learning with convolutional neural networks. In: Proceedings of the IEEE Conference on Computer Vision and Pattern Recognition, pp. 685–694 (2015)
10. Song, H.O., Girshick, R.B., Jegelka, S., Mairal, J., Harchaoui, Z., Darrell, T.: On learning to localize objects with minimal supervision. In: ICML, pp. 1611–1619 (2014)
11. Chen, Y., Wang, J.Z.: Image categorization by learning and reasoning with regions. J. Mach. Learn. Res. **5**, 913–939 (2004)
12. Sapienza, M., Cuzzolin, F., Torr, P.H.: Learning discriminative space–time action parts from weakly labelled videos. Int. J. Comput. Vis. **110**(1), 30–47 (2014)
13. Hong, R., Wang, M., Gao, Y., Tao, D., Li, X., Wu, X.: Image annotation by multiple-instance learning with discriminative feature mapping and selection. IEEE Trans. Cybern. **44**(5), 669–680 (2014)
14. Achanta, R., Shaji, A., Smith, K., Lucchi, A., Fua, P., Süsstrunk, S.: Slic superpixels (2010)
15. Zitnick, C.L., Dollár, P.: Edge boxes: locating object proposals from edges. In: Fleet, D., Pajdla, T., Schiele, B., Tuytelaars, T. (eds.) ECCV 2014. LNCS, vol. 8693, pp. 391–405. Springer, Cham (2014). doi:10.1007/978-3-319-10602-1_26
16. Maron, O., Lozano-Pérez, T.: A framework for multiple-instance learning. In: Advances in Neural Information Processing Systems, pp. 570–576 (1998)
17. Zhang, Q., Goldman, S.A.: EM-DD: an improved multiple-instance learning technique. In: Advances in Neural Information Processing Systems, pp. 1073–1080 (2001)

18. Xu, X., Frank, E.: Logistic regression and boosting for labeled bags of instances. In: Dai, H., Srikant, R., Zhang, C. (eds.) PAKDD 2004. LNCS, vol. 3056, pp. 272–281. Springer, Heidelberg (2004). doi:10.1007/978-3-540-24775-3_35
19. Andrews, S., Tsochantaridis, I., Hofmann, T.: Support vector machines for multiple-instance learning. In: Advances in Neural Information Processing Systems, pp. 561–568 (2002)
20. Zhou, Z.-H., Sun, Y.-Y., Li, Y.-F.: Multi-instance learning by treating instances as non-iid samples. In: Proceedings of the 26th Annual International Conference on Machine Learning, pp. 1249–1256. ACM (2009)
21. Berg, T.L., Berg, A.C., Edwards, J., Maire, M., White, R., Teh, Y.-W., Learned-Miller, E., Forsyth. D.A.: Names and faces in the news. In: 2004 Proceedings of the 2004 IEEE Computer Society Conference on Computer Vision and Pattern Recognition, CVPR, vol. 842, pp. II-848–II-854. IEEE (2004)
22. Felzenszwalb, P.F., Girshick, R.B., McAllester, D., Ramanan, D.: Object detection with discriminatively trained part-based models. IEEE Trans. Pattern Anal. Mach. Iintell. **32**(9), 1627–1645 (2010)
23. Gu, C., Arbeláez, P., Lin, Y., Yu, K., Malik, J.: Multi-component models for object detection. In: Fitzgibbon, A., Lazebnik, S., Perona, P., Sato, Y., Schmid, C. (eds.) ECCV 2012. LNCS, vol. 7575, pp. 445–458. Springer, Heidelberg (2012). doi:10.1007/978-3-642-33765-9_32
24. Deselaers, T., Alexe, B., Ferrari, V.: Weakly supervised localization and learning with generic knowledge. Int. J. Comput. Vis. **100**(3), 275–293 (2012)
25. Wang, X.-F., Huang, D.S., Xu, H.: An efficient local Chan-Vese model for image segmentation. Pattern Recogn. **43**(3), 603–618 (2010)
26. Huang, D.S.: Systematic Theory of Neural Networks for Pattern Recognition (in Chinese). Publishing House of Electronic Industry of China, Beijing, May 1996
27. Li, B., Huang, D.S.: Locally linear discriminant embedding: An efficient method for face recognition. Pattern Recogn. **41**(12), 3813–3821 (2008)
28. Huang, D.S., Du, J.-X.: A constructive hybrid structure optimization methodology for radial basis probabilistic neural networks. IEEE Trans. Neural Networks **19**(12), 2099–2115 (2008)
29. Huang, D.S.: Radial basis probabilistic neural networks: model and application. Int. J. Pattern Recogn. Artif. Intell. **13**(7), 1083–1101 (1999)
30. Wang, X.-F., Huang, D.S.: A novel density-based clustering framework by using level set method. IEEE Trans. Knowl. Data Eng. **21**(11), 1515–1531 (2009)
31. Huang, D.S., Ip, H.H.S., Chi, Z.: A neural root finder of polynomials based on root moments. Neural Comput. **16**(8), 1721–1762 (2004)

A Novel Method for Generating Benchmark Functions Using Recurrent Neural Network

Fengyang Sun, Lin Wang(✉), Bo Yang, Jin Zhou, and Zhenxiang Chen

Shandong Provincial Key Laboratory of Network Based Intelligent Computing, University of Jinan, Jinan 250022, China
wangplanet@gmail.com

Abstract. In recent years numerous evolutionary algorithms have been proposed to optimize multi–modal problems. These algorithms test the performance by benchmark functions for simulating real-world problems. However, the benchmark functions don't have enough similarity and complexity compared to real world. Thus, Recurrent Benchmark Generator (RBG) is proposed in this paper to generate complex and different benchmark functions. This generator obtains a mass of modals by recurrent neural network, which are added various fluctuations of normal benchmark functions to keep a balance between complexity and gradient. The experimental results indicate that the novel approach produces more complex benchmark functions which are more conformed to real world problems.

Keywords: Benchmark function · Recurrent neural network · Random · Probability density

1 Introduction

There has been a growing interest in studying soft computing in recent years due to its fault tolerance and robustness. Soft computing, including Artificial Neural Network, Genetic Algorithm [1], etc. solves problems effectively by simulating the operating mechanism of natural intelligent systems like human sense, evolution, etc. These evolutionary algorithms (EA) [2–5] have been widely utilized in various applications due to the ability of self-adaption, self-organization and self-learning. All the EAs may face a problem in premature convergence into local minima, thus benchmark functions are required to test the performance of them.

However, existing benchmark functions are not similar enough to the landscape of real world problems. Although the algorithms to generate benchmark functions have been developed and improved in recent years, these approaches [6–9] are too simple and have too many repeated structures, which is unappeasable for testing of EAs.

In view of the problem, we proposed Recurrent Benchmark Generator (RBG), which uses Recurrent Neural Network (RNN) to generate a mass of peaks and valleys. Because the weights of the recurrent neural network are generated randomly in certain probability density, theoretically it can produce infinite variety of complex benchmark functions. Moreover, the obstacles and difficulties of benchmark functions: dimensionality, multimodality, deceptiveness, bias and flatness [10] are well solved.

D.-S. Huang et al. (Eds.): ICIC 2017, Part I, LNCS 10361, pp. 768–773, 2017.
DOI: 10.1007/978-3-319-63309-1_68

The rest of the paper is organized as follows. Section 2 presents the details and analyzes RBG. Then some experiments are carried out to present difficulty of RBG in Sect. 3. Finally, Sect. 4 concludes the paper.

2 Methodology

Recurrent Benchmark Generator (RBG) is a totally different approach to generate benchmark functions. Firstly, the algorithm generates a one-dimension weight array randomly based on neurons' number in a certain probability density to put into the RNN. Secondly, after getting the network, input the data, run iteration part based on the recurrent steps. Then, compute the weighted output result N_{out}. Finally, add fluctuations of normal benchmark function to N_{out} as the final fitness.

2.1 Recurrent Neural Network

Classical feed-forward neural network, including input layer, hidden layer and output layer, is fully-connected between layers and non-connected inside each layer. Other than the classical neural network the Recurrent Neural Network (RNN) [11] introduces recurrence, which is that the neurons in hidden layer is connected and the inputs of hidden layer not only include the outputs of the input layer but also include the outputs of the hidden layer in the last iteration (clock).

The RNN we use is a bit similar to Echo State Network, which varies widely from classical RNNs. There are 3 characteristics of it:

- Its nuclear structure is a reservoir which is generated randomly and remains the same. The reservoir is a large-scale, random and sparse-connected (we change it to fully-connected, but small weights are more and large weights are less based on probability density) iterative structure.
- The weight matrix from reservoir to output layer is the only part to be adjusted.
- Simple liner regression can finish the training of network (we don't need to do this part due to have set the distribution of weights in advance).

The weight range is expanded from original $\left[-\sqrt{\frac{2}{\pi}}, \sqrt{\frac{2}{\pi}}\right]$ to $[-6, 6]$ for magnifying the extent of variation. The classical activation function Eq. (1) is Sigmoid function:

$$f(x) = \frac{1}{1 + e^{-x}} \tag{1}$$

Its expanded version Eq. (2) is presented as follows:

$$f(x) = \frac{2}{1 + e^{-x}} - 1 \tag{2}$$

We change the bounded function to a new unbounded activation function Eq. (3) to make more complex landscape, noted Bi-Log function:

$$f(x) = \begin{cases} -\ln(1-x), & x < 0 \\ \ln(1+x), & x \geq 0 \end{cases} \quad (3)$$

2.2 Probability Density

Probability Density Function (PDF) is a function that describes the probability of the output value of a random variable nearby a certain point. For instance, the random variable *X*'s probability density value is 0.798 when *X* is equal to 0, which means the probability of *X*= 0 is 0.798. In other word, if we generate a bunch of random numbers, the number of nearby the point 0 in proportion to amount random numbers will be approximately 0.798. The final distribution of network's weights is determined by the above definition, which is the Eq. (4):

$$f(x) = \sqrt{r^2 - x^2}, x \in \left[-\sqrt{\frac{2}{\pi}}, \sqrt{\frac{2}{\pi}}\right], r = \sqrt{\frac{2}{\pi}} \quad (4)$$

3 Experiments

To examine the complexity of the proposed RBG algorithm, several experiments are carried out and the results are added the fluctuations of other well-known existing benchmark functions. All experiments were conducted in MATLAB 8.4 using the same

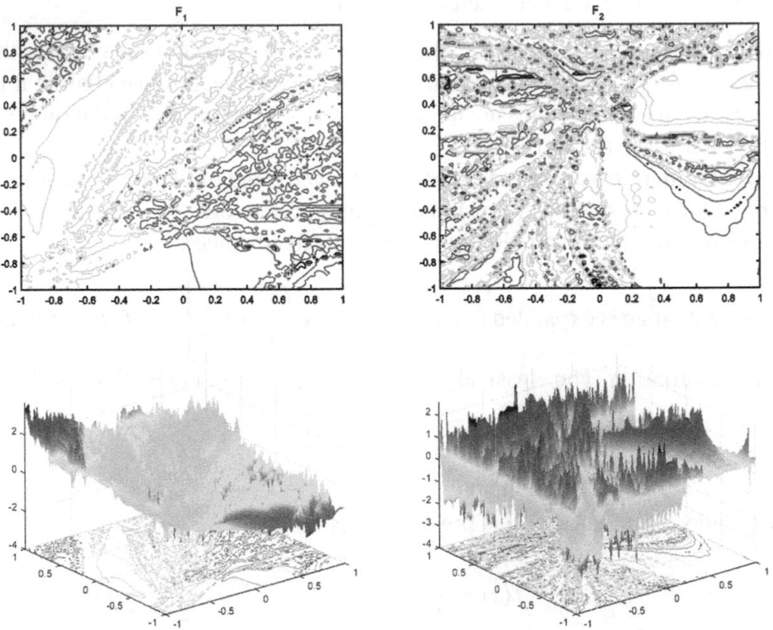

Fig. 1. Contour maps and matching 3-D colored surface maps for two different 2-D functions (Bi-Log) plus fluctuations of shifted sphere and shifted ackley

machine with a Core (TM) i5-3230 M 2.60 GHz CPU, 8 GB of Ram, and Windows 7 ultimate operating system.

3.1 Analysis and Discussion

Experiments were conducted to present RBG plus the fluctuations of 4 benchmark functions [7], which are shifted sphere, shifted ackley, shifted rotated rastrigin and shifted rotated griewank without bounds.

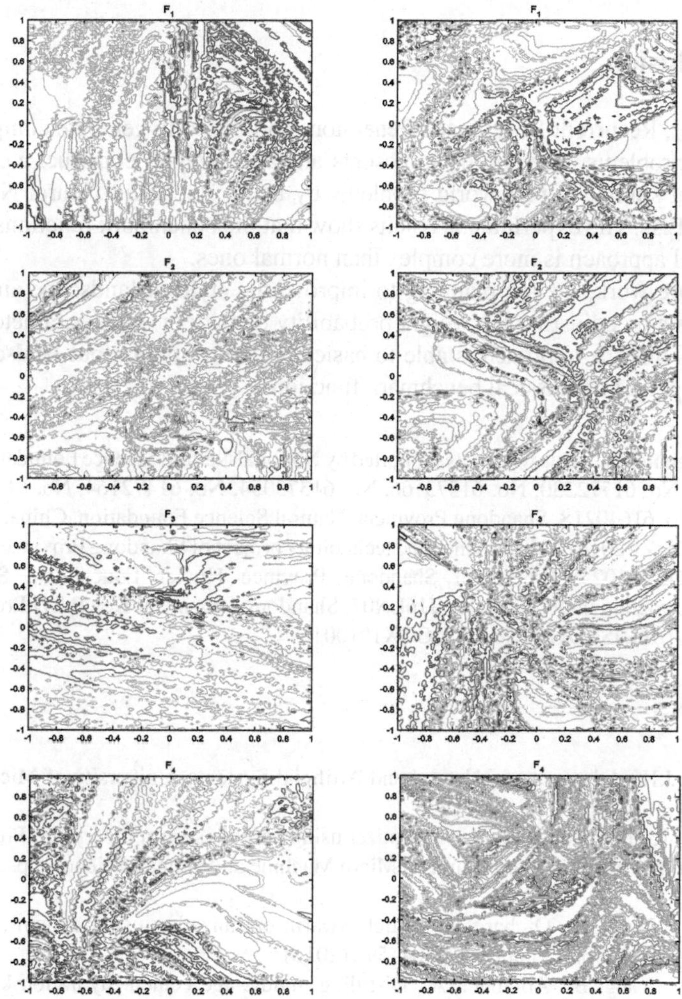

Fig. 2. Contour maps for 8 different 2-D functions plus fluctuations of shifted sphere, shifted ackley, shifted rotated rastrigin and shifted rotated griewank without bounds, Bi-Log as the activation function

Figure 1 presents the contour map and 3-D surface map for two different functions with fluctuations of shifted sphere and shifted ackley. As can be seen from the figure, the novel approach obviously generates more local optima. It also generates some deceptive areas which reveal misleading information of global optimum. The blank part in the contour map illustrates flatness of the landscape, which provides very little information of finding possible best solution.

Figure 2 illustrates contour maps of several RBG functions with different activation function. The Bi-Log-version benchmark functions in Fig. 2 generate more complex, chaotic and fickle landscape.

4 Conclusion

In this paper, Recurrent Benchmark Generator is proposed to generate complex multi-modal test problems. The approach presents an easier way to construct a challenging environment similar to real world problems by using Recurrent Neural Network for comparing EAs. The experimental results show that the benchmark functions generated by the novel approach is more complex than normal ones.

For future work, we are planning to improve the style of landscape and generate larger modals by adjusting the weight probability density or other parameters. One of our goals is that the approach is able to basically control the output landscape of the network to get rid of classical benchmark functions.

Acknowledgements. This work was supported by National Natural Science Foundation of China under Grant No. 61572230, No. 61573166, No. 61373054, No. 61472164, No. 61472163, No. 61672262, No. 61640218, Shandong Provincial Natural Science Foundation, China, under Grant ZR2015JL025, ZR2014JL042. Science and technology project of Shandong Province under Grant No. 2015GGX101025, Project of Shandong Province Higher Educational Science and Technology Program under Grant no. J16LN07. Shandong Provincial Key R&D Program under Grant No. 2016ZDJS01A12, No. 2016GGX101001.

References

1. Holland, J.H.: Adaptation in Natural and Artificial Systems. University of Michigan Press, Ann Arbor (1975)
2. Eberhart, R., Kennedy, J.: A new optimizer using particle swarm theory. In: Proceedings of the Sixth International Symposium on Micro Machine and Human Science, pp. 39–43. IEEE (1995)
3. Wang, L., Yang, B., Orchard, J.: Particle swarm optimization using dynamic tournament topology. Appl. Soft Comput. **48**, 584–596 (2016)
4. Wang, L., Yang, B., Abraham, A.: Distilling middle-age cement hydration kinetics from observed data using phased hybrid evolution. Soft. Comput. **20**(9), 3637–3656 (2016)
5. Wang, L., Yang, B., Chen, Y., Zhang, X., Orchard, J.: Improving neural-network classifiers using nearest neighbor partitioning. IEEE Trans. Neural Netw. Learn. Syst. (2016, in Press). doi:10.1109/TNNLS.2016.2580570

6. Qu, B.Y., Liang, J.J., Wang, Z.Y., Chen, Q., Suganthan, P.N.: Novel benchmark functions for continuous multimodal optimization with comparative results. Swarm Evol. Comput. **26**, 23–34 (2016)
7. Suganthan, P.N., Hansen, N., Liang, J.J., Deb, K., Chen, Y.P., Auger, A., Tiwari, S.: Problem definitions and evaluation criteria for the CEC 2005 special session on real-parameter optimization. KanGAL report, 2005005 (2005)
8. Li, T., Rogovchenko, Y.V.: Oscillation criteria for even-order neutral differential equations. Appl. Math. Lett. **61**, 35–41 (2016)
9. Li, T., Rogovchenko, Y.V.: Oscillation of second-order neutral differential equations. Math. Nachr. **288**(10), 1150–1162 (2015)
10. Mirjalili, S., Lewis, A.: Obstacles and difficulties for robust benchmark problems: a novel penalty-based robust optimisation method. Inf. Sci. **328**, 485–509 (2016)
11. Pineda, F.J.: Generalization of back-propagation to recurrent neural networks. Phys. Rev. Lett. **59**(19), 2229 (1987)

Optimization of Neural Tree Based on Good Point Set

Hao Teng[1(✉)], Yuehui Chen[2], and Shixian Wang[1]

[1] School of Information Science and Engineering, University of Jinan, Jinan 250022, China
ise_tengh@ujn.edu.cn

[2] Shandong Provincial Key Laboratory of Network Based Intelligent Computing, Jinan 250022, China

Abstract. Particle swarm optimization algorithm is often used in the optimization of parameters in the model of flexible neural tree. Using the principle of good point set in number theory, a novel algorithm of creating good point is presented to modify initial population setting, which is combined with the disturbance, so as to be used to overcome the shortcomings of particle swarm optimization algorithm, such as being slow, easily trapped in local optimal solution. In this paper, flexible neural tree, using hyperbolic tangent as transfer function, with the novel particle swarm optimization algorithm is applied to provide a reliable and effective forecast framework for stock market indices. The Nasdaq-100 index of Nasdaq Stock Market and the S&P CNX NIFTY stock index are analyzed. Several years of Nasdaq 100 main-index values and NIFTY index values are considered using several intelligent computing techniques. The result shows that the proposed algorithm could represent the stock indices behavior very accurately.

Keywords: Flexible neural tree · Particle swarm optimization · Good point set · Hyperbolic tangent

1 Introduction

Flexible neural tree [1–7] means a novel forward artificial neural network, which is presented by Chen. In this model, instead of a fully connected NN classifier, the connection between layers should not have to be completely. Connection cross between layers is allowed. According to the applied environment, this kind of flexible neural tree can reduce or select the features or the function of the input variable automatically, and nonlinear neurons of hidden and output layer have different transfer function, there into it the nonlinear neurons of hidden is variable.

There are two important steps to build the model of FNT. The structures of FNT are usually optimized by probabilistic incremental program evolution (PIPE) [8], and parameter by particle swarm optimization algorithm (PSO) [9, 10]. Due to the traditional particle swarm optimization algorithm has some shortcomings, such as slow and easy to trapped into local optimal solution and so on, this paper presents an improved

D.-S. Huang et al. (Eds.): ICIC 2017, Part I, LNCS 10361, pp. 774–785, 2017.
DOI: 10.1007/978-3-319-63309-1_69

particle swarm optimization algorithm based on good-point set and used it to optimize the parameters of FNT.

As an application, we picked the Nasdaq-100 index from 11 January 1995 to 11 January 2002 and the NIFTY index from 01 January 1998 to 03 December 2001 and analyzed the trend.

All data is roughly divided into two equal parts. For the two indexes, there are no other special rules when selecting the training set except guarantying a reasonable parameter space for the representation of the problem domain.

The rest of this paper is as follows. A brief introduce about the good point set theory is given in Sect. 2. A basically introduce about the new particle swarm optimization and PIPE is given in Sects. 3 and 4. In Sect. 5 the FNT and the FNT based on the new particle swarm optimization has been briefed. Section 6 explains the experiments about the stock indices of that has been carried out and others. Finally, Sect. 7 shows the conclusions of all works.

2 Good-Point Set

2.1 Good Point Generation

Let $\gamma = f(\gamma_1, \ldots, \gamma_s) \in c^s$.

If p_n forms the first n points of

$$\gamma = \{(\{\gamma_1 k\}, \cdots, \{\gamma_s k\}), k = 1, 2 \cdots\} \tag{1}$$

Where $\{\gamma\}$ denotes the fractional part of γ, with discrepancy $D(n, p_n) = o(\mathrm{n}^{-1+\varepsilon})$ as $n \to \infty$, then the set p_n is called a good point set (*gp* set) and γ is a good point.

For simplicity, the method using this good set is called *gp* method [11, 12]. The following are some useful good points:

Let $p_1, \ldots, p_s$ be the first s primes. Take

$$r = \left(\sqrt{p_1}, \cdots, \sqrt{p_s}\right) \tag{2}$$

Let p be a prime and $\mathrm{p} \geq 2\mathrm{s} + 3$. Take

$$r = \left(\left\{2\cos\frac{2\pi}{p}\right\}, \left\{2\cos\frac{4\pi}{p}\right\}, \cdots, \left\{2\cos\frac{2\pi s}{p}\right\}\right) \tag{3}$$

2.2 Advantages of Good Point Set

If the good point is used to calculate the approximate integral, the order of the error is only related to the number of the points and has nothing to do with the dimension t of the space, which provides a very good algorithm for the approximate calculation of high dimension [13, 14].

Zhang [13] uses a random method to get n points for an unknown distribution objects, and the deviation is $O(n^{-1/2})$. Using good point set method takes n points, the deviation is $O(n^{(-1+E)})$, the difference between the square times. If $n = 10000$, then the deference is between $O(10^{-2})$ and $O(10^{-4})$. That is why the good point set method converges faster.

3 PSO and Its Improvements

3.1 Standard PSO

Firstly, n samples of the population are generated in the initial generation randomly. Then the flying of each particle are decided by its own flying experience and its companions flying experience in PSO. The ith particle is represented as: $x_i = (x_{i1}, x_{i2}, \cdots, x_{id})^T, i = 1, 2, \cdots, m$, the best previous position of any particle is recorded and represented as $p_{ibest} = (p_{i1}, p_{i2}, \cdots, p_{id})^T$. g_i is the position of the best particle in all the particles among the population (means global best position, *gbest*). The velocity for particle i^{th} is $v_i = (v_{i1}, v_{i2}, \cdots, v_{id})^T$. The speed and position of the particles are described as Eqs. 4 and 5:

$$v_{ik} = \omega v_{ik} + c_1 rand_1 (p_{ik} - x_{ik}) + c_2 rand_2 (g_k - x_{ik})\, i = 1, \cdots, m;\ k = 1, \cdots d \tag{4}$$

$$x_{ik+1} = x_{ik} + v_{ik+!}\, i = 1, \cdots, m;\ k = 1, \cdots d \tag{5}$$

where $c_1, c_2 > 0$, $rand_1$ and $rand_2$ are two random functions which are in the range [0,1].

3.2 PSO with Gaussian Disturbance

Because of the nature of PSO algorithm, all the particles will converge to a collective point the particles rarely search further before the next iteration and the diversity of population is not enough. In response to it, an improved step is to add Gaussian disturbance in PSO algorithm so as to improve the diversity and performance of the standard algorithm [15].

Three strategies can often be used as follow:

(A) Focus on the average of the best position, join the disturbance with obey Gauss distribution in the average of the best position.
(B) Focus on the global best position, join the disturbance with obey Gauss distribution in the global best position.
(C) Comprehensive consideration the both, join the disturbance with obey Gauss distribution in both A and B.

The strategy A or strategy B is usually adopted.

Here, the Gaussian disturbance can be joined in the average of the best position, as Eq. 6, where e is a preset parameter, R is a random numbers which satisfy the Gaussian disturbance, set $e = 0.005$.

$$Mbest = mbest + e * R \tag{6}$$

3.3 The Improved Particle Swarm Optimization

The new improving PSO algorithm is combined with the good point and disturbance which have been described as preamble, the particular steps show as follow.

(1) Create an initial population by good point set, as Eq. 6; n samples are generated based on the good point set according to the distribution of each dimension instead of randomly initialized population. It can make the initialized population distribution more uniform.
(2) Evaluation: calculate the fitness of each particle according to the fitness function;
(3) Find the Pbest: Compare the fitness of each particle to search the best previous position of each particle, Pbest.
(4) Find the Gbest: Compare the fitness of all the particles to search the global best position of the population, Gbest.
(5) Update the Velocity: updated the speed and position of the particles.
(6) If satisfactory solution is found, the search procedure stopped; otherwise go to (2).

4 Probabilistic Incremental Program Evolution

Probabilistic incremental program evolution is adopted to get a better tree structure in FNT model here. Its proposed to find an optimal or near-optimal neural tree. A basic flow of PIPE is given as below which is based on Population based Incremental Learning (PBIL).

(1) Initialize probabilistic prototype tree;
(2) Repeat until termination condition is met:
    (2.1) Create the population of programs. The probabilistic prototype tree can be grows if required;
    (2.2) Evaluate population. Smaller programs is preference when all sub trees is equal;
    (2.3) Update and mutate probabilistic prototype tree;
    (2.4) Prune probabilistic prototype tree;
(3) If a better structure has been found, then return, otherwise go to (2).

5 FNT Based on the Improved Particle Swarm Optimization

5.1 Instructions of FNT

FNT and its function instruction set and terminal instruction set can be represented as follows [1–3]:

Terminal instruction set is: $T = \{x_1, x_2, \ldots, x_n\}$, where $+_i\ (i = 2, 3, \ldots, n)$, Function instruction set is: $F = \{+_2, +_3, \ldots, +_N\}$. It means i arguments is taken. Inputs, $x_1, x_2, \ldots, x_n$ are the instructions of leaf nodes, instructions, $+_i$ are non-leaf nodes instructions. The instruction $+_i$ is also called a flexible neuron operator which has i inputs. In the creation process of neural tree, when a non-terminal instruction being selected, i real values are generated randomly and used to represent the connection strength between the node $+_i$ and its children.

5.2 Transfer Function

Gauss's kernel function, Sigmoid, hyperbolic tangent can be used as transfer function.

Papers usually take Gauss's kernel function as transfer function. Two adjustable parameters a_i and b_i are randomly created as flexible activation function parameters, the following flexible activation function is usually used as Eq. 7:

$$f(a_i, b_i, x) = \exp\left(-\left(\frac{x - a_i}{b_i}\right)^2\right) \quad (7)$$

The output of a flexible neuron $+_n$ can be calculated as Eq. 8. The total excitation of $+_n$ as:

$$net_n = \sum_{j=1}^{N} w_j * x_j \quad (8)$$

Where $x_j (j = 1, 2, \cdots, n)$ are the inputs to node $+_n$. The output of the node $+_n$ can be calculated by Eq. 9.

$$out_n = f(a_i, b_i, net_n) = \exp\left(-\left(\frac{net_n - a_i}{b_i}\right)^2\right) \quad (9)$$

A FNT and its function instruction set and terminal instruction set is as Fig. 1.

In our idea, FNT is essentially a kind of a forward artificial neural network, can choose one of Sigmoid and hyperbolic tangent. In order to determine the function for the researches are analyzed and compared respectively on the two functions. They have several different: Sigmoid can only take positive values, hyperbolic tangent can be positive or negative. Due to the complexity of the data, if there a negative correlation between objects exists, hyperbolic tangent is better. The convergence rate is different. Within (−1, 1) the guide function of hyperbolic tangent shows faster than the guide

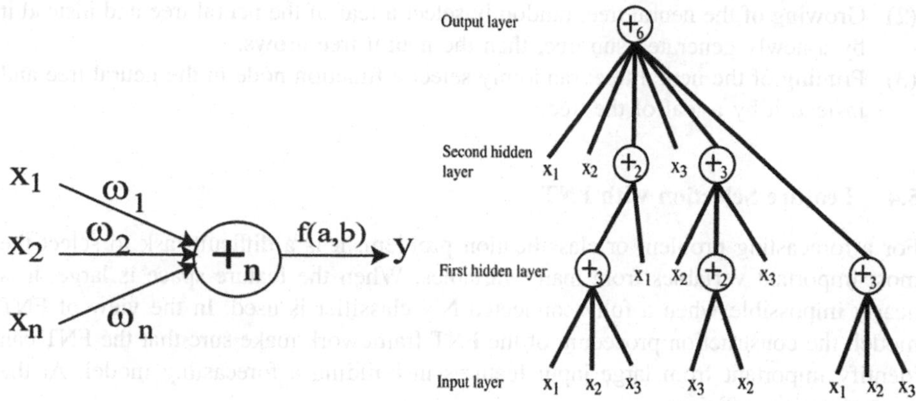

Fig. 1. A FNT and its function instruction set and terminal instruction set

function of Sigmoid function, error correction to the larger, so that the convergence speed of neural network is faster. The network convergence speed of hyperbolic tangent function than the sigmoid function, it is easier to avoid maximum of local search. Therefore, this study determined the hyperbolic tangent function for transfer function.

In this paper, the following flexible activation function is used as Eq. 10 instead of Eq. 7, the output of the node $+_n$ can be calculated by Eq. 11, only one adjustable parameters a_i is needed here:

$$f(a_i, x) == \frac{\exp(a_i * x) - \exp(-a_i * x)}{\exp(a_i * x) + \exp(-a_i * x)} \tag{10}$$

$$out_n = f(a_i, net_n) = \frac{\exp(a_i * net_n) - \exp(-a_i * net_n)}{\exp(a_i * net_n) + \exp(-a_i * net_n)} \tag{11}$$

5.3 Optimization of the Tree Structure

First an optimal neural tree or an approximate optimal neural tree must be found by PIPE. Then the neural tree which was found can be evaluated, several methods such as the crossover and selection used as same as those of standard GP. A number of neural tree mutation operators are defined as follows [1–3]:

(1) Terminal nodes Changed: Single terminal node changed or all the terminal nodes changed can be done. For a single terminal node, select a node in the tree and replace it with another terminal node randomly. For all the terminal nodes, select each terminal node in the neural tree and replace everyone with another terminal node.

(2) Growing of the neural tree: randomly select a leaf of the neural tree and instead it by a newly generated sub tree, then the neural tree grows.
(3) Pruning of the neural tree: randomly select a function node in the neural tree and instead it by a leaf of the tree.

5.4 Feature Selection with FNT

For a forecasting problem or classification problem, it is a difficult task to select the most important variablcs from many variables. When the feature space is large, it is nearly impossible when a fully connected NN classifier is used. In the view of FNT model, the construction procedure of the FNT framework make sure that the FNT can identify important from large input features in building a forecasting model. At the same time it is efficient.

The procedures of input selection in the FNT are defined as follows.

(1) Initialization. The input variables are selected to build the FNT model initially, at start the input variables are with same probabilities;
(2) Update the Probabilities. It was based on the survival of the fittest. By an evolutionary procedure the variables which shows more importance to the objective function will be changed and have high probabilities to survive in the next generation. At the same time the variables which show less importance will be changed with less probabilities.
(3) Evolution. To choice input selection methods used by the neural tree to select appropriate variables, some standard evolutionary operators can be adopted, such as cross over or mutation [1–3].

5.5 FNT Based on the Improved PSO

FNT based on the improved particle swarm optimization produced by the process as below:

(1) Create an initial population randomly;
(2) PIPE is adopted to get an optimal tree structure or an approximate optimal tree structure;
(3) If a better structure has been found, then go to (4), otherwise go to (2), the normalized mean squared error (RMSE) is adopted to find an optimal FNT;
(4) When the structure is gotten, parameter optimization is achieved by an improved PSO algorithm combined with the good point and disturbance which has been described above is used;
(5) If the maximum of local search reach, or no better parameter sets appears for enough long time then go to step (6);
(6) If satisfactory solution is found, the algorithm stopped; otherwise go to (2).

6 Experimental Results and Analysis

6.1 Empirical Comparison of RMSE Results for Several Learning Methods

The experiment is to investigate the performance analysis of the model. Here the parameters of the FNT model are optimized by the improved PSO.

Large data was analyzed including 7 years stock data for the Nasdaq-100 index and 4 years for the NIFTY index. The target was to find an efficient forecast models that could predict the index value of the following trading day according to the opening, closing, and maximum values on any given day.

For the two indexes, we divided the entire data into two parts.

In literature [2], several algorithms are used in the predicting indexes.

FNT algorithm was used with the instruction sets = $\{+_2, +_3, \ldots, +_{10}, x_0, x_1, x_2\}$ forecasting the Nasdaq-100 index and instruction sets = $\{+_2, +_3, \ldots, +_{10}, x_0, x_1, x_2, x_3, x_4\}$ forecasting the NIFTY index.

A feed-forward neural network includes three input nodes and a single hidden layer with 10 neurons was used for modeling the Nasdaq-100 index. A feed-forward neural network includes five input nodes and a single hidden layer with 10 neurons was used for modeling the NIFTY index.

WNN-Prediction Model uses the standard form of a wavelet neural network. A WNN includes three input nodes and a single hidden layer with 10 neurons was used for modeling the Nasdaq-100 index. A WNN includes five input nodes and a single hidden layer with 10 neurons was used for modeling the NIFTY index.

A Local Linear WNN Prediction Model (LLWNN) includes three input nodes and a hidden layer with five neurons was used for modeling Nasdaq-100 index. A LLWNN includes five inputs and a single hidden layer consisting of five neurons was used for forecasting NIFTY index.

Our algorithm uses a set of parameters similar to the FNT. The comparison of our method with these methods above shows below.

The Nasdaq-100 index value between 11 January 1995 and 11 January 2002 was analyzed. For this index, we divided the entire data into almost two equal parts (449 in train dataset and 450 in test dataset).

The results of experiment data are shown in Table 1, including the experimental results of different methods. The smaller RMSE means the difference between the predicted value and the true value is smaller, the higher the prediction accuracy is. According to the empirical results, in terms of RMSE values, for the Nasdaq-100 index, our new method performed better than other models such as standard FNT, FNT combined with PSO, WNN combined with PSO, LLWNN combined with PSO.

4-year for the NIFTY index, between 01 January 1998 and 03 December 2001 was analyzed. For this index, we divided the entire data into almost two equal parts (400 in train dataset and 384 in test dataset). The results of experiment data are shown in Table 2, including the experimental results of our method and others. According to the empirical results, in terms of RMSE values, for the NIFTY index, the new method greatly superior than other models such as standard FNT, FNT combined with PSO, WNN combined with PSO, LLWNN combined with PSO.

Table 1. The RMSE results of this method and others for Nasdaq-100 index

Methods	RMSE
FNT [2]	0.01882
WNN-PSO [2]	0.01789
LLWNN-PSO [2]	0.01968
Our method	0.01656

Table 2. The RMSE results of this method and others for NIFTY index

Methods	RMSE
FNT [2]	0.01428
WNN-PSO [2]	0.01426
LLWNN-PSO [2]	0.01564
FNT based on QPSO [16]	0.01377
Our method	0.009028

The parameters of experiment are showed in Table 3.

Table 3. The parameter of experiment

Parameter	Set
Function instruction set	$+_2$, $+_3$, $+_4$
Terminal instruction set	x_0, x_1, x_2, x_3
Init terminal probability	0.5
Mutation probability	0.8
Mutation rate	0.4
Number of particle	20
e	0.005

6.2 Empirical Comparison of RMSE Results for FNT Ensemble Technique

CHEN shows that the flexible neural tree (FNT) ensemble technique could be well represented used in the stock markets. Literature [17] investigates the development of novel reliable and efficient techniques to model the behavior of stock markets. The structure and parameters of FNT are optimized using genetic programming (GP) like tree structure-based evolutionary algorithm and particle swarm optimization (PSO) algorithms, respectively.

In literature [17], a basic ensemble method (BEM) and a general ensemble method (GEM) are defined.

A simple method to combining network outputs is to simply average them together as the BEM. The output of BEM is defined as Eq. 12:

$$f_{BEM} = \frac{1}{n}\sum_{i=1}^{n} f_i(x) \quad (12)$$

This approach by itself can lead to improved performance, but does not take into account the fact that some FNTs may be more accurate than others. It has the advantage that it is easy to understand and implement and can be shown not to increase the expected error.

A generalization to the BEM method (GEM) is to find weights for each output that minimize the positive and negative classification rates of the ensemble. The output of GEM is defined as Eq. 13:

$$f_{BEM} = \sum_{i=1}^{n} \alpha_i f_i(x) \quad (13)$$

where the a_i are chosen to minimize the root mean square error between the FNT outputs and the desired values. For comparison purpose, the optimal weights of the ensemble predictor are optimized by using PSO algorithm.

Literature [17] shows FNT ensemble learning paradigm could be well represented for the behavior of stock indices. Results on the two data sets using FNT ensemble models clearly reveal it is better than single FNT. In terms of RMSE values, for the Nasdaq-100 index and the NIFTY index, Our method performed better than other models. Our method is based on single FNT. For this reason, FNT ensemble learning paradigm based on our FNT Would show better than our single FNT (Table 4).

Table 4. Empirical comparison of RMSE results for four learning methods

Methods	Nasdaq-100	NIFTY
Best-FNT [17]	0.01854	0.01315
BEM [17]	0.01824	0.01258
GEM [17]	0.01635	0.01222
Our method	0.01656	0.009028

7 Conclusion

In the optimization process, the novel algorithm combined with good point set is to improve the distribution of the initial population, which makes them more uniform.

In the course of evolution, population can run out of local minima by introducing the Gaussian disturbance. The network characteristics are improved by using the hyperbolic tangent function as transfer function. So the performance of FNT is improved.

The result shows the difference between the proposed method and others in forecast models and the proposed method shows the advantage. So it will have wide prospect of the application in the future.

Acknowledgements. The research work was supported by National Natural Science Foundation of China under Grant No. 60573065.

References

1. Chen, Y., Yang, B., Dong, J.: Nonlinear system modeling via optimal design of neural trees. Int. J. Neural Syst. **14**(02), 125–137 (2004). PMID: 15112370
2. Chen, Y., Abraham, A.: Hybrid-Learning Methods for Stock Index Modeling. Artificial Neural Network Finance and Manufacturing, pp. 63–78 (2006)
3. Chen, Y., Abraham, A., Yang, B.: Feature selection and classification using flexible neural tree. Neurocomputing **70**(13), 305–313 (2006). Neural Networks Selected Papers from the 7th Brazilian Symposium on Neural Networks (SBRN '04) 7th Brazilian Symposium on Neural Networks
4. Bao, W., Chen, Y., Wang, D.: Prediction of protein structure classes with flexible neural tree. Bio-Med. Mater. Eng. **24**(6), 3797–3806 (2014)
5. Yang, B., Chen, Y., Jiang, M.: Reverse engineering of gene regulatory networks using flexible neural tree models. Neurocomputing **99**, 458–466 (2013)
6. Ammar, M., Bouaziz, S., Alimi, A.M., Abraham, A.: Recurrent flexible neural tree model for time series prediction. In: Proceedings of the 16th International Conference on Hybrid Intelligent Systems (HIS 2016), 21–23 November 2016, Marrakech, Morocco, pp. 58–67 (2016)
7. Ojha, V.K., Abraham, A., Snasel, V.: Ensemble of heterogeneous flexible neural tree for the approximation and feature-selection of poly (Lactic-co-glycolic Acid) micro- and nanoparticle. In: Abraham, A., Wegrzyn-Wolska, K., Hassanien, A.E., Snasel, V., Alimi, A.M. (eds.) Proceedings of the Second International Afro-European Conference for Industrial Advancement AECIA 2015. AISC, vol. 427, pp. 155–165. Springer, Cham (2016). doi:10.1007/978-3-319-29504-6_16
8. Salustowicz, R., Schmidhuber, J.: Probabilistic incremental program evolution. Evol. Comput. **5**(2), 123–141 (1997)
9. Eberhart, R., Kennedy, J.: A new optimizer using particle swarm theory. In: Proceedings of the Sixth International Symposium on Micro Machine and Human Science, MHS 1995, October 1995, pp. 39–43 (1995)
10. Kennedy, J.: Particle Swarm Optimization. Springer, Heidelberg (2011)
11. Wang, Y., Hua, L.K.: Applications of Number Theory to Numerical Analysis. Springer, Heidelberg (2012)
12. Yan, H., Cao, Y., Yang, J.: Statistical tolerance analysis based on good point set and homogeneous transform matrix. Procedia CIRP **43**, 178–183 (2016)
13. Zhang, L., Zhang, B.: Good point set based genetic algorithm. Chin. J. Comput. **24**(9), 917–922 (2001)
14. Liu, X., Xuan, S.-B., Liu, F.: An advanced particle swarm optimization based on good-point set and application to motion estimation. In: Proceedings Intelligent Computing Theories and Technology - 9th International Conference, ICIC 2013, 28–31 July 2013, Nanning, China, pp. 494–502 (2013)
15. Wang, X., Long, H., Sun, J.: Quantum-behaved particle swarm optimization based on gaussian disturbance. Appl. Res. Comput. **27**(6), 2093–2096 (2006)

16. Chen, Y.H., Teng, H., Liu, S.H.: Optimization of neural tree based on an improved quantum particle swarm optimization. In: Sensors, Measurement and Intelligent Materials II. Applied Mechanics and Materials, vol. 475, pp. 956–959. Trans Tech Publications 3 (2014)
17. Chen, Y., Yang, B., Abraham, A.: Flexible neural trees ensemble for stock index modeling. Neurocomputing **70**(4–6), 697–703 (2007)

A Three-Step Deep Neural Network Methodology for Exchange Rate Forecasting

Juan Carlos Figueroa-García[1(✉)], Eduyn López-Santana[1], and Carlos Franco-Franco[2]

[1] Universidad Distrital Francisco José de Caldas, Bogotá, Colombia
{jcfigueroag, eduynl}@udistrital.edu.co
[2] Management Department, Universidad del Rosario, Bogotá, Colombia
carlosa.franco@urosario.edu.co

Abstract. We present a methodology for volatile time series forecasting using deep learning. We use a three-step methodology in order to remove trend and nonlinearities from data before applying two parallel deep neural networks to forecast two main features from processed data: absolute value and sign. The proposal is successfully applied to a volatile exchange rate time series problem.

1 Introduction and Motivation

Volatility in volatile time series is an interesting problem since many statistical methods are not good, so the use of Computational Intelligence (CI) methods have gained popularity due to its capability to deal with uncertainties and nonlinearities. This way, we can decompose a time series into different features in order to better be computed using different CI methods.

Our main goal is to eliminate trend and cyclical effects of a volatile time series to decompose it into different signals to be forecasted using specialized deep neural nets. In particular, trend and (at least partially) heteroscedasticity in exchange rate series can be removed before decompose it into two features: its absolute value, and sign whose can be forecasted using multilayer perceptrons and backpropagation networks.

The paper is organized as follows: Sect. 1 is the introduction. Section 2 presents some basics on time series; Sect. 3 describes the proposal. Section 4 presents an application example, and Sect. 5 presents the concluding remarks of the study.

2 Basics in Time Series

A *time series* is a set of observations x_t each one recorded at a specific instant $t \in T$ where T is the set of instants for which x_t is observed (usually discrete and finite). Thus, $x \in X$ can be considered as a random variable for which we want to find a model to specify some of their properties. A *time series model* for x_t is a specification of the joint distribution of X_t, or at least its expected value and variance structure. The most popular time series model is the following linear equation:

D.-S. Huang et al. (Eds.): ICIC 2017, Part I, LNCS 10361, pp. 786–795, 2017.
DOI: 10.1007/978-3-319-63309-1_70

$$X_t = m_t + s_t + Y_t, \quad t = 1, 2, \cdots, n \tag{1}$$

where m_t is a trend component, s_t is a cyclic/seasonal component, and y_t is a random noise component independent from $Y_t = X_t - (m_t + s_t), \quad t = 1, 2, \cdots, n$ with zero autocorrelation.

A time series model is considered to be appropriate if Y_t is a zero-mean gaussian process which ensures symmetry and estimability. In general, if X_t has a linear structure then the model (1) provides good results. Thus, the random variable Y_t is better known as the *residuals* of the model, so we can rewrite Eq. (1) as follows:

$$Y_t = X_t - (m_t + s_t), \quad t = 1, 2, \cdots, n \tag{2}$$

$$Y_t = X_t - \hat{X}_t, \quad t = 1, 2, \cdots, n \tag{3}$$

where $\hat{X}_t$ is the estimate of X_t through m_t and s_t.

This way, a time series model $\hat{X}_t$ is tested to contrast four desirable properties via its residuals Y_t: (i) zero mean, (ii) zero autocorrelation, (iii) independence, and (iv) normality. While it is somehow easy to obtain conditions (i) and (ii), many linear models do not provide conditions (iii) and (iv). Now, in most of cases is easy to remove the effect of trend m_t using differentiation, as shown as follows:

$$\nabla X_t = X_t - X_{t-1} = (1 - B)X_t, \tag{4}$$

where B is the backward shift (lag) operator:

$$BX_t = X_{t-1}. \tag{5}$$

From which we can obtain as lags as desired.

2.1 General Discussion over Volatile Time Series

The model described in (1) has been extensively analyzed from a statistical perspective by many authors (see Brockwell [1]), and it is often called *time series linear model*. Most of available models try to predict future values of x_t, this is x_{t+h}, $h \leq n$ using autoregressions and/or moving averages, so our idea is to estimate the effect of trend and seasonal components using statistical methods before using neural nets.

Many financial/weather/economic time series have a hard to estimate nonlinear behavior, so classical methods do not provide good forecasts for nonlinear problems. Another kind of complex problems are the self-called volatile time series (see Fig. 1) whose variance structures are highly nonlinear hard problems to estimate using structural equations. Any random variable can be expressed in the following way:

$$X_t = \mu_t + \delta_t, \quad t = 1, 2, \cdots, n$$

where μ_t is the mean of X_t and δ_t is its variance.

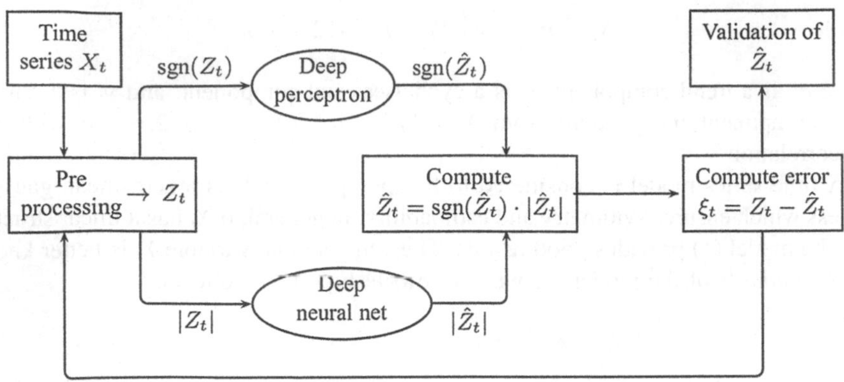

Fig. 1. Proposed three-step deep neural network

A stationary process has $\mu_t = \mu$ and $\delta_t = \delta$ which means homoscedasticity, so we consider *volatility* as the effect of nonlinear δ_t (see Engels [2], Figueroa-García & Soriano [3], Figueroa-García [4]). In this paper, we attempt to predict x_{t+h} using a three-step methodology based on statistical analysis, deep neural networks, and series decomposition (a divide-and-conquer strategy).

3 The Proposed Methodology

Based on ideas of deep learning (see Bengio et al. [5], Bengio [6], Ciresan et al. [7], Deng & Yu [8], Bengio, LeCun & Hinton [9–11], Hinton et al. [12], Schmidhuber [13], Deng et al. [14], Lee [15], and Panchal et al. [16]), we decompose X_t into three main subproblems: a *complexity reduction* of X_t, and a two-neural net structure to extract two basic features namely *value of the series* and its *sign* which can be easily treated by classical deep neural networks.

Throughout different exchange rate examples we found that statistical pre-processing plus a decomposition of the series into easy-to-observe features improve its overall performance (computational efficiency and accuracy). Particularly, the first difference ∇X_t plus a standardization method can remove trend m_t from Eq. (1), and signal processing decomposition such as wavelets, kalman filter, etc. improves performance. The proposed methodology (see Fig. 1) is described as follows:

i. **Statistical pre-processing:** This first step attempts to remove as many components of Eq. (1) as possible using the following transformations based on Eq. (4):

$$\dot{X}_t = \frac{\nabla X_t - \nabla \bar{X}_t}{\nabla s_t}, \tag{6}$$

$$Z_t = \frac{\dot{X}_t}{|\max_x\{\dot{X}_t\}|}, \tag{7}$$

$$Z_t = \text{sgn}(Z_t) \cdot |Z_t|, \tag{8}$$

where $\nabla \bar{X}_t$ is the average of $\dot{X}_t$ and ∇s_t is the standard deviation of $\dot{X}_t$.

In most cases, those transformations obtain normally distributed zero mean, unity standard deviation, with no trend effect m_t series, this is $\dot{X} : X \to N(0,1)$ and $Z : X \to N(0, S^2)$ where S^2 can be easily computed from Z_t.

ii. **Deep perceptron:** We use a deep perceptron to compute sgn (Z_t) of Eq. (8). The following sets of inputs and outputs are recommended:

$$\text{Inputs} = \begin{cases} \text{sgn}\{Z_{t-1}, Z_{t-2}, \cdots, Z_{t-h}\}, \\ \{\dot{X}_{t-1}, \dot{X}_{t-2}, \cdots, \dot{X}_{t-p}\}, \end{cases} ; \quad \text{Outputs} = \{\text{sgn}(Z_t),$$

where $h \in \mathbb{N}^+$ and $p \in \mathbb{N}^+$ are lag operators.

A deep-perceptron is a deep backpropagation net whose output layer is a perceptron with hardlim activation function. All its weights $\mathbb{W}$ are updated using the errors ξ, a gradient strategy $\nabla\mathbb{W}$, and a learning rate σ to improve convergence.

iii. **Deep neural net:** Now, the component $|Z_t|$ of Eq. (8) can be computed using any deep neural network. We recommend to use multi-layer backpropagation or time delay networks using the following sets of inputs and outputs:

$$\text{Inputs} = \begin{cases} \{|Z_{t-1}|, |Z_{t-2}|, \cdots, |Z_{t-q}|\}, \\ \{\dot{X}_{t-1}, \dot{X}_{t-2}, \cdots, \dot{X}_{t-r}\}, \end{cases} ; \quad \text{Outputs} = \{|Z_t|,$$

where $q \in \mathbb{N}^+$ and $r \in \mathbb{N}^+$ are lag operators.

All weights $\mathbb{W}$ of this deep back-propagation neural net are updated using the errors ξ, a gradient strategy $\nabla\mathbb{W}$, and a learning rate σ to improve convergence.

The quantity of hidden layers and the shapes of the activation functions depend on data complexity. We recommend to use sigmoid/linear activation functions for continuous inputs, hardlim activation functions for $\{-1, 1\}$ inputs and the output sgn (Z_t), and sigmoid activation function for continuous outputs. At least 10 neurons per input in the input layer, and two (or more) additional hidden layers are recommended.

As always, the amount of neurons per hidden layer is a degree of freedom that depends on the available data (usually the number of inputs should be larger than the total neurons in the net). The difference between a deep perceptron and a deep multi-layer neural net lies in the outputs of a deep perceptron since they are in the domain $\{-1, 1\}$ (the values of sgn Z_t) while the outputs of a multilayer neural net are in the domain $[-\infty, \infty]$ (which are bounded to $[0, \infty]$).

The selection of h, p, q, r is a combinatorial problem that affects the forecasts. Some recommendations have been provided by Figueroa-García & Soriano [3], Figueroa-García [4], Figueroa-García et al. [17–19], Connor et al. [20], and Huang

[21]. Some successful deep neural models applied to automatic trading were reported by Arévalo et al. [22], and Niño-Peña & Hernández-Pérez [23].

4 Application Example

The selected example is the COP/USD exchange rate which shows volatility (understood either as highly nonlinear variance or heteroscedasticity). A total of 544 observations were used in this example, 489 for training and 55 for validation.

COP/USD exhibits a nonlinear behavior (see Fig. 2) even when applied ∇X_t to remove trend m_t (see Fig. 3), and pre-processing $\dot{X}_t$ and Z_t as shown in Eqs. (4) to (8).

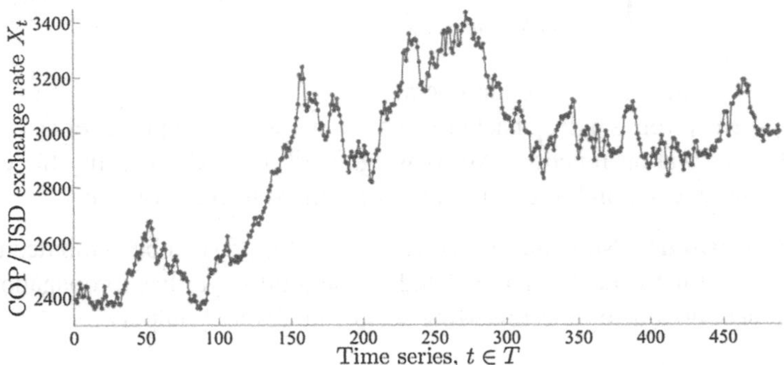

Fig. 2. COP/USD exchange rate X_t

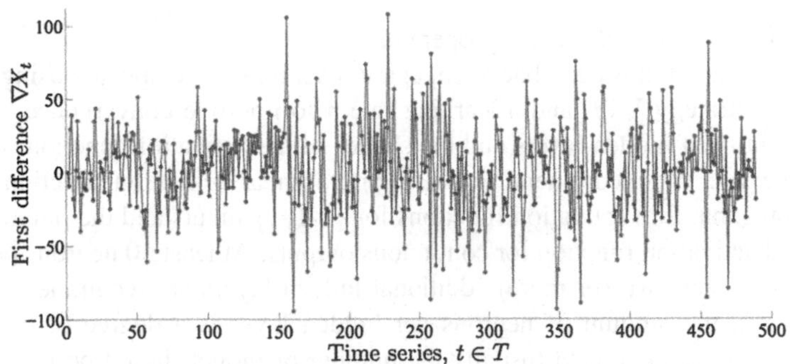

Fig. 3. First difference of COP/USD exchange rate X_t

Table 1. Goldfeld-Quant results over Z_t

Time period	F-test	p-value
t_1/t_2	0.566	0.001
t_1/t_3	0.707	0.028
t_2/t_3	1.248	0.158

Some heteroscedasticity tests are performed to contrast the goodness of Z_t. The Goldfeld-Quant test (see Table 1) is based on an *f*-test over variances over three equally length time periods t_1, t_2, t_3, not that t_2/t_3 shows homoscedasticity.

Table 2 shows that Z_t does not have ARCH effect, a god sign of homoscedasticity. Preprocessed and decomposed series are shown in Figs. 4 and 5.

Table 2. ARCH test results

Original series X_t				Pre-processed series Z_t.		
Lag	ARCH stat	Critical Value	p-value	ARCH stat	Critical value	p-value
1	928.44	3.8415	≈0	1.167	3.8415	0.2799
2	927.45	5.9915	≈0	1.948	5.9915	0.377
3	926.45	7.8147	≈0	5.435	7.8147	0.143
4	925.45	9,4877	≈0	6.830	9.4877	0.145
5	924.45	11.07	≈0	7.144	11.07	0.210

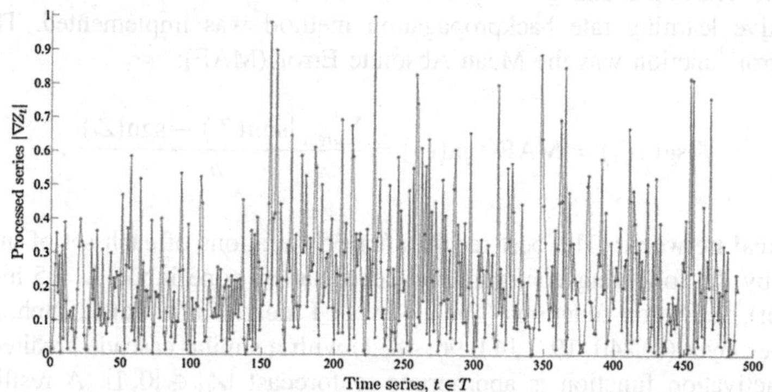

Fig. 4. Pre-processed COP/USD exchange rate $|Z_t|$

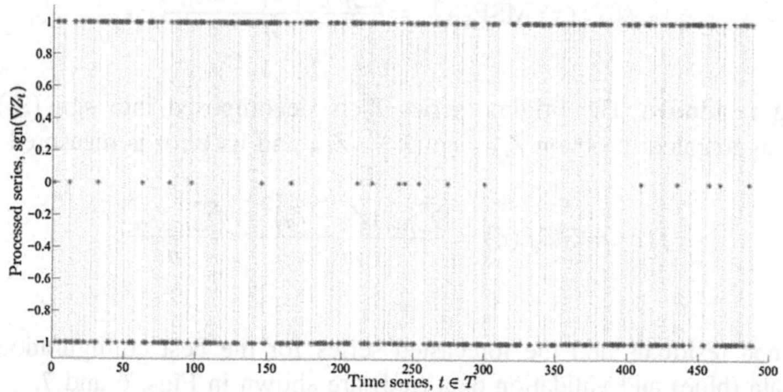

Fig. 5. Sign of $|Z_t|$

4.1 Configuration of the Networks

Pre-processing: First, we have decomposed X_t into Z_t using Eqs. (6), (7) and (8), so in order to select an initial set of inputs we have computed autocorrelations of sgn (Z_t) and $|Z_t|$. After 12 iterations of the methodology, the final set of inputs is:

$$\text{Deep perceptron inputs} = \left\{ \begin{array}{l} \text{sgn}\{Z_{t-1}, Z_{t-2}, Z_{t-4}, Z_{t-5}\}, \\ \{\dot{X}_{t-1}, \dot{X}_{t-2}, \dot{X}_{t-5}\}, \end{array} \right. ; \quad \text{Outputs} = \{\text{sgn}(Z_t),$$

$$\text{Deep neural net inputs} = \left\{ \begin{array}{l} \{|Z_{t-1}|, |Z_{t-2}|, |Z_{t-5}|, |Z_{t-7}|\}, \\ \{\dot{X}_{t-1}, \dot{X}_{t-2}, \dot{X}_{t-3}, \dot{X}_{t-5}\}, \end{array} \right. ; \quad \text{Outputs} = \{|Z_t|.$$

Deep perceptron: After 25 iterations of each inputs set, the best configuration is as follows: five layers (one input, $L = 3$ hidden, and one output), activation functions (Tanh, Tanh, Linear, Hardlims), neurons per layer (7, 210, 140, 70, 1) in a layer cascade architecture (see Arévalo et al. [22], and Niño-Peña & Hernández-Pérez [23]). It is clear that sgn $(Z_t) \in \{-1, 1\}$ so Hardlims activation function is the most appropriate for forecasting exchange rate direction. Also a Gradient descent with momentum and adaptive learning rate backpropagation method was implemented. The implemented error function was the Mean Absolute Error (MAE):

$$f(\text{sgn}(\xi)) = \text{MAE}(\text{sgn}(\xi)) = \frac{\sum_{t \in n} |\text{sgn}(Z_t) - \text{sgn}(\hat{Z}_t)|}{n}.$$

Deep neural network: The best results after 35 iterations of each set of inputs were provided by the following configuration: seven layers (one input, $L = 5$ hidden, and one output), activation functions (Tanh, Linear, Linear, Tanh, Tanh, Tanh, Sigmoid), neurons per layer (8, 240, 192, 144, 96, 48, 1) with a similar cascade architecture. The sigmoid activation function is appropriate to forecast $|Z_t| \in [0, 1]$. A resilient backpropagation method was implemented in order to speed up memory. The implemented error function was the Mean Squared Error (MSE):

$$f(|\xi|) = \text{MSE}(|\xi|) = \frac{\sum_{t \in n} (|Z_t| - |\hat{Z}_t|)^2}{n}.$$

Analyzing residuals: The original series Z_t is decomposed into sgn (Z_t) and $|Z_t|$ which are aggregated to obtain $Z_t = \text{sgn}(Z_t) \cdot |Z_t|$, and its error is measured via MSE:

$$f(\xi) = \text{MSE}(\xi) = \frac{\sum_{t \in n} (Z_t - \hat{Z}_t)^2}{n} = \frac{\sum_{t \in n} \xi_t^2}{n}.$$

The obtained residuals and the forecasted series for the best configuration for both training data (blue) and validation data (red) are shown in Figs. 6 and 7.

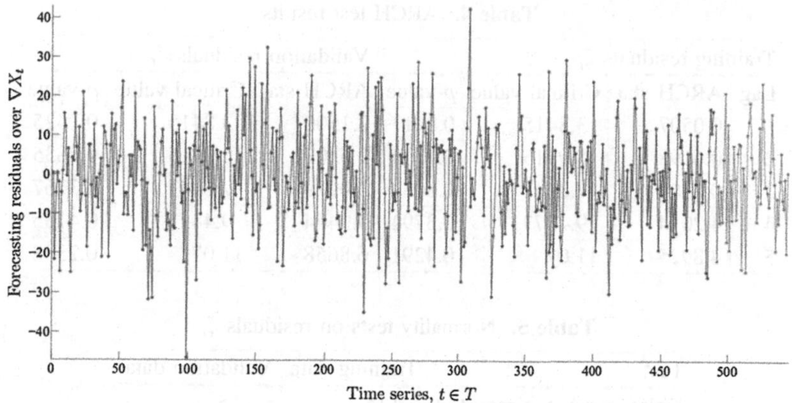

Fig. 6. Residuals ξ_t (Color figure online)

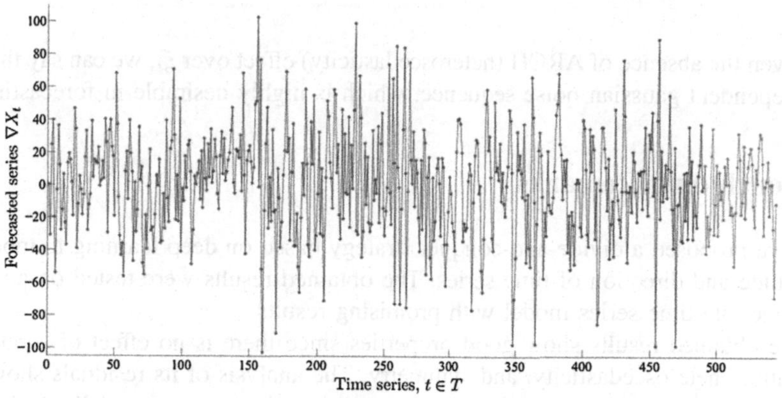

Fig. 7. Forecasted series $\hat{Z}_t$ (Color figure online)

Some statistics are collected in Table 3, and the results if the ARCH test over training and validation errors ξ_t are shown in Table 4.

Table 3. Error measures

Dataset	MAE for sgn (Z_t)	MSEfor $\|Z_t\|$	MSE for $\hat{Z}_t$
Training	0.4089	0.0031	161.13
Validation	0.4	0.0025	129.13

Several normality tests are performed over either training and validation errors to verify if either the model is appropriate or not (see Table 5). All tests suggest that the residuals of both training and validation datasets are normally distributed.

Table 4. ARCH test results

Training residuals ξ_t				Validation residuals ξ_t		
Lag	ARCH stat	Critical value	p-value	ARCH stat	Critical value	p-value
1	0.0507	3.8415	0.8217	2.1500	3.8415	0.1425
2	0.6198	5.9915	0.7334	2.5207	5.9915	0.2835
3	0.7307	7.8147	0.8659	3.0979	7.8147	0.3767
4	4.2014	9.4877	0.3794	3.9444	9.4877	0.4135
5	4.8925	11.07	0.4291	6.8658	11.07	0.2308

Table 5. Normality tests on residuals ξ_t

Test	Training data	Validation data
Kolmogorov-Smirnov test	>0.15	>0.15
Anderson-Darling test	0.833	0.498
d'Agostino-Pearson test	0.385	0.769
Shapiro-Wilks test	0.707	0.664

Given the absence of ARCH (heteroscedasticity) effect over ξ_t, we can say that ξ_t is an independent gaussian noise sequence, which is highly desirable in forecasting.

5 Concluding Remarks

We have proposed a divide-and-conquer strategy based on deep learning to model the magnitude and direction of time series. The obtained results were tested on a volatile exchange rate time series model with promising results.

The obtained results show good properties since there is no effect of serial autocorrelation, heteroscedasticity, and symmetry. The analysis of its residuals shows that the model is appropriate for forecasting, according to time series modeling principles.

Finally, we have computed two specialized deep learning neural networks in order to distribute computing using less resources. Instead of designing a bigger network, we propose to decompose the time series into less complex signals which can be better forecasted using specialized networks. We recall that every time series should be analyzed before using our results since it can be decomposed into different signals, and consequently different neural networks should be applied to.

References

1. Brockwell, P.J.: Time Series: Theory and Methods. Springer, Heidelberg (2000)
2. Engel, R.: Autoregressive conditional heteroscedasticity with estimates of the variance of United Kingdom inflation. Econometrica **1**(50), 987–1007 (1982)
3. Figueroa-García, J.C., Soriano, J.J.: A comparison of ANFIS, ANN and DBR systems on volatile time series identification. In: Annual Meeting of the North American Fuzzy Information Processing Society, vol. 26, pp. 321–326. IEEE (2007)

4. Figueroa-García, J.C.: An evolutive interval type-2 TSK fuzzy logic system for volatile time series identification. In: Conference on Systems, Man and Cybernetics, pp. 1–6. IEEE (2009)
5. Bengio, Y., Lamblin, P., Popovici, D., Larochelle, H.: Greedy layer-wise training of deep networks. In: Proceedings of NIPS 2006 (2006)
6. Bengio, Y.: Learning Deep Architectures for AI. Now Publishers, Hanover (2009)
7. Ciresan, D., Meier, U., Masci, J., Schmidhuber, J.: Multi-column deep neural network for traffic sign classification. Neural Netw. **32**(1), 333–338 (2012)
8. Deng, L., Yu, D.: Deep learning: Methods and applications. Found. Trends Sig. Process. **7**(3–4), 197–387 (2014)
9. Schmidhuber, J.: Deep learning in neural networks: an overview. Neural Netw. **61**, 85–117 (2015)
10. Hinton, G.E.: Learning multiple layers of representation. Trends Cogn. Sci. **11**(1), 428–434 (2007)
11. Hinton, G.E.: Deep belief networks. Scholarpedia **4**(5), 5947 (2009)
12. Hinton, G.E., Osindero, S., Teh, Y.W.: A fast learning algorithm for deep belief nets. Neural Comput. **18**(7), 1527–1554 (2006)
13. Bengio, Y., LeCun, Y., Hinton, G.: Deep learning. Nature **521**, 436–444 (2015)
14. Deng, L., Abdel-Hamid, O., Yu, D.: A deep convolutional neural network using heterogeneous pooling for trading acoustic invariance with phonetic confusion. In: Proceedings of ICASSP 2013 (2006)
15. Lee, H., Grosse, R., Ranganath, R., Ng, A.Y.: Convolutional deep belief networks for scalable unsupervised learning of hierarchical representations. ICML Trans. ACM **1**, 609–616 (2009)
16. Panchal, G., Ganatra, A., Kosta, Y., Panchal, D.: Behaviour analysis of multilayer perceptrons with multiple hidden neurons and hidden layers. Int. J. Comput. Theor. Eng. **3**(2), 332–337 (2011)
17. Kalenatic, D., Figueroa-García, J.C., Lopez, C.A.: A neuro-evolutive interval type-2 TSK fuzzy system for volatile weather forecasting. In: Huang, D.-S., Zhao, Z., Bevilacqua, V., Figueroa, J.C. (eds.) ICIC 2010. LNCS, vol. 6215, pp. 142–149. Springer, Heidelberg (2010). doi:10.1007/978-3-642-14922-1_19
18. Figueroa-García, J.C., Kalenatic, D., López-Bello, C.A.: An evolutionary approach for imputing missing data in time series. J. Syst. Circ. Comput. **19**(1), 107 (2010)
19. Figueroa-García, J.C., Kalenatic, D., López-Bello, C.A.: Missing data imputation in multivariate data by evolutionary algorithms. Comput. Hum. Behav. **27**(5), 1468–1474 (2011)
20. Connor, J., Martin, R., Atlas, L.: Recurring neural networks and robust time series prediction. IEEE Trans. Neural Netw. **5**(2), 240–254 (1994)
21. Huang, G.B., Zhu, Q.Y., Siew, C.K.: Extreme learning machine: theory and applications. Neurocomputing **704**(1), 489–501 (2005)
22. Niño Peña, J.N., Hernández-Pérez, G.: Price direction prediction on high frequency data using deep belief networks. Commun. Comput. Inf. Sci. **657**(1), 69–79 (2016)
23. Arévalo, A., Niño, J., Hernández, G., Sandoval, J.: High-frequency trading strategy based on deep neural networks. In: Huang, D.-S., Han, K., Hussain, A. (eds.) ICIC 2016. LNCS, vol. 9773, pp. 424–436. Springer, Cham (2016). doi:10.1007/978-3-319-42297-8_40

Android Malware Clustering Analysis on Network-Level Behavior

Shanshan Wang[1], Zhenxiang Chen[1(✉)], Xiaomei Li[2], Lin Wang[1], Ke Ji[1], and Chuan Zhao[1]

[1] School of Information Science and Engineering, University of Jinan, Jinan 250022, China
czx@ujn.edu.cn

[2] Department of Political Theory, Shandong Sport University, Jinan 250102, China

Abstract. Android becomes the most popular operating system for smart phones today. However, malicious application proposes a huge threat on Android platform. Many malware are designed to steal personal information of user or control the device of user through the network. In this paper, we show how to efficiently cluster network behavior by analyzing the statistical information of HTTP flow at the network level. To do so, we observe the specific statistical information on HTTP flow generated by more than 8,000 malware. In the end, we separate malware's malicious network into seven different clusters using clustering technology. Our evaluation experiments show that HTTP flows in the same cluster have similar network behavior and there are big differences between the different clusters. This similarity and variability are manifested at some specific network-level statistical characteristics. In addition, in order to show the results of the study more intuitively, we reduce the dimensionality of the original features, and show the final clustering results in two-dimensional space.

Keywords: Android malware · Clustering · Dimensionality reduction

1 Introduction

In today, Android platform becomes one of the most popular operating systems on smart phones. There are millions of applications on different application markets which provide users with a rich of functionality. These huge numbers of applications are convenient to the daily life of user, but also provide users with more entertainment services. A recent report [1] shows that the number of apps in the Google Play Store has risen from 16 thousand in December 2009 to more than 2 million in February 2016. Unfortunately, the smartphones running on Android platform have gradually become the target of the attacker and are infected by malicious applications. Comparing with other platforms, Android allows users to install applications from a variety of ways, such as from application markets and from forums, which makes the malware easy to spread. According to the latest report [2], the number of malicious installation packages hit more than 8.5 million in 2016, three times more than 2015.

The Android platform offers several security measures to harden malware's installation, such as Android permission system. In order to perform a definite task on the

D.-S. Huang et al. (Eds.): ICIC 2017, Part I, LNCS 10361, pp. 796–807, 2017.
DOI: 10.1007/978-3-319-63309-1_71

device, such as making a call, each app must request the explicit permission from the user in the process of installing app. However, many users tend to ignore the meaning of this permission so weaken the original purpose of permission system. In practice, malicious applications are difficult to be controlled by Android permissions system.

In this paper, we focus on network-level behavioral cluster of HTTP flow generated by malware. Because HTTP protocol is the main means of malware communicating with the attacker, we pay attention on the statistical information of HTTP flows. In fact, Zhou et al. [10] have shown that more than 90% of malware-infected mobile terminals are controlled by botnets through network or SMS commands. Their observations give us insights on exploring apps' network behaviors. Different variants of the same malware or different malware in same family will exhibit similar malicious activities. We cluster these network flows according to the similarity of network behavior, and then analyze the common network behavior of each cluster. To sum up, the main contribution of this paper are summarized as follows:

- We analyze a large number of malicious network traffic data and then extract seven statistical features that can effectively characterize each HTTP flow.
- K-Means clustering algorithm is used to separate the network traffic into seven clusters. Finally, the flows with similar network behavior are gathered into the same cluster.
- We use TSNE [21] tool to reduce the dimension of network-level features, and visualize cluster result of flows in the low-dimensional space.

The paper is organized as follows. Section 2 introduces the details of this solution. Section 3 presents evaluation and Sect. 4 discusses the related work with our solution. Finally, Sect. 5 concludes the paper.

2 Methodology

We propose a solution for clustering and analyzing the network-level behavior of malware. The proposed approach undergoes several stages of implementation. The detailed discussion on the stages of proposed work outlines as follows:

(a) Traffic Collection. We design an Android traffic collection platform specifically to generate the required network traffic data (Sect. 2.1).
(b) Traffic Pre-process. A flow extraction algorithm is designed to help extract individual HTTP flow from mixed traffic file (Sect. 2.2).
(c) Feature Analysis. We extract seven network-level features to characterize each HTTP network flow. At the same time, the features are analyzed and processed (Sect. 2.3).
(d) Traffic clustering. We use the K-Means clustering algorithm to cluster the HTTP flows with similar network behavior and analyze the common network behavior of each cluster (Sect. 2.4).

2.1 Traffic Collection

We design the Android traffic collection platform which is used for collecting the traffic data generated by a large number of malware and deployed in the University of Jinan campus network. In order to ensure the safety of the traffic collection framework, a firewall and NAT server are deployed at the gateway of the campus. As shown in Fig. 1, the platform consists of three parts, namely, control center, app storage & traffic management server, and traffic generation & collection. These three parts work together to complete the generation and acquisition of Android network traffic. The control center connects with the traffic generation & collection part and the app storage & traffic management via LAN switch. The control center is responsible for scheduling task. Each Android app in the app storage server is assigned to a traffic collection machine in the traffic generation & collection part by the control center. All spare machines can be used to generate and collect network traffic. Furthermore, collected traffic files will be transferred to the traffic management server. Each traffic collection machine typically consists of two process. They are executing an app and collecting network traffic data generated by the app respectively. We execute the apps on several simulators, which are automatically driven by the Android tool, monkey. The monkey tool can send some events to the Android device randomly when the app is running. To ensure the network traffic data we collected is generated by the specific app, we only execute one app on one simulator device each time. The collected traffic data is save as PCAP format files and the details of this Android traffic collection platform are described in paper [11].

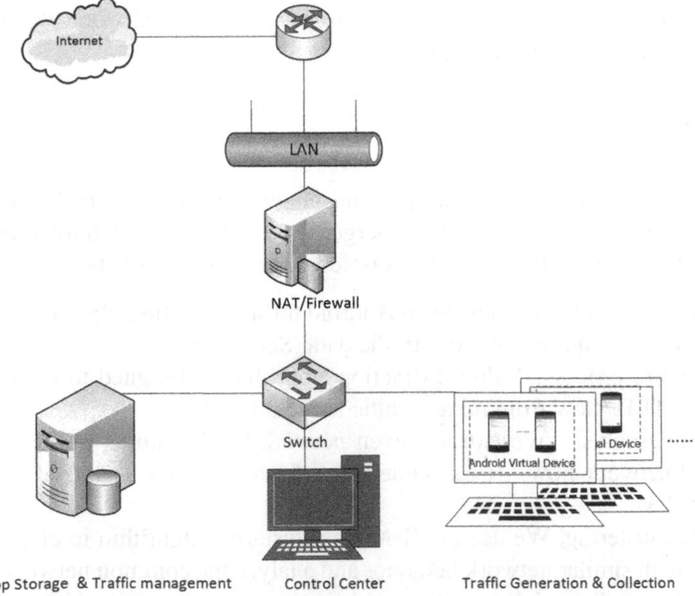

Fig. 1. The architecture of Android traffic collection platform

2.2 Traffic Pre-process

As we all known, the network flow is a basic unit that the app interacts with the network. The original network traffic data we obtain from the traffic collection platform is generated by an app in the first few minutes. However, numerous HTTP flows generated by the app are mixed with other HTTP flows, and we should separate each complete flow from the mixed traffic. We have proposed an algorithm that can extract each complete HTTP flow. The specific process is described in Algorithm 1. The input of this algorithm are the collected traffic (original traffic files), and the outputs are individual HTTP flows. This algorithm is implemented by combination Python script with T-shark command.

Data: Network traffic data (a pcap file that generated by one app in a few minutes.
Result: Individual and complete HTTP flows.
Initialize the index id as 0 on searching HTTP flow;
while *true* **do**
  Use T-shark command to extract HTTP flow which is pointed by the current index;
  if *the bytes number of the HTTP flow is smaller than a fixed number* **then**
  break;
  else
    Write the HTTP flow to a new pcap file from memory;
    Increase the index id of HTTP flow;
  end
end

Algorithm 1. The algorithm of obtaining individual HTTP flows

2.3 Features Analysis

Feature Selection. Many of the network traffic features are used to detect malicious activity or identify particular network behavior in the area of cyber security. The features we select in our proposed approach are based on different malware behaviors. These features come from paper [12]. The authors of paper [12] also point out that seven features are particularly representative. At the same time, they prove that these seven features can completely distinguish between benign and malicious network traffic with their experimental results. The reason is that these seven features can completely describe the characteristics of an HTTP network flow. In this paper, we use these seven features to analyze malicious network traffic data generated by malware. The seven features are summarized in Table 1.

Data Normalization. Data normalization is a basic work in data mining. Different evaluation indicators have different dimensions and the difference between the values may be large, so analyzing the original data directly may effect the final results. In order

Table 1. Network-level features that can effectively characterize network flow

ID	Traffic feature
1	Average packet size (in bytes)
2	Ratio of incoming to outgoing bytes
3	Average number of bytes received per second
4	Average number of packets sent per flow
5	Average number of packets received per flow
6	Average number of bytes sent per flow
7	Average number of bytes received per flow

to eliminate the influence of the differences between the dimensions and the range of values, it is necessary to carry out standardized processing, that is, to scale the data in proportion to a specific area. Minimum - maximum standardization is a popular data normalization method and is used to process traffic feature data. It performs a linear change on the original data, and the value is mapped to between 0 and 1. The conversion formula is shown in Eq. 1.

$$x^{*} = \frac{x - min}{max - min} \quad (1)$$

Where *max* is the maximum value of the sample data and *min* is the minimum value of the sample data. The x^* represents the normalized data x. Minimum - maximum standardization preserves the relationships that exist in the original data and is the easiest way to eliminate the effects of dimensions and range of data.

2.4 Traffic Cluster

There are many malware family classification methods, many of which are based on static scanning of the source code of malware to divide malware to different families. However, this method is not applied to the network level. That is, although the malware from same family have similar malicious code fragments, the malicious network behavior is not necessarily similar. Based on this consideration, we plan to cluster the malicious network behavior from the network traffic level.

Clustering techniques are widely used in malware detection over the past many years and can also be applied to detect unknown malicious network traffic through online learning without labeling. There are many well-known clustering algorithms, such as K-Means clustering algorithm, hierarchical clustering, density-based clustering algorithm. Each algorithm has its own most suitable application scenarios. In order to pick out the best clustering algorithm for our data set, we reduced the network traffic data using principal component analysis (PCA) algorithm and visualize the data after dimensionality reduction and finally we choose K-Means algorithm to cluster the traffic flows.

K-Means is one of the simplest unsupervised learning algorithm that solves the well known clustering problem. This algorithm requires the number of clusters to be specified. It scales well to large number of samples and has been used across a large range of application areas in many different fields. In our experiment, the total 8960 malicious HTTP flows are clustered into seven clusters with K-Means algorithm.

Data Visualization. We use TSNE [21], one famous visualization tools on clustering results to visualize the clustering result. The design of TSNE is to visualize high-dimensional data. We always like to show the results of research in a intuitive way, clustering is no exception. However, it is generally difficult to display the clustering results directly with the original features when the number of features is usually high (more than 3). While TSNE provides an effective way to reduce data, so that we can show clustering results in low-dimensional space. Figure 2 shows the cluster result in the two-dimensional space. Different colors represent different cluster and samples share common color within one cluster. Because K-Means is based on the distance of samples to perform division, the number of samples per cluster is not consistent in Fig. 2.

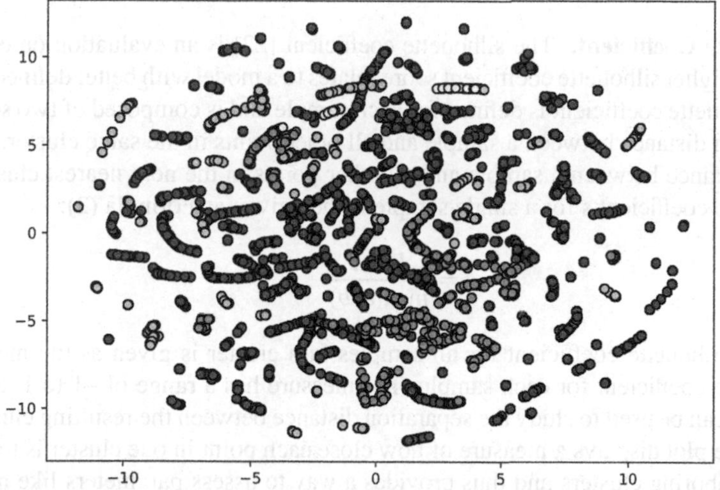

Fig. 2. Data visualization of K-Means clustering result

3 Evaluation

3.1 Data Sets

Our initial malicious app set is obtained from VirusShare.com website [13] which is a repository of malware samples to provide security researchers. These malicious malware contain a total of 8203 samples and are shared on VirusShare.com website from 2015 to 2017. Using our own designed traffic collection platform, the malware generate 14.2 GB network traffic data. We extract 113,735 HTTP flows from these traffic data.

There is one thing to be noted that: Not all network traffic generated by malicious app are malicious traffic. Many malware are from the repackaged benign apps, so the malware also contain the basic functions of benign apps. The network traffic they generate mixes benign and malicious network traffic together, even most of the traffic is benign and only a small part of them is the malicious network traffic. In order to ensure label accuracy of training set, we perform a selection operation on the network traffic generated by malicious apps. We extract the host or target IP field of each flow and upload it to Virustotal. If the Host or target IP is malicious, the flow is considered as a malicious. This malicious network traffic flow will be added to our malicious traffic data set as one sample of training set. At last, we get 8960 malicious flows that have exact label from malwares traffic.

3.2 Cluster Evaluation

In our data set, the ground truth label of each sample is not known, so the evaluation must be performed using the model itself. There are two well known cluster evaluation metric without labels of dataset and they are Silhouette Coefficient and Calinski-Harabaz Index respectively.

Silhouette Coefficient. The silhouette coefficient [22] is an evaluation on clustering, where a higher silhouette coefficient score relates to a model with better defined clusters. The silhouette coefficient is defined for each sample and is composed of two scores: {a: The mean distance between a sample and all other points in the same cluster.} {b: The mean distance between a sample and all other points in the next nearest cluster.} The silhouette coefficient s for a single sample is then given as Formula (2):

$$s = \frac{b - a}{\max(a, b)} \tag{2}$$

The silhouette coefficient for all samples in a cluster is given as the mean of the silhouette coefficient for each sample. The measure has a range of –1 to 1. Silhouette analysis can be used to study the separation distance between the resulting clusters. The silhouette plot displays a measure of how close each point in one cluster is to points in the neighboring clusters and thus provides a way to assess parameters like number of clusters visually. Silhouette coefficients (as these values are referred to as) near +1 indicate that the sample is far away from the neighboring clusters. A value of 0 indicates that the sample is on or very close to the decision boundary between two neighboring clusters and negative values indicate that those samples might have been assigned to the wrong cluster.

In our evaluation experiment, the silhouette analysis is used to choose an optimal value for the number of cluster. Figure 3 shows the different silhouette coefficient values when the cluster number k is set different values. In Fig. 3, the silhouette plot shows that the silhouette coefficient at different cluster number k are every close. The cluster number we set as 5, 6, 7 and 8 respectively and corresponding silhouette coefficient at different cluster number are 0.50, 0.50, 0.52 and 0.48. The optimal value for the number of cluster is set as 7 because the silhouette coefficient is maximum. The cluster size from the thickness of the silhouette plot also can be visualized. The number of samples assigned to cluster 3 is the largest while cluster 6 owns a few samples when cluster number is 7.

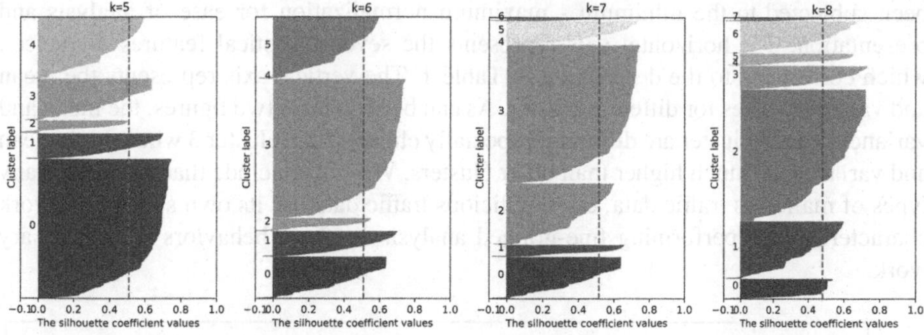

Fig. 3. The silhouette coefficient and cluster size at different cluster number

Calinski-Harabaz Index. If the ground truth labels are not known, the calinski-harabaz index also can be used to evaluate the model, where a higher calinski-harabaz score relates to a model with better defined clusters. For *k* clusters, the calinski-harabaz score s is given as the ratio of the between-clusters dispersion mean and the within-cluster dispersion:

$$\mathrm{s}(k) = \frac{T_r(B_k)}{T_r(w_k)} \times \frac{N-k}{k-1} \tag{3}$$

where B_K is the between group dispersion matrix and W_K is the within-cluster dispersion matrix defined by:

$$W_k = \sum\nolimits_{q=1}^{k} \sum\nolimits_{x \in C_q} (x - c_q)(\mathrm{x} - c_q)^T B_k = \sum\nolimits_{q} n_q (c_q - c)(c_q - c)^T \tag{4}$$

with N be the number of points in our data, C_q be the set of points in cluster q, c_q be the center of cluster q, c be the center of E, n_q be the number of points in cluster q.

In the previous step, we have determined the value of K according to silhouette coefficient. When the K number is set as 7, the corresponding calinski-harabaz Index reaches to 10145.46.

3.3 Network Behavior Analysis

We use seven network-level statistical features to separate malicious network traffic into seven clusters and cluster labels are cluster 0, cluster 1, cluster 2… and cluster 6 respectively. The number of HTTP flows in each cluster is 1303, 122, 2070, 4351, 2, 839, 3 respectively. Among them, we can find most of HTTP flows are separated into cluster 0, cluster 2 and cluster 3, and the proportion accounts for 15.54%, 23.10% and 48.56% respectively.

In addition, we calculate the mean and variance of all the features in each cluster in Figs. 4 and 5. Because there are big differences between the different indicators and values in individual indicators in the original data, the data we show in Figs. 4 and 5 has been subjected to the minimum - maximum normalization for ease of analysis and presentation. The horizontal axis represents the seven statistical features we select, which correspond to the description in Table 1. The vertical axis represents the mean and variance values for different cluster. As can be seen from two figures, the mean and variance of each cluster are different, especially cluster 0 and cluster 3 where both mean and variance are much higher than other clusters. We can conclude that there are many types of malicious traffic data, each malicious traffic data has its own specific network characteristic, so performing fine-grained analysis on these behaviors is a necessary work.

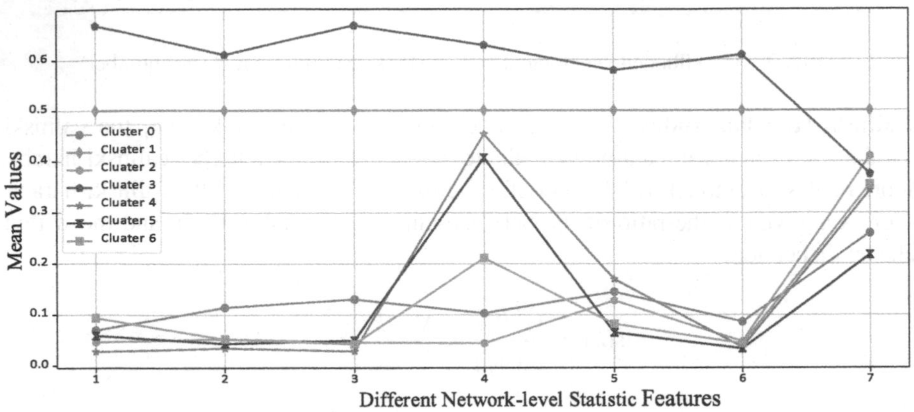

Fig. 4. The mean values of all features in per cluster

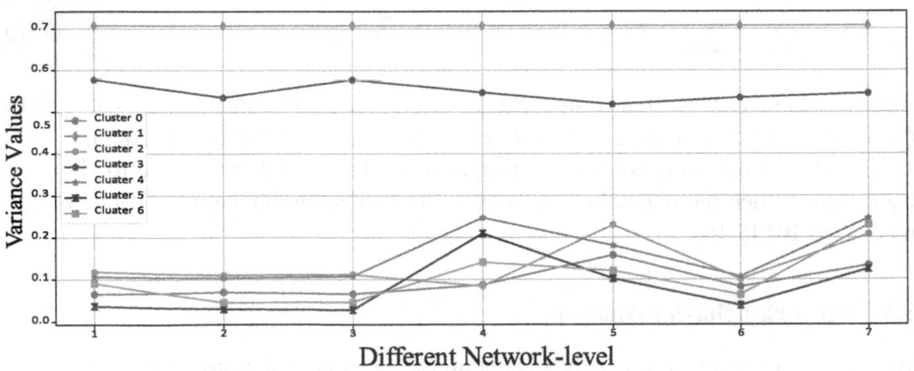

Fig. 5. The variance values of all features in per cluster

4 Related Work

A large body of research methods on Android malware analysis and detection in recent years. These methods can be roughly divided into static analysis, dynamic analysis and network-based traffic analysis. Static analysis method is popular in malware detection, such as [3–5]. However, static analysis is challenged by malware's code polymorphism and code obfuscation [6], which are used to generate variants of malware to evade detections. Dynamic analysis method, such as [7–9] can monitor the behavior of an application at running-time. Although this approach is very effective on identifying malicious behavior, the monitoring at app running-time has to bear a lot of overhead, so this method is not well applied to the mobile device.

The work [14, 15] study the clustering of malware at the system level. Specifically, the study [15] proposes a scalable malware clustering algorithm based on malware system events. They first perform dynamic analysis to obtain the execution traces of malware programs. These execution traces are then generalized into behavioral profiles, which characterize the activity of a program in more abstract terms. By contrast, the method based on network traffic is always smaller overhead and effectively. There are some explorations on automatic network signatures generated by malware in previous works [16–18]. In particular, [17] uses clustering technology to split the worm flow to different worm classes, but this clustering algorithm does not apply to a large number of traffic data sets. In the paper [19], the authors propose FLOWR, a system that automatically identifies mobile apps by continually learning the apps distinguishing features via traffic analysis. FLOWR focuses solely on key-value pairs in HTTP headers and intelligently identifies the pairs suitable for app signatures. The study [20] shows an effective method that group mobile botnets families by analyzing the HTTP traffic they generate. They also extracts a few statistical features just from HTTP headers for the task of Android malware detection.

Some malware are more dangerous than the others because of the fact that they connect with some malicious remote servers in the background to obtain commands or leak the privacy data of user. Due to the malicious performance of each malware is different, the malicious network behaviors they produce have big differences. Different malware can be separated based on these differences. Keeping the seriousness of malware network-level behavior cluster in mind, this paper has analyzed their network traffic behavior and observed the deviations in the network behavior of malware.

5 Conclusion

In this paper, we perform analysis on a malicious network traffic that many malware generate. We extract seven statistical features that can describe the network behavior of malware. By clustering network flows into different clusters, we observe that the network flows belonging to the same cluster have similar statistical information, that is, they have similar network behavior. In addition, the TSNE which is a visualization tool on clustering results is used to data dimensionality reduction, and finally we visualize the flow data in two-dimensional space.

Acknowledgement. This work was supported by the National Natural Science Foundation of China under Grants No. 61672262, No. 61573166, No. 61472164 and No. 61572230, the Natural Science Foundation of Shandong Province under Grants No. ZR2014JL042 and No. ZR2012FM010, the Shandong Provincial Key R&D Program under Grants No. 2016GGX101001.

References

1. Google play: number of available apps 2009–2016. http://www.statista.com/statistics/266210/
2. Report: 2016 saw 8.5 million mobile malware attacks, ransomware and IoT threats on the rise. http://www.techrepublic.com/article/report-2016-saw-8-5-millionmobile-malware-attacks-ransomware-and-iot-threats-on-the-rise/
3. Enck, W., et al.: On lightweight mobile phone application certification (2009)
4. Felt, A.P., Chin, E., Hanna, S., et al.: Android permissions demystified. In: ACM Conference on Computer and Communications Security, pp. 627–638. ACM (2011)
5. Grace, M., Zhou, Y., Zhang, Q., et al.: RiskRanker: scalable and accurate zero-day android malware detection. In: International Conference on Mobile Systems, Applications, and Services, pp. 281–294. ACM (2012)
6. Moser, A., Kruegel, C., Kirda, E.: Limits of static analysis for malware detection. In: Twenty-Third Annual Conference on Computer Security Applications, ACSAC 2007, pp. 421–430. IEEE Xplore (2008)
7. Enck, W., Gilbert, P., Han, S., et al.: TaintDroid: an information-flow tracking system for realtime privacy monitoring on smartphones. In: Usenix Symposium on Operating Systems Design and Implementation, OSDI 2010 Proceedings, 4–6 October 2010, Vancouver, BC, Canada, pp. 393–407. DBLP (2010)
8. Zhou, Y., Wang, Z., Zhou, W., et al.: Hey, you, get off of my market: detecting malicious apps in official and alternative android markets. In: Proceedings of Annual Network & Distributed System Security Symposium (2012)
9. Yan, L.K., Yin, H.: DroidScope: seamlessly reconstructing the OS and dalvik semantic views for dynamic android malware analysis. In: Usenix Security Symposium, pp. 569–584 (2012)
10. Zhou, Y., Jiang, X.: Dissecting android malware: characterization and evolution. In: Security and Privacy, pp. 95–109. IEEE (2012)
11. Cao, D., Wang, S., Li, Q., Cheny, Z., Yan, Q., Peng, L., Yang, B.: Droidcollector: a high performance framework for high quality android traffic collection. In: 2016 IEEE Trustcom/BigDataSE/ISPA, August 2016, pp. 1753–1758 (2016)
12. Arora, A., Garg, S., Peddoju, S.K.: Malware detection using network traffic analysis in android based mobile devices. In: Eighth International Conference on Next Generation Mobile Apps, Services and Technologies, pp. 66–71. IEEE (2014)
13. Virusshare.com. https://virusshare.com/
14. Bailey, M., Oberheide, J., Andersen, J., et al.: Automated classification and analysis of internet malware. In: International Conference on Recent Advances in Intrusion Detection, pp. 178–197. Springer-Verlag (2007)
15. Bayer, U., Comparetti, P.M., Hlauschek, C., et al.: Scalable, behavior-based malware clustering. In: Network and Distributed System Security Symposium, NDSS 2009, February 2009, San Diego, California, USA. DBLP (2009)
16. Li, Z., Sanghi, M., Chen, Y., et al.: Hamsa: Fast Signature Generation for Zero-day PolymorphicWorms with Provable Attack Resilience, p. 47 (2006)
17. Newsome, J., Karp, B., Song, D.: Polygraph: automatically generating signatures for polymorphic worms. In: IEEE Symposium on Security & Privacy, pp. 226–241. IEEE (2005)
18. Xie, Y., Yu, F., Achan, K., et al.: Spamming botnets: signatures and characteristics. In: ACM SIGCOMM Computer Communication Review, pp. 171–182 (2008)
19. Xu, Q., Liao, Y., Miskovic, S., et al.: Automatic generation of mobile app signatures from traffic observations. In: Computer Communications, pp. 1481–1489. IEEE (2015)

20. Aresu, M., Ariu, D., Ahmadi, M., et al.: Clustering android malware families by http traffic. In: International Conference on Malicious and Unwanted Software, pp. 128–135 (2015)
21. TSNE. http://scikit-learn.org/stable/modules/generated/sklearn.manifold.TSNE.html
22. Luan, S., Kong, X., Wang, B., et al.: Silhouette coefficient based approach on cellphone classification for unknown source imagee. In: IEEE International Conference on Communications, pp. 6744–6747. IEEE (2012)

Author Index

Printed in the United States
By Bookmasters